AF385814

Latin American Mathematics Series

Latin American Mathematics Series – UFSCar subseries

Published under the Latin American Mathematics Series, which was created to showcase the new, vibrant mathematical output that is emerging from this region, this subseries aims to gather high-quality monographs, graduate textbooks, and contributing volumes based on mathematical research conducted at/with the Federal University of São Carlos, a technological pole located in the State of São Paulo, Brazil. Submissions are evaluated by an international editorial board and undergo a rigorous peer review before acceptance.

Alessandro Arsie • Igor Mencattini

Geometry of Integrable Systems

An Introduction

Alessandro Arsie
Natural Sciences and Mathematics
University of Toledo
Toledo, OH, USA

Igor Mencattini
Inst. de Ciências Mat. e de Computação
Universidade de São Paulo
São Carlos, São Paulo, Brazil

ISSN 2524-6755   ISSN 2524-6763   (electronic)
Latin American Mathematics Series
ISSN 2524-6755   ISSN 2524-6763   (electronic)
Latin American Mathematics Series – UFSCar subseries
ISBN 978-3-031-96281-3   ISBN 978-3-031-96282-0   (eBook)
https://doi.org/10.1007/978-3-031-96282-0

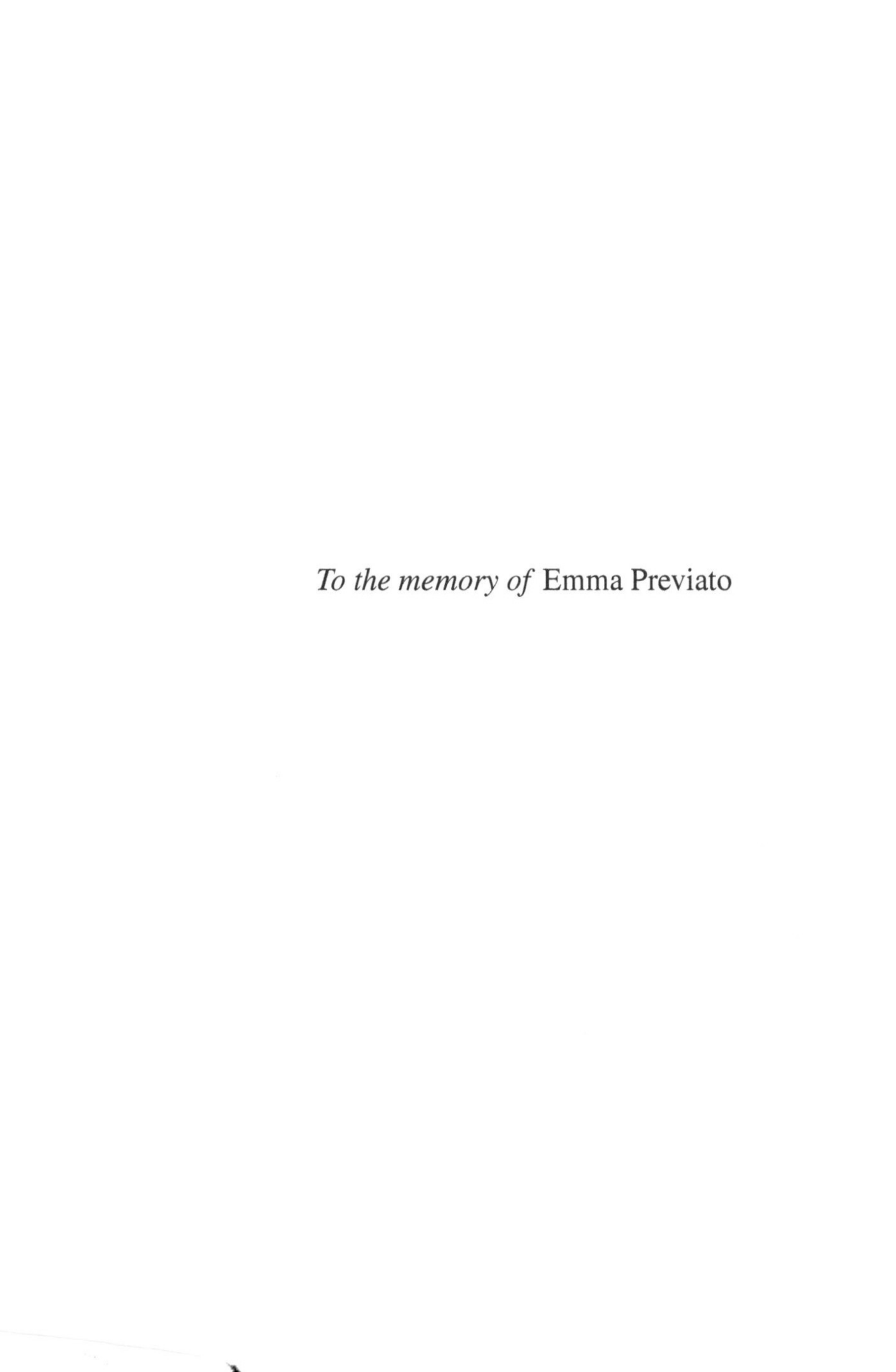

To the memory of Emma Previato

Preface

Complete integrability, in a broader sense, is a subtle property. At some extent one could say that a (dynamical) system is integrable if its equations are *solvable* and if it exhibits classes of globally defined and well-behaved solutions. In spite of this and of similar proposals, at this level of generality, it seems easier to exhibit examples of integrable systems than to pin down a precise definition of integrability, see [178, Chapter 1]. On the other hand, restricting the attention to the class of finite dimensional Hamiltonian system, a nice and geometrical definition of integrability is available. If one defines an Hamiltonian system to be a tuple (M, ω, H) of a symplectic manifold (M, ω) and a smooth function $H : M \to \mathbb{R}$, one can say that (M, ω, H) is integrable, or completely integrable, if H belong to a suitable maximal commutative subalgebra A of $(C^\infty(M), \{\cdot, \cdot\})$, where $\{\cdot, \cdot\}$ is the Poisson bracket defined by the symplectic form ω. The integrability of the original Hamiltonian system follows from the following observations. If M has dimension $2n$, A is (functionally) generated by n-generically independent functions, which will cut on M a family of Lagrangian submanifolds. Since $H \in A$, once chosen the initial conditions, the flow of the corresponding Hamiltonian vector field is doomed to be trapped into one of these submanifolds, reducing by one-half the number of degrees of freedom of the original dynamical system. This will suffice to provide the sought complete integrability. Note that, in this description, integrability and ergodicity are antithetical properties. It is worth mentioning that behind the previous (for the time being) sketchy description, there is (perhaps not-so) hidden the notion of *symmetry*. In fact all the elements of A are (globally defined) first integrals of H. This short presentation will be greatly expanded in the course of these notes, where all the ingredients introduced above will be carefully defined. For the time being we would like to remark that the above definition of integrability, though phrased in a *different* jargon, is generally attributed to Joseph Liouville (1809–1882) and it is commonly referred to as integrability *à la Liouville*. We refer the reader to [160, Chapter XVI] for a detailed historical reconstruction of the genesis of this result and for the analysis of its relations with the previous work of William Rowan Hamilton (1805–1865) and Carl Gustav Jacob Jacobi (1804–1851). Anyway, as underlined above, this result as many more in the field of classical mechanics, was originally stated with little (if none) geometric flavor.[1] In fact, as highlighted by Alain Weinstein in [246], although

Classical mechanics in the time of Huygens (1629–1695) and Newton (1642–1727) was very geometrical...... After Newton, there came a period of "mécanique analytique," during which Lagrange (1736–1813) could boast that his treatise on mechanics contained no pictures. Following the path of Euler (1707–1783) and Lagrange, Jacobi (1804–1851) and Hamilton (1805–1865) continued the development of analytic techniques for the explicit solution of the differential equations describing mechanical systems. Finally, geometry has taken a new role in mechanics through the contributions of Poincaré (1854–1912) and Birkhoff (1884–1944).

Without entering in the intricate details of the interaction between classical mechanics and geometry, see for example [161], we cannot avoid to mention that the geometric approach to integrability followed in these notes is due to Vladimir Igorevich Arnold (1937–2010). In fact, borrowing the words of Alexander Givental, [106]

Among many concepts owing Arnold their existence......one is the geometric notion of integrability in Hamiltonian systems. There is a lot of controversy over which of the known integrability mechanisms is most fundamental, but there is a consensus that integrability means a complete set of Poisson-commuting first integrals. This definition and "Liouville's Theorem" on geometric consequences of the integrability property (namely, foliation of the phase space by Lagrangian tori) are in fact Arnold's original inventions.

This theoretical framework suggests that the first step in the attempt to prove the complete integrability of a given Hamiltonian system is to show that it possesses a sufficient number of Poisson commuting first integrals. For this reason, it is natural to seek for geometrical recipes which provide this plethora of distinguished functions.

These observations will lead us to the theory of bi-Hamiltonian structures and recursion operators. The latter were introduced by Peter Olver in [204] as generators of infinitesimal symmetries of evolution equations, while the birth of the theory of the bi-Hamiltonian structures, i.e. pairs of *compatible* Poisson structures, can be traced back to the fundamental work of Franco Magri [163], see also [98], whose aim was, in the author words,

...to suggest a constructive approach to the infinite-dimensional integrable Hamiltonian equations, i.e., to the evolution equations possessing an infinite sequence of independent integrals which are in involution.

Both the Olver's and the Magri's work are part of the so-called *soliton revolution*, [255, 256], started with the groundbreaking work of Martin Kruskal and Norman Zabusky [257] on the KdV equation and for this reason they belong to the realm of the theory of nonlinear integrable PDEs. In spite of their infinite dimensional

---

[1] With at least one notable exception, the work of Louis Poinsot (1777–1859).

origin, these works opened the door to a new differential geometry topic, the theory of bi-Hamiltonian manifolds, which, hereafter, will receive considerable attention.

Our aim is to present an introduction to the theory of completely integrable Hamiltonian systems from the point of view of differential geometry. A particular attention will be given to the bi-Hamiltonian aspects of this theory.

What Is in This Book

We divided the content of these notes in two parts. The first one contains a detailed, though elementary, exposition of the topics we think necessary to start a serious geometrical analysis of complete integrability, while the second one is devoted to the analysis of three classical examples of integrable systems. More precisely, Chaps. 1 and 2 provide a background of symplectic and Poisson geometry. In the third chapter we set up the stage for the study of the Hamiltonian systems with symmetry. The fourth chapter is a primer on the theory of completely integrable systems, with emphasis on their geometric description. Finally, Chap. 5 is devoted to a (quite detailed) presentation of bi-Hamiltonian geometry, highlighting the role played by the recursion operators, i.e. emphasizing the theory of the so-called Poisson-Nijenhuis structures. The second part of these notes starts in Chap. 6 with the description of the (free) n-dimensional rigid-body, while Chaps. 7 and 8 are devoted to the rational Calogero-Moser and, respectively, to the open Toda system. In each case first we describe the system, then we discuss its integrability, and finally we present (at least) one of its (known) bi-Hamiltonian descriptions. In Chap. 6, before delving in the rigid-body theory, we outline a few aspects of the geometry of (co)tangent bundles of Lie groups and we introduce the notion of a Lax representation. The book has four appendices, where we collected basics concepts of symplectic linear algebra, differential geometry, and Lie theory, whose presence we thought could facilitate the reading the main text. Moreover, each chapter closes with a section titled a *Concluding Remarks and Further Topics*. In each of them we collect a few items which, in spite of their more advanced nature and because of their importance, we thought could at least be mentioned in an elementary presentation. The format of these eight sections is quite different compared with the main sections, for example proofs are mostly absent. For this reason these sections are meant to be short introductory surveys to be used as starting points for further readings.

What Is Not in This Book

The list of omitted topics is, obviously, very long. Here we mention just two of them, which, because of their nature, we think very much related to the main theme of these notes. First, in these notes there is no mention to algebraic geometry and to its (many) relations with integrability, which can be traced back to the very beginning of this theory and which selects the class of the so-called *algebraically completely integrable* systems, a.c.i. systems from now on. A point of entrance in this theory is the Lax representation with spectral parameter. From such a representation one gains the so-called spectral curve, whose Jacobian, or more generally its Prym variety, can be used to linearize the dynamics of the original system. A nice introduction to this topic can be found in [19]. The second (almost complete) omission is the theory of the integrable PDE, which, as we have mentioned above, flourished after the publication of [257] where the KdV equation *was solved numerically.*[2] This story, and much more than this, is beautifully told in [197]. Two other references we would like to mention are [198] and [79], both addressing the theory of the *Inverse Scattering Method*, a sophisticated nonlinear analogue of the Fourier transform. There are many points of contact between the theory of the integrable PDE and the subject of this book. One of those is briefly presented at the end of Chap. 8, where it is described a famous correspondence between the solutions of the rational Calogero-Moser system and the equations of the Kadomtsev-Petviashvili hierarchy. Another one, already mentioned before, see [98, 163], is the fact that these infinite dimensional dynamical systems often admit a multi-Hamiltonian formulation. Both the theory of a.c.i. systems and the one of integrable PDE can easily occupy more than one volume of the same size as the present one. For example, the Kruskal-Zabusky breakthrough gave birth to a new research area in mathematical-physics, which is nowadays still very active and which have ramifications in many, apparently unrelated, areas of mathematics like algebraic geometry, combinatorics, and probability [22, 134, 146, 148]. Furthermore, any discussion of integrable discrete systems (difference and partial difference equations, billiards, maps, etc.) is completely missing.

Suggestions for the Reader

This book is addressed to an advanced undergraduate or a beginning graduate student with a strong interest in the geometrical methods of mathematical physics. The topics discussed hereafter have as prerequisites an introductory course in differential geometry at the level of [227, 244] and some familiarity with Hamiltonian and Lagrangian mechanics as presented in [10, 84]. Chapters 1–3 could form the core of an introductory one-semester course in symplectic-Poisson geometry. Starting from a more advanced level, after skimming through these first three chapters, one could focus on Chaps. 4 and 5, which together with one of the examples presented in the last three chapters could complete a one-semester course on the geometry of finite dimensional integrable systems.

---

[2] [256]

Notations and Mathematical Conventions

(i) In what follows every manifold will be a *smooth manifold*, i.e., a *locally Euclidean, second countable and Hausdorff* topological space N, endowed with a smooth structure $\mathscr{A}$, i.e., a *maximal smooth atlas*. In spite of the fact that a manifold is a pair $(N, \mathscr{A})$, we will never mention the smooth structure and we will simply write N.

(ii) Unless explicitly stated, all manifolds will be assumed to be connected. In particular all Lie groups will be connected.

(iii) We will generically work over the field of the real numbers, so that when we write *"Let M be a manifold...."* or *"Consider the vector space V....,"* M will be a *real* manifold and V will be a *real* vector space. Said that, we could leave open the possibility to have a different ground field (which will be typically the field of complex numbers $\mathbb{C}$), and for this reason will agree that when it is better to leave this possibility open, we will denote the ground field with the letter $\mathbb{K}$.

(iv) All tensor fields will be considered *smooth*.

(v) The letters N, M, P will be used to denote *manifolds*, while the letters X, Y, Z will denote *vector fields*. We will use Π to denote Poisson tensors and ω to denote symplectic forms, with the only exception of the *canonical symplectic* form defined on the cotangent bundles, which will be denoted with the letter Ω. The *canonical* or *Liouville* one-form will be instead denoted with the letter Θ.

(vi) $\Omega^k(M)$ will denote the $C^\infty(M)$-module of k-differential forms on the manifold M, $\mathscr{D}(M)$ the group of diffeomorphisms of M, $\mathfrak{X}(M)$ the Lie algebra of vector fields on M, equipped with the Lie bracket $[\cdot, \cdot]$.

(vii) We will also adopt the two following conventions:

   (a) $\Omega = d\Theta$, instead of $\Omega = -d\Theta$.

   (b) When we come to Lie algebras (Lie groups) morphisms, we will always prefer the homomorphisms to the anti-homomorphisms. This choice will be systematically adopted and, in particular, will be reflected in the following ones:

      (i) We will define a *Hamiltonian vector field* X_f with Hamiltonian function f as a vector field such that:

$$df = -i_{X_f}\omega \text{ instead of } df = i_{X_f}\omega.$$

      (ii) Given a *left G-action* on N, the *fundamental vector field* X_x associated to $x \in \mathfrak{g}$ will be defined by:

$$X_x(n) = \left.\frac{d}{dt}\right|_{t=0} \varphi_{\exp(-tx)}(n) \text{ instead of } X_x(n) = \left.\frac{d}{dt}\right|_{t=0} \varphi_{\exp(tx)}(n).$$

In case of a *right G-action* we will define the *fundamental vector field* associated to $x \in \mathfrak{g}$ via:

$$X_x(n) = \left.\frac{d}{dt}\right|_{t=0} \varphi_{\exp(tx)}(n) \text{ instead of } X_x(n) = \left.\frac{d}{dt}\right|_{t=0} \varphi_{\exp(-tx)}(n).$$

This choice will imply that, in both cases, the *infinitesimal* $\mathfrak{g}$-*action* $\mathfrak{g} \to \mathscr{X}(N)$, $x \rightsquigarrow X_x$, is a *homomorphism* of Lie algebras.

(viii) *Generally* the differential of a map $f : M \to N$ at the point $m \in M$ will be denoted using the "*-*notation*," i.e. the differential of f at m will be denoted by $f_{*,m} : T_m M \to T_{f(m)} N$. Anyway we should warn the reader that this general rule is not strict and sometimes it will be broken in favor of the notation $df_m : T_m M \to T_{f(m)} N$. In particular the $*$-notation will be preferred every time we will need to handle the differential of a map (or function) together with its *dual* or *transpose* map. In this case the *transpose map* of $f_{*,m} : T_m M \to T_{f(m)} N$ will be denoted by $f^*_{,m}$ and it will be a linear map $T^*_{f(m)} N \to T^*_m M$.

(ix) In particular, if $f : M \to M$ is a diffeomorphism, then $(f^{-1})_{*,f(m)} : T_{f(m)} M \to T_m M$ and its transpose is $(f^{-1})^*_{f(m)} : T^*_m M \to T^*_{f(m)} M$.

Acknowledgments We are pleased to thank our friends, collaborators, and teachers who helped and inspired us during the preparation of this text: Claudio Bartocci, Ugo Bruzzo, Henrique Bursztyn, Franco Cardin, Gregorio Falqui, Pedro Frejlich, Letterio Gatto, Rui Loja Fernandes, Luen-Chau Li, Paolo Lorenzoni, Franco Magri, Volodya Rubtsov, and Farid Tari.

We are very grateful to Marco Pedroni, who carefully read a preliminary version of the manuscript, and to Washington Marar, who drew the pictures and was always available to help with many good advices.

We also acknowledge the Springer team, in particular our Series Editor, César R. de Oliveira, and our Publishing Editor, Robinson dos Santos.

<table>
<tr><td>Toledo, OH, USA</td><td>Alessandro Arsie</td></tr>
<tr><td>São Carlos, SP, Brazil</td><td>Igor Mencattini</td></tr>
</table>

Contents

Part I
Preliminary Material

Chapter 1
Symplectic Geometry

Symplectic geometry is the branch of differential geometry which studies the properties of manifolds endowed with a *symplectic structure*, i.e., a closed and nondegenerate two-form. Two, not mutually exclusive, classes of examples of these manifolds are cotangent bundles and coadjoint orbits. Symplectic manifolds appear in classical mechanics as the *phase spaces* of classical dynamical systems. For example, if N is the configuration space of a mechanical system, its cotangent bundle T^*N plays the role of its phase space. It is worth noticing that symplectic geometry is not only the geometric apparatus of Hamiltonian dynamics, but it also finds plenty of applications in many other areas of mathematics, such as representation theory, Fourier integral operators, geometric optics, and control theory, just to cite a few of them. Since the 1980s, with the fundamental works of A. Floer, M. Gromov, V. Arnol'd, A. Givental, H. Hofer, and many others, this area has expanded also into symplectic topology. Indeed, even though, unlike Riemannian manifolds, all symplectic manifolds of a given dimension locally look alike (Darboux theorem), they exhibit unexpected global rigidity properties that have been started to be fully studied in the last 40 years. In this chapter, after introducing the notion of symplectic manifold together with a few of its properties, we shall briefly discuss the *symmetries* of a symplectic structure. This will be taken as the starting point to introduce the Hamilton-Jacobi theory, a classical topic of mathematical physics. The chapter will end with the definition of a classical integrable system, which is the main topic of these lectures.

1 Symplectic Manifolds: Definition, Main Properties, and Examples

We start this chapter introducing one of the main characters of our narrative.

© The Author(s), under exclusive license to Springer Nature Switzerland AG 2026

A. Arsie, I. Mencattini, *Geometry of Integrable Systems*, Latin American Mathematics Series – UFSCar subseries,

https://doi.org/10.1007/978-3-031-96282-0_1

Definition 1.1 A two-form ω is called *symplectic* if it is *closed* and *nondegenerate*. A *symplectic manifold* is a pair (M, ω) of a manifold M and of a symplectic form $\omega \in \Omega^2(M)$. A symplectic form is said to be, or to define, a *symplectic structure* on the underlying manifold. $\triangle$

Remark 1.2 In the above definition, the requirement that ω is nondegenerate is equivalent to claim that for all $m \in M$, $(T_m^* M, \omega_m)$ is a symplectic vector space, i.e., for all $m \in M$,

$$\omega_m : T_m M \times T_m M \longrightarrow \mathbb{R}$$

is a *skew-symmetric, nondegenerate* bilinear form, [30, Section 1.1] $\triangle$

The most basic example of a symplectic manifold is the following.

Example 1.3 (Standard Symplectic Manifold) If $M = \mathbb{R}^{2n}$ is endowed with coordinates $(p, x) = (p_1, \ldots, p_n, x_1, \ldots, x_n)$, then the two-form $\omega_{st} = \sum_{i=1}^{n} dp_i \wedge dx_i$ is symplectic. Hereafter the pair $(\mathbb{R}^{2n}, \omega_{st})$ is dubbed the *standard symplectic manifold*, while ω_{st} and (p, x) as above is called the *standard symplectic form* and a set of *standard symplectic* or of *Darboux coordinates*, respectively; see Sect. 3.3. Note that $(\mathbb{R}^{2n}, \omega_{st})$ should be thought as the *skew analogue* of the standard Euclidean space $\mathbb{R}^n$ endowed with its Euclidean metric. In particular it represents the model space for symplectic geometry; see Sect. 3.3. Observe also that $\omega_{st}^{\wedge n}$ fixes one of the two orientations in $\mathbb{R}^{2n}$. $\triangle$

Remark 1.4 If $v, u \in \mathbb{R}^{2n}$ and $v = \sum_{i=1}^{n} v_i \frac{\partial}{\partial x_i} + \sum_{i=1}^{n} V_i \frac{\partial}{\partial p_i}$ and $u = \sum_{i=1}^{n} u_i \frac{\partial}{\partial x_i} + \sum_{i=1}^{n} U_i \frac{\partial}{\partial p_i}$, a simple computation shows that

$$\omega_{st}(v, u) = \sum_{i=1}^{n} \det \begin{pmatrix} V_i & v_i \\ U_i & u_i \end{pmatrix}.$$

In other words, $\omega_{st}(v, u)$ is the sum of the *oriented areas* of the *parallelograms* spanned by the vectors (v_i, V_i) and (u_i, U_i) on the planes (x_i, p_i), for all $i = 1, \ldots, n$. By "oriented areas" we mean that the areas have a sign, depending if the ordered pair of spanning vectors is oriented positively or negatively with respect to the induced orientation given by $\omega_{st}^{\wedge n}$. $\triangle$

A generalization of Example 1.3 is the content of the following.

Example 1.5 Let V be a finite-dimensional vector space, and let V^* be its dual. The vector space $M = V^* \oplus V$ is symplectic. In fact a simple computation shows that for each $(\phi, x), (\psi, y) \in M$ the bilinear form $\tilde{\omega}\big((\phi, x), (\psi, y)\big) = \phi(y) - \psi(x)$ is skew-symmetric and nondegenerate and it defines a symplectic form on M. The symplectic manifold $(M, \tilde{\omega})$ is the coordinate-free version of Example 1.3. $\triangle$

Remark 1.6 The previous examples belong to the realm of *linear* symplectic geometry. For more information about this topic, we refer to [30, Chapter 1], where in particular the notion of symplectic vector space, linear symplectic maps and distinguished subspaces of a symplectic vector space are fully analyzed. See also Appendix A for a shorter introduction. △

One way to get new symplectic manifolds from old ones is presented in the following

Example 1.7 (Cartesian Product of Symplectic Manifolds) Given two symplectic manifolds (M_1, ω_1), (M_2, ω_2), one can consider the following diagram:

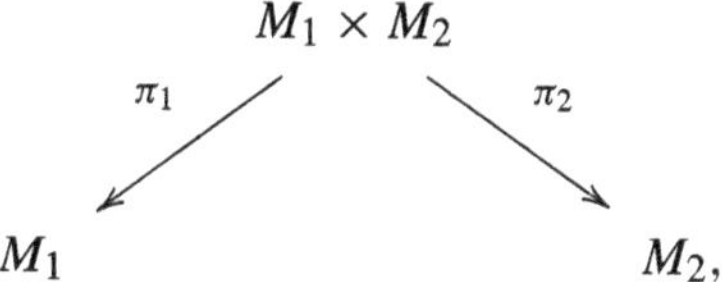

where π_1, π_2 are the projections on the first and on the second factor, respectively. Let $\omega = \pi_1^* \omega_1 - \pi_2^* \omega_2$ be the two-form on $M_1 \times M_2$ defined by the formula

$$\omega_{(m_1,m_2)}((v_1, v_2), (w_1, w_2)) = (\omega_1)_{m_1}(v_1, w_1) - (\omega_2)_{m_2}(v_2, w_2), \tag{1.1}$$

for all $(m_1, m_2) \in M_1 \times M_2$ and for all $(v_1, v_2) \in T_{m_1} M_1 \oplus T_{m_2} M_2 \simeq T_{(m_1,m_2)}(M_1 \times M_2)$. Then $(M_1 \times M_2, \omega)$ is a symplectic manifold, called the Cartesian product of (M_1, ω_2) with (M_2, ω_2). The symplectic form ω defined above is sometimes denoted by $\omega_1 \ominus \omega_2$. △

Problem 1.8 Prove that the two-form (1.1) is symplectic, i.e., it is closed and nondegenerate. What about $\omega_1 \oplus \omega_2 := \pi_1^* \omega_1 + \pi_2^* \omega_2$? △

Example 1.9 The two-dimensional sphere S^2 is a symplectic manifold. In fact if $\alpha = (\alpha_1, \alpha_2, \alpha_3)$ is any point of S^2, and v, w are two tangent vectors to the sphere at α, the area of the parallelogram spanned by v and w is given by the determinant of the matrix

$$\mathrm{vol}(v, w) = \begin{pmatrix} \alpha_1 & \alpha_2 & \alpha_3 \\ v_1 & v_2 & v_3 \\ w_1 & w_2 & w_3 \end{pmatrix}. \tag{1.2}$$

The right-hand side of (1.2) can be written as $\alpha \cdot (v \times w)$, where $\cdot$ and $\times$ denote the standard *scalar* and *cross product* in $\mathbb{R}^3$, respectively. In this way

$$\omega_\alpha(v, w) = \alpha \cdot (v \times w) \tag{1.3}$$

defines a bilinear, nondegenerate, and skew-symmetric form in the arguments v and w, which depends smoothly on α (Fig. 1.1).

In other words, (1.3) defines a symplectic form on S^2. Note that (1.2), or equivalently (1.3), is a volume form of the sphere, which, if written in Cartesian

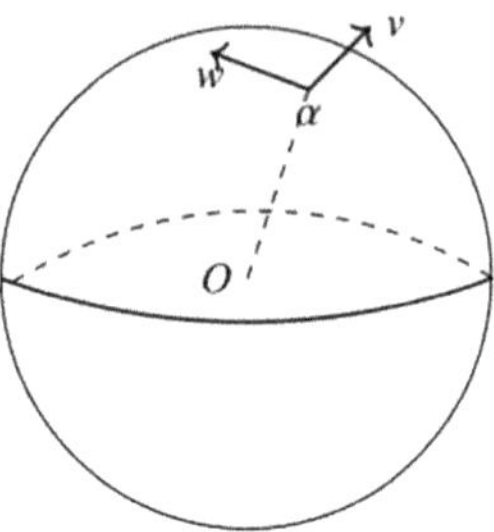

Fig. 1.1 Tangent space of S^2 at α generated by w and v

coordinates, assumes the following expression:

$$\omega = x\,dy \wedge dz + y\,dz \wedge dx + z\,dx \wedge dy, \tag{1.4}$$

as one can check by computing $\langle \omega, v \wedge w \rangle(\alpha)$, where $v = v_1 \frac{\partial}{\partial x} + v_2 \frac{\partial}{\partial y} + v_3 \frac{\partial}{\partial z}$, $w = w_1 \frac{\partial}{\partial x} + w_2 \frac{\partial}{\partial y} + w_3 \frac{\partial}{\partial z}$ and $\alpha = (\alpha_1, \alpha_2, \alpha_3)$. $\triangle$

The following problem is borrowed from [20].

Problem 1.10 Let $(\theta, z) \in [\pi, -\pi) \times (-1, 1)$ be the *cylindrical* coordinates on $S^2 \setminus \{(0, 0, 1), (0, 0, -1)\}$ defined by $x = \sqrt{1 - z^2} \cos \theta$, $y = \sqrt{1 - z^2} \sin \theta$, and $z = z$. Show that the symplectic form (1.4), if written in these coordinates, becomes $\omega = d\theta \wedge dz$. Prove that the total volume of (1.4) is 4π. $\triangle$

Example 1.11 The two-form $\omega = \frac{dx \wedge dy}{y^2}$ defined on $\{(x, y) \in \mathbb{R}^2 \mid y > 0\}$ is closed for dimensional reason and nondegenerate, as it follows by a direct inspection. For this reason it defines a symplectic form. Note that ω is the volume form on the *upper-half*, or *Poincaré*, plane $\mathbb{H}$, i.e., the subset of the two-dimensional (real) plane with positive ordinate endowed with the so-called *Poincaré* metric $g = \frac{dx^2 + dy^2}{y^2}$. $\triangle$

Remark 1.12 A few remarks are in order:

(i) As the attentive reader will have noticed, there is a basic obstruction for a manifold M to carry a symplectic structure. In fact, the condition of nondegeneracy of the symplectic form forces the underlying manifold to be *even-dimensional*.

(ii) Every symplectic manifold (M, ω) is *orientable*, i.e., it carries a never-vanishing differential form of maximal degree $\mathrm{Vol}_\omega = \frac{1}{n!} \underbrace{\omega \wedge \cdots \wedge \omega}_{n-\text{times}}$, where $2n$ is the dimension of M. Recall that a differential form of maximal degree is called a volume form.

(iii) Note also that since a symplectic form ω is closed, it defines a class $[\omega]$ in $H^2_{dR}(M)$.

$\triangle$

Example 1.13 Every orientable two-dimensional compact real surface is a symplectic manifold, whose symplectic form is, up to a nonzero constant, a volume form on the underlying manifold. The simplest example, next to the two-sphere discussed in Example 1.9, is the two-dimensional torus $\mathbb{T}^2$. This is diffeomorphic to the product $\mathbb{S}^1 \times \mathbb{S}^1$ and can be obtained as a quotient of the two-plane $\mathbb{R}^2$ by a lattice isomorphic to $\mathbb{Z}^2$. In this case, the symplectic form on $\mathbb{T}^2$ comes from the symplectic form $\omega = dp \wedge dx$ defined on $\mathbb{R}^2$, since that is translation invariant. $\triangle$

The symplectic manifolds whose symplectic form defines the trivial class in $H^2_{dR}(M)$ deserve a particular attention.

Definition 1.14 A symplectic manifold (M, ω) is called *exact* if $\omega = d\theta$ for some $\theta \in \Omega^1(M)$, called a *symplectic potential* for ω or for (M, ω). $\triangle$

Note that not every symplectic manifold is exact. In fact

Proposition 1.15 *A closed (i.e., compact and without boundary) symplectic manifold (M, ω) cannot be exact.*

Proof Suppose there exists a one-form θ such that $d\theta = \omega$. Then, $\mathrm{Vol}_\omega = \underbrace{\omega \wedge \cdots \wedge \omega}_{n-times} = d\eta$, where $\eta = \theta \wedge \underbrace{\omega \wedge \cdots \wedge \omega}_{(n-1)-times}$. By Stokes theorem we get

$$vol(M) = c \int_M \mathrm{Vol}_\omega = \int_{\partial M} d\eta = 0,$$

for some nonzero constant c. This is not possible since Ω is a volume form on M. $\square$

As a consequence,

Corollary 1.16 *An m-dimensional sphere S^m is symplectic if and only if $m = 2$.*

Proof We already know that S^2 is symplectic. Suppose now that $m > 2$, that S^m is symplectic, and recall that the de Rham cohomology of S^m decomposes as follows (see, for example, [36]):

$$H^\bullet_{dR}(S^m) = H^0_{dR}(S^m) \oplus H^m_{dR}(S^m).$$

Since we suppose that $m \neq 2$, $H^2_{dR}(S^m) = 0$, and for this reason, if ω is symplectic on S^m, ω is exact. But this, by Proposition 1.15, is not possible. $\square$

This shows in particular that there are orientable, even-dimensional manifolds which admit no symplectic structure.

An important example of exact symplectic manifold is the cotangent bundle of a manifold. The importance of this example, as already mentioned in the introduction to this chapter, stems from the prominent role that this class of manifold plays in the theory of the *Hamiltonian dynamical systems*. We shall discuss in more details this class of symplectic manifolds in the following subsection.

1.1 Cotangent Bundles

Let N be a manifold, $M = T^*N$ its cotangent bundle, and $\pi : M \longrightarrow N$ the canonical projection. We aim to prove that M carries a *canonical exact* symplectic structure, where canonical means depending only the geometry of M. To this end, we start defining $\Theta : M \to \mathbb{R}$ by letting, for all $\alpha \in M$ and $v \in T_\alpha M$,

$$\langle \Theta_\alpha, v \rangle = \langle \alpha, \pi_{*,\alpha} v \rangle.$$

Note that α is an element in $T^*_{\pi(\alpha)} N$ and $\pi_{*,\alpha} v \in T_{\pi(\alpha)} N$, so that both sides of the above equation are well defined. Furthermore, observe that Θ_α is a linear function on $T_\alpha M$. We claim that Θ is a one-form on M and that $d\Theta = \Omega$ is symplectic. To prove these statements we work with local coordinates on M. Let $(U, x_1, \ldots, x_n)$ be a local chart around the point $\pi(\alpha) \in N$ and let $(x_1, \ldots, x_n, p_1, \ldots, p_n)$ the induced coordinates on $V = \pi^{-1}(U)$, $\alpha \in V$. Then

$$\alpha = \sum_{i=1}^{n} p_i(\alpha) dx_i \quad \text{and}$$

$$\Theta_\alpha = \sum_{i=1}^{n} \left\langle \Theta_\alpha, \frac{\partial}{\partial x^i} \right\rangle dx_i + \sum_{i=1}^{n} \left\langle \Theta_\alpha, \frac{\partial}{\partial p_i} \right\rangle dp_i = \sum_{i=1}^{n} p_i(\alpha) dx_i,$$

implying $\Theta = \sum_{i=1}^{n} p_i dx_i$, showing that Θ is a one-form on M. Settled this first point, since

$$\Omega = d\Theta = \sum_{i=1}^{n} dp_i \wedge dx_i,$$

one can conclude that Ω is nondegenerate and, clearly, closed.

Definition 1.17 The one-form Θ is called the *Liouville, tautological,* or *canonical* one-form, and its differential $\Omega = d\Theta$ is called the *canonical symplectic form* of T^*N. $\quad\triangle$

As it was already mentioned above, the term *canonical* refers to the fact that Θ and Ω are intrinsic to the geometry of T^*N. Here below, we comment more on this point. First, we observe that if $\alpha \in \Omega^1(N)$, then for every $n \in N$

$$(\pi^*\alpha)_{\alpha_n} = \Theta_{\alpha_n}.$$

This follows directly from the definition of Θ. In fact, for all $v \in T_{\alpha_n} M$

$$\langle (\pi^*\alpha)_{\alpha_n}, v \rangle = \langle \alpha_{\pi(\alpha_n)}, \pi_{*,\alpha_n} v \rangle \overset{\pi(\alpha_n)=n}{=\!=\!=} \langle \alpha_n, \pi_{*,\alpha_n} v \rangle = \langle \Theta_\alpha, v \rangle.$$

On the other hand, since $\alpha \in \Omega^1(N)$ is a (smooth) section of $\pi : M \to N$, one can pull back Θ from M to N via α. In this case one discovers that

Proposition 1.18 *For every one-form α*

$$\alpha^* \Theta = \alpha. \tag{1.5}$$

Proof In fact, if $n \in N$ and $v \in T_n N$,

$$\langle (\alpha^* \Theta)_n, v \rangle = \langle \Theta_{\alpha_n}, \alpha_{*,n} v \rangle = \langle \alpha_n, \pi_{*,\alpha_n}(\alpha_{*,n} v) \rangle = \langle \alpha_n, (\pi \circ \alpha)_{*,n} v \rangle = \langle \alpha_n, v \rangle.$$

Comparing now the first and the last term in the previous computation, we have that

$$(\alpha^* \Theta)_n = \alpha_n, \ \forall n \in N,$$

proving (1.5). $\qquad\square$

The *zero section* of T^*N is the smooth map $\underline{0} : N \to T^*N$ defined by

$$\underline{0}(n) = (n, 0), \ \forall n \in N.$$

It defines a smooth embedding of N into T^*N and, for all $(n, 0)$ belonging to its image, one has that

$$T_{\underline{0}(n)} T^*N \simeq Vert_n N \oplus T_{\underline{0}(n)} \underline{0}(N), \tag{1.6}$$

where $Vert_n N = \ker(\pi_{*,\underline{0}(n)})$. In particular, the differential at n of $\underline{0}$ is a linear isomorphism between $T_n N$ and $T_{\underline{0}(n)} \underline{0}(N)$, for all $n \in N$. On the other hand, the canonical symplectic form defines an isomorphism between $Vert_n N$ and $T_n^* N$, which maps the vector $v \in Vert_n N$ to the covector $i_v \omega_{\underline{0}(n)} = \omega_{\underline{0}(n)}(v, \cdot) \in T_n^* N$. Written in local coordinates (p, x), one has that $v = \sum_{i=1}^n v_i \frac{\partial}{\partial p_i}$ and

$$i_v \omega_{\underline{0}(n)} = \sum_{i=1}^n v_i dx_i.$$

In other words, one can rewrite isomorphism (1.6) as

$$T_{\underline{0}(n)} T^*N \simeq T_n^* N \oplus T_n N. \tag{1.7}$$

Since T^*N is a symplectic manifold, the tangent space at any of its points is a symplectic vector space. For the notion of symplectic vector space and linear symplectomorphism, see Appendix A and [30, Sections 1.1,1.2]. A linear map $\phi : (V_1, \omega_1) \to (V_2, \omega_2)$ is called anti-symplectic or an anti-symplectomorphism if $\omega_2(\phi(u), \phi(v)) = -\omega_1(u, v)$ for all $uv, \in V_1$. In particular

Proposition 1.19 *For each $n \in N$, $T_{\underline{0}(n)} T^* N$, with its structure of a symplectic vector space defined by ω is anti-symplectomorphic to $T_n^* N \oplus T_n N$ endowed with the linear symplectic structure defined in Example 1.5.*

Proof We prove that the isomorphism in (1.7) is indeed an anti-symplectomorphism. To this end, let $v, u \in T_{\underline{0}(n)} T^* N$ which, in local coordinates, can be written as

$$v = \sum_{i=1}^{n} v_i \frac{\partial}{\partial x_i} + \sum_{i=1}^{n} \tilde{v}_i \frac{\partial}{\partial p_i} \quad \text{and} \quad u = \sum_{i=1}^{n} u_i \frac{\partial}{\partial x_i} + \sum_{i=1}^{n} \tilde{u}_i \frac{\partial}{\partial p_i}.$$

Under isomorphism (1.7), they correspond to

$$(v_0, v_1) = \sum_{i=1}^{n} \left(\tilde{v}_i dx_i, v_i \frac{\partial}{\partial x_i} \right) \quad \text{and} \quad (u_0, u_1) = \sum_{i=1}^{n} \left(\tilde{u}_i dx_i, u_i \frac{\partial}{\partial x_i} \right).$$

Then, on one side, one has

$$\tilde{\omega}\big((v_0, v_1), (u_0, u_1)\big) = \langle u_0, v_1 \rangle - \langle v_0, u_1 \rangle = \sum_{i=1}^{n} (\tilde{u}_i v_i - \tilde{v}_i u_i),$$

while on the other side

$$\Omega_{\underline{0}(n)}(v, u) = i_v \Omega_{\underline{0}(n)}(u) = \sum_{i=1}^{n} \left\langle \tilde{v}_i dx_i - v_i dp_i, u_i \frac{\partial}{\partial x_i} + \sum_{i=1}^{n} \tilde{u}_i \frac{\partial}{\partial p_i} \right\rangle$$

$$= \sum_{i=1}^{n} (\tilde{v}_i u_i - \tilde{u}_i v_i),$$

proving the statement. $\qquad\square$

In the next example we present a few results which will be proved in full details in Chap. 6.

Example 1.20 (Cotangent Bundle of G) Let G be a Lie group. Identifying $T^* G$ with $G \times \mathfrak{g}^*$ via left translations (see Formulas (6.2)), the forms Θ and Ω define on $G \times \mathfrak{g}^*$ the one-form ϑ and the two-form ω, respectively. More precisely

$$(1) \quad \langle \vartheta_{(g,\alpha)}, (v, \beta) \rangle = \langle \alpha, x \rangle \text{ and}$$

$$(2) \quad \omega_{(g,\alpha)}\big((v, \beta), (u, \gamma)\big) = \langle \beta, y \rangle - \langle \gamma, x \rangle - \langle \alpha, [x, y] \rangle$$

for all $(g, \alpha) \in G \times \mathfrak{g}^*$ and $(v, \beta) \in T_g G \oplus \mathfrak{g}^*$; see Formulas (6.7) and (6.8), Proposition 6.8. $\qquad\triangle$

Problem 1.21 Let N be a manifold and let T^*N be its the cotangent bundle equipped with the canonical symplectic form Ω; see Sect. 1.17. Prove the following statements.

(i) For every $\sigma \in \Omega^2(N)$, $\omega = \Omega + \pi^*\sigma$ is a nondegenerate two-form on T^*N.
(ii) The form ω so defined is symplectic if and only if σ is closed and, if written in coordinates, it assumes the following expression:

$$\omega = \sum_{i=1}^{n} dp_i \wedge dx_i + \frac{1}{2} \sum_{i,j=1}^{n} \sigma_{ij} dx_i \wedge dx_j.$$

When σ is closed, the symplectic form $\omega = \Omega + \pi^*\sigma$ is called a *magnetic* symplectic form, or a *magnetic extension* of Ω, and the pair (T^*N, ω) is called a *magnetic cotangent bundle*. $\triangle$

In the following subsection we introduce another important class of symplectic manifolds. We refer the reader to Appendix C for the relevant background material connected to the following example.

1.2 Coadjoint Orbits as Symplectic Manifolds

For the following important class of symplectic manifolds, it might be useful for the reader to review the relevant notions in Appendix C.

Let G be a Lie group, $\mathfrak{g}$ its Lie algebra, and $\mathfrak{g}^*$ the dual vector space of $\mathfrak{g}$. Given $\alpha \in \mathfrak{g}^*$, let $\mathcal{O}_\alpha$ be its orbit under the coadjoint action of G.

Recall that $\mathcal{O}_\alpha \simeq G/G_\alpha$, where G_α is the stabilizer of α in G and that the fundamental vector field associated to $x \in \mathfrak{g}$ by coadjoint action is

$$X_x(\alpha) = \frac{d}{dt}\bigg|_{t=0} \mathrm{Ad}^\sharp_{\exp(-tx)}(\alpha) = -\mathrm{ad}^\sharp_x \alpha, \text{ for all } \alpha \in \mathfrak{g}^* \text{ and } x \in \mathfrak{g};$$

see Sect. 1.3 in Appendix C. The identification $T_\alpha \mathcal{O}_\alpha \simeq \mathfrak{g}/\mathfrak{g}_\alpha$ implies that the tangent space at each point of the orbit is spanned by the values assumed by the fundamental vector fields at that point. More precisely, since $\mathfrak{g}_\alpha$ acts trivially on $\mathcal{O}_\alpha$, one has the following isomorphism of vector spaces:

$$i : T_\alpha \mathcal{O} \xrightarrow{\; v = X_x(\alpha) \rightsquigarrow \bar{x} = x \bmod \mathfrak{g}_\alpha \;} \mathfrak{g}/\mathfrak{g}_\alpha. \tag{1.8}$$

This isomorphism implies that if $v \in T_\alpha \mathcal{O}_\alpha$ and $x, y \in \mathfrak{g}$ are such that $X_x(\alpha) = v = X_y(\alpha)$, then $x - y \in \mathfrak{g}_\alpha$.

Starting from these preliminary remarks, we prove that each coadjoint orbit is a symplectic manifold. To this end, let $\omega_\alpha : T_\alpha \mathcal{O}_\alpha \times T_\alpha \mathcal{O}_\alpha \to \mathbb{K}$ be the bilinear form

defined by

$$\omega_\alpha(u, v) = \langle \alpha, [x, y] \rangle, \tag{1.9}$$

where $x, y \in \mathfrak{g}$ are any two Lie algebra elements such that $X_x(\alpha) = v$ and $X_y(\alpha) = u$. Then ω_α is (i) well defined, (ii) skew-symmetric, and (iii) nondegenerate. Property (i) follows from the previous observation: if $x, x' \in \mathfrak{g}$ are such that $X_x(\alpha) = u = X_{x'}(\alpha)$, then $x - x' \in \mathfrak{g}_\alpha$ which implies that $\langle \alpha, [x', y] \rangle = \langle \alpha, [x, y] \rangle$ for all $y \in \mathfrak{g}$. Property (ii) follows from the skew-symmetry of the Lie brackets. Finally, property (iii): suppose that there is $v \in T_\alpha \mathscr{O}_\alpha$ such that $\omega_\alpha(v, u) = 0$ for all $u \in T_\alpha \mathscr{O}_\alpha$. Then there exists $x \in \mathfrak{g}$ such that $X_x(\alpha) = v$ such that $\langle \alpha, [x, y] \rangle = 0$ for all $y \in \mathfrak{g}$. But this implies that $\mathrm{ad}_x^\sharp(\alpha) = 0$, i.e., that $x \in \mathfrak{g}_\alpha$ which yields that $0 = X_x(\alpha) = v$. Clearly, the choice of the point $\alpha \in \mathscr{O}_\alpha$ is totally arbitrary: we could choose a different $\alpha' \in \mathscr{O}_\alpha$ and define the corresponding form $\omega_{\alpha'}$, which will have the same properties of ω_α (in this case we should consider $T_{\alpha'}\mathscr{O}_\alpha \simeq \mathfrak{g}/\mathfrak{g}_{\alpha'}$). In other words, given an orbit $\mathscr{O}_\alpha$, we have defined a map $\mathscr{O}_\alpha \ni \alpha' \rightsquigarrow \omega_{\alpha'} \in \Lambda^2 T_{\alpha'}^* \mathscr{O}_\alpha$. The form ω given in Formula (1.9) is *smooth* since it is the *restriction* to $\mathscr{O}_\alpha$ of a bilinear form on $\mathfrak{g}$ which depends *linearly* on α. In this way we proved that:

Lemma 1.22 *Formula* (1.9) *defines a two-form on* $\mathscr{O}_\alpha$.

We show that this form is *closed*. Let θ be the *(left) Maurer-Cartan* form of G; see Example C.104. Given $\alpha \in \mathfrak{g}^*$, the orbit $\mathscr{O}_\alpha$ is diffeomorphic to G/G_α via

$$j : \mathscr{O}_\alpha \xrightarrow{\mathrm{Ad}_g^\sharp \alpha \rightsquigarrow \overline{g}} G/G_\alpha,$$

where $\overline{g}$ is the class of g in G/G_α. Under this isomorphism, calling e the identity of G, the class $[e] \in G/G_\alpha$ corresponds to $\alpha \in \mathscr{O}_\alpha$. Moreover, the isomorphism j factors the canonical projection

$$p : G \xrightarrow{g \rightsquigarrow \overline{g}} G/G_\alpha$$

as follows:

$$
\begin{array}{ccc}
G & \xrightarrow{\;\;p\;\;} & G/G_\alpha \\
{\scriptstyle \widetilde{p}}\Big\downarrow & \nearrow_{\;j} & \\
\mathscr{O}_\alpha & &
\end{array}
$$

The map $\widetilde{p} : G \to \mathscr{O}_\alpha$ is clearly surjective and, since $T_\alpha \mathscr{O}_\alpha \simeq \mathfrak{g}/\mathfrak{g}_\alpha$ (see 1.8), it is also submersive. Let Ω be the form given by

$$\Omega := \widetilde{p}^*(\omega). \tag{1.10}$$

Since Ω is a left invariant two-form on G (see Problem 1.23), it is completely determined by its value at $e \in G$. Let $x, y \in T_e G \simeq \mathfrak{g}$ and let $\widetilde{p}_{*,e} : T_e G \longrightarrow T_\alpha \mathcal{O}_\alpha$ be the differential of $\widetilde{p}$ calculated at the identity of G. Then

$$(\widetilde{p}^* \omega)_e(x, y) = \omega_{\widetilde{p}(e)}(\widetilde{p}_{*,e} x, \widetilde{p}_{*,e} y) = \omega_\alpha(v, u) = \langle \alpha, [x, y] \rangle,$$

where we denoted by v and u the tangent vectors at α corresponding to the elements x and y, respectively. Let $\beta_\alpha \in \Omega^1(G)$ be defined by

$$\beta_\alpha = -\langle \alpha, \theta \rangle,$$

and let $X, Y \in \mathfrak{G}$ be the (unique) left invariant such that $X(e) = x$ and $Y(e) = y$. Then

$$(d\beta_\alpha)_e(x, y) = \langle \alpha, [x, y] \rangle = \langle \alpha, \theta_e([X, Y])(e) \rangle. \tag{1.11}$$

In fact

$$(d\beta_\alpha)_e(x, y) = (d\beta_\alpha)_e(X(e), Y(e))$$

$$= X(e)\big(\beta_\alpha(Y)\big) - Y(e)\big(\beta_\alpha(X)\big) - (\beta_\alpha)_e\big([X, Y](e)\big)$$

$$= -(\beta_\alpha)_e([x, y]) \tag{1.12}$$

$$= \langle \alpha, [x, y] \rangle. \tag{1.13}$$

Here we used the fact that, if $X, Y \in \mathfrak{X}^L(G)$, $X(e) = x$ and $Y(e) = y$, then

$$\beta_\alpha(X)(g) = \langle (\beta_\alpha)_g, X(g) \rangle = -\langle \alpha, \theta_g\big(X(g)\big) \rangle$$

$$= -\langle \alpha, \big(L_{g^{-1}}\big)_{*,g}(X(g)) \rangle = -\langle \alpha, x \rangle,$$

as it follows from the definition of the one-form β_α, from the left invariance of X and, finally, from the condition that $X(e) = x$. From this, one deduces that $\beta_\alpha(X)$ and $\beta_\alpha(Y)$ are constant functions on G. This observation, together with the definition of the Lie bracket of $\mathfrak{g}$ obtained using the left invariant vector fields (see Formula C.3), gives us the equality in line (1.12). On the other hand, the equality in line (1.13) follows from the following computation:

$$-(\beta_\alpha)_e([x, y]) = \langle \alpha, \theta_e([x, y]) \rangle = \langle \alpha, [x, y] \rangle,$$

which follows from the definition of β and θ. From these computations, Formula (1.11) follows. In turn, this formula implies

$$\Omega_e(\eta, \xi) = (d\beta_\alpha)_e(\eta, \xi) \quad \forall \eta, \xi \in \mathfrak{g}. \tag{1.14}$$

Since Ω and β_α are left invariant two-forms which coincide at e (see 1.11 and 1.14), we conclude that $\widetilde{p}^*\omega = \Omega = d\beta_\alpha$. Then

$$d(\tilde{p}^*\omega) = d^2\beta_\alpha = 0. \tag{1.15}$$

To conclude, it suffices to observe that, since $\widetilde{p} : G \to \mathcal{O}_\alpha$ is submersive, the map $\widetilde{p}^*$ is injective which, together with (1.15), implies that $d\omega = 0$.

Problem 1.23 Prove that the two-form Ω defined in Formula (1.10) is left invariant. $\triangle$

In this way one arrives to the following important result.

Theorem 1.24 *For every $\alpha \in \mathfrak{g}^*$, the corresponding coadjoint orbit $\mathcal{O}_\alpha$ is a symplectic manifold.*

It is worth making a couple of observations.

(i) One simple consequence of Theorem 1.23 is that coadjoint orbits are always *even-dimensional* manifolds. On the other hand, as it is shown in Example 2.69, adjoint orbits in general *are not even-dimensional*, and for this reason, in general, they do not carry a symplectic form.
(ii) Coadjoint orbits are homogeneous manifolds carrying a symplectic form. In Sect. 3 of Chap. 3 it will be shown that that these two properties blend together providing a characterization of this class of manifolds.

For more information about the nature of the symplectic structure defined on the coadjoints orbit, we refer the reader to Sect. 3 in Chap. 2.

2 The Lie Algebras Defined by a Symplectic Structure

In this section we introduce several Lie algebras that are naturally defined on every symplectic manifold. Our presentation starts with the following simple observation, which is an immediate consequence of the nondegeneracy of the symplectic form ω. For every $m \in M$, the symplectic form ω defines a (linear) isomorphism $\omega_m^\flat : T_m M \to T_m^* M$, by

$$v \rightsquigarrow \iota_v \omega_m = \omega_m(v, \cdot). \tag{1.16}$$

This application induces a (bundle) isomorphism between the tangent and the cotangent bundle of M, which, in turn, defines an isomorphism between $\mathfrak{X}(M)$ and $\Omega^1(M)$, $X \rightsquigarrow i_X\omega = \omega(X, \cdot)$. By means of the inverse of this isomorphism, one can define the linear map

$$\Lambda : C^\infty(M) \to \mathfrak{X}(M)$$

$$f \rightsquigarrow X_f, \tag{1.17}$$

where, by definition, X_f is the unique vector field such that $df = -\iota_{X_f}\omega$, i.e.,

$$X_f(m) = -(\omega_m^\flat)^{-1}(df(m)) \tag{1.18}$$

Remark 1.25 Note that Λ is *not* an isomorphism. In particular, since M is connected, the kernel of Λ is $\mathbb{R}$. $\triangle$

Definition 1.26 The vector field $X_f = \Lambda(f)$, $f \in C^\infty(M)$ is called the Hamiltonian vector field associated to f and the function f is called the *Hamiltonian function* of X_f. $\triangle$

We also give the following:

Definition 1.27 (Hamiltonian System) A Hamiltonian system is a triple (M, ω, H) where (M, ω) is a symplectic manifold and $H : M \to \mathbb{R}$ is a smooth function, called *the* Hamiltonian function of the Hamiltonian system. $\triangle$

Note that two different Hamiltonian functions of a given Hamiltonian vector field differ at most by a constant function (remind that we always assume that M is connected). For this reason in Definition 1.27 we should better call H *a* Hamiltonian function of the given Hamiltonian system. In other words, from a geometric viewpoint a faithful representation of a Hamiltonian system is provided by the corresponding Hamiltonian vector field and not by the Hamiltonian H.

It is worth spending a few words to explain the origin of the adjective *Hamiltonian* in Definition 1.26.

Remark 1.28 (Hamiltonian Equations) Consider the standard symplectic manifold $(\mathbb{R}^{2n}, \omega_{st})$ with standard symplectic coordinates (p, x). Let $H \in C^\infty(M)$ and let $dH = \sum_{i=1}^{n} \frac{\partial H}{\partial x_i}dx_i + \sum_{i=1}^{n} \frac{\partial H}{\partial p_i}dp_i$ be its differential. The vector field $X = \sum_{i=1}^{n} X_i\frac{\partial}{\partial x_i} + Y_i\frac{\partial}{\partial p_i}$ having as Hamiltonian function H is defined by the condition $dH = -\iota_X\omega$, which, written in the coordinates (p, x), becomes $X_i = \frac{\partial H}{\partial p_i}$, $Y_i = -\frac{\partial H}{\partial x_i}$ $\forall i = 1, \ldots, n$. The adjective Hamiltonian in Definition 1.26 is due to the following observation. The integral curves of this vector field satisfy the following system of differential equations:

$$\dot{x}_i = \frac{\partial H}{\partial p_i} \quad \text{and} \quad \dot{p}_i = -\frac{\partial H}{\partial x_i}, \ \forall i = 1, \ldots, n, \tag{1.19}$$

which are classically known as the *Hamiltonian equations* of the mechanical system described H; see [10, 84]. Note that, as already remarked above, these equations are invariant under the transformation $H \rightsquigarrow H + c$, for every $c \in \mathbb{R}$. $\triangle$

Let us now introduce the following important notion (Fig. 1.2).

Fig. 1.2 Level set of a first
integral of H

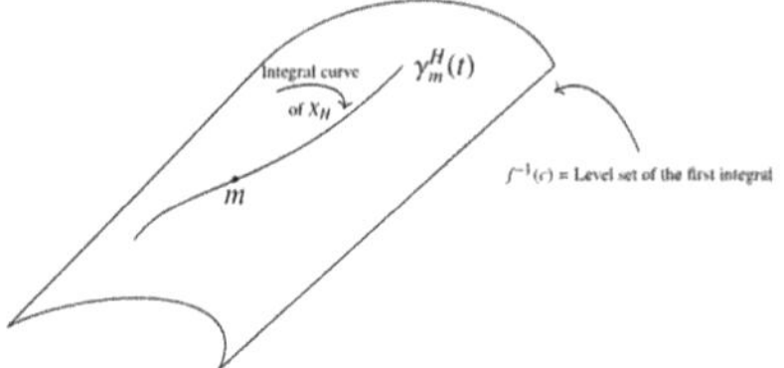

Definition 1.29 (First Integral)   A function $f \in C^\infty(M)$ is called a *first integral* of the Hamiltonian system (M, ω, H), if

$$df(X_H) = X_H(f) = \mathscr{L}_{X_H} f = 0.$$

$\triangle$

Observe that $f \in C^\infty(M)$ is a first integral of (M, ω, H) *if and only if* the integral curves of X_H are contained in its level sets. In particular H is always a first integral of (M, ω, H) since $\mathscr{L}_{X_H} H = \omega(X_H, X_H) = 0$.

The notion of first integral is not limited to dynamical systems of Hamiltonian type. In fact, more generally, if $X \in \mathfrak{X}(M)$, one can say that $f \in C^\infty(M)$ is a first integral of X if $\mathscr{L}_X f = 0$. As before, this condition is equivalent to say that f is constant along the flow generated by X or, equivalently, that the integral curves of X are contained in the level sets of f. The relevance of the first integrals in mechanics stems from the fact that they can be used to reduce the number of degrees of freedom of the mechanical system, simplifying the process of integration of the equation of motion. At the level of a system of ODEs this entails the fact that one can reduce the number of equations of the original system, reducing the complexity of the original problem.

Note that in a neighborhood of a point $m \in M$ where $X(m) \neq 0$ the Frobenius theorem (see Theorem B.5), guarantees the existence of $\dim M - 1$ *local* first integrals for X. On the other hand, the existence of *globally* defined first integrals is a nontrivial property of a dynamical system. When the latter admits a Hamiltonian formulation, i.e., a local representation in terms of a system of $2n$ ODEs like the one in (1.19), it turns out that the existence of *only n* first integrals (instead of $2n - 1$) suffices to integrate the original system using elementary methods. The Hamiltonian systems having this property are quite rare, they are called *completely integrable* (see Definition 1.120), and they are the main object of study of the present work.

Note that if a vector field X is Hamiltonian, i.e., if $X = X_f$ for some $f \in C^\infty(M)$, then $\mathscr{L}_X \omega = 0$. In fact,

$$\mathscr{L}_X \omega = d(\iota_X \omega) + \iota_X(d\omega) = -d^2 f = 0.$$

In view of this observation it makes sense to give a definition that generalizes the one of Hamiltonian vector field.

Definition 1.30 A vector field X is called *locally Hamiltonian* if $\mathscr{L}_X \omega = 0$, i.e., if the one-form $\iota_X \omega$ is closed. $\triangle$

The locally Hamiltonian vector fields on (M, ω) form a vector space which we denote with $\mathscr{H}_l(M)$. Similarly, the Hamiltonian vector fields form a vector space which is denoted with $\mathscr{H}(M)$. Clearly,

$$\mathscr{H}(M) = \Lambda(C^\infty(M)) \subset \mathscr{H}_l(M).$$

Example 1.31 (Locally Hamiltonian But Not Hamiltonian) Let $(\mathbb{R}^2 \backslash \{(0, 0)\}, \omega = r d\phi \wedge dr)$, where (r, ϕ) are the polar coordinates on $\mathbb{R}^2 \backslash \{(0, 0)\}$. Then

$$X = \frac{1}{r} \frac{\partial}{\partial \phi}$$

is a *Hamiltonian* vector field, while

$$Y = \frac{1}{r} \frac{\partial}{\partial r}$$

is a *locally Hamiltonian* vector field which is *not* Hamiltonian. $\triangle$

Observe that $\mathscr{Z}^1(M) \simeq \mathscr{H}_l(M)$ and $\mathscr{B}^1(M) \simeq \mathscr{H}(M)$, where $\mathscr{Z}^1(M)$ and $\mathscr{B}^1(M)$ are the spaces of closed and exact one-forms, respectively. From this, the isomorphism

$$H^1_{dR}(M) \simeq \frac{\mathscr{H}_l(M)}{\mathscr{H}(M)}$$

follows.

Problem 1.32 Let

$$X_1 = -p \frac{\partial}{\partial x}, \quad X_2 = p \frac{\partial}{\partial p} - x \frac{\partial}{\partial x}, \quad X_3 = -x \frac{\partial}{\partial p}$$

be three vector fields defined on $(\mathbb{R}^2, dp \wedge dx)$. Show that:

(i) $\{X_1, X_2, X_3\}$ generates a Lie subalgebra of $\mathfrak{X}(\mathbb{R}^2)$.
(ii) X_1, X_2, X_3 are Hamiltonian vector fields and find a corresponding Hamiltonian function for each of them.

$\triangle$

Before moving to the next result, we need to introduce the following important concept. On the ring of smooth functions $C^\infty(M)$ of (M, ω), one can define the following bilinear, skew-symmetric bracket:

$$\{f, g\} = \omega(X_f, X_g) \, \forall f, g \in C^\infty(M). \tag{1.20}$$

The proof of Proposition 1.34 is based on the following preliminary:

Lemma 1.33 $\mathcal{H}_l(M)$ *is a Lie algebra with respect to the usual commutator of* $\mathfrak{X}(M)$, *and*

$$[\mathcal{H}_l(M), \mathcal{H}_l(M)] \subset \mathcal{H}(M).$$

Moreover, for each $f, g \in C^\infty(M)$ *one has*

$$[X_f, X_g] = X_{\{f,g\}}. \tag{1.21}$$

Proof It suffices to show that if $X, Y \in \mathcal{H}_l$, then $\iota_{[X,Y]}\omega = df$ for some $f \in C^\infty(M)$. Indeed, one has

$$\iota_{[X,Y]}\omega = \mathcal{L}_X(\iota_Y\omega) = d(\iota_X\iota_Y\omega).$$

The proof of the second part follows from the same computation, applied to the case $X = X_f$ and $Y = X_g$. $\qquad\square$

The previous lemma shows that on every symplectic manifold (M, ω) one can select a distinguished Lie subalgebra $\mathcal{H}_l(M)$ of the Lie algebra $\mathfrak{X}(M)$ and that $\mathcal{H}(M)$ is an ideal of such a Lie algebra. We can now prove the following important

Proposition 1.34 *The bracket* $\{\cdot, \cdot\}$ *defined in* (1.20) *fulfills the Jacobi identity, i.e.,*

$$\{\{f, g\}, h\} + \{\{h, f\}, g\} + +\{\{g, h\}, f\} = 0,$$

for all $f, g, h \in C^\infty(M)$.

Proof Using the identity (1.21), we can write

$$\{\{f, g\}, h\} = \omega(X_{\{f,g\}}, X_h) = \omega([X_f, X_g], X_h).$$

Then Jacobi identity becomes

$$\omega([X_f, X_g], X_h) + \omega([X_h, X_f], X_g) + \omega([X_g, X_h], X_f) = 0. \tag{1.22}$$

For every two-form η and every $X, Y, Z \in \mathfrak{X}(M)$, one has

$$3d\eta(X, Y, Z) = X\eta(Y, Z) + Z\eta(X, Y) + Y\eta(Z, X) + \eta([X, Y], Z)$$
$$+ \eta([Z, X], Y) + \eta([Y, Z], X). \tag{1.23}$$

To prove (1.22) it suffices to apply (1.23) when η is the symplectic form ω and when X, Y, Z are Hamiltonian vector fields. $\qquad\square$

Remark 1.35 Note that every two-form on M defines a skew-symmetric bracket $\{\cdot, \cdot\}$ via Formula 1.20. Proposition 1.34 shows that this bracket fulfills the Jacobi identity *if and only if* the defining two-form is closed. $\triangle$

Beside being a Lie bracket, the bilinear form defined in (1.20) interacts nicely with the commutative product of $C^\infty(M)$. More precisely

Lemma 1.36 *Lie bracket defined in* (1.20) *is a bi-derivation of* $C^\infty(M)$.

Proof Since $\{\cdot, \cdot\}$ defined in (1.20) is skew-symmetric, it suffices to show that $\{fh, g\} = f\{h, g\} + h\{f, g\}$ for all $f, g, h \in C^\infty(M)$. This follows from the definition of the bracket. In fact

$$\{fh, g\} = \omega(X_{fh}, X_g) = f\omega(X_h, X_g) + h\omega(X_f, X_g) = f\{h, g\} + h\{f, g\}.$$

$\square$

Remark 1.37 Since for each $f \in C^\infty(M)$ $\{f, \cdot\}$ is a derivation of $C^\infty(M)$, this corresponds to a vector field on M, which, as one can easily check, coincides with the Hamiltonian vector field X_f. In fact, for all $g \in C^\infty(M)$, calling X the vector field corresponding to the derivation $\{f, \cdot\}$, one has

$$\langle dg, X \rangle = \{f, g\} = \omega(X_f, X_g) = -\omega(X_g, X_f) = \langle dg, X_f \rangle.$$

$\triangle$

Definition 1.38 (Poisson Bracket Defined by the Symplectic Form) The bracket (1.20) is called the *Poisson bracket* defined by the symplectic form ω. The pair $(C^\infty(M), \{\cdot, \cdot\})$ is called a *Poisson algebra*. $\triangle$

We will study more in depth the general theory of Poisson manifolds and of the corresponding Poisson algebras in Chap. 2. For the time being we would like to observe that since a Poisson bracket is a bi-derivation of $C^\infty(M)$, it is completely determined by the so-called *elementary brackets*, or *elementary Poisson brackets*, i.e., the values of the Poisson bracket on an arbitrary set of coordinate functions.

Remark 1.39 We would like to stress that the notion of a Poisson algebra is a genuinely algebraic concept that can be defined regardless of the existence of an underlying manifold. More precisely, if $\mathbb{K}$ is a field, a Poisson $\mathbb{K}$-algebra is a commutative and associative $\mathbb{K}$-algebra A endowed with a $\mathbb{K}$-bilinear bracket $\{\cdot, \cdot\} : A \times A \to A$ which is (i) skew-symmetric, (ii) a bi-derivation of the product of A, and (iii) satisfies the Jacobi identity. In spite of this possible more general perspective, hereafter a Poisson algebra will be always the algebra of functions of a manifold endowed with the additional structure of a Poisson bracket. $\triangle$

Example 1.40 (Standard Poisson Structure)   The Poisson bracket $\{\cdot,\cdot\}$ defined on $C^\infty(\mathbb{R}^{2n})$ by the standard symplectic form $\omega_{st} = \sum_{i=1}^n dp_i \wedge dx_i$ is called *the standard Poisson bracket*. A simple computation shows that

$$\{f, g\} = \sum_{i=1}^n \frac{\partial f}{\partial p_i} \frac{\partial g}{\partial x_i} - \frac{\partial g}{\partial p_i} \frac{\partial f}{\partial x_i}. \tag{1.24}$$

The elementary Poisson brackets assume a particularly nice form if expressed in terms of a set of standard coordinates (p, x), i.e.,

$$\{p_i, p_j\} = 0, \quad \{x_i, x_j\} = 0 \quad \text{and} \quad \{p_i, x_j\} = \delta_{ij}, \ \forall i, j = 1, \ldots, n,$$

as follows from (1.24). On the other hand, a simple application of the chain rule shows that given any set of local coordinates $(x_1, \ldots, x_{2n})$ on U open subset of M, the knowledge of the elementary Poisson brackets $\pi_{ij} = \{x_i, x_j\}$ for all i, j fixes uniquely the value $\{f, g\}$ for all smooth functions f, g defined on U. $\triangle$

Problem 1.41   Write an explicit formula for the Poisson bracket defined on $C^\infty(\mathcal{H})$ by the symplectic form $\omega = dx \wedge \frac{dy}{y^2}$; see Example 1.11. $\triangle$

Summarizing, one can say that to every symplectic manifold (M, ω) can be associated with four Lie algebras, $\mathcal{H}(M) \subset \mathcal{H}_l(M) \subset \mathfrak{X}(M)$ and $C^\infty(M)$. The first three are Lie algebras of vector fields, endowed with the corresponding canonical bracket between vector fields; the fourth one is the Lie algebra of smooth functions on M, whose Lie bracket is Poisson bracket defined by ω. Note that since $[\mathcal{H}_l(M), \mathcal{H}_l(M)] \subset \mathcal{H}(M)$, $\mathcal{H}(M)$ is Lie ideal of $\mathcal{H}_l(M)$. The relation between these Lie algebras of vector fields and $(C^\infty(M), \{\cdot, \cdot\})$ is enclosed in the following:

Corollary 1.42   *The map* $\Lambda : C^\infty(M) \longrightarrow \mathcal{H}(M)$ *defined in Formula* (1.17) *is a homomorphism of Lie algebras.*

In other words, one can say that on every symplectic manifold (M, ω) the following exact sequence of Lie algebras

$$0 \to \mathbb{R} \to C^\infty(M) \xrightarrow{\Lambda} \mathcal{H}(M) \to 0 \tag{1.25}$$

is defined. Of course, this exact sequence splits as a sequence of vector spaces, albeit *noncanonically*. More precisely, the splittings of (1.25) are in one-to-one correspondence with the points of M. In fact, given $m \in M$, the map $C^\infty(M) \to \mathcal{H}(M) \oplus \mathbb{R}$, $f \rightsquigarrow (X_f, f(m))$ is an isomorphism of vector spaces, whose inverse is given by the map $(X, a) \rightsquigarrow f - f(m) + a$, where f is *any* Hamiltonian of $X \in \mathcal{H}(M)$. However, (1.25) does *not split* as an exact sequence of Lie algebras; see item (ii) of Example C.80. More precisely, suppose that to $\mathbb{R}$ is given the structure of a *trivial* $\mathscr{H}(M)$-module, i.e., $X.a = 0$, for every $X \in \mathscr{H}(M)$ and $a \in \mathbb{R}$. After choosing $m \in M$, define $c_m : \mathscr{H}(M) \times \mathscr{H}(M) \to \mathbb{R}$ as

$$c_m(X, Y) = \omega(X, Y)(m), \forall X, Y \in \mathscr{H}(M). \tag{1.26}$$

Then

Proposition 1.43 *The bilinear form c_m defined in Formula (1.26) is two-cocycle of $\mathscr{H}(M)$ with values in $\mathbb{R}$. The resulting extension of $\mathscr{H}(M)$ by $\mathbb{R}$ is isomorphic, as a Lie algebra, to $(C^\infty(M), \{\cdot, \cdot\})$.*

Proof This is just a direct check. If f, g, h are functions on $C^\infty(M)$, we have

$$c_m([X_f, X_g], X_h) = \omega([X_f, X_g], X_h) = \{\{f, g\}, h\}$$

so that the cocycle condition (see Formula C.19) is equivalent to the Jacobi identity for the Poisson bracket. For the second claim, let $f, g \in C^\infty(M)$. From (1.25) one deduces that

$$[(X_f, f(m)), (X_g, g(m))] = \big([X_f, X_g], \omega(X_f, X_g)(m)\big) = \big(X_{\{f,g\}}, \{f, g\}(m)\big).$$

$\square$

As a consequence, the family of cocycles $\{c_m\}_{m \in M}$ define equivalent extensions of $\mathscr{H}(M)$ by $\mathbb{R}$. Extension (1.25) is said to be *central* because $\mathbb{R}$ maps into the center of $(C^\infty(M), \{\cdot, \cdot\})$.

Problem 1.44 Let m be a point of M and consider the vector space $\mathscr{H}(M)_m$ consisting of all the Hamiltonian vector fields which are zero at m.

(i) Show that $\mathscr{H}(M)_m$ is a Lie subalgebra of $\mathscr{H}(M)$.
(ii) Show that the cocycle c_m restricted to $\mathscr{H}(M)_m$ vanishes (so that the resulting central extension of $\mathscr{H}(M)_m$ by $\mathbb{R}$ splits). $\triangle$

Problem 1.45 On $(\mathbb{R}^2, \omega_{st})$, endowed with its standard Poisson structure $\{\cdot, \cdot\}$ defined in Formula (1.24), consider the quadratic functions $f_1 = -\frac{p^2}{2}$, $f_2 = -px$ and $f_3 = \frac{x^2}{2}$. Show that:

(i) $\{f_1, f_2, f_3\}$ generate a Lie subalgebra of $(C^\infty(\mathbb{R}^2), \{\cdot, \cdot\})$.
(ii) $\Lambda : C^\infty(\mathbb{R}^2) \to \mathfrak{X}(\mathbb{R}^2)$ maps isomorphically the Lie algebra generated by $\{f_1, f_2, f_3\}$ onto the one generated by $\{X_1, X_2, X_3\}$ defined in Problem 1.32.

See also Example 3.29. $\triangle$

3 From Symplectomorphisms to Hamilton-Jacobi Theory and Back

The goal of this section is to introduce the concepts of *symplectic morphism* and of *Lagrangian submanifold*. We will see how these two are related to each other and how they relate to the classical notion of *generating function* via the theory of Hamilton and Jacobi.

3.1 Some Tools from Differential Geometry

Before proceeding, we recall some essential tools from differential geometry.

3.1.1 Vector Fields and Diffeomorphisms

Vector fields on a manifold M are differential operators of order one acting on functions or, equivalently, derivations of $C^\infty(M)$. More precisely, every $X \in \mathfrak{X}(M)$ is the application from $C^\infty(M)$ to itself tacking f to Xf, where $(Xf)(m) = X(m)f = \langle df, X \rangle(m)$ is the *derivative* of f at the point $m \in M$ *along the direction* $X(m)$. Moreover, if $X, Y \in \mathfrak{X}(M)$, then

$$[X, Y](m)f = X(m)(Yf) - Y(m)(Xf). \tag{1.27}$$

Given a manifold M, its diffeomorphisms form a group denoted $\mathscr{D}(M)$. Such a group acts on M and on the full tensor algebra associated to it. In particular if $\varphi \in \mathscr{D}(M)$, then

$$(\varphi f)(m) = f(\varphi^{-1}(m)), \ \forall f \in C^\infty(M),$$

$$(\varphi X)(m) = \varphi_{*,\varphi^{-1}(m)} X(\varphi^{-1}(m)), \ \forall X \in \mathfrak{X}(M), \tag{1.28}$$

$$(\varphi \alpha)(m) = (\varphi^{-1})^*_{,m} \alpha(\varphi^{-1}(m)), \ \forall \alpha \in \Omega^1(M),$$

for all $m \in M$. The previous formulas define a *left action* of $\mathscr{D}(M)$ on $C^\infty(M)$, $\mathfrak{X}(M)$, and $\Omega^1(M)$, respectively.

If ξ is any tensor, we say that ξ is φ-invariant if $(\varphi\xi)(m) = \xi(m)$ for all $m \in M$ and we say that ξ is $\mathscr{D}(M)$-invariant if it is invariant for all $\varphi \in \mathscr{D}(M)$. In particular, X is φ-invariant if and only if

$$\varphi_{*,m} X(m) = X(\varphi(m)), \tag{1.29}$$

for all $m \in M$, i.e., if and only if X is φ-related to itself. Recall that if $f : M \to N$ is a smooth map and X, Y are two vector fields on M and N, respectively, one says that X is f-related to Y, $X \sim_f Y$, if for all $m \in M$,

$$f_{*,m} X(m) = Y(f(m)). \tag{1.30}$$

3.1.2 Integral Curves, Flows, and Lie Derivative

To every vector field X and to every point $m \in M$, one can associate a curve $\gamma_m^X :$ $(a(m), b(m)) \to M$, where $(a(m), b(m))$ is an open interval of $\mathbb{R}$ depending on the point m and containing 0. Such a curve is smooth, it satisfies $\gamma_m^X(0) = m$, and it is such that $X(\gamma_m^X(t)) = \dot{\gamma}_m^X(t)$ for all $t \in (a, b)$; in other words, the curve γ_m^X is the unique curve passing through the point $m \in M$ and such that at each of its points it has as a tangent vector the value of X at that point. The *maximal* among the curves having all such properties is called the *integral curve* of X passing through m. In what follows γ_m^X denotes the maximal integral curve of X passing through the point $m \in M$. The existence and unicity of such a curve follows from the existence and unicity theorem for the systems of ODE. In fact, γ_m^X is the solution of the following system of ordinary differential equations:

$$\begin{cases} \dfrac{dx_i(\gamma_m^X(t))}{dt} = X_i(x_1(t), \ldots, x_n(t)) \\[2mm] x_i(\gamma_m^X(0)) = x_i(m), \end{cases}$$

for all $i = 1, \ldots, n$. Let X be a vector field. For each $t \in \mathbb{R}$, let

$$\mathcal{D}_t^X = \{m \in M \,|\, t \in (a(m), b(m))\}$$

and let φ_t^X be the application defined by $\varphi_t^X(m) = \gamma_m^X(t)$, for all $m \in \mathcal{D}_t^X$. Then

(i) For each $m \in M$, there exists an open neighborhood U of m and an $\epsilon > 0$ such that the map $(t, m') \rightsquigarrow \varphi_t^X(m')$ is a smooth map from $(-\epsilon, \epsilon) \times U$ to M.
(ii) For each $t \in \mathbb{R}_{>0}$, $\mathcal{D}_t^X$ is open and $M = \bigcup_{t \geq 0} \mathcal{D}_t^X$.
(iii) $\varphi_t^X : \mathcal{D}_t^X \to \mathcal{D}_{-t}^X$ is a diffeomorphism with inverse φ_{-t}.
(iv) Given $s, t \in \mathbb{R}$, on the domain of $\varphi_t^X \circ \varphi_s^X$ we have $\varphi_t^X \circ \varphi_s^X = \varphi_{s+t}^X$.

For a proof of these statements see [244]. What we have just recalled can be summarized saying that $\{\varphi^X\}_{t \in \mathbb{R}}$ is a *one-parameter group of local diffeomorphisms*. The vector field X is called complete if $\mathcal{D}_t = M$ for all $t \in \mathbb{R}$. In this case $\{\varphi_t^X\}_{t \in \mathbb{R}}$ is a *one-parameter group of diffeomorphisms*. This can be rephrased saying that $\varphi : \mathbb{R} \times M \to M$ is a (left) action of the group $(\mathbb{R}, +)$ on M, or that $t \rightsquigarrow \varphi_t^X$ is a representation of $(\mathbb{R}, +)$ on the group of diffeomorphisms of M.

Let $X \in \mathscr{X}(M)$ and $\{\varphi_t^X\}_{t \in \mathbb{R}}$ be the corresponding one-parameter group of local diffeomorphisms. If ξ is any tensor field, we define its *Lie derivative* at the point

$m \in M$ with respect to X to be

$$(\mathscr{L}_X \xi)(m) = \left.\frac{d}{dt}\right|_{t=0} (\varphi_{-t}^X)_{*,\varphi_t^X(m)} \xi(\varphi_t^X(m)), \tag{1.31}$$

for all $m \in M$. In particular:

(i) $(\mathscr{L} f)(m) = \langle df, X \rangle(m)$, for all $f \in C^\infty(M)$
(ii) $(\mathscr{L}_X Y)(m) = [X, Y](m)$, for all $Y \in \mathscr{X}(M)$

for all $m \in M$. Here $[X, Y]$ is defined in (1.27). Finally, we want to record the following result. Let $X, Y \in \mathfrak{X}(M)$ and let $\{\varphi_t^X\}_{t \in \mathbb{R}}$, $\{\varphi_t^Y\}_{t \in \mathbb{R}}$ be the corresponding one-parameter groups of (local) diffeomorphisms.

Proposition 1.46 *Then:*

(i) *The two vector fields commute, i.e., $[X, Y] = 0$ if and only if $\varphi_t^X \circ \varphi_s^Y = \varphi_s^Y \circ \varphi_t^X$, for all $s, t \in \mathbb{R}$ such that the left- and the right-hand sides of the previous formula are defined.*
(ii) *Let $\{\varphi_t^{X+Y}\}_{t \in \mathbb{R}}$ be the one-parameter group of (local) diffeomorphisms defined by the vector field $X + Y$. Then*

$$\varphi_t^{X+Y} = \lim_{n \to \infty} (\varphi_{t/n}^X \circ \varphi_{t/n}^Y)^n$$

where the left-hand side is defined where the right-hand side is. In particular, if $[X, Y] = 0$,

$$\varphi_t^X \circ \varphi_t^Y = \varphi_t^{X+Y} = \varphi_t^Y \circ \varphi_t^X$$

for all $t \in \mathbb{R}$ where the composition(s) make(s) sense.

For the proof see, for example, [2]. We will apply these general remarks and comments to the study of a particular subgroup of the group of the diffeomorphisms of a symplectic manifold, which is introduced below.

3.2 The Group of Symplectomorphisms

We start introducing the notion of symplectic morphism between two symplectic manifolds (M_1, ω_1) and (M_2, ω_2).

Definition 1.47 (Symplectic Morphism) A smooth map $\varphi : M_1 \to M_2$ is called *symplectic*, or a *symplectic morphism*, if

$$\varphi^* \omega_2 = \omega_1,$$

i.e., if for all $m \in M$ and for all $u, v \in T_m M$

$$(\omega_1)_m(v, u) = (\varphi^* \omega_2)_m(v, u) = (\omega_2)_{\varphi(m)}(\varphi_{*,m} v, \varphi_{*,m} u). \tag{1.32}$$

$\triangle$

The following proposition follows at once from the previous definition.

Proposition 1.48 *Every symplectic morphism $\varphi : M_1 \to M_2$ is an immersion.*

Proof We need to show that $\varphi_{*,m} : T_m M_1 \to T_{\varphi(m)} M_2$ is injective for all $m \in M_1$. To this end, let $m \in M_1$ and $v \in T_m M_1$. If $\phi_{*,m}(v) = 0$,

$$0 = (\omega_2)_{\varphi(m)}(\varphi_{*,m}(v), \cdot) = (\varphi^* \omega_2)_m(v, \cdot) = (\omega_1)_m(v, \cdot),$$

which entails $v = 0$ since ω_1 is nondegenerate. $\qquad\square$

In particular if $\varphi : M_1 \to M_2$ is a symplectic morphism, then $\dim M_1 \leq \dim M_2$, and if $\dim M_1 = \dim M_2$, then $\phi : M_1 \to M_2$ is a *local diffeomorphism*. The symplectic morphisms on exact symplectic manifolds are characterized in the following:

Example 1.49 Let (M, ω) be an exact symplectic manifold; see Definition 1.14. If $\varphi : M \to M$ is a symplectic morphism, then $\omega = \varphi^* \omega = \varphi^*(d\theta) = d(\varphi^* \theta)$, implying that the one-form $\theta - \varphi^* \theta$ is closed. $\triangle$

Now suppose $M_1 = M = M_2$ and $\omega_1 = \omega = \omega_2$. Then

Definition 1.50 (Symplectomorphism) A *bijective* symplectic morphism $\varphi : M \to M$ is called a *symplectomorphism*. $\triangle$

The proof of the following statements is left as an exercise for the reader: (i) every symplectomorphism is a diffeomorphism, (ii) the inverse of every symplectomorphism is a symplectomorphism, and (iii) the composition of two symplectomorphisms is still a symplectomorphism. From these its follows that to every symplectic manifold (M, ω), one can associate its *group of symplectomorphisms*, which is denoted by $\mathrm{Symp}(M, \omega)$, or simply by $\mathrm{Symp}(M)$.

The group of symplectomorphisms of (M, ω) is a very complicated mathematical object; see [179, Chapter 10]. On the other hand, it is very common to be interested in smaller subgroups of $\mathrm{Symp}(M, \omega)$, in particular to the ones associated with a Lie group action on the underlying manifold. More precisely, if G is a Lie group acting smoothly M via $\varphi : G \times M \to M$, one says that

Definition 1.51 (Symplectic G-Action) φ is a symplectic action, or that G acts by symplectomorphisms on M, if $\varphi_g^* \omega = \omega$ for all $g \in G$, where $\varphi_g = \varphi(g, \cdot)$. $\triangle$

Note that in this case $\varphi : G \to \mathrm{Symp}(M, \omega)$ is a group homomorphism. We postpone a more in-depth discussion about this class of G-action until Chap. 3. For the time being we consider the following important case. The group $\mathrm{Symp}(\mathbb{R}^{2n}, \omega_{st})$ has a distinguished subgroup, the group of *linear symplectomorphisms* $\mathrm{Sp}_{2n}(\mathbb{R})$; see

[30, Section 1.2]. The relation between $\mathrm{Symp}(\mathbb{R}^{2n}, \omega_{st})$ and $\mathrm{Sp}_{2n}(\mathbb{R})$ is analyzed more in detail in the following.

Example 1.52 If $(M, \omega) = (\mathbb{R}^{2n}, \omega_{st})$, a diffeomorphism $\varphi : \mathbb{R}^{2n} \to \mathbb{R}^{2n}$ is a symplectomorphism if and only if its Jacobian (at each point) belongs $\mathrm{Sp}_{2n}(\mathbb{R})$. Writing $\varphi(p, q) = (y, x)$ this condition becomes

$$J = \mathrm{Jac}_\varphi^t(p, q) J \, \mathrm{Jac}_\varphi(p, q),$$

where

$$\mathrm{Jac}_\varphi(p, q) = \begin{pmatrix} \frac{\partial y}{\partial p} & \frac{\partial y}{\partial q} \\ \frac{\partial x}{\partial p} & \frac{\partial x}{\partial q} \end{pmatrix} (p, q).$$

$\mathrm{Jac}^t(p, q)$ is its transpose and

$$J = \begin{pmatrix} 0 & \mathrm{id} \\ -\mathrm{id} & 0 \end{pmatrix}.$$

In fact using (1.32), writing ω to denote ω_{st}, one has

$$\langle \omega_m^\flat(v), u \rangle = \langle \omega_{\varphi(m)}^\flat(\varphi_{*,m}v), \varphi_{*,m}u \rangle = \langle \varphi_{,m}^* \omega_{\varphi(m)}^\flat \varphi_{*,m}(v), u \rangle$$

(see 1.16), which amounts to say that

$$\omega_m^\flat = \varphi_{,m}^* \omega_{\varphi(m)}^\flat \varphi_{*,m}, \ \forall m.$$

$\triangle$

In the following example we introduce two classes of symplectomorphisms of (T^*N, Ω), playing an important role both in classical mechanics and in the theory of the classical integrable systems. For a more extensive discussion about this topic see Appendix D.

Example 1.53 (Cotangent Lifts and Translations Along the Fibers) If $\varphi \in \mathscr{D}(N)$, its cotangent lift $\widetilde{\varphi} : T^*N \to T^*N$, defined by $\widetilde{\varphi}(n, \alpha) = \left(\varphi(n), (\varphi^{-1})_{,\varphi(n)}^* \alpha \right)$ for all $n \in N$ and $\alpha \in T_n^*N$, is a symplectomorphism of (T^*N, Ω); see Lemma D.1. On the other hand, given any $\eta \in \Omega^1(N)$, the application $\tau_\eta : T^*N \to T^*N$, defined by

$$\tau_\eta(\alpha) = \alpha + \eta_n, \ \forall n \in N, \ \alpha \in T_n^*N,$$

is a diffeomorphism of T^*N, which belongs to $\mathrm{Symp}(T^*N, \Omega)$ *if and only if* $d\eta = 0$; see Lemma D.13. τ_η is called the *translation* along the fibers of T^*N defined by η.

$\triangle$

Remark 1.54 ($\mathcal{D}(M)$, $\mathcal{D}_{\mathbf{vol}}(M)$, **and** $\mathrm{Symp}(M)$) The previous discussion entails that $\mathrm{Symp}(M)$ is a subgroup of $\mathcal{D}(M)$, the group of diffeomorphisms of M. Moreover, since every symplectomorphism preserves the volume form of M (see item (ii) in Remark 1.12), $\mathrm{Symp}(M)$ is also naturally a subgroup of the group of the *volume-preserving diffeomorphisms* of M, $\mathcal{D}_{\mathrm{vol}}(M)$. In this way, to every symplectic manifold (M, ω) the following chain of groups is naturally associated: $\mathrm{Symp}(M) \subset \mathcal{D}_{\mathrm{vol}}(M) \subset \mathcal{D}(M)$. For a thorough discussion about the group of diffeomorphisms of a (closed) symplectic manifold we refer the reader to [179] and to [12] and [131] for a very nice and complete discussion about the applications of the theory of the groups of diffeomorphisms and of symplectomorphisms to hydrodynamics. △

Example 1.55 (Locally Hamiltonian Vector Fields vs Symplectomorphisms) Recall that a vector field X on a manifold is called complete if its integral curves $\gamma_m^X(t)$ are defined for all $t \in \mathbb{R}$. Let X be a *complete*, locally Hamiltonian vector field and let $\{\varphi_t^X\}_{t\in\mathbb{R}}$ be the corresponding one-parameter group of diffeomorphisms; see, for instance, [227, Section 8, Chapter II]. Then, for each $t \in \mathbb{R}$, φ_t^X is a symplectomorphism. In fact,

$$(\varphi_t^X)^*\omega - \omega = \int_0^t \frac{d}{ds}(\varphi_s^X)^*\omega\,ds = \int_0^s (\varphi_s^X)^*\mathcal{L}_X\omega\,ds = 0,$$

since $\mathcal{L}_X\omega = 0$. In other words, if X is locally Hamiltonian, $\{\varphi_t^X\}_{t\in\mathbb{R}}$ is a one-parameter group of symplectomorphisms. Note that if X is not complete, $\{\varphi_t^X\}_{t\in\mathbb{R}}$ is a one-parameter group of *local* symplectomorphisms. Indeed in this case, indicating with $(a(m), b(m))$ the maximal interval of existence of the integral curve of X starting at time $t = 0$ at m, $\mathcal{D}_t^X := \{m \in M | t \in a(m), b(m)\}$ is, in general, a *proper* open subset of M, where the restriction of ω defines a symplectic form with respect to which φ_t^X is a symplectomorphism. For example, on $(\mathbb{R}^2, \omega_{st})$ $X = \frac{\partial}{\partial p}$ is a complete Hamiltonian vector field, whose Hamiltonian function, modulo constants, is $H(p, x) = x$. The corresponding one-parameter group of canonical diffeomorphism is $\varphi_t^X(p, x) = (p, x + t)$, for all $t \in \mathbb{R}$ and all $(p, x) \in \mathbb{R}^2$. △

We start discussing the relation between symplectomorphisms and classical mechanics next.

Remark 1.56 (Symplectomorphisms and Classical Mechanics) Symplectomorphisms are one of the main characters in classical Hamilton-Jacobi theory and in classical mechanics, where they are usually dubbed *canonical transformations*. For the time being, we restrict ourselves to a few general comments, referring the reader to Sect. 3.5 for a short introduction to this beautiful topic and to the references [1, 10] for a complete account. The problem of integrating the equations

$$\dot{x}_i = \frac{\partial H}{\partial p_i} \quad \text{and} \quad \dot{p}_i = -\frac{\partial H}{\partial x_i}, \ \forall i = 1, \dots, n \tag{1.33}$$

describing the evolution of the Hamiltonian system $(\mathbb{R}^{2n}, \omega_{st}, H)$ can be rephrased as follows.

Find a symplectomorphism $\varphi : \mathbb{R}^{2n} \to \mathbb{R}^{2n}$, whose components are

$$a_i = a_i(p, x) \quad \text{and} \quad b_i = b_i(p, x), \ \forall i = 1, \ldots, n,$$

such that the new Hamiltonian function $\tilde{H}(b, a) = H\big(\varphi^{-1}(b, a)\big)$ does not depend, for example, on the coordinates $(b_1, \ldots, b_n)$.

The knowledge of such a symplectomorphism is sufficient to solve explicitly (1.33). In fact if $\tilde{H}(b, a) = \tilde{H}(a)$, these equations become

$$\dot{a}_i = 0 \quad \text{and} \quad \dot{b}_i = -\frac{\partial \tilde{H}}{\partial a_i}, \ \forall i = 1, \ldots, n,$$

which are easily integrated as

$$a_i(t) = a_i(0) \quad \text{and} \quad b_i(t) = -\frac{\partial \tilde{H}}{\partial a_i}(a_1(0), \ldots, a_n(0))t + b_i(0), \ \forall i = 1, \ldots, n,$$

$$(1.34)$$

which are the equations for the integral curves of the Hamiltonian vector field $X_{\tilde{H}}$. To find the solutions of the original system (1.33) it suffices to apply φ^{-1} to (1.34).

In other words, the problem of integrating system (1.19) is solved finding a suitable symplectomorphism of the phase space. This strategy is at the heart of the classical Hamilton-Jacobi theory and it will be further discussed in Sect. 3.5 below; see also Definitions 1.85 and 1.92 and Remark 1.86.

We close this preliminary discussion proving a result characterizing the symplectomorphisms in terms of their action on the Hamiltonian vector fields. More precisely, if (M, ω) and (N, Ω) are two symplectic manifolds of the same dimension, then

Proposition 1.57 *A diffeomorphism $\varphi : M \to N$ is a symplectomorphism, i.e., $\varphi^*\Omega = \omega$, if and only if*

$$\varphi X_H = X_{H \circ \varphi^{-1}},$$

for all $H \in C^\infty(M)$, where φX_H is defined in (1.28), i.e., via the following formula:

$$(\varphi X_H)(m) = \varphi_{*, \varphi^{-1}(m)} X_H(\varphi^{-1}(m)), \ \forall m \in M. \qquad (1.35)$$

Proof Let $n \in N$ and $v \in T_n N$a and suppose that φ is a symplectomorphism.

$$\big(i_{X_{H \circ \varphi^{-1}}} \Omega\big)_n(v) = \Omega_n(X_{H \circ \varphi^{-1}}(n), v)$$

$$= -\langle d(H \circ \varphi^{-1})_n, v \rangle$$

$$
= -\langle dH_{\varphi^{-1}(n)}, (\varphi^{-1})_{*,n} v \rangle
$$

$$
= \omega_{\varphi^{-1}(n)}\Big(X_H(\varphi^{-1}(n)), (\varphi^{-1})_{*,n} v \Big)
$$

$$
= \omega_{\varphi^{-1}(n)}\Big((\varphi^{-1})_{*,n}\varphi X_H(n), (\varphi^{-1})_{*,n} v \Big) \, see\,(1.35)
$$

$$
= \Omega_n(\varphi X_H(n), v), \ since \ \varphi^*\Omega = \omega
$$

$$
= (i_{\varphi X_H}\Omega)_n(v),
$$

which proves that $\varphi X_H = X_{H\circ\varphi^{-1}}$. Since for each $m \in M$

$$
T_m M = \mathrm{span}\langle X_H(m) \mid H \in C^\infty(M) \rangle,
$$

to prove the converse statement it suffices to show that $(\varphi^*\Omega)_m(X_H(m), \cdot) = \omega_m(X_H(m), \cdot)$, for all $m \in M$ and $H \in C^\infty(M)$. Choosing $m \in M$, $H \in C^\infty(M)$, and $v \in T_m M$

$$
(\varphi^*\Omega)_m(X_H(m), v) \overset{n=\varphi(m)}{=} \Omega_n(\varphi_{*,\varphi^{-1}(n)} X_H(\varphi^{-1}(n)), \varphi_{*,\varphi^{-1}(n)} v)
$$

$$
= \Omega_n\big((\varphi X_H)(n), \varphi_{*,\varphi^{-1}(n)} v\big), \ see\,(1.35)
$$

$$
= \Omega_n(X_{H\circ\varphi^{-1}}(n), \varphi_{*,\varphi^{-1}(n)} v), \ since \ \varphi X_H = X_{H\circ\varphi^{-1}}
$$

$$
= -\langle d(H \circ \varphi^{-1})_n, \varphi_{*,\varphi^{-1}(n)} v \rangle
$$

$$
= -\langle \varphi^*_{,\varphi^{-1}(n)} d(H \circ \varphi^{-1})_n, v \rangle
$$

$$
= -\langle dH_m, v \rangle
$$

$$
= \omega_m(X_H(n), v).
$$

$$\square$$

Note that the previous proposition can be stated equivalently saying that $\varphi : (M, \omega) \to (N, \Omega)$ is an anti-symplectomorphism, i.e., $\varphi^*\Omega = -\omega$ if and only if

$$
\varphi X_H = -X_{H\circ\varphi^{-1}}, \ \forall H \in C^\infty(M). \tag{1.36}
$$

$$\triangle$$

After all these comments about the class of symplectic maps called symplectomorphisms, we say that two symplectic manifolds (M_1, ω_1) and (M_2, ω_2) are *symplectomorphic* if there is a symplectomorphism $\phi : M_1 \to M_2$. In this case $\dim M_1 = \dim M_2$. On the other hand, two symplectic manifolds having the same dimension in general are not symplectomorphic. For example, a two-dimensional sphere cannot be symplectomorphic to $(\mathbb{R}^2, \omega_{st})$, since the former is compact, while the latter is not. In spite of this, in the next subsection we will prove an important result stating that, at least *locally*, any two symplectic manifolds having the same dimension are *symplectically indistinguishable*.

3.3 The Darboux Theorem

In this section we prove the so-called *Darboux theorem*, a very important result stating that on a neighborhood of every point of a symplectic manifold one can define coordinates which reduce the symplectic form to the standard one ω_{st}. The resulting atlas is called a *symplectic atlas*, while a neighborhood of a point where such a system of coordinates is defined is called a *Darboux neighborhood*. Let M be a manifold.

Definition 1.58 An atlas $\mathcal{S} = \{(U_i, \phi_i)\}_{i \in I}$ on M is called *symplectic*, or a *symplectic atlas*, if, for all U_i, U_j such that $U_i \cap U_j \neq \emptyset$,

$$\phi_i \circ \phi_j^{-1} : (\phi_j(U_j), \omega_{st}) \to (\phi_i(U_i), \omega_{st})$$

is a symplectomorphism. $\triangle$

Lemma 1.59 *A manifold M that admits a symplectic atlas is a symplectic manifold.*

Proof Let (U_i, ϕ_i) be a chart of $\mathcal{S}$ and define $\omega_i = \phi_i^*(\omega_{st}) \in \Omega^2(U_i)$. This two-form is closed, as $d\omega_i = d\phi_i^*(\omega_{st}) = \phi_i^*(d\omega_{st}) = 0$, and it is nondegenerate. So it defines a symplectic form on the open set U_i. If (U_j, ϕ_j) is another chart in $\mathcal{S}$, such that $U_i \cap U_j \neq \emptyset$, then on the intersection $U_i \cap U_j$, we have

$$\omega_j = \phi_j^*(\omega_{st}) = \phi_i^* \circ (\phi_i^{-1})^* \circ \phi_j^*(\omega_{st}) = \phi_i^* \circ (\phi_j \circ \phi^{-1})^*(\omega_{st}) = \omega_i.$$

So a closed and nondegenerate two-form is defined on the entire manifold M. $\square$

Conversely, we have the following important result.

Theorem 1.60 (Darboux Theorem) *Every symplectic manifold has a symplectic atlas.*

Remark 1.61 A remarkable consequence of the Darboux theorem is that for every point p of a $2n$-dimensional symplectic manifold (M, ω), there exists an open neighborhood U of p and local coordinates (p, x) on U, such that $\omega_{|U} = \sum_{i=1}^n dp_i \wedge dx_i$. In other words, every $2n$-dimensional symplectic manifold is *locally symplectomorphic* to $(\mathbb{R}^{2n}, \omega_{st})$. Thus, symplectic manifolds are, from the viewpoint of their local behavior, very different from the Riemannian ones. In fact, a Riemannian metric can be cast in *standard form*, i.e., described by the identity matrix, at every point of the manifold, but, in general, it cannot be put in standard form in any neighborhood of any point of the manifold where the curvature of the metric is different from zero. On the contrary, the Darboux theorem states that on a suitable neighborhood of each point of a symplectic manifold one can define coordinates that reduce the symplectic form to the standard one. In other words *symplectic manifolds have no local invariants.* $\triangle$

The proof of the Darboux theorem that we are going to present is based on the following preliminary result.

Lemma 1.62 *Let ω_1 and ω_0 be two symplectic forms on M and let $Q \subset M$ be a closed and embedded submanifold such that*

$$\omega_1|_Q = \omega_0|_Q.$$

Then there exists a neighborhood U_0 of Q and a diffeomorphism $\Psi : M \to M$ with the following properties:

$$\Psi|_Q = \mathrm{id}_Q \quad and \quad \Psi^*\omega_1 = \omega_0.$$

In our proof of Lemma 1.62 the following two results will be used.

Lemma 1.63 *Let ω_t be a one-parameter family of symplectic forms, with $t \in I = [0, 1]$. The existence of a family of diffeomorphisms $\{\psi_t\}_{t\in I}$ such that $\psi_t^*\omega_t = \omega_0$ for each $t \in I$ is equivalent to the existence of a one-parameter family of one-forms $\{\sigma_t\}_{t\in I}$ with the property*

$$\frac{d\omega_t}{dt} = -d\sigma_t, \ \forall t \in I. \tag{1.37}$$

Proof Let $\{\psi_t\}_{t\in I}$ be a family of diffeomorphisms such that $\psi_t^*\omega_t = \omega_0$ for each $t \in I$. Then

$$0 = \frac{d(\psi_t^*\omega_t)}{dt} = \psi_t^*\left[\frac{d\omega_t}{dt} + d(\iota_{X_t}\omega_t) + \iota_{X_t}(d\omega_t)\right];$$

see, for example, [114, p. 158–160] for a proof of the second equality. Since ω_t is closed and ψ_t is a diffeomorphism for each t, one has

$$\frac{d\omega_t}{dt} = -d(\iota_{X_t}\omega_t),$$

which proves one implication if we define $\sigma_t = \iota_{X_t}\omega_t$. Conversely, let us suppose that the condition (1.37) holds true. Since ω_t is nondegenerate for each t, one can define X_t as the only vector field such that $\iota_{X_t}\omega_t = \sigma_t$. We can now define the family $\{\psi_t\}_{t\in I}$ as the one-parameter family of diffeomorphism generated by $\{X_t\}_{t\in I}$. Then

$$\frac{d(\psi_t^*\omega_t)}{dt} = \psi_t^*\left[\frac{d\omega_t}{dt} + d(\iota_{X_t}\omega_t)\right].$$

The right-hand side of this equality is zero by the definition of σ_t, which implies that $\psi_t^* \omega_t$ does not depend on t, i.e., $\psi_t^* \omega_t = \omega_0$. $\qquad\square$

Lemma 1.64 (Relative Poincaré Lemma) *Let Q be an embedded submanifold of M. Suppose that η is a p-form on M which is closed on an open neighborhood of Q and identically zero on Q. There exists an open neighborhood U_0 of Q and a $(p-1)$-form θ defined on U_0 such that $\eta|_{U_0} = d\theta$.*

Proof By the tubular neighborhood theorem (see, e.g., [226, p. 465]) we can find an open neighborhood U_0 of Q and a retraction of U_0 onto Q given by a one-parameter family of smooth maps $\{\psi_t\}_{t \in I}$, such that ψ_t is a diffeomorphism for $t > 0$, $\psi_1 = \mathrm{id}_{U_0}$, $\psi_0 : U_0 \longrightarrow Q$ and $\psi_t|_Q = \mathrm{id}_Q$, for all $t \in [0, 1]$. Then one can write

$$\psi_1^*(\eta) - \psi_0^*(\eta) = \int_0^1 \frac{d(\psi_t^* \eta)}{dt} = \int_0^1 \psi_t^* d(\iota_{X_t} \eta) dt + \int_0^1 \psi_t^* (\iota_{X_t} d\eta) dt = d\theta,$$

$$\tag{1.38}$$

where $\theta = \int_0^1 \psi_t{}^*(\iota_{X_t} \eta) dt$. Since η vanishes on Q, one has $\eta|_{U_0} = d\theta$. $\qquad\square$

Remark 1.65 (Poincaré Lemma) If Q is a point $m \in M$ and U_0 a small (open) ball around m, the previous result reduces to the usual Poincaré lemma; see also Remark B.21. $\qquad\triangle$

Proof of Lemma 1.62 Let $\omega_t = (1 - t)\omega_0 + t\omega_1 = \omega_0 + t\alpha$ be a one-parameter family of closed two-forms, where $\alpha = \omega_1 - \omega_0$. From the hypothesis, $\alpha|_Q \equiv 0$ and $d\alpha = 0$. Moreover, if $\{\psi_t\}_{t \in I}$ is the retraction of the tubular neighborhood of Q onto Q itself as in Lemma 1.64, we have $\psi_0^* \alpha = 0$. The form α fulfills the hypothesis of Lemma 1.64, so that one can find a neighborhood of Q and a one-form $\beta = \int_0^1 \psi_t^*(\iota_{X_t} \alpha) dt$, such that $\alpha = d\beta$. Moreover, from the proof of Lemma 1.64 one can deduce that the restriction of β to Q is identically zero. Since $\omega_t|_Q = (\omega_0 + t\alpha)|_Q = \omega_i|_Q, i = 0, 1$, then there exists an open neighborhood of Q where ω_t is nondegenerate (nondegeneracy is an open condition and Q is a closed set). Now let us take an open neighborhood of Q where the one-form β is defined and ω_t is nondegenerate for each $t \in [0, 1]$. Define $\{X_t\}_{t \in I}$ via the condition

$$\beta = -\iota_{X_t} \omega_t \tag{1.39}$$

and let $\{\varphi_t\}_{t \in I}$ be the family of diffeomorphism defined by $\{X_t\}_{t \in I}$. Then

$$\frac{d(\varphi_t^* \omega_t)}{dt} = \psi_t^* \left[\frac{d\omega_t}{dt} + \mathscr{L}_{X_t} \omega_t \right] + \psi_t^* [\alpha + d(\iota_{X_t} \omega_t)] = 0.$$

Then $\varphi_t^* \omega_t$ does not depend on t, and since $\varphi_0 = id$,

$$\varphi_1^* \omega_1 = \varphi_0^* \omega_0 = \omega_0.$$

The first part of the statement follows defining $\Psi = \psi_1$. To conclude it suffices to note that since $\beta|_Q \equiv 0$, $X_t|_Q \equiv 0$ for each $t \in [0, 1]$, one has $\varphi_t(Q) = Q$ for each $t \in [0, 1]$. $\qquad\square$

Finally, we can give the long promised proof of Darboux theorem.

Proof of Darboux Theorem Let (M, ω) be a symplectic manifold and $Q = \{m\}$ be the embedded submanifold of Lemma 1.62, where m is a point of M. The restriction of ω to m defines a (linear) symplectic form on the tangent space $T_m M$. Without loss of generality, one can assume that $\omega_m = \sum_{i=1}^n dp_i \wedge dx_i$, since in every symplectic vector space, we can always find a symplectic basis. Now fix a Riemannian metric on M and consider the exponential map associated to such a metric. This is a diffeomorphism (see [227]):

$$\rho : V_0 \longrightarrow U_m,$$

such that $\rho(0) = m$, and V_0 and U_m are suitable neighborhoods of the origin in $T_m M$ and of the point $m \in M$, respectively. The chosen symplectic basis in $T_m M$ defines the exponential coordinates on U_m. With respect to such coordinates, the symplectic form $\omega_1 = (\rho^{-1})^*(\omega_m)$ is written in canonical form. Moreover, the symplectic form $\omega_0 = \omega$ has the following property: $\omega_0|_m = \omega_1|_m$. Then, all the hypotheses of Lemma 1.62 are satisfied. So we can find an open neighborhood U of the point $m \in M$ and a diffeomorphism $\Psi : U \longrightarrow U$ such that $\Psi^*\omega_1 = \omega_0$ and $\Psi(m) = m$. $\qquad\square$

3.4 Special Submanifolds and Generating Functions

The main goal of this subsection is to discuss the interplay between Lagrangian submanifolds and the classical theory of generating functions. Since our presentation is aimed at introducing the Hamilton-Jacobi theory, we start discussing more in general the *sub-geometry* of a symplectic manifold. We shall say that (N, Ω) is a symplectic submanifold of (M, ω) if:

(i) (N, i) is an immersed submanifold of M.
(ii) $\Omega = i^*\omega$.

In general a submanifold of M is *not* a symplectic submanifold of (M, ω), and it may very well not to be a symplectic manifold at all (for example, if its dimension is odd). This is one of the basic differences between symplectic and Riemannian geometry, since every submanifold of a Riemannian manifold is a Riemannian submanifold of the ambient manifold.

Problem 1.66 Construct a nontrivial even-dimensional linear subspace of $(\mathbb{R}^{2n}, \omega_{st})$ which is not symplectic. $\qquad\triangle$

Problem 1.67 Prove that (N, i) is a symplectic submanifold of (M, ω) if and only if for all $m \in N$, $T_m N$ is a symplectic vector subspace of $(T_m M, \omega_m)$. For the definition of symplectic subspace see Appendix A and also [30, Chapter 1]. △

In spite of the previous observation, symplectic manifolds possess a very rich *sub-geometry*, i.e., every symplectic manifold admits several classes of *distinguished submanifolds*, as the following definition shows. If (M, ω) be a symplectic manifold, then

Definition 1.68 An immersed submanifold $N \subset M$ is called *coisotropic* (*isotropic*) if, for each $n \in N$, $T_n N$ is a *coisotropic* (*isotropic*) vector subspace of $T_n M$; see Appendix A and [30, Chapter 1]. A coisotropic (isotropic) submanifold of M of *minimal* (*maximal*) dimension is called a *Lagrangian* submanifold. △

Problem 1.69 Consider the symplectic vector space $(\mathbb{R}^{2n}, \omega_{st})$. Let $n \geq 4$. Construct an explicit coisotropic subspace of codimension greater than one. Construct an explicit isotropic subspace of dimension greater than one. Construct an explicit Lagrangian subspace. Are there subspaces which are not Lagrangian, isotropic, coisotropic, nor symplectic? △

Example 1.70 Every smooth hypersurface $X \subset M$ is a coisotropic submanifold of M. Every smooth one-dimensional submanifold (curve) $\gamma \subset M$ is an isotropic submanifold of M. △

Before analyzing in some more detail the geometry of coisotropic submanifolds, we present a result concerning quotient of coisotropic subspaces, which, despite being elementary, is the prototypical mechanism in the various incarnations of a reduction procedure which will play an important role in many parts of these notes. In particular, the following algebraic result will be used to prove the so-called Marsden-Weinstein-Meyer reduction; see Theorem 3.81.

Theorem 1.71 (Linear Symplectic Reduction) *Let $U \subset V$ be a coisotropic vector subspace and U^ω its symplectic orthogonal; see Definition A.6. Then the vector space $V' = U/U^\omega$ is canonically a symplectic vector space.*

Proof Since U is coisotropic, U^ω is a subspace of U. Let $j : U \hookrightarrow V$ be the inclusion and denote by $\pi : U \longrightarrow U' = U/U^\omega$ the canonical projection and by $[v]$ the class of the element $v \in U$. Then, define $\omega' : U' \times U' \longrightarrow \mathbb{R}$ via the following condition:

$$\omega'([v], [u]) = \omega(u, v) \quad \text{for all} \quad u, v \in U.$$

Note that ω' is well defined. In fact, if $u \sim u'$ and $v \sim v'$, then $\omega'([v'], [u']) = \omega(v', u') = \omega(v + t, u + z)$ for some $t, z \in U^\omega$. Since $v, u \in U$, we have $\omega(v + t, u + z) = \omega(v, u) = \omega'([v], [u])$. The form ω' is skew-symmetric and bilinear. Let us prove that ω is nondegenerate. Suppose there exists $[v] \in U'$ such that $\omega'([v], [u]) = 0$ for all $[u] \in U'$. Then $\omega(v, u) = 0$ for all $u \in U$, implying that $v \in U^\omega$, i.e., $[v] = 0$. □

We divide what remains in this subsection in three parts: in the first and in the second we collect a few properties of the coisotropic and of the Lagrangian submanifolds, respectively, while the last one is devoted to an introduction to the classical theory of the generating functions.

3.4.1 Coisotropic Submanifolds

Let $C \subset M$ be a coisotropic submanifold of codimension k and let $\omega^C = i^*\omega$, where $i : C \to M$ is the canonical injection. In this case, for all $m \in C$, $T_m C^\omega = \ker \omega_m^C$ is a k-dimensional linear subspace of $T_m C$.

Problem 1.72 Prove the previous claim. (Hint: See the proof of Proposition 1.74.)

$\triangle$

In particular the application that to every point $m \in C$ associates $T_m C^\omega$ defines a regular distribution of dimension k, called the *characteristic distribution* of the coisotropic submanifold C. Moreover, an element $X \in \mathfrak{X}(M)$, tangent to C, is called a *characteristic vector field* for C if $X_m \in T_m C^\omega$ for all $m \in C$. As a warm-up for the proof of the following proposition, one can prove that the Hamiltonian vector field X_H is characteristic for C *if and only if*, for all $m \in C$, there exists an open neighborhood $U \subset M$ of m such that H is constant on $C \cap U$.

Problem 1.73 Prove the previous statement. (Hint: Recall that $dH = -i_{X_H}\omega$ and that, for all m, $TC_m^\omega = \{v \in T_m C \mid i_v \omega_m^C = 0\}$.) $\triangle$

Proposition 1.74 *The characteristic distribution is completely integrable; see Definition B.4.*

Proof First, observe that the characteristic distribution of C is smooth, since it is locally generated by vector fields. In fact, if C has codimension k, then for all $m \in C$ one can find an open neighborhood U of m in M and k independent functions $f_1, \ldots, f_k \in C_M^\infty(U)$ such that $C \cap U = \{m' \in U \mid f_i(m') = 0\}$, whose corresponding Hamiltonian vector fields X_{f_j}s generate the characteristic distribution on $C \cap U$. To conclude the proof it suffices to observe that since $d\omega^C = 0$, if X, Y vector fields such that $i_X \omega^C = 0 = i_Y \omega^C$, then $i_{[X,Y]}\omega^C = 0$. Now the proof follows from the Frobenius theorem; see Theorem B.5. $\square$

Remark 1.75 Note that if functions $f_1, \ldots, f_k$ are as in the proof of the previous proposition, then $\{f_i, f_j\}_{|C} = 0$ for all i, j, where the Poisson bracket is the one defined by the ambient symplectic structure. In fact, recalling that $\omega^C = i^*\omega$ where $i : C \to M$ is the standard injection, one has that

$$(\omega^C([X_{f_i}, X_{f_j}], \cdot))_m = \omega_{i(m)}(i_{*,m}([X_{f_i}, X_{f_j}](m)), \cdot)$$

$$\overset{i(m)=m}{=} \omega_m(i_{*,m} X_{\{f_i, f_j\}}(m), \cdot).$$

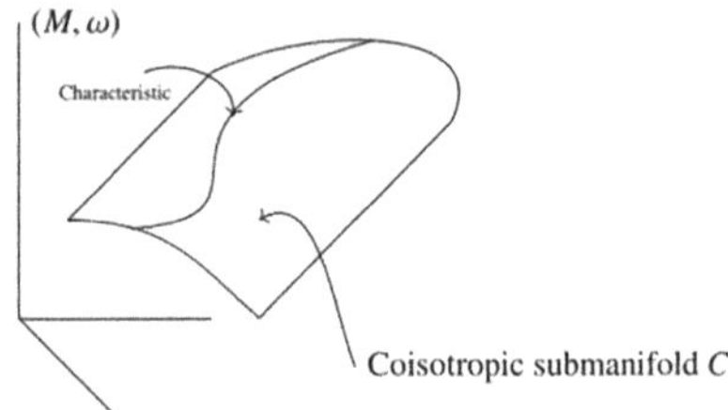

Fig. 1.3 A coisotropic submanifold C and one of its characteristics

Since $\omega^C([X_{f_i}, X_{f_j}], \cdot))_m = 0$ for all $m \in C$ and ω is nondegenerate, the hypothesis that i is an immersion entails that the restriction to C of $\{f_i, f_j\}$ is identically zero.

$\triangle$

On a coisotropic submanifold C the integral submanifolds defined by the characteristic distribution are called the *characteristic submanifolds* or, simply, the *characteristics* of C. The following statement complements the one before Proposition 1.74; see Problems 1.72 and 1.73 (Fig. 1.3).

Problem 1.76 Show that a Hamiltonian vector field X_H is tangent to the coisotropic submanifold C *if and only if* H is constant on the characteristics of C. (Hint: Using the notations introduced in the proof of Proposition 1.74, note that X_H is tangent to C if and only if $\langle df_j, X_H \rangle = 0$ for all $j = 1, \ldots, k$, which is equivalent to the condition $\omega(X_H, X_{f_j}) = \langle dH, X_{f_j} \rangle = 0$ for all $j = 1, \ldots, k$. Now it suffices to observe that the vector fields X_{f_j} generate, locally, the characteristic distribution.)

$\triangle$

A first consequence of Proposition 1.74 is enclosed in the following:

Corollary 1.77 *Every coisotropic submanifold C of codimension k is foliated by k-dimensional isotropic submanifolds.*

Proof If $m \in C$ and S is the characteristic of C passing through m, one has $T_m S = T_m C^\omega \subset T_m C \subset T_m M$ implying that $T_m S$ is an isotropic subspace of $T_m M$. In fact, since $T_m S \subset T_m C$, $T_m C^\omega \subset T_m S^\omega$, but $T_m C^\omega = T_m S$. $\square$

Example 1.78 Let (M, ω) be a symplectic manifold and let $H \in C^\infty(M)$. Then if $h \in \mathbb{R}$ is a regular value of H, $W_h = H^{-1}(h)$ is a smooth (embedded) hypersurface of M, i.e., it is a coisotropic submanifold of (M, ω) of maximal dimension. The Hamiltonian vector field X_H is tangent to W_h, and for each $m \in W_h$, the characteristic of W_h going through this point is the integral curve of X_H passing through m.

$\triangle$

The last property of coisotropic submanifolds we want to discuss is the so-called *absorption principle*, [28, 243], which, roughly speaking, states that the Lagrangian submanifolds *contained* in a coisotropic one *absorb* its characteristics.

More precisely, let $C \subset M$ be a coisotropic submanifold of (M, ω) and let L be a Lagrangian submanifold of (M, ω) immersed in C. Then

Proposition 1.79 (Absorption Principle) *For each $m \in C$ such that $m \in L$, if S is the characteristic of C passing through m, then $T_m S \subset T_m L$. In particular every Lagrangian submanifold of (M, ω) which is a submanifold of C is a union of characteristics.*

Proof The first statement follows noticing that since $T_m L \subset T_m C$, $T_m S = T_m C^\omega \subset T_m L^\omega = T_m L$. The second statement follows from the first one. $\square$

Example 1.80 The absorption principle applied to Example 1.78 yields the following conclusion: the Hamiltonian vector field X_H is tangent to every Lagrangian submanifold of (M, ω) contained in W_h. In particular, if an integral curve of X_H hits L at a given $t = t_0$, it is trapped in L for all t in its definition domain. $\triangle$

Remark 1.81 The absorption principle plays an important role in the *coordinate-free* formulation of Hamilton-Jacobi theory; see the above cited references [28, 243] and Sect. 3.5 below. $\triangle$

3.4.2 Lagrangian Submanifolds

We now move on to discuss a few properties shared by the Lagrangian submanifolds. The first one, which is elementary, is the content of the following.

Lemma 1.82 *Every Lagrangian submanifold N of (M, ω) has dimension equal to $\frac{1}{2} \dim M$.*

Proof Since N is a Lagrangian submanifold of M, for every point $n \in N$ the tangent space $T_n N$ is a Lagrangian subspace of $T_n M$. Then the lemma follows from the fact that in a symplectic vector space of dimension $2n$, an isotropic subspace has dimension $k \leq n$, while a coisotropic subspace has dimension $k \geq n$. $\square$

Example 1.83

(i) Let V be a vector space, let V^* be its dual, and let $M = V \oplus V^*$ with the symplectic structure ω defined in Example 1.5. Then, $V, V^* \subset M$ are Lagrangian submanifolds of (M, ω).

(ii) Both the fibers of $\pi : T^*N \to N$ and the zero section of π are Lagrangian submanifolds of (T^*N, Ω).

$\triangle$

A more interesting result states that

Proposition 1.84 *Every Lagrangian submanifold of (T^*N, Ω), transversal to the canonical projection, is the image of a closed one-form $\alpha \in \Omega^1(N)$.*

Proof Since every submanifold of M transversal to π is the image of some $\alpha \in \Omega^1(N)$ and since $\alpha^*(\Theta) = \alpha$, where Θ is the Liouville one-form (see

Proposition 1.18), one can conclude that the restriction of Θ to $\alpha(N)$ coincides with α. Then

$$d\alpha = d\alpha^*(\Theta) = \alpha^*(\omega) = \omega_{|\operatorname{im}(\alpha)}.$$

So $d\alpha = 0$ if and only if $\alpha(N)$ is Lagrangian. $\square$

Applying the Poincaré lemma to Proposition 1.84, one concludes that if $L \subset (T^*N, \Omega)$ is a Lagrangian submanifold *transversal* to the canonical projection π, for each $p \in N$ there exists an open neighborhood $U \subset N$ of p and a function $S \in C_N^\infty(U)$ such that $dS(U) = L|_U$. Clearly, S is unique up to an arbitrary constant and, over U, L is described by the equations $p_i = \frac{\partial S}{\partial x_i}$, $i = 1, \ldots, n$.

Definition 1.85 (Local Generating Function for L) The function S so defined is called a *local generating function* for the Lagrangian submanifold L. Assuming U connected, S is defined uniquely only up to a constant, compare with Definition 1.92 below. $\triangle$

Remark 1.86 (Exact Lagrangian Submanifold) Note that every $f \in C^\infty(N)$ defines a Lagrangian submanifold of T^*N transversal to π. Transversal Lagrangian submanifolds defined as the image of exact one-forms are called *exact*. Note that (local) generating functions and the corresponding exact Lagrangian submanifolds play a crucial role in the *Hamilton-Jacobi* theory; see Sect. 3.5. $\triangle$

Example 1.87 Let us consider a couple of examples:

(i) Every $\alpha \in \Omega^1(\mathbb{R})$ defines an *exact* Lagrangian submanifold of $(\mathbb{R}^2, \omega_{st})$. More precisely, the Lagrangian submanifold L defined by $\alpha = f\,dx$ corresponds to the graph of f and every primitive of f defines a generating function for L.
(ii) On $N = \mathbb{R}^2 \backslash \{(0, 0)\}$ the form $\alpha = \frac{x\,dy - y\,dx}{x^2 + y^2}$ is closed but *not* exact and for this reason the corresponding Lagrangian submanifold of (T^*N, Ω) is not exact. On the other hand, on every simply connected subset of N, any primitive of α is local generating function for the corresponding Lagrangian submanifold. For example, on $\mathbb{R}^2 \backslash \{(x, y) \mid y = 0, \ x \geq 0\}$ one can choose

$$S(x, y) = \begin{cases} \arctan\left(\frac{y}{x}\right) - \pi & \text{if } x > 0, \ y > 0, \\[4pt] \arctan\left(\frac{y}{x}\right) & \text{if } x < 0, \\[4pt] \arctan\left(\frac{y}{x}\right) + \pi & \text{if } x > 0, \ y < 0, \\[4pt] \qquad -\frac{\pi}{2} & \text{if } y > 0, \ x = 0, \\[4pt] \qquad \frac{\pi}{2} & \text{if } y < 0, \ x = 0. \end{cases}$$

$$\triangle$$

Another important class of Lagrangian submanifolds is described in the following.

Example 1.88 (Conormal Bundles) Let $P \subset N$ be a submanifold of N. The *conormal bundle* of P is

$$CP = \{(x, \alpha) \mid x \in P, \ \alpha \in T_x^*N \text{ s.t. } \langle \alpha, v \rangle = 0 \ \forall \, v \in T_xP\}.$$

In other words, CP is the vector bundle on P defined as the kernel of the application $(T^*N)_{|P} \to T^*P$ defined by the restriction to TP of the one-forms on N. Another way to look at CP is to consider the exact sequence defining the normal bundle to P in N:

$$0 \to TP \to TN_{|P} \to NP \to 0.$$

Dualizing it one gets

$$0 \to N^*P \to T^*N_{|P} \to T^*P \to 0,$$

from which CP is identified with N^*P.

CP is a Lagrangian submanifold of T^*N since

$$\langle \Theta_\alpha, w \rangle = \langle \alpha, \pi_{*,\alpha}(w) \rangle = 0,$$

for all $w \in T_\alpha CP$, as it follows from the definitions of the canonical one-form Θ and of CP. Thus, the restriction of the canonical symplectic form $\Omega = d\Theta$ to CP vanishes. To prove the claim it suffices to show that $\dim CP = \dim N$; see Problem 1.89. $\triangle$

Problem 1.89 Prove that CP is a vector subbundle of $(T^*N)_{|P}$ of rank equal to $\dim N - \dim P$ and conclude that $\dim CP = \dim N$. $\triangle$

Problem 1.90 Let N, P be as above. Describe the conormal bundle of P if (i) $P = \{n\}$, i.e n is a point of N and (ii) if $P = N$. $\triangle$

As we promised, we discuss now the relation between symplectomorphisms and Lagrangian submanifolds. An important and useful characterization of symplectic morphisms is the content of the next proposition. Let (M_1, ω_1) and (M_2, ω_2) be two symplectic manifolds and let (M, ω) be the product symplectic manifold; see Example 1.7. Then

Proposition 1.91 *A map $\varphi : M_1 \to M_2$ is symplectic if and only if its graph*

$$\Gamma_\varphi = \{(x, y) \mid y = \varphi(x)\} \subset M$$

is a Lagrangian submanifold of (M, ω).

Proof First, note that if $(m_1, m_2) \in \Gamma_\varphi$ and $U \in T_{(m_1,m_2)}\Gamma_\varphi$, then $U = (u_1, u_2)$, where $v_1 \in T_{m_1}M_1$, $v_2 \in T_{m_2}M_2$, $m_2 = \varphi(m_1)$ and $v_2 = \varphi_{*,m_1}(v_1)$. Furthermore, note that Γ_φ is Lagrangian if and only if $i_\varphi^*\omega = 0$ where $i_\varphi : \Gamma_\varphi \to M_1 \times M_2$ is the canonical immersion. Then

$$(i_\varphi^* \omega)_{(m_1, m_2)}(U, V) = \omega_{(m_1, \varphi(m_1))}(U, V)$$

$$\overset{(1.1)}{=} (\omega_1)_{m_1}(u_1, v_1) - (\omega_2)_{m_2}(u_2, v_2)$$

$$= (\omega_1)_{m_1}(u_1, v_1) - (\omega_2)_{\varphi(m_1)}(\varphi_{*,m_1} u_1, \varphi_{*,m_1} v_1)$$

$$= (\omega_1)_{m_1}(u_1, v_1) - (\varphi^* \omega_2)_{m_1}(u_1, v_1),$$

i.e.,

$$i_\varphi^* \omega = \omega_1 - \varphi^* \omega_2, \tag{1.40}$$

which proves the statement. $\square$

We now discuss how Lagrangian submanifolds enter in the classical theory of generating functions.

3.4.3 Generating Functions of Symplectic Morphisms

Beside its intrinsic interest, the previous result plays a fundamental role in the theory of Hamilton-Jacobi. In view of these applications and to simplify the exposition, hereafter, without loss of generality, we use Proposition 1.91 under the assumptions that $M_1 = M_2 = \tilde{M}$ and $\omega_1 = \omega_2 = d\theta$. This is because most of the issues involved in Hamilton-Jacobi theory and, more generally, in the theory of generating functions are of a local nature. Now keeping the notations introduced in the proof of the proposition above, Formula (1.40) entails that φ is symplectic if and only if

$$i_\varphi^* \omega = d(\theta - \varphi^* \theta) = 0.$$

In other words, $\varphi : (\tilde{M}, d\theta) \to (\tilde{M}, d\theta)$ is symplectic *if and only if* each $p \in M = \tilde{M} \times \tilde{M}$ has a neighborhood U where it is defined as a smooth function S such that

$$(\theta - \varphi^* \theta)|_U = dS.$$

Definition 1.92 (Generating Function for φ) The function S so defined is called a (local) *generating function* for φ; compare with Definition 1.85 above. $\triangle$

We apply the previous comments to the generating functions of the symplecto-morphisms of the cotangent bundle of a given manifold N. Since the existence of these functions is a local matter, we can restrict ourselves to the case $N = \mathbb{R}^n$. Let $\varphi : (\mathbb{R}^{2n}, (p, q)) \to (\mathbb{R}^{2n}, (y, x))$ be a diffeomorphism which we write in coordinates as

$$(y, x) = \varphi(p, q) = (G(p, q), F(p, q)).$$

Since $\omega_{st} = d(\sum_{i=1}^{n} p_i dq_i)$, and $\mathbb{R}^{2n}$ is simply connected, if φ is a symplectomorphism, then

$$dS = \sum_{i=1}^{n} p_i dq_i - \varphi^*\left(\sum_{i=1}^{n} y_i dx_i\right),$$

for some smooth function $S = S(x, q)$. The first problem we want to address is how to get S from φ. To this end one can try to invert the relation $x = F(p, q)$, find $p = f(x, q)$, and insert the latter in $y = G(p, q)$. One can achieve the first step in the neighborhood of every point (p_0, q_0) where $\det\left(\frac{\partial x}{\partial p}(p_0, q_0)\right) \neq 0$. Where this condition is verified, one gets

$$p_i = f_i(x, q) \quad \text{and} \quad y_i = g_i(x, q) = G_i(f(x, q), q)$$

which yields the following systems of differential equations for S:

$$\frac{\partial S}{\partial x_i} = -g_i \quad \text{and} \quad \frac{\partial S}{\partial q_i} = f_i, \ \forall i = 1, \ldots, n,$$

Note that, by construction,

$$\det\left(\frac{\partial^2 S}{\partial x \partial q}\right) \neq 0, \tag{1.41}$$

in a suitable neighborhood of (x_0, q_0), where $x_0 = F(p_0, q_0)$. Observe further that S is neither a function on $(\mathbb{R}^{2n}, (p, q))$ nor on $(\mathbb{R}^{2n}, (y, x))$, but it is a function defined on the *auxiliary space* $(\mathbb{R}^{2n}, (x, q))$, where the coordinates of the original source and target spaces of φ are shuffled. The usefulness of the previous construction is due to the fact that it can be reversed. More precisely, if $S = S(x, q)$ is such that $\det\left(\frac{\partial^2 S}{\partial x \partial q}\right)(x_0, q_0) \neq 0$, one can invert the n-relations $p_i := \frac{\partial S}{\partial q_i}$ on suitable neighborhood of (p_0, q_0), to get a new functional relation $x = x(p, q)$, where $x = (x_1(p, q), \ldots, x_n(p, q))$. Defining $y_i = -\frac{\partial S}{\partial x_i}(x(p, q), q)$ for all $i = 1, \ldots, n$, the (locally defined) application $\varphi : (\mathbb{R}^{2n}, (p, q)) \to (\mathbb{R}^{2n}, (y, x))$ is a symplectomorphism of $(\mathbb{R}^{2n}, \omega_{st})$, as it can be shown with a simple computation.

Definition 1.93 The function S introduced above is called a *free* generating function of φ. △

Example 1.94 Let $\varphi : \mathbb{R}^2 \to \mathbb{R}^2$, $\varphi(p, q) = (y, x)$, defined by $y = ap + bq$ and $x = cp + dq$, where $a, b, c, d \in \mathbb{R}$. This application is symplectic if and only if the matrix

$$A = \left(\begin{array}{c|c} a & b \\ \hline c & d \end{array}\right)$$

belongs to $\mathrm{Sp}_1(\mathbb{R}) = \mathrm{SL}_2(\mathbb{R})$, i.e., if and only if $\det A = ad - bc = 1$. If $c \neq 0$, one can invert *in p* the relation $x = cp + dq$ to get $p = c^{-1}x - c^{-1}dq$, which inserted in $y = y(p, q)$ yields $y = ac^{-1}x - ac^{-1}dq + bq$, i.e.,

$$\begin{cases} y = ac^{-1}x + (b - ac^{-1}d)q = ac^{-1}x - c^{-1}q \\ p = c^{-1}x - dc^{-1}q. \end{cases}$$

Writing $\frac{\partial S}{\partial q} = c^{-1}x - dc^{-1}q$ and $\frac{\partial S}{\partial x} = c^{-1}q - ac^{-1}x$, one obtains after a first integration in x,

$$S(x, q) = c^{-1}qx - \frac{ac^{-1}}{2}x^2 + V(q),$$

which, together with the first relation, implies

$$\frac{\partial S}{\partial q} = V'(q) + c^{-1}x = c^{-1}x - dc^{-1}q,$$

i.e., $V(q) = -\frac{dc^{-1}}{2}q^2$ (plus an omitted arbitrary constant). In other words, a *free* generating function for the linear symplectomorphism defined by A is

$$S(x, q) = -\frac{ac^{-1}}{2}x^2 + c^{-1}qx - \frac{dc^{-1}}{2}q^2.$$

△

One thing that this simple example is teaching us is that *not* all symplectomorphisms admit a free generating function, i.e., a generating function depending on the *new* and *old positions* (x, q). For example, a linear symplectomorphism of $(\mathbb{R}^2, \omega_{st})$ of the form $\varphi(p, q) = (ap + bq, dq)$, $a = d^{-1}$, does *not* admit a free generating function. More in general, a symplectomorphism of $(\mathbb{R}^{2n}, \omega_{st})$ of the form $\varphi(p, q) = (y(p, q), x(q))$ does not admit a free generating function.

Problem 1.95 Generalize Example 1.94; more precisely, find a formula for a free generating function for the linear symplectomorphism of $(\mathbb{R}^{2n}, \omega_{st})$ defined by the matrix

$$A = \left(\begin{array}{c|c} a & b \\ \hline c & d \end{array}\right)$$

in $\mathrm{Sp}_{2n}(\mathbb{R})$. △

Problem 1.96 Find a general form for a generating function of a symplectomorphism of the type $\varphi(p, q) = (y(p, q), x(q))$. Observe that the cotangent lift of a diffeomorphism of $\mathbb{R}^n$ is of that form. Find a general expression for a generating function for these symplectomorphisms. △

In the construction of the free generating function of a symplectomorphism discussed above, the first step was to invert the functional relation $x = F(p, q)$ in a neighborhood of a point where $\det(\frac{\partial x}{\partial p}) \neq 0$; see Example 1.94. As noticed in this example and in the discussion that follows it, this condition is not always verified. Moreover, when such a condition is fulfilled, the generating function is, by construction, a function of the new and old positions, which is not always desirable. Without entering the intricacy of the general theory of the generating functions, for which we refer the reader to the classical texts [10, 84, 158], we introduce one more important class of generating functions depending on the *old positions* and on the *new momenta*, i.e., $S = S(q, y)$ keeping up the notation introduced above. Is this case the nondegeneracy condition (1.41) becomes

$$\det \left(\frac{\partial^2 S}{\partial q \, \partial y} \right) \neq 0. \tag{1.42}$$

In particular, one can prove that if $(q_0, y_0) \in \mathbb{R}^{2n}$ is a point where (1.42) is verified, if

$$p_i := \frac{\partial S}{\partial q_i} \quad \text{and} \quad x_i := \frac{\partial S}{\partial y_i}, \ \forall i = 1, \ldots, n, \tag{1.43}$$

then $\varphi : \mathbb{R}^{2n} \to \mathbb{R}^{2n}$, defined by $\varphi(p, q) = (y, x)$, is a symplectomorphism, i.e., $\varphi^*(\sum_{i=1}^n dy_i \wedge dx_i) = \sum_{i=1}^n dp_i \wedge dq_i$.

The following simple example exemplifies the previous observations.

Example 1.97 Let us write h and k to denote y and x, respectively, and let $\varphi : (\mathbb{R}^2, (p, q)) \to (\mathbb{R}^2, (h, k))$, defined by

$$h(p, q) = \frac{p^2 + \varpi^2 q^2}{2} \quad \text{and} \quad k(p, q) = \frac{1}{\varpi} \arccos \left(\frac{\varpi q}{\sqrt{p^2 + \varpi^2 q^2}} \right), \tag{1.44}$$

where ϖ is a positive constant. A simple computation shows that $\varphi^*(dk \wedge dh) = dp \wedge dq$, i.e., that φ is a symplectomorphism. Choosing the *negative* determination of the square root in the formula $p = \pm\sqrt{2h - \varpi^2 q^2}$, (1.43) entails

$$\frac{\partial S}{\partial q} = -\sqrt{2h - \varpi^2 q^2}$$

$$\frac{\partial S}{\partial h} = \frac{1}{\varpi} \arccos \left(\frac{\varpi q}{\sqrt{2h}} \right), \tag{1.45}$$

implying

$$S(q, h) = -\int^q \sqrt{2h - \varpi^2 u^2} \, du + C(h) \stackrel{(1.45)}{\Rightarrow} \frac{\partial C(h)}{\partial h} = 0$$

i.e.,

$$S(q, h) = - \int^q \sqrt{2h - \varpi^2 u^2} du,$$

which, up to an arbitrary (omitted) constant, is the *generating function* of the symplectomorphism (1.44). Note that already in this simple example the function S is only locally defined as a consequence of fact that only locally one can invert the functional relation between h, p, and q. $\triangle$

In the next subsection we will see how coisotropic submanifolds, Lagrangian submanifolds, and generating functions enter in the theory of Hamilton and Jacobi to integrate the Hamilton equations (1.19).

3.5 Hamilton-Jacobi and a Theorem of Carathéodory-Jacobi-Lie

In the previous subsections we saw that:

(i) To solve the Hamilton equations defined by a Hamiltonian function H, one could wish to find a symplectomorphism of the ambient symplectic manifold which transforms the old Hamiltonian in a new one depending only on a suitable subset of coordinates. More precisely, if we use Darboux coordinates, this subset of coordinates must span at most a Lagrangian subspace to solve automatically Hamilton equations.

(ii) The symplectomorphisms of (M, ω) are associated to suitable Lagrangian submanifolds of $(M \times M, \omega \ominus \omega)$ which locally can be described by a single function, called the generating function of the original canonical transformation, a.k.a. of the original symplectomorphism.

The aim of this subsection is to present a short introduction to the classical Hamilton-Jacobi theory. As it will be shortly discussed in more details, the main aim of this theory is to provide a systematic way to produce, given a Hamiltonian function, a generating function of a symplectomorphism to accomplish the proposal of point (i) above. From an operative viewpoint, in the Hamilton-Jacobi theory one trades a system of ODE, the Hamilton equations, for a single PDE, the Hamilton-Jacobi equation, which belongs to a very well-known class of partial differential equations, for which several methods of solution are available. Before delving in a more formal discussion, we present the idea behind this method. Hereafter we assume that the underlying symplectic manifold is (T^*N, Ω), i.e., a cotangent bundle endowed with its canonical symplectic form. Let $H \in C^\infty(T^*N)$ and suppose that we are looking for a solution of the corresponding system of Hamiltonian ODE in a neighborhood of a nonsingular point of H, i.e., in a neighborhood of a point $m \in T^*N$ such that $dH_m \neq 0$. The idea behind the

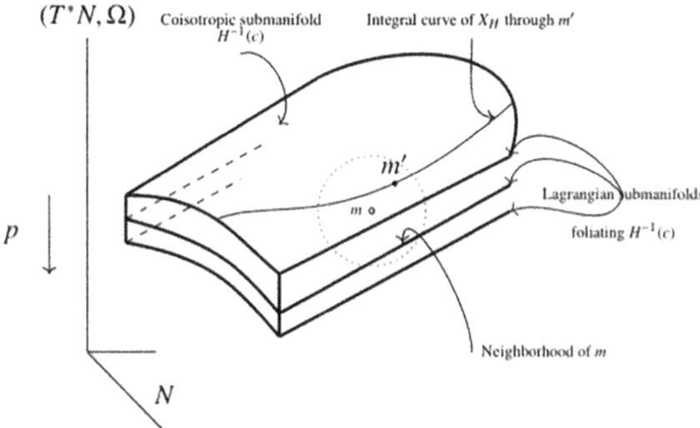

Fig. 1.4 A Lagrangian family

Hamilton-Jacobi method is to find a family of Lagrangian submanifolds, or a *Lagrangian family*, of (T^*N, Ω), foliating a suitable neighborhood of m and such that:

(1) Every submanifold of the Lagrangian family is transversal to the canonical projection p of T^*N.
(2) The Lagrangian family defines a foliation on each level set of H, which, if smooth, is a coisotropic submanifold of (T^*N, Ω), see (Fig. 1.4).

Under these assumptions every integral curve of X_H starting at a point m' sufficiently closed to m is trapped into the Lagrangian submanifold going through m'; see Examples 1.78 and 1.80. In this way one reduces by one-half the number of degrees of freedom of the initial problem: the ODE describing the integral curves of X_H depends only on the parameters describing the Lagrangian fibration.

To formalize these introductory comments we let (T^*N, Ω, H) be a Hamiltonian system and $(U, q_1, \ldots, q_n)$ be a local chart on N. Under these assumptions and keeping the notations introduced above, we give the following.

Definition 1.98 (Hamilton-Jacobi Equation) The Hamilton-Jacobi equation (HJ-equation from now on) defined on U by the Hamiltonian system (T^*N, Ω, H) is

$$H\left(\frac{\partial S}{\partial q}, q\right) = H\left(\frac{\partial S}{\partial q_1}, \ldots, \frac{\partial S}{\partial q_n}, q_1, \ldots, q_n\right) = h, \qquad (1.46)$$

where h is a regular value of H. A *local*, *smooth* solution of the HJ-equation is any $S \in C_N^\infty(U)$ satisfying (1.46). $\triangle$

Remark 1.99 We will see shortly that one needs to consider a more general class of solutions of (1.46) which depends on a certain number of *additional parameters* $a_1, \ldots, a_k$; see the discussion in Sect. 3.5.2. $\triangle$

From the analytical viewpoint (1.46) is a first-order nonlinear PDE, which does not depend explicitly on the unknown function S. On the other hand, from a geometrical viewpoint the HJ-equation is described by the one-codimensional, coisotropic submanifold $W_h = H^{-1}(h)$ of (T^*N, Ω) and its solutions correspond to the Lagrangian submanifolds of W_h, transversal to the canonical projection. Since each of these Lagrangian submanifolds is the image of a closed one-form (see Proposition 1.84), a local solution of (1.46) is a (local) generating function of a Lagrangian submanifold L contained in W_h. Note that these Lagrangian submanifolds are invariant with respect to the Hamiltonian flow generated by H; see (1.80). These comments can be summarized with the following synthetic representation of (1.46):

$$H \circ dS = h.$$

Example 1.100 Note that even the existence of local solutions of (1.46) is not guaranteed. For example, let $N = \mathbb{R}$ and $(T^*N, \Omega) = (\mathbb{R}^2, \omega_{st})$.

(i) If $H(p, q) = \frac{p^2}{2} + \frac{\varpi^2 q^2}{2}$, where ϖ is a nonzero constant, and $h = \frac{1}{2}$, which is a regular value of H, (1.46) reads as

$$\left(\frac{\partial S}{\partial q}\right)^2 + \varpi^2 q^2 = 1.$$

It is simple to check that this equation does not admit solutions in any neighborhood of the points $\left(\pm\frac{1}{\varpi}, 0\right)$.

(ii) If $H(p, q) = q$, every $h \in \mathbb{R}$ is a regular value of H, but the equation $H\left(\frac{\partial S}{\partial q}, q\right) = h$ has no solution if $h \neq 0$. If $h = 0$, then $S(q) = 0$ is a global, but obviously, trivial solution of (1.46).

(iii) On the opposite side of the spectrum, if $N = \mathbb{R}^n$, $(M, \omega) = (T^*N, \Omega)$ and $H(p, q) = p_n$, then every $h \in \mathbb{R}$ is a regular value for H and $H\left(\frac{\partial S}{\partial q}, q\right) = 0$ becomes $\frac{\partial S}{\partial q_n} = 0$. In other words, every $S \in C^\infty(N)$ which does not depend on q_n is a solution of (1.46).

△

We study in more detail the interplay between the solutions of the HJ-equation (1.46) and the solutions of the Hamilton equations defined by H. As we have already remarked, the Hamilton-Jacobi theory is a two-way street: (1) it prescribes how to obtain a solution of the Hamilton equations starting from a solution of the HJ-equation and (2) it provides a solution of the HJ-equation starting from a solution of the Hamilton equations. Both ways relay on the properties of the coisotropic submanifolds discussed in Sect. 3.4, in particular on the absorption principle; see Proposition 1.79. Note that (1) is the direction commonly used in analytical mechanics, while (2) is the direction used in the theory of PDEs and, in this context, it is known as the *method of the characteristics*; see, for example, [112, Section 4, Chapter 2].

Hereafter our presentation will be informal, and it will be mainly, if not only, concerned with the geometrical aspects of the HJ-theory. For an exhaustive account of this beautiful topic from the perspective of analytical mechanics, we refer the reader to the classical texts [1, 10]. For a more analytical point of view we suggest to consult [53, 104, 105], while for a more geometrical perspective we refer the reader to [158], [243], [28], and [50]. Hereafter we follow the presentation of [158, 243].

3.5.1   From Hamilton to Hamilton-Jacobi

First, we see how to use the solutions of the Hamilton equations defined by H to find a *solution* of the corresponding HJ-equation; see Remark 1.104. To this end we need to introduce the following.

Definition 1.101 (Noncharacteristic Cauchy Data)   A set of *Cauchy data* for the HJ-equation $H = h$ is any $n - 1$-dimensional, connected, and isotropic submanifold $J \subset W_h = H^{-1}(h)$. A set of Cauchy data J is called *noncharacteristic* if

$$T_m J \cap T_m W_h^\omega = \{0\}, \ \forall m \in J.$$

$\triangle$

Remark 1.102   If h is a regular value of H, then, for each $m \in W_h$, $T_m W_h^\omega$ is generated by $X_H(m)$. This entails that, for h regular, J is noncharacteristic *if and only if* the integral curves of X_H are *transversal* to it at every point. $\triangle$

We can now state the following important result. Let $\varphi^H : \mathbb{R} \times T^*N \to T^*N$ be the *flow* associated to the Hamiltonian vector field X_H and recall that its domain of definition is an open subset of $\mathbb{R} \times T^*N$.

Theorem 1.103   *Let $h \in \mathbb{R}$ be a regular value of $H \in C^\infty(T^*N)$ and let $J \subset W_h$ a set of noncharacteristic Cauchy data for the HJ-equation $H = h$. Let $\Gamma \subset \mathbb{R} \times J$ be the intersection of $\mathbb{R} \times J$ with the domain of definition of the flow φ^H. Then the restriction of φ^H to Γ spans a solution of the HJ-equation $H = h$; see Remark 1.104.*

Proof *(Sketch)* The idea of the proof is relatively simple and it goes as follows. Since J is isotropic, for each $m \in J$, $\omega_m(u, v) = 0$ for all $u, v \in T_m J$. Moreover, since J is a set of noncharacteristic Cauchy data (see Definition 1.101), the vector space $\mathbb{R}X_H(m) \oplus T_m J$ is Lagrangian. Finally, since the flow φ^H defines a one-parameter group of local symplectomorphisms (see Example 1.55), the vector subspace $\mathbb{R}X_H(\varphi^H(m)) \oplus \varphi^H_{*,m}(T_m J)$ is a Lagrangian subspace of $T_{\varphi^H(m)} W_h$ for each $m \in J$. This reasoning shows that the flow of φ^H spans, starting from J, a Lagrangian submanifold of W_h; see the picture below. $\square$

A *generalized or geometrical* solution of HJ-equation is a Lagrangian submanifold of (T^*N, Ω), which is obtained flowing-out a set of noncharacteristic Cauchy data with the flow defined by the Hamiltonian vector field X_H. In general it is not everywhere transversal to the canonical projection p (Fig. 1.5).

Fig. 1.5 A generalized
solution of the
Hamilton-Jacobi equation

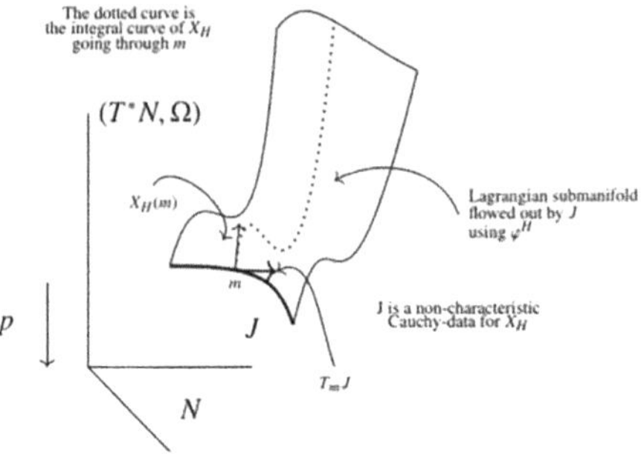

Remark 1.104 (Generalized or Geometrical Solution) The solutions of the
HJ-equation described on the previous theorem should be called *generalized or
geometrical* solutions. In fact, as should be clear from their construction, they are
Lagrangian submanifolds of T^*N which in general fail to be everywhere transversal
to the canonical projection p. At those *singular* points a local generating function
is not available and their generalized solutions cease to be useful to solve the
Hamilton equations. To save the day one needs to consider the more general notion
of *generating family*; see Sect. 4.2.   △

   For more details about the *method of characteristics*, summarized in the theorem
above, we refer the reader to the monographs [112] and [53]; see also [11].

3.5.2   From Hamilton-Jacobi to Hamilton

We see now how to recover solutions of (1.19) starting from (a particular class
of) solutions of the HJ-equation. To this end it will be necessary to consider
solutions of the HJ-equation smoothly depending on additional parameters. From
a geometrical viewpoint this generalization yields the introduction of families
of Lagrangian submanifolds of (T^*N, Ω). Let us start formalizing these ideas
introducing the notion of complete integral of the HJ-equation. In this framework
it is convenient to think of an HJ-equation as a *general coisotropic submanifold*
$C \subset (T^*N, \Omega)$, i.e., not necessarily one-codimensional; see [28, 158], and [50].

Definition 1.105 (Complete Integral) Let A be a manifold. A *complete integral*
of an HJ-equation $C \subset T^*N$ is a smooth function $S : N \times A \to \mathbb{R}$, such that:

   (i) For each $a \in A$, $S_a = S \circ i_a$ is a generating function of a solution of the
       Hamilton-Jacobi equation associated to C, where $i_a : N \to N \times A$ is defined
       by $i_a(n) = (n, a)$ for all $n \in N$.
  (ii) The family $\{\Lambda_a\}_{a \in A}$ of the solutions of the Hamilton-Jacobi equation associ-
       ated to C parametrized by A defines a (Lagrangian) foliation of C. In particular,
       for every $n \in C$ there exists a unique $a \in A$ such that $n \in \Lambda_a$.
 (iii) The application $\pi : C \to A$ which to every $n \in C$ associates (the unique)
       $a \in A$ such that $n \in \Lambda_a$ is smooth; see the picture below (Fig. 1.6).

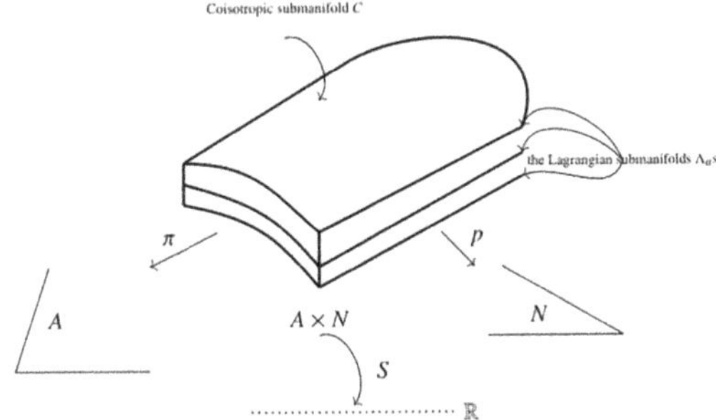

Fig. 1.6 A complete integral of the Hamilton-Jacobi equation

$\triangle$

The application π defines a fibration whose total space is the coisotropic submanifold C and whose fibers are Lagrangian submanifolds of (T^*N, Ω). These Lagrangian submanifolds are by definition transversal to the canonical projection p, since they are defined as the image of dS where $S : A \times N \to \mathbb{R}$ is a smooth function, generating the fibration.

Remark 1.106 The definition above is taken from [28]. One could introduce the slightly more general notion of a *local* complete integral of an HJ-equation. In this case the family of Lagrangian submanifolds $\{\Lambda_a\}_{a \in A}$ would define only a local foliation of the coisotropic submanifold C. Obviously, there is no need to think of S as a globally defined function on $N \times A$; rather one could consider functions defined on suitable open subsets of this Cartesian product; see [158]. $\triangle$

The previous definition implies that π has constant rank equal to $n - k$, where $n = \dim N$ and $k = \operatorname{codim} C$. In particular if $a_1, \ldots, a_{n-k}$ are local coordinates on A and $(p_1, \ldots, p_n, q_1, \ldots, q_n)$ are local coordinates on T^*N, one has that:

(i) Each Λ_a is locally described by the relations $p_i = \frac{\partial S_a}{\partial q_i}$, for $i = 1, \ldots, n$.

(ii) The $n \times (n-k)$ matrix $\left(\frac{\partial^2 S}{\partial q_i \partial a_j}\right)_{i=1,\ldots,n; \, j=1,\ldots,n-k}$ has everywhere maximal rank, called *completeness condition*.

We discuss in detail two cases: (i) $k = 1$, when the coisotropic submanifold C is (locally) described as the level set of a single Hamiltonian function, and (2) $k = 0$, i.e., when C is an open subset of T^*N. Note that in this case C is *not* a coisotropic submanifold of (T^*N, Ω_N). In the first case we show how to get a solution of system (1.19) if it is known a complete integral of the HJ-equation of level energy h, where h is a regular value of the Hamiltonian H.

3.5.3 Codimension One

Since we will need to consider at the same time the cotangent bundle of more than one manifold, hereafter we will use the notation Ω_N to denote the canonical symplectic form of T^*N. A formal statement enshrining the relation between a complete integral of the HJ-equation and system (1.19) is enclosed in the following.

Theorem 1.107 (Jacobi) *The knowledge of a complete integral of the HJ-equation associated to (T^*N, Ω_N, H) and to a given regular value h of H yields the determination, by means of elementary operations, of the trajectories of the Hamiltonian vector field X_H contained in the level set $W_h = H^{-1}(h)$.*

Proof *(Sketch)* First of all we would like to clarify that by *elementary operations* is meant integration, partial differentiation, and elimination.Then the proof of the theorem consists in showing that if a complete integral is known, one can find an explicit description of the characteristics of the coisotropic submanifold W_h. This is accomplished showing that the completeness condition above is sufficient to define a submersive smooth map φ from a neighborhood of W_h in T^*N with values in T^*A such that the connected component of its level sets are exactly the maximal arcs of trajectory of X_H contained in C. We conclude observing that the submersive map above mentioned is a symplectic map from a suitable open set of (T^*N, Ω_N) to (T^*A, Ω_A), whose graph, therefore, is a Lagrangian submanifold of $(T^*N \times T^*A, \Omega_N \ominus \Omega_A)$. Using this map one defines the change of coordinates which lets us integrate the Hamiltonian equations defined by X_H using only the elementary operations mentioned above. For a detailed treatment we refer the reader to the references [28] and [158]. $\square$

So far we considered only the problem of the integration of the Hamilton equation for a *fixed* regular value h of the total energy, i.e., of the Hamiltonian function. From a more geometric point of view, this entails to consider a fixed level set W_h of the Hamiltonian function, i.e., a one-codimensional coisotropic submanifold of the cotangent bundle of N. In this case a complete integral of the HJ-equation produces a fibration of W_h, whose fibers are Lagrangian submanifolds of (T^*N, Ω). Suppose now one needed to find a symplectic change of coordinates to integrate the Hamiltonian equations (1.19) *simultaneously* for different values of the Hamiltonian. This case forces us to consider not just one level set of H, but rather a family of level sets $\{W_h\}_{h \in I}$, where $I \subset \mathbb{R}$ is an open interval made of regular values of H. From a geometrical point of view the family $\{W_h\}_{h \in I}$ sweeps a domain in T^*N which looks like a tubular neighborhood of a given W_{h_0}, with $h_0 \in I$. In this case the geometric intuition suggests that a complete integral in the sense of Definition 1.105 should define a Lagrangian foliation in each of the W_hs. In this *relative* framework the constant h is traded for a function $h = K(a) = K(a_1, \ldots, a_n)$, i.e., h is allowed to depend smoothly on the parameters $a = (a_1, \ldots, a_n)$.

Remark 1.108 (About the Parameters Involved) What we called relative framework consists of a one-parameter family, $\{W_h\}_{h \in I}$, of one-codimensional (coisotropic) submanifolds of (T^*N, Ω), each of them foliated by a family of Lagrangian submanifolds of the ambient symplectic manifold. If dim $N = n$, to get a parametric description of the geometric objects involved, one needs (in total)

n-independent parameters, one to label the family of the coisotropic submanifolds, the remaining $n - 1$ labelling the Lagrangian submanifolds. Since the coisotropic submanifolds are (regular) level sets of the Hamiltonian H, setting its value to a giving h yields the relation $h = K(a_1, \ldots, a_n)$ for some K. In particular, for each regular value h of H, of the n-parameters $a_1, \ldots, a_n$, only $n - 1$ remain independent. In this way, one recovers the description given in Definition 1.105, where, for $k = 1$, the dimension of the parameter space A is $n - 1$. On the other hand, from these observations it follows that h is *a* natural candidate to be one of the n-parameters, say $a_n = h$. With this choice $A = B \times I$, where B is a $n - 1$-dimensional manifold (say an open subset of $\mathbb{R}^{n-1}$), parametrizing the Lagrangian submanifolds of the foliations and $I \subset \mathbb{R}$ is an open subset made of regular values of H. In this way, fixing $a_n = h \in I$ determines the coisotropic submanifold, W_h, whose Lagrangian foliation is completely determined by fixing the values of the remaining $n - 1$-parameters, $a_1, \ldots, a_{n-1}$. Note that in both cases S is a smooth function on (an open subset of) $A \times N$. △

Thus, one arrives to the *classical* formulation of the HJ-problem. Find a smooth function defined on (an open subset of) N such that (i) it smoothly depends on $n = \dim N$ parameters, (ii) it satisfies the conditions in Definition 1.105, and (iii) it satisfies

$$H\left(\frac{\partial S}{\partial q}, q\right) = K(a). \tag{1.47}$$

Remark 1.109 It is a quite common usage, in particular in the classical text about analytical mechanics, to refer to (1.47) as to the HJ-equation and to a function S satisfying the properties above recalled as to a complete integral of this equation. In this context the Completeness Condition 3.5.2 reads as

$$\det\left(\frac{\partial^2 S}{\partial q_i \partial a_j}\right)_{i,j=1,\ldots,n} \neq 0.$$

△

In complete analogy with 1.107, one has

Theorem 1.110 (Jacobi) *Given a complete integral of* (1.47) *one can solve, by means of elementary operations, system* (1.19) *defined by the Hamiltonian H.*

Proof *(Sketch)* Under the above assumptions, S is the generating functions of a symplectomorphism $\varphi : \mathbb{R}^{2n} \to \mathbb{R}^{2n}$, $\varphi(p, q) = (a, b)$ where

$$p_i = \frac{\partial S}{\partial q_i}(q, a) \quad \text{and} \quad b_i = \frac{\partial S}{\partial a_i}(q, a), \ \forall i = 1, \ldots, n,$$

and if $\tilde{H} = H \circ \varphi^{-1}(a, b)$, then

$$\dot{a}_i = 0 \quad \text{and} \quad \dot{b}_i = \frac{\partial \tilde{H}}{\partial a_i}, \ \forall i = 1, \ldots, n \tag{1.48}$$

which yields

$$a_i(t) = a_i(0) \quad \text{and} \quad b_i(t) = \eta_i(a(0))t + b_i(0), \ \forall i = 1, \ldots, n.$$

In other words, the dynamics linearizes in the new variables. Note that the *new* Hamiltonian $\tilde{H}$ depends only on the a's variables, i.e., $\tilde{H} = \tilde{H}(a)$, and that $\eta_i(a(0)) = \frac{\partial \tilde{H}}{\partial a_i}(a_1(0), \ldots, a_n(0))$, for all i. From the previous observations it follows that the complete integrals of the HJ-equation as defined above are generating functions of the type introduced in (1.43). $\qquad\square$

The example below is the follow-up of Example 1.97.

Example 1.111 (One-Dimensional Harmonic Oscillator) Let $N = \mathbb{R}$ so that $(T^*N, \Omega) = (\mathbb{R}^2, \omega_{st})$ and let $H(p, q) = \frac{p^2}{2} + \frac{\varpi^2 q^2}{2}$, where ϖ is a nonzero constant. In this case $h = 0$ is the only nonregular value for H. So let $h \neq 0$. The differential equation

$$\left(\frac{\partial S}{\partial q}\right)^2 + \varpi^2 q^2 = 2h$$

has as solutions

$$S_{\pm}(q, h) = \pm \int^q \sqrt{2h - \varpi^2 u^2} du, \tag{1.49}$$

where the $\pm$ sign depends on the choice of the branch of the square root. The latter are the generating functions for the canonical transformations $\varphi_{\pm}(p, q) = (h, k_{\pm})$, where $h = \frac{1}{2}(p^2 + \varpi^2 q^2)$ and

$$k_+ = \frac{1}{\varpi}\arcsin\left(\frac{\varpi q}{\sqrt{2h}}\right) \quad \text{and} \quad k_- = \frac{1}{\varpi}\arccos\left(\frac{\varpi q}{\sqrt{2h}}\right);$$

see Example 1.97. To fix the ideas, let us work on the case defined by φ_+. One has $\tilde{H} = H\left(\varphi_+^{-1}(h, k_+)\right) = h$, yielding

$$\dot{h} = 0 \quad \text{and} \quad \dot{k}_+ = 1, \tag{1.50}$$

as follows from (1.48) with $a = h$ and $b = k_+$. (1.50) are readily integrated:

$$h(t) = h(0) \quad \text{and} \quad k_+(t) = t + k_+(0). \tag{1.51}$$

Note that the previous equations imply that the *conjugate* variable to the energy is the time. Moreover, inverting the relation between k_+ and q and using (1.51) one can easily deduce

$$q(t) = \frac{\sqrt{2h}}{\varpi} \sin\left(\varpi t + \varpi k_+(0)\right) \quad \text{and} \quad p(t) = \sqrt{2h} \cos\left(\varpi t + \varpi k_+(0)\right).$$

$$(1.52)$$

We leave to the reader to check that choosing k_- yields the same solutions. More generally, if the system under examination were described by a Hamiltonian of the form

$$H(p,q) = \frac{p^2}{2} + V(q),$$

where $V = V(x)$ is a smooth function, using the total energy of the system as parameter, the HJ-equation becomes

$$\left(\frac{\partial S}{\partial x}\right)^2 + 2V(q) = 2h.$$

A solution of this equation can be written in the implicit form

$$S(q,h) = \pm\sqrt{2} \int^q \sqrt{h - V(u)}\,du,$$

which defined the variable canonically conjugate to h via the formula

$$k = \frac{\partial S}{\partial h} = \pm \int^q \frac{du}{\sqrt{2h - 2V(q)}}.$$

Clearly, an explicit expression for such a function is available as soon as one is able to find an explicit formula for a primitive of $\frac{1}{\sqrt{2h-2V(q)}}$. $\triangle$

To summarize in a few lines the previous discussion, one could say that at the heart of the HJ-theory stands the notion of complete integral depending on auxiliary parameters, which is used to define, locally on the phase space (T^*N, Ω), a *Lagrangian fibration* on each regular level set of the Hamiltonian H, the fibration being given by the parameters themselves. This Lagrangian fibration is *adapted* to the Hamiltonian system (T^*N, Ω, H) in the sense that the integral curves of the latter are trapped into its leaves, which end up straight to the Hamiltonian flow.

Hereafter we will discuss, in the setting of a general symplectic manifold, (M, ω), which consequences one can draw from the existence of a (local) Lagrangian fibration. As we will see, this notion yields the one of *integrable Hamiltonian system* (see 1.120), which is the main object of study of the present work.

3.5.4 Codimension Zero

In our previous discussion we were concerned with the HJ-problem for a single or, more in general, for a family of one-codimensional coisotropic submanifolds of a cotangent bundle. We now turn our attention to the case of an open subset of T^*N, where we want to define a Lagrangian foliation transversal to the fibers of the canonical projection. To this end, as in the one-codimensional case, one can introduce a smooth function $S : N \times A \to \mathbb{R}$, where dim $A=$dim N. As in the proof of Theorem 1.107, dS defines a Lagrangian submanifold of $(T^*N \times T^*A, \Omega_N \ominus \Omega_A)$, which turns out to be the graph of a symplectomorphism φ defined from an open set $U_N \subset T^*N$ to an open set $U_A \subset T^*A$. To frame our exposition in a more general context which will be used in the next chapter, consider a $2n$-dimensional symplectic manifold (M, ω) and a *submersive* map $\pi : M \to A$, where dim $A = n$, such that for all $a \in A$ in the image of π, $M_a = \pi^{-1}(a)$ is a Lagrangian submanifold of (M, ω). Let $a_1, \ldots, a_n$ be local coordinates on the open set $U \subset \mathrm{Im}(\pi)$ and let $f_1, \ldots, f_n$ be the functions defined by $f_i = a_i \circ \pi$, for $i = 1, \ldots, n$. Under these assumptions

Lemma 1.112

 (i) $f_1, \ldots, f_n$ *Poisson commute with respect to the Poisson bracket defined by the restriction of ω to $\pi^{-1}(U)$.*
 (ii) *For each $a' \in U$, there exists a neighborhood U', such that the restrictions of $f_1, \ldots, f_n$ to U' are independent.*

Proof The proof of (i) follows from the following two observations. First, the fibers of (the submersive map) π are (locally) described as the intersection of the level sets of the f_is. Second, since the fibers of π are Lagrangian submanifolds of (M, ω), $[X_{f_i}, X_{f_j}] = 0$, for all $i, j = 1, \ldots, n$; see also Lemma 4.3. The proof of part (ii) follows from the definition of the f_is and from the definition of a submersion. $\square$

The following result is a crucial point in the proof of the local integrability of the Hamilton equations defined by a Hamiltonian which belongs to a maximal family of (locally defined) Poisson commuting functions.

We first need a preliminary lemma, whose proof can be found, for example, in [10, 49]:

Lemma 1.113 *Let (V, ω) be a symplectic vector space of dimension $2n$ and let $\{e_1, \ldots, e_n, f_1, \ldots, f_n\}$ be a symplectic basis of V, i.e., $\omega(e_i, e_j) = \omega(f_i, f_j) = 0$ and $\omega(e_i, f_j) = \delta_{ij}$. Let $L \subset V$ be a Lagrangian vector subspace of V. Then there exists a subset $I \subset \{1, \ldots, n\}$ such that if $J = \{1, \ldots, n\}\backslash I$, then the subspace $L' = span\langle \{e_i\}_{i \in I}, \{f_j\}_{j \in J} \rangle$ is Lagrangian in V and $L \oplus L' = V$.*

Theorem 1.114 (Carathéodory-Jacobi-Lie Theorem) *Let (M, ω) be as above, let $m \in M$, and let U be a Darboux neighborhood of m; see Sect. 3.3 above. Let $f_1, \ldots, f_n \in C_M^\infty(U)$ such that:*

(i) They are functionally independent, *i.e., $df_1 \wedge \cdots \wedge df_n(p) \neq 0$ for all $p \in U$.*
(ii) They are in involution, *i.e., $\{f_i, f_j\} = 0$ for all $i, j = 1, \ldots, n$.*

Then there exists a neighborhood V of m, eventually properly contained in U, and $g_1, \ldots, g_n \in C_M^\infty(V)$ such that

$$\omega|_V = \sum_{i=1}^n dg_i \wedge df_i.$$

Before proving it, let us remark that Carathéodory-Jacobi-Lie theorem basically says that n functionally independent Poisson commuting functions can be taken to be part of a system of Darboux coordinates.

Proof Let $(p_1, \ldots, p_n, x_1, \ldots, x_n)$ be a set of Darboux coordinates on U (see Sect. 3.3), such that $x_i(m) = 0$ for all $i = 1, \ldots, n$. Notice that this condition can be always achieved summing to each of the x_i a suitable constant. Since $f_1, \ldots, f_n$ are independent and satisfy $\{f_i, f_j\} = 0$ for all $i, j = 1, \ldots, n$, Lemma 1.113 implies that $(x_1, \ldots, x_n, f_1, \ldots, f_n)$ is a coordinate system in a neighborhood of m, eventually properly contained in U. Since $\omega(X_{f_i}, X_{f_j}) = 0$ for all i, j, the restriction of ω to this neighborhood belongs to the differential ideal generated by $df_1, \ldots, df_n$; see Theorem B.9. For this reason, without loss of generality, we can assume that the restriction of the symplectic form to U can be written as

$$\omega = \omega|_U = \sum_{i=1}^n \alpha_i \wedge df_i, \tag{1.53}$$

where $\alpha_1, \ldots, \alpha_n \in \Omega_M^1(U)$ are such that $\alpha_i = \sum_{i=1}^n \gamma_{ik}(x, f)dx_k, i = 1, \ldots, n$. Let $\{\psi_t\}_{t \in [0,1]}$ be a one-parameter family of smooth maps, where $\psi_t : U \to U$ for all $t \in [0, 1]$, is defined by

$$\psi_t(x_1, \ldots, x_n, f_1, \ldots, f_n) = ((1-t)x_1, \ldots, (1-t)x_n, f_1, \ldots, f_n) := ((1-t)x, f).$$

Note that, as t changes from 0 to 1, $\{\psi_t\}_t$ shrinks U onto the slice defined by the equations $x_i = 0$, for $i = 1, \ldots, n$, that $\psi_0 = \mathrm{id}_U$ and that $\psi_1(x_1, \ldots, x_1, f_1, \ldots, f_n) = (0, \ldots, 0, f_1, \ldots, f_n)$. If $\psi : [0, 1] \times U \to U$ is the smooth map defined by $\psi(t, m) = \psi_t(m)$, for all $t \in [0, 1]$ and $m \in U$, one can define a *homotopy operator* $\mathcal{H} : \Omega^k(U) \to \Omega^{k-1}(U), k \geq 1$; see Proposition B.20 and Formula (B.6). Since $d\omega = 0$, $\psi_0^*\omega = \omega$ and $\psi_1^*\omega = 0$, one has that

$$d\mathcal{H}(\omega) = -\omega;$$

see Formula (B.4). To conclude the proof it suffices to prove that $\mathcal{H}(\omega)$ belongs to the differential ideal generated by $f_1, \ldots, f_n$. In fact, in this case, one could write

$$\mathcal{H}(\omega) = -\sum_{i=1}^{n} g_i df_i,$$

with $g_1, \ldots, g_n$ smooth functions defined on a suitable neighborhood V of the point m, eventually contained in U. The proof of this last claim is obtained by a direct computation. In fact, the first one computes

$$\psi^* \omega \overset{(1.53)}{=} \psi^* \Big(\sum_{i,k=1}^{n} \gamma_{ik}(x, f) \, dx_k \wedge df_i \Big)$$

$$= \sum_{i,k} (1 - t)\gamma_{ik}((1-t)x, f) \, dx_k \wedge df_i - \sum_{i,k=1}^{n} x_k \gamma_{ik}((1-t)x, f) \, dt \wedge df_i,$$

which yields

$$\mathcal{H}(\omega) \overset{(B.6)}{=} \int_0^1 (i_t^* \circ i_{\frac{\partial}{\partial t}} \circ \psi^*)\omega \, dt = \sum_{i=1}^{n} - \Big(\sum_{k=1}^{n} x_k \int_0^1 \gamma_{ik}((1-t)x, f) \, dt \Big) df_i,$$

proving the claim. $\square$

Remark 1.115 The previous proof is constructive, at least at some extent. In fact the functions $g_1, \ldots, g_n$ are defined by

$$g_i(x, f) = \sum_{k=1}^{n} x_k \int_0^1 \gamma_{ik}((1-t)x, f) \, dt, \ i = 1, \ldots, n.$$

$\triangle$

Remark 1.116 The proof of Theorem 1.114 shows that the *complementary* set of functions $(g_1, \ldots, g_n)$ can be found using only *elementary operations*, i.e., *integration, partial differentiation, and elimination*. For this reason sometimes it is said that the $g_1, \ldots, g_n$ are found by *quadrature*. $\triangle$

As already mentioned, the previous theorem implies the existence of a (local) symplectomorphism which transforms the original Hamiltonian H into a *new* Hamiltonian depending only on a half of the Darboux coordinates. More precisely, one can prove the following.

Corollary 1.117 (Local Integrability) *Let $H : M \to \mathbb{R}$ be a smooth function and suppose that $m \in M$ is a nonsingular point of the Hamiltonian vector field X_H. If there exists a Darboux neighborhood U of m and $f_1, \ldots, f_n \in C_M^\infty(U)$ satisfying*

(i) and (ii) of Theorem 1.114 and the conditions

$$\{H, f_i\} = 0, \ \forall i = 1, \ldots, n,$$

then one can find local symplectic coordinates on a suitable neighborhood of m such that, if written in these coordinates, the Hamiltonian flow defined by H is linear.

Proof Using the result contained in Theorem 1.114, given $m \in M$ such that $X_H(m) \neq 0$, we can find a neighborhood V of m and $(g_1, \ldots, g_n)$ smooth functions defined on V such that

$$\omega|_V = \sum_{i=1}^{n} dg_i \wedge df_i.$$

Since $\{f_i, H\} = 0$ for all $i = 1, \ldots, n$, the restriction of the function H to V cannot depend on the g_i, i.e., $H = H(f_1, \ldots, f_n)$. Writing now the Hamilton equations for the function H expressed in terms of the (local) symplectic coordinates $(f_1, \ldots, f_n, g_1, \ldots, g_n)$, one gets

$$\dot{f_i} = \frac{\partial H}{\partial g_i} = 0 \quad \text{and} \quad \dot{g_i} = -\frac{\partial H}{\partial f_i}, \ \forall i = 1, \ldots, n.$$

$\square$

In this way we close the circle of thoughts we started in Remark 1.56 and formalized with Jacobi theorem 1.107: a complete integral of the HJ-equation defines a symplectomorphism that transforms the *old Hamiltonian* into a *new one* depending only on a half of the canonical coordinates. With respect to this new Hamiltonian, the Hamiltonian equations are readily integrated as explained in Remark 1.56 and in Corollary 1.117. Before introducing the notion of completely integrable system, whose properties will be further investigated in Chap. 4, we would like to make a final comment about the Hamilton-Jacobi theory we sketched above. In particular we would like to stress that there is an extension of the HJ-theory to the *time-dependent* Hamiltonian formalism. In this enhanced setting (1.46) becomes

$$\frac{\partial S}{\partial t} + H\left(\frac{\partial S}{\partial q}, q, t\right) = 0, \tag{1.54}$$

where $S = S(q, t) : U \times I \to \mathbb{R}$ is a smooth function defined on the Cartesian product of an open set $U \subset N$ with the open interval $I \subset \mathbb{R}$. We will not comment much more on this topic, referring the reader to the reference [1] for a complete analysis of this general case. We would like only to add the following three observations.

(PR1) The time-dependent formalism is not only necessary when the Hamiltonian is time-dependent, but it surges also naturally when one needs to apply the theory of generating functions to the one-parameter group of symplecto-morphisms defined by a (complete locally) Hamiltonian vector field; see Example 1.55. One interesting case is the following. Let $H \in C^\infty(T^*N)$ and let L be a Lagrangian submanifold of (T^*N, Ω) transversal to the canonical projection. In this case one can prove that if X_H is transversal to L, the flow generated by X_H, for each $t \in I_\epsilon = [0, \epsilon]$, $\epsilon << 1$, will move the L to $L_t \subset T^*N$, which is still a Lagrangian submanifold, transversal to the canonical projection. The family of submanifolds so defined will be described by a one-parameter family of smooth functions $\{S_t\}_{t \in I_\epsilon}$, locally defined on N. One can show that this family is a solution of (1.54).

(PR2) The second observation is that the right geometrical formalism to frame the time-dependent Hamilton-Jacobi theory is *not* symplectic but rather *contact geometry*. We will not even introduce the basic structures populating this beautiful subject since this would lead us very far from our main goals. We refer the reader to [1, 10] for the applications of contact geometry to the problems arising in analytical mechanics and to [97] for a very complete introduction to the geometrical and topological subtleties of the contact world.

(PR3) Finally, as we have already mentioned the HJ-method is a two-way process which produces solutions of a system of ODEs from a solution of a nonlinear PDE and vice-versa. In spite of its elegance, this theory would be of little use if special methods for solving the classes of PDEs/ODEs involved were not be available. In this respect we would like to mention that for the HJ-equation the so-called *separation of variables* (SoV) has been proved to be a very powerful method of solution. Also in this case we will make only very superficial comments, referring the reader to the classical text [53] for a detailed exposition and to Remark 5.116 for some comments about the relation between the SoV and the theory of bi-Hamiltonian manifolds. In its simplest form, the method of SoV goes as follows. One looks for a system of (local) coordinates $(p_1, \ldots, p_n, q_1, \ldots, q_n)$ which *separate* the original Hamiltonian in the sum of functions each of which depends only on a conjugate pair (p_i, q_i). Written in these coordinates the HJ-equation

$$H\left(\frac{\partial S}{\partial q}, q\right) = K(a)$$

assumes the form

$$\sum_{i=1}^{n} H_i\left(\frac{\partial S}{\partial q_i}, q_i\right) = K(a) \tag{1.55}$$

where $K(a) = \sum_{i=1}^{n} K_i(a_i)$ and the K_is are suitable functions. This suggests that one could look for a solution of the HJ-equation of the form

$$S(q_1, \ldots, q_n, a_1, \ldots, a_n) = \sum_{i=1}^{n} S_i(q_i, a_i).$$

In this case (1.55) splits into a sum of n independent equations:

$$H_i\left(\frac{\partial S_i}{\partial q_i}, q_i\right) = K_i(a_i), \ i = 1, \ldots, n,$$

whose solutions, in general, are easier to be found.

A simple application of this method is presented in the following two examples.

Example 1.118 (n-Dimensional Harmonic Oscillator) Let $(M, \omega) = (T^*\mathbb{R}^n, \Omega) = (\mathbb{R}^{2n}, \omega_{st})$, and let

$$H(p, q) = \frac{1}{2} \sum_{i=1}^{n} p_i^2 + \frac{1}{2} \sum_{i=1}^{n} \varpi_i^2 q_i^2 \tag{1.56}$$

be the Hamiltonian of the *n-dimensional harmonic oscillator* where the ϖ_is are positive constants; see also Example 1.111. In this case the physical coordinates (p, q) are already separation variables for the Hamiltonian (1.56); in fact one can write $H(p, q) = \sum_{i=1}^{n} f_i(p_i, q_i)$ where

$$f_i(p_i, q_i) = \frac{p_i^2}{2} + \frac{\varpi_i^2 q_i^2}{2}, \ i = 1, \ldots, n. \tag{1.57}$$

In other words the Hamiltonian (1.56) uncouples as the superposition n harmonic oscillators each of them described by the Hamiltonian (1.57). Choosing $K(h_i) = h_i$ for all $i = 1, \ldots, n$, one obtains

$$\left(\frac{\partial S_i}{\partial q_i}\right)^2 + \varpi_i^2 q_i^2 = 2h_i,$$

a solution to which is

$$S_i(q_i, h_i) = \int^{q_i} \sqrt{2h_i - \varpi_i^2 u^2}\, du = \sqrt{2h_i - \varpi_i^2 q_i^2}\, h_i q_i + \frac{h_i}{\varpi_i} \arcsin\left(\frac{\varpi_i q_i}{\sqrt{2h_i}}\right)$$

(see (1.49)), which implies

$$S(q, h) = \sum_{i=1}^{n} \left[\sqrt{2h_i - \varpi_i^2 q_i^2}\, h_i q_i + \frac{h_i}{\varpi_i} \arcsin\left(\frac{\varpi_i q_i}{\sqrt{2h_i}}\right)\right].$$

Note that if the ϖ_is are all the same, the harmonic oscillator is called isotropic.　△

Example 1.119 (Central Potential) Let us consider a mechanical system made of
a point of mass equal to one moving in the Euclidean three-dimensional space under
a *central potential*. The Hamiltonian of such system, in Cartesian coordinates, is

$$H(p, x) = \frac{p_x^2 + p_y^2 + p_z^2}{2} + V(x), \tag{1.58}$$

where $V = V(\sqrt{x^2 + y^2 + z^2})$. If written in spherical coordinates (r, θ, ϕ) (1.58)
assumes the following form:

$$H(p_r, p_\theta, p_\phi, r, \theta, \phi) = \frac{1}{2}\left(p_r^2 + \frac{p_\theta^2}{r^2} + \frac{p_\phi^2}{r^2 \sin^2 \theta} \right) + \mathcal{V}(r). \tag{1.59}$$

Recall that $r \in [0, +\infty)$, $\theta \in (0, \pi)$, and $\phi \in [0, 2\pi]$. The corresponding Hamilton-
Jacobi equation is

$$\frac{1}{2}\left[\left(\frac{\partial S}{\partial r}\right)^2 + \frac{1}{r^2}\left(\frac{\partial S}{\partial \theta}\right)^2 + \frac{1}{r^2 \sin^2 \theta}\left(\frac{\partial S}{\partial \phi}\right)^2 \right] + \mathcal{V}(r) = h. \tag{1.60}$$

The *nested* structure of (1.59)

$$H(p_r, p_\theta, p_\phi, r, \theta, \phi) = \frac{1}{2}\left(p_r^2 + \frac{1}{r^2}\left(p_\theta^2 + \left(\frac{p_\phi^2}{\sin^2 \theta}\right) \right) \right) + \mathcal{V}(r)$$

suggests that one can separate (1.60) using the following ansatz:

$$S(p_r, p_\theta, p_\phi, r, \theta, \phi) = S_1(\phi, h_1) + S_2(\theta, h_1, h_2) + S_3(r, h_1, h_2, h)$$

which inserted in (1.60) gives

$$\frac{1}{2}\left(\frac{\partial S_3}{\partial r}\right)^2 + \mathcal{V}(r) + \frac{1}{2r^2}\left[\left(\frac{\partial S_2}{\partial \theta}\right)^2 + \frac{1}{\sin^2 \theta}\left(\frac{\partial S_1}{\partial \phi}\right)^2 \right] = h,$$

which separates as

$$\left(\frac{\partial S_1}{\partial \phi}\right)^2 = h_1,$$

$$\left(\frac{\partial S_2}{\partial \theta}\right)^2 + \frac{h_1}{\sin^2 \theta} = h_2,$$

$$\frac{1}{2}\left(\frac{\partial S_3}{\partial r}\right)^2 + \mathcal{V}(r) + \frac{h_2}{2r^2} = h.$$

Since $p_\phi = \frac{\partial S_1}{\partial \phi}$ is constant along the motion of the system described by (1.59) (note that ϕ is a cyclic coordinate), the first equation integrates trivially

$$S_1(\phi, h_1) = p_\phi \phi = \sqrt{h_1}\,\phi.$$

On the other hand, the second equation and the third equations are solved by

$$S_2(\theta, h_1, h_2) = \pm \int^\theta \sqrt{h_2 - \frac{h_1}{\sin^2 u}}\,du \quad \text{and}$$

$$S_3(r, h_1, h_2, h) = \pm \int^r \sqrt{2h - 2\mathcal{V}(u) - \frac{h_2}{u^2}}\,du.$$

The case $\mathcal{V}(r) = -\frac{\kappa}{r}$, where κ is a positive constant, is of particular interest, describing the so-called *Kepler system* . $\triangle$

3.5.5 Integrable Systems

We close this section introducing the concept of completely integrable system, whose geometric analogue, the one of Lagrangian fibration, will be analyzed in depth in Chap. 4. More generally, integrable systems on Poisson manifolds will be briefly dealt with in Sect. 3.1. Let (M, ω, H) be a Hamiltonian system. Suppose $\dim M = 2n$.

Definition 1.120 The triple (M, ω, H) is called a *completely integrable system* if there exist $f_1, \ldots, f_n \in C^\infty(M)$ such that:

(i) $f_1, \ldots, f_n$ are generically independent on M.
(ii) $\{f_i, f_j\} = 0$ for all $i, j = 1, \ldots, n$.
(iii) $\{H, f_i\} = 0$ for all $i = 1, \ldots, n$.

$\triangle$

A few observations are in order. Condition (i) in the above definition means that for every m in a dense open subset $M^\circ \subset M$

$$df_1 \wedge \cdots \wedge df_n(m) \neq 0.$$

Observe further that in view of $\dim M = 2n$, since $f_1, \ldots, f_n$ are generically independent and pairwise Poisson commuting, they generate a *maximal commutative* Poisson subalgebra of $(C^\infty(M), \{\cdot, \cdot\})$. This means that if $f \in C^\infty(M)$ is such that $\{f, f_i\} = 0$ for all $i = 1, \ldots, n$, then there exists a smooth $G : \mathbb{R}^n \to \mathbb{R}$ such that

$$f = G(f_1, \ldots, f_n).$$

This observation, by virtue of (iii), can be applied to H which, obviously, Poisson commute with every function $G = G(f_1, \ldots, f_n)$. Recall that the functions which Poisson commute with H are called *first integrals* or *conserved quantities* of the Hamiltonian system defined by H; see Definition 1.29. From these observations it follows that the complete integrability of the Hamiltonian system (M, ω, H) *is not* defined by the maximal Poisson commuting set $(f_1, \ldots, f_n)$, but rather by the *maximal Poisson commuting algebra* generated by these functions to which H belongs. For this reason we can always assume that the Hamiltonian H is one of $(f_1, \ldots, f_n)$, for example, $f_1 = H$. We should also stress that our definition of completely integrable system can be extended in several directions, for example, allowing the phase space being a Poisson (non-symplectic) manifold. We will not pursue this matter any further, referring the interested reader, for example, to [150, Chapter 12].

In Chap. 4 we will explore the geometrical consequence of the definition above. For the time being we present a list of (more or less) elementary examples.

Example 1.121 (Systems with One and Two Degrees of Freedom) Every Hamiltonian system with a one degree of freedom is completely integrable everywhere its Hamiltonian has nonzero differential. For example, if $(M, \omega) = (\mathbb{R}^2, \omega_{st})$ and $H : M \to \mathbb{R}$ is a smooth function of the form

$$H = \frac{p^2}{2} + V(q), \tag{1.61}$$

on $M^0 \subset M$ where $(\frac{dV}{dq}, p) \neq (0, 0)$, H is the Hamiltonian of a completely integrable system. Note that the Hamiltonian systems defined by (1.61) have a *periodic* behavior when we can find $q_1, q_2 \in \mathbb{R}$ such that:

(i) $q_1 \leq q_0 \leq q_2$.
(ii) $V(q) \leq h$ for all $q \in (q_1, q_2)$.
(iii) $V(q_1) = h = V(q_2)$.
(iv) q_1, q_2 are *not* critical points of the function $V(q)$.

Under these assumptions one can solve (implicitly) the equation of motion:

$$t = \pm\sqrt{\frac{1}{2}} \int_{q_0}^{q} \frac{du}{\sqrt{h - V(u)}}.$$

The conditions (i), (ii), (iii), and (iv) imply that the velocity of the system is zero only when it reaches the positions labelled by q_1 and q_2, where the motion is inverted; see the picture below (Fig. 1.7).

Note that these points are the abscissas of two consecutive intersection points of the graph of the potential function with the horizontal line at ordinate equal to h. The period of motion is

$$T(h) = \sqrt{2} \int_{q_1}^{q_2} \frac{dq}{\sqrt{h - V(q)}}. \tag{1.62}$$

Fig. 1.7 When $q = q_1, q_2$
the velocity of the particle is
zero

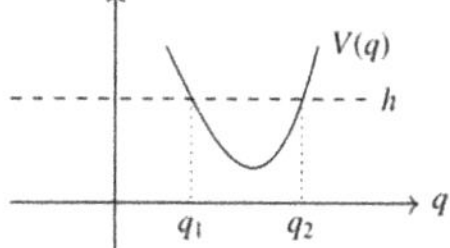

In fact, choosing $p = +\sqrt{h - V(q)}$

$$T(h) = \sqrt{\frac{1}{2}} \int_{q_0}^{q_2} \frac{dq}{\sqrt{h - V(q)}} - \sqrt{\frac{1}{2}} \int_{q_2}^{q_1} \frac{dq}{\sqrt{h - V(q)}} + \sqrt{\frac{1}{2}} \int_{q_1}^{q_0} \frac{dq}{\sqrt{h - V(q)}}.$$

Note that choosing the negative determination of the function p yields the same result. It is important to note that $T(h)$ is, in general, a nonconstant function of the energy of the system. Moreover, once the initial condition (p_0, q_0) is chosen, the periodic motion takes place on the curve

$$C_h = \{(p, q) \mid H(p, q) = h\}.$$

Remark 1.122 A couple of comments are in order. The motion described above is called *libration* or *oscillation*. The condition $(\frac{dV}{dq}, p) \neq (0, 0)$, i.e., $(\frac{\partial H}{\partial q}, \frac{\partial H}{\partial p}) \neq (0, 0)$, guarantees that the curve implicitly defined by $H(p, q) = h$ is smooth. Note that the motions of oscillation occur in a neighborhood of a point of *stable equilibrium*; see, for example, [84]. $\qquad\triangle$

On the other hand, every Hamiltonian system with two degrees of freedom, possessing two independent first integrals in involution is completely integrable. One example in this class of systems is the called *spherical pendulum* which will be analyzed more in depth in Chap. 4. $\qquad\triangle$

Among the integrable systems with three or more degrees of freedom, we just mention the systems of point particles in a central potential and the so-called spinning tops.

Example 1.123 (Central Potential) Let us consider the case of a point particle moving in a central potential. Using the notations introduced in Example 1.119, and writing $\{\cdot, \cdot\}$ to denote the Poisson bracket defined by the standard symplectic form in $T^*\mathbb{R}^3$, one can show that

$$\{p_\phi, H\} = 0, \quad \left\{ p_\theta^2 + \frac{p_\phi^2}{\sin^2 \theta}, H \right\} = 0 \text{ and } \left\{ p_\phi, p_\theta^2 + \frac{p_\phi^2}{\sin^2 \theta} \right\} = 0,$$

and that the functions H, p_ϕ and $p_\theta^2 + \frac{p_\phi^2}{\sin^2 \theta}$ are independent, implying that $(T^*\mathbb{R}^3, \Omega, H)$ is a complete integrable system. Note that H is the total energy of the system, while p_ϕ and $p_\theta^2 + \frac{p_\phi^2}{\sin^2 \theta}$ are the component of the angular momentum along the z-axis and the squared of the module of the angular momentum of the system, respectively. $\triangle$

Example 1.124 (Spinning Tops) Following [19] and [6], we say that a *spinning top* is a (three-dimensional) rigid body, with a fixed point, in a gravitational field. We will discuss the bi-Hamiltonian aspects of the n-dimensional free rigid-body dynamics in Chap. 6. For the time being we just make a few comments about these very important dynamical systems in the three-dimensional case; see also Sect. 4.2 in Chap. 6. For (many) more details we refer the reader to the classical monographs [84] and [10], where both the kinematical and the dynamical aspects of the rigid-body mechanics are treated; to [19], where the algebro-geometrical aspects of the dynamics of these systems are presented; and to [59] for a deep analysis of the Liouville integrability of the *Euler-Poinsot* and of the *Lagrange* spinning tops, from the viewpoint of differential geometry. Finally, we refer the reader to the beautiful, unfortunately unpublished, lecture notes [85], where the reader will find a nice survey about *noncommutative integrability* and its application to the rigid-body dynamics.

Below, we will limit ourselves recalling that a rigid body is a dynamical system whose points are subjected to *the rigidity constraint*, i.e., the prescription that the relative distance between any two points of the system does not change during its temporal evolution. This condition implies that the configuration space of a three-dimensional rigid body can be identified with $\mathbb{R}^3 \times SO(3)$ that reduces further to $SO(3)$ if the rigid body has a *fixed point*, i.e., a point which is fixed with respect to an inertial reference system. As it will be discussed in Sect. 4.1 of Chap. 6, the equations describing the time evolution of a spinning top are equivalent to the Euler equations on the (co)adjoint orbits of maximal dimension of the three-dimensional Euclidean group. Since these orbits are four-dimensional, the complete integrability of the spinning top follows once a first integral is known, independent of the Hamiltonian of the system. The existence of such an integral has been proven only for the following cases:

(Euler-Poinsot, 1758) In the *Euler-Poinsot* case the fixed point O coincides with the center of mass G of the rigid body. In this case the motion of the system is equivalent to the free motion (absence of the gravitational force) around O. In this case the three components of the angular momentum (with respect to an inertial reference system) are first integrals.

(Lagrange, 1788) The *Lagrange* spinning top is a *symmetric top*. More precisely, in this case there is one *moving frame* with respect to which two principal moments of inertia are equal and the line connecting the fixed point O with the center of mass G of the body is a revolution axis.

(Kowalevski, 1889) In the *Kowalevski* case there is a moving frame with respect to which the principal momenta of inertia are $I_1 = I_2 = 2I_3$ and with respect to which the center of mass G lies on the coordinate plane generated by the principal axis with equal eigenvalues.

$$\triangle$$

Finally, as examples of integrable systems, or better of classes of integrable systems, we mention the so-called integrable lattices and the n-dimensional harmonic oscillator.

Example 1.125 (Integrable Lattices) The class of the integrable lattices provides many examples of highly sophisticated integrable systems. These are dynamical systems constituted by n point masses interacting via a potential depending on the distance of the particles. Among the many classes, we recall here the so-called *Toda* and *Calogero-Moser* lattices. In Chaps. 8 and 7 the case of the so-called *rational Calogero-Moser* and of the *open Toda* system, respectively, will be analyzed carefully. $\triangle$

Example 1.126 (Harmonic n-Dimensional Oscillator) As a final example of integrable system with more than one degree of freedom, we consider again the n-dimensional harmonic oscillator; see Example 1.118. Recall that the phase space of this dynamical system is $\mathbb{R}^{2n}$ with its canonical symplectic form and its Hamiltonian is $H = \frac{1}{2}\sum_{i=1}^{n}(p_i^2 + \varpi^2 q_i^2)$, where the ϖ_is are nonnegative constants. To ascertain the complete integrability of this Hamiltonian system it suffices to observe that, if the f_is are as in (1.57), then

$$\sum_{i=1}^{n} \frac{\partial f_k}{\partial p_i}\frac{\partial f_j}{\partial q_i} - \frac{\partial f_k}{\partial p_i}\frac{\partial f_j}{\partial q_i} = 0, \ \forall k, j = 1, \ldots, n,$$

and that $df_1 \wedge \cdots \wedge df_n \neq 0$ on an open and dense subset of $\mathbb{R}^{2n}$, i.e., the rank of the matrix

$$\begin{pmatrix} \varpi_1^2 q_1 & 0 & \cdots & 0 & p_1 & 0 & \cdots & 0 \\ 0 & \varpi_2^2 q_2 & \cdots & 0 & 0 & p_2 & \cdots & 0 \\ \vdots & \vdots & \ddots & \vdots & \vdots & \vdots & \ddots & \vdots \\ 0 & 0 & \cdots & \varpi_n^2 q_n & 0 & \cdots & 0 & p_n \end{pmatrix}$$

is maximal (equal to n) on a dense and open subset of $\mathbb{R}^{2n}$. $\triangle$

4 Concluding Remarks and Further Topics

4.1 The Moser Argument

The argument we used to prove the Darboux theorem goes back to the paper [188], where the author showed that on any compact, connected, and oriented manifold M, two normalized volume forms σ_1 and σ_2 are *diffeomorphism equivalent*, i.e., there exists a diffeomorphism $\psi : M \to M$ such that $\psi^*\sigma_1 = \sigma_2$. In the same paper it was shown that such a diffeomorphism ψ can be chosen to be *isotopic* to the identity of M. More precisely:

Theorem 1.127 (Moser, 1965) *Let M be a compact, connected, and oriented manifold and let σ_1 and σ_2 be two volume forms such that*

$$\int_M \sigma_1 = \int_M \sigma_2.$$

Then, there exists an isotopy $\psi : I \times M \to M$, such that $\psi_1^*\sigma_1 = \sigma_0$; see Definition B.19.

The proof of the theorem above is based on the so-called *Moser trick*, which consists in trading the search for the isotopy $\{\psi\}_{t\in I}$ with the one for a *time-dependent* vector field $\{X_t\}_{t\in I}$ such that

$$\left. \frac{d\psi_s}{ds} \right|_{s=t} (m) = X_t(\psi_t(m)),$$

for all $t \in I$ and $m \in M$. This change of viewpoint simplifies the problem since while equations which are satisfied by the members of $\{\psi_t\}_{t\in I}$ are nonlinear, the ones satisfied by the time-dependent vector field $\{X_t\}_{t\in I}$ are linear; see [188, p. 292].

For example, the proof of Lemma 1.62 followed exactly this path: we found a time-dependent vector field X_t such that the condition in Formula (1.39) would be satisfied, and from this, using the result in Proposition B.17, we could deduce the existence of an isotopy would prove the statement of the lemma.

4.2 Generating Families and the Maslov-Hörmander Theorem

In our exposition of the geometric approach to Hamilton-Jacobi theory, we considered only Lagrangian submanifolds of (T^*N, Ω) *transversal* to the canonical projection $p : T^*N \to N$, i.e., such that, for all $\alpha \in L$,

$$T_\alpha L \cap V_\alpha = \{0\}, \tag{1.63}$$

where $V_\alpha = \ker(p_{*,\alpha} : T_\alpha(T^*N) \to T_{p(\alpha)}N)$.

Problem 1.128 Prove that if L is Lagrangian (1.63) is equivalent to $T_\alpha L + V_\alpha = T_\alpha(T^*N)$. $\triangle$

For a general Lagrangian submanifold L, i.e., non-transversal to the canonical projection, a point $\alpha \in L$ is called *regular* if (1.63) is satisfied and *singular* or *critical* otherwise. The image of the singular set, i.e., the set of all singular points of L, under the canonical projection is called the *caustic* of L. Recall that a Lagrangian submanifold is regular, i.e., formed only by regular points, if and only if it is the image of a closed one-form; see Proposition 1.84. Note that the regular points of a Lagrangian submanifold are the points where the rank of the differential of the canonical projection is maximal, equal to the dimension of N. In spite that singular points are not in the image of a *smooth* closed one-form, they could be in the image of a one-form belonging to a *weaker* class of differentiability.

Example 1.129 Let $(M, \omega) = (T^*\mathbb{R}, \Omega)$, and let (p, q) be global coordinates (p as usual is thought as the fibered coordinate defined by q). Then the (global) generating function $S(q) = \frac{3}{4} q^{\frac{4}{3}}$ defines a Lagrangian submanifold of (M, ω) whose points are all regular except that the point $(0, 0)$. In fact at $q = 0$ the function S does not admit a second derivative. $\triangle$

In spite of the Lagrangian submanifold of the previous example not admitting a representation as the image of a C^∞ generating function, we show below that it admits a description in terms of what is called a *smooth generating family*. To introduce this concept let $\pi : Z \to N$ be a smooth submersive map and suppose that $\dim Z = n + k$ and $\dim N = n$. Consider the subbundle C of T^*Z whose fiber at a given point is the vector space formed by the covectors annihilating the vertical space of π at the same point, i.e., C is the union of the conormal bundles of the fibers of π; see Example 1.88. One can prove that C is a coisotropic submanifold of (T^*Z, Ω_Z) of codimension k, fibered over T^*N with k-dimensional isotropic fibers. Let $\tilde{\pi} : C \to T^*N$ be the projection map.

Example 1.130 Let $Z = \mathbb{R}^2$ with coordinates (q, x) and $N = \mathbb{R}$ with coordinate q. Consider the map $\pi : Z \to N$ defined by $\pi(q, x) = q$. Identifying T^*Z with $\mathbb{R}^4$ with coordinates (q, x, p, y), C can be identified with $\mathbb{R}^3$ with coordinates (q, x, p). In this case the isotropic distribution is generated by the constant vector field $\frac{\partial}{\partial x}$ and the *quotient* of C by such distribution can be identified with $T^*\mathbb{R} = \mathbb{R}^2$ with coordinates (q, p). Note that $\tilde{\pi} : C \to T^*\mathbb{R}$ is the quotient map defined by $\tilde{\pi}(q, x, p) = (q, p)$ and $i_0 : C \to T^*\mathbb{R}^2$ is the injection map defined by $i_0(q, x, p) = (q, x, p, 0)$, $i^*\Omega_Z = \tilde{\pi}^*\Omega_N$. Clearly, the same would be true if we traded i_0 with i_a for any $a \in \mathbb{R}$. $\triangle$

Going back to the general case, let $S \in C^\infty(Z)$ and let L_S be the Lagrangian submanifold of (T^*Z, Ω_Z) obtained as the image of dS. Sometimes the pair (π, S) is called a *smooth generating family*; see [28]. Then one can prove that

Theorem 1.131 (Maslov-Hörmander) *If L_S is transversal to C, the image L of $L_S \cap C$ under the $\tilde{\pi}$ is (in general) an immersed Lagrangian submanifold of*

(T^*N, Ω_N). *On the other hand, one has that* every *Lagrangian submanifold L of*
(T^*N, Ω_N) *is* locally *the image of a smooth generating function $S \in C^\infty(Z)$ for a
suitable fibration $\pi : Z \to N$.*

For a proof of the previous result we refer the reader, for example, to [50]. Note that
the transversality hypothesis in the first part of the previous theorem is a sufficient
but not necessary condition. Let us introduce another piece of notation. The *critical
set* of (π, S) is the set

$$\mathrm{Crit}(Z) = \{\xi \in Z \mid \langle dS_\xi, v \rangle = 0 \; \forall v \in T_\xi Z \text{ s.t. } \pi_{*,\xi} v = 0\},$$

i.e., $\mathrm{Crit}(Z)$ are the points where the differential of S, restricted to the fibers of π,
vanishes; see Definition 4.5 in [28].

Example 1.132 Let Z, N, and π as in 1.130. Given $S \in C^\infty(Z)$, one sees that

$$\mathrm{Crit}(Z) = \left\{ (q, x) \mid \frac{\partial S}{\partial x}(q, x) = 0 \right\}.$$

In fact, $dS = \frac{\partial S}{\partial q} dq + \frac{\partial S}{\partial x} dx$ and the kernel of the differential of π, at every point, is
generated by $\frac{\partial}{\partial x}$. $\triangle$

Let $(q, x) = (q_1, \ldots, q_n, x_1, \ldots, x_k)$ local coordinates on Z adapted to the fibration
$\pi : Z \to N$ and let (q, x, p, y) be the corresponding local coordinates on T^*Z.
Then one can show that the Lagrangian submanifold L of (T^*N, Ω_N) is described
(locally) by

$$p_i = \frac{\partial S}{\partial q_i} \; \forall i = 1, \ldots, n \quad \text{and} \quad 0 = \frac{\partial S}{\partial x_j} \; \forall j = 1, \ldots, k. \tag{1.64}$$

Moreover, the generating family (π, S) is *Morse*, i.e., L_S and C intersect transversally, if and only if the matrix

$$\left[\frac{\partial^2 S}{\partial x_i \partial q_j} \middle| \frac{\partial^2 S}{\partial x_i \partial x_j} \right] \tag{1.65}$$

has maximal rank at the point of the $\mathrm{Crit}(Z)$; see [28] Theorems 4.6 and 4.7.

Example 1.133 Going back to the setup of Examples 1.129 and 1.130, let

$$S(q, x) = xq - \frac{1}{4}x^4.$$

Applying the criterium expressed in (1.65) one sees that (π, S) is Morse, i.e., the L_S and C intersect transversally. The Lagrangian submanifold of (T^*N, Ω_N) is described by (1.64), which in this case reads as

$$p = x \quad \text{and} \quad 0 = q - x^3,$$

which imply that $q = p^3$, which is the expression of L in the coordinates (q, p). In other words we showed that in spite the Lagrangian submanifold of Example 1.129 does not admit a representation in terms of a C^∞ generating function, which can be (globally) represented in terms of C^∞ (Morse) generating family. $\triangle$

5   Bibliographical Notes

There are several books and survey articles where the material of this chapter is discussed. In particular the book [114] gives a beautiful treatment of symplectic geometry with several applications to physical problem. One can find there a very nice discussion about the applications of symplectic geometry to harmonic analysis, geometric optics, and integrable systems. Our treatment of Darboux theorem is actually taken from there. Other very useful references are the books [20, 21, 30] and the survey article [13]. Another nice introduction to symplectic geometry can be also found in the book [48]. Our short presentation to the Hamilton-Jacobi theory in Sect. 3 was greatly influenced by review paper [243]. Other two very nice, complete, and readable summaries of this very important theory can be found in [28] and [50]. For a vast array of applications of generating families to a plethora of physical problems, see [50] and the literature quoted therein.

Chapter 2
Elements of Poisson Geometry

This chapter aims to be a short introduction to Poisson geometry which generalizes, in a sense clarified below, the theory of symplectic structures studied in Chap. 1. More precisely, while every symplectic manifold carries a Poisson bracket (see Formula (1.20)), not every Poisson structure comes from an associated symplectic one. Moreover, it will be shown that every Poisson manifold turns out to be a kind of collage of symplectic manifolds, the ones defining its *symplectic foliation*.

We would like to stress that Poisson geometry is a very active and important branch of differential geometry which sits at the crossroad of many area of mathematics and physics like Lie theory, foliation theory, and classical and quantum mechanics, just to name a few.

1 Poisson Brackets and Poisson Manifolds

First, let us introduce the following notion. A *biderivation* of $C^\infty(M)$ is a bilinear map $\delta : C^\infty(M) \times C^\infty(M) \to C^\infty(M)$ such that

$$\delta(fg, h) = f\delta(g, h) + g\delta(f, h) \quad \delta(f, gh)$$
$$= g\delta(f, h) + h\delta(f, g), \forall f, g, h \in C^\infty(M).$$

A biderivation will be called skew-symmetric if

$$\delta(f, g) = -\delta(g, f), \forall f, g \in C^\infty(M).$$

As the space of derivations of $C^\infty(M)$ is in one-to-one correspondence with the Lie algebra of vector fields $\mathfrak{X}(M)$, the biderivations of $C^\infty(M)$ are in one-to-one correspondence with the elements of $\Gamma(\Lambda^2 TM)$, the vector space of (global)

© The Author(s), under exclusive license to Springer Nature Switzerland AG 2026

A. Arsie, I. Mencattini, *Geometry of Integrable Systems*, Latin American Mathematics Series – UFSCar subseries,

https://doi.org/10.1007/978-3-031-96282-0_2

smooth sections of the second exterior power of TM. Hereafter, a smooth section of $\Lambda^2 TM$ will be called a *bivector field*. The correspondence between bivector fields and biderivations of $C^\infty(M)$ goes as follows. If δ is a skew-symmetric biderivation of $C^\infty(M)$ and $x_1, \ldots, x_n$ are (local) coordinates on M, the corresponding $\Pi \in \Gamma(\Lambda^2 TM)$ assumes the following form:

$$\Pi = \frac{1}{2} \sum_{i,j=1}^{n} \delta(x_i, x_j) \frac{\partial}{\partial x_i} \wedge \frac{\partial}{\partial x_j} = \sum_{i<j} \delta(x_i, x_j) \frac{\partial}{\partial x_i} \wedge \frac{\partial}{\partial x_j}. \tag{2.1}$$

On the other hand, if Π is a bivector field on M, the bilinear map δ defined by

$$\delta(f, g) := \langle \Pi, df \wedge dg \rangle$$

defines a skew-symmetric biderivation of $C^\infty(M)$, whose associated bivector field is Π. We can now introduce the following important notion.

Definition 2.1 (Poisson Structure, Poisson Bracket, and Poisson Bivector Field)
A *Poisson structure*, or a *Poisson bracket*, on M is a skew-symmetric biderivation $\{\cdot, \cdot\}$ of $C^\infty(M)$ which satisfies the Jacobi identity, i.e.,

$$\{\{f, g\}, h\} + \{\{h, f\}, g\} + \{\{g, h\}, f\} = 0, \, \forall f, g, h \in C^\infty(M).$$

A bivector field Π whose corresponding skew-symmetric biderivation is a Poisson structure will be called a *Poisson bivector field* and a *Poisson manifold* will be a pair $(M, \{\cdot, \cdot\})$ (or (M, Π)) consisting of a manifold M and a Poisson bracket $\{\cdot, \cdot\}$ (or of a manifold and a Poisson bivector field). $\triangle$

In a local chart $(U, x_1, \cdots, x_n)$ a Poisson bivector field is completely determined by the set of (local) functions $\{x_i, x_j\}$, $i, j = 1, \ldots, n$ (see Formula (2.1)), which are usually called the *elementary Poisson brackets*. For all $f, g \in C^\infty(M)$,

$$\{f, g\} = \frac{1}{2} \sum_{i,j=1}^{n} \Pi_{ij} \frac{\partial f}{\partial x_i} \frac{\partial g}{\partial x_j} = \sum_{j=1}^{n} \{f, x_j\} \frac{\partial g}{\partial x_j}, \tag{2.2}$$

since, for all $j = 1, \ldots, n$,

$$\{f, x_j\} = \frac{1}{2} \sum_{i=1}^{n} \Pi_{ij} \frac{\partial f}{\partial x_i}. \tag{2.3}$$

Let us present a few elementary examples of Poisson manifolds together with the corresponding Poisson structures.

Example 2.2

(i) Every manifold is a Poisson manifold if endowed with the trivial Poisson bracket, i.e., with the bracket such that

$$\{f, g\} = 0, \ \forall f, g \in C^\infty(M).$$

(ii) Let $M = \mathbb{R}^2$ with coordinates (x, y) and let $f \in C^\infty(M)$. Then

$$\Pi = f(x, y)\frac{\partial}{\partial x} \wedge \frac{\partial}{\partial y}$$

is a Poisson tensor. More in general one can show that every bivector field on a two-dimensional manifold defines a Poisson structure; see Sect. 1.1. On the other hand, already in dimension three it is very easy to find a bivector field which does not define a Poisson structure. For example, on $M = \mathbb{R}^3$, choosing coordinates x, y, z, a simple computation shows that

$$\Pi = x\frac{\partial}{\partial x} \wedge \frac{\partial}{\partial y} + y\frac{\partial}{\partial y} \wedge \frac{\partial}{\partial z} + z\frac{\partial}{\partial x} \wedge \frac{\partial}{\partial z}$$

is not Poisson.
(iii) If (M, Π) is a Poisson manifold, it is (M, Π^-), where Π^- is defined by

$$\Pi^-(df, dg) = -\Pi(df, dg), \ \forall f, g \in C^\infty(M).$$

The Poisson structure Π^- is called the *opposite* of Π.

$$\triangle$$

In the following example we recall an important class of Poisson structures which was introduced in Chap. 1; see Formula (1.20).

Example 2.3 Every symplectic manifold (M, ω) is a Poisson manifold whose Poisson bracket is defined by

$$\{f, g\} = \omega(X_f, X_g),$$

for all $f, g \in C^\infty(M)$. In local coordinates $x_1, \ldots, x_m$, $m = 2n$, the corresponding Poisson bivector field assumes the following form:

$$\Pi = \sum_{i<j}(\omega^{-1})_{ji}\frac{\partial}{\partial x_i} \wedge \frac{\partial}{\partial x_j} \tag{2.4}$$

if, in the chosen local coordinates,

$$\omega = \sum_{i<j} \omega_{ij} dx_i \wedge dx_j.$$

To comment more on this point, fix one of the coordinate functions, say x_i, and let $X \in \mathfrak{X}(M)$ be the corresponding Hamiltonian vector field, i.e., X is such that $i_X \omega = -dx_i$. Writing $X = \sum_s \xi_s \frac{\partial}{\partial x_s}$, one can compute

$$i_X \omega = \frac{1}{2} \left(\sum_{k,l,s} \omega_{kl} \xi_s \delta_{ks} dx_l - \sum_{k,l,s} \omega_{kl} \xi_s \delta_{ls} dx_k \right)$$

$$= \frac{1}{2} \left(\sum_{l,s} \omega_{sl} \xi_s dx_l - \sum_{k,s} \omega_{ks} \xi_s dx_k \right)$$

$$= \sum_{l,s} \omega_{sl} \xi_s dx_l,$$

i.e.,

$$dx_i = - \sum_{l,s} \omega_{sl} \xi_s dx_l = - \sum_l \left(\sum_s \omega_{sl} \xi_s \right) dx_l = \sum_l \left(\sum_s \omega_{ls} \xi_s \right) dx_l,$$

which corresponds to the following linear system:

$$\begin{cases} \sum_s \omega_{ls} \xi_s = 0 \text{ if } l \neq i, \\ \sum_s \omega_{is} \xi_s = 1. \end{cases}$$

The (unique) solution of this system yields

$$X = X_{x_i} = \sum_s (\omega^{-1})_{si} \frac{\partial}{\partial x_s}. \tag{2.5}$$

To prove (2.4) it suffices to shows that $\{x_i, x_j\} = (\omega^{-1})_{ji}$. To this end

$$\{x_i, x_j\} = \omega(X_{x_i}, X_{x_j}) = -\langle dx_i, X_{x_j} \rangle \overset{(2.5)}{=} -\sum_s (\omega^{-1})_{sj} \left\langle dx_i, \frac{\partial}{\partial x_s} \right\rangle$$

$$= -\sum_s (\omega^{-1})_{sj} \delta_{is} = -(\omega^{-1})_{ij} = (\omega^{-1})_{ji}.$$

The Poisson bivector field defined by a symplectic form will be called *invertible* or *nondegenerate*.   △

Other classes of Poisson structures are introduced in the following two examples. Both have their origin in the theory of Lie algebras.

Example 2.4 (Linear Poisson Structures) Let $\mathfrak{g}$ be a finite-dimensional Lie algebra, whose Lie bracket is $[\cdot, \cdot]$ and let $\mathfrak{g}^*$ be the dual vector space of $\mathfrak{g}$ endowed with its canonical structure of differentiable manifold. For all $\alpha \in \mathfrak{g}^*$, the canonical isomorphism $T_\alpha \mathfrak{g}^* \simeq \mathfrak{g}^*$ implies that $\mathrm{Hom}(T_\alpha \mathfrak{g}^*, \mathbb{R}) \simeq \mathfrak{g}$. Under this identification, for every $f \in C^\infty(\mathfrak{g}^*)$, df_α can be thought as an element in $\mathfrak{g}$. Now, for all $f, g \in C^\infty(\mathfrak{g}^*)$, one can define

$$\{f, g\}(\alpha) = \langle \alpha, [df_\alpha, dg_\alpha] \rangle, \forall \alpha \in \mathfrak{g}^*. \tag{2.6}$$

One can easily prove that (2.6) is bilinear and skew-symmetric. Moreover, it fulfills the Jacobi identity since the Lie bracket does. Finally, since $d(fg)_\alpha = f(\alpha)dg_\alpha + g(\alpha)df_\alpha$ for all $\alpha \in \mathfrak{g}^*$ and all $f, g \in C^\infty(\mathfrak{g}^*)$, (2.6) is a biderivation of $C^\infty(\mathfrak{g}^*)$, showing that this bracket defines a Poisson structure on $\mathfrak{g}^*$. To find out the local expression of the corresponding Poisson bivector field, let $e_1, \ldots, e_n$ be a basis of $\mathfrak{g}$ and let $x_1, \ldots, x_n$ be the corresponding (global) *linear coordinates* on $\mathfrak{g}^*$, defined by

$$x_i(\alpha) = \langle \alpha, e_i \rangle, \forall i = 1, \ldots, n, \ \alpha \in \mathfrak{g}^*. \tag{2.7}$$

Then, for each $\alpha \in \mathfrak{g}^*$

$$\{x_i, x_j\}(\alpha) = \langle \alpha, [e_i, e_j] \rangle = \sum_{k=1}^{n} c_{ij}^k \langle \alpha, e_k \rangle = \sum_{k=1}^{n} c_{ij}^k x_k(\alpha),$$

i.e.,

$$\{x_i, x_j\} = \sum_{k=1}^{n} c_{ij}^k x_k, \forall i, j = 1, \ldots, n. \tag{2.8}$$

In other words, the Poisson bivector field defined by (2.6), if written in the coordinates (2.7), assumes the following form:

$$\Pi = \sum_{i<j} \sum_{k=1}^{n} c_{ij}^k x_k \frac{\partial}{\partial x_i} \wedge \frac{\partial}{\partial x_j}.$$

Hereafter we will refer to (2.6) as to a *linear* Poisson bracket, since the bracket of two linear functions is still a linear function; see (2.8). Because of this, we will often denote this bracket by $\{\cdot, \cdot\}_{LP}$ and the corresponding Poisson tensor by Π_{LP}. $\triangle$

Remark 2.5 The Poisson bracket (2.6) is sometimes called *Kostant-Kirillov bracket* or *Kostant-Kirillov-Souriau bracket*, after the mathematicians Bertram

Kostant, Alexandre Aleksandrovic Kirillov, and Jean-Marie Souriau, who gave many and very important contributions to both representation theory and symplectic geometry.   △

Before dealing with the next example, we introduce the following:

Definition 2.6 (Quadratic Vector Spaces and Quadratic Lie Algebras) A vector space V is called *quadratic* if it carries a bilinear form:

$$B : V \otimes V \to \mathbb{K},$$

 (i) *Nondegenerate*, i.e., $B(x, y) = 0$ for all $y \in V$ implies $x = 0$
 (ii) *Symmetric*, i.e., $B(x, y) = B(y, x)$ for all $x, y \in V$

We will say that $\mathfrak{g}$ is a quadratic Lie algebra if $\mathfrak{g}$ is a quadratic vector space whose bilinear form is $\mathfrak{g}$-invariant, i.e., $(\mathrm{ad}_x y \,|\, z) + (y \,|\, \mathrm{ad}_x z) = 0$ for all $x, y, z \in \mathfrak{g}$. Moreover, a vector subspace $W \subset V$ will be called *isotropic* if $(v \,|\, u) = 0$ for all $u, v \in U$.   △

Remark 2.7 Note that if a vector space V is endowed with a nondegenerate bilinear form $B : V \otimes V \to \mathbb{K}$, then V can be identified with its dual via the linear map $\check{B} : V \to B^*$:

$$\check{B} : v \rightsquigarrow \alpha_v$$

where $\langle \alpha_v, u \rangle = B(v, u)$ for all $u \in V$.   △

Since B is nondegenerate, $\check{B}$ is invertible and its inverse $(\check{B})^{-1} : V^* \to V$ takes each $\alpha \rightsquigarrow v_\alpha$, where $B(v_\alpha, u) = \langle \alpha, u \rangle$ for all $u \in V$.

With this notion at our disposal we have:

Example 2.8 (Linear Poisson Structure on $\mathfrak{g}$) Every quadratic Lie algebra $(\mathfrak{g}, B)$ carries a linear Poisson structure defined as the pullback via $\check{B}$ of the linear Poisson structure of $\mathfrak{g}^*$. More precisely, one defines $\{\cdot, \cdot\} : C^\infty(\mathfrak{g}) \times C^\infty(\mathfrak{g}) \to C^\infty(\mathfrak{g})$ by

$$\{f, g\}(x) = B(x, [\check{B}^{-1}(df_x), \check{B}^{-1}(dg_x)]), \ \forall f, g \in C^\infty(\mathfrak{g}), \ x \in \mathfrak{g}. \tag{2.9}$$

The proof that the formula above defines a Poisson structure on $\mathfrak{g}$ follows at once from the following identity:

$$\{f, g\}(x) = \{f \circ \check{B}^{-1}, g \circ \check{B}^{-1}\}_{LP}(\check{B}(x)),$$

which holds for all $f, g \in C^\infty(\mathfrak{g})$ and $x \in \mathfrak{g}$. Defining the *gradient* of $f \in C^\infty(\mathfrak{g})$ with respect to B by

$$\nabla_x f := \check{B}^{-1}(df_x), \ \forall x \in \mathfrak{g}, \tag{2.10}$$

(2.9) becomes

$$\{f, g\}(x) = B(x, [\nabla_x f, \nabla_x g]). \tag{2.11}$$

An important example of this construction is provided by $(\mathfrak{g}, B) = (\mathfrak{gl}_n(\mathbb{K}), \mathrm{tr})$, where $\mathrm{tr} : \mathfrak{gl}_n(\mathbb{K}) \to \mathbb{K}$ is usual *trace*, i.e., $\mathrm{tr}(x) = \sum_{i=1}^{n} x_{ii}$. We leave to the reader to verify that the trace defines a nondegenerate, bilinear form on $\mathfrak{gl}_n(\mathbb{K})$, which is invariant with respect to the adjoint action. If $E_{ij} : \mathfrak{gl}_n(\mathbb{K}) \to \mathbb{K}$ is the *linear* function defined by $E_{ij}(x) = x_{ij}$, for all $i, j = 1, \ldots, n$, one observes that

$$\nabla_x E_{ij} \overset{(2.10)}{=} \check{B}^{-1}(dE_{ij,x}) = \check{B}^{-1}(E_{ij}) = e_{ji}, \ \forall i, j = 1, \ldots, n, \tag{2.12}$$

where e_{ji} is the $n \times n$ elementary matrix with a 1 in the position (j, i). In the previous formula, the second equality follows from the linearity of E_{ij}, while the third one follows from the properties of the trace. After this preliminary observation, one can compute the *elementary* Poisson brackets:

$$
\begin{aligned}
\{E_{ij}, E_{kl}\}(x) \ &\overset{(2.11)}{=} \ B(x, [\nabla_x E_{ij}, \nabla_x E_{kl}]) \\
&\overset{(2.12)}{=} \ B(x, [e_{ji}, e_{lk}]) \\
&= \ \mathrm{tr}(x[e_{ji}, e_{lk}]) \\
&= \ \delta_{il} \, \mathrm{tr}(x e_{jk}) - \delta_{kj} \, \mathrm{tr}(x e_{li}) \\
&= \ \delta_{il} \sum_{r,s=1}^{n} x_{rs} \, \mathrm{tr}(e_{rs} e_{jk}) - \delta_{kj} \sum_{r,s=1}^{n} x_{rs} \, \mathrm{tr}(e_{rs} e_{li}) \\
&= \ \delta_{il} \sum_{r,s=1}^{n} x_{rs} \delta_{sj} \delta_{rk} - \delta_{kj} \sum_{r,s=1}^{n} x_{rs} \delta_{sl} \delta_{ri} \\
&= \ \delta_{il} E_{kj}(x) - \delta_{kj} E_{il}(x),
\end{aligned}
$$

for all $x \in \mathfrak{gl}_n(\mathbb{K})$. In other words

$$\{E_{ij}, E_{kl}\} = \delta_{il} E_{kj} - \delta_{kj} E_{il}, \ \forall i, j, k, l = 1, \ldots, n.$$

$$\triangle$$

Example 2.9 (Frozen Argument Bracket) Let $\mathfrak{g}$ be a finite-dimensional Lie algebra and let $A \in \mathrm{Hom}(\Lambda^2 \mathfrak{g}, \mathbb{K})$. Then $\{\cdot, \cdot\}_A : C^\infty(\mathfrak{g}^*) \times C^\infty(\mathfrak{g}^*) \to C^\infty(\mathfrak{g}^*)$ defined by

$$\{f, g\}_A(\beta) = A(df_\beta, dg_\beta), \tag{2.13}$$

for all $f, g \in C^\infty(\mathfrak{g}^\infty)$ and for all $\beta \in \mathfrak{g}^*$, is automatically a skew-symmetric biderivation. It is also Poisson bracket. In fact if $x_1, \ldots, x_n$ are as in (2.7), for all i, j

$$\Pi_{ij}^A = \{x_i, x_j\}_A = A(e_i, e_j)$$

is a constant function on $\mathfrak{g}^*$ and for this reason

$$\{\{x_i, x_j\}_A, x_k\}_A = 0, \ \forall i, j, k,$$

which implies that (2.13) satisfies the Jacobi identity.

As a particular case one can consider A_α defined by

$$A_\alpha(x, y) = \langle \alpha, [x, y] \rangle, \ \forall x, y \in \mathfrak{g},$$

where $\alpha \in \mathfrak{g}^*$. In this case (2.13) becomes

$$\{f, g\}_\alpha(\beta) = \langle \alpha, [df_\beta, dg_\beta] \rangle, \ \forall \beta \in \mathfrak{g}^*, \tag{2.14}$$

and it is called the *frozen argument bracket* with argument α. $\triangle$

In the next example we show that the Cartesian product of two Poisson manifolds can be endowed, in a natural way, with a structure of a Poisson manifold. First, let us introduce a piece of notation. Let (M, Π_M), (N, Π_N) be two Poisson manifolds and let $m_0 \in M$ and $n_0 \in N$. Consider the maps $i_M : M \to M \times N$, $i_N : N \to M \times N$ defined by $i_M(m) = (m, n_0)$ and $i_N(n) = (m_0, n)$, for all $m \in M$ and $n \in N$. Moreover, let $\pi_M : M \times N \to M$ and $\pi_N : M \times N \to N$ be the canonical projections. Then one has that $\pi_M \circ i_M = \mathrm{id}_M$ and $\pi_N \circ i_N = \mathrm{id}_N$, while $\pi_M \circ i_N$ and $\pi_N \circ i_M$ are constant maps. In the example below, we will denote the coordinates on M with x and the ones on N with y. In particular, every $f \in C^\infty(M \times N)$ will be thought as a function of (x, y) and the Cartan differential d on $M \times N$ will split as $d = d_x + d_y$, where d_x denotes the differential *only* with respect to the variables x, i.e., keeping y constant, while d_y will denote the differential *only* with respect to the variables y.

Example 2.10 (Product of Poisson Manifolds) Given (M, Π_M), (N, Π_N) as above, $M \times N$ is a Poisson manifold with Poisson tensor Π defined in the following formula:

$$\Pi(df, dg) = \Pi_M\big(i_M^*(d_x f), i_M^*(d_x g)\big) + \Pi_N\big(i_N^*(d_y f), i_N^*(d_y g)\big), \tag{2.15}$$

for all $f, g \in C^\infty(M \times N)$. $\triangle$

Problem 2.11 Prove that the bivector field defined in Formula (2.15) is indeed a Poisson tensor field whose corresponding Poisson bracket is

$$\{f, g\}_\Pi(x, y) = \{i_M^* f, i_M^* g\}_{\Pi_M}(x) + \{i_N^* f, i_N^* g\}_{\Pi_N}(y),$$

for all $f, g \in C^\infty(M \times N)$.

Hint: Note that $i_M^*(d_x f) = d_M(i_M^* f)$ and $i_N^*(d_y f) = d_N(i_N^* f)$, for all $f \in C^\infty(M \times N)$, where d_M and d_N denote the Cartan differential on $C^\infty(M)$ and $C^\infty(N)$, respectively.   △

As we have already mentioned above, the biderivations of $C^\infty(M)$ are in one-to-one correspondence with the sections of $\Lambda^2 TM$. On the other hand, not every biderivation of $C^\infty(M)$ fulfills the Jacobi identity and for this reason not every biderivation defines a Poisson structure on M. The constraint imposed by the Jacobi identity, if read at the level of the sections of $\Lambda^2 TM$, becomes the following statement:

Proposition 2.12 *A bivector field Π on M is Poisson if and only if there exists an atlas $\{(U_\alpha, x_1^\alpha, \ldots, x_n^\alpha)\}_{\alpha \in \mathcal{A}}$, such that, for each $\alpha \in \mathcal{A}$:*

$$\sum_{r=1}^{n} \left(\Pi_{kr}^\alpha \frac{\partial \Pi_{ij}^\alpha}{\partial x_r^\alpha} + \Pi_{jr}^\alpha \frac{\partial \Pi_{ki}^\alpha}{\partial x_r^\alpha} + \Pi_{ir}^\alpha \frac{\partial \Pi_{jk}^\alpha}{\partial x_r^\alpha} \right) = 0, \forall i, j, k = 1, \ldots, n, \qquad (2.16)$$

where the $\{\Pi_{ij}^\alpha\}_{i,j=1,\cdots,n}$ are the components of the bivector field Π when restricted to the open subset U_α, i.e.,

$$\Pi|_{U_\alpha} = \sum_{i<j} \Pi_{ij}^\alpha \frac{\partial}{\partial x_i^\alpha} \wedge \frac{\partial}{\partial x_j^\alpha}.$$

Proof Let $(U, x_1, \ldots, x_n)$ be a chart on M (we drop the index of the chart to keep at a minimum the notation). Since $\Pi_{ij} = \{x_i, x_j\}$, one can write

$$\{\{x_i, x_j\}, x_k\} = \{\Pi_{ij}, x_k\} = \sum_{l,m=1}^{n} \Pi_{lm} \frac{\partial \Pi_{ij}}{\partial x_l} \frac{\partial x_k}{\partial x_m} = \sum_{l=1}^{n} \Pi_{lk} \frac{\partial \Pi_{ij}}{\partial x_l}.$$

In this way, taking cyclic permutations over i, j, k, one has

$$\{\{x_i, x_j\}, x_k\} + \{\{x_k, x_i\}, x_j\} + \{\{x_j, x_k\}, x_i\}$$

$$= \sum_{l=1}^{n} \left(\Pi_{lk} \frac{\partial \Pi_{ij}}{\partial x_l} + \Pi_{lj} \frac{\partial \Pi_{ki}}{\partial x_l} + \Pi_{li} \frac{\partial \Pi_{jk}}{\partial x_l} \right).$$

From this computation it follows that if the bracket $\{f, g\} = \langle \Pi, df \wedge dg \rangle$ satisfies the Jacobi identity, then (2.16) holds. On the other hand, let $f, g, h \in C^\infty(M)$. Then

$$\{f, \{g, h\}\}$$

$$= \sum_{k,l=1}^{n} \Pi_{kl} \frac{\partial f}{\partial x_k} \sum_{i,j=1}^{n} \left(\frac{\partial \Pi_{ij}}{\partial x_l} \frac{\partial g}{\partial x_i} \frac{\partial h}{\partial x_j} + \Pi_{ij} \frac{\partial^2 g}{\partial x_l \partial x_i} \frac{\partial h}{\partial x_j} + \Pi_{ij} \frac{\partial g}{\partial x_i} \frac{\partial^2 h}{\partial x_l \partial x_j} \right).$$

The sum of the r.h.s of the previous equality over the cyclic permutations of f, g, h is composed of the following four terms:

$$(1) \quad \sum_{i,j,k,l=1}^{n} \Pi_{kl} \frac{\partial \Pi_{ij}}{\partial x_l} \left(\frac{\partial f}{\partial x_k} \frac{\partial g}{\partial x_i} \frac{\partial h}{\partial x_j} + \frac{\partial f}{\partial x_i} \frac{\partial g}{\partial x_j} \frac{\partial h}{\partial x_k} + \frac{\partial f}{\partial x_j} \frac{\partial g}{\partial x_k} \frac{\partial h}{\partial x_i} \right),$$

$$(2) \quad \sum_{i,j,k,l=1}^{n} \Pi_{kl} \Pi_{ij} \left(\frac{\partial f}{\partial x_k} \frac{\partial h}{\partial x_j} \frac{\partial^2 g}{\partial x_l \partial x_i} + \frac{\partial f}{\partial x_i} \frac{\partial h}{\partial x_k} \frac{\partial^2 g}{\partial x_l \partial x_j} \right),$$

$$(3) \quad \sum_{i,j,k,l=1}^{n} \Pi_{kl} \Pi_{ij} \left(\frac{\partial f}{\partial x_j} \frac{\partial g}{\partial x_k} \frac{\partial^2 h}{\partial x_l \partial x_i} + \frac{\partial f}{\partial x_k} \frac{\partial g}{\partial x_i} \frac{\partial^2 h}{\partial x_l \partial x_j} \right),$$

$$(4) \quad \sum_{i,j,k,l=1}^{n} \Pi_{kl} \Pi_{ij} \left(\frac{\partial g}{\partial x_j} \frac{\partial h}{\partial x_k} \frac{\partial^2 f}{\partial x_l \partial x_i} + \frac{\partial g}{\partial x_k} \frac{\partial h}{\partial x_i} \frac{\partial^2 f}{\partial x_l \partial x_j} \right).$$

After the substitution $i \rightsquigarrow k, \ j \rightsquigarrow i, \ k \rightsquigarrow j$, and $j \rightsquigarrow k, \ k \rightsquigarrow i, \ i \rightsquigarrow j$ in the second and in the third summand of (1), respectively, the latter becomes

$$\sum_{i,j,k,l=1}^{n} \left(\Pi_{kl} \frac{\partial \Pi_{ij}}{\partial x_l} + \Pi_{jl} \frac{\partial \Pi_{ki}}{\partial x_l} + \Pi_{il} \frac{\partial \Pi_{jk}}{\partial x_l} \right) \frac{\partial f}{\partial x_k} \frac{\partial g}{\partial x_j} \frac{\partial h}{\partial x_i},$$

which, if (2.16) holds, is equal to zero for all $f, g, h \in C^\infty(M)$. On the other hand, (2), (3), and (4) are identically equal to zero. Let us check this statement for (2). This can be rewritten as

$$\sum_{k,l=1}^{n} \Pi_{kl} \left(\frac{\partial f}{\partial x_k} \sum_{i,j=1}^{n} \Pi_{ij} \frac{\partial h}{\partial x_j} \frac{\partial^2 g}{\partial x_l \partial x_i} + \frac{\partial h}{\partial x_k} \sum_{i,j=1}^{n} \Pi_{ij} \frac{\partial f}{\partial x_i} \frac{\partial^2 g}{\partial x_l \partial x_j} \right).$$

Now, fixing the values of k and l, for $i = k$ and $j = l$, one has the following contribution:

$$\Pi_{kl}\Pi_{kl}\frac{\partial f}{\partial x_k}\frac{\partial^2 g}{\partial x_l \partial x_k}\frac{\partial h}{\partial x_l} + \Pi_{kl}\Pi_{kl}\frac{\partial h}{\partial x_k}\frac{\partial^2 g}{\partial x_l \partial x_l}\frac{\partial f}{\partial x_k} + \Pi_{lk}\Pi_{kl}\frac{\partial f}{\partial x_l}\frac{\partial^2 g}{\partial x_k \partial x_k}\frac{\partial h}{\partial x_l}$$

$$+ \Pi_{lk}\Pi_{kl}\frac{\partial f}{\partial x_k}\frac{\partial^2 g}{\partial x_l \partial x_k}\frac{\partial h}{\partial x_l},$$

while, for $i = l$ and $j = k$, one gets

$$\Pi_{kl}\Pi_{lk}\frac{\partial f}{\partial x_k}\frac{\partial^2 g}{\partial x_l \partial x_l}\frac{\partial h}{\partial x_k} + \Pi_{kl}\Pi_{lk}\frac{\partial h}{\partial x_k}\frac{\partial^2 g}{\partial x_l \partial x_l}\frac{\partial f}{\partial x_l} + \Pi_{lk}\Pi_{lk}\frac{\partial f}{\partial x_l}\frac{\partial^2 g}{\partial x_k \partial x_l}\frac{\partial h}{\partial x_k}$$

$$+ \Pi_{lk}\Pi_{lk}\frac{\partial f}{\partial x_l}\frac{\partial^2 g}{\partial x_k \partial x_k}\frac{\partial h}{\partial x_l},$$

which, because of the skew-symmetry of the components of Π, sum up to zero, proving the statement. Similarly, (3) and (4) can be checked to be identically zero.

$$\square$$

1.1   The Schouten Bracket

To contextualize the previous result one needs to introduce the so-called *Schouten* or *Schouten-Nijenhuis* bracket. To this end let $\Gamma(\Lambda^\bullet T\mathbb{R}^n) = \oplus_{k\geq 0}\Gamma(\Lambda^k T\mathbb{R}^n)$ be the space of global sections of the exterior algebra of the tangent bundle of $\mathbb{R}^n$. The element $\eta \in \Gamma(\Lambda^\bullet T\mathbb{R}^n)$ is called homogeneous of degree k if $\eta \in \Gamma(\Lambda^k T\mathbb{R}^n)$. In the coordinates $x_1, \ldots, x_n$,

$$\eta = \sum_{i_1 < \cdots < i_k} \eta_{i_1 \cdots i_k} \frac{\partial}{\partial x_{i_1}} \wedge \cdots \wedge \frac{\partial}{\partial x_{i_k}},$$

where $\eta_{i_1 \ldots i_k} \in C^\infty(\mathbb{R}^n)$ for all multi-index $(i_1, \ldots, i_k)$. In other words, η can be regarded as a *homogeneous polynomial* of degree k with coefficients in $C^\infty(\mathbb{R}^n)$ in the *noncommutative variables* $\frac{\partial}{\partial x_1}, \ldots, \frac{\partial}{\partial x_n}$. Renaming $\frac{\partial}{\partial x_i}$ by ξ_i for all i and dropping the wedge product, one can write

$$\eta = \sum_{i_1 < \cdots < i_k} \eta_{i_1 \ldots i_k} \xi_{i_1} \cdots \xi_{i_k}.$$

For all l let us define the noncommutative derivation $\frac{\partial}{\partial \xi_{i_l}}$ via

$$\frac{\partial(\xi_{i_1} \cdots \xi_{i_k})}{\partial \xi_{i_l}} = (-1)^{k-l}\xi_{i_1} \cdots \hat{\xi}_{i_l} \cdots \xi_{i_k}.$$

In this way $\Gamma(\Lambda^\bullet T\mathbb{R}^n)$ can be endowed with a remarkable $\mathbb{R}$-bilinear bracket $[\![\cdot,\cdot]\!]$, defined on its homogeneous elements by

$$[\![\eta_1,\eta_2]\!] = \sum_{i=1}^{n} \frac{\partial\eta_1}{\partial\xi_i}\frac{\partial\eta_2}{\partial x_i} - (-1)^{(k_1-1)(k_2-1)}\sum_{i=1}^{n}\frac{\partial\eta_2}{\partial\xi_i}\frac{\partial\eta_1}{\partial x_i}, \tag{2.17}$$

for all $\eta_1 \in \Gamma(\Lambda^{k_1}T\mathbb{R}^n)$ and $\eta_2 \in \Gamma(\Lambda^{k_2}T\mathbb{R}^n)$. Note that for all k_1, k_2

$$[\![\cdot,\cdot]\!] : \Gamma(\Lambda^{k_1}T\mathbb{R}^n) \times \Gamma(\Lambda^{k_2}T\mathbb{R}^n) \to \Gamma(\Lambda^{k_1+k_2-1}\mathbb{R}^n). \tag{2.18}$$

Definition 2.13 (Schouten-Nijenhuis Bracket) The bracket defined in (2.17) is called the Schouten-Nijenhuis bracket (SN-bracket from now on). $\triangle$

The proof of the following theorem is left to the reader as an exercise; see, for example, [69, Theorem 1.8.1].

Theorem 2.14 *The SN-bracket satisfies the following properties. For all* $\eta_1 \in \Gamma(\Lambda^{k_1}T\mathbb{R}^n)$, $\eta_2 \in \Gamma(\Lambda^{k_2}\mathbb{R}^n)$, *and* $\eta_3 \in \Gamma(\Lambda^{k_3}\mathbb{R}^n)$:

(1) $[\![\eta_1,\eta_2]\!] = -(-1)^{(k_1-1)(k_2-1)}[\![\eta_2,\eta_1]\!]$.

(2.a) $[\![\eta_1,\eta_2\wedge\eta_3]\!] = [\![\eta_1,\eta_2]\!]\wedge\eta_3 + (-1)^{(k_1-1)k_2}\eta_2\wedge[\![\eta_1,\eta_3]\!]$.

(2.b) $[\![\eta_1\wedge\eta_2,\eta_3]\!] = \eta_1\wedge[\![\eta_2,\eta_3]\!] + (-1)^{(k_3-1)k_2}[\![\eta_1,\eta_3]\!]\wedge\eta_2.$ $\qquad$ (2.19)

(3) $(-1)^{(k_1-1)(k_3-1)}[\![\eta_1,[\![\eta_2,\eta_3]\!]]\!] + (-1)^{(k_2-1)(k_1-1)}[\![\eta_2,[\![\eta_3,\eta_1]\!]]\!]$

$\quad + (-1)^{(k_3-1)(k_2-1)}[\![\eta_3,[\![\eta_1,\eta_2]\!]]\!] = 0.$

(4) *If* $X \in \Gamma(\Lambda^1 T\mathbb{R}^n)$, *then*

$$[\![X,\eta]\!] = \mathscr{L}_X\eta, \ \forall\eta\in\Gamma(\Lambda^k T\mathbb{R}^n), \ \forall k, \tag{2.20}$$

where $\mathscr{L}_X$ *is the Lie derivative with respect to* X.

Note that if $k_1 = 2 = k_2$, then $[\![\eta_1,\eta_2]\!] \in \Gamma(\Lambda^3\mathbb{R}^n)$. In particular one can prove that

Theorem 2.15 $\Pi \in \Gamma(\Lambda^2\mathbb{R}^n)$ *is a Poisson bivector if and only if*

$$[\![\Pi,\Pi]\!] = 0. \tag{2.21}$$

Proof It suffices to show that the condition in (2.21) is equivalent to the one in (2.16). Suppose that $\Pi = \frac{1}{2}\sum_{i<j}\Pi_{ij}(x)\,\xi_i\xi_j$. Then

$$\frac{\partial\Pi}{\partial\xi_k}\frac{\partial\Pi}{\partial x_k} = \left(-\frac{1}{2}\sum_{k,j}\Pi_{kj}\xi_j + \frac{1}{2}\sum_{k,i}\Pi_{ik}\xi_i\right)\left(\frac{1}{2}\sum_{l,m}\frac{\partial\Pi_{lm}}{\partial x_k}\xi_l\xi_m\right)$$

$$= \frac{1}{2} \left(\sum_{k,i} \Pi_{ki} \xi_i \right) \left(\sum_{l,m} \frac{\partial \Pi_{lm}}{\partial x_k} \xi_l \xi_m \right)$$

$$= \frac{1}{2} \sum_{i,k,l,m} \Pi_{ik} \frac{\partial \Pi_{lm}}{\partial x_k} \xi_i \xi_l \xi_m .$$

Since $\xi_i \xi_l \xi_m = \xi_m \xi_i \xi_l = \xi_l \xi_m \xi_i$, using the previous computation one can write

$$\frac{\partial \Pi}{\partial \xi_k} \frac{\partial \Pi}{\partial x_k} = \frac{1}{2} \sum_{i,l,m} \left(\sum_k \Pi_{ik} \frac{\partial \Pi_{lm}}{\partial x_k} + \Pi_{mk} \frac{\partial \Pi_{il}}{\partial x_k} + \Pi_{lk} \frac{\partial \Pi_{mi}}{\partial x_k} \right) \xi_i \xi_l \xi_m .$$

On the other hand, from (2.18) it follows at once that

$$[\![\Pi, \Pi]\!] = 2 \sum_k \frac{\partial \Pi}{\partial \xi_k} \frac{\partial \Pi}{\partial x_k} ,$$

which together with the previous computation implies that

$$[\![\Pi, \Pi]\!] = \sum_{i,l,m} \left(\sum_k \Pi_{ik} \frac{\partial \Pi_{lm}}{\partial x_k} + \Pi_{mk} \frac{\partial \Pi_{il}}{\partial x_k} + \Pi_{lk} \frac{\partial \Pi_{mi}}{\partial x_k} \right) \xi_i \xi_l \xi_m ,$$

which proves the statement. $\qquad\square$

In other words, the condition expressed in Formula (2.16) is nothing else than the one expressed in Formula (2.21). The previous result can be extended without any effort to a general manifold (we proved it for $M = \mathbb{R}^n$). In particular, it should be clear now why on a two-dimensional manifold M every bivector Π field defines a Poisson structure (see (ii) in Example 2.2): in fact, under this assumption, $[\![\Pi, \Pi]\!] = 0$ for all $\Pi \in \Gamma(\Lambda^2 TM)$ for dimensional reasons.

Problem 2.16

(*i*) Let $M = \mathbb{R}^n$ with coordinates $x_1, \ldots, x_n$ and let A be a skew-symmetric $n \times n$ matrix. Define $\Pi_{ij} = A_{ij} x_i x_j$ for all $i, j = 1, \ldots, n$. Show that the bivector field

$$\Pi = \frac{1}{2} \sum_{i,j=1}^n \Pi_{ij} \frac{\partial}{\partial x_i} \wedge \frac{\partial}{\partial x_j} \tag{2.22}$$

is Poisson. Note that the Poisson bracket defined by (2.22) is *quadratic*, i.e., the bracket of any two linear functions is a *quadratic* function. For this reason (2.22) is said to define a quadratic Poisson structure.

(ii) Let $\alpha \in \Omega^2(M)$ and define $\Pi \in \Gamma(\Lambda^2 T^*M)$ via

$$\Pi = \sum_{i=1}^{n} \frac{\partial}{\partial p_i} \wedge \frac{\partial}{\partial x_i} + \sum_{i<j} \alpha_{ij}(x) \frac{\partial}{\partial p_i} \wedge \frac{\partial}{\partial p_j}, \qquad (2.23)$$

where (p, x) are (local) Darboux coordinates with respect to the canonical symplectic form of T^*M and α_{ij} denotes coefficients of the pullback of α with respect to the canonical projection $p : T^*M \to M$. Prove that (2.23) is Poisson if and only if α is closed. Note that in this case Π is invertible. $\triangle$

1.2 Hamiltonian Vector Fields

We will now introduce a class of vector fields playing an important role in Poisson geometry. To this end, we start observing that every bivector field Π on M defines a $C^\infty(M)$-linear map:

$$\Pi^\sharp : \Omega^1(M) \to \mathfrak{X}(M),$$

via

$$\langle \beta, \Pi^\sharp(\alpha) \rangle = \Pi(\alpha, \beta) = \langle \alpha \wedge \beta, \Pi \rangle, \qquad (2.24)$$

for all $\alpha, \beta \in \Omega^1(M)$. In the following we will write $X_\alpha = \Pi^\sharp(\alpha)$. Note that $\Pi^\sharp$ is skew-symmetric; in fact

$$\langle \beta, \Pi^\sharp(\alpha) \rangle = \Pi(\alpha, \beta) = -\Pi(\beta, \alpha) = -\langle \alpha, \Pi^\sharp(\beta) \rangle = -\langle \Pi^\sharp(\beta), \alpha \rangle,$$

for all $\alpha, \beta \in \Omega^1(M)$. If $x_1, \ldots, x_n$ are coordinates on $U \subset M$ such that $\alpha_U = \sum_{i=1}^{n} \alpha_i dx_i$ and $\Pi_U = \sum_{i<j} \Pi_{ij} \frac{\partial}{\partial x_i} \wedge \frac{\partial}{\partial x_j}$,

$$X_\alpha = \sum_{j=1}^{n} \left(\sum_{i=1}^{n} \Pi_{ij} \alpha_i \right) \frac{\partial}{\partial x_j}.$$

For all $m \in M$, $\Pi^\sharp$ defines the $\mathbb{K}$-linear map

$$\Pi_m^\sharp : T_m^*M \to T_mM, \qquad (2.25)$$

via

$$\Pi_m^\sharp(\sigma) = \Pi^\sharp(\alpha)(m),$$

where α is any one-form defined in a neighborhood of m such that $\alpha_m = \sigma$. Since $\Pi^\sharp$ is $C^\infty(M)$-linear, its image is *locally* generated by the vector fields of the form X_α, with $\alpha = df$, for some locally defined smooth function f. In analogy to what is discussed in Chap. 1 (see Definitions 1.26 and 1.30), one can introduce the following notions.

Definition 2.17 A vector field $X \in \mathfrak{X}(M)$ is called:

(i) *Locally Hamiltonian* if $X = X_\alpha$, with $d\alpha = 0$.
(ii) *Hamiltonian* if $X = X_{df}$, for some $f \in C^\infty(M)$. In this case we will write, with a slight notational abuse, X_f instead of X_{df}.

$\triangle$

Note that for every $f, g \in C^\infty(M)$,

$$\{f, g\} = \langle df \wedge dg, \Pi \rangle = \langle dg, X_f \rangle. \tag{2.26}$$

Problem 2.18 Prove that $[\![\Pi, f]\!] = -X_f$ for all $f \in C^\infty(M)$. (Hint: Write Π in local coordinates and use (2.19)). $\triangle$

Problem 2.19 Let Π be any bivector on a manifold M and consider for any $m \in M$ the linear map $\Pi_m^\sharp : T_m^* M \to T_m M$. Show that $\ker(\Pi_m^\sharp)$ can be identified with $(\mathrm{im}(\Pi_m^\sharp))^\circ$ where $(\mathrm{im}(\Pi_m^\sharp))^\circ$ is the annihilator of $\mathrm{im}(\Pi_m^\sharp)$, which is the vector subspace of $T_m^* M$ consisting of all covectors ξ such that $\xi(v) = 0$ for all $v \in \mathrm{im}(\Pi_p^\sharp)$. $\triangle$

Remark 2.20 In this remark we will add a few comments about the Poisson tensors defined by a symplectic structure. In this case, if $\omega = \sum_{i<j} \omega_{ij} dx_i \wedge dx_j$, one knows how to write the Poisson tensor defined by the bracket $\{f, g\} = \omega(X_f, X_g)$; see Formula (2.4) in Example 2.3. We want now to show that for every $f \in C^\infty(M)$

$$i_{\Pi^\sharp(df)}\omega = -df,$$

i.e., that the Hamiltonian vector field with respect the Poisson structure (2.4) *is* the Hamiltonian vector field defined by the corresponding symplectic form. To this end, if $X = \sum_s \xi_s \frac{\partial}{\partial x_s}$ and $i_X \omega = -df$, then, for all k,

$$\xi_k = \sum_r (\omega^{-1})_{kr} \frac{\partial f}{\partial x_r},$$

which implies that

$$X = \sum_j \left(\sum_k (\omega^{-1})_{jk} \frac{\partial f}{\partial x_k} \right) \frac{\partial}{\partial x_j};$$

see the computations in Example 2.3. On the other hand, (2.4) implies that

$$
\Pi^\sharp(df) = \frac{1}{2}\left(\sum_{i,j,k}(\omega^{-1})_{ji}\frac{\partial f}{\partial x_k}\delta_{ik}\frac{\partial}{\partial x_j} - (\omega^{-1})_{ji}\frac{\partial f}{\partial x_k}\delta_{jk}\frac{\partial}{\partial x_i}\right)
$$

$$
= \frac{1}{2}\left(\sum_{j,k}(\omega^{-1})_{jk}\frac{\partial f}{\partial x_k}\frac{\partial}{\partial x_j} - \sum_{i,k}(\omega^{-1})_{ki}\frac{\partial f}{\partial x_k}\frac{\partial}{\partial x_i}\right)
$$

$$
= \sum_{j}\left(\sum_{k}(\omega^{-1})_{jk}\frac{\partial f}{\partial x_k}\right)\frac{\partial}{\partial x_j},
$$

which proves our assertion. If (p, x) is a set of Darboux coordinates for ω, i.e., such that $\omega = \sum_i dp_i \wedge dx_i$ the matrix of the map $\omega^\flat$ (see Formula (1.18)), is

$$
\omega^\flat = \left(\begin{array}{c|c} 0 & \mathrm{id} \\ \hline -\,\mathrm{id} & 0 \end{array}\right), \tag{2.27}
$$

where blocks are $n \times n$; see Example 1.52. Note that in this representation we ordered the basis of local one-forms writing first the positions x_is and then the moments p_is. The Poisson tensor defined by ω written in the coordinates (p, x) is

$$
\Pi = \sum_{i,j}\{p_i, x_j\}\frac{\partial}{\partial p_i} \wedge \frac{\partial}{\partial x_j}
$$

$$
= \sum_{i,j}\omega(X_{p_i}, X_{x_j})\frac{\partial}{\partial p_i} \wedge \frac{\partial}{\partial x_j}
$$

$$
= \sum_{i,j}\omega\left(\frac{\partial}{\partial x_i}, -\frac{\partial}{\partial p_j}\right)\frac{\partial}{\partial p_i} \wedge \frac{\partial}{\partial x_j}
$$

$$
= \sum_{i,j}\delta_{ji}\frac{\partial}{\partial p_i} \wedge \frac{\partial}{\partial x_j}
$$

$$
= \sum_{i}\frac{\partial}{\partial p_i} \wedge \frac{\partial}{\partial x_i},
$$

where we used $X_{p_i} = \frac{\partial}{\partial x_i}$ and $X_{x_i} = -\frac{\partial}{\partial p_i}$ for all i. Obviously the previous computation shows that in Darboux coordinates $\Pi_{ij} \overset{(2.4)}{=} (\omega^{-1})_{ji} \overset{(2.27)}{=} \omega_{ij}$. A different, though equivalent, way to look at the last computation goes as follows: the tensor Π associated to ω is the image of ω via the natural extension to $\Omega^2(M)$ of the map $(\omega^\flat)^{-1} : \Omega^1(M) \to \mathfrak{X}(M)$. In fact the latter, in Darboux coordinates, is given by $(\omega^\flat)^{-1}(dp_i) = -\frac{\partial}{\partial x_i}$ and $(\omega^\flat)^{-1}(dx_i) = \frac{\partial}{\partial p_i}$, which implies that its extension $(\omega^\flat)^{-1} : \Omega^2(M) \to \Gamma(\Lambda^2 TM)$ satisfies

$$(\omega^\flat)^{-1} \sum_i dp_i \wedge dx_i = \sum_i \left(-\frac{\partial}{\partial x_i}\right) \wedge \frac{\partial}{\partial p_i} = \sum_i \frac{\partial}{\partial p_i} \wedge \frac{\partial}{\partial x_i}$$

Finally, it is worth observing that with our conventions

$$\Pi^\sharp = -(\omega^\flat)^{-1}, \tag{2.28}$$

as one can easily check using Darboux coordinates. To summarize this discussion one could say that if Π is a Poisson tensor coming from a symplectic structure ω, then:

(i) $i_{\Pi^\sharp(df)}\omega = -df$, for all $f \in C^\infty(M)$, i.e., the Hamiltonian vector field of f with respect to ω coincides with the Hamiltonian vector field of f with respect to the corresponding Poisson structure.

(ii) $\Pi_{ij} = \omega(X_{x_i}, X_{x_j})$ for all i, j, where $x_1, \ldots, x_n$ are local coordinates and X_f is the Hamiltonian vector field of f with respect to ω or, equivalently, with respect to Π. More generally, $\Pi(df, dg) = \omega(X_f, X_g)$, for all $f, g \in C^\infty(M)$, i.e., the Poisson brackets defined by Π and ω are the same.

(iii) $(\omega^\flat)^{-1} = -\Pi^\sharp$.

$\triangle$

Hereafter $\mathcal{H}(M, \Pi)$ will denote the $\mathbb{K}$-vector space of the Hamiltonian vector fields defined on M by the Poisson bivector field Π.

Problem 2.21 Prove that $\mathcal{H}(M, \Pi)$ is a vector space. $\triangle$

Example 2.22 Let $\mathfrak{g}^*$ be endowed with its linear Poisson structure and $f \in C^\infty(\mathfrak{g}^*)$. Then, in the linear coordinates $x_1, \ldots, x_n$, one has

$$X_f = \sum_{j=1}^n \left(\sum_{i,k=1}^n c_{ij}^k x_k \frac{\partial f}{\partial x_i} \right) \frac{\partial}{\partial x_j}. \tag{2.29}$$

$\triangle$

Problem 2.23 Show that, for each $f \in C^\infty(\mathfrak{g}^*)$ and every $\alpha \in \mathfrak{g}^*$, one has that

$$X_f(\alpha) = -\operatorname{ad}_{df_\alpha}^\sharp(\alpha), \tag{2.30}$$

where the X_f on the left-hand side is the Hamiltonian vector field (2.29). In particular, if f is a *linear* function on $\mathfrak{g}^*$, i.e., $f \in \mathfrak{g}$, then (2.29) is the fundamental vector field defined by f via the coadjoint action of corresponding Lie group; see Appendix C. $\triangle$

At a risk of abusing the notation, we will write $\Pi^\sharp$ to denote the $\mathbb{K}$-linear map which to every (locally defined) function f associates the corresponding Hamiltonian vector field. Now one can introduce the following.

Definition 2.24 A function $f \in C^\infty(M)$ is called a *Casimir* of Π if $f \in \ker(\Pi^\sharp)$, i.e., f is a Casimir if and only if $X_f = 0$. $\triangle$

Formula (2.26) implies that f is a Casimir if and only if $\{f, g\} = 0$ for all $g \in C^\infty(M)$. In particular if $f \in C^\infty_M(U)$ is such that $\{f, g\} = 0$ for all $g \in C^\infty_M(U)$, then f is called a *local* Casimir of Poisson structure defined by Π; see also Remark 2.29.

Note that f is a Casimir if and only if there is an atlas $\{(U_\alpha, x_1^\alpha, \ldots, x_n^\alpha)\}_{\alpha \in \mathcal{A}}$ such that, for all $\alpha \in \mathcal{A}$

$$\sum_{i=1}^{n} \Pi_{ij}^\alpha \frac{\partial f}{\partial x_i^\alpha} = 0, \ \forall j = 1, \ldots, n,$$

where $\{\Pi_{ij}^\alpha\}_{i,j=1,\cdots,n}$ are the coefficients of Π on U_α; see Formulas (2.2) and (2.3). For instance, if Π is defined by a symplectic form ω, f is a Casimir if and only if $df = 0$, i.e., if f is constant on M (recall that we always assume that M is connected). On the other hand, if Π is a linear Poisson structure defined by $\Pi_{ij}(x) = \sum_{k=1}^{n} c_{ij}^k x_k$, f is a Casimir of Π if and only if

$$\sum_{i=1}^{n} \sum_{k=1}^{n} c_{ij}^k x_k \frac{\partial f}{\partial x_i} = 0, \ \forall j = 1, \ldots, n,$$

or, equivalently, if and only if

$$X_{x_j}(f) = 0, \ \forall j = 1, \ldots, n,$$

where X_{x_j} is the fundamental vector field defined by $x_j \in \mathfrak{g}$ via the coadjoint action; see Formula (2.30).

Example 2.25 For all $n \geq 1$, $H_k : \mathfrak{gl}_n(\mathbb{K}) \to \mathbb{K}$ defined by $H_k(x) = \dfrac{\mathrm{tr}(x^k)}{k}$ is a Casimir of the Poisson bracket defined in (2.11). In fact

$$B(\nabla_x H_k, y) = \frac{d}{dt}\bigg|_{t=0} H_k(x + ty) = \frac{1}{k} \frac{d}{dt}\bigg|_{t=0} \mathrm{tr}(x + ty)^k = \mathrm{tr}(x^{k-1}y)$$

$$= B(x^{k-1}, y), \ \forall y \in \mathfrak{gl}_n(\mathbb{K}),$$

which implies $\nabla_x H_k = x^{k-1}$, for all $x \in \mathfrak{gl}_n(\mathbb{K})$. Then

$$\{H_k, f\}(x) = B(x, [\nabla_x H_k, \nabla_x f]) = B([\nabla_x H_k, x], \nabla_x f) = 0,$$
$$\forall f \in C^\infty(\mathfrak{g}), \ x \in \mathfrak{g}.$$

$\triangle$

Let $\mathrm{Cas}(M, \Pi)$ be the set of the Casimir functions of the Poisson manifold (M, Π). One easily checks that $\mathrm{Cas}(M, \Pi)$ is vector space which is closed with respect to both the Poisson bracket and the multiplication defined on $C^\infty(M)$. This last statement follows since the Poisson bracket is a biderivation of $C^\infty(M)$, while the proof of the first statement follows at once by the Jacobi identity. To study the relation between Casimir functions and Hamiltonian vector fields, we need the following preliminary result, which is interesting per se.

Lemma 2.26

$$\mathscr{L}_{X_f} \Pi = 0. \tag{2.31}$$

See below for a proof of Lemma 2.26.

Using (2.31) one can prove that

Proposition 2.27 *The vector space $\mathscr{H}(M, \Pi)$ is closed with respect to the restriction of the (canonical) Lie bracket defined on $\mathfrak{X}(M)$, more precisely for all $f, g \in C^\infty(M)$*

$$[X_f, X_g] = X_{\{f,g\}}.$$

Proof In fact

$$[X_f, X_g] = \mathscr{L}_{X_f} X_g = \mathscr{L}_{X_f}\left(\Pi^\sharp(dg)\right) \overset{(2.31)}{=} \cancel{\mathscr{L}_{X_f}\Pi^\sharp(dg)} + \Pi^\sharp\left(d(\mathscr{L}_{X_f} g)\right)$$
$$= \Pi^\sharp(d\{f, g\}) = X_{\{f,g\}}.$$

$\square$

Remark 2.28 Note that the proof of the previous proposition follows at once from the Jacobi identity. In fact, if $f, g, h \in C^\infty(M)$, then

$$[X_f, X_g](h) = X_f(X_g(h)) - X_g(X_f(h)) = X_f(\{g, h\}) - X_g(\{f, h\})$$
$$= \{f, \{g, h\}\} - \{g, \{f, h\}\} = \{f, \{g, h\}\} + \{g, \{h, f\}\}$$
$$= \{\{f, g\}, h\} = X_{\{f,g\}}(h).$$

$\triangle$

We can now, as promised, go back to the relation between Casimir functions and Hamiltonian vector fields. In fact, the previous discussion entails that every Poisson manifold has associated the following sequence of Lie algebras, which should be compared with the exact sequence (1.25):

$$0 \to \mathrm{Cas}(M, \Pi) \to (C^\infty(M), \{\cdot, \cdot\}) \xrightarrow{\Pi^\sharp} \mathcal{H}(M, \Pi) \to 0. \tag{2.32}$$

Note that $\mathrm{Cas}(M, \Pi)$ is an ideal of the Lie algebra $(C^\infty(M), \{\cdot, \cdot\})$, which reduces to the space of constant functions on M if the Poisson structure is invertible. Now let us prove (2.31).

Proof (of Lemma 2.26) To this end let $(U, x_1, \ldots, x_n)$ be a local set of coordinates and let

$$\Pi = \frac{1}{2} \sum_{k,l} \Pi_{kl} \frac{\partial}{\partial x_k} \wedge \frac{\partial}{\partial x_l} \quad \text{and} \quad X_f = \sum_{i,j=1}^{n} \Pi_{ij} \frac{\partial f}{\partial x_i} \frac{\partial}{\partial x_j}.$$

Since the Lie derivative is a derivation of the exterior algebra of TM and $\mathscr{L}_X Y = [X, Y]$ for all $X, Y \in \mathfrak{X}(M)$, one has

$$\mathscr{L}_{X_f} \Pi = \frac{1}{2} \left\{ \sum_{k,l=1}^{n} X_f(\Pi_{kl}) \frac{\partial}{\partial x_k} \wedge \frac{\partial}{\partial x_l} + \sum_{k,l=1}^{n} \Pi_{kl} \left[X_f, \frac{\partial}{\partial x_k} \right] \wedge \frac{\partial}{\partial x_l} \right.$$

$$\left. + \sum_{k,l=1}^{n} \Pi_{kl} \frac{\partial}{\partial x_k} \wedge \left[X_f, \frac{\partial}{\partial x_l} \right] \right\}. \tag{2.33}$$

Since

$$X_f(\Pi_{kl}) = \sum_{i,j=1}^{n} \Pi_{ij} \frac{\partial f}{\partial x_i} \frac{\partial \Pi_{kl}}{\partial x_j}$$

and

$$\left[X_f, \frac{\partial}{\partial x_k} \right] = - \sum_{i,j=1}^{n} \frac{\partial \Pi_{ij}}{\partial x_k} \frac{\partial f}{\partial x_i} \frac{\partial}{\partial x_j} - \sum_{i,j=1}^{n} \Pi_{ij} \frac{\partial^2 f}{\partial x_k \partial x_i} \frac{\partial}{\partial x_j},$$

one concludes that (2.33) can be rewritten as the sum of the following two contributions:

$$\frac{1}{2} \sum_{i,j,k,l=1}^{n} \left(\Pi_{ij} \frac{\partial \Pi_{kl}}{\partial x_j} \frac{\partial f}{\partial x_i} \frac{\partial}{\partial x_k} \wedge \frac{\partial}{\partial x_l} - \underbrace{\Pi_{kl} \frac{\partial \Pi_{ij}}{\partial x_k} \frac{\partial f}{\partial x_i} \frac{\partial}{\partial x_j}}_{(1)} \wedge \frac{\partial}{\partial x_l} \right.$$

$$\left. -\,\Pi_{kl}\underbrace{\frac{\partial \Pi_{ij}}{\partial x_l}\frac{\partial f}{\partial x_i}\frac{\partial}{\partial x_k}}_{(2)}\wedge\frac{\partial}{\partial x_j}\right) \tag{2.34}$$

$$-\frac{1}{2}\sum_{i,j,k,l=1}^{n}\left(\underbrace{\Pi_{kl}\Pi_{ij}\frac{\partial^2 f}{\partial x_k\partial x_i}\frac{\partial}{\partial x_j}\wedge\frac{\partial}{\partial x_l}}_{(3)}+\Pi_{kl}\Pi_{ij}\frac{\partial^2 f}{\partial x_l\partial x_i}\frac{\partial}{\partial x_k}\wedge\frac{\partial}{\partial x_j}\right). \tag{2.35}$$

Changing j with k in (1) and j with l in (2), (2.34) becomes

$$\frac{1}{2}\sum_{i,k,l=1}^{n}\left(\sum_{j=1}^{n}\Pi_{ij}\frac{\partial \Pi_{kl}}{\partial x_j}+\Pi_{lj}\frac{\partial \Pi_{ik}}{\partial x_j}+\Pi_{kj}\frac{\partial \Pi_{li}}{\partial x_j}\right)\frac{\partial f}{\partial x_i}\frac{\partial}{\partial x_k}\wedge\frac{\partial}{\partial x_l}$$

which is equal to zero because of (2.16). On the other hand, changing l with k in (3), (2.35) becomes

$$-\sum_{i,j,k,l=1}^{n}\Pi_{kl}\Pi_{ij}\frac{\partial^2 f}{\partial x_l\partial x_i}\frac{\partial}{\partial x_k}\wedge\frac{\partial}{\partial x_j}$$

which is also equal to zero as one can check changing, in the formula, k with j and l with i. $\square$

Remark 2.29 Observe that it is possible to define a *local* version of the exact sequence (2.32). In fact every open set $U \subset M$ inherits from Π a Poisson structure, whose Poisson bracket will be denoted by $\{\cdot,\cdot\}_U$; see Corollary 2.54. This defines $\mathrm{Cas}_M(U,\Pi) = \{f \in C_M^\infty(U)\,|\,\{f,g\}_U = 0\}$ and $\mathcal{H}_M(U,\Pi) = \{X \in \mathfrak{X}_M(U)\,|\,X = \Pi^\sharp(df)\}$, which turn out to be an ideal of $(C_M^\infty(U),\{\cdot,\cdot\}_U)$ and a Lie subalgebra of $(\mathfrak{X}_M(U),[\cdot,\cdot]_U)$, respectively, and fit in the following exact sequence:

$$0 \to \mathrm{Cas}_M(U,\Pi) \to (C_M^\infty(U),\{\cdot,\cdot\}_U) \xrightarrow{\Pi^\sharp} \mathcal{H}_M(U,\Pi) \to 0,$$

which is the *local* analogue of the (2.32). The elements of $\mathrm{Cas}_M(U,\Pi)$ will be called *local Casimirs* of Π on the open set U; see also the comment below Definition 2.24. Note further that for all $U \subset V \subset M$ open sets the restriction map $C_M^\infty(V) \to C_M^\infty(U)$ restricts to a map $\mathrm{Cas}_M(V,\Pi) \to \mathrm{Cas}_M(U,\Pi)$; in particular every Casimir $f \in \mathrm{Cas}(M,\Pi)$ defines an element in $\mathrm{Cas}_M(U,\Pi)$, for all U open in M. $\triangle$

The Lie algebras entering in the exact sequence (2.32) are not the only ones that can be associated to every Poisson manifold. In fact, every Poisson bivector field on M induces a Lie bracket on $\Omega^1(M)$, extending the Poisson bracket on $C^\infty(M)$.

More precisely, let Π be a bivector field on M and let $\{\cdot,\cdot\}_\Pi : \Omega^1(M) \times \Omega^1(M) \to \Omega^1(M)$ defined by

$$\{\alpha,\beta\}_\Pi = \mathscr{L}_{\Pi^\sharp(\alpha)}\beta - \mathscr{L}_{\Pi^\sharp(\beta)}\alpha + d\langle \alpha, \Pi^\sharp(\beta)\rangle. \tag{2.36}$$

First, one observes

Lemma 2.30

$$\{df, dg\}_\Pi = d\{f, g\}, \quad \forall f, g \in C^\infty(M). \tag{2.37}$$

Problem 2.31 Prove Formula (2.37). $\triangle$

Armed with this result, one can prove the following important.

Lemma 2.32 Π *is a Poisson tensor if and only if*

$$\Pi^\sharp(\{\alpha,\beta\}_\Pi) = [\Pi^\sharp(\alpha), \Pi^\sharp(\beta)], \tag{2.38}$$

for all $\alpha, \beta \in \Omega^1(M)$.

Proof First, note that the $\mathbb{K}$-bilinear form $C : \Omega^1(M) \times \Omega^1(M) \to \mathfrak{X}(M)$ defined by

$$C(\alpha,\beta) = \Pi^\sharp(\{\alpha,\beta\}_\Pi) - [\Pi^\sharp(\alpha), \Pi^\sharp(\beta)], \quad \forall \alpha, \beta \in \Omega^1(M),$$

is $C^\infty(M)$-linear. For this reason it suffices to check (2.38) on the exact forms. Then, on one side

$$\langle df, \Pi^\sharp(\{dg, dh\}_\Pi\rangle \overset{(2.37)}{=} \langle df, \Pi^\sharp(d\{g,h\})\rangle = \{\{g,h\}, f\}, \tag{2.39}$$

while on the other side

$$\begin{aligned}
\langle df, [\Pi^\sharp(dg), \Pi^\sharp(dh)]\rangle &= \langle d\langle df, \Pi^\sharp(dh)\rangle, \Pi^\sharp(dg)\rangle - \langle d\langle df, \Pi^\sharp(dg)\rangle, \Pi^\sharp(h)\rangle \\
&= \langle d\Pi(dh, df), \Pi^\sharp(dg)\rangle - \langle d\Pi(dg, df), \Pi^\sharp(h)\rangle \\
&= \{g, \{h, f\}\} + \{h, \{f, g\}\}
\end{aligned}$$

which, together with (2.39), gives

$$\begin{aligned}
\langle df, \Pi^\sharp(\{dg, dh\}_\Pi) - [\Pi^\sharp(dg), \Pi^\sharp(dh)]\rangle &= \{\{g,h\}, f\} + \{\{h, f\}, g\} \\
&\quad + \{\{f, g\}, h\}, \quad \forall f, g, h \in C^\infty(M).
\end{aligned}$$

$\square$

Finally, we can prove the following:

Proposition 2.33 *A bivector P is Poisson if and only if* (2.36) *is a Lie bracket.*

Proof Suppose first that Π is Poisson. Since the bracket in (2.36) is a skew-symmetric $\mathbb{K}$-bilinear form on $\Omega^1(M)$, to prove the first part of the statement we are left to show that it satisfies the Jacobi identity. Let $\alpha, \beta, \gamma \in \Omega^1(M)$ and let us compute

$$\{\{\alpha, \beta\}_\Pi, \gamma\}_\Pi$$
$$= \mathscr{L}_{\Pi^\sharp(\{\alpha,\beta\}_\Pi)}\gamma - \mathscr{L}_{\Pi^\sharp(\gamma)}\{\alpha, \beta\}_\Pi + d\langle\{\alpha, \beta\}_\Pi, \Pi^\sharp(\gamma)\rangle$$
$$\overset{(2.38)\text{ and }(2.36)}{=} \mathscr{L}_{\Pi^\sharp(\alpha)}\mathscr{L}_{\Pi^\sharp(\beta)}\gamma - \mathscr{L}_{\Pi^\sharp(\beta)}\mathscr{L}_{\Pi^\sharp(\alpha)}\gamma - \mathscr{L}_{\Pi^\sharp(\gamma)}\mathscr{L}_{\Pi^\sharp(\alpha)}\beta$$
$$+ \mathscr{L}_{\Pi^\sharp(\gamma)}\mathscr{L}_{\Pi^\sharp(\beta)}\alpha - \mathscr{L}_{\Pi^\sharp(\gamma)}\langle\alpha, \Pi^\sharp(\beta)\rangle$$
$$- d\langle\Pi^\sharp(\{\alpha, \beta\}_\Pi), \gamma\rangle.$$

This computation entails that

$$\{\{\alpha, \beta\}_\Pi, \gamma\}_\Pi + \{\{\gamma, \alpha\}_\Pi, \beta\}_\Pi + \{\{\beta, \gamma\}_\Pi, \alpha\}_\Pi = d\beth(\alpha, \beta, \gamma), \qquad (2.40)$$

where $\beth : \Omega^1(M) \times \Omega^1(M) \times \Omega^1(M) \to C^\infty(M)$ is the $\mathbb{K}$-trilinear form defined by

$$\beth(\alpha, \beta, \gamma) = \Pi^\sharp(\gamma)\big(\Pi(\alpha, \beta)\big) + \Pi^\sharp(\beta)\big(\Pi(\gamma, \alpha)\big) + \Pi^\sharp(\alpha)\big(\Pi(\beta, \gamma)\big)$$
$$+ \Pi(\gamma, \{\alpha, \beta\}) + \Pi(\beta, \{\gamma, \alpha\}) + \Pi(\alpha, \{\beta, \gamma\}).$$

To conclude the first part of the proof, it suffices to observe that $\beth$ is $C^\infty(M)$-trilinear and

$$\beth(df, dg, dh) = 2(\{f, \{g, h\}\} + \{h, \{f, g\}\} + \{g, \{h, f\}\}), \quad \forall f, g, h \in C^\infty(M). \tag{2.41}$$

On the other hand, if (2.36) satisfies the Jacobi identity, then

$$0 \overset{(2.40)}{=} d\beth(df, dg, dh) \overset{(2.41)}{=} 2d(\{f, \{g, h\}\} + \{h, \{f, g\}\} + \{g, \{h, f\}\}),$$
$$\forall f, g, h \in C^\infty(M),$$

which implies that $\{\cdot, \cdot\}$ satisfies the Jacobi identity; see Problem 2.34. $\qquad\square$

Problem 2.34 Prove that if $d(\{f, \{g, h\}\} + \{h, \{f, g\}\} + \{g, \{h, f\}\}) = 0$ for all $f, g, h \in C^\infty(M)$, then $\{\cdot, \cdot\}$ satisfies the Jacobi identity. (Hint: We know that $\beth(df, dg, dh) = c_{f,g,h}$ where $c_{f,g,h}$ is constant depending on f, g, h. Thus,

$\beth(d(fk), dg, dh) = c_{fk,g,h}$, but $d(fk) = f\,dk + k\,df$ and $\beth$ is $C^\infty(M)$-trilinear, so...) $\qquad\qquad\qquad\qquad\qquad\qquad\qquad\qquad\qquad\qquad\qquad\qquad\qquad\qquad\qquad \triangle$

Note in particular that if Π is a Poisson tensor, then

Corollary 2.35 $\Pi^\sharp$ *is a Lie algebra morphism between* $(\Omega^1(M), \{\cdot, \cdot\}_\Pi)$ *and* $(\mathfrak{X}(M), [\cdot, \cdot])$.

We close this section listing a set of conditions equivalent to the existence of a Poisson structure. More precisely, let Π be a bivector field on M and let $\{\cdot, \cdot\}$ be the bracket defined by Π via Formula (2.26). The following conditions are equivalent to each other and any of them defines a Poisson structure on M.

(EP1) $[\![\Pi, \Pi]\!] = 0$ (see Theorem 2.15).
(EP2) The $\mathbb{K}$-bilinear form $\{\cdot, \cdot\}$ on $C^\infty(M)$ defined by $\{f, g\} = \Pi(df, dg)$ satisfies the Jacobi identity (see Definition 2.1).
(EP3) $[X_f, X_g] = X_{\{f,g\}}$, for all $f, g \in C^\infty(M)$, where $X_f = \Pi^\sharp(df)$, for all $f \in C^\infty(M)$ (see Proposition 2.27).
(EP4) Finally, $\Pi^\sharp\{\alpha, \beta\}_\Pi = [\Pi^\sharp(\alpha), \Pi^\sharp(\beta)]$ for all $\alpha, \beta \in \Omega^1(M)$ (see Lemma 2.32).

Remark 2.36 We borrowed the following historical remark from [140], to which we refer for more details and for many other very interesting information. The bracket defined in (2.36) was introduced in the first edition of [1], where it was defined in the case of a nondegenerate Poisson structure. Later, it appeared in the work of several authors working in the theory of integrable systems, [95, 99, 167], and, about 20 years later after its first appearance, its relevance in the theory of Lie algebroids and Lie groupoids was stressed in [52]; see also Sect. 4.3 at the end of this chapter. An in-depth analysis of the properties of the bracket (2.36) and, in particular, a proof of Lemma 2.32 and of Proposition 2.33 can be found in [141, Section 3.2]. Equation (2.36) has been extended on one-forms on infinite jet spaces of formal loops; see [16]. $\qquad\qquad\qquad\qquad\qquad\qquad\qquad\qquad\qquad\qquad\qquad\qquad\qquad \triangle$

2 Poisson Morphisms and Poisson Submanifolds

Let (N, Π_N) and (P, Π_P) be two Poisson manifolds and let $\{\cdot, \cdot\}_N, \{\cdot, \cdot\}_P$ be their corresponding Poisson brackets. In this section we will introduce and then analyze the concept of Poisson morphism. We start our discussion introducing the following:

Definition 2.37 (Poisson Morphism) A smooth map $\psi : N \to P$ is called a *Poisson morphism* if

$$\{f, g\}_P \circ \psi = \{f \circ \psi, g \circ \psi\}_N$$

or equivalently

$$\psi^*\{f, g\}_P = \{\psi^* f, \psi^* g\}_N,$$

for all $f, g \in C^\infty(P)$. $\triangle$

Example 2.38 Let $\mathfrak{g}$ and $\mathfrak{h}$ be two finite-dimensional Lie algebras. Every Lie algebra homomorphism $l : \mathfrak{g} \to \mathfrak{h}$ defines a linear Poisson morphism $l^* : (\mathfrak{h}^*, \{\cdot, \cdot\}_{LP}) \to (\mathfrak{g}^*, \{\cdot, \cdot\}_{LP})$, where $\langle l^*(\alpha), x \rangle = \langle \alpha, l(x) \rangle$ for all $\alpha \in \mathfrak{h}^*$ and $x \in \mathfrak{g}$. $\triangle$

Poisson morphisms can be characterized in terms of their action on Hamiltonian vector fields and on Poisson bivectors. More precisely,

Proposition 2.39 *A smooth map $\psi : N \to P$ is a Poisson morphism between (N, Π_N) and (P, Π_P) if and only if one of the following two equivalent conditions hold true:*

(i) For all $f \in C^\infty(P)$ the Hamiltonian vector fields X_f and $X_{f \circ \psi}$ are ψ-related, i.e.,

$$\psi_{*,n}\big(X_{f \circ \psi}(n)\big) = X_f\big(\psi(n)\big) \tag{2.42}$$

for all $n \in N$.
(ii) ψ intertwines $\Pi_N^\sharp$ with $\Pi_P^\sharp$, i.e.,

$$
\begin{array}{ccc}
T_n^* N & \xrightarrow{\;\Pi_{N,\,n}^\sharp\;} & T_n N \\[4pt]
{\scriptstyle \psi_{,n}^*}\uparrow & & \downarrow{\scriptstyle \psi_{*,n}} \\[4pt]
T_{\psi(n)}^* P & \xrightarrow[\;\Pi_{P,\,\psi(n)}^\sharp\;]{} & T_{\psi(n)} P
\end{array}
$$

commutes for every $n \in N$.

Proof Suppose that ψ is a Poisson morphism. Then

$$\{f, g\}_P(\psi(n)) = \Big\langle dg_{\psi(n)}, \Pi_{P,\,\psi(n)}^\sharp df_{\psi(n)} \Big\rangle = \Big\langle dg_{\psi(n)}, X_f^P(\psi(n)) \Big\rangle,$$

where we denoted by X_f^P the Hamiltonian vector field with Hamiltonian function f and defined by the Poisson structure Π_P. On the other hand,

$$\{\psi^* f, \psi^* g\}_N(n) = \langle d(\psi^* g)_n, \Pi_{N,\,n}^\sharp d(\psi^* f)_n \rangle = \langle d(\psi^* g)_n, X_{\psi^* f}^N(n) \rangle$$

$$= \langle \psi_{,n}^*(dg_{\psi(n)}), X_{\psi^* f}^N(n) \rangle = \langle dg_{\psi(n)}, \psi_{*,n}(X_{\psi^* f}^N(n)) \rangle$$

implying that

$$\psi_{*,n}\left(X^N_{\psi^*f}(n)\right) = X^P_f(\psi(n)),$$

for all $n \in N$, $f, g \in C^\infty(P)$. On the other hand, if $\psi_{*,n}(X^N_{\psi^*f}(n)) = X^P_f(\psi(n))$ for all $n \in N$ and $f \in C^\infty(P)$, then

$$\Pi^\sharp_{P,\,\psi(n)}df_{\psi(n)} = X^P_f(\psi(n)) = \psi_{*,n}(X^N_{\psi^*f}(n))$$

$$= \psi_{*,n}(\Pi^\sharp_{N,\,n}d(\psi^*f)_n) = \psi_{*,n}\left(\Pi^\sharp_{N,\,n}(\psi^*_{,n}df_{\psi(n)})\right)$$

$$= \psi_{*,n} \circ \Pi^\sharp_{N,\,n} \circ \psi^*_{,n}(df_{\psi(n)}),$$

implying

$$\Pi^\sharp_{P,\,\psi(n)} = \psi_{*,n} \circ \Pi^\sharp_{N,\,n} \circ \psi^*_{,n}, \quad \forall n \in N. \tag{2.43}$$

On the other hand, if (2.43) holds for all $n \in N$,

$$\{f, g\}_P(\psi(n)) = \Pi_{P,\,\psi(n)}(df_{\psi(n)}, dg_{\psi(n)}) \overset{(2.43)}{=} \langle \psi_{*,n}\left(\Pi^\sharp_{N,\,n}(\psi^*_{,n}df_{\psi(n)})\right), dg_{\psi(n)}\rangle$$

$$= \langle \Pi^\sharp_{N,\,n}(\psi^*_{,n}df_{\psi(n)}), \psi^*_{,n}dg_{\psi(n)}\rangle = \langle \Pi^\sharp_{N,\,n}d(\psi^*f)_n, d(\psi^*g)_n\rangle$$

$$= \{\psi^*f, \psi^*g\}_N(n),$$

proving the statement. $\square$

Remark 2.40 Another equivalent condition to the ones in Proposition 2.39 states that $\psi : N \to P$ is a Poisson morphism if and only if Π_N and Π_P are ψ-related, i.e., $\psi_{*,n}\Pi_{N,n} = \Pi_{M,\psi(n)}$. $\triangle$

It is worth mentioning the following property of Poisson morphisms.

Corollary 2.41 *Let (N, Π_N) and (P, Π_P) be two Poisson manifolds and let $\psi : N \to P$ be a Poisson morphism. Then, for all $f \in C^\infty(P)$, ψ intertwines the local flows of X_f and $X_{\psi \circ f}$, i.e.,*

$$\varphi^{\psi \circ f}_t \circ \psi = \psi \circ \varphi^f_t,$$

for all t such that both sides are defined.

Proof Let $n \in N$, $f \in C^\infty(P)$ and $\gamma^{f \circ \psi}_n(t)$ be the integral curve of $X_{f \circ \psi}$ passing through n when $t = 0$. Computing the time derivative of the curve $(\psi \circ \gamma^{f \circ \psi}_n)(t)$, one obtains

$$\frac{d}{dt}\left(\psi \circ \gamma^{f \circ \psi}_n\right)(t) = \psi_{*,\gamma^{f \circ \psi}_n(t)}\frac{d}{dt}\gamma^{f \circ \psi}_n(t)$$

$$= \psi_{*,\gamma_n^{f\circ\psi}(t)} X_{f\circ\psi}(\gamma_n^{f\circ\psi}(t))$$

$$\overset{(2.42)}{=} X_f\big(\psi \circ \gamma_n^{f\circ\psi}(t)\big), \ \forall t,$$

which proves that $\psi \circ \gamma_n^{f\circ\psi}(t)$ is the integral curve of X_f, passing through $\psi(n)$ when $t = 0$. In other words the previous computation showed that for all $n \in N$

$$\gamma_{\psi(n)}^f(t) = \psi \circ \gamma_n^{\psi\circ f}(t), \ \forall t.$$

To conclude it suffices to recall that for every vector field X, $\varphi_t^X(n) = \gamma_n^X(t), \forall n,$ and t. $\qquad\square$

Remark 2.42 Let (N, Π_N) and (P, Π_P) be Poisson manifolds and let (N, i) be a submanifold of P. Then, the inclusion map $i : N \to P$ is a Poisson morphism if and only if

$$i_{*,n}\big(X_{f\circ i}(n)\big) = X_f(i(n)), \ \forall n \in N;$$

see Formula (2.42) in Proposition 2.39. Since i is the restriction to $N \subset P$ of the identity map $\mathrm{id} : P \to P$, the condition expressed in Formula (2.42) is equivalent to the following statement: *i is a Poisson morphism if and only if, for all $f \in C^\infty(P)$, the restriction to N of X_f coincides with the Hamiltonian vector field, for Π_N, with Hamiltonian $i^* f$. In formulas*

$$X_{f|_N} = X_f|_N, \ \forall f \in C^\infty(P).$$

In other words, if $i : N \to P$ is a Poisson morphism, then *the restriction to N of every Hamiltonian vector field X_f is a vector field tangent to N.* We also observe that, at the level of Poisson brackets, i is a Poisson morphism if and only if *the Poisson bracket of the restriction to N of any two functions is equal to the restriction to N of the Poisson bracket of the same functions.* In formulas

$$\{f|_N, g|_N\}_{\Pi_N} = \{f, g\}_{\Pi_P}|_N, \ \forall f, g \in C^\infty(P);$$

see also the Remark 2.52. Finally, note that all these conditions make sense locally, in a neighborhood of each point $n \in N \subset P$, so that $f \in C^\infty(P)$ can be replaced by $f \in C_P^\infty(U)$, for all U open subset of P with non-empty intersection with N. $\triangle$

Here is another interesting and important characterization of a Poisson morphism. To present it, in analogy with Definition 1.68 (see also [30, Definition 1.25]), we need to introduce the following.

Definition 2.43 (Coisotropic Submanifold) A vector subspace $V \subset T_m M$ of the Poisson manifold (M, Π) is called *coisotropic* if for any $\alpha, \beta \in T_m^* M$ such that $\langle \alpha, v \rangle = \langle \beta, v \rangle = 0$ for all $v \in V$, one has $\langle \alpha \wedge \beta, \Pi_m \rangle = 0$. A submanifold

(N, i) of (M, Π) is called *coisotropic* if for all $n \in N$, $T_n N \subset T_n M$ is a coisotropic subspace. $\triangle$

Equivalently, $V \subset T_m M$ is coisotropic if and only if its annihilator

$$V^\circ = \{\alpha \in T_m^* M \mid \langle \alpha, v \rangle = 0, \ \forall v \in V\}$$

is a vector subspace of the *Poisson orthogonal* of V°, i.e., of

$$(V^\circ)^\perp = \{\beta \in T_m^* M \mid \langle \alpha \wedge \beta, \Pi_m \rangle = 0, \ \forall \alpha \in V^\circ\}.$$

From the definition above it follows that (C, i) is a coisotropic submanifold of (M, Π) if and only if one of the following equivalent conditions holds true:

(i)

$$\Pi^\sharp(NC) \subset TC, \tag{2.44}$$

where NC is the *conormal bundle* of C; see Example 1.88.

(ii)

$$\{I_C, I_C\}_M \subset I_C, \tag{2.45}$$

where $I_C = \{f \in C^\infty(M) \mid i^* f = 0\}$.

Problem 2.44 Prove that Conditions (2.44) and (2.45) are equivalent to C and coisotropic. $\triangle$

Finally, we can state the sought characterization of Poisson morphisms. If (M, Π_M) and (P, Π_P) are two Poisson manifolds

Proposition 2.45 *A smooth map* $\phi : M \to P$ *is a Poisson morphism if and only if its graph is a coisotropic submanifold of* $(M \times P, \Pi_{M \times P})$, *where* $\Pi_{M \times P} = \Pi_M \times \Pi_P^-$ *and* Π_P^- *is the opposite of* Π_P, *defined in Example 2.2.*

Proof Let $(m, \phi(m)) \in \Gamma_\phi \subset M \times P$, where Γ_ϕ is the graph of ϕ. Then

$$T_{(m,\phi(m))}\Gamma_\phi = \{(v, u) \in T_m M \times T_{\phi(m)} P \mid u = \phi_{*,m}(v)\}.$$

On the other hand, if $\alpha \in T_m^* M$ and $\beta \in T_{\phi(m)}^* P$,

$$\langle (\alpha, \beta), (v, \phi_{*,m}(v)) \rangle = \langle \alpha, v \rangle + \langle \beta, \phi_{*,m}(v) \rangle = \langle \alpha, v \rangle + \langle \phi_{,m}^* \beta, v \rangle$$

$$= \langle \alpha + \phi_{,m}^* \beta, v \rangle,$$

implying that

$$N_{(m,\phi(m))}\Gamma_\phi = \{(\alpha, \beta) \in T_m^* M \times T_{\phi(m)}^* P \mid \alpha = -\phi_{,m}^* \beta\}.$$

Suppose first that ϕ is a Poisson morphism; then

$$
\begin{aligned}
\Pi^\sharp_{M\times P,\,(m,\phi(m))}(-\phi^*_{,m}\beta,\beta) &= \left(-\Pi^\sharp_{M,\,m}\phi^*_{,m}\beta,\ \Pi^\sharp_{P^-,\,\phi(m)}\beta\right) \\
&= \left(-\Pi^\sharp_M\phi^*_{,m}\beta,\ -\Pi^\sharp_{P,\phi(m)}\beta\right) \\
&\overset{(2.43)}{=} \left(-\Pi^\sharp_M\phi^*_{,m}\beta,\ -\phi_{*,m}\Pi^\sharp_M\phi^*_{,m}\beta\right),
\end{aligned}
$$

proving that (2.44) is satisfied, i.e., that the graph of ϕ is a coisotropic submanifold.
On the other hand, if one supposes that Γ_ϕ is coisotropic, then

$$
\Pi^\sharp_{M\times P,(m,\phi(m))} = (-\phi^*_{,m}\beta,\beta) \in T_{(m,\phi(m))}\Gamma_\phi,\ \ \forall m \in M,\ \ \beta \in T^*_{\phi(m)}P,
$$

implying that for all $m \in M$ and $\beta \in T^*_{\phi(m)}P$ there exists $v \in T_m M$ such that

$$
\Pi^\sharp_{M\times P,(m,\phi(m))}(-\phi^*_{,m}\beta,\beta) = (v,\phi_{*,m}v),
$$

which, once spelled out, yields that for all $\beta \in T^*_{\phi(m)}P$ there exists $v \in T_m M$ such
that

$$
\begin{cases}
-\Pi^\sharp_{M,\,m}\phi^*_m\beta = v, \\
\Pi^\sharp_{P^-,\,\phi(m)}\beta = \phi_{*,m}v
\end{cases}
$$

which implies

$$
\Pi^\sharp_{P,\,\phi(m)}\beta = \phi_{*,m}\Pi^\sharp_{M,\,m}\phi^*_m\beta,
$$

for all $m \in M$ and all $\beta \in T^*_{\phi(m)}P$, i.e., if Γ_ϕ is coisotropic, then (2.43) holds and ϕ
is Poisson. $\qquad\square$

To conclude this discussion we would like to notice that the definition of
coisotropic submanifold in the category of the Poisson manifolds is the natural
generalization of the analogue notion introduced in Chap. 1; see Definition 1.68.
In fact if (C,i) is a coisotropic submanifold of a symplectic manifold (M,ω), then,
for all $m \in C$, $T_m C$ is a coisotropic subspace of $T_m M$, i.e., if for all $m \in M$,
$T_m C^\omega \subset T_m C$, where $T_m C^\omega$ is the symplectic orthogonal of $T_m C$. Note that, since
ω is nondegenerate, $\omega^\flat_m$ restricts to a linear isomorphism between $T_m C^\omega$ and $T_m C^\circ$.
Since for all $m \in M$, $\Pi^\sharp_m$ is, *up to the sign*, the inverse of $\omega^\flat_m$ (see (2.28) in
Remark 2.20), one recovers Condition (2.44), which implies that C is coisotropic
as a submanifold of the Poisson manifold (M,Π) where Π is the Poisson structure
defined by ω. The notion of coisotropic submanifold in Poisson geometry can also
be seen as a generalization of the notion of Lagrangian submanifold in symplectic
geometry, in the following sense. A diffeomorphism between two manifolds is

symplectic precisely when its graph is a Lagrangian submanifold of the product symplectic manifold (see Proposition 1.91); likewise a morphism between Poisson manifolds is Poisson precisely when its graph is coisotropic in the product Poisson manifold.

Problem 2.46 Check that if (M, Π) is a Poisson manifold whose Poisson tensor is invertible, then every coisotropic submanifold (C, i) is coisotropic also as a submanifold of the corresponding symplectic manifold; see Definition 1.68. $\triangle$

We will have now a short look at symplectic manifolds and their morphisms from the viewpoint of Poisson geometry. Let (M_1, ω_1) and (M_2, ω_2) be two symplectic manifolds endowed with their corresponding Poisson structures and recall that $\psi :$ $M_1 \to M_2$ is called symplectic if $\psi^* \omega_2 = \omega_1$; see Proposition 1.48. Looking superficially at the relation between symplectic and Poisson manifolds, we could be tempted to conclude that every symplectic map between two symplectic manifold is necessarily a Poisson morphism between the corresponding Poisson manifolds. But this in general is false, as the following example shows.

Example 2.47 Let (M, ω) be a symplectic manifold and let m be a zero-dimensional manifold which we think as a *trivial* symplectic manifold, i.e., endowed with the zero symplectic structure. Let us denote with $\{\cdot, \cdot\}_0$ the corresponding Poisson bracket (which is obviously identically equal to zero). Then any smooth map between m and M is automatically a symplectomorphism (the pullback of a two-form on a zero-dimensional manifold is the a zero two-form). On the other hand, if ψ is such a map, one has

$$\psi^* \{f, g\}_\omega(m) = \{f, g\}_\omega \psi(m) \neq 0$$

for some $f, g \in C^\infty(M)$, while

$$\{\psi^* f, \psi^* g\}_0(m) = 0, \ \forall f, g \in C^\infty(M),$$

showing that there is no Poisson morphism between the Poisson manifolds $(m, \{\cdot, \cdot\}_0)$ and $(M, \{\cdot, \cdot\}_\omega)$. $\triangle$

More generally, one has that

Proposition 2.48 *If (M_1, ω_1) and (M_2, ω_2) are two symplectic manifolds, then any Poisson morphism $\psi : (M_1, \{\cdot, \cdot\}_{\omega_1}) \to (M_2, \{\cdot, \cdot\}_{\omega_2})$ is a submersion.*

Proof Since the map ψ is a Poisson morphism, we have that

$$\psi_{*, m}\left(X_{f \circ \psi}(m)\right) = X_f\left(\psi(m)\right), \ \forall f \in C^\infty(M_2) \text{ and } m \in M_1;$$

see Formula (2.42) in Proposition 2.39. Now, given $m \in M_1$ let us choose a vector $v \in T_{\psi(m)}M_2$. Since M_2 is a symplectic manifold, we can find a function f, such that the corresponding Hamiltonian vector field X_f satisfies

$$X_f(\psi(m)) = v;$$

see Problem 2.51. The previous formula and Formula (2.42) yield the proof. □

Remark 2.49 Proposition 2.48 can be generalized as follows. Let (P, Π) and (M, ω) be a Poisson and a symplectic manifold, respectively, and let $\psi : (P, \{\cdot, \cdot\}_\Pi) \to (M, \{\cdot, \cdot\}_\omega)$ be a morphism of Poisson manifolds. Then ψ is a submersion. △

Remark 2.50 (Symplectic vs Poisson Morphisms) As one has already observed, for symplectic manifolds, the concepts of symplectic and Poisson morphism are not equivalent. More precisely, since a symplectic morphism between symplectic manifold is always an immersion (see Proposition 1.48) and since a Poisson morphism between symplectic manifold is always a submersion, see Proposition 2.48, we can conclude that a symplectic (Poisson) morphism which is also a Poisson (symplectic) morphism is forced to be a local diffeomorphism. This in particular says that a necessary condition for a symplectic (Poisson) morphism between two symplectic manifolds to be a Poisson (symplectic) morphism is that the dimension of the two symplectic manifolds would be the same. △

Problem 2.51 Let (M, ω) be a symplectic manifold. Show that the tangent space at each point of M is generated by Hamiltonian vector fields. (Hint: You can use Proposition 2.48). △

Before presenting a few consequences of the previous proposition, we introduce the notion of a *Poisson submanifold*.

Definition 2.52 A Poisson manifold (N, Π_N) is called a *Poisson submanifold* of (P, Π_P) if:

(i) $N \subset P$ is a submanifold of P.
(ii) The inclusion $i : N \to P$ is a Poisson morphism, i.e.,

$$i^*\{f, g\}_P = \{i^*f, i^*g\}_N, \ \forall f, g \in C^\infty(P).$$

△

Recall that on a Poisson manifold (M, Π) a $C^\infty(M)$-linear map $\Pi^\sharp$ is defined (see Formula (2.24)), which, for every $m \in M$, restricts to a $\mathbb{K}$-linear map:

$$\Pi_m^\sharp : T_m^*M \to T_mM;$$

see Formula (2.25). For every $m \in M$,

$$\mathscr{C}_m = \mathrm{im}\,(\Pi_m^\sharp : T_m^*M \to T_mM) \tag{2.46}$$

is called the *characteristic* vector space of Π at m.

After having introduced this notion, a useful characterization of the Poisson submanifolds goes as follows.

Proposition 2.53 *Let (M, Π_M) be a Poisson manifold and let (N, i) be an immersed submanifold of M. Then the following three conditions are equivalent:*

(i) There exists a Poisson bivector $\Pi_N \in \Gamma(\Lambda^2 TN)$ such that (N, Π_N) is a Poisson submanifold of (M, Π_M).
(ii) $\mathscr{C}_n \subset T_n N$ for all $n \in N$.
(iii) For any $U \subset M$ open subset and for all $f \in C_M^\infty(U)$, the Hamiltonian vector field X_f is tangent to N.

If one of the previous conditions holds true, the corresponding Poisson bivector field Π_N is unique.

Proof The proof follows from Proposition 2.39 and Corollary 2.41. We leave the details as an exercise for the reader. $\square$

In particular

Corollary 2.54 *Every open subset of a Poisson manifold inherits a Poisson structure making it into a Poisson submanifold of ambient Poisson manifold.*

Let (M, ω) be a symplectic manifold, and let (N, i) be a submanifold of M. Recall that N is a symplectic submanifold of (M, ω) if and only if $\omega_N = i^*\omega$ is symplectic. Note that this condition is equivalent to

$$T_n N \cap T_n N^\omega = \{0\}, \ \forall n \in N,$$

where $T_n N^\omega$ is the symplectic orthogonal of $T_n N$ in $T_n M$. Moving from the symplectic to the Poisson framework, one can prove that

Proposition 2.55 *$(N, \{\cdot, \cdot\}_{\omega_N})$ is a Poisson submanifold of $(M, \{\cdot, \cdot\}_{\omega})$ if and only if N is an open submanifold of M.*

Proof In fact if $(N, \{\cdot, \cdot\}_{\omega_N})$ is a Poisson submanifold of $(M, \{\cdot, \cdot\}_{\omega})$, then the immersion $i : N \to M$ is a submersion (see Proposition 2.48), which implies that i is a local diffeomorphism. Now the statement of the proposition follows observing that i is the restriction to N of the identity map $id : M \to M$. More precisely, since i is a local diffeomorphism, for each $n \in N$, there exists an open subset of N containing n such that $i|_{U_n} : U_n \to i(U_n)$ is a diffeomorphism. In particular, for every $n \in N$, $i(U_n)$ is open in M. On the other hand, since $U_n = i(U_n)$ and $N = \cup_{n \in N} U_n$, $N = \cup_{n \in N} i(U_n)$, i.e., $N \subset M$ can be written as a union of open subsets of M. $\square$

We conclude this section with the following important result:

Theorem 2.56 *Let (P, Π) be a Poisson manifold and let $\psi : P \to N$ be a surjective submersion. Then the following two conditions are equivalent:*

(i) For every $f, g \in C^\infty(N)$, $\{\psi^* f, \psi^* g\}$ is constant on the fibers of ψ.

(ii) There exists a unique Poisson structure on N for which ψ is a Poisson morphism.

Proof It suffices to observe that since ψ is submersive and surjective, the map

$$\psi^* : C^\infty(N) \xrightarrow{\ f \rightsquigarrow f \circ \psi\ } C^\infty(P)$$

is injective and its image is the subalgebra of $C^\infty(P)$ of all constant functions along the fibers of ψ. Then, given (ii),

$$\{\psi^* f, \psi^* g\}_P = \psi^*\{f, g\}_N,$$

which implies that $\{\psi^* f, \psi^* g\}_P$ is constant along the fibers of ψ. Vice versa, if (i) holds, defining $F = \psi^* f$, $G = \psi^* g$, and

$$\{f, g\}_N = \{F, G\}_P, \ \forall f, g \in C^\infty(N),$$

we obtain a well-defined Poisson structure on N with respect to which the map ψ is Poisson morphism. The uniqueness of $\{\cdot, \cdot\}_N$ follows from its definition. $\qquad\square$

Here we mention just one application of the previous result which will be generalized in Sect. 4.2; see also Sect. 4.

Corollary 2.57 *Let (P, Π) be a Poisson manifold and let $\varphi : G \times P \to P$ be a Poisson G-action. This means that for every $g \in G$, $\varphi_g : P \to P$ is a Poisson morphism, i.e.,*

$$\varphi_g^*\{f_1, f_2\} = \{\varphi_g^* f_1, \varphi^* f_2\}, \ \forall f_1, f_2 \in C^\infty(P).$$

Furthermore, suppose that the G-action is proper and that the resulting projection $\pi : P \to P/G$ is a submersion. Then, there is a unique Poisson structure on P/G which makes π into a Poisson morphism.

Proof The hypotheses made on the G-action are sufficient to guarantee that (1) a structure of a smooth manifold can be defined on the quotient space P/G and that (2) the canonical projection $\pi : P \to P/G$ is a smooth map. Since the Poisson structure is G-invariant, the algebra $C^\infty(P)^G$ of the G-invariant functions is a Poisson subalgebra of $(C^\infty(P), \{\cdot, \cdot\}_\Pi)$. The proof now follows at once from Theorem 2.56. In fact, since π is submersive, π^* injects $C^\infty(P/G)$ into $C^\infty(P)^G$, which coincides with the algebra of smooth functions on P whose elements are constant along the projection π. $\qquad\square$

3 Characteristic Distribution and Symplectic Foliation

Let (M, Π) be a Poisson manifold and let $\{\cdot, \cdot\}$ be the Poisson bracket defined by Π. In this section we aim to analyze in more detail the properties of the characteristic distribution defined on M by Π. To this end it will be relevant to consider the notion of *generalized distribution*, for which we refer the reader to Sect. 1 in Appendix B and to the references therein.

Recall that, in the previous section, we introduced $\mathscr{C}_m$, the characteristic vector space of Π defined at $m \in M$; see Formula (2.46). As m changes in M, $m \rightsquigarrow \mathscr{C}_m$ defines a *distribution* of vector subspaces on M, which plays a prominent role in the interplay between Poisson and symplectic geometry. To make this observation more precise, we start introducing the following concepts.

Definition 2.58 The *rank* of the Poisson tensor at $m \in M$ is the dimension of the vector space $\mathscr{C}_m$ and it will be denoted by rk (Π_m). The rank of Π is rk $\Pi :=$ $\max_{m \in M}$rk (Π_m). The application $\mathscr{C}$ that to every point $m \in M$ associates the vector space $\mathscr{C}_m$ defines a *generalized distribution* on M, called the *characteristic distribution* of Π. If rk (Π_m) does not depend on m, then Π is called a *regular* Poisson tensor and (M, Π) a regular Poisson manifold. $\qquad\qquad \triangle$

Remark 2.59 Note that, sometimes, the rank of Π at m is called the rank of the Poisson manifold at the point m and, accordingly, the rank of Π is called the rank of the Poisson manifold. $\qquad\qquad \triangle$

Using the above definition and the results from the previous section, one can prove the following.

Corollary 2.60 *If ψ is a Poisson morphism between two Poisson manifolds (N, Π_N) and (P, Π_P), then for all $n \in N$*

$$rk\,(\Pi_{N,\,n}) \geq rk\,(\Pi_{P,\,\psi(n)}). \tag{2.47}$$

In particular, if ψ is an immersion, i.e., the differential $\psi_{*,n}$ is injective for every $n \in N$,

$$rk\,(\Pi_{N,\,n}) = rk\,(\Pi_{P,\,\psi(n)}), \ \forall n \in N.$$

Proof Inequality (2.47) follows almost immediately from part (ii) of Proposition 2.39. In fact, since

$$\Pi^{\sharp}_{P,\,\psi(n)} = \psi_{*,n} \circ \Pi^{\sharp}_{N,\,n} \circ \psi^{*}_{,n},$$

one can conclude that for all $n \in N$

$$\mathrm{im}\,\big(\Pi^{\sharp}_{P,\,\psi(n)}\big) \subset \psi_{*,n}(\mathrm{im}\,\Pi^{\sharp}_{N,\,n}).$$

The proof of the second part of the corollary follows from (2.47) and from the definition of immersion. □

Note that $\mathrm{rk}\,(\Pi_m)$ is always *even*; in fact $\Pi_m^\sharp$ induces a skew-symmetric, *nondegenerate*, linear isomorphism between $T_m^*M/\ker(\Pi_m^\sharp) \simeq \mathscr{C}_m$ and $im(\Pi_m^\sharp)$, which forces $\dim\mathscr{C}_m$ to be even.

Furthermore, since the characteristic distribution is generated by the (locally) Hamiltonian vector fields, it is *smooth* and *invariant* under their flows. These observations, together with the so-called Stefan-Sussmann theorem (see Theorem B.13), yield the following important result (see [205] and also Sect. 1 in Appendix B).

Theorem 2.61 (Symplectic Foliation Theorem) *The characteristic distribution $\mathscr{C}$ of a Poisson manifold (M, Π) is integrable, i.e., there is a (unique) generalized foliation whose tangent distribution is $\mathscr{C}$. Each maximal leaf L of such a foliation is an immersed submanifold of M carrying a symplectic form ω_L, such that the canonical immersion $i : L \to M$ is a Poisson morphism between $(L, \{\cdot, \cdot\}_{\omega_L})$ and $(M, \{\cdot, \cdot\})$.*

Proof For a detailed proof of this result we refer the reader to the monograph [205]. Here we want just remark that the symplectic structure ω_L defined on the leaf L is completely determined by the Hamiltonian vector fields tangent to L. More precisely, all $m \in L$ and for all $u, v \in T_m L$

$$\omega_{L, m}(v, u) = \{f, g\}(m), \qquad (2.48)$$

where f and g, any two smooth functions defined on an open neighborhood of m in M, are such that $X_f(m) = u$ and $X_g(m) = v$ and where $\{\cdot, \cdot\}$ is the ambient Poisson bracket. □

Problem 2.62 Show that (2.48) is a well-defined two-form on L. △

We conclude this discussion noticing that the Stefan-Sussmann theorem and (many) other results characterizing integrability of generalized distributions play an important role in *geometric control theory*, where the maximal connected integral submanifolds are called *accessible, reachable*, or, sometimes, *attainable sets*; see [7] and [128].

Before presenting a few examples of symplectic foliations, we would like to make two final and technical comments. The first will concern the rank of a Poisson tensor as a *function* of the point m, i.e., $m \rightsquigarrow \mathrm{rk}\,\Pi_m$, while the second one will be about the Casimir functions of a Poisson manifold.

Let us start with the one about the rank. This, as a function of the point m, is *lower semicontinuous*, i.e., every $m_0 \in M$ has a neighborhood U such that $\mathrm{rk}\,\Pi_m \geq \mathrm{rk}\,\Pi_{m_0}$ for all $m \in U$. In fact, if $x_1, \ldots, x_d$ is a set of local coordinates around m_0, $\mathrm{rk}\,\Pi_{m_0} = 2n$ if and only the matrix $(\Pi_{ij}(m_0))_{i,j} = (\{x_i, x_j\}(m_0))_{i,j}$ is such that (1) it has a minor of size $2n \times 2n$ with nonzero determinant and (2) all the minors of size greater than $2n$ have zero determinant. Since the determinant is a continuous (indeed a smooth) function of the coordinates, if it is nonzero at m_0, it will stay nonzero in some neighborhood of m_0, proving the statement. A point $m_0 \in M$ which belongs to a neighborhood U such that rk $(\Pi_m) = $ rk (Π_{m_0}) for all $m \in U$ is called *regular*; otherwise, it is called a *singular* point. The set of the regular points of a Poisson manifold is open and dense. In fact, every $m \in M$ has a(n) open) neighborhood U which intersects nontrivially the set of regular points of the underlying Poisson manifold. On the contrary, such a neighborhood should contains only singular points, which is impossible since U is infinite and at each of its point (which is singular by assumption) the rank is not constant, contradicting the fact that rk Π can assume on U only a finite number of values, bounded by 0 and the dimension of M.

The next comment points to the relation between the Casimir functions and the symplectic foliation of a given Poisson structure, which is summarized in the following.

Proposition 2.63 $f \in \mathrm{Cas}(\Pi)$ *if and only if it is constant on every symplectic leaf. Furthermore, let* (M, Π) *be an n-dimensional Poisson manifold,* $U \subset M$ *an open subset such that:*

(i) rk $(\Pi_m) = n - s$ for all $m \in U$.
(ii) $f_1, \ldots, f_s \in \mathrm{Cas}(U, \Pi)$ satisfying $df_1 \wedge \cdots \wedge df_s(m) \neq 0$ for all $m \in U$.

Then the symplectic foliation of $\Pi_{|U}$ coincides with the foliation defined by $F = (f_1, \ldots, f_s) : U \to \mathbb{R}^s$. In other words, in a neighborhood of a regular point the symplectic foliation is a zero set of Casimirs of the Poisson structure.

Instead of giving a detailed proof of the proposition above, for which we refer the reader to the monograph [150, Proposition 1.32], we make a few comments to clarify its content. First, note that if $f \in \mathrm{Cas}(M, \Pi)$, then $df_m \in \ker \Pi_m^{\sharp}$, for all $m \in M$. More in general, if $f_1, \ldots, f_k \in \mathrm{Cas}_M(U, \Pi)$ are functionally independent, i.e., $df_1 \wedge \cdots \wedge df_k(m) \neq 0$ for all $m \in U$, then their differentials form an independent subset of $\ker \Pi_m^{\sharp}$, which, nevertheless, in general is not generated by the differentials evaluated at m of the (local) Casimirs of Π. For example, $(\mathbb{R}^2, \Pi)$, where $\Pi = (x^2 + y^2)\frac{\partial}{\partial x} \wedge \frac{\partial}{\partial y}$, does not possess nonconstant Casimirs, neither globally nor locally defined, but the kernel of $\Pi^{\sharp}$ at the origin is a two-dimensional vector space. In other words if $f_1, \ldots, f_k \in \mathrm{Cas}_M(U, \Pi)$ are independent, then $k \leq \dim \ker \Pi_m^{\sharp}$ for all $m \in U$. Now, let L_m be the leaf of the symplectic foliation containing the point m, let U be an open neighborhood of m, and let $f_1, \ldots, f_k$ be k-independent elements in $\mathrm{Cas}_M(U, \Pi)$. Then

$$\dim L_m = \mathrm{rk}\ (\Pi_m) = \dim M - \dim \ker \Pi_m^{\sharp} \leq \dim M - k, \qquad (2.49)$$

which implies that the dimension of L_m is less than or equal to the dimension of the submanifold (locally) defined by the equations $F^{-1}(F(m))$ where $F = (f_1, \ldots, f_k)$. These observations should give an idea about the proof of the previous proposition and about the relations between Casimirs and symplectic leaves of a Poisson manifold. These relations will be further investigated in the following examples.

Example 2.64 (Regular Poisson Structures) If the Poisson manifold is symplectic, then the rank of the Poisson tensor is equal to the dimension of the underlying manifold. In this case the symplectic foliation has only one leaf, of maximal dimension, the manifold itself. More in general, let $M = \mathbb{R}^n$ with coordinates $x_1, \ldots, x_n$ and let $2s \leq n$. Then

$$\Pi = \sum_{i=1}^{s} \frac{\partial}{\partial x_i} \wedge \frac{\partial}{\partial x_{i+s}} \tag{2.50}$$

is a Poisson bivector field of constant rank equal to $2s$. The invertible case is recovered if $2s = n$. On the other hand, if $2s < n$, then the symplectic foliation is no longer trivial: each leaf has dimension equal to $2s$ and it is (globally) defined by the equation $x_{2s+1} = c_{2s+1}, \ldots, x_n = c_n$, where $(c_{2s+1}, \ldots, c_n)$ are constants. Note that the functions $x_{2s+1}, \ldots, x_n$ are independent and globally defined Casimirs of (2.50). This is the prototypical example of a regular Poisson manifold; see Definition 2.58 $\triangle$

Example 2.65 The symplectic foliation of a linear Poisson structure defined on $\mathfrak{g}^*$ (see Example 2.2) coincides with the foliation defined by the coadjoint action of the (unique up to isomorphism) connected and simply connected Lie group G whose Lie algebra is $\mathfrak{g}$. In fact the characteristic distribution of a linear Poisson structure is generated by the *fundamental* vector fields associated to the coadjoint action; see Problem 2.23. In this case Formula (2.48) reduces to

$$\omega_{L,\,\alpha}(u, v) = \{x, y\}(\alpha) = \langle \alpha, [x, y] \rangle, \;\; \forall \alpha \in \mathfrak{g}^* \text{ and } u, v \in T_\alpha L, \tag{2.51}$$

where x, y are any two linear functions (i.e., elements of $\mathfrak{g}$) such that $X_x(\alpha) = u$ and $X_y(\alpha) = v$. In other words, the symplectic form defined on the symplectic leaves coincides with the one defined on the coadjoint orbits introduced in Sect. 1.2 of Chap. 1; see Formula (1.9). $\triangle$

Here below we present a few explicit instances of the previous general example.

Example 2.66 ($SO_3(\mathbb{R})$) Let $M = \mathbb{R}^3$ endowed with coordinates x, y, z. The skew-symmetric bracket $\{\cdot, \cdot\} : C^\infty(\mathbb{R}^3) \times C^\infty(\mathbb{R}^3) \to C^\infty(\mathbb{R}^3)$ defined on the coordinate functions by

$$\{x, y\} = z, \quad \{z, x\} = y \quad \text{and} \quad \{y, z\} = x \tag{2.52}$$

and extended to $C^\infty(\mathbb{R}^3)$ as a biderivation is a linear Poisson bracket on $\mathbb{R}^3$ whose corresponding Poisson bivector field is

$$\Pi = z \frac{\partial}{\partial x} \wedge \frac{\partial}{\partial y} + y \frac{\partial}{\partial z} \wedge \frac{\partial}{\partial x} + x \frac{\partial}{\partial y} \wedge \frac{\partial}{\partial z}. \tag{2.53}$$

The matrix of the sharp map $\Pi^\sharp$, written in terms of the above coordinates, is

$$\Pi^\sharp = \begin{bmatrix} 0 & -z & y \\ z & 0 & -x \\ -y & x & 0 \end{bmatrix}.$$

The rank of (2.53) is equal to two at every point $(x, y, z) \neq (0, 0, 0)$, where it drops to zero. This observation implies that the symplectic leaf passing through the origin is a singleton, i.e., it is a zero-dimensional submanifold, while the one going through $(x, y, z) \neq (0, 0, 0)$ is a two-dimensional submanifold of $\mathbb{R}^3$. To make this statement more clear notice that the function $f(x, y, z) = x^2 + y^2 + z^2$ is a Casimir of Π. In fact $df = 2xdx + 2ydy + 2zdz$ and

$$\Pi^\sharp(df) = \begin{bmatrix} 0 & -z & y \\ z & 0 & -x \\ -y & x & 0 \end{bmatrix} \begin{bmatrix} 2x \\ 2y \\ 2z \end{bmatrix} = \begin{bmatrix} 0 \\ 0 \\ 0 \end{bmatrix}.$$

The knowledge of the rank of (2.53), together with the previous simple computation and Proposition 2.63, entails that the symplectic foliation of (2.53) is defined by the level sets of the function $f = x^2 + y^2 + z^2$, which are spheres centered at the origin of $\mathbb{R}^3$. Every sphere of radius greater than zero is a regular leaf, i.e., formed by regular points, and the sphere of radius zero is the only *singular* leaf of the foliation. It is worth noticing that the Poisson structure defined by (2.52) is the linear Poisson structure of the Lie algebra $\mathfrak{so}_3(\mathbb{R})$, the Lie algebra of the (real) special orthogonal group. For this reason the symplectic foliation just described is nothing else than the foliation defined on the dual of $\mathfrak{so}_3(\mathbb{R})$ by the coadjoint action. To look at the symplectic foliation from the viewpoint of Lie theory first, one identifies $\mathfrak{so}_3(\mathbb{R})^*$ with $\mathfrak{so}_3(\mathbb{R})$, which, in turn, is identified with $(\mathbb{R}^3, \times)$, where $\times$ is the standard cross-product. The first step yields the identification of the coadjoint orbits with the adjoint ones, while the second one identifies each adjoint orbit with a sphere in $\mathbb{R}^3$ centered at the origin. In this case the symplectic form defined on the coadjoint orbit of the point $\alpha \in \mathbb{R}^3$ is

$$\omega_\alpha(v, u) = \alpha \cdot (v \times u), \ \forall u, v, w \in \mathbb{R}^3;$$

see Formula (2.51) in Example 2.65. $\triangle$

Problem 2.67 ($SL_2(\mathbb{R})$) Let $M = \mathbb{R}^3$ with coordinates x, y, and z and define the skew-symmetric bracket $\{\cdot, \cdot\} : C^\infty(\mathbb{R}^3) \times C^\infty(\mathbb{R}^3) \to C^\infty(\mathbb{R}^3)$ first on the linear functions by

$$\{x, y\} = z, \quad \{z, x\} = 2x \quad \text{and} \quad \{z, y\} = -2y, \tag{2.54}$$

and then extending it to $C^\infty(\mathbb{R}^3)$ as a biderivation (Fig. 2.1).

Fig. 2.1 Coadjoint orbits of
$\mathfrak{sl}_2(\mathbb{R})$

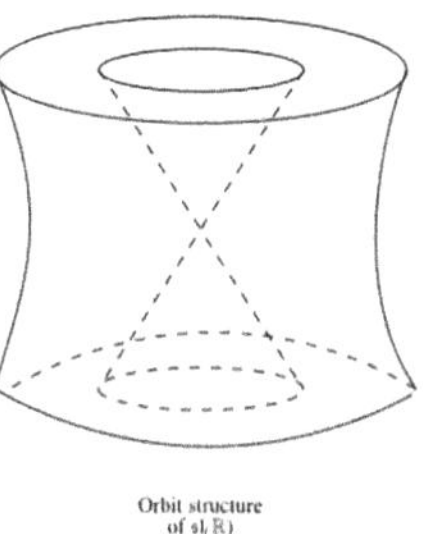

(i) Show that $\{\cdot, \cdot\}$ is a Poisson bracket.
(ii) Describe the symplectic foliation and find a formula for the symplectic form on
the leaves.

The Lie algebra corresponding to the linear Poisson bracket defined in (2.54) and its corresponding Lie group can be (faithfully) represented in terms of 3×3 real matrices with trace equal to zero and in terms of 3×3 real matrices with determinant equal to one, respectively. They represent the (nontrivial) lowest-dimensional examples of the so-called real *special linear* Lie algebras and, respectively, special linear Lie groups and they are generally denoted by $\mathfrak{sl}_2(\mathbb{R})$ and $SL_2(\mathbb{R})$, respectively. $\triangle$

Going back to $SO_3(\mathbb{R})$, its coadjoint orbits are examples of symplectic manifolds not diffeomorphic to cotangent bundles. On the other hand, there are Lie groups whose coadjoint orbits are (diffeomorphic to) cotangent bundle, as the next two examples show.

Example 2.68 ($\mathbb{H}_3$) Let consider again $\mathbb{R}^3$ with coordinates (x, y, z) and let $\{\cdot, \cdot\} : C^\infty(\mathbb{R}^3) \times C^\infty(\mathbb{R}^3) \to C^\infty(\mathbb{R}^3)$ be the skew-symmetric bracket defined on the coordinate functions by

$$\{x, y\} = z, \quad \{z, x\} = 0 \quad \text{and} \quad \{y, z\} = 0 \tag{2.55}$$

and extended to $C^\infty(\mathbb{R}^3)$ as a biderivation. A simple computation shows that (2.55) is a (linear) Poisson bracket on $\mathbb{R}^3$, whose corresponding Poisson bivector field is

$$\Pi = z \frac{\partial}{\partial x} \wedge \frac{\partial}{\partial y}. \tag{2.56}$$

The matrix of the sharp map $\Pi^\sharp$, written in terms of the above coordinates, is

$$\Pi^\sharp = \begin{bmatrix} 0 & -z & 0 \\ z & 0 & 0 \\ 0 & 0 & 0 \end{bmatrix},$$

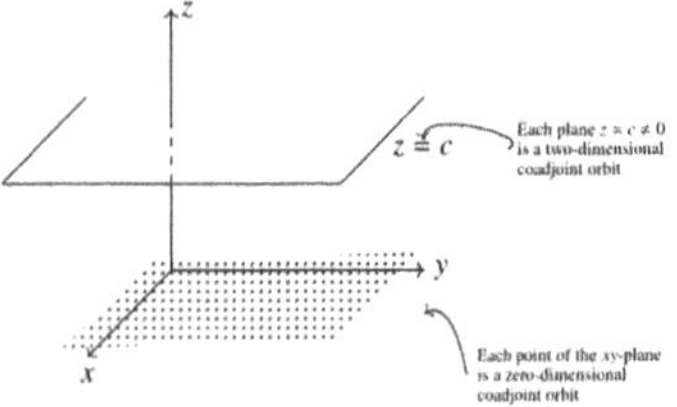

Fig. 2.2 Coadjoint orbits of the three-dimensional Heisenberg group

which implies that the rank of the Poisson structure is two everywhere in $\mathbb{R}^3$ but on the plane (x, y) $(z = 0)$, where it is equal to zero at every point; see the picture above (Fig. 2.2).

The function $f(x, y, z) = z$ is a Casimir of (2.56) and its level sets are two-dimensional planes orthogonal to the z-axis. The level sets $z = c$, with $c \neq 0$, are symplectic leaves, while the zero-level set is not. In fact, since the rank of Π at every point of the plane $z = 0$ is equal to zero, every point of this plane is a zero-dimensional symplectic leaf. In other words, the symplectic foliation of (2.56) has the following description: the regular leaves are the two-dimensional planes defined by the equations $z = c$, with $c \neq 0$, while every point of the plane $z = 0$ is a zero-dimensional singular leaf. A slightly different way to recover this result is to notice that if $g \in C^\infty(\mathbb{R}^3)$, the corresponding Hamiltonian vector field is

$$X_g = \Pi^\sharp(dg) = \begin{bmatrix} 0 & -z & 0 \\ z & 0 & 0 \\ 0 & 0 & 0 \end{bmatrix} \begin{bmatrix} \frac{\partial g}{\partial x} \\ \frac{\partial g}{\partial y} \\ \frac{\partial g}{\partial z} \end{bmatrix} = \begin{bmatrix} -z\frac{\partial g}{\partial y} \\ z\frac{\partial g}{\partial x} \\ 0 \end{bmatrix},$$

i.e.,

$$X_g = -z\frac{\partial g}{\partial y}\frac{\partial}{\partial x} + z\frac{\partial g}{\partial x}\frac{\partial}{\partial y}. \tag{2.57}$$

This entails that every point of the plane $z = 0$ is *stationary* for X_g and for this reason the integral curve of any Hamiltonian vector field going through any point of this plane will reduce to a singleton. On the other hand, (2.57) implies that flowing out any point of the form (x, y, c), $c \neq 0$, with all possible Hamiltonian vector fields one can reach any other point on the plane $z = c$. The induced symplectic form on the plane $z = c$, $c \neq 0$, is the standard symplectic form on $\mathbb{R}^2$ scaled by the factor c, i.e.,

$$\omega_\alpha(v, u) = c,$$

for all α in the plane $z = c$ and for all v, u such that $v = X_x(\alpha)$ and $u = X_y(\alpha)$. If $c = 0$, ω is the zero form. The Lie algebra defined by the linear Poisson bracket (2.55) is the so-called three-dimensional *Heisenberg Lie algebra*, and it can be obtained as a one-dimensional central extension of a two-dimensional abelian Lie algebra. For a nice and complete discussion about the theory of the Heisenberg Lie algebras and Lie groups, we refer the reader to [132]. Note the regular leaves of these Poisson structures are symplectomorphic to the cotangent bundle of $\mathbb{R}$ endowed with its canonical symplectic structure. $\triangle$

Finally, it is worth mentioning that while coadjoint orbits carry always a symplectic structure, the adjoint orbits in general do not, as it is shown in the following example.

Example 2.69 ($\mathrm{Aff}_2(\mathbb{R})$) Let $G = \mathbb{R}^* \times \mathbb{R}$ endowed with the following multiplication:

$$(a_1, b_1)(a_2, b_2) = (a_1 a_2, b_1 + a_1 b_2), \ \forall (a_1, b_1), \ (a_2, b_2) \in G. \tag{2.58}$$

This defines a group structure on G, whose unit is $(1, 0)$ and whose inverse map is defined by $(a, b)^{-1} = (a^{-1}, -a^{-1}b)$, for all (a, b). The corresponding group, denoted by $\mathrm{Aff}_2(\mathbb{R})$, is called the two-dimensional (real) *affine group*. It is an example of semi-direct product (see Example C.65), obtained from the action of $\mathbb{R}^*$ on $\mathbb{R}$ defined by the (usual) multiplication between real numbers. In other words, $\mathrm{Aff}_2(\mathbb{R}) = \mathbb{R}^* \ltimes \mathbb{R}$. It becomes a Lie group if one endows $\mathbb{R}$ and $\mathbb{R}^*$ with their natural structure of smooth manifolds. It is worth noting that $\mathrm{Aff}_2(\mathbb{R})$ has a nice representation in terms of 2×2 real matrices. More precisely to the element $(a, b) \in \mathrm{Aff}_2(\mathbb{R})$, one associates the matrix

$$M_{(a,b)} = \begin{bmatrix} a & b \\ 0 & 1 \end{bmatrix}. \tag{2.59}$$

One can check that under this map (2.58) becomes the standard matrix multiplication row-by-column. The Lie algebra of $\mathrm{Aff}_2(\mathbb{R})$ is the semidirect product of a one-dimensional Lie algebra $\mathbb{R}$ with a one-dimensional vector space $\mathbb{R}$, it is denoted by $\mathfrak{aff}_2(\mathbb{R}) = \mathbb{R} \times \mathbb{R}$, and its Lie bracket has the following form:

$$[(x_1, y_1), (x_2, y_2)] = (0, x_1 y_2 - x_2 y_1), \tag{2.60}$$

for all $(x_1, y_1), (x_2, y_2) \in \mathfrak{aff}_2(\mathbb{R})$. In analogy with (2.59), $\mathfrak{aff}_2(\mathbb{R})$ has the following matrix representation:

$$m_{(x,y)} = \begin{bmatrix} x & y \\ 0 & 0 \end{bmatrix},$$

and as in the group case, under the map $(x, y) \rightsquigarrow m_{(x,y)}$, (2.60) becomes the standard Lie bracket between matrices. Looking at the adjoint orbits of $\mathrm{Aff}_2(\mathbb{R})$ one discovers the following structure. There is one zero-dimensional orbit, the one of the point $(0, 0)$. All the other orbits are one-dimensional. More precisely, the orbit of every point $(x, 0)$, $x \neq 0$, is the line parallel to $y = 0$ going through $(x, 0)$ and the

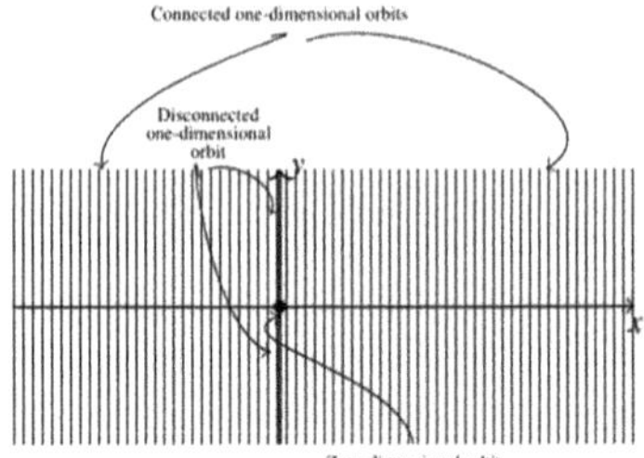

Fig. 2.3 Coadjoint orbits of the two-dimensional affine real group

orbit of $(0, y)$, $y \neq 0$, is the y-axis minus the origin. Since all the nontrivial adjoint orbits are 1-dimensional, they cannot carry a symplectic form; see the picture above (Fig. 2.3).

$\triangle$

Problem 2.70 Double-check the unproven statements contained in the previous example. Analyze the coadjoint orbit structure of $\mathrm{Aff}_2(\mathbb{R})$ and find an explicit form for the symplectic form defined on the corresponding symplectic leaves. $\triangle$

As a final example we discuss the case of the three-dimensional *Euclidean group*.

Example 2.71 ($E_3(\mathbb{R})$) Another example of a semi-direct product (see Example C.65) is provided by the so-called Euclidean group. In the three-dimensional case this is the group defined on $SO_3(\mathbb{R}) \times \mathbb{R}^3$ by the multiplication law:

$$(g_1, v_1) \cdot (g_2, v_2) = (g_1 g_2, v_1 + g_1 v_2), \tag{2.61}$$

for all $(g_i, v_i) \in SO_3(\mathbb{R})$. In analogy to the previous example, one can prove that the identity element is $(\mathrm{Id}, 0)$ and that $(g, v)^{-1} = (g^{-1}, -g^{-1}v)$, for all (g, v). Moreover, endowing $SO_3(\mathbb{R}) \times \mathbb{R}$ with the structure of a product manifold, the group law defined in (2.61) turns $SO_3(\mathbb{R}) \times \mathbb{R}^3$ into a Lie group $E_3(\mathbb{R}) = SO_3(\mathbb{R}) \ltimes \mathbb{R}^3$, called the three-dimensional *Euclidean Lie group*. The Lie algebra of $E_3(\mathbb{R})$ is $\mathfrak{e}(3) = \mathfrak{so}_3(\mathbb{R}) \ltimes \mathbb{R}^3$, the semidirect product of the Lie algebra $\mathfrak{so}_3(\mathbb{R})$ with $\mathbb{R}^3$, whose bracket is

$$[(A_1, v_1), (A_2, v_2)] = ([A_1, A_2], A_1 v_2 - A_2 v_1), \tag{2.62}$$

for all $(A_i, v_i) \in \mathfrak{so}_3(\mathbb{R}) \times \mathbb{R}^3$. The first entry in the right-hand side of (2.62) is the Lie bracket of $\mathfrak{so}_3(\mathbb{R}^3)$, while the second one is obtained via the standard action of $\mathfrak{so}_3(\mathbb{R})$ on $\mathbb{R}^3$, i.e., the standard multiplication of a matrix by a vector. Using the usual identification of $\mathfrak{so}_3(\mathbb{R})$ with $(\mathbb{R}^3, \times)$ (see Example C.11), one can write the Lie-Poisson structure of $\mathfrak{e}(3)$ as follows:

$$\{m_i, m_j\} = \epsilon_{ijk} m_k, \quad \{m_i, f_j\} = \epsilon_{ijk} f_k \quad \text{and} \quad \{f_i, f_j\} = 0, \tag{2.63}$$

where ϵ_{ijk} is the totally antisymmetric Levi-Civita symbol, i.e., ϵ_{ijk} is the signature of the permutation $(1, 2, 3) \mapsto (i, j, k)$, where both (m_1, m_2, m_3) and (f_1, f_2, f_3) denote the canonical basis of $\mathbb{R}^3$. Note that the last set of equations of (2.63) reflects the fact that $\mathbb{R}^3$ embeds in $\mathfrak{e}(3)$ as an abelian Lie subalgebra. The Poisson bivector field defined by (2.63)

$$\Pi = \frac{1}{2} \sum_{i,j=1}^{3} \epsilon_{ijk} m_k \frac{\partial}{\partial m_i} \wedge \frac{\partial}{\partial m_j} + \frac{1}{2} \sum_{i,j=1}^{3} \epsilon_{ijk} f_k \frac{\partial}{\partial m_i} \wedge \frac{\partial}{\partial f_j},$$

whose $\Pi^\sharp$, in the basis $(dm_1, dm_2, dm_3, df_1, df_2, df_3)$, is represented by the matrix

$$\Pi^\sharp = \begin{bmatrix} 0 & -m_3 & m_2 & 0 & -f_3 & f_2 \\ m_3 & 0 & -m_1 & f_3 & 0 & -f_1 \\ -m_2 & m_1 & 0 & -f_2 & f_1 & 0 \\ 0 & -f_3 & f_2 & 0 & 0 & 0 \\ f_3 & 0 & -f_1 & 0 & 0 & 0 \\ -f_2 & f_1 & 0 & 0 & 0 & 0 \end{bmatrix}. \tag{2.64}$$

This Poisson structure has two quadratic Casimir functions $C_1 = f_1^2 + f_2^2 + f_3^2$ and $C_2 = m_1 f_1 + m_2 f_2 + m_3 f_3$. The symplectic foliation of this linear Poisson structure turns out to be more complicated compared to those described in the previous examples. More precisely, there are two one-parameter families of symplectic leaves and one singleton. The latter is the zero-dimensional coadjoint orbit (the orbit of the zero element in $\mathfrak{e}(3)$). The two nontrivial families are composed of two- and four-dimensional orbits. The two-dimensional orbits are diffeomorphic to spheres, while each of the four-dimensional ones is diffeomorphic to the cotangent bundle of a two-dimensional sphere. The symplectic form of each member of the first family is (necessarily) a multiple of the volume form of the underlying manifold, while the symplectic form of each member of the four-dimensional family turns out to coincide, up the sign, with its canonical symplectic form. For the proofs of these last statements we refer the reader to [114] and to [175]. $\triangle$

4 Concluding Remarks and Further Topics

This final section is divided in three parts, each one devoted to a topic related to the main subject of this chapter. The first is an important theorem proved in [247] which generalizes the classical theorem of Darboux to the context of Poisson manifolds. In the second subsection we present a result giving sufficient conditions to *reduce* a Poisson structure, generalizing the Marsden-Weinstein reduction presented in Chap. 3 in the framework of the Hamiltonian actions of Lie groups. The Poisson case is technically more complicated and it was proved in [174]. In the last subsection we will show that the cotangent bundle of every Poisson manifold carries a structure of a Lie algebroid whose cohomology is the Poisson cohomology of the underlying Poisson manifold. In this section the discussion will be more informal and mainly without proofs, for which we refer the reader to the cited references.

4.1   The Darboux-Weinstein Theorem

In this first subsection we will present a remarkable result which extends to the Poisson setting the Darboux theorem presented in Sect. 3.3 of Chap. 1, providing the existence of a normal form of a Poisson tensor in a neighborhood of every point of the underlying manifold.

Theorem 2.72 (Darboux-Weinstein Theorem) *Let (P, Π) be a Poisson manifold of dimension m and let p be a point of P where the Poisson tensor Π has rank equal to $2k$. Then there exist an open neighborhood U of p in P and local coordinates centered at p, $(p_1, \ldots, p_k, x_1, \ldots, x_k, z_1, \ldots, z_{m-2k})$ such that:*

1. $\{x_i, x_j\} = \{p_i, p_j\} = 0 \ \forall \, i, j = 1, \ldots, s.$
2. $\{z_l, x_i\} = \{z_l, p_i\} = 0 \ \forall \, l = 1, \ldots, k, \ i = 1, \ldots, s.$
3. $\{p_i, x_j\} = \delta_{ij} \ \forall \, i, j = 1, \ldots, s.$
4. $\{z_l, z_r\}(p) = 0 \, \forall \, l, r = 1, \ldots, k.$

In particular, using these coordinates, the Poisson tensor Π assumes the following form:

$$\Pi = \sum_{i=1}^{k} \frac{\partial}{\partial p_i} \wedge \frac{\partial}{\partial x_k} + \sum_{1 \leq i < j \leq m-2k} \{z_i, z_j\} \frac{\partial}{\partial z_i} \wedge \frac{\partial}{\partial z_j},$$

or, equivalently,

$$\Pi^{\sharp} = \begin{bmatrix} 0 & id & 0 \\ -id & 0 & 0 \\ 0 & 0 & \Lambda_{lr}(z) \end{bmatrix},$$

where id represents the $k \times k$ identity matrix and $\Lambda_{lr}(z)$ is a $m - 2k \times m - 2k$ block whose entries are the functions $\Lambda_{lr}(z) = \sum_{i=1}^{k} c_{lr}^i z_i + O(z^2).$

The previous theorem can be rephrased as follows. *For every point p of an m-dimensional Poisson manifold P where the rank of the Poisson tensor is equal to $2k$, there exists a pair of local Poisson submanifolds (S, N) of complementary dimension $2k$ and $m - 2k$, transverse to each other and intersecting at p, such that the Poisson structure induced on S by Π is of maximal rank (i.e., the induced Poisson tensor is invertible), while the Poisson tensor defined on N is equal to zero at p. In other words, a Poisson manifold admits, at every point, a* splitting *into a* pair of local (Poisson) submanifolds (S, N), where S is symplectic, while N is a Poisson manifold whose Poisson tensor is equal to zero at p. It is clear from the statement of Theorem 2.72 that the coordinates $(p_1, \ldots, p_k, x_1, \ldots, x_k)$ are local *Darboux coordinates* for the symplectic structure induced on S. In particular, if the Poisson tensor is regular of maximal rank, i.e., it is invertible and defines a symplectic structure on P, then Theorem 2.72 reduces to the Darboux theorem; see Sect. 3.3 in Chap. 1. In the case when the Poisson tensor has rank equal to zero at the point p, the symplectic part of the splitting will be absent. Finally, note that the cartesian product $N \times S$ of the Poisson manifolds described in Theorem 2.72 is isomorphic in the sense of Lemma 2.10, i.e., as a Poisson manifold, to a suitable open neighborhood of (P, Π).

4.2 Poisson Reduction

In this short subsection we would like to mention an important result which extends to the case of a general Poisson manifold (P, Π) the Marsden-Weinstein theorem; see Theorem 3.81. To this end we first introduce the following notions. Let $M \subset P$ be a submanifold and let $i : M \to P$ be the canonical immersion. Denote by $T_M P$ the *restriction* of the tangent bundle of P to M, i.e., $(T_M P)_m = T P_m$ for all $m \in M$. Let $E \subset T_M P$ be a vector subbundle. Think of $E \cap T M$ as a distribution on M which associates to every $m \in M$ the vector space $E_m \cap T M_m$ and suppose that such a distribution is *integrable* and that the corresponding space of leaves is a smooth manifold $M/\sim$ such that $\pi : M \to M/\sim$ is a (smooth) submersion. Under these assumptions

Definition 2.73 The tuple (P, M, E) is called *Poisson reducible* if $M/\sim$ can be endowed with a Poisson structure $\{\cdot, \cdot\}_\sim$ such that for all $F_1, F_2 \in C^\infty(M/\sim)$

$$\{f_1, f_2\}_\Pi \circ i = \{F_1, F_2\}_\sim \circ \pi, \qquad (2.65)$$

for all $f_1, f_2 \in C^\infty(P)$ smooth extensions of $F_1 \circ \pi$ and of $F_2 \circ \pi$, respectively, such that $d(F_1 \circ \pi)|_E = 0 = d(F_2 \circ \pi)|_E$. $\triangle$

Using the same notations as above, let us list the following conditions:

(PR1) The distribution $E \cap T M$ is integrable.
(PR2) The space of the leaves $M/\sim$ of the foliation defined by $E \cap T M$ is a smooth manifold and the corresponding projection map $\pi : M \to M/\sim$ is a smooth submersion.
(PR3) For all $f_1, f_2 \in C^\infty(P)$ such that $df_1|_E = 0$ and $df_2|_E = 0$, $d\{f_1, f_2\}_P|_E = 0$ as well.

One can now state the following result (see [174]).

Theorem 2.74 (Marsden-Ratiu) *If (PR1),(PR2) and (PR3) hold, the tuple (P, M, E) is reducible if and only if*

$$\Pi^\sharp(E^\circ) \subset TM + E, \tag{2.66}$$

where $E^\circ \subset T^*P$ is the annihilator of E, i.e., the vector bundle whose fiber $E_p^\circ = \{\alpha \in T_p^*P \mid \langle \alpha, v \rangle = 0, \ \forall v \in E\}$.

For the proof of this important result we refer the reader to the above cited reference. Here below we discuss a few examples.

Example 2.75 A particular instance of the previous theorem was already discussed in Corollary 2.57. We recall that in case a Lie group G acts on a Poisson manifold via Poisson morphisms, i.e., for all $g \in G$, the corresponding diffeomorphism $\varphi_g : P \to P$ satisfies $\varphi_g^*\{f_1, f_2\}_\Pi = \{\varphi_g^* f_1, \varphi_g^* f_2\}_\Pi$, for all $f_1, f_2 \in C^\infty(P)$. Under the assumptions that (1) the orbit space P/G is a smooth manifold and (2) the projection $\pi : P \to P/G$ is a (smooth) submersion, Corollary 2.57 ensures that on such a manifold a unique Poisson structure $\{\cdot, \cdot\}_\sim$ is defined such that

$$\pi^*\{F_1, F_2\}_\sim = \{\pi^* F_1, \pi^* F_2\}_\Pi, \tag{2.67}$$

for all $F_1, F_2 \in C^\infty(P/G)$. To recover this result via Theorem 2.74 one chooses $M = P$ and E equal to the tangent bundle to the orbits of the G-action, i.e., the so-called vertical bundle. In this way (PR1) and (PR2) are obviously verified. On the other hand, if $f_1, f_2 \in C^\infty(P)$ are in the annihilator of the vertical distribution, i.e., $\mathscr{L}_{X_x} f_1 = 0 = \mathscr{L}_{X_x} f_2$ for all $x \in \mathfrak{g}$, then $\mathscr{L}_{X_x}\{f_1, f_2\}_\Pi = 0$ since $\mathscr{L}_{X_x}\{f_1, f_2\}_\Pi = \{\mathscr{L}_{X_x} f_1, f_2\}_\Pi + \{f_1, \mathscr{L}_{X_x} f_2\}_\Pi$ because of the G-invariance of $\{\cdot, \cdot\}_\Pi$. In this way Conditions (PR1), (PR2), and (PR3) are fulfilled. On the other hand, (2.66) is also obviously satisfied and for this reason (P, M, E) is reducible, i.e., P/G carries a unique Poisson structure fulfilling (2.65) which, in the present case, is (2.67). In particular if $Z \in \mathfrak{X}(P)$ is complete and (1) $\mathscr{L}_Z \Pi = 0$ and (2) the space of its integral curves is a smooth manifold, then it inherits a (unique) Poisson structure $\{\cdot, \cdot\}_\sim$ satisfying (2.67). $\triangle$

As a second example we discuss the result in 3.81 from the viewpoint of 2.74.

Example 2.76 To this end recall that the input data of 3.81 are a symplectic manifold (P, ω) and a strongly Hamiltonian action of a Lie group G whose corresponding moment map is $\mu : P \to \mathfrak{g}^*$. Under suitable assumptions, for which we refer the reader to Sect. 4, given $\alpha \in \mathfrak{g}^*$, on $\mu^{-1}(\alpha)/G_\alpha$ a unique symplectic structure ω_α is defined such that (3.45) is satisfied. To recover this result from Theorem 2.74 it suffices to choose $M = \mu^{-1}(\alpha)$ and E equal to the tangent bundle to the G_α-orbits. Note that with these choices $\Pi^\sharp(E^\circ) \subset TM$. On the other hand, (PR1) and (PR2) are trivially satisfied while, as far as (PR3) is concerned, one notes that the G-invariance of the symplectic form entails the G-invariance of the corresponding Poisson bracket and the argument used in the previous example implies that also in this case (PR3) holds true. In other words the tuple (P, M, E)

is reducible and a simple check shows that the reduced Poisson bracket defined via Theorem 2.74 is the one associated to the reduced symplectic form ω_α. $\triangle$

We close this subsection stating the following result which extends Theorem 3.86 to the present setting. To this end let (P, Π) be a Poisson manifold, $H \in C^\infty(P)$, and let $M \subset P$ be a submanifold. Following [174], M will be called *conserved* for H if $X_H(m) \in T_m M$ for all $m \in M$, where $X_H = \Pi^\sharp(dH)$.

Example 2.77 If E is as above and $\Pi^\sharp(E^\circ) \subset TM$, then M is conserved for all $H \in C^\infty(P)$ whose differential (at each point) belongs to the annihilator of E (at the given point). $\triangle$

After these preliminary observations, one can state the following important result

Theorem 2.78 (Dynamic Reduction) *Let (P, M, E) be Poisson reducible and let $H \in C^\infty(P)$ for which M is conserved. Then the one-parameter group of (local) diffeomorphisms $\{\varphi_t^H\}_t$ induces on $M/\sim$ a one-parameter group of local diffeomorphisms $\{\tilde{\varphi_t}\}_t$ such that $\tilde{\varphi_t}$ is Poisson for every t. Moreover:*

(i) The vector field on $M_\sim$ whose flow is $\{\tilde{\varphi_t}\}_t$ is X_K, where $K \in C^\infty(M_\sim)$ is defined by $K \circ \pi = H \circ i$ and $i : M \to P$ is the canonical immersion.
(ii) The vector fields $X_H|_M$ and X_K are π-related.

For the proof of this theorem we refer the reader to [174].

4.3 Lie Algebroids and Poisson Cohomology

In this last subsection we discuss, very briefly, how Poisson structures enter in the theory of the Lie algebroids. For a general reference about Lie algebroids we refer the reader to [162] and also the very well-written [54] for a shorter treatment. A Lie algebroid is a vector bundle $p : E \to M$ whose space of sections $\Gamma(E)$ is endowed with a Lie bracket $[\cdot, \cdot]_E$ and $C^\infty(M)$-linear morphism $a : \Gamma(E) \to \mathfrak{X}(M)$, called the anchor of the Lie algebroid, such that

$$[s_1, fs_2]_E = f[s_1, s_2]_E + (a(s_1)f)s_2, \ \forall s_1, s_2 \in \Gamma(E), \ f \in C^\infty(B). \tag{2.68}$$

One can show that (2.68) implies $a[s_1, s_2]_E = [a(s_1), a(s_2)]$, for all $s_1, s_2 \in \Gamma(E)$, i.e., that a is a morphism of Lie algebras. We will denote by $(E, a, [\cdot, \cdot]_E)$ the Lie algebroid defined on E and whose anchor and Lie bracket are a and $[\cdot, \cdot]_E$, respectively. Note that $[\cdot, \cdot]_E$ is not $C^\infty(M)$-bilinear unless $a(s) = 0$ for all $s \in \Gamma(E)$. In this case the bracket $[\cdot, \cdot]_E$ restricts to each fiber of the vector bundle to a well-defined Lie bracket, making it into a bundle of Lie algebras. Among the many examples of Lie algebroids we mention the following ones.

Example 2.79 The tangent bundle of every manifold is a Lie algebroid, called the tangent Lie algebroid, whose anchor is the identity and whose Lie bracket is the

Lie bracket between vector fields. This Lie algebroid will be denoted by $TM_{\mathrm{id}} = (TM, \mathrm{id}, [\cdot, \cdot]_{TM})$. $\triangle$

Example 2.80 Every (finite-dimensional) Lie algebra is a Lie algebroid over a point. In this case the Lie bracket is the Lie bracket of the Lie algebra and the anchor is (necessarily) the trivial application. $\triangle$

A more important case for our purposes is enclosed in the following.

Example 2.81 Let (M, Π) be a Poisson manifold. Then $p : T^*M \to M$ is a Lie algebroid whose Lie bracket is defined in Formula (2.36) and whose anchor is $\Pi^\sharp : \Omega^1(M) \to \mathfrak{X}(M)$. This Lie algebroid will be denoted with $T^*M_\Pi = (T^*M, \Pi^\sharp, \{\cdot, \cdot\}_\Pi)$. $\triangle$

To every Lie algebroid is associated with a cohomology as follows. Let $(E, [\cdot, \cdot]_E, a)$ be a Lie algebroid over B and let E^* the dual vector bundle of E, i.e., the vector bundle over B such that $E_b^* = \mathrm{Hom}_{\mathbb{K}}(E_b, \mathbb{K})$, for all $b \in B$. The $\mathbb{K}$-linear map induced on $\Gamma_E = \oplus_{k \geq 0} \Gamma(\Lambda^k E^*)$ by $d_E : \Gamma(\Lambda^k E^*) \to \Gamma(\Lambda^{k+1} E^*)$, defined by

$$(d_a \alpha)(s_1, \ldots, s_{k+1}) = \sum_{i=1}^{k+1} (-1)^{i+1} a(s_i).\alpha(s_1, \ldots, \hat{s}_i, \ldots, s_{k+1})$$

$$= \sum_{i<j} (-1)^{i+j} \alpha([s_i, s_j]_E, s_1, \ldots, \hat{s}_i, \ldots, \hat{s}_j, \ldots, s_{k+1}),$$

$$\tag{2.69}$$

for all $s_1, \ldots, s_{k+1} \in \Gamma(E)$ and all $\alpha \in \Gamma(\Lambda^k E^*)$, squares to zero, defining the complex (Γ_E, d_E). The cohomology of this complex is called the cohomology of the Lie algebroid $(E, [\cdot, \cdot]_E, a)$. For example, the cohomology of the tangent Lie algebroid 2.79 is the de Rham cohomology of M, while the Lie algebroid cohomology of the Lie algebroid defined in 2.80 is the cohomology of the Lie algebra $\mathfrak{g}$; see Sect. 1.6 in Appendix C. Looking at our last Example 2.81, one deduces that to every Poisson manifold (M, Π) is associated with a cohomology, called the Poisson cohomology of (M, Π), which coincides with the cohomology of the Lie algebroid defined on T^*M. The complex defining this cohomology is introduced accordingly with the general procedure sketched above. The relevant graded vector space is $\Gamma_{T*M} = \oplus_{k \geq 0} \Gamma(\Lambda^k TM)$, while the differential will be defined as in (2.69), using $a = \Pi^\sharp$ and $[\cdot, \cdot]_E = \{\cdot, \cdot\}_\Pi$; see Formula (2.36). Note that $\Gamma(\Lambda^0 TM) = C^\infty(M)$, $\Gamma(\Lambda^1 TM) = \mathfrak{X}(M)$ and $\Gamma(\Lambda^2 TM)$ is the vector space of the bivector fields on M. Moreover,

$$(d_\Pi f)(\xi) = \Pi^\sharp(\xi).f = \langle \Pi^\sharp(\xi), df \rangle = -\langle \xi, \Pi^\sharp(df) \rangle = -\langle \xi, X_f \rangle$$

$$\overset{\text{see Exercise 2.18}}{=} [\![\Pi, f]\!](\xi),$$

while

$$(d_\Pi X)(\xi_1, \xi_2) = \Pi^\sharp(\xi_1).\langle X, \xi_2 \rangle - \Pi^\sharp(\xi_2).\langle X, \xi_1 \rangle - \langle X, \{\xi_1, \xi_2\}_\Pi \rangle$$

$$\stackrel{(2.36)}{=} \langle \xi_2, [\Pi^\sharp(\xi_1), X] \rangle - \langle \xi_1, [\Pi^\sharp(\xi_2), X] \rangle + \langle d\Pi(\xi_1, \xi_2), X \rangle$$

$$= -\mathscr{L}_X(\Pi(\xi_1, \xi_2)) + \Pi(\mathscr{L}_X \xi_1, \xi_2) + \Pi(\xi_1, \mathscr{L}_X \xi_2)$$

$$= -(\mathscr{L}_X \Pi)(\xi_1, \xi_2)$$

$$\stackrel{(2.20)}{=} [\![\Pi, X]\!](\xi_1, \xi_2).$$

More in general it can be shown that

$$d_\Pi = [\![\Pi, \cdot]\!]. \tag{2.70}$$

Summarizing, to every Poisson manifold (M, Π) is associated a complex (Γ_{T^*M}, d_Π) where d_Π can be represented in terms of Π and of the Schouten brackets as in Formula (2.70). This complex coincides with the one associated to the Lie algebroid naturally defined on the cotangent bundle of M.

5 Bibliographical Notes

There are many references which discuss the topics presented in this chapter. We refer the reader to the monographs [69, 150, 240] and in particular to the very-well written recent graduate textbook [54]. Other two very rich sources of information are the seminal paper [247] and its companion [248]. In particular, [240] and [69] contain a nice and very readable introduction to the theory of the Lie algebroids (and Lie groupoids) and its relation with Poisson geometry. An encyclopedic discussion of these results can be found in [162]. Finally, a very readable, comprehensive, and up-to-date reference is [54].

Chapter 3
Hamiltonian G-Actions and the Marsden-Weinstein-Meyer Reduction

In this chapter, which aims to set a convenient framework for the study of the class of Hamiltonian systems with symmetries, the definition of Hamiltonian G-action and of moment map will be introduced. The former provides a geometrical model for the notion of symmetry in Hamiltonian mechanics, the latter is a broad generalization of the concepts of linear and angular momenta, and it provides a geometrical setting where the *symmetries* become an efficient a tool to solve the Hamilton equations. In particular, as it will be emphasized hereafter, the moment map is the main ingredient in the theory of the Hamiltonian reduction which, besides its applications to dynamics, is one of the more, if not the most, efficient tool to produce new symplectic manifolds from old ones.

1   Cotangent Lift of a G-Action

In this section, we introduce the moment map for a particular class of G-actions defined on cotangent bundles, the so-called cotangent lifts; see Example 1.53 and Appendix D. More precisely, we show that to this class of G-actions, it is naturally associated with a remarkable Poisson morphism from the cotangent bundle to the dual of the Lie algebra of G, which intertwines the G-action on the cotangent bundle with the coadjoint action on $\mathfrak{g}^*$. The notions introduced in this context will be extended to a more general class of symplectic G-manifolds in the next section, where the reader will find an in-depth discussion about the properties of the moment map defined in this broader framework. To keep up with our plan and for the reader's convenience, we start our discussion with the following reminder; see Appendix D for more details.

Let N be a manifold and let $\mathscr{D}(N)$, $\mathscr{D}(T^*N)$ be the groups of diffeomorphisms of N and of its cotangent bundle, respectively. The *cotangent lift* $\mathscr{D}(N) \to \mathscr{D}(T^*N)$, $\varphi \rightsquigarrow \widetilde{\varphi}$, is a group homomorphism. Moreover, for all $\varphi \in \mathscr{D}(N)$, $\widetilde{\varphi}$ is a bundle map

A. Arsie, I. Mencattini, *Geometry of Integrable Systems*, Latin American Mathematics Series – UFSCar subseries,
https://doi.org/10.1007/978-3-031-96282-0_3

covering φ which preserves the canonical one-form Θ; see Theorem D.4. This last property, in particular, implies that $\tilde{\varphi}$ is a *symplectomorphism* of (T^*N, Ω), where $\Omega = d\Theta$.

The *infinitesimal* analogue of the cotangent lift above recalled induces a Lie algebra homomorphism between $\mathfrak{X}(N)$ and $(C^\infty_{\mathrm{lin}}(T^*N), \{\cdot, \cdot\})$, where $\{\cdot, \cdot\}$ is the Poisson bracket defined by the canonical symplectic form Ω; see Lemma D.10.

Now let G be a Lie group acting on N via a left action φ. Then

Proposition 3.1

(i) φ can be lifted to a G-action on the cotangent bundle T^*N called the cotangent lift of φ, which preserves the canonical one-form Θ. In particular, the cotangent lift of any G-action on N defines a symplectic G-action on (T^*N, Ω); see Definition 1.51.

(ii) The cotangent lift of φ defines a (nontrivial) homomorphism of Lie algebras $\mathfrak{g} \to (C^\infty_{\mathrm{lin}}(T^*N), \{\cdot, \cdot\})$.

Proof (i) Recall that a G-action on N is the same thing as a group homomorphism $\Psi : G \to \mathscr{D}(N)$; see Lemma C.40. The proof of (i) follows composing this homomorphism with the one defined by the cotangent lift of the $\mathscr{D}(N)$-action. More precisely, it is the homomorphism

$$ G \xrightarrow{\; g \rightsquigarrow \tilde{\varphi}_g \;} \mathscr{D}\big(T^*(N)\big) $$

defined by the composition

$$ G \xrightarrow{\; g \rightsquigarrow \varphi_g \;} \mathscr{D}(N) \xrightarrow{\; \varphi \rightsquigarrow \tilde{\varphi} \;} \mathscr{D}(T^*N), $$

where

$$ \tilde{\varphi}_g(\alpha) = (\varphi_g^{-1})^*_{\varphi(n)}(\alpha) \in T^*_{\varphi(n)}N, \quad \forall \alpha \in T^*_n N; \tag{3.1} $$

see Formula (D.1).

(ii) Recall that to every G-action corresponds a $\mathfrak{g}$-action defined as the Lie algebra homomorphism $\mathfrak{g} \to \mathfrak{X}(N)$, $x \rightsquigarrow X_x$; see Sect. 1.3 in Appendix C. The homomorphism of Lie algebras mentioned in (ii) is the one closing

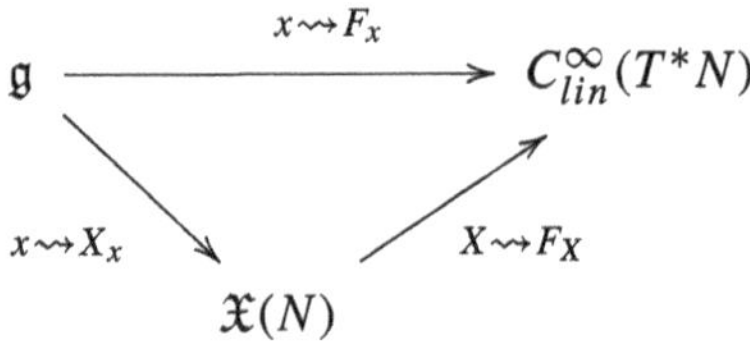

to a commutative diagram of Lie algebras. $\square$

Since

$$\mathfrak{g} \xrightarrow{\ x \rightsquigarrow F_x\ } C^\infty(T^*N)$$

is linear, one can define the smooth map

$$\mu : T^*N \xrightarrow{\ \alpha \rightsquigarrow F_{(\cdot)}(\alpha)\ } \mathfrak{g}^* \tag{3.2}$$

where $\langle \mu(\alpha), x \rangle = F_x(\alpha) = \langle \alpha, X_x(\pi(\alpha)) \rangle$, for all $x \in \mathfrak{g}$; see also Formula (D.5).

Definition 3.2 The map μ defined in Eq. (3.2) is called the *moment*, or the *momentum map* associated to the cotangent lift of the G-action. △

We record below two important properties of the moment map associated to the cotangent lift of a Lie group action. The first one is enclosed in the following.

Proposition 3.3 *The map in (3.2) defines a Poisson morphism between* $(T^*N, \{\cdot, \cdot\})$ *and* $(\mathfrak{g}^*, \{\cdot, \cdot\}_{LP})$, *where* $\{\cdot, \cdot\}$ *is the Poisson bracket defined by* Ω *and* $\{\cdot, \cdot\}_{LP}$ *is the Lie-Poisson bracket on* $\mathfrak{g}^*$.

Proof First, we show that the statement holds for every pair of linear functions $x, y \in \mathfrak{g}$, i.e., that

$$\{\mu^* x, \mu^* y\}(\alpha) = \{x, y\}_{LP}(\mu(\alpha)),$$

for all $\alpha \in T^*N$. The right-hand side of the previous formula is

$$\{x, y\}_{LP}(\mu(\alpha)) = \langle F_{(\cdot)}(\alpha), [dx_{F_{(\cdot)}(\alpha)}, dy_{F_{(\cdot)}(\alpha)}] \rangle.$$

Since x, y are linear functions on $\mathfrak{g}^*$, $dx_{F_{(\cdot)}(\alpha)} = x, dy_{F_{(\cdot)}(\alpha)} = y$ for every α. Then

$$\begin{aligned}
\{x, y\}_{LP}(\mu(\alpha)) &= \langle F_{(\cdot)}(\alpha), [dx_{F_{(\cdot)}(\alpha)}, dy_{F_{(\cdot)}(\alpha)}] \rangle \\
&= \langle F_{(\cdot)}(\alpha), [x, y] \rangle \\
&= F_{[x,y]}(\alpha) \\
&= \{F_x, F_y\}(\alpha); \text{ see (ii) in Proposition 3.1.}
\end{aligned}$$

On the other hand, since $\mu^* x(\alpha) = F_x(\alpha)$ and $\mu^* y(\alpha) = F_y(\alpha)$, for all $\alpha \in T^*N$, $\{\mu^* x, \mu^* y\}(\alpha) = \{F_x, F_y\}(\alpha)$, showing that

$$\{\mu^* x, \mu^* y\} = \mu^*\{x, y\}_{LP}, \ \forall\ x, y \in \mathfrak{g}^*.$$

Now let $f, g \in C^\infty(\mathfrak{g}^*)$. Using the previous result one can write

$$\{\mu^*(df_{\mu(\alpha)}), \mu^*(dg_{\mu(\alpha)})\}(\alpha) = \{df_{\mu(\alpha)}, dg_{\mu(\alpha)}\}_{LP}\mu(\alpha)$$
$$= \langle \mu(\alpha), [df_{\mu(\alpha)}, dg_{\mu(\alpha)}]\rangle = \{f, g\}_{LP}(\mu(\alpha)).$$

On the other hand, since $df_{\mu(\alpha)}$ is a *linear function* on $\mathfrak{g}^*$ and the differential of linear function coincides with the linear function itself, one has

$$d\big(\mu^*(df_{\mu(\alpha)})\big) = \mu^*(df_{\mu(\alpha)}). \tag{3.3}$$

Moreover, if $v \in T_\alpha(T^*N)$

$$\langle d(f \circ \mu)_\alpha, v\rangle = \langle (d\mu^* f)_\alpha, v\rangle = \langle (\mu^* df)_\alpha, v\rangle = \langle df_{\mu(\alpha)}, \mu_{*,\alpha}v\rangle = \langle \mu^*(df_{\mu(\alpha)}), v\rangle,$$

meaning that for all $\alpha \in T^*N$

$$d(f \circ \mu)_\alpha = \mu^*(df_{\mu(\alpha)}). \tag{3.4}$$

This yields

$$i_{X_{f\circ\mu}(\alpha)}\Omega_\alpha = -d(f \circ \mu)_\alpha \overset{(3.4)}{=} -\mu^* df_{\mu(\alpha)} \overset{(3.3)}{=} -d\big(\mu^*(df_{\mu(\alpha)})\big)$$
$$= i_{X_{\mu^*(df_{\mu(\alpha)})}(\alpha)}\Omega_\alpha, \ \forall \alpha \in T^*N,$$

which entails $i_{X_{f\circ\mu}}\Omega = i_{X_{\mu^*(df_{\mu(\alpha)})}}\Omega$. This implies that $f \circ \mu$, as a function on T^*N, is equal to $\mu^*(df_{\mu(\alpha)})$ up to a constant. In this way one arrives to the following identity:

$$\{\mu^*(df_{\mu(\alpha)}), \mu^*(dg_{\mu(\alpha)})\}(\alpha) = \{\mu^* f, \mu^* g\}(\alpha), \ \forall \alpha \in T^*N,$$

which, compared with

$$\{\mu^*(df_{\mu(\alpha)}), \mu^*(dg_{\mu(\alpha)})\}(\alpha) = \{f, g\}_{LP}(\mu(\alpha)), \ \forall \alpha \in T^*N,$$

concludes the proof. $\qquad\qquad\qquad\qquad\qquad\qquad\qquad\qquad\qquad\qquad\square$

The second remarkable property of the map μ defined in Formula (3.2) is contained in the following.

Proposition 3.4 *The moment map μ interwines the G-action on T^*N defined in Formula (3.1), with the coadjoint action of G on the dual of its Lie algebra $\mathfrak{g}^*$, i.e.,*

$$
\begin{array}{ccc}
T^*N & \xrightarrow{\ \mu=F_{(\cdot)}\ } & \mathfrak{g}^* \\[2pt]
\tilde{\varphi}_g \downarrow & & \downarrow \mathrm{Ad}_g^{\sharp} \\[2pt]
T^*N & \xrightarrow{\ \mu=F_{(\cdot)}\ } & \mathfrak{g}^*
\end{array}
$$

is a commutative diagram for every $g \in G$.

Proof First, recall that, for all $x \in \mathfrak{g}$ and $g \in G$, $\varphi_g X_x = X_{\mathrm{Ad}_{g^{-1}x}}$ (see Proposition C.43), and that if $\varphi \in \mathscr{D}(N)$ and $\tilde{\varphi} \in \mathscr{D}(T^*N)$ is its cotangent lift, then for all $X \in \mathfrak{X}(N)$, one has

$$
F_{\varphi X} = \tilde{\varphi}^{-1} F_X;
$$

see Lemma D.1 and Proposition D.11. Writing F_{X_x} to denote the function F_x and applying the previous formula to this case, we obtain

$$
F_{X_{\mathrm{Ad}_{g^{-1}x}}} = \tilde{\varphi}_g^{-1} F_{X_x}.
$$

Furthermore,

$$
\langle \mathrm{Ad}_g^{\sharp}\big(F_{(\cdot)}(\alpha)\big), x \rangle = \langle F_{(\cdot)}(\alpha), \mathrm{Ad}_{g^{-1}} x \rangle = F_{X_{\mathrm{Ad}_{g^{-1}x}}}(\alpha) = \big(\tilde{\varphi}_{g^{-1}} F_{X_x}\big)(\alpha)
$$

$$
= F_{X_x}\big(\tilde{\varphi}_g(\alpha)\big) = \langle F_{(\cdot)}\big(\tilde{\varphi}_g(\alpha)\big), x \rangle,
$$

for all $x \in \mathfrak{g}$, $\alpha \in T^*N$ and $g \in G$. In other words,

$$
\mathrm{Ad}_g^{\sharp} \circ F_{(\cdot)} = F_{(\cdot)} \circ \tilde{\varphi}_g, \ \forall g \in G.
$$

$\square$

Before discussing some examples we make the following observation.

Remark 3.5 As it was shown above, the cotangent lift of a G-action on N is naturally associated with a Poisson morphism:

$$
F_{(\cdot)} = \mu : T^*N \to \mathfrak{g}^*,
$$

called the moment map. However, as it should be clear from the previous discussion, the existence of such a map is an *infinitesimal* issue. In other words, the existence of a Poisson morphism from the cotangent bundle of a manifold N to the dual of a Lie algebra $\mathfrak{g}$ is *equivalent* to the existence of an infinitesimal action of the same Lie algebra on the manifold N. $\triangle$

Besides its intrinsic importance, our first example aims to explain the origin of the name *moment map*.

Example 3.6 (Linear and Angular Momentum) Let $V = \mathbb{R}^3$ and $T^*V \simeq \mathbb{R}^6$ be its cotangent bundle with its standard symplectic form $\omega = \sum_{i=1}^{3} dp_i \wedge dx_i$.

(i) *Linear momentum.* Let $G = (\mathbb{R}^3, +)$ and let us consider the action $G \times \mathbb{R}^3 \ni (u, v) \rightsquigarrow u + v \in \mathbb{R}^3$. For every $v \in \mathbb{R}^3 \simeq \mathfrak{g}$, note that the corresponding fundamental vector field is $X_v(u) = \frac{d}{dt}\big|_{t=0} \left(u + \exp(tv)\right) = v$. Using Formula (3.2) one can write $F_v(u, \alpha) = \langle \alpha, v \rangle$ and consequently $\mu(u, \alpha) = \alpha$. After identifying V with its dual V^* using, for example, the Euclidean inner product, the previous formula assumes the following form:

$$\mu : \mathbb{R}^6 \xrightarrow{\;(\vec{p},\vec{x}) \rightsquigarrow \langle \vec{p}, \cdot \rangle\;} \mathbb{R}^3.$$

In particular, for all i, $\langle \mu(p, x), e_i \rangle = p_i$, which can be recognized as the ith component of the *linear momentum*.

(ii) *Angular momentum.* With V as above but $G = SO_3(\mathbb{R})$ acting on V by multiplication on the left, given $A \in \mathfrak{so}_3(\mathbb{R})$, the corresponding fundamental vector field is $X_A(v) = \frac{d}{dt}\big|_{t=0} \exp(-tA)v = -Av$. Then, using again Formula (3.2), one can write

$$F_A(v, \alpha) = \langle \alpha, X_A(v) \rangle = \langle \alpha, -Av \rangle.$$

Using again the Euclidean inner product to identify V with V^* and identifying the Lie algebra $\mathfrak{so}_3(\mathbb{R})$ with $(\mathbb{R}^3, \times)$, the previous formula becomes

$$F_A : T^*V \xrightarrow{\;(v,\alpha) \rightsquigarrow (v_\alpha, -Av) = (v_\alpha, v \times v_A)\;} \mathbb{R},$$

where $v_A \in V$ corresponds to $A \in \mathfrak{so}_3(\mathbb{R})$. On the other hand, if $(\cdot, \cdot)$ is the Euclidean scalar product, then $(v_\alpha, v \times v_A) = (v_A, v_\alpha \times v)$, which yields

$$\mu(v, \alpha) = v_\alpha \times v.$$

In particular for $i = 1, 2, 3$

$$\langle F_{(\cdot)}(v, \alpha), e_i \rangle = \epsilon_{ijk} p_j x_k,$$

whose right-hand side, where the sum over the repeated indices in understood, is, up to the sign, the ith component of *the angular momentum*;

see [10, 84]. $\triangle$

Remark 3.7 In point (ii) of the previous example the *minus sign* in the formula for the fundamental vector field comes from our definition of X_x, i.e.,

$$X_x(n) = \left.\frac{d}{dt}\right|_{t=0} \varphi_{\exp(-tx)} n;$$

see Formula (C.5). Obviously this minus sign could be avoided defining

$$X_x(n) = \left.\frac{d}{dt}\right|_{t=0} \varphi_{\exp(tx)} n;$$

see also the part *Notations and mathematical conventions* in the Introduction. $\triangle$

Problem 3.8 Under the *standard* identification of $\mathfrak{so}_3(\mathbb{R})$ with $(\mathbb{R}^3, \times)$ (see Example C.11), one has

$$e_1 \rightsquigarrow A_1 = \begin{pmatrix} 0 & 0 & 0 \\ 0 & 0 & -1 \\ 0 & 1 & 0 \end{pmatrix}, \quad e_2 \rightsquigarrow A_2 = \begin{pmatrix} 0 & 0 & 1 \\ 0 & 0 & 0 \\ -1 & 0 & 0 \end{pmatrix} \text{ and } e_3 \rightsquigarrow A_3 = \begin{pmatrix} 0 & -1 & 0 \\ 1 & 0 & 0 \\ 0 & 0 & 0 \end{pmatrix},$$

where (e_1, e_2, e_3) is the standard orthonormal basis of $\mathbb{R}^3$. If (x_1, x_2, x_3) are the coordinates on $\mathbb{R}^3$ by defined by (e_1, e_2, e_3), and ϵ_{ijk} is the Levi-Civita symbol (see Example 2.71), show that:

(i) The fundamental vector fields defined by the elements $A_i \in \mathfrak{so}_3(\mathbb{R})$ are

$$X_{A_i} = \epsilon_{ijk} x_k \frac{\partial}{\partial x_j}.$$

(ii) The cotangent lift of the vector fields X_{A_i} are defined by the following formula:

$$\widetilde{X}_{A_i} = \epsilon_{ijk}\left(x_k \frac{\partial}{\partial x_j} + p_k \frac{\partial}{\partial p_j}\right).$$

(iii) Given the symplectic form $\omega = \sum_{i=i}^{3} dp_i \wedge dx_i$, show that the vector fields $\widetilde{X}_{A_i}$, $i = 1, \ldots, 3$, are Hamiltonian and that the corresponding Hamiltonian functions are equal, up to a constant, to

$$F_{A_i} = \epsilon_{ijk} p_j x_k.$$

$\triangle$

Remark 3.9 In Example 3.6 it is explained the reason why (3.2) is called moment map: it generalizes the classical linear and angular momentum. In the following section we will study a further generalization of this construction. $\triangle$

In the following example we compute the moment map for the cotangent lift of the adjoint action; see Lemma C.66.

Example 3.10 (Moment Map for the Adjoint Action) Let G be a Lie group acting on its Lie algebra $\mathfrak{g}$ via the *adjoint action* $(g, x) \rightsquigarrow \mathrm{Ad}_g x$. Then, for each $x \in \mathfrak{g}$, the corresponding fundamental vector field is

$$X_x(y) = \frac{d}{dt}\bigg|_{t=0} \mathrm{Ad}_{\exp(-tx)} \, y = -\,\mathrm{ad}_x(y) = [y, x].$$

After identifying $T^*\mathfrak{g}$ with $\mathfrak{g} \oplus \mathfrak{g}^*$, since $\langle \alpha, [y, x] \rangle = -\langle \mathrm{ad}_y^\sharp(\alpha), x \rangle$, given

$$F_x : T^*\mathfrak{g} \xrightarrow{\;(y,\alpha)\rightsquigarrow F_x(y,\alpha)=\langle \alpha, X_x(y)\rangle=\langle \alpha,[y,x]\rangle\;} \mathbb{R},$$

one can define $F_{(\cdot)} : T^*\mathfrak{g} \simeq \mathfrak{g} \oplus \mathfrak{g}^* \to \mathfrak{g}^*$ such that $F_{(\cdot)}(y, \alpha) = -\,\mathrm{ad}_y^\sharp(\alpha)$, for all $(y, \alpha) \in \mathfrak{g} \oplus \mathfrak{g}^*$. In this way one discovers that the moment map associated to the cotangent lift of the adjoint action is the map:

$$F_{(\cdot)} : \mathfrak{g} \oplus \mathfrak{g}^* \xrightarrow{\;(x,\alpha)\rightsquigarrow -\,\mathrm{ad}_x^\sharp \alpha\;} \mathfrak{g}^*.$$

$$\triangle$$

Hints about the following problems can be found in Sect. 1 in Chap. 6.

Problem 3.11 (Cotangent Lift of Left and Right Translations of G) Let G be a Lie group. Show that:

(i) For all $g \in G$, the fundamental vector fields X_x^l, X_x^r, associated to $x \in \mathfrak{g}$ and defined by the *left* and the *right*, translations are given by

$$X_x^l(g) = -X_x^R(g) \quad \text{and} \quad X_x^r(g) = X_x^L(g), \, respectively.$$

In the previous formulas $X_x^L(g) = (L_g)_{*,e}(x)$ and $X_x^R(g) = (R_g)_{*,e}(x)$ are the *left* and *right invariant* vector fields, respectively, defined by the element $x \in \mathfrak{g}$.

(ii) The moment maps μ^l and μ^r defined by the cotangent lifts of the left and right translations, respectively, are given by

$$\mu^l : \alpha \rightsquigarrow -(R_g)^*_{,e}\alpha \quad \text{and} \quad \mu^r : \alpha \rightsquigarrow (L_g)^*_{,e}\alpha,$$

for all $g \in G$ and $\alpha \in T_g^*G$.

(iii) Trivialize T^*G by means of the left translations, i.e., consider the diffeomorphism $T^*G \simeq G \times \mathfrak{g}^*$ defined by $t_l : \alpha \rightsquigarrow (g, (L_g)^*_{,e}\alpha)$; see Sect. 1.1 in Chap. 6. Show that, for all $\alpha \in T_g^*G$ and for all $g \in G$, in this trivialization the moment maps for the left and right translations becomes

$$\mu^l : \alpha \rightsquigarrow - \operatorname{Ad}^\sharp_g \alpha \ \text{ and } \ \mu^r : \alpha \rightsquigarrow \alpha.$$

$\triangle$

Example 3.12 (Translations)   The action of $(\mathbb{R}^n, +)$ on $(\mathbb{R}^{2n}, \omega_{st})$ defined by $\varphi_t(p, x) = (p + t, x + t)$, for all $t \in \mathbb{R}^n$ and for all $(p, x) \in \mathbb{R}^{2n}$, is Hamiltonian. This is easily checked on the generators. In fact, a basis of infinitesimal generators of this action is $X_i = \frac{\partial}{\partial x_i}$ $P_i = \frac{\partial}{\partial p_i}$. Then one can write $i_{X_i}\omega = -dp_i$ and $i_{P_i}\omega = dx_i$. In other words, for every $i = 1, \ldots, n$, the vector field X_i is Hamiltonian with Hamiltonian function $-p_i$, while the vector field P_i is Hamiltonian, with Hamiltonian function x_i. One says that $(\mathbb{R}^n, +)$ acts by *translations* on $\mathbb{R}^{2n}$. $\triangle$

The following problem deals with a generalization of Example 3.6.

Problem 3.13 Let V be a real, n-dimensional vector space and let $G = \mathrm{GL}(V)$ be the group of the invertible endomorphisms of V. Let $\varphi : G \times V \to V$ be the G–action defined by $\varphi(g, v) = gv$, for all $g \in G$ and $v \in V$.

(i) Prove that the cotangent lift of φ defines (and it is defined by) the group homomorphism $\Phi : \mathrm{GL}_n(\mathbb{R}) \to \mathrm{Sp}_n(\mathbb{R})$, where

$$\Phi(g) = \begin{pmatrix} g & 0 \\ 0 & (g^{-1})^T \end{pmatrix};$$

see Sect. 1.2 in Appendix A for the relevant definitions. Note that the restriction of Φ to $\mathrm{SO}_n(\mathbb{R})$ defines the diagonal action of latter group on $\mathbb{R}^{2n} = \mathbb{R}^n \oplus \mathbb{R}^n$; see also Example 3.108.

(ii) Let $A = \sum_{i,j} A_{ij} E_{ij} \in \mathfrak{gl}_n(\mathbb{R})$, where for all $i, j = 1, \ldots, n$, $A_{ij} \in \mathbb{R}$ and $E_{ij} \in \mathfrak{gl}_n(\mathbb{R})$ is the $n \times n$ *elementary matrix* with zero everywhere but in the position (i, j) where there is 1. Show that the moment map $\mu : \mathbb{R}^{2n} \to \mathfrak{gl}_n(\mathbb{R})^*$ corresponding to the cotangent lift of the $\mathrm{GL}_n(\mathbb{R})$-action described in (i) can be represented by the map

$$\mu : (p, x) \rightsquigarrow L,$$

where $L \in \mathfrak{gl}_n(\mathbb{R})$ is the matrix such that $L_{ij} = p_i x_j$. Hint: First, identify $\mathfrak{gl}_n(\mathbb{R})$ with its dual vector space using the nondegenerate bilinear form $(A, B) = -\operatorname{tr}(AB^T)$. Then note that for each $A \in \mathfrak{gl}_n(\mathbb{R})$, $F_A(p, x) = -\sum_{i=1}^n A_{ij} p_i x_j$, for all $(p, x) \in \mathbb{R}^{2n}$. Finally, observe that

$$F_A(p, x) = \operatorname{tr}(AL^T),$$

where L is the $n \times n$-matrix defined above. Note that the minus sign in front of the trace form defined above is needed to compensate the minus sign present in front of the fundamental vector field X_A defining the function F_A.

$\triangle$

2 Hamiltonian Actions and Moment Maps

In the previous section it was shown that:

(i) The cotangent lift of a G-action on N is a symplectic G-action on (T^*N, Ω).
(ii) The cotangent lift $\widetilde{X}_x$ of every fundamental vector field X_x is a Hamiltonian vector field on T^*N, whose Hamiltonian function is F_x.
(iii) The map $F_{(\cdot)} : \mathfrak{g} \to C^\infty(T^*M)$ is a homomorphism of Lie algebras which closes

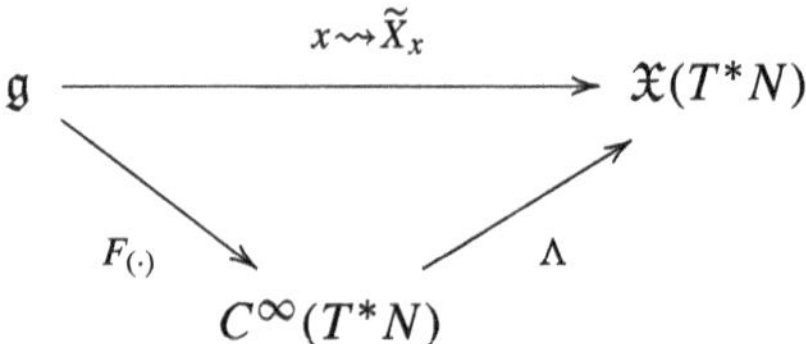

to a commutative diagram of Lie algebras, where Λ was defined in Formula (1.17).

For these statements see Proposition 3.1. Moreover, it was shown that the cotangent lift of a G-action defines a map $\mu : T^*N \to \mathfrak{g}^*$ (see Formula (3.2)), such that:

(iv) It is a Poisson morphism (see Proposition 3.3).
(v) It intertwines the G-action on T^*N with the coadjoint action on $\mathfrak{g}^*$ (see Proposition 3.4).

The main goal of this section is to introduce and study a class of symplectic actions defined on general symplectic manifolds (M, ω), which share many of the properties owned by the cotangent lift of a G-action. To this end, we start introducing the following concept.

Definition 3.14 A symplectic G-action on (M, ω) is called *Hamiltonian* if all the associated fundamental vector fields are Hamiltonian. $\triangle$

Example 3.15 (Hamiltonian vs Non-Hamiltonian) If $H^1_{dR}(M) = 0$, every symplectic action is Hamiltonian. In fact, under this assumption, every locally Hamiltonian vector field is globally Hamiltonian; see also Example 1.31. $\triangle$

On the other hand,

Example 3.16 (Exact Symplectic Manifolds) If (M, ω) is exact, $\theta \in \Omega^1(M)$ is a symplectic potential for ω (see Definition 1.14) and $\varphi : G \times M \to M$ is such that $\varphi_g^* \theta = \theta$ for all $g \in G$, then φ is Hamiltonian. In fact, since $\mathscr{L}_{X_x} \theta = 0$ for every fundamental vector field X_x, $d i_{X_x} \theta = -i_{X_x} \omega$; see also Proposition D.8. Note that this is the case of the G-actions on T^*N defined by cotangent lift. $\triangle$

In the following lemma it is enclosed an important property of Hamiltonian G-actions, which will play a key role in what follows.

Lemma 3.17 *Every Hamiltonian G-action on (M, ω) defines a linear map:*

$$\mu^\sharp : \mathfrak{g} \to C^\infty(M)$$

which closes the following to a commutative diagram of vector spaces:

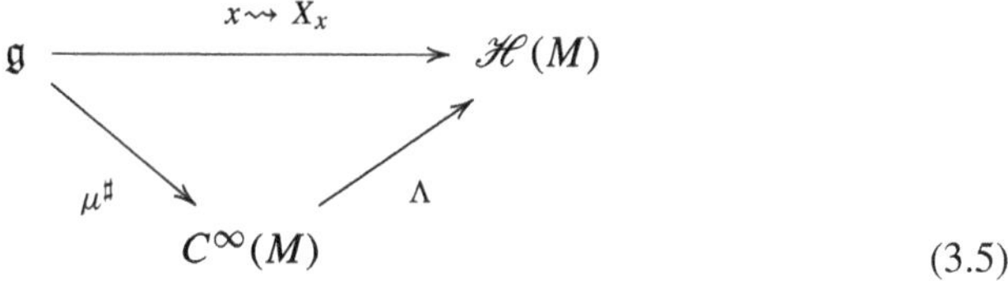

$$\tag{3.5}$$

Vice versa, if there exists a linear map $\mu^\sharp : \mathfrak{g} \to C^\infty(M)$ which closes (3.5) to a commutative diagram, then the symplectic G-action is Hamiltonian.

Proof If φ is Hamiltonian, X_x is Hamiltonian for each $x \in \mathfrak{g}$. Then, for each $x \in \mathfrak{g}$, one can define $\mu^\sharp(x) \in C^\infty(M)$ such that $d\mu^\sharp(x) = -i_{X_x}\omega$. To prove the first part of statement it suffices to show that the map $\mu^\sharp$ can be chosen to be linear. To this end recall that the exact sequence of Lie algebras

$$0 \longrightarrow \mathbb{R} \longrightarrow C^\infty(M) \xrightarrow{f \rightsquigarrow X_f} \mathscr{H}(M) \longrightarrow 0$$

splits, *noncanonically*, as exact sequence of vector spaces. More precisely, for each $m \in M$ one can define the isomorphism $f \rightsquigarrow (X_f, f(m))$, whose inverse is given by the map $(X, a) \rightsquigarrow f - f(m) + a$, where f is *any* Hamiltonian function for X. Moreover, this isomorphism of vector space becomes an isomorphism of Lie algebras if on $\mathscr{H}(M) \oplus \mathbb{R}$ the Lie bracket $\left[(X_f, a), (X_g, b) \right] = (X_{\{f,g\}}, \{f, g\}(m))$ is defined; see the discussion after Corollary 1.42. Anyway, after choosing $m \in M$ defining the above isomorphism between $C^\infty(M)$ and $\mathscr{H}(M) \oplus \mathbb{R}$, one can define

$$\mu^\sharp(x) = f_x - f_x(m), \tag{3.6}$$

where f_x is any Hamiltonian function for X_x. Note that this map is the composition of the map $\mathscr{H}(M) \to \mathscr{H}(M) \oplus \mathbb{R}$, $X_x \rightsquigarrow (X_x, 0)$, with the isomorphism $\mathscr{H}(M) \oplus$

$\mathbb{R} \to C^\infty(M), (X, a) \rightsquigarrow f - f(m) + a$, recalled above. Note that the map defined in Formula (3.6) depends on the choice of the point $m \in M$. Since the map $\mathfrak{g} \to \mathfrak{X}(M)$, $x \rightsquigarrow X_x$, is linear,

$$d\mu^\sharp(x + y) = -i_{X_{x+y}}\omega = -i_{X_x}\omega - i_{X_y}\omega = d\mu^\sharp(x) + d\mu^\sharp(y)$$

yields

$$\mu^\sharp(x + y) - \mu^\sharp(x) - \mu^\sharp(y) = c \in \mathbb{R}.$$

On the other hand, (3.6) entails $\mu^\sharp(z)(m) = 0$ for all $z \in \mathfrak{g}$, which forces $c = 0$, proving that $\mu^\sharp(x + y) = \mu^\sharp(x) + \mu^\sharp(y)$, for all $x, y \in \mathfrak{g}$. Using a similar argument, one can show that $\mu^\sharp(ax) = a\mu^\sharp(x)$, for all $a \in \mathbb{R}$ and $x \in \mathfrak{g}$, proving that the map $\mu^\sharp$ defined in Formula (3.6) is linear. The vice versa is clear. $\qquad\square$

Remark 3.18 Note that the map $X_x \rightsquigarrow (X_x, 0)$, defining a splitting of the exact sequence of vector spaces recalled above, in general is *not* a morphism of Lie algebras. In fact, under such a map $[X_x, X_y] = X_{[x,y]} \rightsquigarrow (X_{[x,y]}, 0)$, but $\left[(X_x, 0), (X_y, 0)\right] = \left(X_{[x,y]}, \omega(X_x, X_y)(m)\right)$. If this map were a Lie algebra morphism, then $\mu^\sharp$ would be so. We will see below that, while it is always possible to close (3.5) to a commutative diagram of vector spaces, it is in general *impossible* to close it to a commutative diagram of Lie algebras. $\qquad\triangle$

Remark 3.19 Since a Hamiltonian vector field defines its Hamiltonian function only up to a constant, the map $\mu^\sharp$ is not uniquely defined by the corresponding Hamiltonian G-action. In particular, if $\mu_1^\sharp, \mu_2^\sharp$ are any two maps defined by the Hamiltonian G-action as in Lemma 3.17, then $d\mu_1^\sharp(x) = -i_{X_x}\omega = d\mu_2^\sharp(x)$ for all $x \in \mathfrak{g}$, which implies that $\mu_1^\sharp - \mu_2^\sharp = C \in \mathfrak{g}^*$. $\qquad\triangle$

After all these preliminary comments and remarks, given a Hamiltonian G-action and a corresponding linear map $\mu^\sharp : \mathfrak{g} \to C^\infty(M)$, one can define

$$\mu : M \xrightarrow{\;m \rightsquigarrow \mu^\sharp(\cdot)(m)\;} \mathfrak{g}^* \tag{3.7}$$

such that for every $x \in \mathfrak{g}$, $\langle \mu(m), x \rangle$ is the value of the function $\mu^\sharp(x)$ at the point m. Accordingly with Remark 3.19, (3.7) is defined by the Hamiltonian G-action *only* up to an additive constant. With a *slight abuse of language* (see Remark 3.21 below), we can now introduce the following.

Definition 3.20 (Co-momentum and Moment Map) Any of the linear maps $\mu^\sharp$ defined as in Formula (3.6) is called a *co-momentum*, or *co-moment map*, for the corresponding Hamiltonian G-action. Given a co-momentum map, the application defined in (3.7) is called a *moment*, or a *momentum map* associated to the Hamiltonian G-action. $\qquad\triangle$

Remark 3.21 It is worth comparing the previous discussion with the content of Sect. 1 where it was discussed the particular case of the cotangent lift of a G-action. In particular note that the moment map associated to the cotangent lift (see Formula (3.2)) is a Poisson morphism (see Proposition 3.3), while the moment map defined in Formula (3.7) is not required to have this property. $\triangle$

2.1 Obstructions

The main outcome of the previous discussion can be summarized saying that given a Hamiltonian G-action, every $m \in M$ induces a linear map $\mu^\sharp$ which closes (3.5) to a commutative diagram of vector spaces. Aiming to pin down the conditions guaranteeing the existence of a morphism of Lie algebras among the $\mu^\sharp$s, we now prove that every co-momentum map defines a class in $H^2(\mathfrak{g}, \mathbb{R})$. Furthermore, we prove that such a class is independent of the chosen $\mu^\sharp$. Hereafter we adopt the following convention.

Convention 3.22 *All the Lie algebras are defined over the field of the real numbers and all cohomology groups of $\mathfrak{g}$ are calculated with respect to the trivial $\mathfrak{g}$-module $\mathbb{R}$. For this reason, we write $H^k(\mathfrak{g})$, $\mathscr{C}^K(\mathfrak{g})$, $\mathscr{Z}^k(\mathfrak{g})$, and $\mathscr{B}^k(\mathfrak{g})$ instead of $H^k(\mathfrak{g}, \mathbb{R})$, $\mathscr{C}^k(\mathfrak{g}, \mathbb{R})$, $\mathscr{Z}^k(\mathfrak{g}, \mathbb{R})$, and $\mathscr{B}^k(\mathfrak{g}, \mathbb{R})$. For the basic Lie algebra and Lie group cohomolgy, we refer the reader to Appendix C.*

To achieve our goal:

(i) First, we show that every co-momentum map $\mu^\sharp$ defined by an Hamiltonian G-action defines an element in $\mathscr{Z}^2(\mathfrak{g})$.
(ii) Then we prove that the difference between any two such classes is an element in $\mathscr{B}^2(\mathfrak{g})$.

We start our discussion proving the following.

Proposition 3.23 *Let $\mu^\sharp : \mathfrak{g} \to C^\infty(M)$ be a co-momentum map corresponding to a Hamiltonian G-action on (M, ω). Then, the bilinear map $d : \mathfrak{g} \times \mathfrak{g} \to C^\infty(M)$ defined by*

$$d(x, y) = \mu^\sharp([x, y]) - \{\mu^\sharp(x), \mu^\sharp(y)\}, \ \forall x, y \in \mathfrak{g}, \tag{3.8}$$

is such that, for all $x, y, z \in \mathfrak{g}$:

(i) $d(x, y) = -d(y, x)$.
(ii) $d([x, y], z) + d([z, x], y) + d([y, z], x) = 0$.

Moreover, for all $x, y \in \mathfrak{g}$, $d(x, y) \in C^\infty(M)$ is a constant function.

Proof The first identity follows from the linearity of $\mu^\sharp$ and from the skew-symmetry of the Lie and of the Poisson bracket. To prove (ii), first, note that for every $x, y \in \mathfrak{g}$, $d(x, y) : M \to \mathbb{R}$ is constant. In fact

$$d\mu^{\sharp}([x, y]) = -i_{X_{[x,y]}}\omega = -i_{[X_x, X_y]}\omega = -\mathscr{L}_{X_x}(i_{X_y}\omega) = -d(i_{X_x}i_{X_y}\omega)$$
$$= d(\omega(X_x, X_y)) = d(\{\mu^{\sharp}(x), \mu^{\sharp}(y)\}).$$

Then

$$d([x, y], z) = \mu^{\sharp}([[x, y], z]) - \{\mu^{\sharp}([x, y]), \mu^{\sharp}(z)\}$$
$$= \mu^{\sharp}([[x, y], z]) - \{\{\mu^{\sharp}(x), \mu^{\sharp}(y)\} + d(x, y), \mu^{\sharp}(z)\}$$
$$= \mu^{\sharp}([[x, y], z]) - \{\{\mu^{\sharp}(x), \mu^{\sharp}(y)\}, \mu^{\sharp}(z)\},$$

since $d(x, y) : M \to \mathbb{R}$ is constant. Taking now cyclic permutations in (x, y, z), the linearity $\mu^{\sharp}$ entails the proof. $\qquad\square$

Since we are interested to attach a class in $H^2(\mathfrak{g})$ to every Hamiltonian G-action, we need to explain how the map d in (3.8) depends on the choice of the co-momentum map used to define it. To this end we prove the following.

Proposition 3.24 *Let $\mu_1^{\sharp}$ and $\mu_2^{\sharp}$ be two co-momentum maps associated to a given Hamiltonian G-action; see Lemma 3.17. If d_1, d_2 are the corresponding bilinear maps defined in (3.8), then there exists $C \in \mathfrak{g}^*$ such that*

$$d_1(x, y) - d_2(x, y) = C([x, y]),$$

for every $x, y \in \mathfrak{g}$.

Proof Since $\mu_1^{\sharp}$ and $\mu_2^{\sharp}$ are associated to the same Hamiltonian G-action, they differ by a constant function C on M, which defines a linear form on $\mathfrak{g}$; see Remark 3.19. If $\mu_1^{\sharp}(x) = \mu_2^{\sharp}(x) + C(x)$ for all $x \in \mathfrak{g}$,

$$\mu_1^{\sharp}([x, y]) - \{\mu_1^{\sharp}(x), \mu_1^{\sharp}(y)\} = \mu_2^{\sharp}([x, y]) + C([x, y]) - \{\mu_2^{\sharp}(x) + C(x),$$
$$\mu_2^{\sharp}(y) + C(y)\}$$
$$= \mu_2^{\sharp}([x, y]) + C([x, y]) - \{\mu_2^{\sharp}(x), \mu_2^{\sharp}(y))\},$$

which yields the result. $\qquad\square$

Before moving on with our discussion it is worth making a summary of the results obtained so far. First, Proposition 3.23 could be rephrased saying that to every co-momentum map $\mu^{\sharp}$ defined by an Hamiltonian G-action is associated an element in $\mathscr{Z}^2(\mathfrak{g})$. In fact, for every $x, y \in \mathfrak{g}$, $d(x, y)$ is a constant function on M and for this reason it can be identified with a bilinear form on $\mathfrak{g}$ which is skew-symmetric and satisfies the cocycle condition; see (i) and (ii) in Proposition 3.23. On the other hand, in Proposition 3.24 is enclosed the statement that two such cocycles differ by a coboundary, i.e., by an element $\xi \in \mathscr{L}^2(\mathfrak{g})$ such that $\xi(x, y) = (\delta C)(x, y)$ for some $C \in \mathfrak{g}^*$; see Sect. 1.6 in Appendix C. In this way we can state the following result.

Theorem 3.25 *Every Hamiltonian G-action defines a class in $H^2(\mathfrak{g})$.*

Proof In fact, by Lemma 3.17, every Hamiltonian G-action defines a co-momentum map $\mu_1^\sharp : \mathfrak{g} \to C^\infty(M)$, which defines an element $d_1 \in \mathscr{L}^2(\mathfrak{g})$; see Proposition 3.23. Now the proof follows, since any other co-momentum map $\mu_2^\sharp : \mathfrak{g} \to C^\infty(M)$ defined by the same Hamiltonian G-action, defines another $d_2 \in \mathscr{L}^2(\mathfrak{g})$ equivalent to d_1. $\qquad\square$

Remark 3.26 The attentive reader will have noticed that the two-cocycle d defined in Formula (3.8) *measures* how $\mu^\sharp$ is far from being a homomorphism of Lie algebras. More precisely, $\mu^\sharp$ is a homomorphism of Lie algebras if and only if d is identically zero. In this case, to any other $\widetilde{\mu}$ stemming from the same Hamiltonian G-action, there corresponds a $\widetilde{d} \in \Lambda^2\mathfrak{g}^*$ defined by $\widetilde{d}(x, y) = C([x, y])$, for all $x, y \in \mathfrak{g}$. $\qquad\triangle$

To move forward with our discussion we note that in Diagram (3.5) both $\Lambda : C^\infty(M) \to \mathscr{H}(M)$ and $\Phi : \mathfrak{g} \to \mathscr{H}(M)$ are homomorphisms of Lie algebras. Starting from this observation it seems natural to focus the attention on the Hamiltonian G-actions having associated a co-momentum map with this same property.

Finally, one can introduce the following notion.

Definition 3.27 (Strongly Hamiltonian G-Actions) A Hamiltonian G-action on (M, ω) is called *strongly Hamiltonian*, if, among the co-momentum maps associated to it, there is one which closes (3.5) to a commutative diagram of Lie algebras. $\quad\triangle$

It is probably worth observing that the class of the strongly Hamiltonian G-actions is not empty. For example, the cotangent lift of a G-action is strongly Hamiltonian.

The next result gives a cohomological characterization of the strongly Hamiltonian G-actions.

Theorem 3.28 *A Hamiltonian G-action is strongly Hamiltonian if and only if its associated cohomology class in $H^2(\mathfrak{g})$ is trivial; see Theorem 3.25.*

Proof If the cohomology class corresponding to the Hamiltonian G-action is trivial then, for every associated co-momentum map $\mu^\sharp : \mathfrak{g} \to C^\infty(M)$, the corresponding two-cocycle d (see (3.8)) is a coboundary, i.e., $d(x, y) = (\delta C)(x, y)$ for all $x, y \in \mathfrak{g}$ and for some $C \in \mathscr{C}^1(\mathfrak{g})$. Under these assumptions one can check that $\mu_1^\sharp := \mu^\sharp - C$ is a homomorphism of Lie algebras. Vice versa, if there exists a co-momentum map $\mu^\sharp : \mathfrak{g} \to C^\infty(M)$ associated to the Hamiltonian G-action that is a homomorphism of Lie algebras, then the corresponding two-cocycle is zero. $\qquad\square$

A simple, though important, example of strongly Hamiltonian G-action is discussed in the following.

Example 3.29 (Quadratic Hamiltonians) Let (V, ω) be a symplectic vector space and let $G = \mathrm{Sp}(V)$ the corresponding symplectic linear group; see Sect. 1.2 in Appendix A. Defining $\varphi : \mathrm{Sp}(V) \times V \to V$ as $\varphi(g, v) = gv$, for all $v \in V$ and $g \in G$, one obtains a symplectic action of G on (V, ω). In fact, $\forall u, v, w \in V$ and $g \in G$

$$(\varphi_g^* \omega)_w (v, u) = \omega_{\varphi_g w}(\varphi_{*, w} u, \varphi_{*, w} v)$$

$$= \omega(gu, gv) \text{ since } \omega \text{ is constant and } \varphi \text{ is } linear$$

$$= \omega(u, v) \text{ since } g \in \mathrm{Sp}(V).$$

Let us show that this action is strongly Hamiltonian. First of all, if $A \in \mathfrak{sp}(V)$, the corresponding fundamental vector field is

$$X_A(u) = \left. \frac{d}{dt} \right|_{t=0} \exp(-tA)u = -Au,$$

for all $u \in V$. Then

$$(i_{X_A} \omega)_u(v) = \omega\big(X_A(u), v\big) = -\omega(Au, v). \tag{3.9}$$

On the other hand, if one defines $\mu_A : V \to \mathbb{R}$ via

$$\mu_A(v) = \frac{1}{2}\omega(Av, v), \ \forall v \in V, \tag{3.10}$$

then, for all $u, v \in V$

$$\begin{aligned}
(d\mu)_{*, u}(v) &= \left. \frac{d}{dt} \right|_{t=0} \mu(u + tv) \\
&= \frac{1}{2} \left. \frac{d}{dt} \right|_{t=0} \omega(Au + tAv, u + tv) \\
&= \frac{1}{2}\big(\omega(Av, u) + \omega(Au, v)\big) \\
&= \omega(Au, v), \tag{3.11}
\end{aligned}$$

since, for every choice of v and u, $\omega(Av, u) + \omega(v, Au) = 0$; see Sect. 1.2 in Appendix A. Comparing now (3.9) with (3.11), one concludes that

$$d\mu_A = -i_{X_A}\omega, \ \forall A \in \mathfrak{sp}(V),$$

proving that each X_A is Hamiltonian. In other words, the $Sp(V)$-action on V is Hamiltonian and a moment map $\mu : V \to \mathfrak{sp}(V)^*$ for this action is defined by

$$\mu(v) = \frac{1}{2}\omega(\cdot v, v),$$

where $\frac{1}{2}\omega(\cdot v, v)$ is the linear form on $\mathfrak{sp}(V)$ defined by $A \rightsquigarrow \frac{1}{2}\omega(Av, v)$.

To show that this action is strongly Hamiltonian, it suffices to prove that $\mu^\sharp[A, B] = \{\mu^\sharp A, \mu^\sharp B\}$, for all $A, B \in \mathfrak{sp}(V)$, where $\mu^\sharp$ is the co-momentum map associated to μ, i.e., is the linear map $\mu^\sharp : \mathfrak{sp}(V) \to C^\infty(V)$ such that $\mu^\sharp A = \mu_A$, for all $A \in \mathfrak{sp}(V)$. To this end, first, we note that

$$\mu^\sharp[A, B](v) = \mu_{[A,B]}(v) = \frac{1}{2}\omega([A, B]v, v) = \frac{1}{2}\omega(ABv, v) - \frac{1}{2}\omega(BAv, v)$$

$$= \omega(Av, Bv),$$

since ω is skew-symmetric and $\omega(Av, u) + \omega(v, Au) = 0$ for all $u, v \in V$ and all $A \in \mathfrak{sp}(V)$. Then we observe that, for all $A, B \in \mathfrak{sp}(V)$ and all $v \in V$,

$$\mu^\sharp[A, B](v) = \omega\big(X_A(v), X_B(v)\big) = \{\mu^\sharp A, \mu^\sharp B\}(v),$$

which is what we wanted to show.

Now we come to the title given to this example. It was shown that for to each element $A \in \mathfrak{sp}(V)$ corresponds the function μ_A defined in (3.10), representing a Hamiltonian function of the vector field X_A. It should be already clear from Formula (3.10) that this function is *quadratic* in the variable v. To make this statement more transparent, let us choose a symplectic basis of $(e_1, \ldots, e_n, f_1, \ldots, f_n)$ of (V, ω); see Definition A.8. This basis defines an isomorphism between $\mathfrak{sp}(V)$ and $\mathfrak{sp}_n(\mathbb{R})$ which lets to represent every $A \in \mathfrak{sp}(V)$ as $2n \times 2n$ matrix, whose block form is given in Formula (A.4). Using this expression and recalling that $\omega(u, v) = v^T J u$ for all u, v column vectors (see Formula (A.2)), one arrives to the following formula:

$$\mu_A(v) = \frac{1}{2}\omega(Av, v) = \frac{1}{2}\left(v_1^T c v_1 + v_1^T d v_2 - v_2^T a v_1 - v_2^T b v_2\right) \tag{3.12}$$

for all $v = \begin{pmatrix} v_1 \\ v_2 \end{pmatrix}$, where v_1, v_2 are two columns vector in $\mathbb{R}^n$. It is clear now that for all $A \in \mathfrak{sp}(V)$, formula (3.10) defines a quadratic polynomial in $2n$-variables, the Darboux coordinates induced on V by the chosen symplectic basis.

Now we detail the case $n = 1$, when $\mathfrak{sp}_1(\mathbb{R}) \simeq \mathfrak{sl}_2(\mathbb{R})$, the Lie algebra of all the 2×2 matrices with trace equal to zero. Writing $v \in \mathbb{R}^2$ as $v = \begin{pmatrix} x \\ p \end{pmatrix}$ and writing the generic element of $\mathfrak{sp}_1(\mathbb{R})$ as $A = \begin{pmatrix} a & b \\ c & -a \end{pmatrix}$, Formula (3.12) becomes

$$\mu_A(v) = \frac{1}{2}cx^2 - apx - \frac{1}{2}bp^2.$$

In particular if $A_1 = \begin{pmatrix} 0 & 1 \\ 0 & 0 \end{pmatrix}$, $A_2 = \begin{pmatrix} 1 & 0 \\ 0 & -1 \end{pmatrix}$ and $A_3 = \begin{pmatrix} 0 & 0 \\ 1 & 0 \end{pmatrix}$, then

$$\mu_{A_1}(v) = -\frac{1}{2}p^2, \quad \mu_{A_2}(v) = -px \text{ and } \mu_{A_3}(v) = \frac{1}{2}x^2,$$

which are the components of the moment map defined by the $\mathfrak{sp}_1(\mathbb{R})$-action on $\mathbb{R}^2$ with respect to the basis $\{A_1, A_2, A_3\}$; see also Problem 1.45. △

Going back to the main topic of this section, Theorem 3.28 tells us that the existence of a strongly Hamiltonian action on a symplectic manifold is *controlled by some cohomology class* in $H^2(\mathfrak{g})$. In particular one has that

Corollary 3.30 *If $H^2(\mathfrak{g})$ is trivial, then every Hamiltonian G-action is strongly Hamiltonian.*

The following example is an (almost) immediate consequence of the previous corollary.

Example 3.31 Every Hamiltonian S^1-action is strongly Hamiltonian. Moreover, every Hamiltonian action of a semi-simple Lie group is strongly Hamiltonian; see Theorem C.81. △

In the next example is discussed in some details the case of the Abelian Lie groups.

Example 3.32 (Hamiltonian Actions of Abelian Lie Groups) First, recall a Lie group is Abelian if and only its Lie algebra is so, i.e., if $[x, y] = 0$ for all $x, y \in \mathfrak{g}$. This forces the differential of the corresponding Chevalley-Eilenberg complex to be trivial, i.e., $\delta = 0$, which yields $\mathscr{B}^k(\mathfrak{g}) = 0$ and $\mathscr{L}^k(\mathfrak{g}) = \Lambda^k\mathfrak{g}^*$, for all $k \geq 1$. These last conditions entail $H^k(\mathfrak{g}) = \Lambda^k\mathfrak{g}^*$, for all $k \geq 1$. These observations imply that, in general, the Hamiltonian action of an Abelian Lie group of dimension greater than one is *not* strongly Hamiltonian. In particular note if G is Abelian, every Hamiltonian G-action selects a *unique* element in $\Lambda^2\mathfrak{g}^*$ (see Formula (3.8)), independent of the co-momentum map $\mu^\sharp$ used to define it; see the proof of Theorem 3.24. In this case the Hamiltonian action is strongly Hamiltonian *if and only if* the corresponding bilinear form is trivial. △

Finally, let us go back to the case of the exact symplectic manifolds.

Example 3.33 (Exact Symplectic Manifolds) Recall that if (M, ω) is exact, every symplectic G-action is Hamiltonian; see Example 3.16. We prove now that such an action is strongly Hamiltonian. To this end it suffices to prove that, if $d\theta = \omega$, the map $\mu^\sharp$ defined by $\mu^\sharp(x) = i_{X_x}\theta$ is a homomorphism of Lie algebras. Let us compute

$$\mu^\sharp([x, y]) = i_{X_{[x,y]}}\theta$$

$$= i_{[X_x, X_y]}\theta$$

$$= i_{\mathscr{L}_{X_x} X_y}\theta$$

$$= \mathscr{L}_{X_x}(i_{X_y}\theta) - i_{X_y}\mathscr{L}_{X_x}\theta$$

$$= d\left(i_{X_x} i_{X_y}\theta\right) + i_{X_x}\left(d i_{X_y}\theta\right)$$

$$= i_{X_x}\left(d i_{X_y}\theta\right) \tag{3.13}$$

$$= -i_{X_x} i_{X_y}\omega \tag{3.14}$$

$$= \omega(X_x, X_y) \tag{3.15}$$

which is what we wanted to prove; see also the problem below. $\triangle$

Problem 3.34 Fill the details of the previous computation. More precisely, in the previous computation justify (3.13), (3.14), and (3.15). $\triangle$

One can rephrase the conclusion of the discussion above saying that given a Hamiltonian G-action, the existence of a Lie algebra homomorphism $\mu^\sharp : \mathfrak{g} \to C^\infty(M)$ closing

$$
\begin{array}{ccccccccc}
0 & \longrightarrow & \mathbb{R} & \longrightarrow & C^\infty(M) & \longrightarrow & \mathscr{H}(M) & \longrightarrow & 0 \\
 & & & & \nwarrow & & \uparrow & & \\
 & & & & & \mu^\sharp & & & \\
 & & & & & & \mathfrak{g} & &
\end{array}
\tag{3.16}
$$

to a commutative diagram of Lie algebra is obstructed by a class in $H^2(\mathfrak{g})$; see Theorem 3.28.

We now assume that this obstruction class is trivial, i.e., that the G-action is strongly Hamiltonian, to understand better how much freedom one has to choose the corresponding Lie algebra homomorphism. To this end one can prove the following..

Proposition 3.35 *If $\mu_1^\sharp$ and $\mu_2^\sharp$ are two Lie algebra homomorphisms associated to the same strongly Hamiltonian G-action, then their difference defines a class in $H^1(\mathfrak{g})$.*

Proof In fact $\mu_1^\sharp = \mu_2^\sharp + C$ for some $C \in \mathfrak{g}^*$; see Proposition 3.24. Then, since $\mu_1^\sharp$ and $\mu_2^\sharp$ are two homomorphisms of Lie algebras, for all $x, y \in \mathfrak{g}$

$$\mu_1^\sharp([x, y]) = \{\mu_1^\sharp(x), \mu_1^\sharp(y)\} = \{\mu_1^\sharp(x) + C(x), \mu_1^\sharp(y) + C(y)\} = \{\mu_2^\sharp(x), \mu_2^\sharp(y)\}$$

$$= \mu_2^\sharp([x, y]).$$

On the other hand, $\mu_1^\sharp([x, y]) = \mu_2^\sharp([x, y]) + C([x, y])$ which shows that the $C \in \mathfrak{g}^*$ is trivial (i.e., identically zero) when restricted to $[\mathfrak{g}, \mathfrak{g}]$, i.e., $C \in \mathscr{Z}^1(\mathfrak{g})$, proving the claim. $\qquad \square$

On the other hand,

Proposition 3.36 *If $\mu^\sharp : \mathfrak{g} \to C^\infty(M)$ is a homomorphism of Lie algebras corresponding to a strongly Hamiltonian G-action on (M, ω) (see Lemma 3.17), then for every $C \in \mathscr{Z}^1(\mathfrak{g})$, $\mu^\sharp + C : \mathfrak{g} \to C^\infty(M)$ is also a homomorphism of Lie algebras defining the same strongly Hamiltonian G-action.*

Proof The claim follows by direct computation that we leave to the reader as an exercise. $\qquad \square$

The previous two propositions entail the following result.

Proposition 3.37 *If the Lie algebra $\mathfrak{g}$ is such that $H^1(\mathfrak{g}) = 0$ and if there exists a lift $\mu^\sharp : \mathfrak{g} \to C^\infty(M)$ which makes (3.16) into a commutative diagram of Lie algebras, then it is unique.*

Example 3.38 As a consequence of the Whitehead lemmas, the co-momentum map associated to a strongly Hamiltonian action of a semi-simple Lie group is unique; see Theorem C.81. $\qquad \triangle$

2.2 Infinitesimal vs Global G-Equivariance

Since a moment map defined by a Hamiltonian G-action is a (smooth) application between two G-manifolds, it is natural to ask under what hypothesis such a map intertwines the two G-actions, i.e., under what hypothesis it is G-*equivariant*. Before delving in more technical details, it seems to be worth giving a formal definition of the latter notion. Let N_1 and N_2 be two G-manifolds with respect to the G-actions φ_1 and φ_2.

Definition 3.39 A smooth map $f : N_1 \to N_2$ is called G-equivariant if

$$f(\varphi_{1g}(n)) = \varphi_{2g}(f(n)), \ \forall n \in N_1, \ g \in G. \tag{3.17}$$

$$\triangle$$

Convention 3.40 *In this subsection Lie groups are not assumed to be connected.*

If the f in the definition above is a moment map, φ_1 is the Hamiltonian G-action on the underlying symplectic manifold, and φ_2 is the coadjoint action of G on $\mathfrak{g}^*$, there are at least two important instances, nonmutually exclusive, when (3.17) is fulfilled: (1) when G is a compact Lie group and (2) when G is connected. We will not discuss the proof of the G-equivariance of the moment map in the first case, for which we refer the reader to [175], but we will analyze in some details the case of connected Lie groups. Before doing that, let us make a general comment to clarify the title of this subsection. Let N_1 and N_2 be G-manifolds and let X_x and Y_x be the fundamental vector fields defined by $x \in \mathfrak{g}$ on N_1 and on N_2, respectively, by the corresponding G-actions. Then

Definition 3.41 (Infinitesimal G-Equivariance) A smooth map $f : N_1 \to N_2$ is called $\mathfrak{g}$-*equivariant*, or *infinitesimal G-equivariant*, if for all $n \in N_1$ and all $x \in \mathfrak{g}$,

$$f_{*,n}(X_x(n)) = Y_x(f(n)). \tag{3.18}$$

$\triangle$

Remark 3.42 Note that (3.18) implies that f is $\mathfrak{g}$-equivariant if and only if, for all $x \in \mathfrak{g}$, X_x and Y_x are f-related; see Formula (1.30). $\triangle$

Clearly, if a map $f : N_1 \to N_2$ is G-equivariant, then it is $\mathfrak{g}$-equivariant, i.e., $\mathfrak{g}$-equivariance is a necessary condition for the G-equivariance. After this preliminary remark, let us go back our symplectic manifold (M, ω) carrying a Hamiltonian G-action and let μ ($\mu^\sharp$) be a corresponding moment (co-momentum) map; see Definition 3.20. Then

Proposition 3.43 *The following properties are equivalent:*

(1) $\mu^\sharp : (\mathfrak{g}, [\cdot, \cdot]) \to (C^\infty(M), \{\cdot, \cdot\})$ *is a Lie algebra morphism.*
(2) μ *is $\mathfrak{g}$-equivariant.*
(3) $\mu : (M, \{\cdot, \cdot\}) \to (\mathfrak{g}^*, \{\cdot, \cdot\}_{LP})$ *is a Poisson morphism.*

Proof *(1) $\Rightarrow$(2). We need to prove that (1) implies that for all $m \in M$ and $x \in \mathfrak{g}$,*

$$\mu_{*,m} X_x(m) = -\,\mathrm{ad}^\sharp_x(\mu(m)). \tag{3.19}$$

To prove (3.19), we start computing its left-hand side. To this end if $y \in \mathfrak{g} \subset C^\infty(\mathfrak{g}^*)$, then

$$\begin{aligned}
\langle \mu_{*,m} X_x(m), y \rangle &= \langle X_x(m), \mu^* y \rangle \\
&= \langle X_x(m), \mu^\sharp(y) \rangle \\
&= -i_{X_y}\omega(X_x)(m) \\
&= \omega(X_x, X_y)(m)
\end{aligned}$$

$$= \{\mu^\sharp(x), \mu^\sharp(y)\}(m)$$

$$= \mu^\sharp[x, y](m), \; \textit{since } \mu^\sharp \textit{ is a homomorphism of Lie algebras}$$

$$= \langle \mu(m), [x, y] \rangle$$

$$= -\langle \mathrm{ad}^\sharp_x \mu(m), y \rangle, \; m \in M.$$

(2) $\Rightarrow$(3). We need to prove that (2) implies that

$$\{f, g\}_{LP} \circ \mu = \{f \circ \mu, g \circ \mu\}, \; \forall f, g \in C^\infty(\mathfrak{g}^*).$$

Let us compute

$$
\begin{aligned}
\{f, g\}_{LP}(\mu(m)) &= \langle \mu(m), [df_{\mu(m)}, dg_{\mu(m)}] \rangle \\
&= -\langle \mathrm{ad}^\sharp_{df_{\mu(m)}} \mu(m), dg_{\mu(m)} \rangle \\
&= \langle \mu_{*,m} X_{df_{\mu(m)}}, dg_{\mu(m)} \rangle \; \textit{see (2)} \\
&= \langle d(g \circ \mu)_m, X_{df_{\mu(m)}} \rangle \\
&= -(i_{X_{g\circ\mu}} \omega)(X_{df_{\mu(m)}})(m) \\
&= \omega(X_{df_{\mu(m)}}, X_{g\circ\mu})(m) \\
&= \{\mu^\sharp(df_{\mu(m)}), g \circ \mu\}(m) \\
&= \{f \circ \mu, g \circ \mu\}(m),
\end{aligned}
$$

since $X_{f\circ\mu} = X_{\mu^\sharp(df_{\mu(m)})}$.
(3) $\Rightarrow$(1). We need to show that (3) implies that

$$\mu^\sharp([x, y]) = \{\mu^\sharp(x), \mu^\sharp(y)\}, \; \forall x, y \in \mathfrak{g}.$$

Let $x, y \in \mathfrak{g} \subset C^\infty(\mathfrak{g}^*)$. Then

$$\{x, y\}_{LP} \circ \mu(m) = \langle \mu(m), [x, y] \rangle = \mu^\sharp([x, y])(m), \; \forall m \in M.$$

On the other hand, since for all $m \in M$, $x \circ \mu(m) = \langle \mu(m), x \rangle = \mu^\sharp(x)(m)$ and, similarly, $y \circ \mu(m) = \mu^\sharp(y)(m)$, one can write

$$\{x \circ \mu, y \circ \mu\} = \{\mu^\sharp(x), \mu^\sharp(y)\},$$

which, together with (3), gives the proof of the statement. $\qquad\square$

In this way one arrives to the following.

Corollary 3.44 *A Hamiltonian G-action is strongly Hamiltonian if and only if it has associated a moment map which is $\mathfrak{g}$-equivariant.*

To recap, here below we *draw* a summary of the previous discussion:

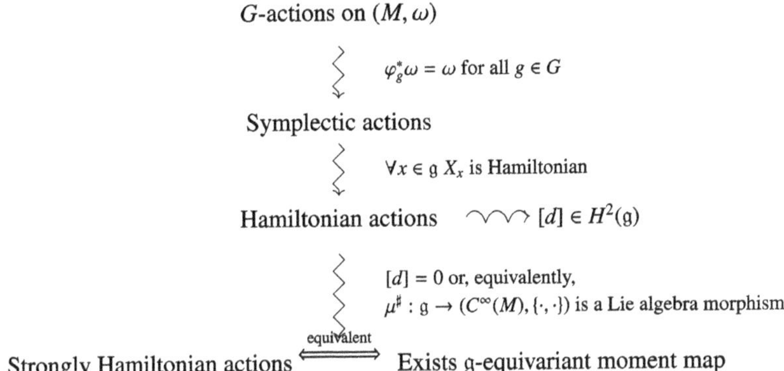

Once the infinitesimal issues are settled, we can turn our attention to the global ones, i.e., from the g-equivariance we move on to investigate the G-equivariance of the moment map. Since a G-equivariant moment map is necessarily g-equivariant, without loss of generality hereafter we will assume that our G-actions are strongly Hamiltonian. Given this assumption, a moment map is G-equivariant if, for all $g \in G$, the following diagram is commutative:

$$
\begin{array}{ccc}
M & \xrightarrow{\ \mu\ } & \mathfrak{g}^* \\
{\scriptstyle \varphi_g}\downarrow & & \downarrow{\scriptstyle \mathrm{Ad}^{\sharp}_g} \\
M & \xrightarrow{\ \mu\ } & \mathfrak{g}^*
\end{array}
$$

or equivalently, if

$$
\mu(\varphi_g(m)) = \mathrm{Ad}^{\sharp}_g(\mu(m)), \ \forall m \in M, \ g \in G;
$$

see Definition 3.39. Finally, we can prove the following.

Theorem 3.45 *If G is a connected Lie group, every infinitesimal G-equivariant moment map is G-equivariant.*

Remark 3.46 It is worth stressing that the previous theorem gives *sufficient* but *not necessary* conditions to ensure the G-equivariance of the moment map. In fact, for example, the moment map associated to the cotangent lift of a Lie group action is always G-equivariant; see Proposition 3.4.   △

Convention 3.47 *In what follows, to simplify the notation we will write $\mu^{\sharp}_x(m)$ instead of $\mu^{\sharp}(x)(m)$ to denote the value of the function $\mu^{\sharp}(x)$ at the point $m \in M$. Furthermore, the result of the G-action on M sometimes will be written as gm instead of $\varphi(g, m)$ or $\varphi_g(m)$.*

We split the proof of Theorem 3.45 in two parts:

(i) First, for every $g \in G$ and $x \in \mathfrak{g}$, we introduce the function $C_{g,x} : M \to \mathbb{R}$ defined by

$$C_{g,x}(m) = \mu_x^{\sharp}(\varphi_g(m)) - \mu_{\mathrm{Ad}_{g^{-1}}x}^{\sharp}(m), \quad \forall m \in M \tag{3.20}$$

and then we show that this is constant on M; see Lemma 3.48 below.

(ii) Second, since for every $(g, x) \in G \times \mathfrak{g}$, $C_{g,x} \in \mathbb{R}$ (see (i) above), and since for each $g \in G$ $C_{g,\cdot} : \mathfrak{g} \xrightarrow{\;x \rightsquigarrow C_{x,g}\;} \mathbb{R}$ is linear, one can define

$$C. : G \xrightarrow{\;g \rightsquigarrow C_{g,\cdot}\;} \mathfrak{g}^*,$$

which turns out to be a one-cocycle of G with value in $\mathfrak{g}^*$; see Lemma 3.49 and Remark 3.50 below and Sect. 1.8 in Appendix C for the relevant definitions of Lie group cohomology.

The cocycle property entails that C is locally constant on G. Since $C_e = 0$, it follows that C is identically zero if restricted to the connected component of the identity of G, proving that μ is G-equivariant if G is connected; see Theorem 3.45. To follow the plan sketched above, we start proving the following:

Lemma 3.48 *For every $(g, x) \in G \times \mathfrak{g}$, $C_{g,x} \in C^{\infty}(M)$ is locally constant.*

Proof Let $m \in M$ and $v \in T_m M$. Then

$$(C_{g,x})_{*,m}(v) = \left(\mu_x^{\sharp} \circ \varphi_g\right)_{*,m}(v) - \left(\mu_{\mathrm{Ad}_{g^{-1}}x}^{\sharp}\right)_{*,m}(v).$$

Let us compute the first term of the right-hand side of the previous identity:

$$
\begin{aligned}
(\mu_x^{\sharp} \circ \varphi_g)_{*,m}(v) &= (\mu_x^{\sharp})_{*,gm} \circ (\varphi_g)_{*,m}(v) \\
&= -\langle i_{X_x(gm)}\omega_{gm}, (\varphi_g)_{*,m}(v)\rangle \\
&= \omega_{gm}\big((\varphi_g)_{*,m}v, X_x(gm)\big) \\
&= (\varphi_g^*\omega)_m\big(v, (\varphi_{g^{-1}})_{*,gm}X_x(gm)\big) \\
&= \omega_m\big(v, (\varphi_g X_x)(m)\big) \\
&\overset{(C.8)}{=} \omega_m\big(v, X_{\mathrm{Ad}_{g^{-1}}x}(m)\big) \\
&= -(i_{X_{\mathrm{Ad}_{g^{-1}}x}}\omega)_m(v) \\
&= \left(\mu_{\mathrm{Ad}_{g^{-1}}x}^{\sharp}\right)_{*,m}(v),
\end{aligned}
$$

for all $m \in M$ and $v \in T_m M$. $\qquad\square$

Since $C_{g,x}$ is constant on M and since it is the difference of two linear forms on $\mathfrak{g}$ depending smoothly on $g \in G$, i.e., $C_{g,x} = \mu_x^\sharp \circ \varphi_g - \mu_{\mathrm{Ad}_{g^{-1}} x}^\sharp$, it defines a (smooth) map $C_. : G \to \mathfrak{g}^*$, $g \rightsquigarrow C_g$, where, for all $x \in \mathfrak{g}$, $C_g(x) = C_{g,x}$. Now one can prove the following.

Lemma 3.49 *For every* $g, h \in G$

$$C_{gh} = \mathrm{Ad}_g^\sharp C_h + C_g.$$

Proof The definition of $C_.$ yields

$$\langle C_{gh}, x \rangle = C_{gh}(x) = \mu_x^\sharp \circ \varphi_{gh} - \mu_{\mathrm{Ad}_{(gh)^{-1}} x}^\sharp = \mu_x^\sharp \circ \varphi_{gh} - \mu_{\mathrm{Ad}_{h^{-1}} \mathrm{Ad}_{g^{-1}} x}^\sharp.$$

On the other hand,

$$\langle \mathrm{Ad}_g^\sharp C_h, x \rangle = \langle C_h, \mathrm{Ad}_{g^{-1}} x \rangle = \mu_{\mathrm{Ad}_{g^{-1}} x}^\sharp \varphi_h - \mu_{\mathrm{Ad}_{h^{-1}} \mathrm{Ad}_{g^{-1}} x}^\sharp.$$

The previous two computations entail that

$$\begin{aligned}
\langle C_{gh} - \mathrm{Ad}_g^\sharp C_h, x \rangle &= \mu_x^\sharp \circ \varphi_{gh} - \mu_{\mathrm{Ad}_{g^{-1}} x}^\sharp \circ \varphi_h \\
&= \left(\mu_x^\sharp \circ \varphi_g - \mu_{\mathrm{Ad}_{g^{-1}} x}^\sharp \right) \circ \varphi_h \\
&= \langle C_g, x \rangle \circ \varphi_h \\
&= \langle C_g, x \rangle,
\end{aligned}$$

since C_g is constant on M. $\qquad\square$

Remark 3.50 Note that Lemma (3.49) tells us that $C_.$ is a one-cocyle of G with values in $\mathfrak{g}^*$, where G acts on $\mathfrak{g}^*$ via coadjoint action; see Sect. 1.8 in Appendix C. $\triangle$

Using the result contained in Lemma 3.49 one can prove the following.

Proposition 3.51 *The function*

$$C_. : G \xrightarrow{\ g \rightsquigarrow C_g\ } \mathfrak{g}^*$$

is locally constant and it is identically equal to 0 *when restricted to the identity component of* G. *In particular, if* G *is connected,* C *is identically zero.*

Proof To prove the statement it suffices to show that, for every $g \in G$ and $\xi \in T_g G$,

$$C_{*,g}(\xi) = 0.$$

That is equivalent to

$$C_{*,g} X_x^R(g) = 0, \ \forall x \in \mathfrak{g}, \ g \in G,$$

where $X_x^R \in \mathfrak{X}(G)$ is the unique right invariant vector field such that $X_x^R(g) = \xi$. Then

$$C_{*,g}(\xi) = C_{*,g} X_x^R(g) = C_{*,g}(R_g)_{*,e}(x) = (C \circ R_g)_{*,e}(x)$$
$$= \frac{d}{dt}\bigg|_{t=0} (C \circ R_g)(\exp(tx)) = \frac{d}{dt}\bigg|_{t=0} C_{\exp(tx)g}.$$

On the other hand, Lemma 3.49 implies that

$$C_{\exp(tx)g} = \mathrm{Ad}^\sharp_{\exp(tx)} C_g + C_{\exp(tx)},$$

or equivalently,

$$\langle C_{\exp(tx)g}, y \rangle = \langle C_g, \mathrm{Ad}_{\exp(-tx)} y \rangle + \langle C_{\exp(tx)}, y \rangle, \tag{3.21}$$

for all $y \in \mathfrak{g}$.

We compute now the time derivative of both terms of the right-hand side of (3.21).

$$\frac{d}{dt}\bigg|_{t=0} \langle C_{\exp(tx)}, y \rangle = \frac{d}{dt}\bigg|_{t=0} \mu_y^\sharp \circ \varphi_{\exp(tx)} - \frac{d}{dt}\bigg|_{t=0} \mu^\sharp_{\mathrm{Ad}_{\exp(-tx)} y}$$
$$= -d\mu_y^\sharp(X_x) + \mu^\sharp_{[x,y]}$$
$$= i_{X_y}\omega(X_x) + \mu^\sharp_{[x,y]}$$
$$= \{\mu_y^\sharp, \mu_x^\sharp\} + \mu^\sharp_{[x,y]}$$
$$= -\mu^\sharp_{[x,y]} + \mu^\sharp_{[x,y]}$$
$$= 0, \tag{3.22}$$

and

$$\frac{d}{dt}\bigg|_{t=0} C_g\big(\mathrm{Ad}_{\exp(-tx)} y\big) = -C_g([x,y]). \tag{3.23}$$

Then, (3.22) and (3.23) yield that for all $g \in G$ and $\xi \in T_g G$

$$\langle C_{*,g}(\xi), y \rangle = \frac{d}{dt}\bigg|_{t=0} C_{\exp(tx)g}(y) = -C_g([x,y]),$$

for all $y \in \mathfrak{g}$, where $x \in \mathfrak{g}$ is such that $X_x^R(g) = \xi$; see above. For this reason, to conclude, it suffices to prove that for every $g \in G$, C_g restricts to 0 to $[\mathfrak{g}, \mathfrak{g}]$. Let us prove this claim.

$$
\begin{aligned}
C_g([x, y]) &= \mu_{[x,y]}^{\sharp} \circ \varphi_g - \mu_{\mathrm{Ad}_{g^{-1}}[x,y]}^{\sharp} \\[2mm]
&= \left\{\mu_x^{\sharp}, \mu_y^{\sharp}\right\} \circ \varphi_g - \mu_{\left[\mathrm{Ad}_{g^{-1}} x, \mathrm{Ad}_{g^{-1}} y\right]}^{\sharp} \\[2mm]
&= \left\{\mu_x^{\sharp}, \mu_y^{\sharp}\right\} \circ \varphi_g - \left\{\mu_{\mathrm{Ad}_{g^{-1}} x}^{\sharp}, \mu_{\mathrm{Ad}_{g^{-1}} y}^{\sharp}\right\} \\[2mm]
&= \left\{\mu_x^{\sharp}, \mu_y^{\sharp}\right\} \circ \varphi_g - \left\{\mu_x^{\sharp} \circ \varphi_g - C_g(x), \mu_y^{\sharp} \circ \varphi_g - C_g(y)\right\} \\[2mm]
&= \left\{\mu_x^{\sharp}, \mu_y^{\sharp}\right\} \circ \varphi_g - \left\{\mu_x^{\sharp} \circ \varphi_g, \mu_y^{\sharp} \circ \varphi\right\} \\[2mm]
&= 0,
\end{aligned}
$$

since φ_g is a Poisson morphism for all $g \in G$, proving that $C. : G \to \mathfrak{g}^*$ is locally constant. Since $C_e = 0$, its restriction to the connected component of the identity is identically zero. $\qquad\square$

Finally, we can present the

Proof (of Theorem 3.45) Let G be a *connected* Lie group and let (M, ω) be a symplectic manifold. We need to show that the moment map $\mu : M \to \mathfrak{g}^*$ of every strongly Hamiltonian G-action, $\varphi : G \times M \to M$, is G-equivariant, i.e.,

$$
\mu(\varphi_g(m)) = \mathrm{Ad}_g^{\sharp}(\mu(m)), \ \forall m \in M \text{ and } g \in G.
$$

To this end recall that the map

$$
C : G \times M \xrightarrow{\ (g,m)\rightsquigarrow C_{g,.}(m)\ } \mathfrak{g}^*
$$

defined in Formula (3.20) is constant over M (see Lemma 3.48), and it is constant and equal to zero with respect to the variable g since we are assuming G-connected; see Proposition 3.51. Then, for all $g \in G$ and $m \in M$, we can write

$$
C_{g,x}(m) = \langle C_{g,.}(m), x \rangle = 0 = \mu_x^{\sharp}(\varphi_g(m)) - \mu_{\mathrm{Ad}_{g^{-1}} x}^{\sharp}(m),
$$

for all $x \in \mathfrak{g}$, which yields

$$
0 = \langle \mu(\varphi_g(m)), x \rangle - \langle \mu(m), \mathrm{Ad}_{g^{-1}}(x) \rangle = \langle \mu(\varphi_g(m)) - \mathrm{Ad}_g^{\sharp}(\mu(m)), x \rangle, \ \forall x \in \mathfrak{g},
$$

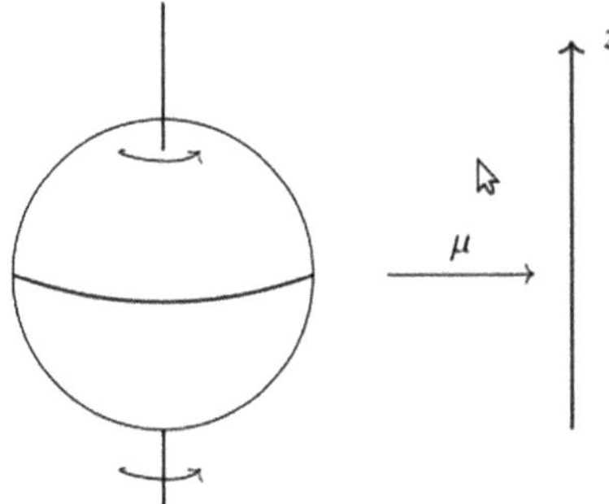

Fig. 3.1 Height function as a moment map

entailing

$$\mu(\varphi_g(m)) = \mathrm{Ad}^{\sharp}_g(\mu(m)), \ \forall g \in G, \ m \in M.$$

$\square$

2.3 *Examples of Hamiltonian G-Actions and of Moment Maps*

In this subsection we present a few examples of strongly Hamiltonian *G*-actions and of their associated moment maps.

Example 3.52 Let $S^2 \subset \mathbb{R}^3$ be the unit sphere centered at the origin and endowed with the its *standard* symplectic form ω; see Example 1.9. Using the *cylindrical* coordinates (θ, z) one finds that $\omega = d\theta \wedge dz$ (see Problem 1.10) and that the map $(\theta, z) \rightsquigarrow (\theta + \eta, z)$ defines a *symplectic* action of the group S^1 on S^2.

Moreover, the fundamental vector field $X = -\frac{\partial}{\partial \theta}$ (unique up to a constant) is *Hamiltonian*, with Hamiltonian function $\mu = z$, see (Fig. 3.1). This function is *the* moment map of the S^1-action described above. Note that in this simple example we could have argued as follows. Since the first de Rham cohomology group of S^2 is trivial, every fundamental vector field is Hamiltonian. Moreover, since the Lie algebra of S^1 is one-dimensional, every map $\mu^{\sharp}$ corresponding to the Hamiltonian S^1-action described above is a Lie algebra homomorphism. $\triangle$

Another interesting case is discussed in the following.

Example 3.53 (Direct Product) Recall that the Lie group G is the direct product of the Lie groups H and K if, as a manifold, $G = H \times K$ and if the product between $g_1 = (h_1, k_1)$ and $g_2 = (h_2, k_2)$ is $g_1 g_2 = (h_1 h_2, k_1 k_2)$ for all $g_1, g_2 \in G$. In this way the canonical immersions $h \rightsquigarrow (h, e_K)$ and $k \rightsquigarrow (e_H, k)$ identify H and K with two Lie subgroups of G which centralize each other. Suppose now that K and H act on (M, ω) in Hamiltonian fashion and let $\mu^K : M \to \mathfrak{k}^*$, $\mu^H : M \to \mathfrak{h}^*$ be two corresponding moment maps. Under these assumptions the induced G-action on (M, ω) is also Hamiltonian and a corresponding moment map $\mu^G : M \to \mathfrak{g}^*$ is $\mu^G = \mu^H + \mu^K$. To prove these assertions observe that if φ^H, φ^K denote the two Hamiltonian actions, and if $g = (h, k)$, then

$$\varphi^G(g, m) = \varphi^G((h, k), m) = \varphi^G\big((h, e_K), \varphi^G((e_H, k), m)\big)$$
$$= \varphi^G\big((e_H, k), \varphi^G((h, e_K), m)\big),$$

which yields

$$\varphi^G(g, m) = \varphi^K(k, \varphi^H(h, m)) = \varphi^H(h, \varphi^K(k, m))$$

for all $g \in G$ such that $g = (h, k)$ and for all $m \in M$. From this observation it follows that the fundamental vector field associated to the element $x = (\xi, \eta) \in \mathfrak{g} \simeq \mathfrak{h} \oplus \mathfrak{k}$ is $X_x = X_\xi + X_\eta$, where X_η and X_ξ are the fundamental vector fields defined by the elements ξ and η via the actions φ^H and φ^K. It follows now that

$$d\mu^{G,x} = -i_{X_x}\omega = -i_{X_\eta}\omega - i_{X_\xi}\omega = d\mu^{H,\xi} + d\mu^{K,\eta},$$

which entails that $\mu^{\sharp,H}(\xi) + \mu^{\sharp,K}(\eta)$ is a Hamiltonian for the fundamental vector field X_x proving the statement. Furthermore, if φ^H and φ^K are strongly Hamiltonian, then φ^G is also strongly Hamiltonian, as one can check using the equivalences in Proposition 3.43. $\triangle$

Before moving to the following example, we would like to present the following nice result (see [20]).

Proposition 3.54 *Let (M, ω) be an exact symplectic manifold and let θ be a symplectic potential for ω. Suppose that:*

(i) M carries a S^1-action $\varphi : S^1 \times M \to M$.
(ii) The corresponding fundamental vector field X satisfies $i_X \mathscr{L}_X \omega = 0$.

Then φ is strongly Hamiltonian with a moment map defined by

$$\mu(m) = \int_{\mathscr{O}_m} \theta, \tag{3.24}$$

where $\mathscr{O}_m$ is the S^1-orbit of the point $m \in M$.

Proof It suffices to show that

$$\langle d\mu_m, v \rangle = \omega_m(v, X(m)), \ \forall m \in M, \ \forall v \in T_m M.$$

To this end let $m \in M$ and $v \in T_m M$ and let Z be the vector field obtained *translating* v along the S^1-orbit going through m, i.e., $Z(\varphi_t(m)) = (\varphi_t)_{*,m}(v)$ for all $t \in S^1$. Finally, let Y be a (smooth) extension of Z to a neighborhood of $\mathscr{O}_m$. Let us now write integral Formula (3.24) using the *parametrization* of $\mathcal{O}_m$ defined by the fundamental vector field X, i.e.,

$$\mu(m) = \int_0^1 \langle \theta_{\varphi_m(t)}, X(\varphi_m(t)) \rangle \, dt$$

and, for each fixed t, let us consider the function $F_t : m \rightsquigarrow \langle \theta_{\varphi_t(m)}, X(\varphi_t(m)) \rangle$. Let us now compute $\langle dF_t, Y \rangle (m)$:

$$\langle dF_t, Y \rangle (m) = \mathscr{L}_Y \langle \theta, X \rangle (\varphi_t(m)) = \underbrace{\langle \mathscr{L}_Y \theta, X \rangle (\varphi_t(m))}_{(1)} + \underbrace{\langle \theta, \mathscr{L}_Y X \rangle (\varphi_t(m))}_{(2)}.$$

The term (2) is equal to zero since, by construction, the restriction of Y to $\mathcal{O}_m$ is X-invariant. The term (1) can be computed using the identity $d\theta(Y, X) = Y \langle \theta, X \rangle - X \langle \theta, Y \rangle - \langle \theta, [Y, X] \rangle$, which the S^1-invariance of (the restriction to $\mathcal{O}_m$ of) Y, reduces to $d\theta(Y, X) = Y \langle \theta, X \rangle - X \langle \theta, Y \rangle$. These observations entail that

$$(1) = \underbrace{d\theta(Y, X)(\varphi_t(m))}_{(3)} + \underbrace{\langle \mathscr{L}_X \theta, Y \rangle (\varphi_t(m))}_{(4)}. \tag{3.26}$$

We now integrate (3) and (4) for t varying in the interval $[0, 1]$. To this end, for a fixed m, let G_m be the function defined by $G_m(t) = \omega(Y, X)(\varphi_m(t))$. We now show that G_m is constant along the flow generated by X computing its t-derivative:

$$\frac{dG_m}{dt} = \underbrace{(\mathscr{L}_X \omega)(Y, X)(\varphi_m(t))}_{(a)} + \underbrace{\omega(\mathscr{L}_X Y, X)(\varphi_m(t))}_{(b)} + \underbrace{\omega(Y, \mathscr{L}_X X)(\varphi_m(t))}_{(c)}.$$

Each of the terms of the right-hand side of the previous equality is zero: (a) is equal to zero since $\mathscr{L}_X \omega$ is the t-derivative of ω and, by hypothesis, $i_X \mathscr{L}_X \omega = 0$, (b) is equal to zero since (the restriction to $\mathcal{O}_m$ of) Y is invariant with respect to the X-flow, and finally, (c) is equal to zero for obvious reasons. From this, it follows that $G_m(t) = \omega(Y, X)(\varphi_m(t))$ does not depend on t, i.e.,

$$d\theta(Y, X)(\varphi_m(t)) = \omega(Y, X)(\varphi_m(t)) = \omega_m(v, X(m)) \, \forall t \in [0, 1].$$

To analyze the term (4) of the right-hand side of (3.26), for each m, let $T_m(t)$ be the function defined by $T_m(t) = \langle \theta, Y \rangle (\varphi_m(t))$ and let us compute its t-derivative:

$$\frac{dT_m}{dt} = \langle \mathscr{L}_X \theta, Y \rangle (\varphi_m(t)) + \langle \theta, \mathscr{L}_X Y \rangle (\varphi_m(t)) = \langle \mathscr{L}_X \theta, Y \rangle (\varphi_m(t)).$$

Then

$$\int_0^1 [d\theta(Y,X)(\varphi_t(m)) + \langle \mathcal{L}_X\theta, Y\rangle(\varphi_t(m))]dt = \int_0^1 \omega_m(v,X)dt$$

$$+ \int_0^1 \frac{d}{dt}[\langle \theta, Y\rangle(\varphi_m(t))]dt$$

$$= \omega_m(v,X) + \langle \theta, Y\rangle(\varphi_m(1)) - \langle \theta, Y\rangle(\varphi_m(0))$$

$$= \omega_m(v,X(m)),$$

since $\varphi_m(1) = \varphi_m(0)$. $\qquad\square$

Remark 3.55 A couple of remarks are in order.

(i) Since $\mathrm{Lie}(S^1)$ is one-dimensional, to every S^1-action is associated a *unique*, up to a scalar multiple, fundamental vector field. For this reason, with a slight abuse of notation, in the statement of the above proposition we wrote *the* fundamental vector field associated to the given S^1-action and we simply denoted it with X.
(ii) Since the moment map for S^1-actions is a (scalar valued) function, Formula (3.24) tells us that its value is constant on the S^1-orbits.

$$\triangle$$

Example 3.56 (Harmonic Oscillator) Let $(\mathbb{R}^{2n}, \omega_{st} = \sum_{i=1}^n dp_i \wedge dx_i)$ be endowed with the diagonal $S^1 \simeq SO_2(\mathbb{R})$-action, whose infinitesimal generator is $X = -\sum_{i=1}^n \left(x_i \frac{\partial}{\partial p_i} - p_i \frac{\partial}{\partial x_i}\right)$. Note that this action is symplectic and $i_X\omega_{st} = -\sum_{i=1}^n (x_i dx_i + p_i dp_i)$, which, together with $-i_X\omega_{st} = dH$, yields

$$H(x,p) = \frac{1}{2}\sum_{i=1}^n (p_i^2 + x_i^2).$$

This function is a Hamiltonian for the n-dimensional (isotropic) harmonic oscillator in $\mathbb{R}^{2n}$ (see Example 1.118) and it is a moment map for S^1-action defined above. $\triangle$

We discuss now in details two important examples of strongly Hamiltonian G-actions. For the relevant definitions about Lie groups and Lie algebras used below, we refer the reader to Appendix C, while for an introduction to the notions of complex geometry used (more or less implicitly) in the next two examples we refer the reader to Sect. 5.2 in this chapter and to the references therein.

Example 3.57 ($U(n)$-Action on $\mathbb{C}^n$) Consider the symplectic vector space $(\mathbb{C}^n, \omega = \frac{1}{2i}\sum_{l=1}^n dz_l \wedge \overline{dz_l})$ endowed with the *usual* $U(n)$-action defined by the product between a matrix and a vector. In particular every $z \in \mathbb{C}^n$ will be thought as a *column vector*. Note that the symplectic form ω above defined is the standard symplectic form of $\mathbb{R}^{2n}$: if $z_k = x_k + \imath y_k$, then

$$\omega = \frac{1}{2\imath} \sum_{k=1}^{n} dz_k \wedge d\bar{z}_k = \frac{1}{2\imath} \sum_{k=1}^{n} (dx_k + \imath dy_k) \wedge (dx_k - \imath dy_k) = \sum_{k=1}^{n} dy_k \wedge dx_k.$$

The $U(n)$-action preserves ω. In fact if $z_k = \sum_{l=1}^{n} g_{kl} w_l$,

$$\begin{aligned}
\sum_{k=1}^{n} dz_k \wedge d\bar{z}_k &= \sum_{k=1}^{n} d\left(\sum_{l=1}^{n} g_{kl} w_l\right) \wedge d\left(\sum_{r=1}^{n} \overline{g_{kr} w_r}\right) \\
&= \sum_{l,r=1}^{n} \sum_{k=1}^{n} g_{kl} \overline{g_{kr}} dw_l \wedge d\bar{w}_r \\
&\stackrel{g^\dagger g = \mathrm{id}}{=} \sum_{k,r=1}^{n} \delta_{rl} dw_l \wedge d\bar{w}_r \\
&= \sum_{r=1}^{n} dw_r \wedge d\bar{w}_r.
\end{aligned}$$

The $U(n)$-action is strongly Hamiltonian and its moment map can be computed as follows. First, for all $\xi \in \mathfrak{u}_n$, the Lie algebra of $U(n)$, one has that

$$X_\xi(z) = -\xi z, \tag{3.27}$$

where the minus sign is due to our convention about the fundamental vector fields; see Proposition C.42. If written in the complex coordinates $(z, \bar{z})$, (3.27) looks like

$$X_\xi(z) = -\sum_{j=1}^{n} \left[(\xi z)_j \frac{\partial}{\partial z_j} + \overline{(\xi z)_j} \frac{\partial}{\partial \bar{z}_j} \right],$$

and

$$\begin{aligned}
i_{X_\xi}\omega(z) &= \frac{1}{2\imath}\left[-\sum_{k=1}^{n} (\xi z)_k d\bar{z}_k + \sum_{k=1}^{n} \overline{(\xi z)_k} dz_k \right] \\
&= \frac{1}{2\imath}\left[-\sum_{k=1}^{n}\sum_{s=1}^{n} (\xi z)_k d\bar{z}_k + \sum_{k=1}^{n}\sum_{s=1}^{n} \bar{\xi}_{ks}\bar{z}_s dz_k \right] \\
&\stackrel{\xi^\dagger = -\xi}{=} \frac{1}{2\imath}\left[-\sum_{k=1}^{n}\sum_{s=1}^{n} (\xi z)_k d\bar{z}_k - \sum_{k=1}^{n}\sum_{s=1}^{n} \xi_{sk}\bar{z}_s dz_k \right]
\end{aligned}$$

$$= -\frac{1}{2\iota}\left[\sum_{k=1}^{n}(\xi z)_k \, d\bar{z}_k + \sum_{s=1}^{n}\bar{z}_s \, d(\xi z)_s\right]$$

$$= -d\left[\frac{1}{2\iota}\sum_{k=1}^{n}(\xi z)_k \bar{z}_k\right].$$

Note that in the last term of the previous computation ξz is a *column* vector and, for this reason, $\bar{z}_k$ should be thought as the kth component of a *row* vector. This observation, together with the previous computation, yields the following:

$$i_{X_\xi}\omega(z) = -d\left(\frac{1}{2\iota}\mathrm{tr}(\xi z z^{\dagger})\right),$$

and since $\mu : \mathbb{C}^n \to \mathfrak{u}_n^*$ is defined by $d\langle\mu(z), \xi\rangle = -i_{X_\xi}\omega(z)$, one concludes

$$\langle\mu(z), \xi\rangle = \frac{1}{2\iota}\mathrm{tr}(\xi z z^{\dagger}) = -\frac{\iota}{2}\mathrm{tr}(\xi z z^{\dagger}) \overset{\xi=-\xi^{\dagger}}{=} \frac{\iota}{2}\mathrm{tr}(\xi^{\dagger} z z^{\dagger}). \tag{3.28}$$

Finally, identifying $\mathfrak{u}_n$ with $\mathfrak{u}_n^*$ via the trace form $(A, B) = \mathrm{tr}(AB^{\dagger})$, (3.28) yields

$$\mu(z) = \frac{\iota}{2}z z^{\dagger}. \tag{3.29}$$

Note that the previous equation says that, for every $z \in \mathbb{C}^n$, the value of the moment map associates to the standard $\mathrm{U}(n)$ is a rank-one skew Hermitian matrix. To comment more about this, thinking of the right-hand side of (3.29) as a linear operator on $\mathbb{C}^n$, one can write

$$\mu(z) = \frac{\iota}{2}z \otimes z^{\dagger}. \tag{3.30}$$

Note that (3.30) acts on $\mathbb{C}^n$ as follows:

$$\mu(z)(u) = \frac{\iota}{2}z \otimes z^{\dagger}(u) = \frac{\iota z}{2}\sum_{k=1}^{n}\bar{z}_k u_k, \quad \forall u \in \mathbb{C}^n.$$

In particular if $\langle u, w\rangle = \sum_{k=1}^{n} u_k \bar{w}_k$ is the standard Hermitian product on $\mathbb{C}^n$,

$$\langle\mu(z)^{\dagger}(u), w\rangle = \langle u, \mu(z)(w)\rangle = \sum_{s=1}^{n} u_s \overline{(\mu(z)w)_s} = \sum_{s=1}^{n} u_s \overline{\frac{\iota z_s}{2}\sum_{k=1}^{n}\bar{z}_k w_k}$$

$$= -\sum_{s=1}^{n} u_s \frac{\iota\bar{z}_s}{2}\sum_{k=1}^{n} z_k \bar{w}_k,$$

which, after reordering, yields

$$\langle \mu(z)^{\dagger}(u), w \rangle = -\sum_{k=1}^{n} \frac{\imath z_k}{2} \left(\sum_{s=1}^{n} \overline{z}_s u_s \right) \overline{w}_k = -\langle \mu(z)(u), w \rangle, \; \forall u, w, z \in \mathbb{C}^n.$$

In other words

$$\mu(z)^{\dagger} = -\mu(z), \; \forall z \in \mathbb{C}^n.$$

$\triangle$

Remark 3.58 It is worth observing that together with what we discussed in the previous example, one can consider the *left* action of $U(n)$ on $\mathbb{C}^n$, thought as the vector space of *row n-vectors*:

$$g.z := zg^{-1}, \; \forall g \in U(n), \; z \in \mathbb{C}^n. \tag{3.31}$$

If $\xi \in \mathfrak{u}_n$, the corresponding fundamental vector field X_ξ is defined to be $X_\xi(z) = z\xi$, for all $z \in \mathbb{C}^n$ which, if written in coordinates, becomes

$$X_\xi(z) = \sum_{k=1}^{n} (z\xi)_k \frac{\partial}{\partial z_k} + \overline{(z\xi)}_k \frac{\partial}{\partial \overline{z}_k},$$

where, in this case,

$$(z\xi)_k = \sum_{l=1}^{n} z_l \xi_{lk}, \; \forall k = 1, \ldots, n. \tag{3.32}$$

Computations completely analogous to the ones in Example 3.57, together with (3.32), yield the conclusion that a moment map associated to (3.31) is

$$\mu(z) = -\frac{\imath}{2} z^{\dagger} \otimes z, \; \forall z \in \mathbb{C}^n,$$

where $z^{\dagger}$, in the previous formula, is the *column* vector whose kth entry is the complex conjugate of the kth entry of the *row* vector z. $\triangle$

In spite of the fact that the next example is, in principle, a particular case of the previous one, we will discuss it in some details.

Example 3.59 Let $V = \mathrm{Mat}_n(\mathbb{C}) \simeq \mathbb{C}^{n^2}$ and observe that

$$\omega = \frac{1}{2\imath} \mathrm{tr}\left(dZ \wedge dZ^{\dagger} \right) = \frac{1}{2\imath} \sum_{i,j=1}^{n} dZ_{ij} \wedge d\overline{Z}_{ij}$$

is the canonical symplectic form on V. To this end it suffices to note that on V the entries Z_{ij} of Z define coordinates and that if $Z_{ij} = X_{ij} + \iota Y_{ij}$, then $\omega = \sum_{i,j=1}^{n} dY_{ij} \wedge dX_{ij}$. Furthermore, observe that $U(n)$ acts on V via the adjoint action

$$g.Z = gZg^{-1}, \ \forall Z \in V, \ g \in U(n),$$

and that this action is symplectic. In fact

$$\mathrm{tr}\big(d(g.Z) \wedge d(g.Z)^{\dagger}\big) = \mathrm{tr}\big(gdZ \wedge dZ^{\dagger}g^{-1}\big) = \mathrm{tr}\big(dZ \wedge dZ^{\dagger}\big), \ \forall g \in U(n),$$

where it was used $g^{\dagger}g = \mathrm{id}$ and the fact that g is a *constant* matrix, i.e., $dg = 0$. Note that $Z \in V$ is a $n \times n$-matrix whose entries $Z_{ij} \in \mathbb{C}$. In this way, still stuck to our conventions, for every $\xi \in \mathfrak{u}_n$, the fundamental vector field X_{ξ} on V defined by the $U(n)$-action is $X_{\xi}(Z) = [\xi, Z]$ for all $Z \in V$, and, once written in the complex coordinates $Z, \overline{Z}$ becomes

$$X_{\xi}(Z) = \sum_{k,l=1}^{n} [Z, \xi]_{lk} \frac{\partial}{\partial Z_{lk}} + \sum_{k,l=1}^{n} \overline{[Z, \xi]}_{lk} \frac{\partial}{\partial \overline{Z}_{lk}},$$

where $[Z, \xi]_{lk} = \sum_{s=1}^{n} (Z_{ls}\xi_{sk} - \xi_{ls} Z_{sk}), k, l = 1, \ldots, n$. One can now compute

$$i_{X_{\xi}}\omega = \frac{1}{2\iota} \sum_{k,l=1}^{n} [Z, \xi]_{lk} d\overline{Z}_{lk} - \overline{[Z, \xi]}_{lk} dZ_{lk}$$

$$= \frac{1}{2\iota} \sum_{k,l=1}^{n} \left(\sum_{s=1}^{n} Z_{ls}\xi_{sk} - \xi_{ls} Z_{sk} \right) d\overline{Z}_{lk} - \left(\overline{\sum_{s=1}^{n} Z_{ls}\xi_{sk} - \xi_{ls} Z_{sk}} \right) dZ_{lk}$$

$$\overset{\xi^{\dagger}=-\xi}{=} \frac{1}{2\iota} \left[\sum_{k,l=1}^{n}\sum_{s=1}^{n} Z_{ls}\xi_{sk} d\overline{Z}_{lk} - \sum_{k,l=1}^{n}\sum_{s=1}^{n} \xi_{ls} Z_{sk} d\overline{Z}_{lk} + \sum_{k,l=1}^{n}\sum_{s=1}^{n} \overline{Z}_{ls}\xi_{ks} dZ_{lk} \right.$$

$$\left. - \sum_{k,l=1}^{n}\sum_{s=1}^{n} \xi_{sl} \overline{Z}_{sk} dZ_{lk} \right]$$

$$
z_{ij}^{\dagger} \overset{=\overline{z}_{ji}}{=} \frac{1}{2\iota} \left[\underbrace{\sum_{k,l=1}^{n}\sum_{s=1}^{n} Z_{ls}\xi_{sk}dZ_{kl}^{\dagger}}_{\textcircled{1}} - \overbrace{\sum_{k,l=1}^{n}\sum_{s=1}^{n} \xi_{ls}Z_{sk}dZ_{kl}^{\dagger}}^{\textcircled{2}} + \underbrace{\sum_{k,l=1}^{n}\sum_{s=1}^{n} Z_{sl}^{\dagger}\xi_{ks}dZ_{lk}}_{\textcircled{3}} \right.
$$

$$
\left. - \overbrace{\sum_{k,l=1}^{n}\sum_{s=1}^{n} \xi_{sl}Z_{ks}^{\dagger}dZ_{lk}}^{\textcircled{4}} \right]
$$

A simple reordering yields

$$
\textcircled{1}-\textcircled{2} = \frac{1}{2\iota}\mathrm{tr}(Z\xi dZ^{\dagger} - \xi Z dZ^{\dagger}) \quad \text{and} \quad \textcircled{3}-\textcircled{4} = \frac{1}{2\iota}\mathrm{tr}(Z^{\dagger}dZ\xi - Z^{\dagger}\xi dZ),
$$

which, using the cyclic property of the trace and $d\xi = 0$, implies

$$
\textcircled{1}-\textcircled{2}+\textcircled{3}-\textcircled{4} = \frac{1}{2\iota}\mathrm{tr}\xi(dZ^{\dagger}Z - ZdZ^{\dagger} + Z^{\dagger}dZ - dZZ^{\dagger})
$$

$$
= \frac{1}{2\iota}\mathrm{tr}\xi d(Z^{\dagger}Z - ZZ^{\dagger})
$$

$$
\overset{\xi^{\dagger}=-\xi}{=} -d\frac{\iota}{2}\mathrm{tr}([Z,Z^{\dagger}]\xi^{\dagger}),
$$

i.e.,

$$
i_{X_{\xi}}\omega(Z) = -d\frac{\iota}{2}\mathrm{tr}([Z,Z^{\dagger}]\xi^{\dagger}) = -d\langle\mu(Z),\xi\rangle.
$$

Identifying $\mathfrak{u}_n^{*}$ with $\mathfrak{u}_n$ via the Hermitian form $(A,B) = \mathrm{tr}(AB^{\dagger})$, one can conclude

$$
\mu(Z) = \frac{\iota}{2}[Z,Z^{\dagger}], \ \forall Z \in V.
$$

$\triangle$

3 Homogeneous Symplectic Manifold

In this section we aim to present a characterization of the class of the so-called *homogeneous symplectic manifolds*, which will be introduced in Definition 3.64. To this end, first, we start show that the coadjoint action of a Lie group G on every of its coadjoint orbits is strongly Hamiltonian.

3.1 *Coadjoint Orbits*

In this subsection we see that the coadjoint action of a Lie group on any of its coadjoint orbits provides an example of a strongly Hamiltonian action. For the definition of the symplectic structure on the coadjoint orbits, we refer the reader to Sect. 1.2 in Chap. 1. Let us just remind that if $\alpha \in \mathfrak{g}^*$ and $\mathscr{O}_\alpha$ is the correspoding coadjoint orbit, then the bilinear form $\omega_\alpha : T_\alpha \mathscr{O} \times T_\alpha \mathscr{O}_\alpha \to \mathbb{R}$, defined by

$$\omega_\alpha(u, v) = \langle \alpha, [x, y] \rangle, \tag{3.33}$$

is a symplectic form on $\mathscr{O}_\alpha$. In the previous formula $x, y \in \mathfrak{g}$ are such that the values at α of the corresponding fundamental vector fields X_x and X_y are equal to u and v, respectively; see Sect. 1.2 in Chap. 1, and see in particular Formula (1.9) where ω is defined. We can now prove the following.

Theorem 3.60 *The symplectic form* (3.33) *is invariant with respect to coadjoint action. Moreover, this action is Hamiltonian.*

Proof First, recall that the tangent space at each point of $\mathscr{O}_\alpha$ is generated by the fundamental vector fields and that the tangent space at α can be identified with $\mathfrak{g}/\mathfrak{g}_\alpha$, using

$$i : T_\alpha \mathscr{O}_\alpha \xrightarrow{\;\; X_x(\alpha) \rightsquigarrow x \bmod \mathfrak{g}_\alpha \;\;} \mathfrak{g}/\mathfrak{g}_\alpha,$$

where $\mathfrak{g}_\alpha$ the Lie algebra of the isotropy subgroup $G_\alpha := \{g \in G$ such that $\mathrm{Ad}_g^\sharp(\alpha) = \alpha\}$.

To prove that the form ω is G-invariant it suffices to show that

$$\left(\mathrm{Ad}_g^\sharp \right)^*_{,\alpha} \omega_{\mathrm{Ad}_g^\sharp \alpha}(v, u) = \omega_\alpha(v, u),$$

for all $g \in G$ and for all $v, u \in T_\alpha \mathscr{O}_\alpha$. Let us compute the left-hand side of the previous formula:

$$\left(\mathrm{Ad}_g^\sharp \right)^*_{,\alpha} \omega_{\mathrm{Ad}_g^\sharp \alpha}(v, u) = \omega_{\mathrm{Ad}_g^\sharp \alpha}\!\left((\mathrm{Ad}_g^\sharp)_{*,\alpha} v, (\mathrm{Ad}_g^\sharp)_{*,\alpha} u \right),$$

where v and u are as above. By virtue of the commutativity of the following diagram,

$$
\begin{array}{ccc}
T_\alpha \mathcal{O}_\alpha & \xrightarrow{\ (\mathrm{Ad}_g^\sharp)_{*,\alpha}\ } & T_{\mathrm{Ad}_g^\sharp \alpha} \mathcal{O}_\alpha \\
{\scriptstyle i}\downarrow & & \downarrow{\scriptstyle i} \\
\mathfrak{g}/\mathfrak{g}_\alpha & \xrightarrow{\ \mathrm{Ad}_g\ } & \mathfrak{g}/\mathfrak{g}_{\mathrm{Ad}_g^\sharp \alpha}
\end{array}
$$

the action of the differential of the *coadjoint map* on u and v can be described as follows:

$$
(\mathrm{Ad}_g^\sharp)_{*,\alpha}(v) \rightsquigarrow (\mathrm{Ad}_g)(x \bmod \mathfrak{g}_\alpha) = \mathrm{Ad}_g(x) \bmod \mathfrak{g}_{\mathrm{Ad}_g^\sharp \alpha}
$$

$$
(\mathrm{Ad}_g^\sharp)_{*,\alpha}(u) \rightsquigarrow (\mathrm{Ad}_g)(y \bmod \mathfrak{g}_\alpha) = \mathrm{Ad}_g(y) \bmod \mathfrak{g}_{\mathrm{Ad}_g^\sharp \alpha},
$$

which yield

$$
\omega_{\mathrm{Ad}_g^\sharp \alpha}\big((\mathrm{Ad}_g^\sharp)_{*,\alpha} v,\, (\mathrm{Ad}_g^\sharp)_{*,\alpha} u\big) = \langle \mathrm{Ad}_g^\sharp \alpha, [\mathrm{Ad}_g\, x, \mathrm{Ad}_g\, y] \rangle = \langle \alpha, [x, y] \rangle = \omega_\alpha(v, u)
$$

as required. To show that the coadjoint action is Hamiltonian we need to prove that each fundamental vector field X_x is Hamiltonian. To this end it suffices to compute the contraction of the symplectic form ω_α with $X_x(\alpha)$. If $v \in T_\alpha \mathcal{O}_\alpha$

$$
i_{X_x(\alpha)}\omega_\alpha(v) = \omega_\alpha(X_x(\alpha), v) = \langle \alpha, [x, y] \rangle = \langle \alpha, [dx_\alpha, y] \rangle = dx_\alpha(v),
$$

where $y \in \mathfrak{g}$ corresponds to the vector $v \in T_\alpha \mathcal{O}_\alpha$ via the isomorphism $T_\alpha \mathcal{O}_\alpha \simeq \mathfrak{g}/\mathfrak{g}_\alpha$; see Problem 3.61. This computation shows that the fundamental vector field X_x is Hamiltonian with Hamiltonian given, up to sign, by the restriction to $\mathcal{O}_\alpha$ of the linear function $x \in \mathfrak{g}$. In formulas, we record this coincidence as follows:

$$
X_x = -X_{x|_{\mathcal{O}}}, \tag{3.36}
$$

where the left-hand side of Formula (3.36) denotes the fundamental vector field associated to the coadjoint action on the coadjoint orbit $\mathcal{O}$ and the right-hand side denotes the Hamiltonian vector field defined on the symplectic manifold $\mathcal{O}$ endowed with the symplectic form (3.33) and whose Hamiltonian is the restriction to $\mathcal{O}$ of (the linear function) $x \in \mathfrak{g} \subset C^\infty(\mathfrak{g}^*)$. $\qquad\square$

Problem 3.61 Let $x \in \mathfrak{g}$, let $\mathcal{O}$ be a coadjoint orbit of G in $\mathfrak{g}^*$ and let $v \in T_\alpha \mathcal{O}_\alpha$. Show that

$$
dx_\alpha(v) = \langle \alpha, [dx_\alpha, y] \rangle,
$$

where $y \in \mathfrak{g}$ is any lifting of v, via the isomorphism $T_\alpha \mathcal{O}_\alpha \simeq \mathfrak{g}/\mathfrak{g}_\alpha$. Hint: Compute

$$dx_\alpha(v) = \left.\frac{d}{dt}\right|_{t=0} x\left(\mathrm{Ad}^\sharp_{\exp(ty)}\alpha\right)$$

and then use the duality between $\mathfrak{g}$ and $\mathfrak{g}^*$. $\triangle$

We show now that the restriction of the coadjoint action of a Lie group on any of its coadjoint orbits is strongly Hamiltonian and that the corresponding moment map is G-equivariant; see Sect. 1.2 in Chap. 1 and Sect. 3.1 in this chapter.

Remark 3.62 Recall that a coadjoint orbit is (in general) only a *injectively immersed* submanifold of $\mathfrak{g}^*$, i.e., in general it is not an embedded submanifolds of $\mathfrak{g}^*$. $\triangle$

To prove that the restriction of the coadjoint action of G to $\mathcal{O}$ is strongly Hamiltonian, we prove that *a* (G-equivariant) moment map associated to this action is the immersion map:

$$i : \mathcal{O} \hookrightarrow \mathfrak{g}^*.$$

Since i is injective,

$$i^* : C^\infty(\mathfrak{g}^*) \to C^\infty(\mathcal{O})$$

is a surjective map which sends $f \in C^\infty(\mathfrak{g}^*)$ to its restriction to $\mathcal{O}$. Since the symplectic structure defined on $\mathcal{O}$ is obtained from the Lie-Poisson structure of $\mathfrak{g}^*$ (see (3.33)), it follows that

$$i^*\{f, g\}_{LP} = \{i^*f, i^*g\}_\omega, \tag{3.37}$$

which proves that $i : \mathcal{O} \to \mathfrak{g}^*$ is a Poisson morphism. In (3.37) we denoted with $\{\cdot, \cdot\}_{LP}$ the Lie-Poisson structure of $\mathfrak{g}^*$ and with $\{\cdot, \cdot\}_\omega$ the Poisson structure on $\mathcal{O}$ induced by the symplectic structure (3.33). In particular this shows that

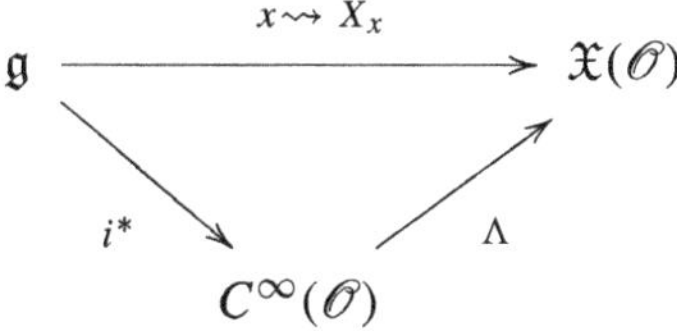

is a commutative diagram of Lie algebras. In fact, since $\mathfrak{g} \subset C^\infty(\mathfrak{g}^*)$ and

$$\{x, y\}_{LP} = [x, y]$$

for all $x, y \in \mathfrak{g}$, Formula (3.37) entails that

$$[x, y]_{|\mathscr{O}} = \{x_{|\mathscr{O}}, y_{|\mathscr{O}}\}_\omega.$$

Since the Hamiltonian vector field whose Hamiltonian function is $[x, y]_{|\mathscr{O}}$ is the fundamental vector field $X_{[x,y]}$ (see Formula (3.36) in the proof of Theorem 3.60), we can write

$$\Lambda(i^*[x, y]) = \Lambda([x, y]_{|\mathscr{O}}) = \Lambda(\{x_{|\mathscr{O}}, y_{|\mathscr{O}}\}_\omega) = X_{\{x_{|\mathscr{O}}, y_{|\mathscr{O}}\}_\omega} = [X_{x_{|\mathscr{O}}}, X_{y_{|\mathscr{O}}}].$$

On the other hand, given $x, y \in \mathfrak{g}$, the fundamental vector field associated to the commutator $[x, y]$ is

$$X_{[x,y]} = [X_x, X_y] = [X_{x_{|\mathscr{O}}}, X_{y_{|\mathscr{O}}}],$$

proving the commutativity of the previous diagram and, equivalently, the fact that the restriction of the coadjoint action to $\mathscr{O}$ is strongly Hamiltonian. To complete the proof we are left to show that

$$i : \mathscr{O} \to \mathfrak{g}^*$$

is G-equivariant. This follows from the definitions of i and of the coadjoint action. We can summarize the above discussion in the following.

Theorem 3.63 *The coadjoint action of G on each of its coadjoint orbits is a strongly Hamiltonian action whose corresponding moment map is G-equivariant.*

3.2 Homogeneous Symplectic Manifolds

To move forward with our discussion we need to introduce the following concept.

Definition 3.64 (Homogeneous Symplectic Manifold) A *homogeneous symplectic manifold* is a symplectic manifold (M, ω) which carries a transitive and strongly Hamiltonian action of a Lie group. $\triangle$

The discussion enclosed in the previous subsection entails that every coadjoint orbit is a homogeneous symplectic manifold. We now show that the coadjoint orbits are, morally speaking, the only examples of homogeneous symplectic manifolds. More precisely, we prove the following.

Theorem 3.65 (Kirillov-Kostant-Souriau) *If (M, ω) is a symplectic manifold carrying a transitive and strongly Hamiltonian G-action φ with an associated G-equivariant moment map μ, then M is locally diffeomorphic to a coadjoint orbit $\mathscr{O} \subset \mathfrak{g}^*$.*

Proof First, we prove that μ maps M to a coadjoint orbit. In fact, since φ is transitive for each $m_1, m_2 \in M$, there exists $g \in G$ such that $\varphi_g(m_1) = m_2$. Then

$$\mu(m_2) = \mu(\varphi_g(m_1)) \overset{G-\text{equivariance}}{=} \mathrm{Ad}_g^\sharp \left(\mu(m_1) \right),$$

which shows that there exists a coadjoint orbit $\mathcal{O}$, such that $\mu(M) \subset \mathcal{O}$. On the other hand, since the restriction of the coadjoint action of G to $\mathcal{O}$ is transitive, given $\alpha_1, \alpha_2 \in \mathcal{O}$ and $m_1 \in M$ such that $\mu(m_1) = \alpha_1$, one can find $g \in G$ such that

$$\alpha_2 = \mathrm{Ad}_g^\sharp(\mu(m_1)) = \mu(\varphi_g(m_1)),$$

proving that μ surjects M to $\mathcal{O}$. We now show that $\mu : M \to \mathcal{O}$ is a local diffeomorphism, i.e., that for every $m \in M$

$$\mu_{*,m} : T_m M \to T_{\mu(m)} \mathcal{O} \tag{3.38}$$

is an isomorphism of vector spaces. To this end, first, we recall that

$$T_{\mu(m)} \mathcal{O} \simeq \mathfrak{g}/\mathfrak{g}_{\mu(m)} \tag{3.39}$$

(see Remark C.51) and that

$$T_m M \simeq \mathfrak{g}/\mathfrak{g}_m. \tag{3.40}$$

Now note that to show that (3.38) is an isomorphism for every $m \in M$ it suffices to prove that it is an isomorphism at some $m \in M$. For this reason, let $m \in M$ and $v \in T_m M$. Hereafter we will use more than once the following argument: since M is a homogeneous manifold, for all $m \in M$ $T_m M$ is spanned by $\{X_x(m) \mid x \in \mathfrak{g}\}$ and, as a result of this, given $v \in T_m M$ there exists $x \in \mathfrak{g}$ such that $v = X_x(m)$. Now first, let us show that $\mu_{*,m}$ is injective. Let v be as above and let us compute

$$\langle \mu_{*,m}(v), y \rangle = -(i_{X_y(m)} \omega_m)(v) = \omega_m(v, X_y(m)).$$

Then, if $\mu_{*,m}(v) = 0$

$$\omega_m(v, X_y(m)) = 0, \ \forall \, y \in \mathfrak{g}.$$

Since $\{X_y(m), \mid y \in \mathfrak{g}\}$ spans $T_m M$ and ω_m in nondegenerate, one can conclude that $v = 0$, proving the injectivity of $\mu_{*,m}$. Then, to show that $\mu_{*,m}$ is a bijection, it suffices to prove that $\dim T_m M = \dim T_{\mu(m)} \mathcal{O}$. To this end recall that since μ is G-equivariant it is $\mathfrak{g}$-equivariant, i.e.,

$$\mu_{*,m}\left(X_x(m)\right) = -\,\mathrm{ad}_x^\sharp \left(\mu(m)\right),$$

for all $m \in M$ and for all $x \in \mathfrak{g}$. Then, since the tangent spaces of M are generated by (the values of) the fundamental vector fields, given $m \in M$ and $v \in T_m M$, there exists $x \in \mathfrak{g}$ such that $v = X_x(m)$ and

$$\mu_{*,m}(v) = \mu_{*,m}\big(X_x(m)\big) = -\operatorname{ad}_x^\sharp\big(\mu(m)\big).$$

Since $\mu_{*,m}$ is injective, one deduces that $x \in \mathfrak{g}_m$, i.e., $X_x(m) = 0$, *if and only if* it belongs to $\mathfrak{g}_{\mu(m)}$. This means in particular that the dimensions of $T_m M$ and $T_{\mu(m)}\mathcal{O}$ are the same (see isomorphisms (3.39) and (3.40)), proving that $\mu_{*,m} : T_m M \to T_{\mu(m)}\mathcal{O}$ is an isomorphism. $\qquad\square$

The content of the previous theorem could be rephrased saying that if (M, ω) is a homogeneous symplectic manifold carrying a strongly Hamiltonian G-action and having a G-equivariant moment map, then M is an *unbranched* covering of a coadjoint orbit.

4 Symplectic Reduction and the Marsden-Weinstein-Meyer Theorem

In this section we describe an important result known as the *Marsden-Weinstein-Meyer reduction* which, as it was already mentioned, is one of the most important methods to construct new symplectic manifolds from old ones. We present this construction under the assumptions of the existence of a strongly Hamiltonian G-action and of an associated G-equivariant moment map, both defined on the initial symplectic manifold. Before stating and proving this important result we need to make a few preliminary general observations, independent on the existence of a G-action on (M, ω), which, hereafter, will be a symplectic manifold of dimension $2n$.

Let (N, j) be a submanifold (either embedded or immersed) of (M, ω) and let $\omega_N = j^*\omega$. Define

$$\ker(\omega_N)_n = T_n N \cap T_n N^\omega,$$

where $T_n N^\omega$ is the symplectic orthogonal complement of $T_n N$ in $T_n M$ (see Definition A.6).

Definition 3.66 The *rank* of ω_N at $n \in N$ is

$$\operatorname{rk}(\omega_N)_n = \dim N - \dim \ker(\omega_N)_n.$$

(N, j) is called of *constant rank* if $\operatorname{rk}(\omega_N)_n$ does not depend on $n \in N$. $\qquad\triangle$

Fig. 3.2 A sketch of
reduction

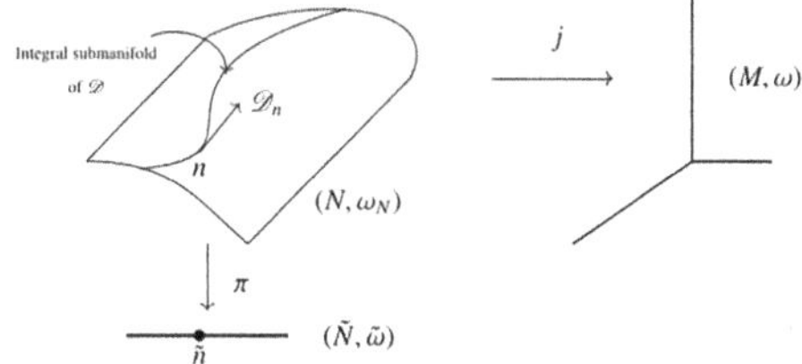

Remark 3.67 Note that (N, j) is of constant rank if and only if the dim ker $(\omega_N)_n$
is constant on N which, in turn, is equivalent to say that dim $T_n N^\omega$ is constant
on N. $\triangle$

Definition 3.68 (The Characteristic Distribution of ω_N) The assignment $\mathscr{D}$:
$n \rightsquigarrow \ker(\omega_N)_n$ defines a generalized distribution on N; see Definition B.10. $\triangle$

Note that the previous remark implies that if (N, j) is of constant rank, the
distribution $\mathscr{D}$ introduced in the previous definition is a constant rank distribution;
see Definition B.1.

Supposing further that $\mathscr{D}$ is *smooth*, if $X, Y \in \mathfrak{X}(N)$ *belong* to $\mathscr{D}$ (see
Definition B.1), one has that

$$i_{[X,Y]}\omega_N = \mathscr{L}_X(i_Y\omega_N) - i_Y(\mathscr{L}_X\omega_N) = -i_Y(i_X d\omega_N + d i_X\omega_N) = 0,$$

where the cancellations follow from the hypothesis that X, Y belong to $\mathscr{D}$, i.e.,
$i_X\omega_N = 0 = i_Y\omega_N$, and from $d\omega_N = 0$. In other words, if $\mathscr{D}$ is a *smooth
and constant rank* distribution, it is *integrable* in the sense of Frobenius; see
Theorem B.5. Under these assumptions the maximal integral manifolds of $\mathscr{D}$ are
called the *null-leaves* of the ω_N. Supposing further that $\tilde{N}$, the *space of the null-
leaves* of ω_N, is a (smooth) manifold and assuming that the canonical map π :
$N \to \tilde{N}$ is a submersion, see (Fig. 3.2), on $\tilde{N}$ there exists a unique symplectic form
$\tilde{\omega}$ such that

$$\pi^*\tilde{\omega} = \omega_N = j^*\omega. \tag{3.41}$$

A quick proof of this statement follows noticing that under these hypothesis the
null-leaves of ω_N are the fibers of the submersion π and that $i_X\omega_N = 0$ (and
$\mathscr{L}_X\omega_N = 0$) for every *vertical* vector field X. All this implies that ω_N is *basic*
(see Appendix C), and, for this reason, there exists a two-form $\tilde{\omega}$ on $\tilde{N}$ such that
$\pi^*\tilde{\omega} = \omega_N$. The facts that $\tilde{\omega}$ is closed and nondegenerate and the second equality in
(3.41) follow from easy arguments and they are left to the reader as exercises.

A more hands-on way to get this proof goes as follows. Under the hypothesis
made above, every $n \in N$ has an open neighborhood endowed with coordinates
$(x_1, \ldots, x_l)$ such that the integral submanifold through n, defined by $\mathscr{D}$, is described
by $x_{k+1} = x_{k+2} = \cdots = x_l = 0$, where $l = \dim N$ and $k = \dim \ker \omega_N$. In these

coordinates the vector fields generating this distribution are

$$X_1 = \frac{\partial}{\partial x_1}, \ X_2 = \frac{\partial}{\partial x_2}, \ldots, \ X_k = \frac{\partial}{\partial x_k}.$$

Writing ω_N in these coordinates, one gets $\omega_N = \sum_{i<j} \omega_{ij} dx_i \wedge dx_j$, where $\{\omega_{ij}\}_{i,j=1,\ldots,l}$ are suitable (local) functions on N. Since $i_{X_i}\omega_N = 0$ for all $i = 1, \ldots, k$

$$\omega = \sum_{k+1 \leq i < j \leq l} \omega_{ij} dx_i \wedge dx_j$$

and since $d\omega = 0$,

$$\frac{\partial \omega_{ij}}{\partial x_s} = 0, \text{ for all } i, j = k+1, \ldots, l \text{ and } s = 1, \ldots, k.$$

In this way $\widetilde{N}$ inherits local coordinates $y_1, \ldots, y_{k-l}$ and a *unique* two-form

$$\widetilde{\omega} = \sum_{1 \leq i, j \leq k-l} \widetilde{\omega}_{ij} dy_i \wedge dy_j,$$

which is nondegenerate by construction and closed since π^* is injective due to the fact that π is a surjective submersion.

The previous discussion is summarized in the following.

Theorem 3.69 *Let (N, j) be a constant rank submanifold of (M, ω). If $\widetilde{N}$, the space of null-leaves of ω_N, is a manifold and if the canonical projection $\pi : N \to \widetilde{N}$ is a submersion, then on $\widetilde{N}$ is defined a unique symplectic form $\widetilde{\omega}$ satisfying* (3.41).

Definition 3.70 (Symplectic Reduction) The symplectic manifold $(\widetilde{N}, \widetilde{\omega})$ is called the symplectic reduction of (M, ω). $\triangle$

Example 3.71 (Coisotropic Submanifolds of Maximal Dimension) Let $H \in C^\infty(M)$ and let $f(m)$ be a regular value of H. Then $N_m = \{m' \in M \mid H(m') = H(m)\}$ is a smooth submanifold of M. Note that, for dimensional reasons, N_m is coisotropic and for every $n \in N_m$, $\ker(\omega_{N_m})_n = \mathrm{span}_{\mathbb{R}}\langle X_H(n)\rangle$ (note that the Hamiltonian vector field X_H is tangent to N_m). Under the assumptions made in the previous theorem, one concludes that $\widetilde{N}_m$, which is the space of the maximal integral curves of the restriction of X_H to N_m, inherits a symplectic form. Note that $\dim \widetilde{N}_m$ is equal to $\dim M - 2$ and that on such a symplectic manifold is defined a *unique* smooth function $\widetilde{H}$ such that $\pi^*\widetilde{H} = j^*H$, where $j : N_m \to M$ is the canonical immersion. $\widetilde{H}$ is called the *reduced Hamiltonian* and the corresponding Hamiltonian equations are, in principle, easier to integrate, if compared to the original ones. The process described in this example is the simplest example of Hamiltonian reduction. $\triangle$

The previous example is generalized as follows.

Example 3.72 (General Coisotropic Submanifolds) An important example of constant rank submanifolds (N, i) of (M, ω) are the coisotropic ones; see Sect. 3.4.1. In this case the distribution $\mathscr{D} : m \rightsquigarrow \ker(\omega_N)_n$ is called the characteristic distribution of the coisotropic submanifold and the null-leaf foliation is made of isotropic submanifolds of (M, ω); see again Sect. 3.4.1. $\triangle$

From now on, we let φ be a strongly Hamiltonian G-action on (M, ω) and $\mu : M \to \mathfrak{g}^*$ an associated G-equivariant moment map. Starting from these ingredients and for each α regular value of μ, we prove that the pair $N_\alpha = \mu^{-1}(\alpha)$, $j_\alpha : N_\alpha \to M$, the canonical immersion, forms a constant rank submanifold of (M, ω) to which one can apply Theorem 3.69. The final result of this process is a new symplectic manifold $(\tilde{N}_\alpha, \tilde{\omega}_\alpha)$ called *Marsden-Weinstein-Meyer* reduction, at $\alpha \in \mathfrak{g}^*$, of the symplectic manifold (M, ω).

Remark 3.73 As we mentioned at the beginning of this chapter, the Marsden-Weinstein-Meyer reduction is the one of the canonical ways to produce *new* symplectic manifolds from *old* ones. This construction generalizes the classical results about the reduction of the number of degree of freedom of mechanical systems with symmetries; see [10, 84, 250]. $\triangle$

Going back to our proposal, first, we prove that for all α regular value of μ, (N_α, j_α) is a constant rank submanifold of (M, ω). To this end we start to note that for each $m \in M$, the transpose of $\mu_{*,m} : T_m M \longrightarrow \mathfrak{g}^*$ is the linear map that associates to each $x \in \mathfrak{g}$ the covector at m defined by

$$T_m M \xrightarrow{v \rightsquigarrow \langle \mu_{*,m}(v), x \rangle} \mathbb{R}.$$

In other words, the transpose of $\mu_{*,m}$ is the linear map defined by

$$\mu^*_m : \mathfrak{g} \xrightarrow{x \rightsquigarrow (\mu_x^\sharp)_{*,m}} T_m^* M, \tag{3.42}$$

where $\mu_x^\sharp = \mu^\sharp(x) \in C^\infty(M)$; see Convention 3.47. Since

$$\langle \mu(\cdot), x \rangle_{*,m} = (\mu_x^\sharp)_{*,m} = -i_{X_x(m)}\omega_m,$$

one deduces that

Lemma 3.74 *For each $m \in M$, $\ker \mu^*_m = \mathfrak{g}_m$, where $\mathfrak{g}_m$ is the Lie algebra of G_m, the isotropy group of m.*

Proof In fact, for each $m \in M$ and $x \in \mathfrak{g}$

$$\langle \mu^*_m(x), v \rangle \overset{(3.42)}{=} \langle x, \mu_{*,m}(v) \rangle = \omega_m\big(v, X_x(m)\big)$$

for each $v \in T_m M$. Then if $x \in \ker \mu^*_m$, $\omega_m\big(v, X_x(m)\big) = 0$ for all $v \in T_m M$ which implies, by the nondegeneracy of ω_m, that $X_x(m) = 0$, i.e., that $x \in \mathfrak{g}_m$. On the other hand, if $x \in \mathfrak{g}_m$, $\omega_m\big(v, X_x(m)\big) = 0$ for all $v \in T_m M$, implying that $\langle \mu^*_m(x), v \rangle = 0$ for all $v \in T_m M$, forcing $x \in \ker \mu^*_m$. $\qquad\square$

This observation yields the following.

Lemma 3.75 *For each $m \in M$, im $\mu_{*,m} = (\mathfrak{g}_m)^\circ$, where $(\mathfrak{g}_m)^\circ$ is the annihilator of $\mathfrak{g}_m$ in $\mathfrak{g}^*$.*

Proof Note that if V and W are finite-dimensional vector spaces and $f : V \to W$ is a linear map, and $f^* : W^* \to V^*$ is its transpose, then $\ker f^* = (\text{im } f)^\circ$. Moreover, $U \subset V$ is e vector subspace, then $(U^\circ)^\circ = U$, where as usual, $U^\circ \subset V^*$ is the annihilator of U. The proof of the lemma follows at once applying these observations to the result in Lemma 3.74. $\qquad\square$

A simple consequence of the previous lemmata is enclosed in the following.

Corollary 3.76

$$dim \big(im \; \mu_{*,m}\big) = dim \; \mathcal{O}_m,$$

where $\mathcal{O}_m$ is the G-orbit of the point $m \in M$. In other words

$$rk \; \mu_{m,*} = dim \, G - dim \, G_m.$$

This, in turn, implies that

Proposition 3.77 *The moment map is a submersion at the point m if and only if $dim \, \mathfrak{g}_m = 0$, i.e., if and only if G_m, the stabilizer of m in G, is a discrete subgroup of G.*

We can now give the following characterization of the kernel of the differential $\mu_{*,m}$ of the moment map.

Proposition 3.78

$$ker \; \mu_{*,m} = (T_m \mathcal{O}_m)^\omega.$$

Proof First, note that

$$\mu_{*,m}(v) = 0 \iff (\mu^\sharp_x)_{*,m}(v) = 0, \; \forall x \in \mathfrak{g}.$$

On the other hand, $\omega_m(X_x(m), v) = 0$ if and only if v is orthogonal to the vector space spanned by $X_x(m)$ for all $x \in \mathfrak{g}$, proving the statement. $\qquad\square$

We collect here below a few simple observations.

- Since $\alpha \in \mathfrak{g}^*$ is a regular value for the moment map, i.e., μ is a submersion at each $m \in N_\alpha = \mu^{-1}(\alpha)$, N_α is an embedded submanifold of M.

- Moreover, since μ is G-equivariant, N_α is a G_α-manifold, where $G_\alpha \subset G$ is the stabilizer of $\alpha \in \mathfrak{g}^*$.
- $\omega_\alpha = j_\alpha^*(\omega) \in \Omega^2(N_\alpha)$, where $j_\alpha : N_\alpha \hookrightarrow M$, is closed.

We can prove the following.

Proposition 3.79 *If $m \in N_\alpha$, then:*

(i) *$\ker(\omega_\alpha)_m = T_m \mathscr{O}_m^{G_\alpha}$, where $\mathscr{O}_m^{G_\alpha}$ is the G_α-orbit of the point $m \in M$.*
(ii) *Moreover, $rk(\omega_\alpha) = 2\dim N_\alpha + \dim \mathscr{O}_\alpha - \dim M$. In particular, the rank of ω_α is constant on N_α.*

Proof First, observe that

$$\ker(\omega_\alpha)_m = T_m N_\alpha \cap (T_m N_\alpha)^\omega$$

and that

$$T_m N_\alpha = \ker \mu_{*,m} = (T_m \mathscr{O}_m)^\omega,$$

where the last equality is Proposition 3.78. Then

$$\ker(\omega_\alpha)_m = \ker \mu_{*,m} \cap T_m \mathscr{O}_m,$$

since $\left((T_m \mathscr{O}_m)^\omega\right)^\omega = T_m \mathscr{O}_m$. Now observe that the tangent space at the point m of $\mathscr{O}_m$ is generated by $\{X_x(m) \mid x \in \mathfrak{g}\}$. Since μ is G-equivariant, it is $\mathfrak{g}$-equivariant, i.e.,

$$\mu_{*,m} X_x(m) = -\operatorname{ad}_x^\sharp(\mu(m)), \forall m \in M \text{ and } x \in \mathfrak{g};$$

see Definition 3.41. This identity implies that $X_x(m) \in \ker \mu_{*,m}$ *if and only if $x \in$* $\mathfrak{g}_\alpha$. But since $T_m \mathscr{O}_m^{G_\alpha}$ is spanned by $\{X_x(m) \mid x \in \mathfrak{g}_\alpha\}$, $x \in \mathfrak{g}_\alpha$ *if and only if $X_x(m)$* belongs to $T_m \mathscr{O}_m^{G_\alpha}$, proving (i). For (ii), observe that since α is a regular value of μ, G_m is a discrete subgroup of G; see Proposition 3.77. Moreover, since μ is G-equivariant, $G_m \subset G_\alpha$. Then

$$\dim T_m \mathscr{O}_m^{G_\alpha} = \dim G_\alpha - \dim G_m = \dim G_\alpha. \tag{3.43}$$

Using again that G_m is a discrete subgroup of G yields

$$\dim T_m \mathscr{O}_m = \dim G,$$

implying that

$$\dim (T_m \mathscr{O}_m)^\omega = \dim M - \dim G.$$

So one can conlude

$$\mathrm{rk}\,(\omega_\alpha)_m \overset{(i)+(3.43)}{=} \dim N_\alpha - \dim G_\alpha$$

$$= \dim N_\alpha - \dim G_\alpha + \dim G - \dim M + \dim N_\alpha$$

$$= \dim N_\alpha + \dim \mathcal{O}_\alpha - \dim M + \dim N_\alpha$$

$$= 2\dim N_\alpha + \dim \mathcal{O}_\alpha - \dim M,$$

where the second equality follows from $\dim N_\alpha = \dim M - \dim G$. This computation shows that dimension of the kernel of the closed two-form ω_α is constant on N_α, completing the proof of (ii). $\qquad\square$

To summarize the previous discussion, one can state that

Proposition 3.80 *For every regular value $\alpha \in \mathfrak{g}^*$ of the G-equivariant moment map $\mu : M \to \mathfrak{g}^*$ associated to a strongly Hamiltonian action, (N_α, j_α), where $N_\alpha = \mu^{-1}(\alpha)$, is a constant rank submanifold of (M, ω) and the null-leaves of $\omega_\alpha = j_\alpha^*(\omega)$ are the G_α°-orbits on N_α, where G_α° is the connected component of the identity of the stabilizer in G of α.*

After all these preliminaries, we can finally state the following important.

Theorem 3.81 (Marsden-Weinstein-Meyer Theorem) *Let $\alpha \in \mathfrak{g}^*$ be a regular value of the G-equivariant moment map $\mu : M \to \mathfrak{g}^*$. Let G_α° be the connected component of the identity of G_α. If G_α° acts freely and properly on N_α, the quotient $\tilde{N}_\alpha = N_\alpha/G_\alpha^\circ$ is a smooth manifold endowed with the symplectic form $\tilde{\omega}_\alpha$, which is the unique symplectic form on $\tilde{N}_\alpha$, such that*

$$\pi_\alpha^*(\tilde{\omega}_\alpha) = j_\alpha^*(\omega), \tag{3.45}$$

where $\pi_\alpha : N_\alpha \to \tilde{N}_\alpha$ is the natural projection.

Proof The proof follows at once from Theorem 3.69 and Proposition 3.80. Note that the hypothesis made on the G_α°-action guarantee that N_α/G_α°, the space of the null-leaves of ω_α, is a smooth manifold. $\qquad\square$

Remark 3.82 The proof of Theorem 3.81 is based on the observation that the infinitesimal action on N_α of the stabilizer G_α generates the distribution defined by the kernel of the two-form $\omega_\alpha = j_\alpha^*\omega$. On the other hand, the orbits of the group G_α, in general, do not coincide with the maximal connected integral manifolds defined by this distribution. This is because the group G_α is not connected in general. For this reason, if we want to use Theorem 3.69 to prove Theorem 3.81 we need to consider the group G_α° instead of the group G_α. $\qquad\triangle$

For the notions borrowed from complex geometry that will be used in the following example and in Problem 3.84, we refer the reader to Sect. 5.2 in this chapter and to the references therein.

Example 3.83 (Complex Projective Space) For all $u, v \in \mathbb{C}^n$, let $h : \mathbb{C}^n \times \mathbb{C}^n \to \mathbb{C}$ defined by $h(u, v) = \sum_{k=1}^n u_k \overline{v}_k$, for all $u = (u_1, \ldots, u_n), v = (v_1, \ldots, v_n) \in \mathbb{C}^n$: h so defined is a *nondegenerate Hermitian* form, sometimes called the standard Hermitian form of $\mathbb{C}^n$. Let $\omega : \mathbb{C}^n \times \mathbb{C}^n \to \mathbb{R}$ be the imaginary part of h, i.e.,

$$\omega(u, v) = \frac{\sum_{k=1}^n u_k \overline{v}_k - \sum_{k=1}^n \overline{u}_k v_k}{2\iota}, \ \forall u, v \in \mathbb{C}^n. \tag{3.46}$$

A simple computation shows that the ω so defined is a nondegenerate, skew-symmetric $\mathbb{R}$-bilinear form, i.e., it is a symplectic form defined on the *real* vector space underlying $\mathbb{C}^n$. Furthermore, if $(z_1, \ldots, z_n)$ are standard coordinates of $\mathbb{C}^n$, another simple computation shows that

$$\omega = \frac{1}{2\iota} \sum_{k=1}^n dz_k \wedge d\overline{z}_k. \tag{3.47}$$

Identifying $\mathbb{C}^n$ with $\mathbb{R}^{2n}$ writing $z_k = x_k + \iota y_k$ for all $k = 1, \ldots, n$, one can prove that ω is the standard symplectic form of $\mathbb{R}^{2n}$, i.e., $\omega = \sum_{k=1}^n dy_k \wedge dx_k$. Now fix $\lambda \in \mathbb{R}$ different from zero and let $\varphi : S^1 \times \mathbb{C}^n \to \mathbb{C}^n$ be the left S^1-action defined by

$$\varphi(e^{\iota t \lambda}, (z_1, \ldots, z_n)) = (e^{-\iota t \lambda} z_1, \ldots, e^{-\iota t \lambda} z_n), \ \forall t \in \mathbb{R}, (z_1, \ldots, z_n) \in \mathbb{C}^n. \tag{3.48}$$

Here λ is a fixed nonzero element of the one dimensional real Lie algebra $\mathbb{R}$, which we identify with the Lie algebra of S^1. A direct inspection shows that φ is symplectic. The fundamental vector field defined by φ and by $\lambda \in \mathbb{R}$ is the vector field $X_\lambda := \frac{d}{dt} \varphi(e^{-\iota t \lambda}, (z_1, \ldots, z_n))|_{t=0}$ so that

$$X_\lambda(u) = \iota \lambda u, \ \forall u \in \mathbb{C}^n.$$

Let us compute

$$(i_{X_\lambda} \omega)_u(v) = \omega\big(X_\lambda(u), v\big) = \omega(\iota \lambda u, v) \overset{(3.46)}{=} \lambda \frac{\sum_{k=1}^n u_k \overline{v}_k + \sum_{k=1}^n \overline{u}_k v_k}{2}$$

$$= \lambda \, Re\left(\sum_{k=1}^n u_k \overline{v}_k\right). \tag{3.49}$$

On the other hand, if $\mu : \mathbb{C}^n \to \mathbb{R}$ is defined by

$$\mu(z_1, \ldots, x_n) = -\sum_{k=1}^n |z_k|^2, \tag{3.50}$$

then

$$(d\mu)_u(v) = \frac{d}{dt}\bigg|_{t=0} \mu(u + tv) = -\frac{d}{dt}\bigg|_{t=0} \sum_{k=1}^{n}(u_k + tv_k)(\overline{u}_k + t\overline{v}_k)$$

$$= -\sum_{k=1}^{n}(u_k\overline{v}_k + \overline{u}_k v_k) = -\mathrm{Re}\left(\sum_{k=1}^{n} u_k\overline{v}_k\right).$$

$$(3.51)$$

Comparing (3.49) with (3.51), one can conclude that the map μ, defined in Formula (3.50), is a moment map for the S^1-action (3.48) (note that we used the obvious identification between $\mathbb{R}$ and $\mathbb{R}^*$). In particular, this computation shows that this action is Hamiltonian and, since S^1 is one-dimensional, implies that φ is strongly Hamiltonian; see Example 3.32 and also Example 3.31. Furthermore, by direct inspection, one can check that μ is S^1-equivariant. Since for every $u \in \mathbb{C}^n$,

$$(d\mu)_u(v) = -Re\left(\sum_{k=1}^{n} u_k\overline{v}_k\right), \ v \in \mathbb{C}^n,$$

one can conclude that if $u \neq 0$, $(d\mu)_u(v) = 0$ *if and only if* $v = 0$, implying that each $c \in \mathbb{R}$, $c \neq 0$, is a regular value of μ. In particular, choosing $c = -1$, $\mu^{-1}(-1) \simeq S^{2n-1}$, i.e., the chosen level set is the unit sphere in $\mathbb{R}^{2n}$. Since S^1 is Abelian and connected, the connected component of the stabilizer of every $\lambda \in \mathbb{R}$ coincides with S^1. Moreover, since the S^1-action on $\mu^{-1}(-1)$ is *proper* (since S^1 is compact) and *free* (as a simple check shows), one can apply Theorem 3.81 to obtain the reduced symplectic manifold which is $\mu^{-1}(-1)/S^1 \simeq \mathbb{P}^{n-1}(\mathbb{C})$, the $n - 1$-dimensional *complex projective space*.   △

Note that the $\mathbb{C}^n$ is not only a symplectic manifold but it is also endowed with a Riemannian metric and a complex structure and these three structures interact to define on $\mathbb{C}^n$ a structure of a *Kähler* manifold. As explained shortly in Sect. 5.2 in this chapter, the Marsden-Weinstein-Meyer reduction can be extended to this class of manifolds to define the so-called *Kähler reduction*, of which 3.83 is a simple, nevertheless, very important example. In particular the complex projective space is a Kähler manifold whose Hermitian form is called the *Fubini-Study* metric and whose corresponding imaginary part is exactly the symplectic form defined on $\mathbb{P}^{n-1}(\mathbb{C})$ via the Marsden-Weinstein-Meyer reduction as in Example 3.48. Examples 3.57 and 3.83 can be generalized as follows.

Problem 3.84 Let $\mathrm{Mat}_{n \times k}(\mathbb{C})$ be the complex vector space of the $n \times k$ matrices with complex entries. Define

$$\omega = \frac{1}{2\imath}\mathrm{tr}(dZ \wedge dZ^\dagger),$$

$$(3.52)$$

and let $\varphi : \mathrm{U}(k) \times \mathrm{Mat}_{n \times k}(\mathbb{C}) \to \mathrm{Mat}_{n \times k}(\mathbb{C})$ be defined by

$$\varphi(g, Z) = Zg^{-1}, \ \forall Z \in \mathbb{C}^{n \times k}, \ g \in \mathrm{U}(k). \tag{3.53}$$

Show that:

(i) The two-form (3.52) is the standard symplectic form of $\mathrm{Mat}_{n \times k}(\mathbb{C})$ and that (3.53) is a symplectic action.

(ii) Show that (3.53) is strongly Hamiltonian. More precisely for all $\xi \in \mathfrak{u}(k)$ write

$$X_\xi(Z) = \sum_{l,k}(Z\xi)_{l,s}\frac{\partial}{\partial Z_{ls}} + \sum_{l,k}\overline{(Z\xi)}_{l,s}\frac{\partial}{\partial \overline{Z}_{ls}},$$

and prove that

$$i_{X_\xi}\omega(Z) = \frac{1}{2\iota}d\mathrm{tr}(Z^\dagger Z\xi). \tag{3.54}$$

(iii) Using (3.54) deduce that the $\mathrm{U}(k)$-action (3.53) admits a moment map $\mu : \mathrm{Mat}_{n \times k}(\mathbb{C}) \to \mathfrak{u}(k)$ defined by

$$\mu(Z) = \frac{1}{2\iota}Z^\dagger Z, \ \forall Z \in \mathrm{Mat}_{n \times k}(\mathbb{C}).$$

(iv) Conclude that the Grassmannian of the k-planes in $\mathbb{C}^n$ is a symplectic manifold.

(Hint: Note that $\mu^{-1}(\frac{\mathrm{id}}{2\iota}) = \{Z \in \mathrm{Mat}_{n \times k}(\mathbb{C}) \mid Z^\dagger Z = \mathrm{id}\}$, i.e., $Z \in \mu^{-1}(\frac{\mathrm{id}}{2\iota})$ *if and only if* the columns of Z are k unitary-orthogonal vectors in $\mathbb{C}^n$. In this way every matrix in the pre-image of $\frac{\mathrm{id}}{2\iota}$ determines a k-plane in $\mathbb{C}^n$. This observation implies that $\mu^{-1}(\frac{\mathrm{id}}{2\iota})/\mathrm{U}(k) \simeq \mathrm{Gr}(n, k)$.) $\triangle$

Remark 3.85 We are going now to link the reduction theorem to dynamics. More precisely, we will prove a Hamiltonian analogue of a theorem more generally stated in the context of the Lagrangian formalism of classical mechanics. This important result, due to Emmy Noether, can be also extended to field theory and it can be resumed saying that the symmetries of the Lagrangian function defining a given mechanical systems correspond to the first integrals of the same mechanical system. For a detailed exposition of this result in the framework of the Lagrangian formalism, we invite the reader to consult, for example, the text [10]. $\triangle$

As we have already announced, we are concerned with the Noether theorem in the framework of the Hamiltonian formalism. Since this theorem gives a link between symmetries and first integrals, its Hamiltonian version is phrased in the jargon of the theory of the Hamiltonian G-actions. More precisely, let $\mu : M \to \mathfrak{g}^*$ be a G-equivariant moment map defined on (M, ω) by a strongly Hamiltonian

G-action. Fix $\alpha \in \mathfrak{g}^*$ a regular value and consider the corresponding level set $N_\alpha = \mu^{-1}(\alpha) \subset M$.

Theorem 3.86 (Noether Theorem) *If H is a G-invariant function on M, then:*

(i) *μ is constant along the trajectories of the vector field X_H, or equivalently, each integral curve of X_H, is completely contained in one level set of the moment map.*

(ii) *Using the same notation as in Theorem 3.81, if $\tilde{N}_\alpha$ is a manifold, $\pi_\alpha : N_\alpha \to \tilde{N}_\alpha$ is a submersion and X_{H_α} is the Hamiltonian vector field on $(\tilde{N}_\alpha, \tilde{\omega}_\alpha)$ whose Hamiltonian is $H_\alpha \in C^\infty(\tilde{N}_\alpha)$ defined by the condition $\pi_\alpha^* H_\alpha = j_\alpha^* H$, then*

$$X_H \sim_{\pi_\alpha} X_{H_\alpha},$$

i.e., X_H is π_α-related to X_{H_α}, that is $\pi_{\alpha,*,m} X_H(m) = X_{H_\alpha}(\pi_\alpha(m))$ for all $m \in M$.

Remark 3.87 Note that in (ii) we wrote X_H to denote the restriction to N_α of the Hamiltonian vector field $X_H \in \mathfrak{X}(M)$. In spite of the fact that, in general, the restriction of a vector field $X \in \mathfrak{X}(M)$ to a submanifold $N \subset M$ does not define a vector field on N, under the hypothesis of the theorem, the restriction of $X_H \in \mathfrak{X}(M)$ to N_α defines a vector field on N_α. $\triangle$

Proof *(of Theorem 3.86)* To prove (i) it suffices to show that $\mathscr{L}_{X_x} H = 0$ for all $x \in \mathfrak{g}$. Let $\gamma_m^H = \gamma_m^H(t)$ the integral curve of X_H going through the point $m \in M$ when $t = 0$ and let $x \in \mathfrak{g}$. Then

$$
\begin{aligned}
\frac{d}{dt} \langle \mu \circ \gamma(t), \xi \rangle &= \langle d\mu_{\gamma(t)}(X_H(\gamma(t))), x \rangle \\
&= \langle X_H(\gamma(t)), (d\mu)^*_{\gamma(t)}(x) \rangle \\
&= \langle X_H(\gamma(t)), (i_{X_x}\omega)_{\gamma(t)} \rangle \\
&= \omega(X_x, X_H)|_{\gamma(t)} \\
&= -dH_{\gamma(t)}(X_x).
\end{aligned}
$$

Since H is G-invariant and X_x is the fundamental vector field defined by x, one has

$$H(\varphi_{\exp(-sx)}\gamma(t)) = (\varphi^*_{\exp(sx)} H)(\gamma(t)) = H(\gamma(t)),$$

so that

$$dH(X_x(\gamma(t)) = \frac{d}{ds}\bigg|_{s=0} H(\varphi_{\exp(-sx)}\gamma(t)) = 0.$$

To prove point (ii) observe first that, since H is G-invariant, it is also G_α-invariant so that, $j_\alpha^* H \in C^\infty(N_\alpha)$ descends to a smooth function $H_\alpha \in C^\infty(\tilde{N}_\alpha)$. More precisely, since π_α is a submersion, π_α^* is injective and H_α is the unique smooth function on $\tilde{N}_\alpha$ such that

$$j_\alpha^* H = \pi_\alpha^* H_\alpha.$$

Taking the differential of both sides, one can write that $j_\alpha^* dH = \pi_\alpha^* dH_\alpha$ or, what is the same, that

$$j_\alpha^* \big(i_{X_H}\omega\big)_m(v) = \pi_\alpha^* \big(i_{X_{H_\alpha}}\omega_\alpha\big)_m(v),$$

for all $m \in N_\alpha$ and for all $v \in T_m N_\alpha$. The previous equality can be rewritten as

$$(\omega_\alpha)_{\pi_\alpha(m)}\big(X_{H_\alpha}(\pi_\alpha(m)), (\pi_\alpha)_{*,m}v\big)$$

$$= \omega_{j_\alpha(m)}\big(X_H(j_\alpha(m)), v\big)$$

$$= (j_\alpha^*\omega)_m\big(X_H(m), v\big) \text{ since } j_\alpha \text{ is the canonical inclusion}$$

$$= (\pi^*\omega_\alpha)_m\big(X_H(m), v\big) \text{ since } \pi^*\omega_\alpha = j_\alpha^*\omega$$

$$= (\omega_\alpha)_{\pi_\alpha(m)}\big((\pi_\alpha)_{*,m}X_H(m), (\pi_\alpha)_{*,m}v\big).$$

Since π_α is a submersion and ω_α is nondegenerate, comparing the first with the last member of the previous computation, one can conclude that

$$(\pi_\alpha)_{*,m}X_H(m) = X_{H_\alpha}(\pi_\alpha(m)),$$

for all $m \in N_\alpha$, proving the assertion. $\square$

Remark 3.88 (Reconstructing the Dynamics) The previous theorem tells us that in the presence of symmetries it is possible to reduce the number of degrees of freedom of the Hamiltonian system (M, ω, X_H) *projecting* it to a *new* Hamiltonian system $(\tilde{N}_\alpha, \tilde{\omega}_\alpha, X_{H_\alpha})$. Since the new system lives on a lower-dimensional manifold, the equations governing its dynamics should be integrated more easily than the ones governing the dynamics of the original system. Once the solutions of the new Hamiltonian equations has been found, one needs to face the problem to get the solutions of the original Hamiltonian equations, i.e., one is left with the problem of *reconstructing the dynamics*; see Sect. 2 in Chap. 6, in particular Remark 6.15. $\triangle$

An important instance of the above theory is discussed in the following subsection.

4.1 Cotangent Bundle Reduction

Let $\varphi : G \times M \to M$ be a proper and free G-action. Then the quotient space B is a manifold and the canonical projection $\pi : M \to B$ is a smooth map. For each $m \in M$, let $\mathrm{Vert}_m = \mathrm{span}\,\langle X_x(m) \mid x \in \mathfrak{g}\rangle \subset T_m M$ the vertical space at m. Note that since the G is free, the infinitesimal action of $\mathfrak{g}$ on M, which takes each $x \in \mathfrak{g}$ to the corresponding fundamental vector field X_x, defines, for each $m \in M$, a linear isomorphism between $\mathfrak{g}$ and Vert_m. Under these hypothesis

$$0 \longrightarrow \mathfrak{g} \xrightarrow{\;X.(m)\;} T_m M \xrightarrow{\;\pi_{*,m}\;} T_{\pi(m)} B \longrightarrow 0$$

is an exact sequence of vector spaces, where $X.(m)$ is the map taking $x \in \mathfrak{g}$ to $X_x(m) \in \mathrm{Vert}_m$. Recall now that φ lifts to a strongly Hamiltonian G-action $\widetilde{\varphi}$ on (T^*M, Ω_M), whose moment map is defined by

$$\langle \mu(\alpha), x \rangle = \langle \alpha, X_x(m) \rangle, \ \forall \alpha \in T_m^* M; \tag{3.55}$$

see (3.2) and Appendix D. Then

Theorem 3.89 (Reduction at 0**)** $0 \in \mathfrak{g}^*$ *is a regular value for the moment map* μ *and the reduced phase space* (T^*M_0, ω_0) *is symplectomorphic to* (T^*B, Ω_B).

Proof First, we prove that T^*M_0 is diffeomorphic to T^*B. To this end, first, note that (3.55) implies that, for each $m \in M$,

$$\mu^{-1}(0) \cap T_m^* M = \mathrm{Vert}_m^\circ . \tag{3.56}$$

Noting this, for all $m \in M$, define $\aleph : \mu^{-1}(0) \to T^*B$ by

$$\langle \aleph(\alpha), \pi_{*,m} v \rangle := \langle \alpha, v \rangle, \tag{3.57}$$

for all $\alpha \in \mu^{-1}(0) \cap T_m^* M$, $v \in T_m M$. Then, (3.56) implies that $\aleph$ is well defined. Moreover, for all $g \in G$ writing $\varphi_g^* \aleph(\alpha) = \aleph\big((\varphi_{g^{-1}})^*_{\varphi_g(m)}\alpha\big)$ (see Formula (3.1) and Appendix D), one has that

$$\langle \varphi_g^* \aleph(\alpha), \pi_{*,\varphi_g(m)} v \rangle = \langle (\varphi_{g^{-1}})^*_{\varphi_g(m)}\alpha, v \rangle$$

$$= \langle \alpha, (\varphi_{g^{-1}})_{*,\varphi_g(m)} v \rangle$$

$$= \langle \aleph(\alpha), \pi_{*,m}\big((\varphi_{g^{-1}})_{*,\varphi_g(m)} v\big) \rangle$$

$$= \langle \aleph(\alpha), (\pi \circ \varphi_{g^{-1}})_{*,\varphi_g(m)} v \rangle$$

$$= \langle \aleph(\alpha), \pi_{*,\varphi_g(m)} v \rangle \ \text{since} \ \pi \circ \varphi_g = \varphi_g, \ \forall g \in G,$$

which implies that $\varphi_g^* \aleph = \aleph$ for all $g \in G$, i.e., $\aleph$ is G-invariant. In this way $\aleph$ descends to a smooth surjective map $\aleph_0 : \mu^{-1}(0)/G \to T^*B$, which closes

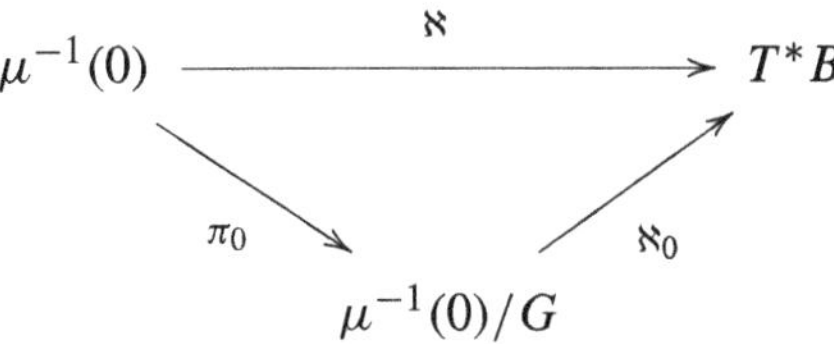

to a commutative diagram, where π_0 is the canonical projection. We prove now that $\aleph_0$ is injective. To this end, note that if $\alpha_1 \in \mu^{-1}(0) \cap T_{m_1}^* M$ and $\alpha_2 \in \mu^{-1}(0) \cap T_{m_2}^* M$ are such that $\aleph_0([\alpha_1]) = \aleph_0([\alpha_2])$, then $\aleph(\alpha_1) = \aleph(\alpha_2)$, which implies that $\pi(m_1) = b = \pi(m_2)$, i.e., there exists a (unique) $g \in G$ such that $\varphi_g(m_1) = m_2$. Given $v \in T_b B$ let $w_1 \in T_{m_1} M$ and $w_2 \in T_{m_2} M$ such that $\pi_{*,m_1} w_1 = v = \pi_{*,m_2} w_2$. From the definition of $\aleph$ it follows that $\langle \alpha_1, w_1 \rangle = \langle \alpha_2, w_2 \rangle$, if $g \in G$ is such that $w_2 = (\varphi_g)_{*,m_1} w_1'$, $w_1' \in T_{m_1} M$,

$$\langle \alpha_1, w_1 \rangle = \langle \alpha_2, (\varphi_g)_{*,m_1} w_1' \rangle. \tag{3.58}$$

The previous identity together with $\aleph(\alpha_1) = \aleph(\alpha_2)$ implies that $\langle \aleph(\alpha_1), \pi_{*,m_1} (w_1 - w_1') \rangle = 0$. In fact

$$\langle \aleph(\alpha_1), \pi_{*,m_1} w_1 \rangle = \langle \aleph(\alpha_2), \pi_{*,m_2} (\varphi_g)_{*,m_1} w_1' \rangle = \langle \aleph(\alpha_2), \pi_{*,m_1} w_1' \rangle.$$

Then $w_1' - w_1 = \xi \in \mathrm{Vert}_{m_1}$, which inserted in (3.58) gives

$$\langle \alpha_1 - (\varphi_g)_{m_1}^* \alpha_2, w_1 \rangle = 0. \tag{3.59}$$

In fact

$$\begin{aligned}
\langle \alpha_1, w_1 \rangle &= \langle \alpha_2, (\varphi_g)_{*,m_1} w_1' \rangle \\
&= \langle \alpha_2, (\varphi_g)_{*,m_1} (w_1 + \xi) \rangle \\
&= \langle \alpha_2, (\varphi_g)_{*,m_1} w_1 \rangle \text{ since } (\varphi_g)_{*,m_1}(\xi) \in \mathrm{Vert}_{m_2} \\
&= \langle (\varphi_g)_{m_1}^* \alpha_2, w_1 \rangle.
\end{aligned}$$

Formula (3.59) suffices to conclude that $\alpha_1 - (\varphi_g)_{m_1}^* \alpha_2 = 0$, or equivalently that $[\alpha_1] = [\alpha_2]$, since for each $m \in M$ $T_m M \simeq T_{\pi(m)} B \oplus \mathrm{Vert}_m$. Since $\aleph_0 : \mu^{-1}(0)/G \to T^*B$ is a smooth bijective map, to prove that it is a diffeomorphism, it is sufficient to show that it is a local diffeomorphism. On the other hand, since $\dim \mu^{-1}(0)/G = \dim T^*B$, to prove that $\aleph_0$ is a local diffeomorphism it is enough to show that $\aleph_0$ is a symplectic map between $(\mu^{-1}(0)/G, \widetilde{\omega}_0)$ and (T^*B, Ω_B); see Proposition 1.48. Summarizing, once we prove that $\aleph_0$ is a symplectic map between the reduced phase space and the cotangent bundle of B, both endowed with their canonical symplectic forms, this will be sufficient to conclude that $\aleph_0$ is a symplectic diffeomorphism, i.e., a symplectomorphism, as stated in the theorem. To prove that $\aleph_0$ is symplectic, it suffices to show that

$$j_0^* \Theta_M = \pi_0^* \aleph_0^* \Theta_B; \tag{3.60}$$

see the diagram (3.61).

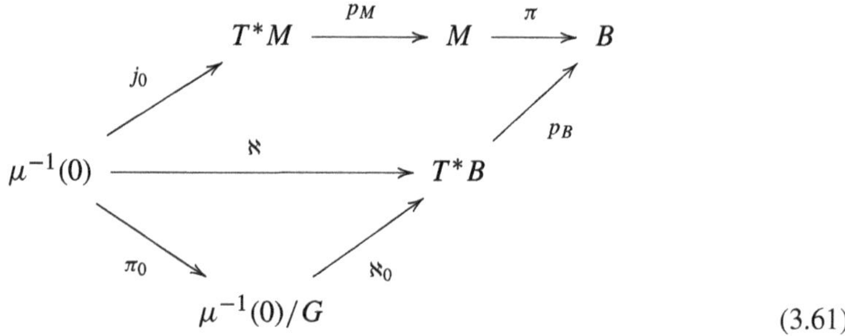

$$\tag{3.61}$$

In fact, if (3.60) holds,

$$j_0^* \Omega_M = d(j_0^* \Theta_M) = d(\pi_0^* \aleph_0^* \Theta_B) = \pi_0^* (\aleph_0^* \Omega_B). \tag{3.62}$$

On the other hand, since the reduced symplectic form $\widetilde{\omega}_0$ is characterized by the property

$$j_0^* \Omega_M = \pi_0^* \widetilde{\omega}_0$$

(see Theorem 3.81), one can conclude that if (3.60) holds, then

$$\pi_0^* \widetilde{\omega}_0 = j_0^* \Omega_M = \pi_0^* (\aleph_0^* \Omega_B),$$

which implies that $\widetilde{\omega}_0 = \aleph_0^* \Omega_B$ being π_0^* injective. To prove (3.60) let $\alpha \in \mu^{-1}(0) \cap T_m^* M$ and let $v \in T_\alpha \mu^{-1}(0)$. Then

$$
\begin{aligned}
\langle (\pi_0^* \aleph_0^* \Theta_B)_\alpha, v \rangle \ &\overset{(3.61)}{=}\ \langle (\aleph^* \Theta_B)_\alpha, v \rangle \\
&=\ \langle (\Theta_B)_{\aleph(\alpha)}, \aleph_{*,\alpha} v \rangle \\
&=\ \langle \aleph(\alpha), (p_B)_{*,\aleph(\alpha)} \circ \aleph_{*,\alpha} v \rangle \\
&=\ \langle \aleph(\alpha), (p_B \circ \aleph)_{*,\alpha} v \rangle \\
&\overset{(3.61)}{=}\ \langle \aleph(\alpha), (\pi \circ p_M \circ j_0)_{*,\alpha} v \rangle \\
&=\ \langle \aleph(\alpha), \pi_{*,m}\big((p_M \circ j_0)_{*,\alpha} v\big) \rangle
\end{aligned}
$$

$$\overset{(3.57)}{=} \ \langle \alpha, (p_M \circ j_0)_{*,\alpha} v \rangle$$

$$= \ \langle \alpha, (p_M)_{*,\alpha}(j_{0\,*,\alpha} v) \rangle$$

$$= \ \langle (\Theta_M)_\alpha, j_{0\,*,\alpha} v \rangle$$

$$= \ \langle (j_0^* \Theta_M)_\alpha, v \rangle \ \text{since } j_0(\alpha) = \alpha.$$

$\square$

Before moving to the next statement we need to make a couple of preliminary observations. We refer the reader to Sect. 2.2 in Appendix C for the basic notions about principal bundles and related structures. The first one is that a connection one-form $\mathcal{A} \in \Omega_M(\mathfrak{g})$ on the G-principal bundle (M, G, B, π) induces, for each $m \in M$, a linear map $\mathcal{A}_m^* : \mathfrak{g}^* \to T_m^* M$ defined by

$$\langle \mathcal{A}_m^*(\xi), v \rangle = \langle \xi, \mathcal{A}_m(v) \rangle = \langle \xi, \langle \mathcal{A}_m, v \rangle \rangle, \ \forall v \in T_m M, \tag{3.63}$$

which, in turn, for each $\xi \in \mathfrak{g}^*$ defines an element $\mathcal{A}^\xi \in \Omega^1(M)$ via

$$\mathcal{A}_m^\xi := \mathcal{A}_m^*(\xi), \ \forall m \in M. \tag{3.64}$$

The second one is that if $\mu : T^*M \to \mathfrak{g}^*$ is the moment map defined by the cotangent lift of the G-action (see Sect. 1 in this chapter), then for each $\xi \in \mathfrak{g}^*$

$$\mu(\mathcal{A}_m^\xi) = \xi, \ \forall m \in M. \tag{3.65}$$

Furthermore,

$$\varphi_g^* \mathcal{A}^\xi = \mathcal{A}^{\mathrm{Ad}_{g^{-1}}^\sharp \xi}, \ \forall g \in G. \tag{3.66}$$

Note that (3.65) says that $\mathcal{A}^\xi$ takes values on the level set $\mu^{-1}(\xi)$, while from (3.66) it follows that if ξ is invariant with respect to the coadjoint action, then $\mathcal{A}^\xi$ is a G-invariant one-form. The proofs of (3.65) and (3.66) follow by a direct computation which we present below for the reader's convenience. For every $x \in \mathfrak{g}$ and every $m \in M$, one has

$$\langle \mu(\mathcal{A}_m^\xi), x \rangle \overset{(3.64)}{=} \langle \mu(\mathcal{A}_m^*(\xi)), x \rangle$$

$$\overset{(3.55)}{=} \langle \mathcal{A}_m^*(\xi), X_x(m) \rangle$$

$$\overset{(3.63)}{=} \langle \xi, \langle \mathcal{A}_m, X_x(m) \rangle \rangle$$

$$\overset{(C.25)}{=} \langle \xi, x \rangle,$$

proving (3.65). On the other hand, for every $m \in M$, $g \in G$ and $v \in T_m M$,

$$
\begin{aligned}
\langle (\varphi_g^* \mathcal{A}^\xi)_m, v \rangle \quad &= \quad \langle \mathcal{A}^\xi_{\varphi_g(m)}, (\varphi_g)_{*,m} v \rangle \\[4pt]
&\overset{(3.64)}{=} \quad \langle \mathcal{A}^*_{\varphi_g(m)}(\xi), (\varphi_g)_{*,m} v \rangle \\[4pt]
&\overset{(3.63)}{=} \quad \langle \xi, \langle \mathcal{A}_{\varphi_g(m)}, (\varphi_g)_{*,m} v \rangle \rangle \\[4pt]
&= \quad \langle \xi, \langle (\varphi_g^* \mathcal{A})_m, v \rangle \rangle \\[4pt]
&\overset{(C.26)}{=} \quad \langle \xi, \mathrm{Ad}_g \langle \mathcal{A}_m, v \rangle \rangle \\[4pt]
&= \quad \langle \mathrm{Ad}^\sharp_{g^{-1}} \xi, \langle \mathcal{A}_m, v \rangle \rangle \\[4pt]
&\overset{(3.63)\text{ and }(3.64)}{=} \quad \langle \mathcal{A}_m^{\mathrm{Ad}^\sharp_{g^{-1}} \xi}, v \rangle
\end{aligned}
$$

proving (3.66). After these preliminary remarks we are ready to generalize Theorem 3.89. Let $\xi \in \mathfrak{g}^*$ and let $\tau_\xi : T^*M \to T^*M$ be defined by

$$
\tau_\xi(\alpha) = \alpha - \mathcal{A}_m^\xi, \quad \forall \alpha \in T_m^* M, \ m \in M;
$$

see Example 1.53. Note that τ_ξ is a *vertical* map, i.e.,

$$
p_M \circ \tau_\xi = p_M. \tag{3.67}
$$

Lemma 3.90 τ_ξ *is a diffeomorphism of T^*M which maps $\mu^{-1}(\xi)$ diffeomorphically to $\mu^{-1}(0)$. Furthermore,*

$$
\tau_\xi^* \Omega_M = \Omega_M - \pi^* d\mathcal{A}^\xi, \tag{3.68}
$$

and

$$
\tau_\xi \circ \widetilde{\varphi}_g = \widetilde{\varphi}_g \circ \tau_{\mathrm{Ad}^\sharp_{g^{-1}} \xi}, \quad \forall g \in G. \tag{3.69}
$$

Proof For the first part of the statement see Example 1.53. On the other hand, if $\alpha \in \mu^{-1}(\xi) \cap T_m^* M$, then for every $x \in \mathfrak{g}$,

$$
\langle \mu(\tau_\xi(\alpha)), x \rangle = \langle \tau_\xi(\alpha), X_x(m) \rangle = \langle \alpha - \mathcal{A}_m^\xi, X_x(m) \rangle = \langle \alpha, X_x(m) \rangle - \langle \xi, x \rangle
$$

$$
= \langle \mu(\alpha) - \xi, x \rangle = 0,
$$

i.e., $\mu(\alpha) \in \mu^{-1}(0) \cap T_m^* M$. Finally, the proof of (3.69) follows from the following computation:

$$\tau_\xi(\widetilde{\varphi}_g \alpha) = \tau_\xi\left((\varphi_{g^{-1}})^*_{\varphi_g(m)}\alpha\right) = (\varphi_{g^{-1}})^*_{\varphi_g(m)}\alpha - \mathcal{A}^\xi_{\varphi_g(m)}. \tag{3.70}$$

On the other hand, for every $v \in T_{\varphi_g(m)}M$

$$\begin{aligned}
\langle \mathcal{A}^\xi_{\varphi_g(m)}, v \rangle &\overset{(3.64)}{=} \langle \mathcal{A}^*_{\varphi_g(m)}(\xi), v \rangle \\[2mm]
&\overset{(3.63)}{=} \langle \xi, \langle \mathcal{A}_{\varphi_g(m)}, v \rangle \rangle \\[2mm]
&= \langle \xi, \langle \mathcal{A}_{\varphi_g(m)}, (\varphi_g)_{*,m}\left((\varphi_{g^{-1}})_{*,\varphi_g(m)}v\right) \rangle \rangle \\[2mm]
&= \langle \xi, \langle (\varphi_g^* \mathcal{A})_m, (\varphi_{g^{-1}})_{*,\varphi_g(m)}v \rangle \rangle \\[2mm]
&\overset{(C.26)}{=} \langle \xi, \mathrm{Ad}_g \langle \mathcal{A}_m, (\varphi_{g^{-1}})_{*,\varphi_g(m)}v \rangle \rangle \\[2mm]
&= \langle \mathrm{Ad}^\sharp_{g^{-1}}\xi, \langle \mathcal{A}_m, (\varphi_{g^{-1}})_{*,\varphi_g(m)}v \rangle \rangle \\[2mm]
&= \langle \mathcal{A}^*_m(\mathrm{Ad}^\sharp_{g^{-1}}\xi), (\varphi_{g^{-1}})_{*,\varphi_g(m)}v \rangle \\[2mm]
&\overset{(3.64)}{=} \langle \mathcal{A}_m^{\mathrm{Ad}^\sharp_{g^{-1}}\xi}, (\varphi_{g^{-1}})_{*,\varphi_g(m)}v \rangle \\[2mm]
&= \langle (\varphi_{g^{-1}})^*_{\varphi_g(m)}\mathcal{A}_m^{\mathrm{Ad}^\sharp_{g^{-1}}\xi}, v \rangle
\end{aligned}$$

proving that

$$\mathcal{A}^\xi_{\varphi_g(m)} = (\varphi_{g^{-1}})^*_{\varphi_g(m)}\mathcal{A}_m^{\mathrm{Ad}^\sharp_{g^{-1}}\xi},$$

which, if inserted in Formula (3.70), yields

$$\tau_\xi(\widetilde{\varphi}_g \alpha) = (\varphi_{g^{-1}})^*_{\varphi_g(m)}\alpha - (\varphi_{g^{-1}})^*_{\varphi_g(m)}\mathcal{A}_m^{\mathrm{Ad}^\sharp_{g^{-1}}\xi} = (\varphi_{g^{-1}})^*_{\varphi_g(m)}\left(\alpha - \mathcal{A}_m^{\mathrm{Ad}^\sharp_{g^{-1}}\xi}\right)$$

$$= \widetilde{\varphi}_g(\tau_\xi \alpha),$$

which is what we wanted to prove. $\qquad\square$

Let $\xi \in \mathfrak{g}^*$ be a G-invariant element, i.e., an element such that $G_\xi = G$. Then

Theorem 3.91 (Reduction at ξ) *The choice of a connection $\mathcal{A} \in \Omega^1_M(\mathfrak{g})$ defines a symplectic diffeomorphism between the reduced phase space $(\mu^{-1}(\xi)/G, \widetilde{\omega}_\xi)$ and $(T^*B, \Omega_B + p_B^* C^\xi)$, where C^ξ is the (unique) two-form on B such that $d\mathcal{A}^\xi = \pi^* C^\xi$.*

Proof Since ξ is G-invariant, τ_ξ intertwines the G-action $\widetilde{\varphi}$ (see Formula (3.69)), and it descends to a diffeomorphism $\widetilde{\tau}_\xi : \mu^{-1}(\xi)/G \to \mu^{-1}(0)/G$. In particular, one can complete (3.61) to the following diagram:

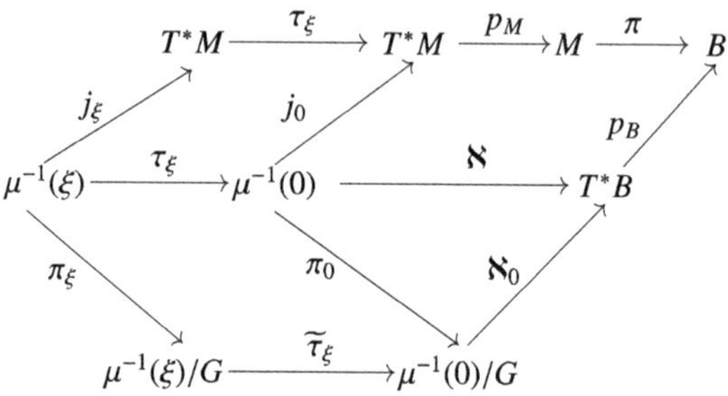

which is also commutative. Defining $\aleph_\xi := \aleph_0 \circ \widetilde{\tau}_\xi$, the proof of the theorem follows if we show that

$$\aleph_\xi^*(\Omega_B + p_B^* C^\xi) = \widetilde{\omega}_\xi, \tag{3.71}$$

where $\widetilde{\omega}_\xi$ is the reduced symplectic form. Using again the defining property of $\widetilde{\omega}_\xi$ (see Theorem 3.81), (3.71) is equivalent to

$$\pi_\xi^*\big(\aleph_\xi^*(\Omega_B + p_B^* C^\xi)\big) = j_\xi^* \Omega_M,$$

which we prove here below. To this end we invite the reader to follow the steps of the computation along the diagram above.

$$
\begin{aligned}
\pi_\xi^*\big(\aleph_\xi^*(\Omega_B + p_B^* C^\xi)\big) &= \pi_\xi^*\big((\aleph_0 \circ \widetilde{\tau}_\xi)^*(\Omega_B + p_B^* C^\xi)\big) \\[4pt]
&= \pi_\xi^*(\aleph_0 \circ \widetilde{\tau}_\xi)^* \Omega_B + (p_B \circ \aleph_0 \circ \widetilde{\tau}_\xi \circ \pi_\xi)^* C^\xi \\[4pt]
&= (\aleph \circ \tau_\xi)^* \Omega_B + (p_B \circ \aleph_0 \circ \widetilde{\tau}_\xi \circ \pi_\xi)^* C^\xi \\[4pt]
&= \tau_\xi^*(\aleph_0 \circ \pi_0)^* \Omega_B + (p_B \circ \aleph_0 \circ \widetilde{\tau}_\xi \circ \pi_\xi)^* C^\xi \\[4pt]
&\overset{(3.62)}{=} \tau_\xi^*(j_0^* \Omega_M) + (p_B \circ \aleph_0 \circ \widetilde{\tau}_\xi \circ \pi_\xi)^* C^\xi \\[4pt]
&= (j_0 \circ \tau_\xi)^* \Omega_M + (p_B \circ \aleph_0 \circ \widetilde{\tau}_\xi \circ \pi_\xi)^* C^\xi \\[4pt]
&= (\tau_\xi \circ j_\xi)^* \Omega_M + (p_B \circ \aleph_0 \circ \widetilde{\tau}_\xi \circ \pi_\xi)^* C^\xi \\[4pt]
&= j_\xi^*(\tau_\xi^* \Omega_M) + (p_B \circ \aleph_0 \circ \widetilde{\tau}_\xi \circ \pi_\xi)^* C^\xi \\[4pt]
&\overset{(3.68)}{=} j_\xi^*(\Omega_M - p_M^* d\theta^\xi) + (p_B \circ \aleph_0 \circ \widetilde{\tau}_\xi \circ \pi_\xi)^* C^\xi
\end{aligned}
$$

$$
\begin{aligned}
&= \; j_\xi^*(\Omega_M - p_M^*\pi^*C^\xi) + (p_B \circ \aleph_0 \circ \tilde{\tau}_\xi \circ \pi_\xi)^*C^\xi \\
&= \; j_\xi^*\Omega_M - (\pi \circ p_M \circ j_\xi)^*C^\xi + (p_B \circ \aleph_0 \circ \tilde{\tau}_\xi \circ \pi_\xi)^*C^\xi \\
&= \; j_\xi^*\Omega_M,
\end{aligned}
$$

which is what we wanted to prove. Note that the last equality follows from (3.67).

$\square$

Let now discuss an important example of the previous theory.

4.1.1 The Mechanical Connection and the Amended Potential

Let (M, B, π, G) be a principal bundle (see Appendix C), and let H be a G-invariant metric on M, i.e., a smoothly varying section of the symmetric product $T^*M \odot T^*M$, such that $(\varphi_g^* H)_m(v, u) = H_m(v, u)$, $\forall m \in M$, $u, v \in T_m M$, where $(\varphi^* H)_m(v, u) = H_{\varphi_g(m)}\big((\varphi_g)_{*,m}v, (\varphi_g)_{*,m}u\big)$, for all $m \in M$, $u, v \in T_m M$. Then, for every $m \in M$, one can define $\mathrm{Hor}_m := \mathrm{Vert}_m^\perp$, where $\perp$ denotes the orthogonal with respect to H. The assignment $m \rightsquigarrow \mathrm{Hor}_m$, defines a connection on M. To prove this statement it suffices to show that $(\varphi_g)_{*,m} \mathrm{Hor}_m = \mathrm{Hor}_{\varphi_g(m)}$, for all $g \in G$ and $m \in M$, which follows essentially from the G-invariance of the metric H and since φ_g is a diffeomorphism. We now prove that the connection one-form $\mathcal{A}^H$ defined by this horizontal distribution, is given by

$$
\mathcal{A}_m^H = I(m)^{-1} \circ \mu \circ h_m, \quad \forall m \in M, \tag{3.72}
$$

where $h_m : T_m M \to T_m^* M$ is defined by

$$
\langle h_m(v), u \rangle = H_m(u, v),
$$

for all $m \in M$ and $u, v \in T_m M$, and $I(m) : \mathfrak{g} \to \mathfrak{g}^*$ is defined by

$$
\langle I(m)x, y \rangle = H_m\big(X_x(m), X_y(m)\big),
$$

for all $m \in M$ and $x, y \in \mathfrak{g}$. Note that for each $m \in M$, $I(m)$ is a linear isomorphism, since if $I(m)x = 0$, then $H_m\big(X_x(m), X_x(m)\big) = 0$, implying that $X_x(m) = 0$, which forces $x = 0$ since φ is free. By a direct inspection, it follows that (3.72) defines a one-form on M with values in $\mathfrak{g}$ such that, for all $m \in M$, $\mathcal{A}_m^H(v) = 0$ for all $v \in \mathrm{Hor}_m$. To prove that (3.72) defines a connection on M, one needs to show that the conditions (C.25) and (C.26) are fulfilled. For what concerns (C.25), if $x \in \mathfrak{g}$ is such that $\mathcal{A}_m^H(X_x(m)) = z$, then

$$
z = \mathcal{A}_m^H\big(X_x(m)\big) = I(m)^{-1}\big(\mu(h_m(X_x(m))),
$$

i.e.,

$$H_m(X_z(m), X.(m)) = I(m)(z) = H_m(X_x(m), X.(m)), \tag{3.73}$$

where $\mu\big(h_m(X_x(m))\big) = H_m(X_x(m), X.(m)) : \mathfrak{g} \to \mathbb{R}$ is defined by

$$\langle \mu\big(h_m(X_x(m))\big), y \rangle = H_m(X_x(m), X_y(m)), \ \forall y \in \mathfrak{g}.$$

Using now the nondegeneracy of H, (3.73) implies $z = x$, proving that $\mathcal{A}_m^H(X_x(m)) = x$ for all $m \in M$ and $x \in \mathfrak{g}$. To conclude the proof, we are left to show that $\mathcal{A}^H$ satisfies (C.26). To this end, first, note that if $v \in \mathrm{Hor}_m$, then

$$\langle (\varphi_g^* \mathcal{A}^H)_m, v \rangle = \langle \mathcal{A}_{\varphi_g(m)}^H, (\varphi_g)_{*,m} v \rangle = 0,$$

since $(\varphi_g)_{*,m} v \in \mathrm{Hor}_{\varphi_g(m)}$, showing that (C.26) in this case holds true. On the other hand,

$$\begin{aligned}
\Big\langle (\varphi_g^* \mathcal{A}^H)_m, X_x(m) \Big\rangle &= \Big\langle \mathcal{A}_{\varphi_g(m)}^H, (\varphi_g)_{*,m} X_x(m) \Big\rangle \\[4pt]
&= \Big\langle \mathcal{A}_{\varphi_g(m)}^H, (\varphi_g)_{*,m} \frac{d}{dt}\Big|_{t=0} \varphi_{\exp(-tx)}(m) \Big\rangle \\[4pt]
&= \Big\langle \mathcal{A}_{\varphi_g(m)}^H, \frac{d}{dt}\Big|_{t=0} \varphi_g\big(\varphi_{\exp(-tx)}(m)\big) \Big\rangle \\[4pt]
&= \Big\langle \mathcal{A}_{\varphi_g(m)}^H, \frac{d}{dt}\Big|_{t=0} \big(\varphi_g \varphi_{\exp(-tx)} \varphi_{g^{-1}}\big)(\varphi_g(m)) \Big\rangle \\[4pt]
&= \Big\langle \mathcal{A}_{\varphi_g(m)}^H, \frac{d}{dt}\Big|_{t=0} \varphi_{\exp(-t\,\mathrm{Ad}_g\,x)}(\varphi_g(m)) \Big\rangle \\[4pt]
&= \Big\langle \mathcal{A}_{\varphi_g(m)}^H, X_{\mathrm{Ad}_g\,x}(\varphi_g(m)) \Big\rangle \\[4pt]
&= \mathrm{Ad}_g\,x \\[4pt]
&= \mathrm{Ad}_g \Big\langle \mathcal{A}_m^H, X_x(m) \Big\rangle,
\end{aligned}$$

which proves (C.26).

Definition 3.92 The connection $\mathcal{A}^H$ defined in Formula (3.72) is called the *mechanical connection* defined by the metric H. $\quad\triangle$

Remark 3.93 The mechanical connection is defined by the commutative diagram:

$$
\begin{array}{ccc}
T^*M & \xrightarrow{\ \mu\ } & \mathfrak{g}^* \\
\big\uparrow{\scriptstyle h_{(\cdot)}} & & \big\uparrow{\scriptstyle \mathcal{I}(\cdot)} \\
T M & \xrightarrow[\ \mathcal{A}^H_{(\cdot)}\ ]{} & \mathfrak{g}
\end{array}
$$

Note that the smooth application $h : T M \to T^*M$ is the *Legendre transform* defined by the metric H; see [10, 84]. In the literature $\mathcal{I}(m) : \mathfrak{g} \to \mathfrak{g}^*$ is called the *locked inertia tensor*, and the $\mathcal{A}^H_m(v)$ is called the *angular velocity* of the locked system (see [172]) $\triangle$

Let K be the bilinear, symmetric, and nondegenerate form defined by H on T^*M. More precisely, for all $\alpha_1, \alpha_2 \in T_m^*M$, let

$$
K_m(\alpha_1, \alpha_2) = H_m(v_1, v_2),
$$

where v_1, v_2 are the unique elements in $T_m M$ such that $h_m(v_1) = \alpha_1$ and $h_m(v_2) = \alpha_2$. Given any G-invariant function V on M, let $k \in C^\infty(T^*M)$, defined by

$$
k(\alpha) = \mathcal{K}(\alpha) + p_M^* V(\alpha), \tag{3.74}
$$

where $\mathcal{K} \in C^\infty(T^*M)$ is defined by

$$
\mathcal{K}(\alpha) = \frac{1}{2} K_m(\alpha, \alpha), \ \forall \alpha \in T_m^*M, \ m \in M.
$$

Problem 3.94 Prove that $\mathcal{K}$ is:

(i) Invariant with respect to the cotangent lift of φ
(ii) The *Legendre transform* of the *kinetic energy* $\mathcal{H} \in C^\infty(T M)$ defined by

$$
\mathcal{H}(v) = \frac{1}{2} H_m(v, v), \ \forall v \in T_m M, \ m \in M
$$

(see [10, 84])

$\triangle$

Note that since $V \in C^\infty(M)$ is G invariant, its pullback to T^*M is invariant with respect to the cotangent lift of the G-action. This observation, together with (i) of the previous exercise, entails the invariance of the function k defined in (3.74).

Definition 3.95 (Simple Mechanical System with Symmetries) (T^*M, Ω, k) is called a *simple mechanical system with symmetries*. $\triangle$

Remark 3.96 The notion of simple mechanical system was introduced in [223] and it has as a straightforward generalization the one of a *symplectic G-system*, i.e., a tuple (M, ω, μ, H) of a symplectic manifold (M, ω), a Hamiltonian function H invariant with respect to a strongly Hamiltonian G-action having a G-equivariant moment map μ; see, for example, [172]. Note that every simple mechanical system with symmetries is a symplectic G-system. △

Let $\xi \in \mathfrak{g}^*$ be a G-invariant element and let $\mathcal{A}^{H,\xi}$ be the G-invariant one-form on M defined by ξ and by the mechanical connection $\mathcal{A}^H$; see (3.64). Using $\mathcal{A}^{H,\xi}$ one can pull back k defined in (3.74) to M to obtain

$$V^\xi(m) := (\mathcal{A}^{H,\xi*}k)(m) \overset{(3.74)}{=} \mathcal{K}_m\left(\mathcal{A}_m^{H,\xi}\right) + (p_M \circ \mathcal{A}^{H,\xi})^* V(m)$$

$$= \frac{1}{2} K_m\left(\mathcal{A}_m^{H,\xi}, \mathcal{A}_m^{H,\xi}\right) + V(m), \ \forall m \in M.$$

Remark 3.97 It is worth noting that the term $K_m(\mathcal{A}_m^{H,\xi}, \mathcal{A}_m^{H,\xi})$ is the restriction to the level set $\mu^{-1}(\xi)$ of the kinetic energy of the original simple mechanical system; see (3.65). △

We now prove that

Lemma 3.98 V^ξ is the function on M defined by

$$V^\xi(m) = \frac{1}{2}\langle I(m)^{-1}(\xi), \xi \rangle + V(m), \ \forall m \in M. \tag{3.75}$$

Proof To prove the statement, it suffices to show that for all $\xi \in \mathfrak{g}^*$

$$K_m(\mathcal{A}_m^{H,\xi}, \mathcal{A}_m^{H,\xi}) = \langle I(m)^{-1}(\xi), \xi \rangle, \ \forall m \in M.$$

To this end observe that for all $m \in M$

$$\langle \mathcal{A}_m^{H,\xi}, v \rangle = 0, \ \forall v \in \mathrm{Hor}_m \quad \text{and} \quad \langle \mathcal{A}_m^{H,\xi}, X_x(m) \rangle = \langle \xi, x \rangle, \ \forall x \in \mathfrak{g}.$$

Moreover,

$$\langle h_m\big(X_{I(m)^{-1}(\xi)}(m)\big), v \rangle = 0, \ \forall v \in \mathrm{Hor}_m \quad \text{and}$$

$$\langle h_m\big(X_{I(m)^{-1}(\xi)}(m)\big), X_x(m) \rangle = \langle \xi, x \rangle, \ \forall x \in \mathfrak{g},$$

which imply, by the nondegeneracy of h_m, that

$$\mathcal{A}_m^{H,\xi} = h_m\big(X_{I(m)^{-1}(\xi)}(m)\big), \ \forall m \in M.$$

Then

$$\frac{1}{2}K_m(\mathcal{A}_m^{H,\xi}, \mathcal{A}_m^{H,\xi}) = \frac{1}{2}H_m\big(X_{\mathcal{I}(m)^{-1}(\xi)}(m), X_{\mathcal{I}(m)^{-1}(\xi)}(m)\big) = \frac{1}{2}\langle \xi, \mathcal{I}(m)^{-1}\xi\rangle,$$

which is what we wanted to prove. $\qquad\Box$

Definition 3.99 (Amended Potential) The function V^ξ defined in (3.75) is called the *amended potential* defined by ξ. $\qquad\triangle$

Remark 3.100 In the literature the amended potential is also called *effective* or *reduced* potential. $\qquad\triangle$

Finally, one can prove that

Lemma 3.101 V^ξ *is G-invariant; hence, it defines a function on B.*

Proof Since V is G-invariant by hypothesis, to prove the statement it suffices to show that the function $m \rightsquigarrow \frac{1}{2}\langle \mathcal{I}(m)^{-1}(\xi), \xi\rangle$ is G-invariant, i.e., that

$$\langle \mathcal{I}\big(\varphi_g(m)\big)^{-1}\xi, \xi\rangle = \langle \mathcal{I}(m)^{-1}\xi, \xi\rangle, \ \forall m \in M, g \in G.$$

This will follow from the more general identity

$$\mathcal{I}\big(\varphi_g(m)\big)^{-1} = \mathrm{Ad}_g \circ \mathcal{I}(m)^{-1} \circ \mathrm{Ad}^\sharp_{g^{-1}} \tag{3.76}$$

which we prove here. Indeed,

$$\begin{aligned}
\langle \mathcal{I}\big(\varphi_g(m)\big)x, y\rangle &= H_{\varphi_g(m)}\big(X_x(\varphi_g(m), X_y(\varphi_g(m))\big)\\
&= H_{\varphi_g(m)}\big((\varphi_g)_{*,m}X_{\mathrm{Ad}_{g^{-1}}x}(m), (\varphi_g)_{*,m}X_{\mathrm{Ad}_{g^{-1}}y}(m)\big)\\
&\quad (\text{see Problem } 3.102)\\
&= (\varphi_g^* H)_m\big(X_{\mathrm{Ad}_{g^{-1}}x}(m), X_{\mathrm{Ad}_{g^{-1}}y}(m)\big)\\
&= H_m\big(X_{\mathrm{Ad}_{g^{-1}}x}(m), X_{\mathrm{Ad}_{g^{-1}}y}(m)\big)\\
&= \langle \mathcal{I}(m)\,\mathrm{Ad}_{g^{-1}}x, \mathrm{Ad}_{g^{-1}}y\rangle\\
&= \langle \mathrm{Ad}^\sharp_g\big(\mathcal{I}(m)\,\mathrm{Ad}_{g^{-1}}x\big), y\rangle,
\end{aligned}$$

which implies that

$$\mathcal{I}\big(\varphi_g(m)\big) = \mathrm{Ad}^\sharp_g \circ \mathcal{I}(m) \circ \mathrm{Ad}_{g^{-1}}, \ \forall m \in M, g \in G,$$

proving (3.76). This concludes the proof of the lemma since, by hypothesis, ξ is G-invariant. $\qquad\Box$

Problem 3.102 Prove that, for all $x \in \mathfrak{g}$,

$$X_x\big(\varphi_g(m)\big) = (\varphi_g)_{*,m} X_{\mathrm{Ad}_{g^{-1}} x}(m), \quad \forall m \in M, \, g \in G.$$

$\triangle$

4.1.2 Relative Equilibria

The amended potential plays an important role in the analysis of the so-called *relative equilibria* of an Hamiltonian system with symmetry. Recall that an *equilibrium point* of an Hamiltonian system (M, ω, H) is a point $m \in M$ such that $X_H(m) = 0$ or, equivalently, where $dH_m = 0$. On the other hand, if (M, ω, μ, H) is a symplectic G-system; see Remark 3.96.

Definition 3.103 (Relative Equilibrium Point) $m \in M$ is called a relative equilibrium point for the symplectic G-system if $X_H(m)$ is contained in the tangent space to the G-orbit of m, i.e., if $X_H(m) \in T_m \mathcal{O}_m$. $\triangle$

A characterization of relative equilibria is enclosed in the following.

Theorem 3.104 *Let (M, ω, μ, H) be a symplectic G-system, and let φ_m^H the integral curve of X_H passing through m at $t = 0$ and let $\xi = \mu(m)$. Then the following conditions are equivalent:*

(i) m is a relative equilibrium point.
(ii) $\varphi_m^H(t) \in G_\xi.m \subset G.m$, for all t.
(iii) There is $x \in \mathfrak{g}$ such that $\varphi_m^H(t) = \exp(tx).m$, $\forall t$.
(iv) There is $x \in \mathfrak{g}$ such that m is a critical point for $H_x(\cdot) := H(\cdot) - \langle \mu(\cdot) - \xi, x \rangle$.
(v) m is a critical point of the smooth map $H \times \mu : M \times M \to \mathbb{R} \times \mathfrak{g}^*$.
(vi) m is a critical point of the restriction of H to $\mu^{-1}(\xi)$.
(vii) m is a critical point of the restriction of H to $\mu^{-1}(\mathcal{O}_\xi)$, where $\mathcal{O}_\xi$ is the orbit of ξ with respect to the coadjoint action of G on $\mathfrak{g}^*$.
(viii) $[m]$ in the reduced phase space is a critical point of the reduced Hamiltonian.

We will not present the proof of this result for which we refer the reader to [172] (see also [185]), but we would like to make a few comments about its content. The first two are terminological. The functions in (iv) and (v) are usually called the *augmented Hamiltonian* and the *energy-momentum map*, respectively. In points (vi)–(viii) of the above equivalences it is assumed that ξ is a regular value of the moment map and in (viii) the reduced phase space is supposed to be smooth (see again [172]). Specializing the previous result to the case of a simple mechanical system with symmetry; see Definition 3.95, one has

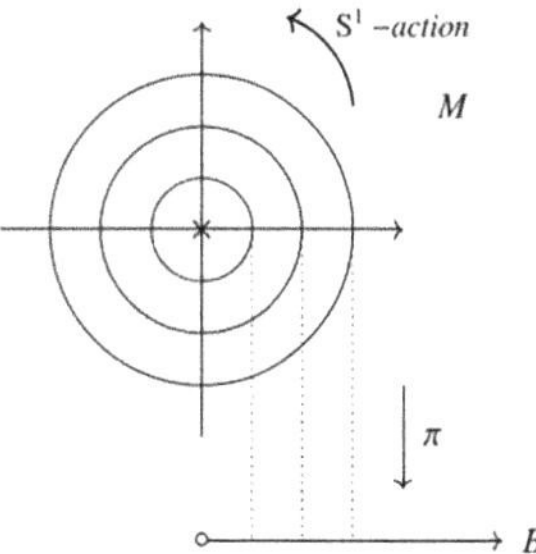

Fig. 3.3 S^1-fibration

Proposition 3.105 *A point* $\alpha \in T_m^* M$ *is a relative equilibrium if and only if there is* $x \in \mathfrak{g}$ *such that:*

(i) $\alpha = h_m(X_x(m))$
(ii) m is a critical point of the emended potential V^ξ where $\xi = \mathcal{I}(m)(x)$

whose proof can be found in the cited reference [172].

An interesting example of mechanical system having relative equilibria is the *spherical pendulum*; see the discussion in 3.2.1 in Sect. 3.2 of Chap. 4.

Example 3.106 (S^1-**Principal Bundle**) Let $(M, B, \pi, S^1) = (\mathbb{R}^2\backslash(0, 0), \mathbb{R}_{>0}, \pi, S^1)$ be the S^1-principal bundle defined by the action

$$\varphi_\theta(x, y) = (x \cos\theta - y \sin\theta, x \sin\theta + y \cos\theta).$$

The S^1-orbits form a one-parameter family of circumferences with center at $(0, 0)$ and radius $R \in \mathbb{R}_{>0}$ and π is the map that takes every point of a given circumference to the value of its radius; see the picture above (Fig. 3.3).

After identifying the Lie algebra of S^1 with $\mathbb{R}$, the fundamental vector field defined by φ and by $\lambda \in \mathbb{R}$ is

$$X_\lambda(x, y) = \lambda \left(-y \frac{\partial}{\partial x} + x \frac{\partial}{\partial y} \right),$$

which, in polar coordinates (r, θ), assumes the form

$$X_\lambda(r, \theta) = \lambda \frac{\partial}{\partial \theta}.$$

The standard Euclidean metric of $\mathbb{R}^2$ is S^1-invariant and in polar coordinates assumes the form

$$H(r, \theta) = dr^2 + r^2 d\theta^2. \tag{3.77}$$

An easy computation shows that

$$\mathcal{I}(r,\theta)(1) = r^2, \quad \mathcal{I}^{-1}(r,\theta)(1) = \frac{1}{r^2}$$

and

$$\mathcal{A}^H_{(r,\theta)} = d\theta. \tag{3.78}$$

Choosing $V = V(r) \in C^\infty(\mathbb{R}^2)$, since every $\xi \in \mathbb{R}$ is S^1-invariant, one can write

$$V^\xi(r) = V(r) + \frac{\xi^2}{2r^2}; \tag{3.79}$$

see Formula (3.75). This function is clearly S^1-invariant and it descends to $\mathbb{R}_{>0}$ to define the amended potential V^ξ, which, written in the coordinate r, has the same analytical expression as the r.h.s of (3.79).

On the other hand, the *kinetic energy* defined by the metric (3.77) is given by the expression

$$\mathcal{H} = \frac{1}{2}\big(\dot{r}^2 + r^2\dot{\theta}^2\big),$$

which, after applying the *Legendre transform*, becomes the function on $T^*\mathbb{R}^2$:

$$\mathcal{K} = \frac{1}{2}\left(p_r^2 + \frac{p_\theta^2}{r^2} \right),$$

where p_θ and p_r are the conjugate momenta of θ and r, respectively; see [10, 84]. Given $V(r) \in C^\infty(\mathbb{R}^2)$, one can define

$$k = \frac{1}{2}\left(p_r^2 + \frac{p_\theta^2}{r^2} \right) + (\pi^*V)(r) \in C^\infty(T^*\mathbb{R}^2), \tag{3.80}$$

which is invariant with respect to the cotangent lift of S^1; see Problem 3.94. Since $\mathcal{A}^H_{(r,\theta)} = d\theta$ (see (3.78)), for all $\xi \in \mathbb{R} = \mathrm{Lie}(S^1)$ and $m = (r,\theta) \in M$, one has

$$\mathcal{A}^{H,\xi}_m = (\mathcal{A}^{H,\xi}_m)^*(\xi) = \xi d\theta,$$

which yields the amended potential

$$V^\xi(r) = \frac{\xi^2}{2r^2} + V(r), \tag{3.81}$$

due to Remark 3.97 and to the fact that the moment map associated to the cotangent lift of the S^1-action $\mu : T^*\mathbb{R}^2 \to \mathbb{R}$ is given by $\mu(r, \theta, p_r, p_\theta) = p_\theta$. To understand better the content of the previous general discussion, let us consider the case of the Keplerian potential $V(r) = -\frac{1}{r}$; see also Sect. 2.2 in Chap. 4. In this case the Hamiltonian equations defined by (3.80) assume the following form:

$$\begin{cases} \dot{r} & = \frac{\partial k}{\partial p_r} = p_r \\ \dot{p}_r & = -\frac{\partial k}{\partial r} = -\frac{p_\theta^2}{r^3} + \frac{1}{r^2} \\ \dot{\theta} & = \frac{\partial k}{\partial p_\theta} = \frac{p_\theta}{r^2} \\ \dot{p}_\theta & = -\frac{\partial k}{\partial \theta} = 0. \end{cases} \tag{3.82}$$

The last of the previous equations entails that p_θ is a conserved quantity which can be used to reduce by two the number of degrees of freedom of the original system. This is achieved by simply fixing its value, for example, choosing $p_\theta = \xi$. In this way, the original system is reduced to a system with a one degree of freedom, whose Hamiltonian is

$$H_\xi(p_r, r) = \frac{1}{2}\left(p_r^2 + \frac{\xi^2}{r^2}\right) - \frac{1}{r} = \frac{p_r^2}{2} + \frac{\xi^2}{2r^2} - \frac{1}{r}, \tag{3.83}$$

i.e., the potential of the *reduced system* is the amended potential. A simple computation shows that the function $V^\xi = \frac{\xi^2}{2r^2} - \frac{1}{r}$ has a unique critical point at $r_\xi = \xi^2$ whose corresponding critical value is $V(r_\xi) = -\frac{\xi^2}{2}$, where V^ξ reaches its (absolute) minimum. Note that $r = r_\xi = \xi^2$ yields the stationary, or equilibrium, point of the reduced system $(p_r, r) = (0, r_\xi)$. In fact the Hamiltonian equations defined by (3.83) are

$$\begin{cases} \dot{r} & = p_r \\ \dot{p}_r & = \frac{r - \xi^2}{r^3} \end{cases}$$

which at the critical point r_ξ become

$$\begin{cases} \dot{r} & = p_r \\ \dot{p}_r & = 0 \end{cases}$$

whose solutions are $(p_r(t), r(t)) = (p_r(0)t + r_\xi, r_\xi)$ which, for $p_r(0) = 0$, yields the desired result. Finally, plugging this solution in the original Hamiltonian system

(3.84) one obtains

$$
\begin{cases}
\dot{r} & = 0 \\
\dot{p}_r & = 0 \\
\dot{\theta} & = \frac{1}{\xi^3} \\
\dot{p}_\theta & = 0,
\end{cases}
\tag{3.84}
$$

whose solution, given the initial conditions $(r_\xi, \theta_0, 0, \xi)$, is $(r(t), \theta(t), p_r(t), p_\theta(t)) = (r_\xi, \frac{t}{\xi^3} + \theta_0, p_r, \xi)$ and it describes a circular motion in the original phase space with constant angular velocity. Note that the corresponding orbit can be described as the integral curve of a fundamental vector field defined by a S^1-action on the original phase space. After identifying the reduced phase space $\mu^{-1}(\xi)/S^1$ with $T^*B \simeq \mathbb{R}^2$ (see Theorem 3.91), (3.81) becomes a function on $\mathbb{R}^2$ with coordinates (r, p_r). $\qquad\triangle$

Problem 3.107 Going through the same steps, analyze the case $V(r) = r^2$. $\qquad\triangle$

The next example can be considered an amplification of the previous one.

Example 3.108 (Central Potentials) Let $(M, \omega) = (\mathbb{R}^{2n}, \omega_{st})$ which we think as the cotangent bundle of $\mathbb{R}^n$ with its canonical symplectic form. Moreover, consider the group $SO_n(\mathbb{R})$ and its *standard* action $\varphi : SO_n(\mathbb{R}) \times \mathbb{R}^n \to \mathbb{R}^n, \varphi : (g, x) \rightsquigarrow gx$. Then, applying Formula (3.1) to the present case, the (cotangent) lift to $\mathbb{R}^{2n}$ of the $SO_n(\mathbb{R})$-action can be written explicitly as

$$
\widetilde{\varphi}_g(x, p) = (gx, gp), \text{ for all } g \in SO_n(\mathbb{R}) \text{ and for all } (x, p) \in \mathbb{R}^{2n};
$$

see also Example 3.13. If $r = \sqrt{x_1^2 + \cdots + x_n^2}$, a smooth function $U : \mathbb{R}^n \to \mathbb{R}$ is a called *a central potential* if $U = U(r)$. The Hamiltonian functions associated to this class of potentials assume the form

$$
H(p, x) = \frac{1}{2} \sum_{i=1}^{n} p_i^2 + U(r),
\tag{3.85}
$$

which can be easily seen to be $SO_n(\mathbb{R})$-invariant with respect to the (cotangent) lift $\widetilde{\varphi}$. Let us now compute the moment map associated to the $SO_n(\mathbb{R})$-action $\widetilde{\varphi}$; see again Example 3.13. To this end observe that the fundamental vector field associated to $v \in \mathfrak{so}_n(\mathbb{R})$ and defined by φ is $X_v, x \rightsquigarrow X_v(x) = vx$, for all $x \in \mathbb{R}^n$. Then, $F_v(x, p) = \langle p, X_v(x) \rangle$, where the components of this map are obtained computing $F_{E_{ij} - E_{ji}}(x, p)$, for all $i, j = 1, \ldots, n$ and all $(x, p) \in \mathbb{R}^{2n}$.

Remark 3.109 Remember that $\mathfrak{so}_n(\mathbb{R})$ is the Lie algebra of all (real) $n \times n$ skew-symmetric matrices and that E_{ij} denotes the elementary $n \times n$-matrix having 1 is the position (i, j). $\qquad\triangle$

After performing this computation, we finally get the $\frac{n(n-1)}{2}$ components of the moment map, which are the functions $F_{ij} = x_i p_j - x_j p_i$, $i, j = 1, \ldots, n$.

Lemma 3.110 *Only $2n - 3$ among the F_{ij} above defined are independent.*

Proof Given $(p, x) \in \mathbb{R}^{2n}$, suppose that $p \neq 0$ and $x \neq 0$ and that p and x are two linearly independent vectors, both thought as vectors of $\mathbb{R}^n$. Then the subgroup of $SO_n(\mathbb{R})$ which fixes the pair (p, x) is isomorphic to $SO_{n-2}(\mathbb{R})$. This observation, together with Corollary 3.76, gives the proof of the lemma. $\qquad \square$

Remark 3.111 Note that $\mathbb{R}^{2n}$ is identified with the cotangent bundle of $\mathbb{R}^n$ via its standard Euclidian inner product. Furthermore, observe that the pairs $(p, x) \in \mathbb{R}^{2n}$ are chosen to be composed by linearly independent vectors. At the points (p, x), with p, x linearly dependent (and both nonzero), one can prove that the rank of the moment map is equal to $n - 1$ and, for this reason, at each of these points among the $\frac{n(n-1)}{2}$ first integrals there would be only $n - 1$-independent. $\qquad \triangle$

A corollary of the previous result is that the level set of any regular value of the moment map is a three-dimensional submanifold of $\mathbb{R}^{2n}$. Once fixed two linearly independent vectors $p, x \in \mathbb{R}^n$, the Hamiltonian flow generated by (3.85) preserve the plane $p \wedge x \in \Lambda^2 \mathbb{R}^n$. In fact, note that the components of this two-vector are the components of the moment map. See also [91] for a different proof of this fact. In this way, we reduced the original system to a two-dimensional one, living on a two-plane with coordinates (z_1, z_2). If ξ_1, ξ_2 are the conjugate momenta to z_1, z_2, the reduced Hamiltonian assumes the form

$$H(\xi, z) = \frac{1}{2} \sum_{i=1}^{2} \xi_i^2 + \widetilde{U}\left(\sqrt{z_1^2 + z_2^2}\right). \tag{3.86}$$

Clearly, this Hamiltonian is S^1-invariant. After introducing polar coordinates, $(z_1, z_2) = (r \cos\theta, r \sin\theta)$, one has $(\dot{z}_1, \dot{z}_2) = (\dot{r}\cos\theta - r\dot{\theta}\sin\theta, \dot{r}\sin\theta + r\dot{\theta}\cos\theta)$. In particular, since

$$L(\theta, \dot{\theta}, r, \dot{r}) = \frac{1}{2}(\dot{r}^2 + r^2\dot{\theta}^2) - V(r),$$

the momenta conjugate to θ and r are $p_\theta = r^2\dot{\theta}$ and $p_r = \dot{r}$, respectively (see, for example, [10]), and the Hamiltonian (3.86) assumes the following form:

$$H = \frac{1}{2}\left(p_r^2 + \frac{p_\theta^2}{r^2}\right) + V(r),$$

whose integrability was analyzed in depth in Example 3.106. To summarize the previous discussion one can state the following.

Proposition 3.112 *Every Hamiltonian system defined on (an open subset of)* $(\mathbb{R}^{2n}, \omega_{st})$ *by an Hamiltonian function of type (3.85) can be reduced to a Hamiltonian system with one degree of freedom. In other words, the Hamiltonian (3.85) is completely integrable.*

Among the central potentials the cases:

(i) $V(r) = kr^2, k > 0$
(ii) $V(r) = -\frac{K}{r}, K > 0$

are particularly important. The one in (i) is the *harmonic potential*, more precisely, it is the potential of the *isotropic harmonic oscillator* (see Example 1.118), while (ii) is the *Keplerian* potential; see Example 3.106 and Sect. 2.2 in Chap. 4. One important, characterizing, property shared by these two potential functions is enclosed in the following classical and beautiful result; see, for example, [108, Section 3.6, Chapter 3].

Theorem 3.113 (Bertrand) *A point mass particle moving in a central potential has all its bounded trajectories closed only if the potential is of the form* $V(r) = -\dfrac{K}{r}$ *or* $V(r) = kr^2$, *where* K, k *are positive constants depending on the physical parameters of the system.*

$$\triangle$$

The next example, borrowed from the beautiful and unpublished [153], discusses the Marsden-Weinstein-Meyer reduction of the cotangent bundle of S^2, acted on by the cotangent lift of the S^1-action defined in Example 3.52.

Example 3.114 (S^1-**Action on** S^2) We remind that the S^1-action on S^2, when expressed in *cylindrical* coordinates (θ, z), $\theta \in [0, 2\pi)$ and $z \in [-1, 1]$, assumes the following form $\gamma.(z, \theta) \rightsquigarrow (z, \theta + \gamma), \forall \gamma \in S^1$. In these coordinates the (parametric) equations of the two-sphere are

$$\begin{cases} x_1 = \sqrt{1 - z^2} \cos \theta \\ x_2 = \sqrt{1 - z^2} \sin \theta \\ x_3 = z, \end{cases}$$

and the round metric on the sphere, i.e the pullback via the canonical *embedding* $S^2 \setminus \{N, S\} \mapsto \mathbb{R}^3$, where $N = (0, 0, 1)$ and $S = (0, 0, -1)$, of the Euclidean metric $g(v, v) = \frac{1}{2} \sum_{i=1}^{2} \dot{x}_i^2$, $v = (\dot{x}_1, \dot{x}_2, \dot{x}_3)$, assumes the following form:

$$L(z, \theta, \dot{z}, \dot{\theta}) = \frac{1}{2}\left(\frac{z^2 \dot{z}^2}{1 - z^2} + \dot{\theta}^2(1 - z^2) + \dot{z}^2 \right). \tag{3.87}$$

The momenta canonically conjugate to (z, θ) are $p_z = \frac{\partial L}{\partial \dot{z}} = \dot{z}\left(1 + \frac{z^2}{1-z^2}\right)$ and $p_\theta = \frac{\partial L}{\partial \dot{\theta}} = \dot{\theta}\,(1 - z^2)$, respectively.

Remark 3.115 Note that since L does not depend on θ, p_θ is a first integral for the mechanical system described by the Lagrangian (3.87). $\triangle$

The canonical symplectic form of $T^* S^2$ now can be written as $\Omega = dp_\theta \wedge d\theta + dp_z \wedge dz = d(p_\theta d\theta + p_z dz)$, and since the fundamental vector field defined by the S^1-action on S^2 is $\frac{\partial}{\partial\theta}$, one can conclude that the moment map associated to the cotangent lift of the S^1-action is

$$\mu(z, \theta, p_z, p_\theta) = p_\theta.$$

The Hamiltonian $H : T^* S^2 \to \mathbb{R}$, obtained from (3.87) via Legendre transform, is

$$H(z, \theta, p_z, p_\theta) = \frac{1}{2}\left((1 - z^2)p_z^2 + \frac{p_\theta^2}{1 - z^2} \right); \tag{3.88}$$

see [10, 84]. As already observed, p_θ is a first integral, i.e.,

$$\{H, p_\theta\} = \frac{\partial H}{\partial p_\theta}\frac{\partial p_\theta}{\partial\theta} - \frac{\partial H}{\partial\theta}\frac{\partial p_\theta}{\partial p_\theta} + \frac{\partial H}{\partial p_z}\frac{\partial p_\theta}{\partial z} - \frac{\partial H}{\partial z}\frac{\partial p_\theta}{\partial p_z} = 0.$$

For what concerns the reduction, note that every $c \in \mathrm{Lie}(S^1)$ is a regular value of the moment map and that its stabilizer is isomorphic to S^1, since S^1 is Abelian. The cotangent lift of the S^1-action is $\gamma.(z, \theta, p_z, p_\theta) \rightsquigarrow (z, \theta + \gamma, p_z, p_\theta)$ and it can be deduced noticing that the infinitesimal action of S^1 on the two sphere is generated by the vector field $\frac{\partial}{\partial\theta}$. After these preliminary observations, one can conclude that if $c \in \mathrm{Lie}(S^1)$,

$$\widetilde{M}_c = \frac{\mu^{-1}(c) = \{(z, \theta, p_z, p_\theta) \mid p_\theta = c)\}}{S^1} \simeq \{(z, p_z) \mid z \in (-1, 1),\ p \in \mathbb{R}\}$$

$$\simeq T^*(-1, 1).$$

If $i_c : \mu^{-1}(c) \to T^* S^2$ is the canonical immersion, $i_c^*\Omega = dp_z \wedge dz$, which is clearly S^1-invariant and, as such, it passes to the quotient as the canonical form of the cotangent bundle $T^*(-1, 1)$. Summarizing, the Marsden-Weinstein-Meyer quotient with respect to the (cotangent lift of the) S^1-action to the cotangent bundle of S^2 with its canonical symplectic form is $(T^*(-1, 1), \widetilde{\Omega})$ where $\widetilde{\Omega}$ is the canonical symplectic form. We conclude this discussion with the following comment. Since the Hamiltonian (3.88) is S^1-invariant, we can apply the Noether theorem, 3.86. In particular, the reduced Hamiltonian vector field is Hamiltonian with respect to the Hamiltonian function $H_c(z, p_z) = \frac{1}{2}\left((1 - z^2)p_z^2 + \frac{c^2}{1-z^2} \right)$. $\triangle$

Remark 3.116 We could generalize the previous discussion considering the Lagrangian function:

$$L = \frac{1}{2}\left(\frac{z^2\dot{z}^2}{1-z^2} + \dot{\theta}^2(1-z^2) + \dot{z}^2 - f(z)\right),$$

where $f : S^2 \to \mathbb{R}$ is a smooth function. With this choice, we would end up with the same reduced phase space, but with a different reduced Hamiltonian, given by

$$H_c(z, p_z) = \frac{1}{2}\left((1-z^2)p_z^2 + \frac{c^2}{1-z^2} + f(z)\right).$$

The case $f(z) = z$ will be discussed in more details in Sect. 3.2 of Chap. 4. We note that in this case, as in the one discussed in the previous Example 3.108, the reduced Hamiltonian function is the Hamiltonian of a one degree of freedom Hamiltonian system having the form

$$H_c(x, p) = P(x, p) + U_c(x),$$

where P is a polynomial function in p of degree two and U_c is a function of x depending on the parameter c. △

5　Concluding Remarks and Further Topics

In this last section we present a few further results related to the Marsden-Weinstein-Meyer reduction. In the first subsection we collect the definition and a few properties of a different approach to the symplectic reduction introduced in [130] and commonly known as the *reduction along an orbit*. In the same subsection we also introduce the concepts of symplectic realization of a Poisson manifolds and the one of dual pair, both introduced in fundamental paper [247]. In the second subsection, the reader will find a few comments about the use of the Marsden-Weinstein-Meyer reduction in the framework of the complex and quaternionic geometry.

5.1　Reduction Along an Orbit and Dual Pairs

In this subsection we discuss an alternative approach to the construction of the reduced phase space, the so-called *symplectic reduction along an orbit* and, at the same time, we make a short detour through the theory of the *dual pairs*. For the proofs of the results stated below, we refer the reader to the monograph [205] and to the references therein.

We start our discussion recalling that the reduction process presented above consists of two steps: first, the original Hamiltonian system is *restricted* to a submanifold of the phase space, then it is *projected* to the reduced phase space. On the other hand, if one is interested in the reduction of the number of degrees of freedom of the original Hamiltonian system, a different strategy is sometime viable. In fact, suppose that on the space of the orbits of the G-action on M is possible to define a structure of a manifold and that the canonical projection $\pi : M \to M/G$ is a submersion, which happens to be true, for example, when the G-action is free and proper. Under these assumptions, since the Hamiltonian function H is G-invariant, one can define $\widetilde{H} \in C^\infty(M/G)$ via the condition $\widetilde{H} \circ \pi = H$. Note that, under the hypothesis we made about the G-action, $C^\infty(M/G)$ is in one-to-one correspondence with $C^\infty(M)^G$, the algebra of the G-invariant function on M. This correspondence is obtained via the projection map π: since the G-action is *proper and free*, π is submersive, so that π^* is injective. Furthermore, since the G-action is symplectic, $\{f_1 \circ \varphi_g, f_2 \circ \varphi_g\} = \{f_1, f_2\} \circ \varphi_g$, for all $g \in G$ and $f_1, f_2 \in C^\infty(M)$, i.e., φ_g is a Poisson morphism for all $g \in G$. From this observation it follows that on the space of the G-orbits inherits a unique Poisson structure $\{\cdot, \cdot\}^\sim$ such that

$$\{F_1 \circ \pi, F_2 \circ \pi\} = \{F_1, F_2\}^\sim \circ \pi, \ \forall F_1, F_2 \in C^\infty(M/G), \tag{3.89}$$

i.e., such that π is a Poisson morphism; see also Corollary 2.57. Since the proofs of these claims are simple, they will be left to the reader as exercises. Anyway, under the hypothesis above, one can define the following diagram:

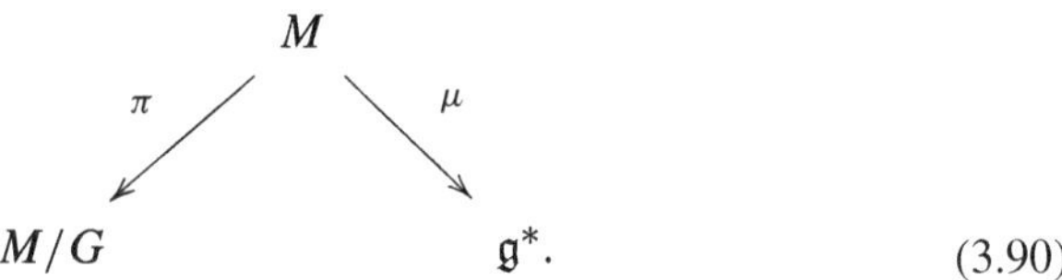

$$\tag{3.90}$$

In this diagram μ and π are two Poisson morphism defined on the same symplectic manifold (M, ω) and taking values in the Poisson manifolds $(\mathfrak{g}^*, \{\cdot, \cdot\}_{LP})$ and $(M/G, \{\cdot, \cdot\}^\sim)$, respectively. Moreover, one can observe that

$$\ker \mu_{*,m} = (\ker \pi_{*,m})^\omega, \ \forall m \in M.$$

This follows from Proposition 3.78, noticing that $\ker \pi_{*,m} = T_m \mathcal{O}_m$. It is now time to introduce the following.

Definition 3.117 (Symplectic Realization) A *symplectic realization* of a Poisson manifold (P, Π) is a pair $\big((M, \omega), \pi\big)$ where (M, ω) is a symplectic manifold and

$$\pi : (M, \omega) \to (P, \Pi)$$

is a Poisson morphism. A symplectic realization is called *full* if the map π is a surjective submersion.   △

Let us present some examples.

Example 3.118 Every symplectic leaf of a Poisson manifold defines a symplectic realization of the Poisson manifold itself via the inclusion map. This, in general, is *not* a full realization.   △

The following examples are particularly important for what is discussed in these notes.

Example 3.119 Let (M, ω) be a symplectic manifold endowed with a strongly Hamiltonian G-action endowed with a surjective moment map μ. Then $\big((M, \omega), \mu\big)$ is a symplectic realization of the Poisson manifold $(\mathfrak{g}^*, \{\cdot, \cdot\}_{LP})$. This realization is *full* if and only if the G-action is *locally free*, i.e., if the Lie algebra $\mathfrak{g}_m$ of the stabilizer G_m of every point $m \in M$ is trivial; see Proposition 3.77.   △

Example 3.120 Let G be a Lie group acting properly and freely on the symplectic manifold (M, ω). Suppose that the canonical projection π is a submersive map. Then, as it was discussed above, the quotient M/G is a manifold and the bracket defined in (3.89) is the unique Poisson structure on M/G which makes π a Poisson morphism. Then $\big((M, \omega), \pi\big)$ is a full symplectic realization of $(M/G, \{\cdot, \cdot\}^{\sim})$.   △

The diagram (3.90) suggests that it is interesting to consider the case of a symplectic manifold (M, ω) defining simultaneously a symplectic realization of two Poisson manifolds (P_i, Π_i), $i = 1, 2$. Diagram (3.90) represents the case of (M, ω) which is endowed with a strongly Hamiltonian G-action. This observation is formalized in the following.

Definition 3.121 (Dual Pairs) The Poisson manifolds (P_1, Π_1) and (P_2, Π_2) form a *dual pair* if there exists a symplectic manifold (M, ω) and two Poisson morphisms π_1, π_2 such that:

(i) $\big((M, \omega), \pi_i\big)$ is a symplectic realization for (P_i, Π_i), for $i = 1, 2$.
(ii) $\mathscr{D}_1 : m \rightsquigarrow \ker \pi_{1*,m}$ and $\mathscr{D}_2 : m \rightsquigarrow \ker \pi_{2*,m}$ are two *symplectically orthogonal* distributions defined on M, namely, they are two distributions such that

$$(\ker \pi_{1*,m})^{\omega} = \ker \pi_{2*,m}, \ \forall m \in M.$$

A dual pair is called *full* if the corresponding symplectic realizations are so.   △

After all these preliminaries remarks, going back to the beginning if this subsection, one can state the following.

Proposition 3.122 *If G is connected and if (M, ω) is endowed with a free, proper, and strongly Hamiltonian G-action, such that the canonical projection $\pi : M \rightarrow$*

M/G is a submersion and the moment map $\mu : M \to \mathfrak{g}^*$ is surjective, then $(\mathfrak{g}^*, \{\cdot, \cdot\}_{LP})$ and $(M/G, \{\cdot, \cdot\}^\sim)$ form a full dual pair.

There are several reasons why (full) dual pairs are mathematical objects worth being studied. One of those is contained in the following.

Proposition 3.123 *If (P_1, Π_1) and (P_2, Π_2) is a full dual pair realized by (M, ω) and such that the fibers of $\pi_1 : M \to P_1$ and $\pi_2 : M \to P_2$ are connected, then there is a one-to-one correspondence between the symplectic leaves of (P_1, Π_1) and the ones of (P_2, Π_2).*

The correspondence between the symplectic leaves of the two Poisson manifolds is realized in the following way: if S_1 is a symplectic leaf of (P_1, Π_1), then $\pi_2(\pi_1^{-1}(S_1))$ is a symplectic leaf of (P_2, π_2), and, vice versa, if S_2 is a symplectic leaf of (P_2, Π_2), then $\pi_1(\pi_2^{-1}(S_2))$ is a symplectic leaf of (P_1, Π_1).

In particular, applying these observations to the full dual pairs described in Proposition 3.122, one can conclude that for all $\alpha \in \mathfrak{g}^*$

$$\pi(\mu^{-1}(\mathscr{O}_\alpha)) = \mu^{-1}(\mathscr{O}_\alpha)/G$$

is a symplectic leaf of $(M/G, \{\cdot, \cdot\}^\sim)$ and that all symplectic leaves of this Poisson manifold are obtained in this way (note that we suppose G to be connected). These observations lead to the following.

Theorem 3.124 (Reduction Along an Orbit) *Under the hypothesis made on G and on the G-action in Proposition 3.122, for all $\alpha \in \mathfrak{g}^*$, $\mu^{-1}(\mathscr{O}_\alpha)$ is a smooth G-manifold and the quotient space $M_{\mathscr{O}_\alpha} = \mu^{-1}(\mathscr{O}_\alpha)/G$ is smooth and it inherits a symplectic structure $\omega_{\mathscr{O}_\alpha}$ so that the symplectic manifold $(M_{\mathscr{O}_\alpha}, \omega_{\mathscr{O}_\alpha})$ is symplectomorphic to the Marsden-Weinstein reduced phase space $(M_\alpha, \omega_\alpha)$.*

We can now conclude this subsection with the following result.

Corollary 3.125 *Let (M, ω) be a symplectic manifold endowed with a free and proper strongly Hamiltonian G-action. Furthermore, assume that the canonical projection $\pi : M \to M/G$ is a submersive map and that the moment map $\mu : M \to \mathfrak{g}^*$ is surjective and with connected fibers. Then the symplectic foliation of the Poisson manifold $(M/G, \{\cdot, \cdot\}^\sim)$ has as symplectic leaves the reduced Marsden-Weinstein-Meyer spaces $(M_\alpha, \omega_\alpha)$.*

This result should clarify the comment made at the very beginning of this subsection: the reduction of the phase space obtained via the Marsden-Weinstein-Meyer prescription and the one obtained via the quotient of the original phase space (M, ω) by the G-action are related as explained in Corollary 3.125. Note that, under the hypothesis of Corollary 3.125, for every $\alpha \in \mathfrak{g}^*$ the reduced Hamiltonian vector field X_{H_α} of Theorem 3.86 is tangent to the symplectic leaf $M_\alpha \subset M/G$.

5.2 Complex and Quaternionic Geometry from the Marsden-Weinstein-Meyer Viewpoint

In this last subsection, we would like to present an extension of the Marsden-Weinstein-Meyer reduction to two important classes of Riemannian manifolds, called *Kähler* and *hyperKähler*, respectively. Because of the nature of this topic, we will necessarily refer the reader to a few references where the relevant background material can be found. Here we start just recalling a few fact about complex geometry. First of all, we recall that an *almost-complex structure* on manifold M is an endomorphism I of TM such that $I^2 = -\,\mathrm{id}$. The existence of such a structure guarantees that every fiber of TM can be endowed with a structure of a complex vector space. In fact, given I, for every $m \in M$ and $v \in T_mM$, one can define $\imath v = Iv$. The existence of such a structure is, for example, guaranteed on every *complex manifold*, i.e., a smooth manifold, locally homeomorphic to $\mathbb{C}^n$, for some n, which admits an atlas whose transition functions are holomorphic. The sufficient and necessary conditions an almost complex structure must satisfy to come from a holomorphic atlas are provided by the fundamental *Newlander-Nirenberg theorem* (after August Newlander and Louis Nirenberg), which states that an almost complex structure comes form a holomorphic atlas *if and only if* its Nijenhuis torsion T_I is trivial; see Sect. 2 in Chap. 5 for the definition of the Nijenhuis torsion. When this condition is satisfied, the almost complex structure I is said to be *integrable* and it is called a *complex structure*. Note that the integrability of an almost complex structure I is equivalent to the existence of a torsion-free linear connection ∇ such that $\nabla I = 0$; see [254, Theorem 3.1]. This condition, in turn, implies that the *holonomy group* of such a connection is contained in $\mathrm{GL}_n(\mathbb{C})$; see [133, 227]. A Riemannian metric g is called *Hermitian* with respect to the complex structure I if $g(IX, IY) = g(X, Y)$ for all $X, Y \in \mathfrak{X}(M)$. Note that in this case

$$g(IY, X) = g(I^2Y, IX) = -g(Y, IX) = -g(IX, Y), \quad \forall X, Y \in \mathfrak{X}(M).$$

In other words, $\omega : \mathfrak{X}(M) \times \mathfrak{X}(M) \to \mathbb{R}$ defined by

$$\omega(X, Y) = g(IX, Y), \quad \forall X, Y \in \mathfrak{X}(M), \tag{3.91}$$

is two-form on M. After these preliminary comments, one can introduce the following.

Definition 3.126 A complex manifold M is called *Kähler* if it admits an Hermitian metric g such that (3.91) is closed. In this case g is called a *Kähler metric* and the pair (g, I) a *Kähler structure*. $\triangle$

It is worth observing that the form h defined by

$$h(X, Y) = g(X, Y) - \imath\omega(X, Y), \forall X, Y \tag{3.92}$$

is Hermitian, i.e., $h(X, Y) = \overline{h(Y, X)}$ for all $X, Y \in \mathfrak{X}(M)$. Finally, note that the existence of a Kähler structure (g, I) amounts to the fact that $\nabla I = 0$, where ∇ is the *Levi-Civita connection* of the metric g, i.e., the unique linear torsion-free connection such that $\nabla g = 0$. This condition, in turn, is equivalent to the fact the holonomy of ∇ is contained in $U(n)$.

Example 3.127 As a very simple, nevertheless, important example consider the complex structure I on $\mathbb{R}^2$ defined as follows. If x_0, x_1 are the standard Euclidean coordinates on $\mathbb{R}^2$, write

$$I \frac{\partial}{\partial x_0} = \frac{\partial}{\partial x_1} \quad \text{and} \quad I \frac{\partial}{\partial x_1} = -\frac{\partial}{\partial x_0},$$

which identifies $(\mathbb{R}^2, I)$ with $\mathbb{C}$. In particular, $z = x_0 + \iota x_1$ is a I-holomorphic coordinate. In fact

$$I \frac{\partial}{\partial z} = \iota \frac{\partial}{\partial z},$$

where $\frac{\partial}{\partial z} = \frac{1}{2}\left(\frac{\partial}{\partial x_0} - \iota \frac{\partial}{\partial x_1} \right)$. Moreover, $h = \frac{1}{2} dz \otimes d\overline{z} = g - \iota\omega$, where

$$g = \frac{1}{2}(dx_0 \otimes dx_0 + dx_1 \otimes dx_1) \quad \text{and} \quad \omega = dx_0 \wedge dx_1;$$

see Formula (3.92). $\triangle$

The generalization of the previous example to $\mathbb{C}^n$ is straightforward and will be left to the reader as an exercise. There are many more examples of Kähler manifolds, for example, for every (n, k), the Grassmannian of the k-planes in $\mathbb{C}^n$ can be shown to possess a Kähler structure. In particular the complex projective space in every dimension is Kähler. More in general, every projective manifold is Kähler; see the references below. It is worth noting that the closed two-form in the definition of a Kähler manifold is nondegenerate. In other words, every Kähler manifold is symplectic. Moreover, as noted in the previous example, on every complex manifold one can define the notion of holomorphic vector field and, more in general, of holomorphic tensor field. In particular, the holomorphic differential forms are characterized as the kernel of the so-called *Del-Bar operator*, commonly denoted by $\overline{\partial}$ and defined, on every (almost) complex manifold, by the decomposition of the Cartan differential as $d = \partial + \overline{\partial}$. Every holomorphic k-form ξ admits the following (local) representation $\xi = \sum_{i_1 < \cdots < i_k} \xi_{i_1 \cdots i_k} dz_{i_1} \wedge \cdots \wedge dz_{i_k}$, where the $\xi_{i_1 \cdots i_k}$s are (locally defined) holomorphic functions and $z_1, \ldots, z_n$ are holomorphic coordinates. As it was recalled above, the integrability of an almost complex structure I is equivalent to the existence of a *torsion-free* linear connection ∇, such that $\nabla I = 0$. From this point of view, the compatibility between metric and complex structure on a Kähler manifold is equivalent to the statement that the complex structure is parallel with respect to the Levi-Civita connection of the underlying metric. For more information about complex and Kähler geometry we refer the reader to [121, 249], and [254]; see also [20]. Before closing this part, we would like to observe that there exists the Kähler analogue of the Marsden-Weinstein-Meyer reduction. In a few words, this goes as follows. Let (M, g, I) be a Kähler manifold and let G be a compact Lie group acting freely on M and preserving both the metric g and the complex structure I. Furthermore, suppose that the G-action is strongly Hamiltonian with respect to the symplectic form (3.91). Then for every G-invariant $\alpha \in \mathfrak{g}^*$, $\mu^{-1}(\alpha)/G$ is smooth and it inherits a unique Kähler structure $(\hat{g}, \hat{I})$ such that $\pi_\alpha^* \hat{\omega} = j_\alpha^* \omega$ and $\hat{g}$ is the quotient metric induced by $j_\alpha^* g$, where $j_\alpha : \mu^{-1}(\alpha) \to M$ and $\pi_\alpha : \mu^{-1}(\alpha) \to \mu^{-1}(\alpha)/G$ are as in Theorem 3.81. A proof of this result can be found in [119]. Before moving on with our presentation, it is worth making a comment about this last result. In many concrete cases, the application of the Marsden-Weinstein-Meyer reduction in the Kähler setting has a purely complex counterpart. More precisely, recall that every compact Lie group G can be embedded in its *complexification* $G_{\mathbb{C}}$,, which is the *unique* complex Lie group whose Lie algebra is the complexification of $\mathfrak{g}$, i.e., whose Lie algebra is $\mathfrak{g}_{\mathbb{C}} = \mathfrak{g} \otimes \mathbb{C}$ with the Lie bracket defined by $[x \otimes z, y \otimes w]_{\mathbb{C}} = [x, y] \otimes zw$ for all $x, y \in \mathfrak{g}$ and $z, w \in \mathbb{C}$, where $\otimes$ is over the real numbers. Note that G embeds in $G_{\mathbb{C}}$ as a maximal compact Lie subgroup, since G is assumed compact.

Example 3.128 For example, if $G = \mathrm{U}(n)$, then $G_{\mathbb{C}} = \mathrm{GL}_n(\mathbb{C})$.   △

Assuming that the G-action φ extends to a *holomorphic* action of $G_{\mathbb{C}}$, i.e., that there exists $\varphi^{\mathbb{C}} : G_{\mathbb{C}} \times M \to M$ such that (i) $\varphi_g^{\mathbb{C}} = \varphi_g$ for all $g \in G$ and (ii) $\varphi_g^{\mathbb{C}} : M \to M$ is a holomorphic map for all $g \in G_{\mathbb{C}}$, one can consider the quotient $M/G_{\mathbb{C}}$. Since $G_{\mathbb{C}}$ is not compact, in general this quotient is not a well-behaved topological space. On the other hand, if one restricts the $G_{\mathbb{C}}$-action to the *open* subset $\tilde{M}$ of the points of M whose $G_{\mathbb{C}}$-orbits intersect (nontrivially) the level set $\mu^{-1}(\alpha)$, then it possible to show that there is a set theoretic bijection between $\mu^{-1}(\alpha)/G$ and $\tilde{M}/G_{\mathbb{C}}$.

Remark 3.129 The points of the open subset $\tilde{M}$ are called *stable*; see (Fig. 3.4) below and, for more information, see [119, 234].   △

Entering in more details of this beautiful theory will drive us too far from the main course of these notes and for this reason we stop here our presentation referring the reader to the lecture notes [234] and to the references therein.

Example 3.130 In Example 3.83 it was shown that the complex projective space can be obtained as a symplectic quotient choosing $M = \mathbb{C}^n$ with its standard symplectic form (3.47), which is just the symplectic form of the Kähler structure defined by the standard Euclidean metric on $\mathbb{R}^{2n}$ and by the complex structure induced by the multiplication by ι; see Example 3.127 above. Recall that the moment map defined by the S^1-action is $\mu(z_1, \ldots, z_n) = -\sum_{i=1}^n |z_i|^2$ and that the quotient $\mu^{-1}(-1)/\mathrm{S}^1$ can be identified with the $n-1$-dimensional complex projective space; see again Example 3.83. On the other hand, note that the complex-

Fig. 3.4 Orbits and stable orbits

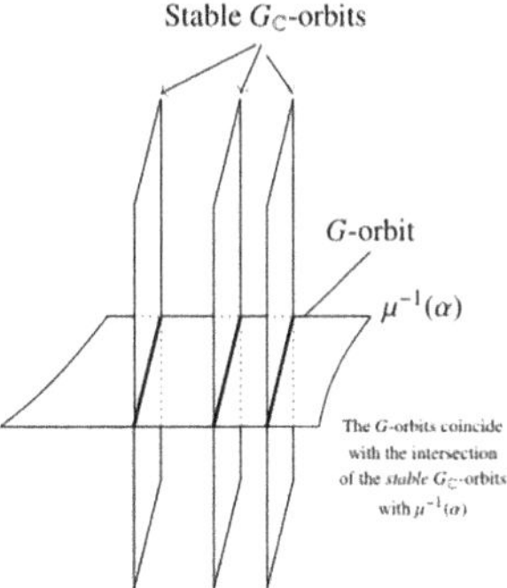

ification of $G = S^1$ is $G_\mathbb{C} = \mathbb{C}^*$ (see Example 3.128), and that (3.48) extends naturally to the holomorphic action of $\mathbb{C}^*$ on $\mathbb{C}^n$ defined by $w.(z_1, \ldots, z_n) = (wz_1, \ldots, wz_n)$, for all $w \in \mathbb{C}^*$ and $(z_1, \ldots, z_n) \in \mathbb{C}^n$. Since the $G_\mathbb{C}$-orbit of $(z_1, \ldots, z_n) \neq (0, \ldots, 0)$ is the line joining this point to the origin of $\mathbb{C}^n$, in this example $\tilde{M} = \mathbb{C}^n \setminus \{(0, \ldots, 0)\}$. In in this way we recover the well known fact that $\mathbb{P}^{n-1}(\mathbb{C}) = \frac{\mathbb{C}^n \setminus \{(0,\ldots,0)\}}{\mathbb{C}^*}$. $\triangle$

In the next subsection we will introduce a *quaternionic* analogue of a Kähler structure and, at the same time, we will discuss the result which extends the Marsden-Weinstein-Meyer reduction to the corresponding class of manifolds.

Remark 3.131 In this chapter we presented a few cases where the interaction between complex and symplectic geometry played an important role; see Examples 3.57, 3.59, and 3.83 and also Problem 3.84. $\triangle$

5.2.1 HyperKähler Manifolds and their Reduction

The quaternionic analogue of a Kähler structure is what is commonly known in the literature as a hyperKähler structure.

Definition 3.132 A hypeKähler structure on a manifold M is tuple (g, I, J) of a Riemannian metric g and two complex structures I, J such that:

(i) g is a Kähler metric with respect to both I and J (see Definition 3.126).
(ii) $IJ = -JI$.

A hyperKähler manifold is a manifold endowed with a hyperKähler structure. $\triangle$

Note that since $I^2 = -\,\mathrm{id} = J^2$, if $K = IJ$, then $K^2 = -\,\mathrm{id}$ as well. Moreover, since g is Kähler with respect to both I and J, $\nabla I = 0 = \nabla J$ where ∇ denotes the Levi-Civita connection of g. This implies that $\nabla K = 0$, i.e., that K is a complex structure and that g is a Kähler metric with respect to it. In this way, from one hand the tangent bundle of a hyperKähler manifold carries an action of the algebra of the quaternions $\mathbb{H}$ (see Example 3.134 below), defined by the three complex structures I, J, K. On the other hand, every hyperKähler manifold is endowed with three symplectic structures, ω_I, ω_J, and ω_K, defined by

$$\omega_I(X, Y) = g(IX, Y), \quad \omega_J(X, Y) = g(JX, Y), \quad \text{and} \quad \omega_K(X, Y)$$
$$= g(KX, Y), \ \forall X, Y \in \mathfrak{X}(M).$$

Observe that if $(a, b, c) \in \mathbb{R}^3$ is a vector of norm equal to 1, then $I_{a,b,c} = aI + bJ + cK$ is also a complex structure compatible with the underlying metric g. In fact

$$I^2_{(a,b,c)} = -(a^2 + b^2 + c^2)\, \mathrm{id} + ab\cancel{(IJ + JI)} + ac\cancel{(IK + KI)} + bc\cancel{(JK + KJ)}$$
$$= -\,\mathrm{id},$$

and $\nabla I_{a,b,c} = 0$. In other words, to every hyperKähler manifold is associated a family of complex structures parametrized by S^2. The existence of this S^2-family of complex structures is encoded in the so-called *twistor space* $(Z, \mathcal{I})$ of the hyperKähler manifold M, where $Z = M \times S^2$ and $\mathcal{I}$ is the almost complex structure defined by $\mathcal{I}_{(m,z)} := I_z \oplus I_0$ where I_0 is standard complex structure of $\mathbb{P}^1(\mathbb{C}) \simeq S^2$ while $I_z = I_{(a,b,c)}$. This almost complex structure turns out to be integrable and it can be shown that the hyperKähler structure of M is completely encoded in the $2m + 1$-dimensional complex manifold $(Z, \mathcal{I})$; see [118, 119]. From the complex geometry viewpoint it is worth noting that

Lemma 3.133 *The complex valued two-form $\Omega = \omega_J + \iota\omega_K$ is holomorphic with the respect to the complex structure I.*

Proof First, one shows that locally $\Omega = \sum_{i<j} \Omega_{ij} dz_i \wedge dz_j$. Since $d\Omega = 0$, one has that

$$0 = d\Omega = \partial\Omega + \bar{\partial}\Omega = \sum_{i<j}\left(\sum_{k=1}^{n} \frac{\partial\Omega_{ij}}{\partial z_k} dz_k\right) \wedge \wedge dz_i \wedge dz_j$$
$$+ \sum_{i<j}\left(\sum_{k=1}^{n} \frac{\partial\Omega_{ij}}{\partial\bar{z}_k} d\bar{z}_k\right) \wedge \wedge dz_i \wedge dz_j,$$

which yields

$$\bar{\partial}\Omega = \sum_{i<j}\left(\sum_{k=1}^{n} \frac{\partial\Omega_{ij}}{\partial\bar{z}_k} d\bar{z}_k\right) \wedge \wedge dz_i \wedge dz_j = 0.$$

Finally, since $\Omega^n \neq 0$, Ω is a holomorphic symplectic form on M, with respect to I.

$\square$

Example 3.134 (Flat HyperKähler Structure on $\mathbb{R}^4$) We now present a simple, though important, example of hyperKähler manifold. To this end recall that the algebra of quaternions $\mathbb{H}$ is an associative, non-commutative algebra generated by $1, \iota, j, k$ with the relations

$$\iota^2 = j^2 = k^2 = -1, \quad \iota j = k = -j\iota, \quad jk = \iota = -kj \quad k\iota = j = -\iota k,$$

1 being central. The quaternionic conjugation, $\bar{\ } : \mathbb{H} \to \mathbb{H}$ defined by $\overline{(a_0 + \iota a_1 + j a_2 + k a_3)} = a_0 - \iota a_1 - j a_2 - k a_3$, defines the quaternionic Hermitian inner product $\langle q_1, q_2 \rangle = q_1 \bar{q}_2$, whose corresponding norm is $|q|^2 = \langle q, q \rangle$, for all $q \in \mathbb{H}$. The obvious identification of $\mathbb{H}$ with $\mathbb{R}^4$ endows the latter with three complex structures, I, J, and K defined by the *left* multiplication of ι, j, and k on $\mathbb{H}$. More precisely, if x_0, x_1, x_2, and x_3 are the standard Euclidean coordinates on $\mathbb{R}^4$,

$$I\frac{\partial}{\partial x_0} = \frac{\partial}{\partial x_1}, \quad I\frac{\partial}{\partial x_1} = -\frac{\partial}{\partial x_0}, \quad I\frac{\partial}{\partial x_2} = \frac{\partial}{\partial x_3} \quad \text{and} \quad I\frac{\partial}{\partial x_3} = -\frac{\partial}{\partial x_2},$$
$$\tag{3.93}$$

$$J\frac{\partial}{\partial x_0} = \frac{\partial}{\partial x_2}, \quad J\frac{\partial}{\partial x_1} = -\frac{\partial}{\partial x_3}, \quad J\frac{\partial}{\partial x_2} = -\frac{\partial}{\partial x_0} \quad \text{and} \quad J\frac{\partial}{\partial x_3} = \frac{\partial}{\partial x_1},$$

$$K\frac{\partial}{\partial x_0} = \frac{\partial}{\partial x_3}, \quad K\frac{\partial}{\partial x_1} = \frac{\partial}{\partial x_2}, \quad K\frac{\partial}{\partial x_2} = -\frac{\partial}{\partial x_1} \quad \text{and} \quad K\frac{\partial}{\partial x_3} = -\frac{\partial}{\partial x_0}.$$

Note that I, J are complex structures on $\mathbb{R}^4$ since they square to minus the identity and they are constant. Observe also that, for the same reasons, $K = IJ$ is a complex structure as well. Moreover, if $q = x_0 + \iota x_1 + j x_2 + k x_3, dq = dx_0 + \iota dx_1 + j dx_2 + k dx_3$ and

$$h = \frac{1}{2} dq \otimes \bar{d}q = g - i\omega_I - j\omega_J - k\omega_K, \tag{3.94}$$

where

$$g = \frac{1}{2} \sum_{i=1}^{3} dx_i \otimes dx_i,$$

$$\omega_I = dx_0 \wedge dx_1 + dx_2 \wedge dx_3, \tag{3.95}$$

$$\omega_J = dx_0 \wedge dx_2 + dx_3 \wedge dx_1, \tag{3.96}$$

$$\omega_K = dx_0 \wedge dx_3 + dx_1 \wedge dx_2. \tag{3.97}$$

A simple computation shows that $\omega_I(X, Y) = g(IX, Y)$, $\omega_J(X, Y) = g(JX, Y)$, and $\omega_K(X, Y) = g(KX, Y)$ for all X, Y vector fields on $\mathbb{R}^4$. In other words, the (g, I, J) is a hyperKähler structure on $\mathbb{R}^4$. Note also that $q = x_0 + \iota x_1 + j x_2 + k x_3 =$

$x_0 + \iota x_1 + (x_2 + \iota x_3)j$ which implies that $z = x_0 + \iota x_1$ and $w = x_3 + \iota x_2$ are holomorphic coordinates with respect to I. In fact if $\frac{\partial}{\partial z} = \frac{1}{2}\left(\frac{\partial}{\partial x_0} - \iota \frac{\partial}{\partial x_1}\right)$ and $\frac{\partial}{\partial w} = \frac{1}{2}\left(\frac{\partial}{\partial x_2} - \iota \frac{\partial}{\partial x_3}\right)$, then (3.95) yields

$$I\frac{\partial}{\partial z} = \iota \frac{\partial}{\partial z} \quad \text{and} \quad I\frac{\partial}{\partial w} = \iota \frac{\partial}{\partial w}.$$

This observation, together with (3.96) and (3.97), entails that $\Omega = \omega_K + \iota \omega_J$ is a I-holomorphic symplectic form on $\mathbb{R}^4$, which, if written in the coordinates (z, w) introduced above, assumes the following form $\Omega = dz \wedge dw$. Note that using these coordinates, (3.94) can be written as

$$h = g - \iota \omega_I - \Omega j, \tag{3.98}$$

which should be compared with (3.92). $\qquad\qquad\qquad\qquad\qquad\qquad\qquad\qquad\triangle$

The generalization of the previous example to $\mathbb{H}^n$ is straightforward and it will be left to the reader as an exercise. On the other end, the following example will be discussed in detail; see [196].

Example 3.135 Let n, s be two integers greater or equal to 1 and let $\text{Mat}_{n,s}(\mathbb{C})$, $\text{Mat}_n(\mathbb{C})$, be the vector space of the $n \times s$ and of the $n \times n$ complex matrices, respectively. On these two vector spaces the following nondegenerate Hermitian product is defined:

$$\langle A, B \rangle = \text{tr}(A B^{\dagger}), \tag{3.99}$$

where $X^{\dagger}$ is the transposed conjugate of X, i.e., $X^{\dagger} = \overline{X}^t$. The $\mathbb{R}$-linear maps I, J defined on

$$V_{n,s} = \text{Mat}_n(\mathbb{C}) \oplus \text{Mat}_n(\mathbb{C}) \oplus \text{Mat}_{n,s}(\mathbb{C}) \oplus \text{Mat}_{s,n}(\mathbb{C})$$

by

$$I(X, Z, v, w) = (\iota X, \iota Z, \iota v, \iota w) \quad \text{and} \quad J(X, Z, v, w)) = (Z^{\dagger}, -X^{\dagger}, w^{\dagger}, -v^{\dagger}) \tag{3.100}$$

satisfy $I^2 = -\text{id} = J^2$ and $IJ = -JI$ and for this reason they define on $V_{n,s}$ two complex structures satisfying the quaternionic commutation relations. Moreover, if $g : V_{n,s} \times V_{n,s} \to \mathbb{R}$ is the real part of the Hermitian inner product (3.99), i.e.,

$$g(x, y) = \frac{1}{2}\text{tr}\left(X_1 X_2^{\dagger} + X_2 X_1^{\dagger} + Z_1 Z_2^{\dagger} + Z_2 Z_1^{\dagger} + v_1 v_2^{\dagger} + v_2 v_1^{\dagger} + w_1 w_2^{\dagger} + w_2 w_1^{\dagger}\right), \tag{3.101}$$

for all $x = (X_1, Z_1, v_1, w_1)$ and $y = (X_2, Z_2, v_2, w_2)$ in $V_{n,s}$, then one can prove by a direct computation that $g(Ix, Iy) = g(x, y)$, $g(Jx, Jy) = g(x, y)$ for all x, y which implies that (g, I, J) is a flat hyperKähler structure on $V_{n,s}$. Another simple computation shows that

$$\omega_I(x, y) = g(Ix, y) = \frac{\iota}{2} \mathrm{tr} \left(X_1 X_2^\dagger - X_2 X_1^\dagger + Z_1 Z_2^\dagger - Z_2 Z_1^\dagger + v_1 v_2^\dagger - v_2 v_1^\dagger \right.$$
$$\left. + w_1 w_2^\dagger - w_2 w_1^\dagger \right),$$

$$\omega_J(x, y) = g(Jx, y) = \frac{1}{2} \mathrm{tr} \left(Z_1^\dagger X_2^\dagger + X_2 Z_1 - X_1^\dagger Z_2^\dagger - Z_2 X_1 + w_1^\dagger v_2^\dagger \right.$$
$$\left. + v_2 w_1 - v_1^\dagger w_2^\dagger - w_2 v_1 \right),$$

which, defining $K = IJ$ as usual, yields

$$\omega_K(x, y) = g(Kx, y) = \frac{\iota}{2} \mathrm{tr} \left(Z_1^\dagger X_2^\dagger - X_2 Z_1 - X_1^\dagger Z_2^\dagger + Z_2 X_1 \right.$$
$$\left. - v_2 w_1 + w_1^\dagger v_2^\dagger - v_1^\dagger w_2^\dagger + w_2 v_1 \right).$$

In particular

$$\omega_I = \frac{\iota}{2} \mathrm{tr}(dX \wedge dX^\dagger + dZ \wedge dZ^\dagger + dv \wedge dv^\dagger - dw^\dagger \wedge dw) \tag{3.102}$$

and, if one defines $\Omega = \omega_J + \iota \omega_K$, i.e.,

$$\Omega(x, y) = \mathrm{tr}(X_2 Z_1 - Z_2 X_1 + v_2 w_1 - w_2 v_1), \ \forall x, y \in V_{n,s},$$

then

$$\Omega = \mathrm{tr}(dZ \wedge dX + dw \wedge dv). \tag{3.103}$$

Note that this is a I-holomorphic symplectic form. $\triangle$

From the viewpoint of the Riemannian holonomy, a manifold admits a hyperKähler structure if and only if its holonomy is contained in the group $\mathrm{Sp}_{\mathbb{H}}(n)$, the so called *quaternionic unitary group*, where

$$\mathrm{Sp}_{\mathbb{H}}(n) = \{g \in \mathfrak{gl}_n(\mathbb{H}) \mid gg^\dagger = \mathrm{id}\},$$

where, in this case, $g^\dagger = \overline{g^t}$ and $\overline{}$ is the quaternionic conjugation. Note that $\mathrm{Sp}_{\mathbb{H}}(n) = U(2n) \cap \mathrm{Sp}_{\mathbb{C}}(2n)$, where $\mathrm{Sp}_{\mathbb{C}}(2n)$ is the Lie group preserving the canonical holomorphic symplectic form of $\mathbb{C}^{2n}$. The lowest-dimensional case is described in the following

Example 3.136 ($\mathrm{Sp}_{\mathbb{H}}(1)$) This is the case of the one-by-one quaternionic matrices such that $gg^{\dagger} = 1$, i.e., it is the set of all quaternions q such that $q\bar{q} = 1$. If $q = a_0 + \imath a_1 + j a_2 + k a_3 \in \mathrm{Sp}_{\mathbb{H}}(1)$, then $a_0^2 + a_1^2 + a_2^2 + a_3^2 = 1$, which shows that there is a (set theoretic) one-to-one correspondence between $\mathrm{Sp}_{\mathbb{H}}(1)$ and $S^3 \subset \mathbb{R}^4$. This set-theoretic correspondence can be easily shown to be a group isomorphism. In particular one has that $\mathrm{Sp}_{\mathbb{H}}(1) \simeq \mathrm{SU}(2)$. $\triangle$

Let us now discuss the Marsden-Weinstein-Meyer reduction in the hyperKähler context. To this end, let G be a compact Lie group acting on M and preserving the metric and the complex structures I and J. This suffices to conclude that this G-action preserves the symplectic forms ω_I, ω_J and ω_K. Suppose that the action of G is strongly Hamiltonian with respect to each of the three symplectic structure and let μ_I, μ_J, and μ_k the corresponding moment maps; see 3.27. Let us write

$$\mu : M \to \mathfrak{g}^* \otimes \mathbb{R}^3,$$

to denote the hyperKähler moment map, i.e., $\mu = (\mu_I, \mu_J, \mu_K)$. In this case one can prove that

Theorem 3.137 *If $\alpha \in \mathfrak{g}^* \otimes \mathbb{R}^3$ is invariant with respect the (component wise) coadjoint action of G and if G acts freely on $\mu^{-1}(\alpha) \subset M$ then the quotient $\mu^{-1}(\alpha)/G$ inherits from M a structure of a hyperKähler manifold of real dimension $\dim M - 4\dim G$.*

This fundamental result was proven in [119] to which we refer the reader for its proof. Here we would like to stress the multifaceted nature of the quotient described in the previous theorem. As carefully explained in [119], the hyperKähler structure on the quotient is obtained from (g, I, J) (the one defined on M) choosing one of the complex structure, say I, and noticing that

(i) $\mu_{\mathbb{C}} : (M, I) \to \mathfrak{g}^* \otimes \mathbb{C}$, defined by $\mu_J + \imath \mu_K$, is a holomorphic map. This implies that, writing $\alpha = (\alpha_{\mathbb{R}}, \alpha_{\mathbb{C}})$, $\mu_{\mathbb{C}}^{-1}(\alpha_{\mathbb{C}})$ is a Kähler manifold, being a complex submanifold of the Kähler manifold (M, g, I).
(ii) $\mu_{\mathbb{C}}^{-1}(\alpha_{\mathbb{C}})$ is G-invariant and $\mu_{\mathbb{R}} = \mu_I$ restricts there to a moment map for the G action.
(iii) Under these assumptions one can perform the *Kähler reduction* (see above), to get a Kähler structure on $\mu_{\mathbb{R}}^{-1}(\alpha_{\mathbb{R}}) \cap \mu_{\mathbb{C}}^{-1}(\alpha_{\mathbb{C}})/G = \mu^{-1}(\alpha)/G$.
(iv) Finally, switching the role of the complex structures, one can conclude that the quotient $\mu^{-1}(\alpha)/G$ is endowed with three complex structures, each of them compatible with the reduced metric, which provides the hyperKähler structure on $\mu^{-1}(\alpha)/G$.

Before we move on with our presentation, it is worth making the following observation. Suppose that the G-action on M extends to an holomorphic action of the complexification $G_{\mathbb{C}}$; see Example 3.128. Under this assumption the latter preserves the holomorphic symplectic form Ω and $\mu_{\mathbb{C}}$ turns out to be a corresponding moment map for this symplectic action. Restricting this holomorphic action to the open subset $\tilde{M}$ of its stable points, one can prove the following bijection:

$$\mu^{-1}(\alpha)/G \simeq \mu_{\mathbb{C}}^{-1}(\alpha_{\mathbb{C}})/G_{\mathbb{C}},$$

which implies that the hyperKähler quotient described in Theorem 3.137 is a symplectic quotient but in the *holomorphic* setting.

In the remaining part of this section we present a few examples of hyperKähler moment maps. The first one is a simple, nevertheless important, case which should be compared with Example 3.29.

Example 3.138 (A Quadratic HyperKähler Moment Map) Let $V = \mathbb{H}^n$ endowed with its structure of flat hyperKähler manifold; see Example 3.134. We write $\vec{q}$ to denote the generic element of V, thought as row vector. The group $\mathrm{Sp}_{\mathbb{H}}(V)$ acts on V by right multiplication, i.e.,

$$g.\vec{q} = \vec{q}\,g^{-1}$$

and this action preserves the hyperKähler structure of V. In fact, it commutes with the action of the complex structures I and J defined by the left multiplication with the unit quaternions $\imath$ and $\jmath$, respectively. Moreover, writing the Hermitian quaternionic product as $\langle \vec{q}_1, \vec{q}_2 \rangle_{\mathbb{H}} = \vec{q}_1 \vec{q}_2^{\dagger}$, where $\vec{q}^{\dagger}$ is the *transposed quaternionic conjugate* of $\vec{q}$, one has

$$\langle g.\vec{q}_1, g.\vec{q}_2 \rangle_{\mathbb{H}} = \langle \vec{q}_1 g^{-1}, \vec{q}_2 g^{-1} \rangle_{\mathbb{H}} = \vec{q}_1 g^{-1} (\vec{q}_2 g^{-1})^{\dagger} \overset{g^{\dagger} = g^{-1}}{=} \vec{q}_1 g^{-1} g \vec{q}_2^{\dagger} = \langle \vec{q}_1, \vec{q}_2 \rangle_{\mathbb{H}}, \tag{3.104}$$

for all $\vec{q}_1, \vec{q}_2 \in V$ and $g \in \mathrm{Sp}_{\mathbb{H}}(V)$. Note that (3.104) implies that the $\mathrm{Sp}_{\mathbb{H}}(V)$-action preserves g, ω_I, ω_J and ω_K (see Formula (3.94)), and that, in particular, it is a symplectic action with respect to both ω_I and the holomorphic symplectic form Ω; see Formula (3.98). The explicit formula for the hyperKähler moment map $\mu : V \to \mathfrak{sp}_{\mathbb{H}}(V)^* \otimes \mathbb{R}^3$ is

$$\langle \mu(\vec{q}), \xi \rangle = \frac{1}{2} \vec{q} \xi \vec{q}^{\dagger}, \tag{3.105}$$

for all $\xi \in \mathfrak{sp}_{\mathbb{H}}(V)$ and $\vec{q} \in V$, where $\mathfrak{sp}_{\mathbb{H}}(V)$, the Lie algebra of $\mathrm{Sp}_{\mathbb{H}}(V)$, consists of all $\xi \in \mathfrak{gl}_{\mathbb{H}}(V)$ such that $\xi + \xi^{\dagger} = 0$. To prove that (3.105) is a moment map for the $\mathrm{Sp}_{\mathbb{H}}(V)$-action, first, observe that

$$(\vec{q} \xi \vec{q}^{\dagger})^{\dagger} = \vec{q} \xi^{\dagger} \vec{q}^{\dagger} \overset{\xi^{\dagger} = -\xi}{=} -\vec{q} \xi \vec{q}^{\dagger},$$

which implies that $\vec{q} \xi \vec{q}^{\dagger} \in Im(\mathbb{H})$, i.e., it is an *imaginary quaternion*. »This observation, together with the standard identification $Im(\mathbb{H}) \simeq \mathbb{R}^3$, implies that the map μ defined via (3.105) takes values in $\mathfrak{sp}_{\mathbb{H}}(V)^* \otimes \mathbb{R}^3$, as it required. Moreover,

$$\langle \mu(g.\vec{q}), \xi \rangle = \frac{1}{2}(g.\vec{q})\xi(g.\vec{q})^{\dagger} = \frac{1}{2}\vec{q}g^{-1}\xi g\vec{q}^{\dagger} = \langle \mu(\vec{q}), \mathrm{Ad}_{g^{-1}}\xi \rangle = \langle \mathrm{Ad}_g^{\sharp} \mu(\vec{q}), \xi \rangle,$$

for all $\xi \in \mathfrak{sp}_{\mathbb{H}}(V)$, $g \in \mathrm{Sp}_{\mathbb{H}}(V)$ and $\vec{q} \in V$, entailing the $\mathrm{Sp}_{\mathbb{H}}(V)$-equivariance of μ. To conclude that (3.105) defines a moment map for the $\mathrm{Sp}_{\mathbb{H}}(V)$ it suffices to prove that μ_I, μ_J and μ_K, defined by

$$\mu_I(\vec{q}) = -\frac{1}{2}\omega_I(\vec{q}X, \vec{q}), \quad \mu_I(\vec{q}) \quad = -\frac{1}{2}\omega_J(\vec{q}X, \vec{q}) \quad \text{and} \quad \mu_K(\vec{q})$$

$$= -\frac{1}{2}\omega_K(\vec{q}X, \vec{q}),$$

are the moment maps for the symplectic forms ω_I, ω_J and ω_K, defined as in (3.94). We leave the details of this computation to the reader, observing that it is completely analogous to the one made in Example 3.29. $\triangle$

Example 3.139 (Eguchi-Hanson Metric) This is a simple but geometrically interesting example which was first discussed in [77]. Let $V = \mathbb{C}^n$ and $T^*V = \mathbb{C}^n \times \mathbb{C}^n$, where the identification of $\mathbb{C}^n$ with its dual is obtained via the Hermitian form (3.92). On the other hand, as already mentioned in Example 3.134, $\mathbb{C}^n \times \mathbb{C}^n$ can be identified with $\mathbb{H}^n$ choosing, for example, the complex structure I defined by the left multiplication by $\imath$, and writing $\vec{q} = \vec{x}_0 + \imath\vec{x}_1 + j\vec{x}_2 + k\vec{x}_3$ as $\vec{q} = \vec{z} + \vec{w}j$, where $\vec{z} = \vec{x}_0 + \imath\vec{x}_1$ and $\vec{w} = \vec{x}_2 + \imath\vec{x}_3$. In this way, one can think of $\mathbb{H}^n$ as the cotangent bundle of $\mathbb{C}^n$. Note that under the above identifications $\Omega = d\vec{w} \wedge d\vec{z} = \sum_{l=1}^{n} dw_l \wedge dz_l$ is the canonical (holomorphic) symplectic form of T^*V and $\omega = \frac{i}{2}\sum_{l=1}^{n}(dz_k \wedge d\bar{z}_k + dw_k \wedge d\bar{w}_k)$ is *minus* the symplectic form ω_I; see (3.95). The S^1-action $e^{\imath\theta}.\vec{z} = e^{\imath\theta}\vec{z}$ on V can be lifted to an S^1-action on T^*V, which, using the above identification, assumes the following form

$$e^{\imath\theta}(\vec{z}, \vec{w}) = (e^{\imath\theta}\vec{z}, e^{-\imath\theta}\vec{w}), \forall\theta \in \mathbb{R}, (\vec{z}, \vec{w}) \in T^*V. \tag{3.106}$$

The S^1-action (3.106) preserves the hyperKähler structure of $\mathbb{H}^n \simeq T^*V$ and it induces on T^*V a hyperKähler moment map. This, accordingly with the identification of $\mathbb{H}^n \simeq \mathbb{C}^n \times \mathbb{C}^n$ made above, splits as $(\mu_{\mathbb{R}}, \mu_{\mathbb{C}})$, where $\mu : T^*V \to \mathrm{Lie}(\mathrm{S}^1) \simeq \imath\mathbb{R}$ and $\mu_{\mathbb{C}} : T^*V \to \mathbb{C}$, where $\mathbb{C} = \mathrm{Lie}(\mathbb{C}^*)$. The latter is I-holomorphic (see (i) below Theorem 3.137), and it can be easily computed from general principles (see Appendix D):

$$\mu_{\mathbb{C}}(\vec{z}, \vec{w}) = -\sum_{l=1}^{n} z_l w_l, \forall(\vec{z}, \vec{w}) \in T^*V.$$

To compute the former observe that, for every $\xi \in \mathrm{Lie}(\mathrm{S}^1)$, the corresponding fundamental vector field, written in the complex coordinates $(z, \overline{z}, w, \overline{w})$, is

$$X_\xi(\vec{z}, \vec{w}) = (-\xi\vec{z}, \xi\vec{w}) = \xi\left(-\sum_{l=1}^n z_l \frac{\partial}{\partial z_l} + \sum_{l=1}^n \overline{z}_l \frac{\partial}{\partial \overline{z}_l} + \sum_{l=1}^n w_l \frac{\partial}{\partial w_l} - \sum_{l=1}^n \overline{w}_l \frac{\partial}{\partial \overline{w}_l}\right).$$

Using this coordinate representation

$$i_{X_\xi}\omega(z, w) = \frac{\iota\xi}{2}d\left(-\sum_{l=1}^l |z_l|^2 + \sum_{l=1}^n |w_l|^2\right),$$

which yields

$$\mu(z, w) = \frac{\iota}{2}\sum_{l=1}^n(|z_l|^2 - |w_l|^2) = \frac{\iota}{2}(||\vec{z}||^2 - ||\vec{w}||^2), \ \forall (\vec{z}, \vec{w}) \in T^*V.$$

In this way, the hyperKähler quotient at the level $\alpha = (\frac{\iota}{2}, 0, 0) \in \iota\mathbb{R} \otimes \mathbb{R}^3$ is

$$\left\{(\vec{z}, \vec{w}) \in T^*V \mid ||\vec{z}||^2 - ||\vec{w}||^2 = 1, \ \sum_{l=1}^n z_l w_l = 0\right\}/\mathrm{S}^1. \tag{3.107}$$

On the other hand, (3.107) can be easily identified with

$$\{(\vec{z}, \vec{w}) \in T^*V \mid \vec{z} \neq 0, \ \sum_{l=1}^n z_l w_l = 0\}/\mathbb{C}^*, \tag{3.108}$$

as one can deduce from the isomorphism $\mathbb{C}^* \simeq \mathbb{R}_{>0} \times \mathrm{S}^1$ and from the observation that for every pair $(\vec{z}, \vec{w}) \in T^*V$ such that $\vec{z} \neq 0$ and $\sum_{l=1}^n z_l w_l = 0$, there exists one and only one $t \in \mathbb{R}_{>0}$ such that $t^2||\vec{z}||^2 - t^{-2}||\vec{w}||^2 = 1$. To get a more geometrical intuition about (3.108), observe that the fiber of $T^*\mathbb{P}(V)$ at a point $[\ell] \in \mathbb{P}(V)$ can be identified with $\ell \otimes \ell^\circ$, where $\ell \subset V$ is a line (through the origin) and $\ell^\circ \in V^*$ is its annihilator, i.e., $\ell^\circ = \{\beta \in V^* \mid \beta(v) = 0 \, \forall v \in \ell\}$. From this observation it follows that the hyperKähler quotient described in this example it the cotangent bundle of $\mathbb{P}(\mathbb{C}^n) = \mathbb{P}^{n-1}(\mathbb{C})$. In particular if $n = 2$ one recovers the so-called *Eguchi-Hanson* hyperKähler structure of the cotangent bundle of $\mathbb{P}^1(\mathbb{C})$, discovered in [77] and then generalized to higher-dimensional projective spaces by Calabi in [45]. $\triangle$

Now we write the explicit form of the (components of the) hyperKähler moment map for a $\mathrm{U}(n)$-action defined on the flat hyperKähler manifold of Example 3.135.

Example 3.140 (Hilbert Schemes of Points on $\mathbb{C}^2$) We start observing that the manifold $V_{n,s}$ introduced in Example 3.135 carries the following (natural) $GL_n(\mathbb{C})$-action

$$g.(X, Z, v, w) = (gXg^{-1}, gZg^{-1}, gv, wg^{-1}), \ \forall (X, Z, v, w) \in V_{n,s}, \ g \in GL_n(\mathbb{C}).$$
$$(3.109)$$

The latter corresponds to the *cotangent lift* of the $GL_n(\mathbb{C})$-action on $\mathrm{Mat}_n(\mathbb{C}) \oplus \mathrm{Mat}_{n,s}(\mathbb{C})$ defined by

$$g.(X, v) = (gXg^{-1}, gv), \ \forall (X, v) \in \mathrm{Mat}_n(\mathbb{C}) \oplus \mathrm{Mat}_{n,s}(\mathbb{C}), \qquad (3.110)$$

after identifying, via the trace form $(A, B) = -\mathrm{tr}(AB)$, $\mathrm{Mat}_n(\mathbb{C})^*$ with $\mathrm{Mat}_n(\mathbb{C})$ and $\mathrm{Mat}_{n,s}(\mathbb{C})^*$, respectively, with $\mathrm{Mat}_{s,n}(\mathbb{C})$. It is worth observing that the restriction of (3.109) to $U(n) \subset GL_n(\mathbb{C})$ preserves the hyperKähler structure of $V_{n,s}$, in particular it preserves ω_I, and Ω; see Formulas (3.102) and (3.103). To compute a hyperKähler moment map defined by this action, one can fix the complex structure I and compute separately $\mu_{\mathbb{R}}$ and $\mu_{\mathbb{C}}$; see the discussion below Theorem 3.137. We start with the moment map $\mu_{\mathbb{C}}$, the one defined by the action of the complex Lie group $GL_n(\mathbb{C})$ on $V_{n,s}$. Since this action is obtained by cotangent lift and Ω is, modulo the identifications already mentioned, the canonical holomorphic form on $T^*V_{n,s}$, $\mu_{\mathbb{C}}$ can be computed from first principles; see Appendix D. More precisely, the fundamental vector field defined on $\mathrm{Mat}_n(\mathbb{C}) \oplus \mathrm{Mat}_{n,s}(\mathbb{C})$ by $\xi \in \mathfrak{gl}_n(\mathbb{C})$ is

$$X_\xi(X, v) = ([X, \xi], -\xi v), \ \forall (X, v) \in \mathrm{Mat}_n(\mathbb{C}) \oplus \mathrm{Mat}_{n,s}(\mathbb{C})$$

(see (3.110)), which entails

$$F_\xi(X, Z, v, w) = -\mathrm{tr}([X, \xi]Z) + \mathrm{tr}(\xi v w) = \mathrm{tr}\xi([X, Z] + vw);$$

see Sect. 1 of this chapter. In other words, $\mu_{\mathbb{C}} : V_{n,s} \to \mathfrak{gl}_n(\mathbb{C})$ is defined by

$$\mu_{\mathbb{C}}(X, Z, v, w) = [X, Z] + vw, \ \forall (X, Z, v, w) \in V_{n,s}. \qquad (3.111)$$

To compute $\mu_{\mathbb{R}}$ one needs to consider (3.109) with $g \in U(n)$ and observe that ω_I in (3.102) is the sum of four contributions: two of them represent the canonical symplectic form on the vector space $\mathrm{Mat}_n(\mathbb{C})$ and the other two are the canonical symplectic form on $\mathbb{C}^n$, seen both as the space of *column* and of *row* n-vectors. Since $U(n)$ acts diagonally via (3.109) on the four components of $V_{n,s}$, the explicit form of $\mu_{\mathbb{R}}$ follows from Examples 3.57 and 3.59; see also Remark 3.58. More precisely, $\mu_{\mathbb{R}} : V_{n,s} \to \mathfrak{u}_n$ is defined by

$$\mu_{\mathbb{R}}(X, Z, v, w) = \frac{\imath}{2}([X, X^\dagger] + [Z, Z^\dagger] + vv^\dagger - w^\dagger w), \ \forall (X, Z, v, w) \in V_{n,s}.$$
$$(3.112)$$

Note that both in (3.111) and (3.112) the Lie algebra involved is identified with its dual via the trace form $(A, B) = -\text{tr}(AB)$. We close this subsection with the following remark. The target space of the above hyperKähler moment map $\mu = (\mu_{\mathbb{R}}, \mu_{\mathbb{C}})$ is $\mathfrak{u}_n^* \otimes \mathbb{R}^3 = \mathfrak{u}_n \otimes (\mathbb{R} \oplus \mathbb{C}) \simeq \mathfrak{u}_n \oplus \mathfrak{gl}_n(\mathbb{C})$. The isomorphism follows from the following facts: first, $\mathfrak{u}_n$ is a real Lie algebra; second, the tensor product is over $\mathbb{R}$, and third, the complexification of $\mathfrak{u}_n$ is $\mathfrak{gl}_n(\mathbb{C})$. If I, J are the complex structures on $V_{n,s}$ defined in (3.100) and $K = IJ$, the map $\varphi : V_{n,s} \to V_{n,s}$, defined by

$$\varphi = \frac{I - K}{\sqrt{2}},$$

preserves the metric (3.101), the sphere of the complex structures of $V_{n,s}$ and it commutes with the $U(n)$-action. Moreover, a simple computation shows that it maps bijectively $\mu_{\mathbb{R}}^{-1}(0) \cap \mu_{\mathbb{C}}(-\text{id})$ onto $\mu_{\mathbb{R}}^{-1}(\iota\,\text{id}) \cap \mu_{\mathbb{C}}(0)$, which can be shown to be smooth submanifolds of $V_{n,s}$ where $U(n)$ acts freely; see, for example, [192, 252]. Now, we would like to mention that it is possible to prove that

$$\frac{\mu_{\mathbb{R}}^{-1}(0) \cap \mu_{\mathbb{C}}(-\text{id})}{U(n)} \simeq \frac{\mu_{\mathbb{R}}^{-1}(\iota\,\text{id}) \cap \mu_{\mathbb{C}}(0)}{U(n)} \tag{3.113}$$

as *hyperKähler* manifolds but they are *not* isomorphic as *complex* manifolds. In other words, the different values chosen to perform the hyperKähler quotient lead to the *same* hyperKähler manifold, but select on it two *non-isomorphic* structures of complex manifolds. We mentioned this example because of its relevance in the theory of *moduli spaces*; see [192]. The case $s = 1$ is very important and will be further discussed in Sect. 5.2. For the time being we only mention that in this case the left-hand side of (3.113) is the (completion of the) phase space of the so-called *rational Calogero-Moser system* (see Chap. 8), while the right-hand side is the so-called *Hilbert scheme* of points on $\mathbb{C}^2$; see [192] for a full-fledged and very accessible account of this beautiful theory. $\triangle$

Remark 3.141 HyperKähler geometry is a very interesting subject which sits at the intersection of (pseudo)Riemannian, symplectic and complex, or better, quaternionic, geometry. Its first appearance can be dated back to the midfifties of the last century, when the unitary quaternionic groups $\text{Sp}_{\mathbb{H}}(n)$, $n \geq 1$, popped up in the Berger's list of the possible holonomy groups of (nonsymmetric and irreducible) Riemannian metrics on simply connected manifolds; see [29] and also [126, 127]. As remarked in the latter two references, the discovery of Riemannian manifolds with unitary quaternionic holonomy happened only many years later the publication of [29]. In particular, the proof of the existence of compact hyperKähler manifolds was obtained as a consequence of the Yau proof of the so-called *Calabi conjecture*; see [126, 127]. In real dimension four, among the examples of hyperKähler manifolds, one finds the so-called *ALE-spaces* (asymptotically locally Euclidean spaces), also known as *gravitational instantons*, of which the Eguchi-

Hanson metric is one instance; see Example 3.139. These hyperKähler structures are defined on the suitable desingularization of the quotients $\mathbb{H}/G$, where G is a finite subgroup of $\mathrm{Sp}_{\mathbb{H}}(1) = \mathrm{SU}(2)$, a so-called platonic group. On the other hand, a class of remarkable examples of (real) four-dimensional compact hyperKähler manifolds is provided by the so-called $K3$-surfaces. These are compact, simply connected two-dimensional complex surfaces, whose holomorphic cotangent bundle is trivial. One example of such a surface is the so-called Fermat quartic, the hypersurface in $\mathbb{P}^3(\mathbb{C})$ defined as the zero-locus of the homogeneous polynomial $z_0^4+z_1^4+z_2^4+z_3^4$. Compact higher-dimensional, i.e., of real dimension $4n$ with $n > 1$, hyperKähler manifolds were constructed by Beauville in [25]; see [121, 122] for nice summaries. It is worth noting that hyperKähler geometry plays an important role in theoretical physics and that many moduli spaces turn out to carry a hyperKähler structure. One example of this phenomenon was mentioned in Example 3.140, where the relevant moduli space was the Hilbert scheme of point on $\mathbb{C}^2$. $\triangle$

6   Bibliographical Notes

Hamiltonian actions and moment maps are crucial ingredients both in symplectic geometry and geometric mechanics. The main result of this chapter, Theorem 3.81, in the geometric framework exposed here, was first proven in [176]. A similar result, in a less geometric form was proved in [182]. There, more specifically, the author proved that if a Hamiltonian system with n degrees of freedom admits p independent integrals *not necessarily in involution*, the system can be reduced to the integration of a Hamiltonian system of $n - p$ degrees of freedom with p parameters and additional quadratures.

Theorem 3.81 underwent many generalizations encompassing topics far beyond the realm of symplectic geometry. For a very extensive and in-depth analysis of these results we refer the reader to the monograph [205] and to the references therein. Other suggested readings are the classical [1, 114, 158] and [175], where the reader will find many applications of the reduction procedure(s) to physical problems. Another very nice reference to be consulted is [20], where the reader will find many nice examples of reduced spaces both of finite and of infinite dimension. For more informations about the cotangent bundle reduction, a topic we sketched in Sect. 4.1, we refer the reader to the monograph [173]. As we mentioned above, the Marsden-Weinstein-Meyer Theorem found many applications and it underwent to many generalizations, one of these summarized in Sect. 5.2.1, where it is sketched the so-called hyperKähler reduction theorem together with few of its applications. Our main reference for this construction was the original paper [119]. Other references about general hyperKähler geometry are the monographs [126, 127] and the references therein, among those we would like to mention the classical review paper [118]. Finally, we would like to mention two more references where the symplectic reduction is presented at a more leisurely pace. The first is [221], a nice introduction to the Marsden-Weinstein-Meyer theorem and to its applications to mechanics. The second is the review paper [44], where the author gives a nice overview of the applications of the symplectic reduction from a viewpoint of a philosopher of physics.

Chapter 4
Lagrangian Fibrations and Integrable Systems

In this chapter we introduce and study in some detail the notion of *Lagrangian fibration with smooth fibers* (Lagrangian fibration from now on) which is the differential geometrical counterpart of the notion of classical integrable system introduced in Chap. 1. A particular emphasis will be given to the case of fibrations with compact, connected, and smooth fibers. For these fibrations, we prove the existence of the so-called *action-angle* coordinates, symplectic coordinates defined on a suitable tubular neighborhood of each fiber. The construction of these coordinates will be presented in the case of a few concrete examples of completely integrable systems and the problem of their global existence will be addressed at the end of the chapter, where we will briefly address also the case of *singular* Lagrangian fibrations, i.e., a generalization of Lagrangian fibration which admits singular fibers.

1 Lagrangian Fibrations, Action-Angle Variables, and Integrable Systems

In this section, first, we introduce the notion of Lagrangian fibration and then we discuss its relation with the theory of the integrable systems. It will be shown that every Lagrangian fibration $\pi : (M, \omega) \to B$ with compact and connected fibers carries a fiber-wise, transitive action of T^*B. This action entails the existence of a Lagrangian covering Λ of B, called the period lattice of the Lagrangian fibration which, for every choice of a Lagrangian local section σ of π over $U \subset B$, defines a local (anti-) simplectomorphism $(T^*U, \Omega_{|T^*U}) \simeq (\pi^{-1}(U), \omega_{|_{\pi^{-1}(U)}})$. This local identification will be a crucial ingredient in our presentation of the construction of the action-angle variables, which are symplectic coordinates defined on suitable tubular neighborhood of every smooth fiber of π. This section will close with the

A. Arsie, I. Mencattini, *Geometry of Integrable Systems*, Latin American Mathematics Series – UFSCar subseries,
https://doi.org/10.1007/978-3-031-96282-0_4

presentation of the so-called *Arnol'd formula*, which, in principle, allows to compute
a set of action-angle variables for the given Lagrangian fibration.

We begin with the following:

Definition 4.1 A Lagrangian fibration is a locally trivial bundle $\pi : M \to B$,
whose fibers are Lagrangian submanifolds of (M, ω).   △

Remark 4.2 Unless differently stated, the Lagrangian fibrations considered hereafter will have connected fibers.   △

One of the main goals of this chapter is to present a proof of Theorem (4.29),
which can be divided conceptually in two parts. In the first one, which is divided
in several steps, one shows that action-angle coordinates can be defined for a
Lagrangian fibration. This is the content of Sect. 1.1. In the second part, it is
shown that an integrable system (M, ω, H) on a symplectic manifold, under suitable
conditions, gives rise to a Lagrangian fibration (possibly with singular fibers) and
away from the singular fibers the results of the first part can be applied. This is
the content of Sect. 1.2. Before proceeding with Sect. 1.1, we detail the various
steps that are necessary to arrive to the existence of the action-angle variables for a
Lagrangian fibration:

- Step 1: for each $b \in B$ there is a transitive action of $T_b^* B$ (viewed as the abelian
  Lie group $(\mathbb{R}^n, +)$) on M_b.
- Step 2: for each $b \in B$ there exists a discrete subgroup $\Lambda_b \subset T_b^* B$ such that
  $M_b \cong T_b^* B / \Lambda_b$. The period lattice is defined as $\Lambda = \cup_{b \in B} \Lambda_b$.
- Step 3: the period lattice Λ is an embedded Lagrangian submanifold of $T_b^* B$
  which is also a smooth covering of B.
- Step 4: since Λ is Lagrangian and a smooth covering, any of its sheets can be
  described via a system of closed 1-forms on B, $\alpha_1, \ldots, \alpha_n$. Restricting them to
  a suitable open set $U \subset B$, $\alpha_i = dI_i$ for $i = 1, \ldots n$. On $T^* U$ we have induced
  Darboux coordinates $(\phi_1, \ldots, \phi_n, I_1, \ldots, I_n)$.
- Step 5: using a suitable section of the Lagrangian fibration over U, one can
  construct a diffeomorphism $\pi^{-1}(U) \cong T^* U$ such that in the induced coordinates
  by $(\phi_1, \ldots, \phi_n, I_1, \ldots, I_n)$ on $\pi^{-1}(U)$ the symplectic form $\omega_{|\pi^{-1}(U)}$ is written
  in action-angle coordinates.

1.1   Existence of Action-Angle Coordinates on a Lagrangian Fibration

Before embarking in the proof of the main result of this subsection, following the
steps enumerated above, we collect few preliminaries.

Let (M, ω) be a symplectic manifold of dimension $2n$ and let $\pi : M \to B$
be a Lagrangian fibration. Let $b \in B$ and $(U, x_1, \ldots, x_n)$ be a local system of

coordinates around b. Without loss of generality we can assume that the system of coordinates is centered at b. Let $f_i = \pi^* x_i$, for all $i = 1, \ldots, n$. Then:

Lemma 4.3 *The functions $(f_1, \ldots, f_n)$ generate a maximal commutative subalgebra of*

$$\left(C_M^\infty(\pi^{-1}(U)), \{\cdot, \cdot\} \right),$$

where $\{\cdot, \cdot\}$ is the restriction of the Poisson bracket defined by ω to the open subset $\pi^{-1}(U)$.

Proof First, observe that since π is a submersion, π^* is injective, which proves that $f_1, \ldots, f_n$ are independent. On the other hand, for all $m \in M_b = \pi^{-1}(b)$ and $i = 1, \ldots, n$, $df_i(m) \in (T_m M_b)^\circ$, the annihilator of $T_m M_b$ in $T_m^* M$. Since M_b is Lagrangian, $(T_m M_b)^\circ \simeq T_m M_b$. More precisely, for each $m \in M_b$, this isomorphism is defined by $\omega_m : T_m M_b \to (T_m M_b)^\circ$, via

$$\omega_m(X_{f_i}(m)) = df_i(m),$$

where $X_{f_i}(m) \in T_m M_b$ is the value at m of the Hamiltonian vector field X_{f_i}. Given f_i, f_j,

$$0 = \langle df_i(m), X_{f_j}(m) \rangle = \langle dx_i(b), \pi_{*,m} X_{f_j}(m) \rangle,$$

for all $m \in M_b$ and for all $b \in U$, which implies that $\{f_i, f_j\} = 0$ for all $i, j = 1, \ldots, n$. Therefore, the functions $f_1, \ldots, f_n$ Poisson commute, and their differentials generate the annihilator of the tangent space at each point of each of the fiber of π. This concludes the proof since the fibers of π are Lagrangian submanifolds of M. The maximality follows from the fact that if there were an additional function, say g, Poisson commuting with the $\{f_i\}_{i=1,\ldots,n}$, g would be functional dependent on the f_is, in particular the differential dg would have to be a linear combination of the differentials df_i, $i = 1, \ldots, n$. $\qquad\square$

Remark 4.4 If instead of considering a Lagrangian fibration, we were considering an *isotropic* (not Lagrangian) fibration $\pi : M \to B$ on a symplectic manifold M of dimension $2n$, then necessarily the dimension of B would be greater than n and therefore the corresponding commutative subalgebra would be generated by more than n functions in involution. $\qquad\triangle$

Given the fibration $\pi : M \to B$, recall that $X \in \mathfrak{X}(M)$ is called *vertical* if $\pi_{*,m} X_m = 0$ for all $m \in M$. This condition is equivalent, say that X is *tangent* to the fibers of π. With this in mind, in the next remark we collect a few consequences of the previous lemma.

Remark 4.5 Let $U \subset B$ be an open subset. Then

(i) For each $f \in C_B^\infty(U)$, $X_{\pi^* f} \in \mathfrak{X}_M(\pi^{-1}(U))$ is a vertical vector field. In fact, for every coordinate function x_i on U

$$
\begin{aligned}
\langle (dx_i)_{\pi(m)}, \pi_{*,m}(X_{\pi^* f}(m)) \rangle &= \langle (\pi^* dx_i)_m, X_{\pi^* f}(m) \rangle \\
&= -\langle (\pi^* df)_m, X_{\pi^* x_i}(m) \rangle \\
&= -\langle (df)_{\pi(m)}, \pi_{*,m} X_{\pi^* x_i}(m) \rangle \\
&= 0
\end{aligned}
$$

since every vector field $X_{\pi^* x_i}$ is vertical; see Lemma 4.3.

(ii) Since $\pi^* : \Omega_B^1(U) \to \Omega_M^1(\pi^{-1}(U))$ is *injective*, each $\beta \in \Omega_B^1(U)$ corresponds to a unique vector field $X_\beta = X_{\pi^* \beta} \in \mathfrak{X}_M(\pi^{-1}(U))$, defined by $i_{X_\beta}\omega = \pi^* \beta$. Moreover,

$$
[X_{\beta_1}, X_{\beta_2}] = 0, \ \forall \beta_1, \beta_2 \in \Omega_B^1(U). \tag{4.1}
$$

To prove this identity it suffices to check it in a coordinate neighborhood $(U, x_1, \ldots, x_n)$, where we can write $\beta_l = \sum_{i=1}^n \beta_{l,i} dx_i$, $l = 1, 2$, $\beta_{l,i} \in C_B^\infty(U)$ for all $i = 1, \ldots, n$ and $l = 1, 2$. To prove that $[X_{\beta_1}, X_{\beta_2}] = 0$, it suffices to observe that $[X_{f_i}, X_{f_j}] = 0$ for all $i, j = 1, \ldots, n$ (here $f_i = \pi^* x_i$) and that $X_{f_i}(\pi^* \beta_{l,j}) = 0$ for all $i, j = 1, \ldots, n$ and $l = 1, 2$, since the X_{f_i}s are *vertical* vector fields. In particular, for each $\beta \in \Omega_B^1(U)$, X_β is a vertical vector field.

△

The next step we undertake is to show that, for each $b \in B$, the fiber M_b of Lagrangian fibration $\pi : M \to B$ carries an *infinitesimal action* of the abelian Lie algebra $T_b^* B$; see Sect. 1.3 in Appendix C. To this end, let $b \in U \subset B$, where U is an open subset.

Lemma 4.6 *If $\beta, \beta' \in \Omega_B(U)$ are such that $\beta(b) = \beta'(b)$, then $X_\beta(m) = X_{\beta'}(m)$ for all $m \in M_b$. In other words, given $\beta \in \Omega_B(U)$, for each $b \in U$, the restriction of the vector field X_β to M_b depends only on $\beta(b) \in T_b^* B$.*

Proof Since $i_{X_\beta}\omega = \pi^* \beta$, one has $i_{X_\beta(m)}\omega_m = \pi^* \beta(m)$ for all $m \in \pi^{-1}(U)$. The proof of the lemma follows now observing that, for each m, $\pi^* \beta(m) = \beta(\pi(m)) = \beta(b)$ (if $m \in M_b$) and from the hypothesis that $\beta(b) = \beta'(b)$. $\qquad\square$

Remark 4.7 This lemma implies that, for each $b \in B$ and each $\xi \in T_b^* B$, one can define a vector field $X_\xi \in \mathfrak{X}_M(M_b)$ via the following procedure. Choose any one-form β (locally defined on a neighborhood of b) such that $\beta(b) = \xi$. Then, define $X_\xi \in \mathfrak{X}_M(M_b)$ as the restriction to M_b of the Hamiltonian vector field X_β (this is possible since X_β is vertical). △

1.1.1 Step 1

After these preliminary observations, one can prove the following.

Proposition 4.8 (Infinitesimal Action) *Each fiber M_b supports an infinitesimal action of $T_b^* B$, equipped the trivial Lie algebra structure ($T_b^* B$ is seen as a commutative Lie algebra).*

Proof Let $U \subset B$ be an open subset containing b. Let $\xi_1, \xi_2 \in T_b^* B$ and let $\beta_1, \beta_2 \in \Omega_B^1(U)$ be such that $\beta_1(b) = \xi_1$ and $\beta_2(b) = \xi_2$. Then the proof of the proposition follows from Formula (4.1) and from Lemma 4.6 (see also Remark 4.7), observing that the map $\xi \rightsquigarrow X_\xi$, from $T_b^* B$ to $\mathfrak{X}_M(M_b)$ is linear, since $\beta \rightsquigarrow X_\beta$ is a linear map from $\Omega_B^1(U)$ to $\mathfrak{X}_M(\pi^{-1}U)$). $\qquad\square$

We now present a sufficient condition to *integrate* the infinitesimal action described in the previous proposition. More precisely,

Lemma 4.9 *If for all $\xi \in T_b^* B$, X_ξ is a complete vector field, then the infinitesimal action of $T_b^* B$ on M_b, defined by $\xi \rightsquigarrow X_\xi$, induces on M_b an action of the abelian Lie group $(T_b^* B, +)$.*

Proof For any $\xi \in T_b^* B$ let $\{\varphi_t^\xi\}_{t \in \mathbb{R}}$ the one-parameter group of diffeomorphisms (of M_b) defined by X_ξ. Let β_1, β_2 be two one-forms extending $\xi_1, \xi_2 \in T_b^* B$ on an open neighborhood U of b and let X_{β_1}, X_{β_2} be the corresponding vector fields; see Remark 4.5. Since X_{β_1}, X_{β_2} commute (see Formula (4.1)), their restrictions X_{ξ_1}, X_{ξ_2} to M_b will commute as well and for this reason

$$\varphi_1^{\xi_1} \circ \varphi_1^{\xi_2} = \varphi_1^{\xi_1 + \xi_2},$$

where both left- and right-hand sides are defined, since X_{ξ_1}, X_{ξ_2}, and $X_{\xi_1 + \xi_2}$ are complete by hypothesis. Here we used two standard results: the flows of two vector fields commute iff the two vector fields have zero Lie bracket (see Proposition 1.46 and also [10, Section 39.E]), and if two vector fields commutes, the flow of their sum is the composition of the corresponding flows; see, for instance, [2, Supplement 4.1.C]. This entails that $\varphi_1 : T_b^* B \to \mathscr{D}(M_b)$ is a group homomorphism defining an action of $(T_b^* B, +)$ on M_b as required by the lemma. $\qquad\square$

Remark 4.10 It is worth observing that under the completeness hypothesis, given $\beta \in \Omega_B^1(U)$ such that $\xi = \beta(b)$, for all $t \in \mathbb{R}$, $\varphi_t^\beta : \pi^{-1}(U) \to \pi^{-1}(U)$ is a fiber-wise diffeomorphism extending the diffeomorphism $\varphi_t^\xi : M_b \to M_b$ to $\pi^{-1}(U)$. In particular one can write

$$\varphi_t^\beta(b') := \varphi_t^{\beta(b')} \ \forall b' \in U,$$

and $\varphi_t^{\beta(b)} = \varphi_t^\xi$. $\qquad\triangle$

Before moving on, recall that if X is a complete vector field, then $\varphi_t^X = \varphi_1^{tX}$ for all $t \in \mathbb{R}$. In particular, if $\xi \in T_b^* B$ is such that $X_\xi \in \mathfrak{X}_M(M_b)$ is complete, then

$$\varphi_1^{t\xi}(m) = \varphi_t^\xi(m), \ \forall t \in \mathbb{R}. \tag{4.2}$$

Now suppose that for each $b \in U$ the action

$$\varphi_1^\cdot : T_b^* B \times M_b \to M_b, \qquad \xi \rightsquigarrow \varphi_1^\xi \tag{4.3}$$

is defined (see Lemma 4.9), and let $(U, x_1, \ldots, x_n)$ be a coordinate neighborhood of b. Then, using Formula (4.2), one can write

$$\varphi_1^\xi = \varphi_1^{t_1 dx_1(b)} \circ \cdots \circ \varphi_1^{t_n dx_n(b)} = \varphi_{t_1}^{dx_1(b)} \circ \cdots \circ \varphi_{t_n}^{dx_n(b)},$$

for each $b \in B$ and for every $\xi \in T_b^* B$, where $\xi = \sum_{i=1}^n t_i dx_i(b)$ for suitable $(t_1, \ldots, t_n) \in \mathbb{R}^n$. Since $\varphi_1^\cdot : T_b^* B \to \mathscr{D}(M_b)$ is a group homomorphism, we can state the following proposition, whose proof is left to the reader.

Proposition 4.11 *Given a coordinate neighborhood* $(U, x_1, \ldots, x_n)$ *of* $b \in B$, *(4.3) induces a smooth action of* $(\mathbb{R}^n, +)$ *on* M_b:

$$\tau : \mathbb{R}^n \times M_b \xrightarrow{((t_1,\ldots,t_n),m) \rightsquigarrow \varphi_{t_1}^{dx^1(b)} \circ \cdots \circ \varphi_{t_n}^{dx^n(b)}(m)} M_b.$$

In particular, for each $m \in M_b$, $\tau_m : \mathbb{R}^n \to M_b$ *defined by*

$$\tau_m : \mathbb{R}^n \xrightarrow{(t_1,\ldots,t_n) \rightsquigarrow \tau((t_1,\ldots,t_n),m)} M_b \tag{4.4}$$

is a smooth map.

We can now prove the following important result.

Lemma 4.12 *Given* $b \in B$ *as above, for each* $m \in M_b$ *the differential at* $(0, \ldots, 0) \in \mathbb{R}^n$ *of (4.4) is a linear isomorphism between* $\mathbb{R}^n$ *and* $T_m M_b$.

Proof The choice of a system of local coordinates around b defines an isomorphism between $T_b^* B$ and $\mathbb{R}^n$. The differential of the map τ_m at the point $(0, \ldots, 0) \in \mathbb{R}^n$ is a linear map $(\tau_m)_{*,(0,\ldots,0)} : \mathbb{R}^n \to T_m M_b$, which to every $\xi \in T_b^* B \simeq \mathbb{R}^n$ associates the value of the vector field X_ξ at the point m. This map is injective and because of dimensional reason it is an isomorphism. $\qquad\square$

Lemma 4.12, together with Lemma C.52, yields the following.

Proposition 4.13 *For every* $b \in B$, *the action of* $T_b^* B$ *on* M_b *defined in (4.3) is* transitive.

1.1.2 Step 2

From Sect. 1.4 in Appendix C and, in particular, from Lemma C.52 and Lemma C.54, we immediately obtain that

Corollary 4.14 *For each $b \in B$ there exists a discrete subgroup $\Lambda_b \subset T_b^* B$ such that $M_b \simeq T_b^* B / \Lambda_b$.*

In particular, if $\pi : M \to B$ is a Lagrangian fibration with *complete* fibers, $\pi^{-1}(b) \simeq \mathbb{R}^{n-k} \times \mathbb{T}^k$ for every $b \in B$; see Proposition C.58. Furthermore, since we assume that the fibration is locally trivial, one can prove the following.

Proposition 4.15 *If (M, ω) is a symplectic manifold and $\pi : M \to B$ is a Lagrangian fibration whose fibers are complete, then for each $b \in B$, there exists an open neighborhood $b \in U \subset B$ such that $\pi^{-1}(U) \simeq U \times \mathbb{R}^{n-k} \times \mathbb{T}^k$, where $0 \le k \le n$.*

Remark 4.16 Since the Lagrangian fibrations we consider are locally trivial, the rank of Λ_b is locally constant (as function of b). Here we are mainly interested in the two opposite cases, $k = n$ and $k = 0$, i.e., the first case is when the fibers are diffeomorphic to n-dimensional *tori*, while the second one concerns the case when each fiber is diffeomorphic to a n-dimensional *affine space*. $\triangle$

Suppose now that $k = n$, i.e., that for every $b \in B$, Λ_b is isomorphic (noncanonically) to $\mathbb{Z}^n$. Let us introduce now a concept which plays a very important role in the classification of the Lagrangian fibrations with compact and connected fibers.

Definition 4.17 (Period Lattice) The *period lattice* of the Lagrangian fibration $\pi : M \to B$ is $\Lambda = \cup_{b \in B} \Lambda_b$. $\triangle$

Remark 4.18 Note that the period lattice can be associated to every Lagrangian fibration whose fibers are complete. In particular to every Lagrangian fibration with compact fibers. Henceforth, we always assume that the Lagrangian fibrations have complete fibers. $\triangle$

Before moving on with our discussion, we define a map between T^*B and M which will play a major role in what has to come. Let $p : T^*B \to B$ be the canonical projection, $U \subset B$ an open subset, and $\sigma : U \to M$ a local section of $\pi : M \to B$ supported in U. We define the map $\chi_\sigma : p^{-1}(U) \to \pi^{-1}(U)$ by

$$\chi_\sigma(\xi) = \varphi_1^\xi\big(\sigma(p(\xi))\big), \ \forall \xi \in p^{-1}(U). \tag{4.5}$$

Note that χ_σ is smooth and

$$\chi_\sigma(\xi) = \sigma\big(p(\xi)\big), \ \forall \xi \in \Lambda(U) = \cup_{b \in U} \Lambda_b.$$

1.1.3 Step 3

In Lemma 4.19 and Proposition 4.21 we prove two important properties of the period lattice. The first one states that the Λ is (locally) generated by local sections of T^*B, while the second claims that Λ is a Lagrangian submanifold of (T^*B, Ω).

Lemma 4.19 *Let $\pi : M \to B$ be a Lagrangian fibration and let Λ be its period lattice. Then, given any $b_0 \in B$ and $\xi_0 \in \Lambda_{b_0}$, there exist an open neighborhood $V \subset B$ containing b_0 and a one-form $\alpha \in \Omega_B^1(V)$ such that $\alpha(b_0) = \xi_0$ and $\alpha(b) \in \Lambda_b$, for all $b \in U$.*

Proof Let $U \subset M$ be an open neighborhood of $b_0 \in B$ and let $\sigma : U \to M$ be a local section of $\pi : M \to B$. Using Formula (4.5) one can define $\Psi_\sigma : p^{-1}(U) \to \pi^{-1}(U)$ as

$$\Psi_\sigma(\xi) = \chi_\sigma(\xi) - \sigma\big(p(\xi)\big),$$

for all $\xi \in T_b^*U$, where the right-hand side of the previous equation is defined in terms of the affine structure of the fiber M_b, i.e., $M_b \simeq T_b^*B/\Lambda_b$. Note that for every $\xi_0 \in \Lambda_b$

$$\Psi_\sigma(\xi_0) = 0.$$

Since φ_1^ξ is a diffeomorphism and σ is a local section of $\pi : M \to B$, the ξ-derivative of Ψ_σ at ξ_0 is different from zero. Therefore, by the *implicit function theorem*, there exists an open neighborhood V of $b_0 = p(\xi_0)$, an open neighborhood $W \subset p^{-1}(U)$ of ξ_0, and a smooth map $\alpha : V \to W$ such that $p \circ \alpha = \mathrm{id}_V$,

$$\alpha(b_0) = \xi_0 \quad \text{and} \quad \Psi_\sigma\big(\alpha(b)\big) = 0, \ \forall b \in V,$$

i.e.,

$$\varphi_1^{\alpha(b)}(\sigma(b)) = \sigma(b), \ \forall b \in V.$$

Since α is smooth, $\alpha \in \Omega_B^1(V)$ as stated. $\qquad\square$

This lemma implies that for any $b \in B$, if $\xi_1, \ldots, \xi_n$ is a system of generators for Λ_b, one can find a suitable open neighborhood V of b and $\alpha_1, \ldots, \alpha_n \in \Omega_B^1(V)$ such that for each $b' \in B$, $\alpha_1(b'), \ldots, \alpha_n(b')$ is a system of generators for $\Lambda_{b'}$, and $\alpha_i(b) = \xi_i$, for all $i = 1, \ldots, n$. This observation entails the following.

Corollary 4.20 *The period lattice Λ of the Lagrangian fibration $\pi : M \to B$ is an embedded submanifold of T^*B. In particular it is a smooth covering of the base manifold B.*

Proof In fact, every point $\xi \in \Lambda$ has a neighborhood in Λ which can be obtained as the graph of a smooth map defined on a neighborhood of $b = p(\xi)$ in B with values in a neighborhood of ξ in T^*B. $\qquad\square$

Finally, we are ready to prove the following important result.

Proposition 4.21 *The period lattice is a Lagrangian submanifold of T^*B.*

Proof It suffices to show that the one-forms $\alpha_1, \ldots, \alpha_n$, locally trivializing the period lattice Λ, are closed (see Proposition 1.84). This follows from the following computation:

$$
\begin{aligned}
(\varphi_1^\alpha)^*\omega - \omega &= \int_0^1 \frac{d}{dt}(\varphi_t^\alpha)^*\omega\, dt = \int_0^1 (\varphi_t^\alpha)^* \mathscr{L}_{X_{\pi^*\alpha}} \omega\, dt \\
&= \int_0^1 (\varphi_t^\alpha)^* d(i_{X_{\pi^*\alpha}}\omega)\, dt = \int_0^1 (\varphi_t^\alpha)^* d(\pi^*\alpha)\, dt \\
&= \int_0^1 d(\pi \circ \varphi_t^\alpha)^*\alpha\, dt = \pi^*(d\alpha),
\end{aligned}
$$

since $\pi \circ \varphi_t^\alpha = \pi$ for all t. Then, since π^* is injective, one deduces that φ_1^α is a symplectomorphism if and only if α is a closed one-form. On the other hand, if α is a one-form locally defining Λ, one knows that $\varphi_1^\alpha = \mathrm{id}$. $\qquad\square$

1.1.4 Step 4

Let $\alpha_1, \ldots, \alpha_n$ be a basis of local sections of Λ. Since the one-forms α_is are closed, one can define (locally) $I_1, \ldots, I_n$ such that

$$
\alpha_i = dI_i, \ \forall i. \tag{4.6}
$$

Moreover, since the α_is are independent, the I_is define a set of local coordinates on B.

Let $(\phi_1, \ldots, \phi_n, I_1, \ldots, I_n)$ be the corresponding Darboux coordinates on T^*U, obtained from the Is eventually shrinking their domain of definition. In particular

$$
\Omega|_{T^*U} = \sum_{k=1}^n d\phi_k \wedge dI_k.
$$

1.1.5 Step 5

The following theorem implies the existence of *special* Darboux coordinates on M, which are induced by the Darboux coordinates (ϕ, I) introduced above. Let (M, ω)

be a symplectic manifold of dimension $2n$ and let $\pi : M \to B$ be a Lagrangian fibration with compact (and connected) fibers.

Theorem 4.22 *Let $q_1, \ldots, q_n$ be a set of local coordinates on $U \subset B$ and let $(p_1, \ldots, p_n, q_1, \ldots, q_n)$ be the corresponding set of Darboux coordinates on T^*U. If $\sigma : U \to M$ is a local section of $\pi : M \to B$, then*

$$\chi_\sigma^*(\omega) = \sum_{i=1}^{n} dq_i \wedge dp_i + \sum_{i,j=1}^{n} A_{ij} dq_i \wedge dq_j, \tag{4.7}$$

where A_{ij} are functions depending only on the q_is, which are zero if σ is Lagrangian.

Proof Since χ_σ is a fiber map (see (4.5)), the restriction of $\chi_\sigma^*(\omega)$ to the fibers of T^*U is identically zero. This implies that

$$\chi_\sigma^*(\omega) = \sum_{i=1}^{n} dp_i \wedge \theta_i + \sum_{i,j=1}^{n} A_{ij} dq_i \wedge dq_j, \tag{4.8}$$

for some one-forms θ_is, independent of the dp_is, and some functions A_{ij}. Furthermore, since χ_σ is a local diffeomorphism, it is locally invertible. Hereafter, to simplify the notation, we write χ, dropping the reference to σ, and χ^{-1} to denote its inverse without specifying its domain; see the diagram below.

$$\begin{array}{ccc} T^*U & \underset{\chi^{-1}}{\overset{\chi}{\rightleftarrows}} & \pi^{-1}(U) \\ & \searrow{\scriptstyle p} \quad \swarrow{\scriptstyle \sigma} \nearrow & \\ & U \subset B & \end{array} \qquad \pi \tag{4.9}$$

Now one notices that

$$p^* dq_i = (\pi \circ \chi)^* dq_i = \chi^*(\pi^* dq_i) = -\chi^*(i_{X_i}\omega), \ \forall i = 1, \ldots, n,$$

where X_i denotes the Hamiltonian vector field $X_{\pi^* dq_i}$. The usual identification between $p^* dq_i$ with dq_i yields

$$dq_i = -\chi^*(i_{X_i}\omega). \tag{4.10}$$

On the other hand, for all $\xi \in T^*U$ and $v \in T_\xi T^*U$, one has

$$(\chi^* i_{X_i}\omega)_\xi(v) = (i_{X_i}\omega)_{\chi(\xi)}(\chi_{*,\xi} v)$$
$$= \omega_{\chi(\xi)}\big(X_i(\chi(\xi)), \chi_{*,\xi}(v)\big)$$

$$= \omega_{\chi(\xi)} \left(\chi_{*,\xi} \underbrace{\chi_{*,\chi(\xi)}^{-1} X_i(\chi(\xi))}_{=\chi^{-1}X_i(\xi)}, \chi_{*,\xi}(v) \right) \quad \text{(see Formula (1.28))}$$

$$= \omega_{\chi(\xi)} \left(\chi_{*,\xi}\left(\chi^{-1}X_i(\xi)\right), \chi_{*,\xi}(v) \right)$$

$$= (\chi^*\omega)_\xi (\chi^{-1}X_i(\xi), v),$$

which, together with (4.10), implies the following identity:

$$i_{\chi^{-1}X_i}(\chi^*\omega) = -dq_i. \tag{4.11}$$

To make the previous formula more transparent, one needs to write the vector field $\chi^{-1}X_i$ in the coordinates (p, q). To this end, we look at its flow. If $\xi \in T^*U$, $\xi = \sum_{i=1}^{n} \xi_i dq_i$

$$\chi^{-1}X_i(\xi) = \chi_{*,\chi(\xi)}^{-1} X_i(\chi(\xi))$$

$$= \chi_{*,\chi(\xi)}^{-1} \left.\frac{d}{dt}\right|_{t=0} \varphi_t^{dq_i}(\chi(\xi))$$

$$= \left.\frac{d}{dt}\right|_{t=0} \chi^{-1}\varphi_t^{dq_i}(\chi(\xi)). \tag{4.12}$$

On the other hand, since $\chi(\xi) \overset{(4.5)}{=} \varphi_1^\xi(\sigma(p(\xi))) = \varphi_1^{\sum_{i=1}^{n}\xi_i dq_i}(\sigma(p(\xi)))$,

$$\varphi_t^{dq_i}(\chi(\xi)) \overset{(4.2)}{=} \varphi_1^{(t+\xi_i)dq_i}\varphi_1^{\sum_{k\neq i}\xi_k dq_k}(\sigma(p(\xi))) = \varphi_1^{(t+\xi_i)dq_i}\varphi_1^{\sum_{k\neq i}\xi_k dq_k}(\sigma(p(\xi'))), \tag{4.13}$$

where $\xi' = (t+\xi_i)dq_i + \sum_{k\neq i}\xi_k dq_k$. But now, it should be clear that the vector field defined in (4.12) is just that $\frac{\partial}{\partial p_i}$, since from (4.13) its flow is the *vertical* translation along the coordinate p_i. Summarizing

$$\chi^{-1}X_i = \frac{\partial}{\partial p_i}. \tag{4.14}$$

Formula (4.7) is now obtained plugging (4.14) and (4.8) into (4.11). To prove the remaining statements, we start to observe that since $d\chi_\sigma^*(\omega) = 0$, the coefficients A_{ij} do not depend on the qs. Finally, if σ is Lagrangian, i.e., if $\sigma^*\omega = 0$, the restriction of $\chi_\sigma^*(\omega)$ to the 0-section of T^*B is zero, which implies that $\sum_{i,j=1}^{n} A_{ij}dq_i \wedge dq_j = 0$, i.e., $A_{ij} = 0$ for all i, j. $\qquad\square$

Remark 4.23 One can make the following *reality check*. If χ_σ were an anti-symplectomorphism, then

$$\chi_\sigma \frac{\partial}{\partial p_i} = -\chi_\sigma X_{p^*q_i} \overset{\text{Formula (1.36)}}{=} X_{(p^*q_i)\circ\chi_\sigma^{-1}} = X_{(p\circ\chi_\sigma^{-1})^*q_i} = X_{\pi^*q_i}. \qquad (4.15)$$

Of course this argument cannot be used in the proof of Theorem 4.22. $\qquad\qquad\triangle$

It is worth making a few comments about the previous theorem:

(AA1) The previous result implies that if σ is Lagrangian, then $\chi_\sigma^* \omega = \sum_{i=1}^n dq_i \wedge dp_i$, i.e., χ_σ is a *local anti-symplectomorphism*.

(AA2) In spite of its appearance, the hypothesis about σ being Lagrangian is not really restrictive. In fact if σ were *not* Lagrangian and if $\omega' = \sum_{i=1}^n dq_i \wedge dp_i$, then

$$\chi_\sigma^* \omega - \omega' \overset{\text{on } U}{=} \sum_{i,j}^n A_{ij} dq_i \wedge dq_j,$$

which is a *basic* closed two-form, i.e., a closed two-form in the image of $p^* : \Omega^2(U) \to \Omega^2(T^*U)$. If $H^2(U, \mathbb{R}) = 0$, which can always be assumed choosing U sufficiently small, then one can find $\gamma \in \Omega^1(U)$ such that $p^*(d\gamma) = \chi_\sigma^* \omega - \omega'$. If $\sigma_\gamma := \chi_\sigma \circ \gamma$, then $\pi \circ \sigma_\gamma = \mathrm{id}_U$ and $\sigma_\gamma^* \omega = 0$, i.e., σ_γ is a *local Lagrangian section* of π. To prove that $\sigma_\gamma^* \omega = 0$, it suffices to compute

$$\sigma_\gamma^* \omega = (\chi_\sigma \circ \gamma)^* \omega$$
$$= \gamma^*(\chi_\sigma^* \omega)$$
$$= \gamma^* \left(\sum_{i=1}^n dq_i \wedge dp_i + \sum_{i,j=1}^n A_{ij} dq_i \wedge dq_j \right)$$
$$= \gamma^*(-d\Theta) + \gamma^*(p^* d\gamma)$$
$$= -d(\gamma^* \Theta) + \underbrace{(p \circ \gamma)^*}_{=\mathrm{id}}(d\gamma)$$
$$= -d\gamma + d\gamma = 0,$$

where the last equality follows from Proposition 1.18.

(AA3) If σ is any section of π over U, writing Λ_U to denote the restriction of the period lattice to U, one has $T^*U/\Lambda_U \overset{\chi_\sigma}{\simeq} \pi^{-1}(U)$. This can be thought as the *relative* version of the isomorphism $T_b^* B/\Lambda_b \simeq M_b$. Since every local section σ can be deformed to a Lagrangian one (see the point (AA2) above), Theorem 4.22 implies that $\pi^{-1}(U)$ can be, *always*, endowed with a symplectic set of coordinates, induced by the Darboux coordinates defined on T^*U. Note that the induced coordinates are *compatible* with the structure of locally trivial bundle of $\pi : M \to B$, in particular the coordinates corresponding to the p_is are *vertical* while the one induced by the q_is are *horizontal*.

Looking at the previous observations and at the proof of Theorem 4.22, one can deduce the following recipe to construct a set of Darboux coordinates on $\pi^{-1}(U)$.

(RDC1) Eventually shrinking U, choose a Lagrangian section of $\pi^{-1}(U) \xrightarrow{\pi} U$.
(RDC2) Choose a set of local coordinates $q_1, \ldots, q_n$ on U and complete it to a set of Darboux coordinates $(p_1, \ldots, p_n, q_1, \ldots, q_n)$ on T^*U.
(RDC3) For each i, $\chi_\sigma \frac{\partial}{\partial p_i}$ is Hamiltonian with Hamiltonian equal to $Q_i := \pi^* q_i$. Moreover, they commute and they are independent, so the corresponding dual one-forms α_i are closed and, by the relative Poincaré lemma, they are locally exact, defining local functions $P_1, \ldots, P_n$.

The functions so obtained form, together with the $Q_1, \ldots, Q_n$, a set of Darboux coordinates on $\pi^{-1}(U)$ for the (restriction of) ω to $\pi^{-1}(U)$, i.e., $\omega|_{\pi^1(U)} \sum_{i=1}^n dQ_i \wedge dP_i$. In fact

$$\delta_{ij} = \langle dP_i, X_{Q_j} \rangle = -\langle i_{X_{P_i}} \omega, X_{Q_j} \rangle = \omega(X_{Q_j}, X_{P_i}) = \{Q_j, P_i\}.$$

It is worth noting the change of the orientation of the form above, which will be stressed writing (Q, P) instead of (P, Q). Note that once such a system of coordinates (Q, P) is fixed, every function of the form $H = \pi^* f$, for $f \in C^\infty(U)$, will not depend on the Ps. The condition $i_{X_H} \omega = -dH$, if written in these coordinates, assumes the following form:

$$\begin{cases} \dot{P}_i = \frac{\partial H}{\partial Q_i} \\ \dot{Q}_i = -\frac{\partial H}{\partial P_i}, \end{cases} \tag{4.16}$$

for all $i = 1, \ldots, n$, which yields

$$\begin{cases} P_i(t) = v_i(Q(0))t + P_i(0) \\ Q_i(t) = Q_i(0), \end{cases} \tag{4.17}$$

where $v_i(Q(0)) = \frac{\partial H}{\partial Q_i}(Q_1(0), \ldots, Q_n(0))$, for all i, i.e., the solutions of the Hamilton equations, if written in the coordinates (P, Q), are linear in time. Note one more time that one obtains (4.16) using ω written in the coordinates (P, Q)s, i.e., $\omega = \sum_{i=1}^n dQ_i \wedge dP_i$.

The previous observations and Theorem 4.22 assume a particular relevance when the coordinates (p, q)s are *related* to the Lagrangian fibration $\pi : M \to B$. More precisely, if $(p, q) = (\phi, I)$s, where the Is are defined in (4.6) and the ϕs are the corresponding conjugate Darboux coordinates, (4.17) are, by construction, not only linear but also *periodic* with period equal to one.

Definition 4.24 (Action-Angle Variables)   The coordinates $Q_1, \ldots, Q_n$ and $P_1, \ldots, P_n$, defined on $\pi^{-1}(U)$ via the construction sketched above starting from the Darboux coordinates (ϕ, I), are called *action-angle variables* and will be denoted by $J_1, \ldots, J_n$ and $\theta_1, \ldots, \theta_n$, respectively. Note that for all i

$$J_i = \pi^* I_i.$$

$\triangle$

1.2   Lagrangian Fibrations and Integrable Systems

We now take a closer look at the relation between completely integrable systems (see Definition 1.120), and Lagrangian fibrations (see Definition 4.1). On the one hand, the latter clearly says that every Lagrangian fibration defines, at least in a suitable neighborhood of every fiber, a completely integrable system. On the other, using the same notations as in Definition 1.120 and writing $\pi : M \to \mathbb{R}^n$ as the application whose components are $f_1, \ldots, f_n$, condition (i) in the same definition implies that for each $b \in \pi(M)$, $M_b = \pi^{-1}(b)$ is an embedded submanifold of M. Furthermore,

Lemma 4.25 M_b *is a Lagrangian submanifold of* (M, ω).

Proof   In fact M_b is a level set of the functions $(f_1, \ldots, f_n)$ which are in involution with respect to the Poisson bracket defined by the symplectic form ω. $\square$

In other words, an integrable system (M, ω, H) defines a fibration $\pi : M \to \mathbb{R}^n$ whose fibers are *embedded*, in general non-connected, *Lagrangian* submanifolds of (M, ω). Anyway, we warn the reader that the issue of deciding when an integrable system defines a *Lagrangian fibration* in the sense of Definition 4.1 is a murky one. In fact, at the level of generality of Definition 1.120, it is not clear at all if the fibration defined by π is a *locally trivial* one. The answer to this question is positive when the level sets of π are (1) diffeomorphic to $\mathbb{R}^n$ and when (2) they are *compact and connected*. In the first scenario, one can appeal to the following result, which we state here without proof, for which we refer the interested reader to the original paper [180].

Theorem 4.26   *Let* $\pi : M \to B$ *be a surjective and submersive smooth map such that for each* $b \in B$, $M_b = \pi^{-1}(b)$ *is diffeomorphic to* $\mathbb{R}^n$, *for a given n. Then* $\pi : M \to B$ *is a locally trivial fibration.*

On the other hand:

Proposition 4.27   *Let* (M, ω, H) *be an integrable system and suppose that the fibers of* $\pi : M \to \mathbb{R}^n$, *where* $\pi = (f_1, \ldots, f_n)$, *are connected and compact. Then* $\pi : M \to \pi(M)$ *is a Lagrangian fibration.*

Proof Clearly $\pi : M \to \pi(M)$ is surjective. Because of (i) and (ii) of Definition 1.120, π is submersive and its level sets are Lagrangian submanifolds of (M, ω). To conclude the proof it suffices to prove that $\pi : M \to \pi(M)$ defines a locally trivial fibration. This follows from the compactness of the fibers of π and from the property of the Ehresmann connection; see, for example, [59]. $\qquad\square$

Remark 4.28

(i) Note that the conclusion of the previous proposition holds true also dropping the assumption about the connectedness of the level sets of the map π. On the other hand, this proposition could be stated under the stronger assumption that the map $\pi : M \to B$ is proper. In this case, it follows from Ehresmann's lemma: if $\pi : M \to B$ is a surjective submersion between smooth manifolds and π is a proper map, then π is a locally trivial fibration.
(ii) Using the same argument used in the proof of the previous proposition, one can show that the set of all $m \in M$ such that $\pi^{-1}(\pi(m))$ is compact and connected is open in M. In other words, if M_b is a compact level set of F, then there exists an open neighborhood $V \subset \mathbb{R}^n$ of b such that $\pi^{-1}(V) \simeq V \times M_b$.

△

The previous discussion implies that Lagrangian fibrations and Hamiltonian integrable systems are two facets of the same object. The same discussion should also clarify the relevance of the Arnold-Liouville-Mineur theorem (see below), as a main tool to analyze both the evolution of an integrable Hamiltonian system and the geometry of the tubular neighborhood of the fiber where this evolution takes place. We are now ready to address the following important result, which is of *semilocal* nature, in the sense that it describes the geometry of the Lagrangian fibration defined by a completely integrable Hamiltonian system on a neighborhood of any of its fibers.

Theorem 4.29 (Arnold-Liouville-Mineur Theorem) *Let (M, ω, H) be an integrable system whose corresponding Lagrangian fibration $\pi : M \to \mathbb{R}^n$ has compact and connected fibers. Let $b \in \pi(M)$ and $M_b = \pi^{-1}(b)$. Then:*

(i) There exists an open neighborhood $U \subset \pi(M)$ of b such that $\pi^{-1}(U) \subset M$ is invariant with respect to the flows generated by the Hamiltonian vector fields of the form X_f, where f belongs to the commutative Poisson algebra generated by $f_1, \ldots, f_n$ (see Proposition 4.27).
(ii) The action-angle variables introduced in Definition 4.24 define a diffeomorphism $(J, \theta) : \pi^{-1}(U) \to U \times \mathbb{T}^n$, such that

$$\omega|_{\pi^{-1}(U)} = \sum_{i=1}^{n} dJ_i \wedge d\theta_i.$$

Recall that $J_i = I_i \circ \pi$ where $I = (I_1, \ldots, I_n) : U \to \mathbb{R}^n$ is a local set of coordinates on U.

Given H as in the theorem above, (4.16) and (4.17), written in the action-angle coordinates, assume the following form:

$$\begin{cases} \dot{\theta}_i = \frac{\partial H}{\partial J_i} \\ \dot{J}_i = -\frac{\partial H}{\partial \theta_i}, \end{cases} \Rightarrow \begin{cases} \theta_i(t) = v_i(J(0)t + \theta_i(0) \\ J_i(t) = J_i(0), \end{cases} \tag{4.18}$$

for all $i = 1, \ldots, n$.

Definition 4.30 The motion of the system (M, ω, H), which takes place on a fixed torus determined by the initial condition, is called *quasi-periodic*. The v_is in (4.18) are called the *frequencies* of the quasi-periodic motion. Moreover, the system (M, ω, H) is called *isochronous* if all its frequencies are constant, i.e., if they do not depend on the action variables Js. On the other hand, if

$$\det\left(\frac{\partial v_i}{\partial J_j}\right) \neq 0,$$

(M, ω, H) is called *anisochronous*.   △

Note that an integrable system is anisochronous if and only if

$$\det\left(\frac{\partial^2 H}{\partial J_i \partial J_j}\right) \neq 0,$$

i.e., if the Hamiltonian function is *nondegenerate* with respect to one (and so with respect to all) system of action-angle coordinates.

Remark 4.31 Here we make a few comments about the statement of Theorem 4.29.

(1) The hypothesis that the fibers of the fibration are compact (and connected) is not restrictive in the following sense. If $M_b = \pi^{-1}(b)$ is compact (and connected), there exists a neighborhood of b such that $M_{b'}$ is diffeomorphic to M_b, and if such a neighborhood is chosen sufficiently small and contractible, then its preimage via π is a trivial fiber bundle with fiber diffeomorphic to M_b; see point (ii) in Remark 4.28. From this observation it follows at once that the pre-image of such a neighborhood is invariant with respect to the flow generated by any of the Hamiltonian vector fields of the form X_f, where f is as in the theorem.

(2) The diffeomorphism mentioned in part (ii) of Theorem 4.29 is obviously a symplectomorphism and the content of the theorem is really about the existence of such a symplectomorphism between a suitable neighborhood $\pi^{-1}(U)$ of M_b and $\pi^{-1}(U) \times \mathbb{T}^n$, endowed with its *canonical* symplectic structure. In other words, the action-angle coordinates are a *sort* of Darboux coordinates on a suitable neighborhood of a compact fiber of the Lagrangian fibration. More precisely, recall that the Darboux theorem states that every $2n$-dimensional

symplectic manifold is locally symplectomorphic to $\mathbb{R}^{2n}$ with its standard symplectic structure, see Sect. 3.3 in Chap. 1. This local equivalence can be rephrased by saying that on (M, ω) there exists a *symplectic atlas*, whose local charts define the local equivalence between (M, ω) and $(\mathbb{R}^{2n}, \omega_0)$, i.e., the symplectic atlas entails a *normal form* for ω. On the other hand, the Arnold-Liouville-Mineur theorem is a statement about the existence of a normal form for a symplectic structure, not in a neighborhood of a point of M, but rather in a *neighborhood of a fiber of a Lagrangian fibration with connected and compact fibers*. For this reason Theorem 4.29 should be thought of as a *Darboux-type* theorem and the action-angle coordinates as *Darboux like* coordinates. Given a Lagrangian fibration with compact and connected fibers, one could ask whether it is possible to put together these coordinates to get an atlas on M, i.e., one could ask if it is possible to define *global action-angle* coordinates on the total space of a Lagrangian fibration with compact fibers. Such an atlas would be the analogue of the symplectic atlas stemming from the Darboux theorem. However, in general the existence of global action-angle variables is obstructed, as we shall discuss later in this chapter; see Sect. 3.2.

$$\triangle$$

Now we go back to the action-angle coordinates to present an important integral formula which, at least in principle, lets us compute a set of action variable for a given integrable system.

1.3 Arnold Formula

Let (M, π, B) be a Lagrangian fibration with compact fibers. We want to discuss an alternative way to define the action variables on $\pi^{-1}(U)$, where U is any contractible open subset of B. We start with a few remarks. First, the exactness of $\omega|_{\pi^{-1}(U)}$ can be proven independently of Theorem 4.29. Indeed, given $b \in U$, since U is contractible and $\pi^{-1}(U) \simeq U \times M_b$, from Künneth formula (see, for example, [36]), one has

$$H^2_{dR}(\pi^{-1}(U)) \simeq \sum_{i=0}^{2} H^i_{dR}(M_b) \otimes_{\mathbb{R}} H^{2-i}_{dR}(U) \simeq H^2_{dR}(M_b) \otimes_{\mathbb{R}} \mathbb{R} \simeq H^2_{dR}(M_b),$$

i.e., that

$$H^2_{dR}(\pi^{-1}(U)) \simeq H^2_{dR}(M_b). \tag{4.19}$$

Note that the above isomorphism is induced by the canonical embedding $j : M_b \to \pi^{-1}(U)$, more precisely, for each $[\alpha] \in H^2_{dR}(\pi^{-1}(U))$ the corresponding class in

$H^2_{dR}(M_b)$ is $[j^*\alpha]$, where j^* is the pull-back map. Writing ω for $\omega|_{\pi^{-1}(U)}$, note that since M_b is Lagrangian $[j^*\omega] = [0] \in H^2_{dR}(M_b)$, which implies that $[\omega] = [0] \in H^2_{dR}(\pi^{-1}(U))$, as (4.19) is an isomorphism. From this one concludes that ω restricted to $\pi^{-1}(U)$ is exact. Keeping the hypothesis that U is contractible, let σ be a Lagrangian section of $\pi : M \to B$ supported on U, and let λ be a primitive of ω on $\pi^{-1}(U)$. Then $\chi_\sigma^* \lambda - \sum_{i=1}^n I_i d\phi_i$ is a closed one-form on $p^{-1}(U)$, and by Poincaré Lemma, there exists $S \in C^\infty\!\left(p^{-1}(U)\right)$ such that

$$\chi_\sigma^* \lambda = \sum_{i=1}^n I_i d\phi_i + dS.$$

We will use this form to define a function on U. This goes as following. Let $\Gamma : U \times [0, 1] \to \pi^{-1}(U)$ be a family of (sufficiently regular) curves smoothly parametrized by U, such that for all $b \in U$, the support of $\gamma^b(\cdot) := \Gamma(b, \cdot)$ is contained in $\pi^{-1}(b)$. Then, denoting with $\gamma(b)$ the support of the curve $\gamma^b(\cdot)$, the function $\mathscr{I} : \pi^{-1}(U) \to \mathbb{R}$ defined by

$$\mathscr{I}(b) := \int_{\gamma(b)} \lambda \tag{4.20}$$

is smooth and is constant along the fibers of $\pi : \pi^{-1}(U) \to U$. Furthermore, since for all $b \in U$, $\chi_\sigma : p^{-1}(U) \to \pi^{-1}(U)$ restricts to a covering map $\chi_\sigma : T_b^* B \to M_b$, given $\gamma^b : [0, 1] \to M_b$, there exists a unique lifting $\tilde{\gamma}^b : [0, 1] \to T_b^* B$ such that the following diagram commutes:

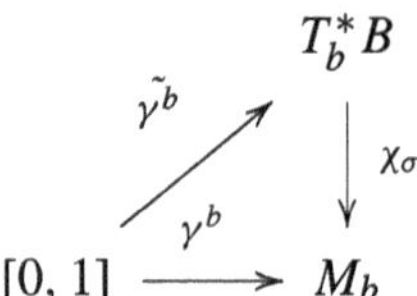

In particular one has

$$\mathscr{I}(b) = \int_{\gamma(b)} \lambda = \int_{\chi_\sigma(\tilde{\gamma}(b))} \lambda = \int_{\tilde{\gamma}(b)} \chi_\sigma^* \lambda = \int_{\tilde{\gamma}(b)} \left(\sum_{i=1}^n I_i d\phi_i + dS \right), \tag{4.21}$$

where, for each $b \in U$, $\tilde{\gamma}(b) \subset \pi^{-1}(b)$ is the support of $\tilde{\gamma}^b$. We come now to the main outcome of this discussion. Let U be an open subset of B such that on $\pi^{-1}(U)$ a set of action-angle coordinates can be defined, and let $\gamma_1, \ldots, \gamma_n$,

$$\gamma_i(b, t) : U \times [0, 1] \to \pi^{-1}(U),$$

be a smooth family parametrized by U of smooth curves such that for each $b \in U$,

$$H_1(M_b, \mathbb{Z}) = \mathrm{span}_{\mathbb{Z}} \langle [\gamma_1(b)], \ldots, [\gamma_n(b)] \rangle.$$

Here, as before, for each $i = 1, \ldots, n$, $\gamma_i(b)$ is the image of $\gamma_i(b, t) : [0, 1] \to M_b$. Then

Theorem 4.32 (Arnold Formula) *If $\lambda \in \Omega^1(\pi^{-1}(U))$ is a potential of $\omega|_{\pi^{-1}(U)}$, the functions $(\mathscr{J}_1, \ldots, \mathscr{J}_n)$, defined by*

$$\mathscr{J}_i(b) = \int_{\gamma_i(b)} \lambda, \ i = 1, \ldots, n \tag{4.22}$$

at most differ by the action variables $J_1, \ldots, J_n$ introduced in Definition 4.24 by an additive constant.

Proof Let U be as above, let $\alpha_1, \ldots, \alpha_n \in \Omega^1_B(U)$ be such that $d\alpha_i = 0$ for all $i = 1, \ldots, n$, and let $\Lambda_b = \mathrm{span}_{\mathbb{Z}} \langle \alpha_1(b), \ldots, \alpha_n(b) \rangle$ for all $b \in U$. Then for every $i = 1, \ldots, n$ one can define a smooth family of smooth curves:

$$\varphi^{\alpha_i(\cdot)}(\cdot) : [0, 1] \times U \to \pi^{-1}(U),$$

where for every $b \in B$, $\varphi^{\alpha_i(b)} : [0, 1] \to M_b$ is the integral curve of the vector field $X_{\alpha_i(b)}$; see Lemma 4.6 and Remark 4.7. Let $(I_1, \ldots, I_n)$ be the action variables defined on U by $\alpha_i = dI_i$, and let $(\phi_1, \ldots, \phi_n)$ be the corresponding (canonically conjugate) fibered coordinates, defined by $\xi = \sum_{i=1}^n \langle \xi, \frac{\partial}{\partial I_i} \rangle dI_i = \sum_{i=1}^n \phi_i(\xi) \, dI_i$. For each i, let $\tilde{\varphi}^{\alpha_i}$ be the lift to $\pi^{-1}(U)$ of the family φ^{α_i}, i.e., the smooth family of smooth curves which closes the following commutative diagram:

$$
\begin{array}{ccc}
 & & p^{-1}(U) \\
 & \overset{\tilde{\varphi}^{\alpha_i}}{\nearrow} & \downarrow {\scriptstyle X_\sigma} \\
[0, 1] \times U & \underset{\varphi^{\alpha_i}}{\longrightarrow} & \pi^{-1}(U)
\end{array}
$$

Note that for every $b \in B$ the previous diagram specializes to

$$
\begin{array}{ccc}
 & & T_b^* B \\
 & \overset{\tilde{\varphi}^{\alpha_i(b)}}{\nearrow} & \downarrow {\scriptstyle X_\sigma} \\
[0, 1] & \underset{\varphi^{\alpha_i(b)}}{\longrightarrow} & M_b
\end{array}
$$

Furthermore, the homotopy class of (the image of) φ^{α_i} is a generator of $\pi_1\big(X_\sigma(\alpha_i(b)), M_b\big)$ and its lift to $T_b^* B$ is such that if $\tilde{\varphi}^{\alpha_i(b)}(0) = \xi$, then

$\tilde{\varphi}^{\alpha_i(b)}(1) = \xi + \alpha_i(b)$. Moreover, since $\varphi^{\alpha_i(b)}$ is the integral curve of the vector field $X_{\alpha_i(b)}$ and since χ_σ is a symplectomorphism, then $\tilde{\varphi}^{\alpha_i(b)}$ is the integral curve of the Hamiltonian vector field defined by $p^*(\alpha(b))$, i.e., the vector field $\frac{\partial}{\partial \phi_i}$. In other words, in the local coordinates (I_i, ϕ_i) one has the following parametric representation:

$$\tilde{\varphi}^{\alpha_i(b)} = \begin{cases} I_j(t) = I_j(b), & \text{for all } j, \\ \phi_j(t) = \phi_j(0) \text{ if } j \neq i \\ \phi_i(t) = t. \end{cases}$$

Then, from (4.21),

$$\mathscr{I}_i(b) = \int_{\varphi^{\alpha_i(b)}} \lambda = \int_{\tilde{\varphi}^{\alpha_i(b)}} \left(\sum_{i=1}^{n} I_i d\phi_i + dS \right) = I_i(b) + \Delta_{\phi_i}(S), \qquad (4.23)$$

where

$$\Delta_{\phi_i}(S) = S(I_1, \ldots, I_n, \phi_1, \ldots, \phi_i+1, \ldots, \phi_n) - S(I_1, \ldots, I_n, \phi_1, \ldots, \phi_i, \ldots, \phi_n).$$

We prove now that $\Delta_{\phi_i}(S)$ is constant. To this end, given $\alpha \in \Omega_B^1(U)$, let τ_α be the corresponding translation along the fibers of $p : T^*B \to B$; see Example 1.53. Then

$$\tau_\alpha^* dS = \tau_\alpha^*(\chi_\sigma^* \lambda) - \tau_\alpha^* \left(\sum_{i=1}^{n} I_i d\phi_i \right) = (\chi_\sigma \circ \tau_\alpha)^* \lambda - \tau_\alpha^* \left(\sum_{i=1}^{n} I_i d\phi_i \right)$$

$$= (\chi_\sigma \circ \tau_\alpha)^* \lambda - \sum_{i=1}^{n} (I_i \circ \tau_\alpha) d(\phi_i \circ \tau_\alpha).$$

Since for all $b \in U$, $\tau_\alpha(\xi) = \xi + \alpha(b)$, $\forall \xi \in T_b^* B$, one has $I_i \circ \tau_\alpha(\xi) = I_i\big(p(\xi + \alpha(b))\big) = I_i(b)$, and

$$\phi_i \circ \tau_\alpha(\xi) = \phi_i\big(\xi + \alpha(b)\big) = \left\langle \xi + \alpha(b), \frac{\partial}{\partial I_i} \right\rangle = \left\langle \xi, \frac{\partial}{\partial I_i} \right\rangle + \left\langle \alpha(b), \frac{\partial}{\partial I_i} \right\rangle$$

$$= \phi_i(\xi) + \alpha_i(b),$$

where $\alpha(b) = \sum_{i=1}^{n} \alpha_i(b) dI_i$. In other words, the functions $\phi_i \circ \tau_\alpha$ and ϕ_i differ by a function which does not depend on ξ, and for this reason

$$d\phi_i = d(\phi_i \circ \tau_\alpha),$$

which, together with the previous computation, implies that

$$\tau_\alpha^* \left(\sum_{i=1}^n I_i d\phi_i \right) = \sum_{i=1}^n I_i d\phi_i.$$

Let us assume now that α is a closed one-form such that $\alpha(b) \in \Lambda_b$, $\forall b \in U$. In this case

$$\chi_\sigma \circ \tau_\alpha(\xi) = \chi_\sigma \big(\tau_\alpha(\xi) \big) = \chi_\sigma \big(\xi + \alpha(b) \big)$$
$$= \varphi_1^{\alpha(b)+\xi} \big(\sigma(b) \big) = \varphi_1^{\alpha(b)} \circ \varphi_1^{\xi} \big(\sigma(b) \big)$$
$$= \varphi_1^{\xi} \big(\sigma(b) \big) \text{ since } \alpha(b) \in \Lambda_b = \chi_\sigma(\xi).$$

In other words, we proved that, for all closed $\alpha \in \Omega_B^1(U)$ such that $\alpha(b) \in \Lambda_b$ for all $b \in U$ one has

$$\tau_\alpha^*(dS) = dS.$$

But this is equivalent to saying that

$$d(\tau_\alpha^* S - S) = 0,$$

i.e.,

$$\tau_\alpha^* S - S = C_\alpha, \tag{4.24}$$

where C is a constant function on $p^{-1}(U)$ depending on α. Let now $\alpha_1, \ldots, \alpha_n$ be closed one-forms such that $\Lambda_b = \mathrm{span}_{\mathbb{Z}} \langle \alpha_1(b), \ldots, \alpha_n(b) \rangle$, $\forall b \in U$. Then,

$$(\tau_{\alpha_i}^* S)\big(I_1, \ldots, I_n, \phi_1, \ldots, \phi_n\big) = S(I_1, \ldots, I_n, \phi_1, \ldots, \phi_i + 1, \ldots, \phi_n)$$

since

$$I_k \circ \tau_{\alpha_i} = I_k, \quad \text{and} \quad \phi_k \circ \tau_{\alpha_i} = \phi_k + \delta_{ik}, \ \forall i, k,$$

and (4.24) implies that

$$\Delta_{\phi_i} S = \tau_{\alpha_i}^* S - S = C_{\alpha_i},$$

as required. To conclude the proof of the theorem we are left to show that the family of cycles smoothly depending on $b \in U$ in (4.23) can be traded for any family generators of the fundamental group of M_b as b changes smoothly in U. This follows from two observations. The first is that $\mathscr{I}(b)$ in (4.20) depends only on the homotopy class of $\gamma(b)$. In fact, if $\gamma(b)$ and $\widetilde{\gamma}(b)$ are two homotopically equivalent curves in M_b and Σ is the two-dimensional surface such that $\partial\Sigma = \gamma(b)\cup(-\widetilde{\gamma}(b))$ (the minus sign keeps track of the orientation of the boundary of Σ), then by Stokes' theorem

$$\int_{\gamma(b)}\lambda - \int_{\widetilde{\gamma}(b)}\lambda = \int_{\Sigma}\omega,$$

and

$$\int_{\Sigma}\omega = 0,$$

since $M_b \subset \pi^{-1}(U)$ is Lagrangian. The second observation is that, by definition, for each $b \in U$, the classes of the curves $\varphi^{\alpha_1(b)}, \ldots, \varphi^{\alpha_n(b)}$ are generators of $H_1(M_b, \mathbb{Z})$. $\qquad\square$

Remark 4.33 Note that in our Formula (4.22) the actions are not normalized by the factor $\frac{1}{2\pi}$. $\qquad\triangle$

In the next section the above constructions will be applied to simple, nevertheless important, integrable systems.

2 Two Examples

In this section the above constructions will be applied to define a set of action-angle variables for two integrable systems: the harmonic oscillator (see Examples 1.111 and 1.118) and the Kepler system (see Example 3.108). In the latter case the action-angle variables will be obtained using the results of Theorem 4.32, while the action-angle variables for the harmonic oscillator will stem from the more geometrical construction presented in the first part of this chapter.

2.1 One-Dimensional Harmonic Oscillator

Recall that in the Darboux coordinates (p, q) the Hamiltonian function of the one-dimensional harmonic oscillator is $H(p, q) = \frac{p^2}{2} + \frac{\varpi^2 q^2}{2}$, where ϖ is a dimensioned positive constant. The phase space of this system can be identified with $\mathbb{R}^2$ with its canonical symplectic form $\omega = dp \wedge dq$, and the Hamiltonian vector field describing the dynamics can be written as

$$X_H = p\frac{\partial}{\partial q} - \varpi^2 q\frac{\partial}{\partial p}.$$

This vector field corresponds to the second-order ODE

$$\ddot{q} + \varpi^2 q = 0, \tag{4.25}$$

describing the motion of the system on the configuration space, which is well known to be *periodic*, with period equal to $T = 2\pi/\omega$. We double-check this claim computing T as follows (see Example 2.1). First, fix the value for the Hamiltonian, i.e., $H = h$, and then solve the equation $\frac{\varpi^2 q^2}{2} = h$, which has solutions

$$q_{\max} = \frac{\sqrt{2h}}{\varpi} \quad \text{and} \quad q_{\min} = -\frac{\sqrt{2h}}{\varpi}. \tag{4.26}$$

Since we set $m = 1$, $H = \frac{\dot{q}^2}{2} + \frac{\varpi^2 q^2}{2}$, which implies

$$dt = \pm \frac{dq}{\sqrt{2h - \varpi^2 q^2}}.$$

Finally, the time elapsed between the positions q_0 and q is

$$t = \pm \int_{q_0}^{q} \frac{du}{\sqrt{2h - \varpi^2 u^2}},$$

where the $\pm$ indetermination is fixed once a branch of the inverse of the square root is chosen. Now a simple integration, which uses (4.26), yields

$$T(h) = \int_{q_{\min}}^{q_{\max}} \frac{dq}{\sqrt{2h - \varpi^2 q^2}} - \int_{q_{\max}}^{q_{\min}} \frac{dq}{\sqrt{2h - \varpi^2 q^2}}$$

$$= 2 \int_{q_{\min}}^{q_{\max}} \frac{dq}{\sqrt{2h - \varpi^2 q^2}} = \frac{2\pi}{\varpi}.$$

This computation shows, in particular, that the period does not depend on the value of H. Given T, the *period lattice* of the harmonic oscillator is $\Lambda = \cup_{h \in \mathbb{R}_{\geq 0}} \Lambda_h$, where

$$\Lambda_h = \left\{ \alpha_n = nT\,dh = \frac{2\pi n}{\varpi} dh \mid n \in \mathbb{Z} \right\} \subset T_h^* \mathbb{R}. \tag{4.27}$$

Note that Λ_h is independent of h and it is generated by the one-form $\alpha_1 = T\,dh = \frac{2\pi}{\varpi} dh$. To compute a set of action-angle variables for our dynamical system we follow the steps indicated in the discussion preceding Definition 4.24. In particular we look first for a (local) coordinate I on the base of the fibration $H : \mathbb{R}^2 \to \mathbb{R}$, such that dI generates the period lattice of the system in the chosen neighborhood $U \subset \mathbb{R}$. In our case the most natural choice is $I = h$. The second step consists in completing h to a symplectic coordinate system on (T^*U, Ω), which is achieved introducing t as the *conjugate* variable of h; see Formula (1.51) in Example 1.111. In this way it remains defined a set of Darboux coordinates (t, h), the so-called *time-energy* coordinates, on the cotangent bundle of the base of the fibration $H :$ $\mathbb{R}^2 \to \mathbb{R}$. Note that, using our conventions, the canonical form of $T^*\mathbb{R}$, written in these coordinates, is $\Omega = dt \wedge dh$. Let us now compute the action-angle variables for the harmonic oscillator. First, note that since $\alpha_1 = T dh$, T is h-independent and $\alpha_1 = dI$ (see (4.6)), one has that

$$J = \frac{2\pi}{\varpi} h. \tag{4.28}$$

Since (t, h) are canonically conjugate, (4.28) implies that the conjugate variable of J is obtained simply *rescaling* t to maintain the symplecticity of the new pair. In other words

$$\theta = \frac{\varpi}{2\pi} t. \tag{4.29}$$

Note that these two functions are defined on the cotangent bundle of the base of the fibration $H : \mathbb{R}^2 \to \mathbb{R}$ which, in this case, coincides with its total space. See also the remarks and observation at the end of this example.

Now let us find, for this simple example, an explicit formula for the map χ defined in Formula (4.5). To this end, let σ be the section of $H(p, q) : \mathbb{R}^2 \to \mathbb{R}$ defined by

$$\sigma(h) = \left(\sqrt{2h}, 0\right). \tag{4.30}$$

Note that for dimensional reasons every section of H is Lagrangian. Then, for every $t \in \mathbb{R}$

$$\chi_\sigma(tdh) \overset{(4.5)}{=} \varphi_1^{tdh}(\sigma(p(tdh))) = \varphi_1^{tdh}(\sigma(h)) \overset{(4.30)}{=} \varphi_1^{tdh}\left(\sqrt{2h}, 0\right) \overset{(4.2)}{=} \varphi_t^{dh}\left(\sqrt{2h}, 0\right).$$

This means that, for every t', $\chi_\sigma(t'dh)$ is the point in $M = \mathbb{R}^2$ obtained *flowing* $(\sqrt{2h}, 0)$ along the integral curve of the Hamiltonian vector field defined by H from $t = 0$ to $t = t'$. As already recalled, the evolution of the oscillator is described by the solutions of (4.25) which, together with the *initial conditions* defined by σ, are

$$\begin{cases} p &= \sqrt{2h}\cos(\varpi t) \\ q &= \frac{\sqrt{2h}}{\varpi}\sin(\varpi t), \end{cases} \tag{4.31}$$

i.e.,

$$\chi_\sigma(tdh) = \left(\sqrt{2h}\cos(\varpi t), \frac{\sqrt{2h}}{\varpi}\sin(\varpi t)\right).$$

As already noticed *en passant*, the choice of the section σ is equivalent to the choice of the initial conditions for (4.25). In particular, (4.30) says that momentum and position at the time $t = 0$ are $p_0 = \sqrt{2h}$ and $q_0 = 0$, respectively. The solutions (4.31) should be compared with Formulas (1.52), with $k_+(0) = 0$, in Example 1.111. Note that (4.31) implies that

$$\chi_\sigma^* \omega = \chi_\sigma^*(dp \wedge dq) = dh \wedge dt = -\Omega,$$

confirming that χ_σ is an anti-symplectomorphism; see Theorem 4.22. Equation (4.31), together with (4.28) and (4.29), defines the following change of variables:

$$\begin{cases} p &= \sqrt{\dfrac{\varpi J}{\pi}} \cos(2\pi\theta) \\ q &= \sqrt{\dfrac{J}{\pi \varpi}} \sin(2\pi\theta), \end{cases} \qquad (4.32)$$

whose inverse expresses the action-angle variables as functions of the physical variables (p, q). Note that $\sigma = \chi_\sigma \circ s$, where s is the 0-section of the canonical fibration $p : T^*\mathbb{R} \to \mathbb{R}$; see the diagram below:

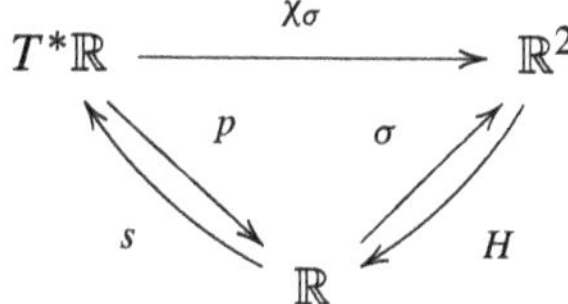

Furthermore, observe that $\chi_\sigma^{-1}(\sigma(h)) = \Lambda_h = \{(h, t) \mid t = nT\}$.

It is probably worth making a few more comments about (4.29). In the previous example the phase space of the dynamical system coincides with the cotangent bundle of the base of the configuration space. For this reason, (4.29) can be obtained observing that since $I'(h) \neq 0$, $J = J(h)$ is a diffeomorphism of $\mathbb{R}$. This can be lifted to its cotangent bundle, defining the symplectomorphism

$$\tilde{J}(k, h) = \left(J(h), \frac{t}{J'(h)} \right)$$

(see Formula (D.1)), which yields (4.29). In this case, the passage from the *energy-time* to the *action-angle* variables is achieved via a linear change of variables. This is possible because the phase space of the system coincides with the cotangent bundle of the Lagrangian fibration. One should also notice that in this example there is no difference between the J and I variables. This is so because the projection π is identified with H and $H^*(I) = \frac{2\pi}{\varpi} H^*(h) = \frac{2\pi}{\varpi} h$. Finally, it is possible to get a clear understanding of the consequences of the change of variables from h to J.

This amounts to a rescaling of the Hamiltonian of the system, which yields the *new* Hamiltonian vector field

$$\tilde{X} = \frac{2\pi p}{\varpi} \frac{\partial}{\partial q} - 2\pi \varpi q \frac{\partial}{\partial p},$$ (4.33)

whose corresponding second-order ODE is $\ddot{q} + 4\pi^2 q = 0$. One can easily check that the solutions of this equation are periodic of period equal to one.

Remark 4.34 A final comment is in order. As already noticed more than once, the construction of the action-angle variables in the example of the one-dimensional harmonic oscillator is simple for two reasons: (1) the phase space of this system is the cotangent bundle of the base of the associated Lagrangian fibration, and (2) the projection of the Lagrangian fibration is the Hamiltonian of the system. This simplified framework makes possible to bypass the recipe summarized in points (RDC1), (RDC2), and (RDC3). On the other hand, following those points would imply to compute $\chi_\sigma \frac{\partial}{\partial \theta}$, where θ and σ are defined in (4.29) and in (4.30), respectively. A simple computation, which uses the change of variable (4.32), shows that χ_σ is the Hamiltonian vector field (4.33); see also (4.15). Following the last step in the recipe for the construction of the action-angle variables, one should find the one-form α such that $\langle \alpha, \chi_\sigma \frac{\partial}{\partial \theta} \rangle = 1$. This one-form is easily guessed being equal to

$$\alpha = \frac{\varpi}{2\pi} \left(\frac{p\,dq - q\,dp}{p^2 + \varpi^2 q^2} \right).$$

This form is closed and, writing it in the energy-time coordinates, one discovers that

$$\alpha = \frac{\varpi}{2\pi} dt,$$

i.e., $\alpha = d\theta$ as it should be; see (4.29). $\triangle$

The previous discussion can be generalized, at least in principle, to the case of Hamiltonian systems with one degree of freedom, whose phase space is as in the previous example and the Hamiltonian is

$$H = \frac{p^2}{2} + V(q).$$ (4.34)

In this case, when the motion is periodic, its period is given by

$$T(h) = \sqrt{2} \int_{q_1}^{q_2} \frac{dq}{\sqrt{h - V(q)}},$$ (4.35)

and the corresponding period lattice is computed as in (4.27). Note that, in general, T is a nonconstant function of h. A concrete example in this more general class of dynamical systems is the so-called *simple pendulum*. This is an (idealized)

Fig. 4.1 Oscillating
pendulum

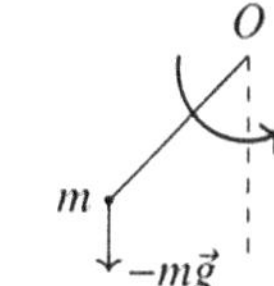

Fig. 4.2 Phase portrait:
Trajectories of the pendulum
in the phase space (p, q)

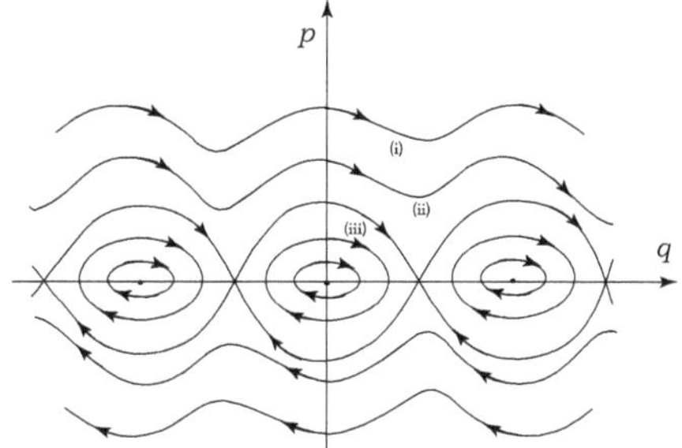

mechanical system constituted by a bob (a small weight) of mass m (which we choose equal to 1), constrained to oscillate in a vertical plane suspended at one end of a rigid rod of negligible weight; the other end of the rod is fixed. The Hamiltonian of this system is of type (4.34), with $V(q) = -\cos q$, i.e.,

$$H = \frac{1}{2}p^2 - \cos q; \tag{4.36}$$

see (Fig. 4.1).

The configuration space of this system is diffeomorphic to S^1, and the phase space is diffeomorphic to the *cylinder* $\mathbb{R} \times S^1$. Note that the image of H is the interval $[-1, +\infty)$. On the dense open subset of $\mathbb{R} \times S^1$ where $dH \neq (0, 0)$, (4.36) is a submersion onto $(-1, +\infty)$. As h varies in this interval, $F_h = H^{-1}(h)$ consists of:

(i) Two connected components if $h > 1$. These are two closed curves wrapping around the cylinder. Also in this case the corresponding motion is periodic.

(ii) Two curves if $h = 1$ called *separatrices*. In this case, the fiber $F_1 = H^{-1}(1)$ corresponds to the so-called *asymptotic* motions: during these motions the system tends to the configuration of the *unstable equilibrium*, without reaching it in any finite time. Of course, the fiber of H contains also the unstable equilibrium point in this case. Observe that the separatrices connect the points $(\pi, 0)$ and $(-\pi, 0)$, which coincide on the cylinder $S^1 \times \mathbb{R}$.

(iii) A closed curve surrounding the origin $(0, 0)$, if $h \in (-1, 1)$. The corresponding motions are *oscillations* (or *librations*) around the position of *stable equilibrium*. Note that for these values of h the fibers F_h are *connected*; see (Fig. 4.2).

Summarizing the previous discussion one can say that the fibers of H are of the following (topological) type:

$$F_h \simeq \begin{cases} S^1 \cup S^1 & \text{if } h \in (1, +\infty), \\ \mathbb{R} \cup \mathbb{R} \cup \{ \text{ the unstable equilibrium point } \} & \text{if } h = 1, \\ S^1 & \text{if } h \in (-1, 1). \end{cases}$$

Note that in this example there are two kinds of closed trajectories: those corresponding to the *oscillations* and those corresponding to the *rotations*. In the first case, the trajectories are homotopically trivial, i.e., they can be continuously deformed to a point, while in the second one, the trajectories wrap around the cylinder, and they are homotopically *not* trivial; see picture above. We will not present the details of the construction of the action-angle variables in this case as one could just proceed as in the example of the harmonic oscillator. The only thing we would like to stress is that in the case of the pendulum the period of the system is a nonconstant function of the energy, i.e., $T(h) = \sqrt{2} \int_{q_1}^{q_2} \frac{dq}{\sqrt{h+\cos q}}$ (see Formula (4.35)), making the computation of the action variable a nontrivial task.

Example 4.35 (n-Dimensional Harmonic Oscillator)   The results enclosed in Sect. 2.1 can be generalized to the case of the harmonic oscillator with n-degrees of freedom; see Examples 1.118 and 1.126. Recall that the phase space of this system is $\mathbb{R}^{2n}$ with its canonical symplectic structure and its Hamiltonian is the function $H = \sum_{i=1}^{n} \frac{p_i^2}{2} + \sum_{i=1}^{n} \frac{\varpi_i^2 q_i^2}{2}$, which is completely separated in the physical coordinates (p_i, q_i) as $H = \sum_{i=1}^{n} f_i$ where $f_i(p_i, q_i) = \frac{1}{2} \sum_{i=1}^{n} p_i^2 + \varpi_i^2 q_i^2$, for all $i = 1, \ldots, n$. As a consequence of these separation relations, the generalization of the results in Sect. 2.1 to the present situation is straightforward. In analogy to (4.28) and (4.29) one introduces

$$J_i = \frac{2\pi}{\varpi_i} h_i \quad \text{and} \quad \theta_i = \frac{\varpi_i}{2\pi} t, \ \forall i = 1, \ldots, n \tag{4.37}$$

(see Example 2.1 and Formula (1.51)), which defines the change of variables (symplectomorphism):

$$\begin{cases} p_i &= \sqrt{\frac{J_i \varpi_i}{\pi}} \cos(2\pi \theta_i) \\ q_i &= \sqrt{\frac{J_i}{\pi \varpi_i}} \sin(2\pi \theta_i); \end{cases} \tag{4.38}$$

see Formula (4.32). It is worth noting, again in analogy with the one-dimensional case, that the canonical transformation (4.38) is *generated* by

$$S(q, h) = \pm \sum_{i=1}^{n} \int^{q_i} \sqrt{2h_i - \varpi_i^2 u_i^2} d\xi_i.$$

Note that the Hamiltonian of the system, written in terms of the coordinates (4.37), does not depend on the angles, i.e.,

$$\tilde{H}(J) = \frac{1}{2\pi} \sum_{i=1}^{n} h_i(J_i),$$

where

$$h_i(J_i) = \frac{\varpi_i}{2\pi} J_i.$$

Furthermore, observe that the h_is, in this case, are the same functions we had in (4.37). These equations show that the frequencies of the system are $v_i = \frac{\varpi_i}{2\pi}$, for all $i = 1, \ldots, n$ (see Definition 4.30), as one could deduce also from (4.37). Finally, observe that the isotropic harmonic oscillator is an anisochronous system; see Definition 4.30. $\triangle$

In the next subsection we analyze the case of the Kepler system.

2.2 *The Kepler System*

In Example 3.108 it was shown that a dynamical system with n-degree of freedom, with central potential (3.85), can be reduced to a system with one-degree of freedom, and for this reason it is completely integrable. As an important, though particular case we discussed the Kepler system, whose potential encapsulates the planetary motion laws in the framework of the classical Newtonian gravitation. This is a two-body interaction whose force is proportional to the *inverse square* of the distance between the two interacting bodies. See also Example 3.106, where the Kepler potential was discussed from the *symmetry reduction* viewpoint. The dynamics of such a system was accurately described by Johannes Kepler at the beginning of the seventeenth century, long before Newton's discovery of the general laws of dynamics made in 1687. As already remarked, the Liouville integrability of this system follows naturally from the reduction process explained in Example 3.108. Here below we present explicit formulas for the action-angle coordinates. To this end, we recall that the motion of the system takes place on the plane orthogonal to the angular momentum, because of conservation of angular momentum itself. This implies that the Hamiltonian of the reduced system, written in polar coordinates (r, θ), looks like

$$H = \frac{p_r^2}{2} + \frac{p_\theta^2}{2r^2} - \frac{K}{r} = \frac{p_r^2}{2} + V, \quad \text{where} \quad V = \frac{p_\theta^2}{2r^2} - \frac{K}{r}, \tag{4.39}$$

where we set $m = 1$ and where K is a positive constant equal to the product of the masses of the two bodies and the universal gravitational constant. At this point the complete integrability of the Kepler system follows observing that p_θ is a first integral for (4.39), i.e., that $\{H, p_\theta\} = 0$, and that H and p_θ are generically independent on the cotangent bundle of $\mathbb{R}^2 \backslash \{(0,0)\}$, the phase space of the system. More precisely, H, p_θ are independent where

$$\begin{pmatrix} \frac{\partial V}{\partial r} & 0 & p_r & \frac{p_\theta}{r^2} \\ 0 & 0 & 0 & 1 \end{pmatrix}$$

has maximal rank, i.e., on the set (here $M = T^*(\mathbb{R}^2 \backslash \{(0,0)\})$)

$$M^\circ = M \backslash \{(r, \theta, 0, \pm\sqrt{Kr}) \mid r > 0,\ \theta \in [0, 2\pi)\}.$$

Now we study the level sets of $\pi|_{M^\circ} : M^\circ \to \mathbb{R}^2$, where $\pi = (H, p_\theta)$. To this end, one fixes

$$p_\theta = \mu \in \mathbb{R}$$

and analyzes the behavior of the level sets of π as a function of the values of H. From (4.39) it follows that

$$H_\mu := \frac{p_r^2}{2} + V_\mu \geq V_\mu \geq v_\mu,$$

where v_μ is the absolute minimum of the amended potential $V_\mu = \frac{\mu^2}{2r^2} - \frac{K}{r}$ (see Sect. 4.1.1), which is achieved for $r = r_{\min} = \frac{\mu^2}{K}$ (Fig. 4.3).

The outcome of these observations is that for every choice of μ, the only values of H_μ *dynamically significant* are those such that $H_\mu \geq v_\mu = -\frac{K^2}{2\mu^2}$.

Remark 4.36 Note that for $H_\mu = v_\mu$, the pre-image of π is diffeomorphic to S^1 but it does not belong to M°. Anyway, the corresponding trajectory is a circle in the physical space (Fig. 4.4). $\triangle$

Fig. 4.3 Plot of the amended potential

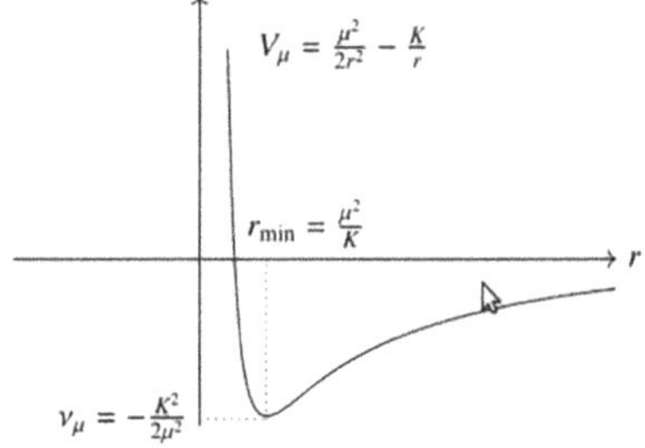

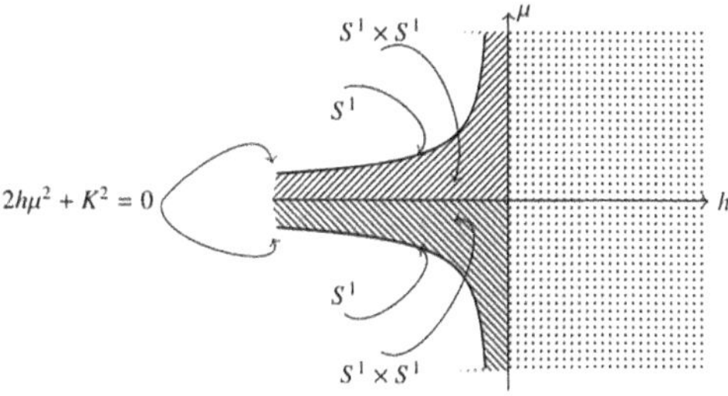

Fig. 4.4 Topological type of the level sets of π

To analyze the topological type of the other pre-images of π, first, we observe that $\pi(M^\circ) \subset \mathbb{R}^2$ is the set

$$B = \{(h, \mu) \mid 2h\mu^2 + K^2 > 0\}. \tag{4.40}$$

Then we observe that for each $b = (h, \mu) \in B$:

(i) If $v_\mu < h < 0$, then $\pi^{-1}(b) = C_+ \cup C_-$, where

$$C_+ = \{(r, \theta, p_r, p_\theta) \mid r \in [r_1(b), r_2(b)], \ \theta \in [0, 2\pi),$$
$$p_r = \sqrt{2(h - V_\mu)} \text{ and } p_\theta = \mu\},$$

$$C_- = \{(r, \theta, p_r, p_\theta) \mid r \in [r_1(b), r_2(b)], \ \theta \in [0, 2\pi),$$
$$p_r = -\sqrt{2(h - V_\mu)} \text{ and } p_\theta = \mu\},$$

(ii) If $h \geq 0$, then $\pi^{-1}(b) = P_+ \cup P_-$, where

$$P_+ = \{(r, \theta, p_r, p_\theta) \mid r \geq r_3(b), \ \theta \in [0, 2\pi), \ p_r = \sqrt{2(h - V_\mu)} \text{ and } p_\theta = \mu\},$$

$$P_- = \{(r, \theta, p_r, p_\theta) \mid r \geq r_3(b), \ \theta \in [0, 2\pi), \ p_r = -\sqrt{2(h - V_\mu)} \text{ and } p_\theta = \mu\}.$$

In particular, case (i) corresponds to *the bounded motions*, i.e., motions constrained to the circular crown $\{(r, \theta) \mid \theta \in [0, 2\pi), \ r \in [r_1(b), r_2(b)]\}$. The values $r_1(b)$ and $r_2(b)$ are called *apsidal* distances, and they correspond to the minimum and the maximum distance from the force center, respectively. Recall that for the Keplerian potential the bounded motions correspond to *closed orbits* (see Theorem 3.113), and these orbits are ellipses with the source of the gravitational force (the Sun) at one of its foci. On the other hand, in the case (ii) the motions are not bounded. From the topological point of view, the fibers of π, as b moves in B, can be two-dimensional *tori*, if $h \in (v_\mu, 0)$, or (infinite) *cylinders*, if $h \in [0, \infty)$. Even though the points for which $h = v_\mu$ do not belong to B, as it was defined here, their inverse images via π correspond to circles, which can be thought of as a degeneration of the two-dimensional tori. The tori are obtained gluing together the circular crowns C_+ and

C_- described in (i), while the cylinders are obtained by gluing together the regions of the plane P_+ and P_-. All the orbits of a Kepler particle are conic sections, whose eccentricity e is a function of h, μ and of the physical parameters of the system. More precisely, still keeping $m = 1$,

$$e = \sqrt{1 + \frac{2h\mu^2}{K^2}}.$$

Note that the orbit is a parabola if and only if $e = 1$, i.e., if $h = 0$, and it is an hyperbola if and only if $e > 1$, i.e., if $h > 0$. These are the only cases corresponding to unbounded motions. The bounded ones correspond to $e < 1$, i.e., $h \in (v_\mu, 0)$, i.e., to elliptic orbits, and to $e = 0$, i.e., to circular orbits, i.e., $h = v_\mu$; see [108, Section 3.7, Chapter 3].

Remark 4.37 The map $\pi = (H, p_\theta) : M^\circ \to B$ is a relatively simple example of *energy-momentum map*. The analysis presented above of the topological type of its fibers can be adapted, without significative changes, to the case of any central potential. This method of analysis was introduced in [223]. $\triangle$

2.3   Action-Angle Variables for the Kepler System

Recall that for all $b = (h, \mu) \in B$ (see (4.40)), such that $v(\mu) < h < 0$, $F_b := F^{-1}(h, \mu) = C_+ \cup C_-$ is two-dimensional torus. Under these assumptions, a set of action-angle variables for the Kepler system can be computed as follows. Choose as generators of the fundamental group of $C_\pm$ the curves

$$\gamma_1(b) = \left\{ (r, \overline{\theta}, \sqrt{2(h - V_\mu)}, \mu) \mid r \in [r_1(b), r_2(b)] \right\} \cup$$

$$\left\{ (r, \overline{\theta}, -\sqrt{2(h - V_\mu)}, \mu) \mid r \in [r_1(b), r_2(b)] \right\}$$

and

$$\gamma_2(b) = \left\{ (\overline{r}, \theta, \sqrt{2(h - V_\mu)}, \mu) \mid \theta \in [0, 2\pi] \right\}$$

for chosen $\overline{r} \in [r_1(b), r_2(b)]$ and $\overline{\theta} \in [0, 2\pi)$. If $\lambda = p_\theta d\theta + p_r dr$ is the Liouville form, the Arnold formula (see Theorem 4.32), yields

$$J_1(b) = \int_{\gamma_1(b)} \lambda = \int_{r_1(b)}^{r_2(b)} \sqrt{2(h - V_\mu)} dr + \int_{r_2(b)}^{r_1(b)} \left(-\sqrt{2(h - V_\mu)} \right) dr$$

$$= 2 \int_{r_1(b)}^{r_2(b)} \sqrt{2(h - V_\mu)} dr \qquad\qquad (4.41)$$

$$= 2\pi \left(-\mu + \frac{K}{\sqrt{-2h}}\right);$$ (4.42)

see Remark 4.38. On the other hand,

$$J_2(b) = \int_{\gamma_2(b)} \lambda = \int_0^{2\pi} \mu \, d\theta = 2\pi \mu.$$ (4.43)

Note that (4.42) and (4.43) yield the following formula expressing the Hamiltonian of the Kepler system in terms of the action variables:

$$H = -\frac{2\pi^2 K^2}{(J_1 + J_2)^2}.$$ (4.44)

From (4.44) one easily computes the frequencies of the system (see Definition 4.30):

$$\nu_i = \frac{\partial H}{\partial J_i} = \frac{4\pi^2 K^2}{(J_1 + J_2)^3}, \quad \text{for } i = 1, 2,$$ (4.45)

which implies that all the orbit are periodic with period

$$T = \frac{1}{\nu_i} = \frac{2\pi K}{(-2H)^{\frac{3}{2}}}.$$

The angles $(\vartheta_1, \vartheta_2)$ corresponding to (J_1, J_2) are found introducing a generating function S, which defines a symplectomorphism mapping the physical variables (r, θ, p_r, p_r) to $(J_1, J_2, \vartheta_1, \vartheta_2)$. To define such an S, first, we write

$$p_\theta \overset{(4.43)}{=} \frac{J_2}{2\pi} \quad \text{and} \quad p_r = \pm\sqrt{-\left(\frac{4\pi^2 K^2}{(J_1 + J_2)^2} + 2V_{J_2}(r)\right)},$$ (4.46)

where V_{J_2} simply means V_μ and where μ has been substituted with its value in terms of J_2, (4.43). To define the generating function we are looking for, we choose the *positive* determination of the square root in (4.46), we fix $P_0 = (r_0, 0)$ and we write

$$S(r, \theta, J_1, J_2) = \int_{P_0}^{P} (p_r dr + p_\theta d\theta) = J_2\theta + \int_{r_0}^{r} \sqrt{-\left(\frac{4\pi^2 K^2}{(J_1 + J_2)^2} + 2V_{J_2}(u)\right)} \, du,$$

which yields

$$\vartheta_1 = \int_{r_0}^{r} \frac{\partial}{\partial J_1}\sqrt{-\left(\frac{4\pi^2 K^2}{(J_1 + J_2)^2} + 2V_{J_2}(u)\right)} \, du,$$

$$\vartheta_2 = \theta + \int_{r_0}^{r} \frac{\partial}{\partial J_2} \sqrt{-\left(\frac{4\pi^2 K^2}{(J_1 + J_2)^2} + 2V_{J_2}(u)\right)} \, du.$$

It is worth making a couple of final comments. First, note that the action variables for the Kepler system can be computed *before* the reduction to the system recalled above; see, for example, [84]. In this case, still setting $m = 1$, the Hamiltonian in spherical coordinates is

$$H = \frac{1}{2}\left(p_r^2 + \frac{p_\theta^2}{r^2} + \frac{p_\varphi^2}{r^2 \sin^2\theta}\right) - \frac{K}{r}.$$

Using the angular variables θ and φ to define two of the three integration cycles, one finds

$$J_2 = 2\pi p_\theta \quad \text{and} \quad J_3 = 2\pi p_\varphi,$$

where J_2 is the same as in (4.43), while J_3 is the action corresponding to the cyclic coordinates φ, i.e., to the first integral p_φ. The last action variable, which in our notation is J_1 and corresponds to r, can be written as $J_1 = -(J_2 + J_3) + 2\pi \frac{K}{\sqrt{-2h}}$ (see (4.42) and Remark 4.38), which yields the following analog of (4.44):

$$H = -\frac{2\pi^2 K^2}{(J_1 + J_2 + J_3)^2},$$

and the frequencies as in (4.45).

We would like to alert again the reader (see Remark 4.33) that in our computations we used Formula (4.22) for the action variables, which differs from the one found in some reference by the normalizing factor $\frac{1}{2\pi}$; see, for example, [84].

Remark 4.38 The integral in (4.41) can be computed using, for example, a trigonometric substitution or via the following argument which we borrowed from the beautiful text [210]. First, observe that

$$\int_{r_1}^{r_2} \sqrt{2(h - V_\mu)} \, dr \overset{(4.39)}{=} \int_{r_1}^{r_2} \sqrt{2h + \frac{2K}{r} - \frac{2\mu^2}{r^2}} \, dr$$

$$= \int_{r_1}^{r_2} \frac{\sqrt{2hr^2 + 2Kr - 2\mu^2}}{r} \, dr.$$

Extending the previous to the complex plane and forgetting, for the time being, the actual values of the constants, one arrives to the following indefinite integral:

$$\int \frac{\sqrt{Az^2 + Bz + C}}{z} \, dz,$$

whose computation now can be performed via the methods of the residues. To this end it is worth noting that the eventual poles of this integral are located at $z = 0$ and at $z = \infty$. A simple check shows that the residue at $z = 0$ is $\sqrt{C}$, while to find out the value of the residue at $z = \infty$, it is convenient to introduce $w = \frac{1}{z}$, which forces $dz = -\frac{1}{w^2}dw$, yielding

$$\int \frac{\sqrt{Az^2 + Bz + C}}{z}dz = -\int \frac{\sqrt{A + Bw + Cw^2}}{w^2}dw.$$

This implies that the residue at $z = \infty$ is $\frac{B}{\sqrt{A}}$ and that

$$\int_{r_1}^{r_2} \frac{\sqrt{Az^2 + Bz + C}}{z}dz = 2\pi\iota \left(\sqrt{C} - \frac{B}{2\sqrt{A}}\right).$$

Substituting $A = 2h$, $B = 2K$, and $C = -\mu^2$ in the previous formula yields

$$\int_{r_1}^{r_2} \frac{\sqrt{2hr^2 + 2Kr - 2\mu^2}}{r}dr = 2\pi\iota \left(\sqrt{-\mu^2} - \frac{2K}{2\sqrt{2h}}\right) = 2\pi \left(-\mu + \frac{K}{\sqrt{-2h}}\right),$$

which proves (4.41). $\triangle$

3 Concluding Remarks and Further Topics

3.1 *Integrable Systems and Action-Angle Coordinates on Poisson Manifolds*

In this work we focus on integrable systems and action-angle coordinate on symplectic manifolds. Now we provide a quick overview of the situation in the case of Poisson manifolds, following [6, Sections 4.2, 4.3] and [149]. Action-angle coordinates have been constructed also on contact and on Dirac manifolds. This is definitely beyond the scope of this book, but the interested reader can find much more about this together with very clear proofs in the insightful [258].

We begin with the following.

Definition 4.39 Let (M, Π) be a Poisson manifold of dimension n and of rank $2r$ (see Definition 2.58), and let $F = (f_1, \ldots, f_s)$ be an s-uple of smooth functions with $s = n - r$ that Poisson commute and are generically functionally independent (meaning their differentials are independent on an open dense subset U_F of M). Then we say that (M, Π, F) is a completely integrable system (in the sense of Liouville). The integer r is the number of degrees of freedom of the completely integrable system and $2r$ is its rank. $\triangle$

In the previous definition, one can think that $n - 2r$ of the s functions $(f_1, \ldots, f_s)$ cut out the symplectic leaves associated to Π and the remaining r cut out Lagrangian submanifolds in the symplectic leaves. Indeed, $2r$ is the dimension of the symplectic leaves of maximal dimension (which are the generic ones for such a Poisson manifold) and r is the number of independent Hamiltonian vector fields on these symplectic leaves. The case of an integrable system on a symplectic manifold is readily obtained by taking $n = 2r$ and the rank of the Poisson tensor equal to $2r$ everywhere. Like in the case of an integrable system on a symplectic manifold, the differential equations representing any of the vector fields X_{f_i} is integrable by quadrature where f_i is one of the independent Hamiltonian vector field on the symplectic leaf (if f_i is a Casimir of course the corresponding vector field is identically zero).

On a Poisson manifold (M, Π) of dimension n and of rank $2r$, denote with $M_{(r)}$ the open subset of M where the rank of Π is $2r$. Then we have the following result that generalizes the Arnold-Liouville-Mineur theorem to the case of integrable systems on Poisson manifolds:

Theorem 4.40 *Let $(M, \Pi, F = (f_1, \ldots, f_s))$ be an integrable system of rank $2r$ on a Poisson manifold of dimension n with $s = n - r$. Then:*

- *The open subset $U_F \cap M_{(r)}$ is preserved by the flows of the vector fields X_{f_i} $i = 1, \ldots, s$. These vector fields define an integrable (in the sense of Frobenius) distribution $\mathcal{D}$ of rank r on $U_F \cap M_{(r)}$.*
- *For a point $m \in U_F \cap M_{(r)}$, the maximal integral submanifold of $\mathcal{D}$ passing through it is called the invariant manifold (of F) through m and denoted with S_m. In particular it is an embedded submanifold of M (see [6]).*
- *(i) If S_m is compact, then there is a diffeomorphism to the torus $\mathbb{T}^r = \mathbb{R}^r / \mathbb{Z}^r$ and through this diffeomorphism the vector fields $X_{f_1}, \ldots, X_{f_s}$ are mapped to linear (i.e., translation invariant) vector fields on the torus.*
  (ii) If S_m is not compact, but the flow of each X_{f_i}, $i = 1, \ldots, s$ is complete, then there is a diffeomorphism mapping S_m to a cylinder $\mathbb{R}^{r-q} \times \mathbb{T}^q$ $(0 \le q < r)$, and through this diffeomorphism the vector fields X_{f_i}, $i = 1, \ldots, s$ are again mapped to linear vector fields.

It is clear that both the notion of integrable system on a Poisson manifold and Theorem 4.40 are a direct generalization of the corresponding notion and results on a symplectic manifold.

Similarly, one can obtain a theorem for existence of action-angle coordinates for integrable systems on Poisson manifolds; see [149] for a detailed proof and [258] for a simpler proof that generalizes also to other cases, including contact manifolds and Dirac manifolds. Here is the result:

Theorem 4.41 *Let (M, Π) be a Poisson manifold of dimension n and rank $2r$. Suppose $F = (f_1, \ldots, f_s)$ is an integrable system on M according to the Definition 4.39. Let m be a point of M such that:*

(i) The differentials of $f_1, \ldots f_s$ are independent at m, i.e., m belongs to U_F.

(ii) The rank of Π at m is $2r$ (thus maximal).

(iii) The integral manifold S_m (see Theorem 4.40) is compact.

Then there exist $\mathbb{R}$-valued smooth functions $(J_1, \ldots, J_r, t_1, \ldots, t_{s-r})$ and $\mathbb{R}/\mathbb{Z}$-valued functions $(\theta_1, \ldots, \theta_r)$ defined in a neighborhood W of S_m in M such that:

(i) The functions $(\theta_1, \theta_r, J_1, \ldots, J_r, t_1, \ldots, t_{s-r})$ define a diffeomorphism $\phi : W \to \mathbb{T}^r \times B^s$ (where B^s is an open s-dimensional ball in $\mathbb{R}^s$).

(ii) The Poisson structure restricted to U in these coordinates assumes the form

$$\Pi = \sum_{i=1}^{r} \frac{\partial}{\partial \theta_i} \wedge \frac{\partial}{\partial J_i},$$

and the functions $t_1, \ldots, t_{s-r}$ are Casimir functions of Π on U. In particular ϕ is a Poisson map.

(iii) The leaves of the surjective submersion $F = (f_1, \ldots, f_s)$ are given by the projection onto the second component of $\mathbb{T}^r \times B^s$, in particular the Hamiltonians depend only on the coordinates $(J_1, \ldots, J_r, t_1, \ldots, t_{s-r})$.

In Theorem 4.41, the functions $(\theta_1, \ldots, \theta_r)$ are called angle coordinates, the functions $(J_1, \ldots, J_r)$ are called action coordinates, and the remaining functions $(t_1, \ldots, t_{s-r})$ are called transverse coordinates. We can visualize part of the content of Theorem 4.41 via the following diagram:

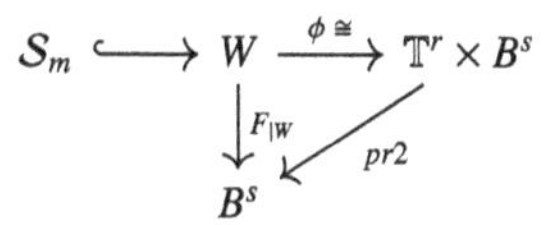

$$S_m \lhook\joinrel\longrightarrow W \xrightarrow{\ \phi\ \cong\ } \mathbb{T}^r \times B^s$$
$$\left\downarrow F_{|W}\right. \qquad \swarrow pr2$$
$$B^s$$

As stated here, Theorem 4.41 applies only to the region of the Poisson manifold where the Poisson tensor is of constant maximal rank. This however can be generalized (see [149]), introducing the notion integrable noncommutative system on a Poisson manifold and developing the corresponding results. We will not pursue this here; we just point out that roughly speaking, an integrable noncommutative system has more constant of motions than expected, so that the tori on which the system linearizes are of smaller dimension; the price to pay is that not all these constant of motion are in involution. Again for a complete development of this line of thought, that at the same time generalizes noncommutative integrable systems on symplectic manifold and Theorem 4.41 we refer to the illuminating [149].

3.1.1   An Example of Smooth Liouville Torus in a Poisson Manifold No Neighborhood of Which Admits Action-Angle Coordinates

Now we rework an example given in [149, Example 3.10] which provides a very simple integrable system on a three dimensional manifold P. This system is trivially integrable and yet it has a continuum of smooth invariant tori, such that for each of them there is no open neighborhood W on which action-angle coordinates are defined. Furthermore, P is completely foliated by smooth Liouville tori and there are no singular tori. Start with $M_0 = [-1, 1] \times (-1, 1)$. Call x the coordinate along $[-1, 1]$ and y the coordinate along $(-1, 1)$. Now we construct out of M_0 a Möbius strip M identifying the points of the form $(-1, y)$ with the points of the form $(1, -y)$: basically the two edges $-1 \times (-1, 1)$ and $1 \times (-1, 1)$ are identified inverting the orientation of one of them.

Observe that the function x descends to M to a smooth function, while the function y does not descend even to a continuous function on M. The function y^2, however, does descend to a smooth function on M, as any even smooth function of y. Notice that the vector field $\mathcal{V} := \frac{\partial}{\partial x}$ is still defined on M, but it has a peculiar property, which is the key in the construction of this example.

Indeed, $\mathcal{V}$ is always periodic, but it has period 2 if one looks at the integral curves going through points of the form $(x_0, 0)$, while it has period 4 if one looks at the integral curves going through points of the form (x_0, y_0) with $y_0 \neq 0$. The period of $\mathcal{V}$ does not depend continuously on the initial condition. The idea is to make these integral curves the tori of an integrable system of a Poisson manifold: then for the torus corresponding to the integral curve going through the points of the form $(x_0, 0)$ (the central circle), there cannot be any action-angle coordinate in a neighborhood of it; the corresponding fibration is not locally trivial around the fiber corresponding to the circle with smallest period.

This is the central idea. Now we develop it further introducing the Poisson manifold P. Define $P_0 = M_0 \times \mathbb{R}$, where we call z coordinate on the $\mathbb{R}$ factor and $P = M \times \mathbb{R}$. On P we consider the Poisson tensor $\Pi = \mathcal{V} \wedge \frac{\partial}{\partial z}$ and similarly we consider on P_0 the Poisson tensor $\Pi_0 = \frac{\partial}{\partial x} \wedge \frac{\partial}{\partial z}$. They are both of rank 2 (for P_0 we do not consider the boundary) and Π is obtained by projecting Π_0 to P. There is however a fundamental difference between the two Poisson tensors. The algebra of Casimir functions for Π_0 is generated by smooth functions of y, while the algebra of Casimir functions for Π is generated by smooth *even* functions of y

Consider now the map $F : P \to \mathbb{R}^2$ given by $F = (f_1, f_2)$ with $f_1 = y^2$ and $f_2 = z$. Observe that f_1 and f_2 Poisson commute everywhere on P and they are functionally independent away from the locus $C_0 = \{y^2 = 0\} \subset P$. Thus (P, Π, F) defines an integrable system on the Poisson manifold P according to the definition 4.39. Let us analyze the Liouville tori in this case and the existence of action-angle variables in a neighborhood of an invariant Liouville torus. Observe that for each $a \geq 0$, the locus $C_a = \{y^2 = a\} \subset P$ is a smooth symplectic manifold (these are exactly the symplectic leaves of Π in P). Moreover, each C_a, with $a \geq 0$ is a cylinder, which is a product of a circle $\mathbb{S}^1 \times \mathbb{R}$.

Observe that $f_2^{-1}(b)$ is just a copy of the Möbius strip. Now we consider $F^{-1}(a, b)$, with $a \geq 0$ and b arbitrary real, which can be thought of intersection of (a copy of) the Möbius strip with the cylinder C_a: this intersection is a circle $S_{a,b}$ which can be easily visualized on M_0: if $a = 0$ it is just the equator of M_0, and if $a > 0$, then it is given in M_0 by two parallel segments located at heights $y = \pm\sqrt{a}$ that in M become a single circle. Notice however that $S_{a,b}$ (for $a > 0$) and $S_{0,b}$ are such that $\pi : S_{a,b} \overset{2:1}{\to} S_{0,b}$ is a two to one unramified covering. These circle are exactly the invariant Liouville tori, since they are the integral curves of the Hamiltonian vector field $\mathcal{V}$ (whose Hamiltonian is, up to a sign, the function f_2). They are one-dimensional as expected (in this case $n = 3$, $2r = 2$, so $r = 1$).

By Theorem 4.41, for each of the tori $S_{a,b}$ (with $a > 0$) there exists a neighborhood $W \subset P$ where action-angle coordinates are defined. However, if we consider one of the invariant tori of the form $S_{0,b}$, they have no neighborhood $W \subset P$ where action-angle variables exist: this is because any such neighborhood W would contain tori $S_{a,b}$ with $a > 0$ arbitrarily close to $S_{0,b}$ and the corresponding Lagrangian fibration is not locally trivial at $a = 0$. The reason is that the period of $\mathcal{V}$ jumps discontinuously from 2 to 4 transitioning from $a = 0$ to $a > 0$ (or if one prefers, all the fibers $S_{a,b}$ ($a > 0$) of the fibration next to $S_{0,b}$ are unramified double covers of $S_{0,b}$. And indeed Theorem 4.41 cannot be applied here since df_1 vanishes identically on $S_{0,b}$.

Therefore, there is a continuum of invariant Liouville tori $S_{0,b}$ ($b \in \mathbb{R}$), none of which has a neighborhood W in P in which action-angle coordinates can be defined.

Next, we introduce a few topics related to the theory of Lagrangian fibrations and not discussed in the main body of this chapter. In what follows all the fibrations considered are with connected fibers.

3.2 Global Action-Angle Variables, Monodromy, and the Spherical Pendulum

One of the main steps in the construction of the (local) action-angle coordinates is the choice of a local section σ of the Lagrangian fibration $\pi : M \to B$. Once such a section is chosen, one can define the map (4.5) which turns out to be a local diffeomorphism and defines the period lattice $\Lambda(U)$ of the Lagrangian fibration over the open set $U \subset M$ supporting σ. All this is summarized in the diagram (4.9) which we propose here again for the reader's and since the hypothesis convenience:

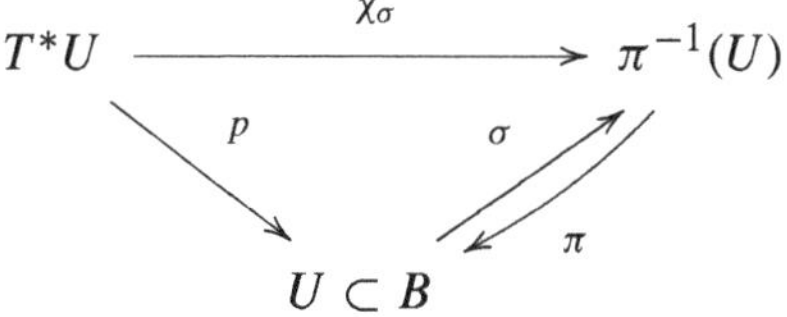

In particular χ_σ is an anti-symplectomorphism if and only if the section σ is Lagrangian; see Theorem 4.22. Under this assumption, a local set of action-angle coordinates can be defined over $\pi^{-1}(U)$, and since the hypothesis about σ of being Lagrangian is not restrictive (see (AA1)), it follows that in a suitable tubular neighborhood of every fiber of $\pi : M \to B$ one can introduce a set of action-angle coordinates. In other words, there is no obstruction to the existence of local action-angle variables. Note that the existence of a set of action-angle variables (ϕ, I) on $\pi^{-1}(U)$ is equivalent to the existence of a smooth map:

$$(\phi, J) : \pi^{-1}(U) \to \mathbb{T}^n \times \mathbb{R}^n$$

such that:

(i) ϕ is injective on the fibers of π.
(ii) J is constant on the fibers of π.
(iii) The restriction of ω to $\pi^{-1}(U)$ is exact, i.e., $\omega_{\pi^{-1}(U)} = \sum_{i=1}^n dJ_i \wedge d\phi_i$.

A set of coordinates (ϕ, J) globally defined on M and satisfying (i), (ii), and (iii) are called *global action-angle coordinates* on M. As the attentive reader may suspect reading the above definition, the existence of *global* action-angle variables for a given Lagrangian fibration with compact fibers is, in general, obstructed. In fact

Theorem 4.42 (Duistermaat) *On the total space of the Lagrangian fibration π : $(M, \omega) \to B$, there exists a global set of action-angle coordinates if and only if the* monodromy *and the* Chern class *of the fibration are trivial and ω is exact.*

For the proof of this important result we refer the reader to the reference [71]. Here we would like to make a few comments about this statement. In particular about the claim that the obstruction to the existence of global action-angle variables is contained in two topological invariant of the fibration, its Chern class and its monodromy, and in the exactness of the symplectic form. While the latter is clear, we feel that it could be worth spending a few words to clarify the meaning of the former two. First, the Chern class of the fibration $\pi : M \to B$ is a cohomology class associated to the period lattice Λ which defines the obstruction to the existence of a global Lagrangian section of $\pi : M \to B$. This means that such a class is zero if and only the Lagrangian fibration admits such a global section. On the other hand, the monodromy of $\pi : M \to B$ is defined as follows. Since the period lattice is a covering of B, for each $b \in B$ one can define a *left action* of $\pi_1(B, b)$, the fundamental group of B at the base point b, on $\pi^{-1}(b) \cap \Lambda := \Lambda_b$ as follows. For each $\xi \in \Lambda_b$ and for each loop $\gamma : [0, 1] \times B \to B$ based at b, we define $[\gamma]\xi = \tilde{\gamma}(1)$, where $\tilde{\gamma} : [0, 1] \to \Lambda$ is the unique lift of γ such that $\tilde{\gamma}(0) = \xi$. One can prove that $[\gamma]\xi$ does not depend on the choice the representative γ, that $[0]\xi = \xi$ and that $[\gamma_1] \cdot [\gamma_2]\xi = [\gamma_1 * \gamma_2]\xi$, for all $[\gamma_1], [\gamma_2] \in \pi_1(B, b)$ and $[0]$ is the class of the trivial, i.e., contractible, loop based at b. In the previous formulas $\gamma_1 * \gamma_2$ is the concatenation of the loops with base at b. This action is equivalent to a representation of $\pi_1(B, b)$ in Λ_b, or, equivalently, to an homomorphism $\rho : \pi_1(B, b) \to \mathrm{Aut}(\Lambda_b)$ which under our topological assumptions turns out to be independent of the choices made and for such a reason is called the *monodromy representation* of the covering Λ and its image in $\mathrm{Aut}(\Lambda_b)$ is called the *monodromy* of Λ. In particular, the monodromy is trivial if the image of ρ consists only of the identity automorphism. Note that the choice of a local basis for Λ_b yields an isomorphism $\mathrm{Aut}(\Lambda_b) \simeq \mathrm{GL}_n(\mathbb{Z})$ and for this reason the monodromy of Λ can be identified with a subgroup of this discrete group. Before moving on with our presentation, we would like to mention that the monodromy representation is equivalent to the existence of a *flat connection* on the (co)tangent bundle of B, whose bases of (locally defined) flat sections are made of (locally defined) closed one-forms on B. The choice of such a basis on the neighborhood of each point defines on B a structure of an affine manifold; see Sect. 3.3.

3.2.1 Spherical Pendulum

Nowadays, there are many known examples of Hamiltonian systems whose corresponding Lagrangian fibration is not topological trivial and for which no global action-angle variables are available; see, for example, [58, 71, 76] and the encyclopedic treatise [59]. To introduce this phenomenon to reader we give a brief and incomplete description of the classical example of the spherical pendulum, which is discussed in every details in [59].

This is a simple dynamical system composed of a bob hanged with a rigid rope at the center of a two-dimensional sphere on which surface can freely oscillate, see (Fig. 4.5). The phase space of this system is identified with (T^*S^2, Ω), the cotangent bundle of the two-dimensional sphere endowed with its canonical symplectic form. Denoting by x, y, and z the standard coordinates of $\mathbb{R}^3$, setting all the physical constants equal to 1, the Lagrangian of the system is

$$\mathcal{L} = T - U = \frac{\dot{x}^2 + \dot{y}^2 + \dot{z}^2}{2} - z \tag{4.47}$$

constrained on the cotangent bundle of S^2, which can be described by $x^2 + y^2 + z^2 = 1$ and $x\dot{x} + y\dot{y} + z\dot{z} = 0$. Note that (4.47) has an S^1-symmetry generated by $p_z = x\dot{y} - y\dot{x}$, the z-component of the angular momentum. To proceed further, we can introduce the energy-momentum map $\mathcal{E}$ of the spherical pendulum (see Remark 4.37), and then analyze how the topological type of its fibers changes

Fig. 4.5 Configuration space
of the spherical pendulum

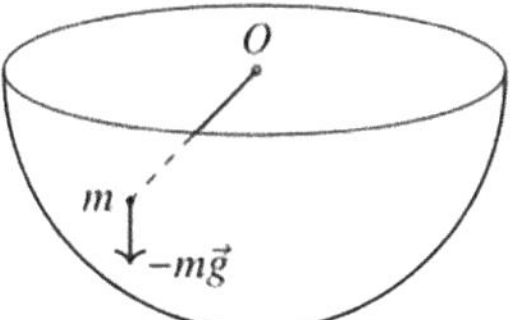

as a function of its image (see Sect. 2.2). To this end, after introducing spherical coordinates $(\theta, \varphi) \in [0, 2\pi) \times [0, \pi]$, (4.47) assumes the following form:

$$L = \frac{\dot{\varphi}^2 + \dot{\theta}^2 \sin^2 \varphi}{2} - \cos \varphi,$$

while the momenta conjugate to the angular coordinates θ and φ becomes

$$p_\theta = \frac{\partial L}{\partial \dot{\theta}} = \dot{\theta} \sin^2 \varphi \quad \text{and} \quad p_\varphi = \frac{\partial L}{\partial \dot{\varphi}} = \dot{\varphi}.$$

The Legendre transform and the previous formulas yield

$$H = \frac{1}{2} \left(p_\varphi^2 + \frac{p_\theta^2}{\sin^2 \varphi} \right) + \cos \varphi \quad \text{and} \quad L = p_\theta.$$

In this way we are left to study the set of the critical values of $\mathcal{E} : T^* S^2 \to \mathbb{R}^2$, where

$$\mathcal{E} = (L, H) = \left(p_\theta, \frac{1}{2} \left(p_\varphi^2 + \frac{p_\theta^2}{\sin^2 \varphi} \right) + \cos \varphi \right). \tag{4.48}$$

Before moving on with our discussion, it is worth noting that:

(i) This analysis needs to take in consideration the constraint involved in the description of the system.
(ii) The spherical coordinates introduce *apparent singularities* in the formulas which force to consider separately the cases $\varphi = 0, \pi$ and $\varphi \in (0, \pi)$. In particular, to analyze the first case it not possible to use this representation. We refer the reader to [59, Chapter IV, Section 3.2] for a detailed analysis of this problem. See also [58, 76].

Now one can prove that the singular set of the energy-momentum map is

$$M_{\text{sing}} = P_N \cup P_S \cup \left\{ (\theta, \varphi, p_\theta, p_\varphi) \mid p_\varphi = 0 \text{ and } p_\theta = \pm \frac{\sin^2 \varphi}{\sqrt{-\cos \varphi}}, \ \varphi \in \left(\frac{\pi}{2}, \pi \right) \right\},$$

where $P_N = ((0, 0, 1), (0, 0, 0))$ and $P_S = ((0, 0, -1), (0, 0, 0))$, i.e., they are the zero (co)vectors above the north and south pole, respectively, of S^2. These two points are the unstable and the stable equilibrium points of the system, respectively, and their image under the energy-momentum map is $\mathcal{E}(P_N) = (1, 0)$ and $\mathcal{E}(P_S) = (-1, 0)$. More in general,

$$\mathcal{E}(M_{\mathrm{sing}})$$

$$= \{(0, 1), (0, -1)\} \cup \left\{ (\mu, h) \in \mathbb{R}^2 \mid h = -\frac{\sin^2 \varphi}{2 \cos \varphi} + \cos \varphi, \ \mu = \pm \frac{\sin^2 \varphi}{\sqrt{-\cos \varphi}} \right\}, \tag{4.49}$$

where $\varphi \in (\frac{\pi}{2}, \pi)$. To get a clearer picture of the set (4.49) it is convenient to introduce the amended potential for the spherical pendulum

$$V = \frac{p_\theta^2}{2 \sin^2 \varphi} + \cos \varphi, \tag{4.50}$$

and write

$$H = \frac{p_\varphi^2}{2} + V.$$

Following the same strategy used to study the Kepler system (see Sect. 2.2), one first proves that, for all values μ of the momentum p_θ, (4.50) has only one critical point which can be shown to be a minimum and is contained in the open interval $(\frac{\pi}{2}, \pi)$. Denoting with $v_0(\mu)$ the value of (4.50) at this point, the image of the energy-moment map can be characterized as the subset of the plane (μ, h) where

$$h \geq v_0(\mu) \tag{4.51}$$

and the image of the singular set is $\{(0, 1), (0, -1)\} \cup \{(\mu, h) \in \mathbb{R}^2 \mid h = v_0(\mu)\}$. Note that the latter set is made of the points $(\mu, h) \in \mathbb{R}^2$ which are solutions of

$$\begin{cases} dV_\mu = 0 \\ h \ \ = V_\mu \end{cases} \tag{4.52}$$

and a simple computation shows that it coincides with

$$\left\{ (\mu, h) \in \mathbb{R}^2 \mid h = -\frac{\sin^2 \varphi}{2 \cos \varphi} + \cos \varphi, \ \mu = \pm \frac{\sin^2 \varphi}{\sqrt{-\cos \varphi}} \right\};$$

see (4.49). In (4.52) V_μ denotes the amended potential (4.50) evaluated at the fixed momentum μ. We close this analysis observing that the curve defined in (4.52) contains the point $(0, -1)$ and it can be parametrized as

$$\begin{cases} h = \frac{u^2}{2} - \frac{3}{2u^2} \\ \mu = u - \frac{1}{u^3} \end{cases}$$

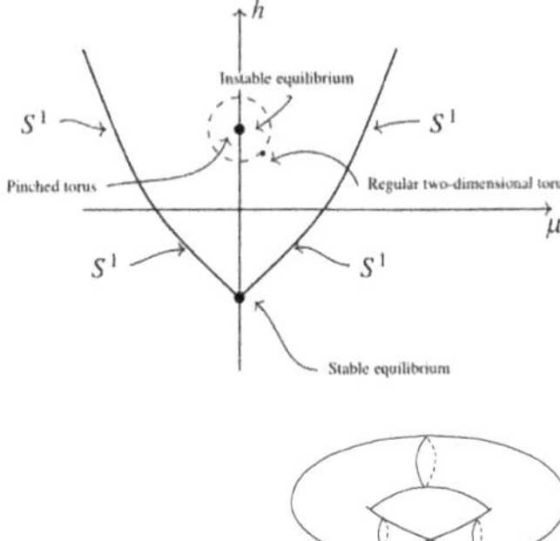

Fig. 4.6 Target space of the energy-momentum map of the spherical pendulum

Fig. 4.7 A torus with a pinched cycle

where $u = \pm\dfrac{1}{\sqrt{-\cos\varphi}}$ with $\varphi \in (\frac{\pi}{2}, \pi)$. The region of the plane (μ, h) defined in (4.51) is represented in (Fig. 4.6).

We discuss now the nature of the different kind of points in (4.51). We start with the points belonging to the region *strictly* above the curve described in (4.52). Each P in this set is a regular value of (4.48). Since $\mathcal{E}^{-1}(P)$ can be shown to be connected and compact, it is a two-dimensional torus; see Sect. 1. On the other hand, every point on curve represented in (4.52), excepted $(0, -1)$, corresponds to a so-called *relative equilibrium*, i.e., it corresponds to an equilibrium point of the system after its reduction by the S^1-symmetry generated by the first integral p_θ. These points correspond to the *horizontal motions* of the pendulum: they take place on circles sitting on planes orthogonal to the z-axis. The point $(0, -1)$, as already mentioned, is an equilibrium point but of different nature: it is the unique stable equilibrium point of the system; see the discussion in 4.1.2 in Sect. 4.1 of Chap. 3. Finally, the point $(0, 1)$, which is the unstable equilibrium point, sits inside the region of the regular points and its pre-image under $\mathcal{E}$ is a so-called *pinched torus* (Fig. 4.7).

At this point the period lattice of the Lagrangian fibration defined by the integrable system *degenerates* and it can be shown that the monodromy of the Lagrangian fibration is nontrivial; see [71]. This means that if one drags a basis (ℓ_1, ℓ_2) of the period lattice of a full turn along a simple loop completely contained in the set of regular values of $\mathcal{E}$ and encircling once the point $(0, 1)$, the basis (ℓ'_1, ℓ'_2) obtained as a final outcome of this process will differ by the original one. The difference will be measured by an invertible 2×2 matrix M, with integer coefficients, which is contained in the image of the monodromy representation. In other words, $\ell'_1 = M_{11}\ell_1 + M_{21}\ell_2$ and $\ell'_2 = M_{21}\ell_1 + M_{22}\ell_2$, where $M_{ij} \in \mathbb{Z}$. Theorem 4.42 implies that it is not possible to introduce global action-angle coordinates on the phase space of the spherical pendulum.

3.3 The Affine Structure Defined by a Lagrangian Fibration

We consider now the geometric structure inherited by the base of a Lagrangian
fibration with compact and connected fibers. More precisely, given such a fibration,
the choice of a set of action-angle coordinates in a suitable neighborhood of its
fibers endows its base with a structure of an *affine manifold*. The details of this
construction are presented below. To this end, recall that one of the first steps of
the construction of the action-angle variables consists in the choice, for any point of
B, of a local basis $\alpha_1, \ldots, \alpha_n$ of sections of the period lattice Λ. The existence of
these local bases is equivalent to the claim that $\pi : \Lambda \to B$ is a locally trivial $\mathbb{Z}^n$-
fiber bundle. In particular, if $(U, \alpha_1, \ldots, \alpha_n)$ and $(V, \beta_1, \ldots, \beta_n)$ are two systems
of locally trivializing charts for Λ with $U \cap V \neq \emptyset$, one has that, for each $b \in U \cap V$,

$$\alpha_i(b) = \sum_{i=1}^{n} m_{ij}(b)\beta_j(b), \ \forall i = 1, \ldots, n,$$

from which it follows that, for each $i, j = 1, \ldots, n$, m_{ij} is a constant function, with
values in $\mathbb{Z}$. Since both the αs and βs are closed, choosing a possibly smaller open
set, one can write

$$dI_i = \sum_{i=1}^{n} m_{ij} dJ_j, \ m_{ij} \in \mathbb{Z}, \ i, j = 1, \ldots, n,$$

which implies that

$$I_i = \sum_{i=1}^{n} m_{ij} J_j + c_i, \ i = 1, \ldots, n,$$

where $c_i, \in \mathbb{R}, \forall i = 1, \ldots, n$.

Recall now that

Definition 4.43 An *affine structure* on a manifold B is an atlas $\{(U_i, \phi_i)\}_{i \in I}$ whose
transition functions are affine, i.e., are such that

$$y_i = \sum_{j=1}^{n} A_{ij} x_j + C_i, \ i = 1, \ldots, n,$$

where $A \in \mathrm{GL}_n(\mathbb{R})$ and $C \in \mathbb{R}^n$. The affine structure is called *integral* if $A \in$
$\mathrm{GL}_n(\mathbb{Z})$. $\triangle$

Therefore,

Proposition 4.44 *The base of a Lagrangian fibration whose fibers are compact and
connected is endowed with an integral affine structure.*

The existence of an affine structure on a manifold is equivalent to the existence of a flat, torsion-free linear connection. In fact, if ∇ is such a connection on B, then for each point $b \in B$ one can find an open neighborhood V and a basis of one-forms $\alpha_1, \ldots, \alpha_n \in \Omega^1_B(V)$ such that $\nabla \alpha_i = 0$, $\forall i = 1, \ldots, n$. If $X_1, \ldots, X_n \in \mathfrak{X}_B(V)$ are such that $\langle \alpha_i, X_j \rangle = \alpha_i(X_j) = \delta_{ij}$ for all $i, j = 1, \ldots, n$, then

$$
\begin{aligned}
d\alpha_i(X_j, X_k) &= X_j \alpha_i(X_k) - X_k \alpha_i(X_j) - \alpha_i([X_j, X_k]) \\
&= -\alpha_i(\nabla_{X_j} X_k - \nabla_{X_k} X_j) \\
&= \nabla_{X_j} \alpha_i(X_k) - \nabla_{X_k} \alpha(X_j) \\
&= 0.
\end{aligned}
$$

In other words, the one-forms $\alpha_1, \ldots, \alpha_n$ are closed. Then, shrinking eventually the open set V, one can find local functions $x_1, \ldots, x_n$ such that $\alpha_i = dx_i$, $i = 1, \ldots, n$. If $\alpha_1, \ldots, \alpha_n \in \Omega^1_B(V)$ and $\beta_1, \ldots, \beta_n \in \Omega^1_B(W)$ are two bases of local flat sections for ∇ and $V \cap W \neq \emptyset$, then on the intersection $V \cap W$ one has $\alpha_i = \sum_{j=1}^n \psi_{ij} \beta_j$, $i = 1, \ldots, n$, and

$$
\nabla \alpha_i = \sum_{j=1}^n d\psi_{ij} \otimes \beta_j + \sum_{j=1}^n \psi_{ij} \nabla \beta_j,
$$

implying that

$$
\sum_{j=1}^n d\psi_{ij} \otimes \beta_j = 0, \forall i = 1, \ldots, n,
$$

which forces $d\psi_{ij} = 0$ for all $i, j = 1, \ldots, n$. In other words, on an open set possibly smaller than $V \cap W$, one has constant transition functions $\{\psi_{ij}\}_{i,j=1,\ldots,n}$ for T^*B. Then, if for all $i = 1, \ldots, n$, $dy_i = \alpha_i$ and $dx_i = \beta_i$, for suitable locally defined local functions $x_1, \ldots, x_n, y_1, \ldots, y_n$, one has

$$
dy_i = \sum_{j=1}^n \psi_{ij} dx_j \Rightarrow d\left(y_i - \sum_{j=1}^n \psi_{ij} x_j \right), \forall i = 1, \ldots, n,
$$

which implies

$$
y_i = \sum_{j=1}^n \psi_{ij} x_j + c_i, \forall i = 1, \ldots, n,
$$

where, $c_1, \ldots, c_n$ are suitable constant functions. On the other hand, if B is endowed with an affine structure, then given a local system of affine coordinates $x_1, \ldots x_n$, one can define a linear connection ∇ simply saying that

$$\nabla dx_i = 0, \ \forall i = 1, \ldots, n.$$

Doing that, if $y_1, \ldots, y_n$ is another system of affine coordinates defined on an open set intersecting nontrivially the one where $x_1, \ldots, x_n$ are defined, then, since

$$y_i = \sum_{j=1}^{n} A_{ij} x_j + c_i, \ \forall i = 1, \ldots, n,$$

for $A \in \mathrm{GL}_n(\mathbb{R})$ and $c \in \mathbb{R}^n$, one has that, for all $i = 1, \ldots, n$,

$$\nabla dy_i = 0,$$

as well. In other words, the flat, torsion-free linear connection associated to the affine structure of B is defined claiming that the local flat section are the differential of the coordinates functions of the affine atlas. In particular one has that

Corollary 4.45 *On the base of a Lagrangian fibration with connected and compact fibers, a flat, torsion-free linear connection is defined.*

In fact, it suffices to note that T^*B, as vector bundle, is isomorphic to $\Lambda \otimes_{\mathbb{Z}} \mathbb{R}$, i.e., T^*B is a vector bundle whose transition functions are constant.

3.4 Admissible Symplectic Forms

Before explaining the meaning of *admissible* in the title, we set up the framework we will work with. Let M be a $2n$-dimensional manifold and let $X \in \mathfrak{X}(M)$, let $X_1, \ldots, X_n \in \mathfrak{X}(M)$ and $a_1, \ldots, a_n \in C^\infty(M)$ such that

$$[X_i, X_j] = 0, \ X_1 \wedge \cdots \wedge X_n(m) \neq 0, \ \langle da_i, X_j \rangle = 0 \text{ and } da_1 \wedge \cdots da_n(m) \neq 0,$$

for all i, j and $m \in M$. These conditions imply that $\pi : M \to \mathbb{R}^n$, where $\pi = (a_1, \ldots, a_n)$, is a fibration whose fibers have trivial tangent bundle, trivialized by the vector fields X_is. If, furthermore, one assumes that the fibers of π are compact, then such a fibration is locally trivial and if $\alpha_1, \ldots, \alpha_n$ are one-forms defined in a suitable neighborhood of a given fiber $F_p = \pi^{-1}(p)$ and they are such that

$$\langle \alpha_i, X_j \rangle = A_{ij}, \quad \det A_{ij} \neq 0 \quad \text{and} \quad A_{ij} = A_{ij}(a_1, \ldots, a_n), \ \forall i, j, \tag{4.53}$$

then one can conclude that, on an eventually smaller tubular neighborhood of F_p, one can find $b_1, \ldots b_n$, $\alpha_i = db_i$, which complete $f_1, \ldots, f_n$ to a set of local coordinates in the given neighborhood of F_p. This is a framework only slightly more general than the one we had in Sect. 1, with the main difference that in this case there is no symplectic form involved. Note that the ODEs defined by any vector field of the form

$$X = \sum_{j=1}^{n} v_i X_i, \tag{4.54}$$

with $v_i = v_i(a_1, \ldots, a_n)$ for all i, is *integrable*. In fact, the system of ODEs corresponding to such a vector field is of the form

$$\begin{cases} \dot{a}_i & = 0 \\ \dot{b}_i & = v_i, \end{cases} \tag{4.55}$$

for all i, whose solutions are readily found to be

$$a_i = a_i(0) \quad \text{and} \quad b_i = v_i(a_1(0), \ldots, a_n(0))t + b_i(0), \ \forall i = 1, \ldots, n,$$

i.e., the v_is are the frequencies of the dynamical system; see Definition 4.30. Obviously this is not very surprising, it is just saying, one more time, that one way to integrate a system of n ODEs on $\mathbb{R}^{2n}$, is to find a new set of variables (a, b) such that the system becomes of the form (4.55). What Hamilton-Jacobi theory does, together with all its declinations, is to provide a theoretical framework where such transformations can take place. Now we come to the meaning of the *admissible* in the title. Starting from the previous observations, one could ask if, under the above assumptions, it is possible to find one symplectic structure, defined in the neighborhood of F_p, with respect to which the above vector field X is Hamiltonian. Such a symplectic form was dubbed *admissible* in [147], where the problem of its existence was studied in depth; see also Chap. 5. To summarize the discussion in the above cited reference, first, we observe that for every $f_1, \ldots, f_n$ such that $f_i = f_i(a_1, \ldots, a_n)$, one can define the two-form:

$$\omega_f = \sum_{i=1}^{n} df_i \wedge \alpha_i, \tag{4.56}$$

which is closed by construction and nondegenerate on the set where $df_1 \wedge \cdots \wedge df_n \neq 0$. Moreover, the vector fields

$$Y_i = \sum_{k=1}^{n} (A^{-1})_{ki} X_k,$$

where the A_{ij}s are as in (4.53), are Hamiltonian, more precisely $i_{Y_i}\omega_f = -df_i$ for all i. Now (4.54) is Hamiltonian with respect to ω_f *if and only if* $i_X\omega_f = -\sum_{i=1}^{n} v_l A_{il} df_i$ is closed, i.e., if and only if

$$\sum_{i=1}^{n} df_i \wedge d\tilde{v}_i = 0, \tag{4.57}$$

where we defined

$$\tilde{v}_i = \sum_{j=1}^{n} A_{ij} v_j, \ \forall i. \tag{4.58}$$

Note that this is a necessary condition for the vector field X to be Hamiltonian. It becomes sufficient as well, because of the relative Poincaré lemma applied to the tubular neighborhood of the fiber. In this way, given X as before, every independent set of solutions of (4.57) defines a symplectic form ω_f *admissible* with respect to X. There are two classes of solutions of (4.57) which will be discussed further in Chap. 5. The first is obtained choosing

$$f_i = \sum_{j=1}^{n} C_{ij}\tilde{v}_j, \tag{4.59}$$

with $C_{ij} \in \mathbb{R}$ and $C_{ij} = C_{ji}$ for all i, j. In fact under this hypothesis

$$\sum_{i=1}^{n} df_i \wedge d\tilde{v}_i \overset{(4.59)}{=} \sum_{i=1}^{n} \left(\sum_{j=1}^{n} C_{ij}d\tilde{v}_j \right) \wedge d\tilde{v}_i = \sum_{i,j=1}^{n} C_{ij}d\tilde{v}_j \wedge d\tilde{v}_i = 0.$$

With this choice one has

$$i_X\omega_f \overset{(4.56)}{=} i_X \left(\sum_{i=1}^{n} df_i \wedge \alpha_i \right) = -\sum_{i=1}^{n} df_i \langle \alpha_i, X \rangle \overset{(4.54)}{=} -\sum_{i=1}^{n} df_i \left\langle \alpha_i, \sum_{k=1}^{n} v_k X_k \right\rangle$$

$$= -\sum_{i,k=1}^{n} v_k \langle \alpha_i, X_k \rangle df_i \overset{(4.53)}{=} -\sum_{i=1}^{n} \left(\sum_{k=1}^{n} A_{ik} v_k \right) df_i$$

$$\overset{(4.58)}{=} -\sum_{i=1}^{n} \tilde{v}_i df_i \overset{(4.59)}{=} -\sum_{i,j=1}^{n} C_{ij}\tilde{v}_i d\tilde{v}_j,$$

i.e.,

$$i_X \omega_f = -d \left(\frac{1}{2} \sum_{i,j=1}^{n} C_{ij} \tilde{v}_i \tilde{v}_j \right).$$

In other words, choosing the f_is as in (4.59), the corresponding ω_f is admissible with respect to (4.54), which is Hamiltonian with Hamiltonian $H = \frac{1}{2} \sum_{i,j=1}^{n} C_{ij} \tilde{v}_i \tilde{v}_j$. The other class of solutions of (4.57) that will be considered further in Chap. 5 is obtained choosing

$$f_i = f_i(\tilde{v}_i), \ \forall i = 1, \ldots, n,$$

i.e., every f_i depends only on the corresponding $\tilde{v}_i$. For example, if one chooses $f_i = \frac{\partial F_i}{\partial \tilde{v}_i}$, where $F_i = F_i(\tilde{v}_i)$ for all i, a computation shows that (4.54) is Hamiltonian with Hamiltonian

$$H = \sum_{i=1}^{n} \frac{\partial F_i}{\partial \tilde{v}_i} \tilde{v}_i - F_i.$$

Note that this function is separated in the variables $\tilde{v}_i$. In the applications we will be interested in Chap. 5, the following case will be important. Suppose than in (4.53) $A_{ij} = \delta_{ij}$, choose $H \in C^\infty(M)$ such that $dH \wedge da_1 \wedge \cdots \wedge da_n = 0$, i.e., $H = H(a_1, \ldots, a_n)$, and define $f_i = \frac{\partial H}{\partial a_i}$ for all i. As already noticed, (4.56) is symplectic if and only if the f_is are independent, i.e., if and only if

$$\det \left(\frac{\partial^2 H}{\partial a_i \partial a_j} \right) \neq 0. \tag{4.60}$$

Moreover, if in (4.54) one chooses $v_i = f_i = \frac{\partial H}{\partial a_i}$, then X is Hamiltonian with respect to the ω_f so defined, i.e., ω_f, is admissible with respect to X. Although this conclusion follows from the discussion above, we prefer to make an explicit computation:

$$i_X \omega_f = -\sum_{i=1}^{n} df_i \langle \alpha_i, X \rangle = -\sum_{i,k=1}^{n} df_i \frac{\partial H}{\partial a_k} \langle \alpha_i, X_k \rangle = -\sum_{i,j=1}^{n} \frac{\partial^2 H}{\partial a_i \partial a_j} \frac{\partial H}{\partial a_i} da_j$$

$$= -\sum_{i=1}^{n} \frac{\partial H}{\partial a_i} d\left(\frac{\partial H}{\partial a_i} \right),$$

i.e., X is Hamiltonian with respect to ω_f with Hamiltonian equal to

$$K = \frac{1}{2} \sum_{i=1}^{n} \left(\frac{\partial H}{\partial a_i} \right)^2. \tag{4.61}$$

Remark 4.46 A few final comments are in order.

 (i) This last example follows from the general one discussed above choosing $A_{ij} = \delta_{ij}$, as already remarked, and $C_{ij} = \delta_{ij}$; see (4.59).
 (ii) The previous construction will be applied to the case of the tubular neighborhood of a fixed fiber of a Lagrangian fibration $\pi : M \to B$ defined on the symplectic manifold (M, ω). In this case $a_i = \pi \circ x_i$ for $i = 1, \ldots n$, where $x_1, \ldots, x_n$ is a set of local coordinates on B; see Sect. 1. In particular, the above discussion implies that, under the suitable hypothesis on H, the vector field X_H, which is Hamiltonian with respect to the background symplectic form, could be endowed with a second Hamiltonian description with respect to an admissible symplectic form. In particular, one could use as local coordinates on the tubular neighborhood of the fixed fiber, the action-angle coordinates. In this case the a_is would be substituted the J_is.
 (iii) The very last example, if looked at in the framework of the Lagrangian fibration, tells us that every Hamiltonian H satisfying (4.60) admits a second Hamiltonian description provided by an admissible symplectic form of the type (4.56). The relation between the two symplectic forms will be discussed in Chap. 5.
 (iv) Still looking at the last example from the enhanced framework of the Lagrangian fibrations, using a set of local action-angle variables (J, θ), one has $\omega = \sum_{i=1}^{n} dJ_i \wedge d\theta_i$, $\omega_f = \sum_{i=1}^{n} dv_i \wedge d\theta_i$, and $K = \frac{1}{2} \sum_{i=1}^{n} v_i^2$, where $v_i := \frac{\partial H}{\partial J_i}$ are the frequencies of the Hamiltonian system, as already observed above; see also Sect. 5 in Chap. 5. In particular, the condition of nondegeneracy of the Hamiltonian H implies that the frequencies can be used as local coordinates.

$$\triangle$$

4 Bibliographical Notes

The results presented in this chapter are central to the theory of the classical integrable systems and they can be found in many classical references like [1, 10], and [84]. Our presentation, which, differs from the one in these references because of its more geometrical flavor, was greatly inspired by the seminal paper [71], where the period lattice was introduced and used to analyze both the local and the global problem of the existence of the action-angle variables. Other two references where this problem is studied in depth are [195] and [64]. The first of these contains a detailed local study, while the second is an in-depth global analysis. Other references treating the problem of the complete integrability of the classical Hamiltonian systems are [19, 21] and [6]. In the first, the reader will find a very nice summary of this theory, with many examples and an introduction to the application of the methods of differential Galois theory and of algebraic geometry to the problem of the complete integrability. The last two references are somehow more advanced and they focus more on the algebraic geometric aspects of integrability. In our opinion the most complete reference treating the problem of integrability from the viewpoint of global differential geometry is [59]. A different and more powerful approach compared to the one introduced here can be found in the enlightening [258]. In this paper, the author presents a construction of action-angle variables for Poisson, pre-symplectic, contact, and Dirac manifolds, using a unifying principle that greatly simplifies the treatment.

Finally, we would like to mention other two very nice introductions to the theory of completely integrable systems. The first is the beautiful, unfortunately unpublished, lecture notes [85] where the reader will also find a nice and very readable introduction to the topic of noncommutative integrability which we did not touch in these lectures. The second one is another very nice, still unpublished, lecture notes in Italian [207], where the reader will find a nice description of the construction of the action-angle variables for both the Kepler and for the spherical pendulum. About our presentation, we would like to mention that the proof of Theorem 4.22 was borrowed from [110], while the discussion about the action-angle variables for the Kepler and the spherical pendulum was borrowed from [207].

Chapter 5
Elements of Bi-Hamiltonian Geometry

In this chapter, we present some rudiments of what is sometimes called multi-Hamiltonian geometry, which, roughly speaking, is the theory of the manifolds endowed with two or more Poisson structures subjected to suitable compatibility conditions. To explain the significance of this class of manifold in the theory of classical integrable systems, we recall that the theorem of Arnold-Liouville-Mineur does not provide any insight into how to find the first integrals of a given Hamiltonian systems, which are necessary to check its complete integrability. For this reason, it becomes important to look for additional structures that could provide a given Hamiltonian vector field with a sufficiently rich supply of first integrals in involution. As we will see in this chapter, the existence of multi-Hamiltonian representation of a dynamical system, induced by a family of compatible Poisson structures, implies the appearance of a set of first integral in involution, which, if further suitable conditions are verified, would imply the integrability of the dynamical system. In our presentation, a particular emphasis will be given to the multi-Hamiltonian structures, called *Poisson-Nijenhuis*, generated by a Poisson bivector field Π and a compatible Nijenhuis tensor N. For this class of structures, a local normal form will be presented, and it will be explained how they relate to master and conformal symmetries. The general notion of symmetry of a vector field still plays a fundamental role in our exposition. In particular, we decided to start the next section introducing the notion of recursion operator, of which the tensor N in the definition of a Poisson-Nijenhuis structure is a particular example. We will see how these operators act on the Lie algebra of infinitesimal symmetries of a given vector field, producing new symmetries from old ones.

A. Arsie, I. Mencattini, *Geometry of Integrable Systems*, Latin American Mathematics Series – UFSCar subseries,
https://doi.org/10.1007/978-3-031-96282-0_5

1 Nijenhuis and Haantjes Torsion of a $(1, 1)$-Tensor Field

The recursion operator N we are going to define is in particular a $(1, 1)$-tensor field. We discuss very briefly some aspects of the geometry of a $(1, 1)$-tensor field N on a manifold M. N can be viewed as a fiber-wise linear map $N : TM \to TM$.

Definition 5.1 Let N be a $(1, 1)$-tensor field on manifold M. Associated to N, there is vector-valued 2-form T_N, which is a $C^\infty(M)$-bilinear map $T_N : \mathfrak{X}(M) \times \mathfrak{X}(M) \to \mathfrak{X}(M)$ defined via

$$T_N(X, Y) = [NX, NY] - N([NX, Y] + [X, NY] - N[X, Y]), \ \forall X, Y \in \mathfrak{X}(M),$$
$$(5.1)$$

called the *Nijenhuis torsion* or simply the *torsion* of N. $\qquad\qquad \triangle$

Problem 5.2 Show that (5.1) is a $C^\infty(M)$-bilinear map, i.e., T_N is a tensor. $\qquad \triangle$

One can introduce a higher analogue of the torsion just defined as follows:

Definition 5.3 Let N be a $(1, 1)$-tensor field on manifold M. Then the *Haantjes torsion* of N is a vector valued 2-form given by the following expression:

$$H_N(X, Y) = T_N(NX, NY) - N\left(T_N(X, NY) + T_N(NX, Y) - NT_N(X, Y)\right),$$

where T_N is the (Nijenhuis) torsion of N $\qquad\qquad\qquad\qquad\qquad\qquad \triangle.$

H_N is simply obtained from the expression of T_N formally substituting the Lie bracket of two vector fields with T_N itself. One introduces the following:

Definition 5.4 $N \in \mathrm{End}(TM)$ is called a Nijenhuis operator if $T_N = 0$, and it is called a Haantjes operator if $H_N = 0$. $\qquad\qquad\qquad \triangle$

Let's briefly discuss the geometric meaning of these two concepts. For each $m \in M$, $N_m : T_m M \to T_m M$ defines a linear map. Suppose this map is diagonalizable, so it is has eigenvalues $\lambda_1(m), \ldots, \lambda_k(m)$ (possible with multiplicities) such that $T_m M = \ker(N_m - \lambda_1(m)\mathrm{Id}) \oplus \cdots \oplus \ker(N_m - \lambda_k(m)\mathrm{Id})$. Suppose this happens for all $m \in M$ (or for all m in an open set $U \subset M$), and furthermore assume that the dimension of any subspace $\ker(N_m - \lambda_i(m)\mathrm{Id})$, $i = 1, \ldots, k$ is constant for all $m \in U$. Then consider a sub-bundle of TM of eigenvector fields associated to the same eigenvalue function: this is called an *eigen-distribution* associated to N. For instance, under the hypotheses just stated for N there are k eigen-distributions $\mathscr{D}_i$ such that $\mathscr{D}_i(m) = \ker(N_m - \lambda_i(m)\mathrm{Id})$, $i = 1, \ldots, k$.

Under the assumptions above, one has $TM_{|U} = \mathscr{D}_1 \oplus \cdots \oplus \mathscr{D}_k$, where each $\mathscr{D}_i$ is a smooth distribution of constant rank. Then saying that each $\mathscr{D}_i$ is separately integrable means that for each of them, one can find an adapted coordinate chart, but not in general a coordinate chart adapted to more than one of them jointly. This means that even if $\mathscr{D}_i$ is integrable for each i, it does not follow in general that the direct sum of two (or more) eigen-distributions is integrable.

However, one has the following result due to Haantjes, for a proof see [103, Theorem 13.28]:

Proposition 5.5 *Let N be a $(1, 1)$-tensor field on M, and assume that for each $m \in U \subset M$, the operator N_m is semi-simple with real eigenvalues $\lambda_i(m)$ and that the multiplicity of each eigenvalue is constant on U. Assume that the eigenvalues $\lambda_i(m)$ are smooth $(i = 1, \ldots, k)$ and the same holds for each eigen-distribution $\mathscr{D}_i$. Then each $\mathscr{D}_i$ and each arbitrary direct sum $\mathscr{D}_{i_1} \oplus \cdots \oplus \mathscr{D}_{i_l}$, where $\{i_1, \ldots, i_l\} \subset \{1, \ldots, k\}$ is integrable (in the sense of Frobenius) if and only if $H_N = 0$ identically on U.*

Of course, $T_N = 0$ implies $H_N = 0$, so the vanishing of the Nijenhuis torsion is a sufficient condition for the integrability (in the sense of Frobenius) of the eigen-distributions mentioned above. With regard to the vanishing of the Nijenhuis torsion T_N, we have the following proposition due to Nijenhuis (for a proof, see [103, Theorem 13.29]):

Proposition 5.6 *Let N be a $(1, 1)$ tensor field on a manifold M and suppose that $T_N = 0$ identically. Suppose that the eigenvalues and the eigen-distributions of N satisfy the same assumptions as in the Proposition 5.5. Then:*

(i) *Each $\mathscr{D}_i$ and each arbitrary direct sum $\mathscr{D}_{i_1} \oplus \cdots \oplus \mathscr{D}_{i_l}$, where $\{i_1, \ldots, i_l\} \subset \{1, \ldots, k\}$ is integrable (in the sense of Frobenius).*
(ii) *If X is a vector field tangent to $\mathscr{D}_i$, then $X(\lambda_j) = 0$ for all $j \neq i$. This means that for each fixed j, the eigenvalue λ_j depends only on the coordinates of the integral submanifold of the distribution $\mathscr{D}_j$.*

In particular, from Proposition 5.6 it follows that if $T_N = 0$ and N is semi-simple with distinct eigenvalues $\lambda_1, \ldots, \lambda_n$, then in a suitable coordinate system, N is diagonal, and each eigenvalue depends only on one coordinate.

One can also ask what happens if N_m is not diagonalizable. In this case, one associates invariant subspaces to N_m coming from the Jordan normal form and can define generalized eigen-distributions. The integrability of these is controlled by *higher torsions* $T_N^{(k)}$, $k = 1, 2, \ldots$ where $T_N^{(1)} = T_N$, $T_N^{(2)} = H_N$. For much more on this, see the beautiful [233].

Problem 5.7 Requiring N to be a (smooth) $(1,1)$ tensor field on an n-dimensional manifold M is equivalent to say that in any coordinate chart with coordinates $\{x^1, \ldots, n^n\}$, N can be expressed as $N = \sum_{i,j=1}^n N_j^i(x) \frac{\partial}{\partial x^i} \otimes dx^j$ where the functions N_j^i are smooth. Prove that $\mathscr{D}_N$, the image distribution of N defined by $\mathscr{D}_N(m) := \mathrm{im}(N_m : T_m M \to T_m M)$, is always smooth, even if its rank is not constant (see Appendix B for the notion of rank and smoothness of a generalized distribution). $\triangle$

2　Recursion Operators

We start our discussion introducing first the concept of symmetry and then the one of a recursion operator of a given $X \in \mathfrak{X}(M)$.

Definition 5.8 A symmetry of a vector field X is a vector field Z such that $[X, Z] = 0$. △

The set $\mathscr{Z}(X)$ of all symmetries of a given vector field is a vector space that, as a consequence of the Jacobi identity, is closed with respect to the restriction of the Lie bracket defined on $\mathfrak{X}(M)$.

Remark 5.9 Note that the vector field of the previous definition should be called, more precisely, an *infinitesimal* symmetry of X. △

Given any $N \in \mathrm{End}(TM)$, its *adjoint* ${}^t N \in \mathrm{End}(T^*M)$ is defined by $\langle {}^t N(\alpha), X \rangle = \langle \alpha, NX \rangle$ for all $X \in \mathfrak{X}(M)$ and $\alpha \in \Omega^1(M)$.

Problem 5.10 Prove that the following conditions are equivalent:

$$T_N = 0,$$

$$\mathscr{L}_{NX} N = N \mathscr{L}_X N, \tag{5.2}$$

$$\mathscr{L}_{NX} {}^t N = \left(\mathscr{L}_X {}^t N \right) {}^t N,$$

$\forall X \in \mathfrak{X}(M)$. △

Problem 5.11 Prove that $\mathscr{L}_X N = 0$ if and only if $\mathscr{L}_X {}^t N = 0$. (Hint: Assuming that $\mathscr{L}_X N = 0$, given $\alpha \in \Omega^1(M)$ and $Y \in \mathfrak{X}(M)$, compute $\langle (\mathscr{L}_X {}^t N)\alpha, Y \rangle$. Then, assuming that $\mathscr{L}_X {}^t N = 0$, compute $\langle \alpha, (\mathscr{L}_X N)Y \rangle$.) △

After these preliminary observations, we can introduce the following:

Definition 5.12 Let $X \in \mathfrak{X}(M)$. A Nijenhuis operator $N \in \mathrm{End}(TM)$ such that $\mathscr{L}_X N = 0$ is called a *recursion operator* of X. △

Note that if $N \in \mathrm{End}(TM)$, $X \in \mathfrak{X}(M)$ is such that $\mathscr{L}_X N = 0$ and $Y \in \mathfrak{X}(M)$ is such that $\mathscr{L}_X Y = 0$, then $\mathscr{L}_X(N^k Y) = 0$ for all $k \geq 1$. In other words, if N is constant with respect to X, its action preserves the Lie algebra of symmetries $\mathscr{Z}(X)$ of X.

By Problem 5.7, every $N \in \mathrm{End}(TM)$ gives rise to a smooth distribution $\mathscr{D}_N$ on M. The condition $T_N = 0$ implies the involutivity of $\mathscr{D}_N$ (see Appendix B). In spite of this, since the rank of N is, in general, non-constant, one cannot use the Frobenius theorem to conclude about the integrability of $\mathscr{D}_N$. On the other hand, one can prove the following integrability results:

Proposition 5.13 *Let $N \in \mathrm{End}(TM)$ be a Nijenhuis operator.*

(i) If $\mathscr{L}_X N = 0$, i.e., if N is a recursion operator of X, then

$$\left[N^k X, N^l X\right] = 0, \ \forall k, l \geq 1. \tag{5.3}$$

(ii) If $\alpha \in \Omega^1(M)$ is such that $d\alpha = 0$ and $d\left({}^t N \alpha\right) = 0$, then

$$d({}^t N^k \alpha) = 0, \ \forall k \geq 1.$$

Proof As far as (i) concerns, one notes that

$$\begin{aligned}
\left[N^k X, N^l X\right] &= T_N\left(N^{k-1} X, N^{l-1} X\right) + N\left(\left[N^k X, N^{l-1} X\right]\right. \\
&\quad \left. + \left[N^{k-1} X, N^l X\right] - N\left[N^{k-1} X, N^{l-1} X\right]\right) \\
&\overset{(5.1)}{=} N\left(\left[N^k X, N^{l-1} X\right] + \left[N^{k-1} X, N^l X\right] - N\left[N^{k-1} X, N^{l-1} X\right]\right),
\end{aligned}$$

which suffices to conclude arguing by induction because $[X, NX] = 0$. On the other hand, to prove (ii), one computes

$$\begin{aligned}
d\left({}^t N^k \alpha\right)(X, Y) &= X\langle {}^t N^k(\alpha), Y\rangle - Y\langle {}^t N^k(\alpha), X\rangle - \langle {}^t N^k(\alpha), [X, Y]\rangle \\
&= X\langle {}^t N^{k-1}(\alpha), NY\rangle - Y\langle {}^t N^{k-1}(\alpha), NX\rangle - \langle {}^t N^{k-1}(\alpha), N[X, Y]\rangle \\
&= \langle d i_X {}^t N^{k-1}(\alpha), NY\rangle + \langle {}^t N^{k-1}(\alpha), [X, NY]\rangle \\
&\quad - \langle d i_Y {}^t N^{k-1}(\alpha), NX\rangle - \langle {}^t N^{k-1}(\alpha), [Y, NX]\rangle \\
&\quad - \langle {}^t N^{k-1}(\alpha), N[X, Y]\rangle \\
&= \langle d i_X {}^t N^{k-1}(\alpha), NY\rangle - \langle d i_Y {}^t N^{k-1}(\alpha), NX\rangle \\
&\quad + \langle {}^t N^{k-1}(\alpha), [NX, Y]\rangle + \langle {}^t N^{k-1}(\alpha), [X, NY]\rangle \\
&\quad - \langle {}^t N^{k-1}(\alpha), N[X, Y]\rangle \\
&= \langle d i_{NX} {}^t N^{k-2}(\alpha), NY\rangle - \langle d i_{NY} {}^t N^{k-2}(\alpha), NX\rangle \\
&\quad + \langle {}^t N^{k-2}(\alpha), [NX, XY]\rangle \\
&= NY\langle {}^t N^{k-2}(\alpha), NX\rangle - \langle d i_{NY} {}^t N^{k-2}(\alpha), NX\rangle \\
&\quad + \langle {}^t N^{k-2}(\alpha), [NX, XY]\rangle \\
&= \langle d(i_{NY} {}^t N^{k-2}(\alpha), NX\rangle + \langle {}^t N^{k-2}(\alpha), [NY, NX]\rangle
\end{aligned}$$

$$- \overline{\langle di_{NY}{}^t N^{k-2}(\alpha), NX \rangle} + \overline{\langle {}^t N^{k-2}(\alpha), [NX, XY] \rangle}$$
$$= 0.$$

$\square$

The previous result clarifies, at least partially, the role of the recursion operators in the search for integrability conditions of a system of ODE. In fact (i) in (5.3) tells that a recursion operator of X gives rise to an abelian Lie subalgebra of $\mathscr{L}(X)$, the one generated by $\{N^k X\}_{k \geq 1}$. On the other hand, (ii) implies that if both α and ${}^t N \alpha$ are closed one-forms such that $i_X \alpha = 0$ and if N is a Nijenhuis recursion operator for X, then one can find a sequence of (locally defined) functions $\{f_k\}_{k \geq 1}$, such that $X(f_k) = 0$ for all k. In fact since α and $\alpha_k = {}^t N^k \alpha$ are exact, they are locally closed, so locally equal to df_k and df, respectively. Then

$$i_X(df_k) = X(f_k) = \mathscr{L}_X(f_k) = \mathscr{L}_X\big({}^t N^k df\big) = \big(\mathscr{L}_X {}^t N^k\big)df + {}^t N^k d\mathscr{L}_X(f) = 0,$$

since N is a recursion operator for X (see Exercise 5.11) and $\mathscr{L}_X(f) = X(f) = \langle X, df \rangle = i_X \alpha = 0$ by hypothesis. In other words, under the above assumptions, one can generate, at least locally, a sequence of first integrals of X, which can be used, if independent, to cut the number of degrees of freedom of the system.

We will now see how recursion operators appear in the realm of Poisson geometry. This will bring us to the main topic of this chapter, i.e., the notion of *Poisson-Nijenhuis* structure (see Sect. 4). Instead of diving in the most general framework, we start our discussion from a (slightly) more stringent hypothesis (see Sect. 4.1 where this topic will be discussed more in depth). General references for this part are [164, 165]. Let (M, ω_0) be a symplectic manifold and let ω_1 be a two-form. Then on $C^\infty(M)$, the following two brackets are defined:

$$\{f, g\}_0 = \omega_0(X_f, X_g) \quad \text{and} \quad \{f, g\}_1 = \omega_1(X_f, X_g), \ \forall f, g \in C^\infty(M), \qquad (5.4)$$

where X_f is the Hamiltonian vector field defined by f with respect to ω_0, i.e., $i_{X_f}\omega_0 = -df$. The first is the Poisson bracket defined by the symplectic form ω_0, while $\{\cdot, \cdot\}_1$ is a skew-symmetric (bi)derivation of $C^\infty(M)$, and for this reason, it corresponds to a bivector field Π_1 defined by the condition

$$\Pi_1(df, dg) = \omega_1(X_f, X_g), \ \forall f, g \in C^\infty(M).$$

Notice that due to the invertibility of ω_0, we can define the tensor $N = \omega_0^{\flat^{-1}} \circ \omega_1^\flat$, where $\omega^\flat(X) = i_X \omega$ for all X (see 1.16). Furthermore, using N, the following relation holds between ω_0 and ω_1:

$$\omega_1(X, Y) = \omega_0(NX, Y), \ \forall X, Y \in \mathfrak{X}(M), \qquad (5.5)$$

as it is immediate to check.

To proceed further with our discussion, we need to introduce the following simple, nevertheless important, general notion.

Definition 5.14 Two bilinear forms $B_0, B_1 : V \times V \to V$ are called *compatible* if

$$B_0(B_1(v, u), w) + B_1(B_0(u, v), w)) + \circlearrowleft (u, v, w) = 0, \ \forall u, v, w \in V. \tag{5.6}$$

where $\circlearrowleft (u, v, w)$ stands for the sum over all the *non-trivial* cyclic permutations of the arguments. $\triangle$

Now let

$$Y_f = \Pi_1^\sharp(df), \ \forall f \in C^\infty(M). \tag{5.7}$$

Since $\langle dg, Y_f \rangle = \langle dg, \Pi_1^\sharp(df) \rangle = \Pi_1(df, dg) = \omega_1(X_f, X_g) \overset{(5.5)}{=} \omega_0(NX_f, X_g) = \langle dg, NX_f \rangle$, one has

$$Y_f = NX_f, \ \forall f \in C^\infty(M).$$

We can now prove that if $\{\cdot, \cdot\}_0$ and $\{\cdot, \cdot\}_1$ are defined as in (5.4), we have the following:

Proposition 5.15 *The following conditions are equivalent:*

(i) $\{\cdot, \cdot\}_0$ *and* $\{\cdot, \cdot\}_1$ *are compatible.*
(ii) $d\omega_1 = 0$.
(iii)

$$[X_f, Y_g] + [Y_f, X_g] = X_{\{f,g\}_1} + Y_{\{f,g\}_0}, \ \forall f, g \in C^\infty(M).$$

Proof First one observes that the compatibility condition (5.6) reads as

$$\big\{f, \{g, h\}_1\big\}_0 + \big\{f, \{g, h\}_0\big\}_1 + \circlearrowleft (f, g, h), \ \forall f, g, h \in C^\infty(M).$$

Then one notes that

$$d\omega_1(X_f, X_g, X_h) = \big\{f, \{g, h\}_1\big\}_0 + \big\{f, \{g, h\}_0\big\}_1 + \circlearrowleft (f, g, h), \ \forall f, g, h \in C^\infty(M).$$

In fact

$$\begin{aligned}
d\omega_1\big(X_f, X_g, X_h\big) &= X_f \omega_1\big(X_g, X_h\big) + X_g \omega_1\big(X_h, X_f\big) + X_h \omega_1\big(X_f, X_g\big) \\
&\quad + \omega_1\big(X_h, [X_f, X_g]\big) + \omega_1\big(X_g, [X_h, X_f]\big) + \omega_1\big(X_f, [X_g, X_h]\big) \\
&= \big\{f, \{g, h\}_1\big\}_0 + \big\{f, \{g, h\}_0\big\}_1 + \circlearrowleft (f, g, h),
\end{aligned}$$

where we used the identities $[X_f, X_g] = X_{\{f,g\}_0}$ and $X_f(g) = \{f, g\}_0$ for all $f, g \in C^\infty(M)$. On the other hand, for all $f, g, h \in C^\infty(M)$,

$$\left\{\{f, g\}_1, h\right\}_0 + \left\{\{f, g\}_0, h\right\}_1 + \left\{\{h, f\}_1, g\right\}_0 + \left\{\{h, f\}_0, g\right\}_1 + \left\{\{g, h\}_1, f\right\}_0$$

$$+ \left\{\{g, h\}_0, f\right\}_1$$

$$= X_{\{f,g\}_1}(h) + Y_{\{f,g\}_0}(h) + X_g(Y_f(h)) + Y_g(X_f(h)) - X_f(Y_g(h)) - Y_f(X_g(h))$$

$$= \left(X_{\{f,g\}_1} + Y_{\{f,g\}_0} + [X_g, Y_f] + [Y_g, X_f]\right)(h).$$

$\square$

Remark 5.16 It is worth noting that if $\omega_1 = \omega_0$, the equivalence between (ii) and (iii) in the previous proposition is the well-known result stating that the bracket defined in (5.4) satisfies Jacobi if and only if $d\omega_0 = 0$ (see Remark 1.35). $\triangle$

Using the previous proposition, we can prove the following important result.

Theorem 5.17 *If $\{\cdot, \cdot\}_1$ is compatible with $\{\cdot, \cdot\}_0$, it satisfies the Jacobi identity if and only if $N \in \mathrm{End}(TM)$, defined by*

$$\omega_1(X, Y) = \omega_0(NX, Y), \quad \forall X, Y \in \mathfrak{X}(M), \tag{5.8}$$

is a Nijenhuis operator.

Proof First note that (5.8) is equivalent to $N = \omega_0^{\flat^{-1}} \circ \omega_1^{\flat}$, where $\omega^\flat(X) = i_X \omega$ for all X (see 1.16). Then observe that since T_N is a tensor, it suffices to check the condition $T_N = 0$ on the Hamiltonian vector fields. Since $\{\cdot, \cdot\}_1$ and $\{\cdot, \cdot\}_0$ are compatible, applying (iii) of Proposition 5.15, we obtain

$$\left[NX_f, X_g\right] + \left[X_f, NX_g\right] - N\left[X_f, X_g\right] = \left[Y_f, X_g\right] + \left[X_f, Y_g\right] - NX_{\{f,g\}_0}$$

$$= \left[Y_f, X_g\right] + \left[X_f, Y_g\right] - Y_{\{f,g\}_0}$$

$$\stackrel{(iii)\ \text{in Proposition 5.15}}{=} X_{\{f,g\}_1},$$

which implies $N([NX_f, X_g] + [X_f, NX_g] - N[X_f, X_g]) = N(X_{\{f,g\}_1}) = Y_{\{f,g\}_1}$. Since $\{\cdot, \cdot\}_1$ satisfies Jacobi if and only if $Y_{\{f,g\}_1} = [Y_f, Y_g]$ (see (EP3) in Chap. 2) and $[Y_f, Y_g] = [NX_f, NX_g]$, one concludes that $\{\cdot, \cdot\}_1$ satisfies the Jacobi identity if and only if $T_N(X_f, X_g) = 0$ for all $f, g \in C^\infty(M)$. $\square$

We summarize the previous discussion in the following:

Theorem 5.18 *Let ω_1 be a two-form on a symplectic manifold (M, ω_0) and let $\{\cdot, \cdot\}_0, \{\cdot, \cdot\}_1$ be the brackets defined in (5.4). Then:*

(i) $\{\cdot, \cdot\}_0$ and $\{\cdot, \cdot\}_1$ are compatible if and only if $d\omega_1 = 0$.

(ii) *If $\{\cdot, \cdot\}_0$ and $\{\cdot, \cdot\}_1$ are compatible, $\{\cdot, \cdot\}_1$ satisfies the Jacobi identity if and only if $N = \omega^{\flat-1} \circ \omega_1^\flat \in \mathrm{End}(TM)$ is a Nijenhuis operator. In this case, N is a recursion operator for every $X \in \mathfrak{X}(M)$ such that $\Pi_1^\sharp(dg) = X = \Pi_0^\sharp(df)$.*

Proof Part (i) and the first statement in (ii) are the content of Proposition 5.15 and Theorem 5.17. To conclude the proof, note that, since Π_1 and Π_0 are Poisson bivector fields, $\mathscr{L}_{Y_g}\Pi_1 = 0 = \mathscr{L}_{X_f}\Pi_0$ (see Lemma 2.26 in Chap. 2). For this reason

$$\mathscr{L}_X N = \mathscr{L}_X \left(\Pi_1^\sharp \Pi_0^{\sharp-1}\right) = \left(\mathscr{L}_{Y_g}\Pi_1^\sharp\right)\Pi_0^\sharp + \Pi_1^\sharp\left(\mathscr{L}_{X_f}\Pi_0^{\sharp-1}\right) = 0.$$

$\square$

We can introduce the following:

Definition 5.19 The pair (ω_1, ω_0) of two-forms where ω_0 is symplectic is called *compatible*, or a *compatible pair*, if (i) ω_1 is closed and (ii) the bracket $\{\cdot, \cdot\}_1$ defined in (5.4) is Poisson. $\triangle$

In other words, a pair of two-forms is compatible if and only if the corresponding brackets (5.4) are compatible and the bracket corresponding to ω_1 is Poisson.

It is worthwhile now to make a few comments about the relations existing among the structures introduced in this chapter up to this point.

(i) If (ω_1, ω_0) is a compatible pair and $\Pi_0^\sharp$ and $\Pi_1^\sharp$ are the Poisson tensors corresponding to $\{\cdot, \cdot\}_0$ and, respectively, to $\{\cdot, \cdot\}_1$, then

$$\Pi_1^\sharp = N\Pi_0^\sharp. \tag{5.9}$$

In fact,

$$\begin{aligned}
\langle dg, \Pi_1^\sharp(df)\rangle &= \{f, g\}_1 = \omega_1(X_f, X_g)\\
&= \omega_0(NX_f, X_g)\\
&= -\omega_0\left(\Pi_0^\sharp(dg), NX_f\right)\\
&\overset{(2.28)}{=} \langle dg, NX_f\rangle\\
&= \langle dg, N\Pi_0^\sharp(df)\rangle,
\end{aligned}$$

proving the statement. Note that the skew-symmetry of Π_1 and Π_0, together with the previous computation, shows that

$$\Pi_0^{\sharp t} N = N\Pi_0^\sharp.$$

(ii) As we have already noticed, if ω_1 and ω_0 are two-forms on M and, for example, ω_0 is non-degenerate, then there exists a unique $N \in \mathrm{End}(TM)$ such that

$\omega_1(X, Y) = \omega_0(NX, Y)$ for all $X, Y \in \mathfrak{X}(M)$. Suppose now that ω_1 is also non-degenerate. In this case, its *inverse* is a bivector field P_1, which can be described in terms of ω_0 and the above-defined N. Notice that the induced operator $P_1^\sharp$ is different from $\Pi_1^\sharp$. More precisely, one has

$$P_1^\sharp \overset{(2.28)}{=} -\left(\omega_1^\flat\right)^{-1} = -\left(\omega_0^\flat N\right)^{-1} = N^{-1}\left(-\omega_0^\flat\right)^{-1} = N^{-1}\Pi_0^\sharp,$$

which should be compared with (5.9). In particular, note that the assumption $d\omega_1 = 0$ implies that P_1 is a Poisson tensor whose Poisson bracket $\{\cdot, \cdot\}_{P_1}$, in general, is *not* compatible with $\{\cdot, \cdot\}_0$. Summarizing, assuming that ω_1 is a non-degenerate two-form and ω_0 is symplectic:

- $d\omega_1 = 0 \iff \{\cdot, \cdot\}_0$ is compatible with $\{\cdot, \cdot\}_1$, as defined in (5.4), but it does not yield the Jacobi identity for $\{\cdot, \cdot\}_1$.
- $d\omega_1 = 0 \Rightarrow$ that $\{\cdot, \cdot\}_{P_1}$ is a Poisson bracket, but it does not imply the compatibility between $\{\cdot, \cdot\}_{P_1}$ and $\{\cdot, \cdot\}_0$.

Remark 5.20 To a pair of compatible two-forms (ω_1, ω_0), it is naturally associated with a Nijenhuis operator N, defined by $N = (\omega_0^\flat)^{-1}\omega_1^\flat$. For this reason, in particular in the literature about integrable systems, a pair of compatible two-forms is more commonly called an ΩN-*structure*. The properties of these structures will be analyzed in more details, from a slightly different viewpoint, in Sect. 4.1. $\triangle$

In the next example, we present an important construction of compatible pairs of Poisson brackets (see [124, 238]).

Example 5.21 (Compatible Pairs on the Cotangent Bundle) Let Q be a manifold and $N \in \mathrm{End}(TQ)$. ^tN defines a smooth map $\tau_N : T^*Q \to T^*Q$ (see D.14), which can be used to pull back the Liouville form $\Theta \in \Omega^1(T^*Q)$ to get a new one-form θ on T^*Q. Defining $\omega = d\theta$ and $\{\cdot, \cdot\}_\omega : C^\infty(T^*Q) \times C^\infty(T^*Q) \to C^\infty(T^*Q)$ as in (5.4), i.e.,

$$\{f, g\}_\omega = \omega(X_f, X_g), \quad \forall f, g \in C^\infty(T^*Q), \tag{5.10}$$

where now X_f, X_g are the Hamiltonian vector fields defined by the canonical symplectic structure $\Omega = d\Theta$, one goes back to the framework introduced above, and it could be asked under which assumptions on N $\{\cdot, \cdot\}_\omega$ and $\{\cdot, \cdot\}$ form a pair of *compatible Poisson brackets* on T^*Q. Note that since ω is exact (because pullback commutes with exterior derivative), $\{\cdot, \cdot\}_\omega$ and $\{\cdot, \cdot\}$ are compatible (see Proposition 5.15), i.e., the compatibility is independent of N. On the other hand, since $\omega = \tau_N^*\Omega$, $\tilde{N} \in \mathrm{End}_{C^\infty(T^*Q)}(TT^*Q)$ defined by

$$\tilde{N} = (\Omega^\flat)^{-1} \circ \omega^\flat,$$

is the *lift* of N to T^*Q, and if $T_N = 0$, then $T_{\tilde{N}} = 0$ (see Sect. 3 in Appendix D). In other words, if N is a Nijenhuis operator, then (5.10) is Poisson, and $\{\cdot, \cdot\}$ and $\{\cdot, \cdot\}_\omega$ form a pair of compatible Poisson brackets (see Theorem 5.17). It is worth to compute explicitly the elementary Poisson brackets of $\{\cdot, \cdot\}_\omega$ in a Darboux coordinates system for Ω. These brackets can be computed from the formula

$$\{f, g\}_\omega = d\theta(X_f, X_g) = X_f \theta(X_g) - X_g \theta(X_f) - \theta([X_f, X_g]),$$

observing that $X_{q_i} = -\frac{\partial}{\partial p_i}$ and $X_{p_i} = \frac{\partial}{\partial q_i}$. More precisely,

$$\{p_i, p_j\}_\omega = \sum_k \left(\frac{\partial N_j^k}{\partial q_i} - \frac{\partial N_i^k}{\partial q_j} \right) p_k, \quad \{q_i, p_j\}_\omega = -N_j^i \quad \text{and}$$

$$\{q_i, q_j\}_\omega = 0, \forall i, j, \tag{5.11}$$

where $N = \sum_{k,l} N_k^l dq_k \otimes \frac{\partial}{\partial q_l}$ and the N_k^ls are local functions on Q. $\triangle$

Problem 5.22 Prove that $\{\cdot, \cdot\}$ and $\{\cdot, \cdot\}_\omega$ form a pair of compatible Poisson bracket by a direct computation using the elementary brackets (5.11). Hint: the only thing to be proven is that $\{\cdot, \cdot\}_\omega$ satisfies the Jacobi identity. Using (5.11), it suffices to prove that $\{\{p_i, p_j\}_\omega, q_k\}_\omega + \circlearrowleft (p_i, p_j, q_k) = 0$ and $\{\{p_i, p_j\}_\omega, p_k\}_\omega + \circlearrowleft (p_i, p_j, p_k) = 0$. These will follow from the hypothesis on N being a Nijenhuis operator.) $\triangle$

In this section, we saw how the recursion operators enter in an apparently *ad hoc* definition of compatibility between pairs of Poisson brackets (see Definition 5.14) when at least one of the two comes from a symplectic structure. In the next section, we will consider this construction in full generality, and we will show how the definition of compatibility proposed is indeed functional to the solution of the problem of the integrability of Hamiltonian systems.

3 Bi-Hamiltonian Manifolds

In this section, we introduce the general notion of bi-Hamiltonian structure (see [165]). Our starting point is the following:

Definition 5.23 Two Poisson tensors Π_0, Π_1 on M are called *compatible* if

$$\Pi = \lambda \Pi_0 + \mu \Pi_1$$

is a Poisson tensor for all $\lambda, \mu \in \mathbb{K}$. $\triangle$

Then

Proposition 5.24 (Π_0, Π_1) *is a pair of compatible Poisson tensors* if and only if

$$[\![\Pi_0, \Pi_1]\!] = 0, \tag{5.12}$$

where $[\![\cdot, \cdot]\!]$ *is the Schouten-Nijenhuis bracket introduced in* (2.17).

Proof In fact, the bivector field Π is a Poisson tensor for all $\lambda, \mu \in \mathbb{K}$, if and only if $[\![\Pi, \Pi]\!] = 0$. On the other hand

$$[\![\Pi, \Pi]\!] = [\![\lambda\Pi_0 + \mu\Pi_1, \lambda\Pi_0 + \mu\Pi_1]\!] = 2\lambda\mu[\![\Pi_0, \Pi_1]\!],$$

since $[\![\Pi_1, \Pi_1]\!] = 0 = [\![\Pi_0, \Pi_0]\!]$, which implies (5.12). $\square$

Rephrased in terms of Poisson brackets, Definition 5.23 reads as:

Definition 5.25 Two Poisson brackets $\{\cdot, \cdot\}_0, \{\cdot, \cdot\}_1$ are called *compatible* if and only if

$$\{\cdot, \cdot\} = \lambda\{\cdot, \cdot\}_0 + \mu\{\cdot, \cdot\}_1 \tag{5.13}$$

is a Poisson bracket for every choice of $\lambda, \mu \in \mathbb{K}$. The Poisson bracket in the left-hand side of (5.13) is called the *Poisson pencil* defined by $\{\cdot, \cdot\}_0, \{\cdot, \cdot\}_1$. $\triangle$

Problem 5.26 Let (Π_0, Π_1) be a pair of compatible Poisson tensor. Show that for each $f \in C^\infty(M)$

$$\mathscr{L}_{X_f}\Pi_1 + \mathscr{L}_{Y_f}\Pi_0 = 0,$$

where $X_f = \Pi_0^\sharp(df)$ and Y_f is defined as in (5.7). $\triangle$

Remark 5.27 Note that the condition in Definition 5.23 is equivalent to say that $\{\cdot, \cdot\}_0, \{\cdot, \cdot\}_1$ are compatible if and only if $\{\cdot, \cdot\} := \{\cdot, \cdot\}_1 + \lambda\{\cdot, \cdot\}_0$ is a Poisson bracket for all $\lambda \in \mathbb{K}$. $\triangle$

Finally, we show that the notions of compatibility introduced in Definition 5.14 and, respectively, in Definition 5.25 are equivalent. More precisely, reading Proposition 5.24 in terms of Poisson brackets, one arrives to

Proposition 5.28 *The two Poisson brackets* $\{\cdot, \cdot\}_0, \{\cdot, \cdot\}_1$ *are compatible in the sense of Definition 5.25 if and only if they are compatible in the sense of Definition 5.14, i.e., if and only if*

$$\{f, \{g, h\}_0\}_1 + \{h, \{f, g\}_0\}_1 + \{g, \{h, f\}_0\}_1 + \{f, \{g, h\}_1\}_0 + \{h, \{f, g\}_1\}_0$$

$$+ \{g, \{h, f\}_1\}_0 = 0, \tag{5.14}$$

$\forall f, g, h \in C^\infty(M)$.

Proof It suffices to show that condition (5.14) is equivalent to $[\![\Pi_0, \Pi_1]\!] = 0$. Since these are both local conditions, it is enough to check their equality in local coordinates. Then let $\Pi_0 = \frac{1}{2} \sum_{i,j} \Pi^0_{ij} \xi_i \xi_j$ and $\Pi_1 = \frac{1}{2} \sum_{m,l} \Pi^1_{lm} \xi_l \xi_m$,

$$
[\![\Pi_0, \Pi_1]\!] = \sum_k \underbrace{\frac{\partial \Pi_0}{\partial \xi_k} \frac{\partial \Pi_1}{\partial x_k}}_{(1)} + \underbrace{\frac{\partial \Pi_0}{\partial \xi_k} \frac{\partial \Pi_1}{\partial x_k}}_{(2)}.
$$

A computation completely analogous to the one done in the proof of Theorem 2.15, whose details will not be presented again here, shows that

$$
(1) = \sum_{i,l,m} \left(\Pi^0_{ik} \frac{\partial \Pi^1_{lm}}{\partial x_k} + \Pi^0_{mk} \frac{\partial \Pi^1_{il}}{\partial x_k} + \Pi^0_{lk} \frac{\partial \Pi^1_{mi}}{\partial x_k} \right) \xi_i \xi_l \xi_m \quad \text{and}
$$

$$
(2) = \sum_{i,l,m} \left(\Pi^1_{ik} \frac{\partial \Pi^0_{lm}}{\partial x_k} + \Pi^1_{mk} \frac{\partial \Pi^1_{il}}{\partial x_k} + \Pi^1_{lk} \frac{\partial \Pi^0_{mi}}{\partial x_k} \right) \xi_i \xi_l \xi_m,
$$

implying that

$$
[\![\Pi_0, \Pi_1]\!] = \sum_{i,l,m} \sum_k \left(\Pi^0_{ik} \frac{\partial \Pi^1_{lm}}{\partial x_k} + \Pi^0_{mk} \frac{\partial \Pi^1_{il}}{\partial x_k} + \Pi^0_{lk} \frac{\partial \Pi^1_{mi}}{\partial x_k} \right.
$$
$$
\left. + \Pi^1_{ik} \frac{\partial \Pi^0_{lm}}{\partial x_k} + \Pi^1_{mk} \frac{\partial \Pi^1_{il}}{\partial x_k} + \Pi^1_{lk} \frac{\partial \Pi^0_{mi}}{\partial x_k} \right) \xi_i \xi_l \xi_m.
$$

Then one concludes that

$$
[\![\Pi_0, \Pi_1]\!] = 0 \text{ iff } \sum_k \left(\Pi^0_{ik} \frac{\partial \Pi^1_{lm}}{\partial x_k} + \Pi^0_{mk} \frac{\partial \Pi^1_{il}}{\partial x_k} + \Pi^0_{lk} \frac{\partial \Pi^1_{mi}}{\partial x_k} \right.
$$
$$
\left. + \Pi^1_{ik} \frac{\partial \Pi^0_{lm}}{\partial x_k} + \Pi^1_{mk} \frac{\partial \Pi^1_{il}}{\partial x_k} + \Pi^1_{lk} \frac{\partial \Pi^0_{mi}}{\partial x_k} \right) = 0,
$$

for all $1 \le i, l, m \le n$. The proof of the proposition follows now from an argument already used in the proof of Proposition 2.12, and for this reason, it will be not reproduced here. $\qquad\square$

We can now introduce the following:

Definition 5.29 A *bi-Hamiltonian structure* on a manifold M is a pair (Π_0, Π_1) of compatible Poisson tensors on M. A *bi-Hamiltonian manifold* is triple (M, Π_0, Π_1), of a manifold M endowed with a bi-Hamiltonian structure (Π_0, Π_1). $\qquad\triangle$

Remark 5.30 Note that one could equivalently define a bi-Hamiltonian structure on M as a pair of *compatible* Poisson brackets on M. $\triangle$

Problem 5.31 Show that if $\mathfrak{g}^*$ is the dual vector space of a finite-dimensional Lie algebra, for all $\alpha \in \mathfrak{g}^*$, $\{\cdot,\cdot\}_{LP}$ and $\{\cdot,\cdot\}_\alpha$ (see 2.14) is a pair of compatible Poisson structures on $\mathfrak{g}^*$. $\triangle$

An interesting class of bi-Hamiltonian structures is obtained in the following way. Let $(M, \{\cdot,\cdot\}_1)$ be a Poisson manifold and let $X \in \mathfrak{X}(M)$ be such that $\{\cdot,\cdot\}_0 : C^\infty(M) \times C^\infty(M) \to C^\infty(M)$, defined by

$$\{f, g\}_0 = \{\mathscr{L}_X f, g\}_1 + \{f, \mathscr{L}_X g\}_1 - \mathscr{L}_X \{f, g\}_1, \tag{5.15}$$

for all $f, g \in C^\infty(M)$, is a Poisson bracket. Then

Proposition 5.32 $\left(\{\cdot,\cdot\}_0, \{\cdot,\cdot\}_1\right)$ *is a bi-Hamiltonian structure on* M.

Proof Note that if $X \in \mathfrak{X}(M)$ and Π_1 is a Poisson tensor, then $\Pi_0 = d_{\Pi_1} X = [\![\Pi_1, X]\!] \in \Gamma(\Lambda^2 TM)$ is such that

$$[\![\Pi_1, [\![\Pi_1, X]\!]]\!] = 0$$

(see Formula 2.70). On the other hand, to get an explicit formula for the bracket defined by $[\![\Pi_1, X]\!]$, it is enough to compute

$$\{f, g\}_0 = [\![\Pi_1, X]\!](df, dg) = -[\![X, \Pi_1]\!](df, dg) = -(\mathscr{L}_X \Pi_1)(df, dg)$$
$$= -\mathscr{L}_X \{f, g\}_1 + \{\mathscr{L}_X f, dg\}_1 + \{df, \mathscr{L}_X g\}_1,$$

for all $f, g \in C^\infty(M)$. $\square$

Definition 5.33 The pair $(\{\cdot,\cdot\}_1, X)$ such that (5.15) holds for all $f, g \in C^\infty(M)$ is called an *exact bi-Hamiltonian structure*, and the corresponding triple $(M, \{\cdot,\cdot\}_1, X)$ is called an *exact bi-Hamiltonian manifold*. $\triangle$

Example 5.34 Note that every Poisson manifold (M, Π_1) can be completed to a bi-Hamiltonian one simply choosing $\Pi_0 = 0$. On the other hand, every $X \in \mathfrak{X}(M)$ such that $\mathscr{L}_X \Pi_1 = 0$ completes $(M, \{\cdot,\cdot\}_1)$ to an exact bi-Hamiltonian manifold $(M, \{\cdot,\cdot\}_1, X)$. $\triangle$

A more interesting example is the following one.

Example 5.35 Let A be a (real) symmetric $n \times n$ matrix. Then the bracket $[\cdot,\cdot]_A : \mathfrak{so}(n) \otimes \mathfrak{so}(n) \to \mathfrak{so}(n)$ defined by

$$[x, y]_A = xAy - yAx, \quad \forall x, y \in \mathfrak{so}(n),$$

is a Lie bracket. In particular,

$$\{f, g\}_A(\alpha) = \langle \alpha, [df_\alpha, dg_\alpha]_A \rangle, \ \forall f, g \in C^\infty(\mathfrak{so}(n)^*), \ \alpha \in \mathfrak{so}(n)^*,$$

is a linear-Poisson structure on $\mathfrak{so}(n)^*$. In fact, since A is symmetric, $[x, y]_A \in \mathfrak{so}(n)$ for all $x, y \in \mathfrak{so}(n)$. The skew-symmetry is evident, while the proof of the Jacobi identity is a simple computation we leave to the reader as an exercise. Using the bilinear form $B : \mathfrak{so}(n) \otimes \mathfrak{so}(n) \to \mathbb{R}$, defined by $B(x, y) = -\frac{1}{2}\operatorname{tr}(xy)$ for all $x, y \in \mathfrak{so}(n)$, one can induce on $\mathfrak{so}(n)$ the following linear-Poisson structure:

$$\begin{aligned}
\{f, g\}_A(x) &= B\big(x, [\nabla_x f, \nabla_x g]_A\big) \\
&= -\frac{1}{2}\operatorname{tr}\big(x[\nabla_x f, \nabla_x g]_A\big), \ \forall x \in \mathfrak{so}(n), \ f, g \in C^\infty(\mathfrak{so}(n))
\end{aligned} \tag{5.16}$$

(see 2.11). If A_1, A_2 are two symmetric matrices $n \times n$, then $A_\lambda = A_1 + \lambda A_2$ is a symmetric matrix of the same size, and one can easily show that $\{\cdot, \cdot\}_{A_\lambda} = \{\cdot, \cdot\}_{A_1} + \lambda\{\cdot, \cdot\}_{A_2}$ is a (linear) Poisson bracket for all $\lambda \in \mathbb{R}$. In other words, the linear Poisson brackets defined via (5.16) by any pair of symmetric $n \times n$ matrices A_1, A_2 are compatible, i.e., $(\{\cdot, \cdot\}_{A_1}, \{\cdot, \cdot\}_{A_2})$ is a bi-Hamiltonian structure on $\mathfrak{so}(n)$. We show now that the linear-Poisson bracket (5.16) is *exact* in the Poisson cohomology defined by the linear Poisson bracket of $\mathfrak{so}(n)$ (see Sect. 4.3 in Chap. 2). Let A be a symmetric matrix and let $X_A \in \mathfrak{X}(\mathfrak{so}(n))$ be defined by $X_A(x) = Ax + xA$ for all $x \in \mathfrak{so}(n)$. Then

$$(\mathscr{L}_{X_A}x)(y) = \frac{d}{dt}\Big|_{t=0} x(y + tX_A(y)) = -\frac{1}{2}\operatorname{tr}(x(Ay + yA)),$$

where we identified $\mathfrak{so}(n)$ with its dual using the trace form $(x, y) = -\frac{1}{2}\operatorname{tr}(xy)$ (see Example C.11 in Appendix C). Using the cyclic property of the trace, the previous formula yields

$$\mathscr{L}_{X_A}x = xA + Ax, \ \forall x \in \mathfrak{so}(n). \tag{5.17}$$

Then

$$\begin{aligned}
(\mathscr{L}_{X_A}\Pi_{LP})(x, y) &= \mathscr{L}_{X_A}(\Pi_{LP}(x, y)) - \Pi_{LP}(\mathscr{L}_{X_A}x, y) - \Pi_{LP}(x, \mathscr{L}_{X_a}y) \\
&= \mathscr{L}_{X_A}[x, y] - \Pi_{LP}(\mathscr{L}_{X_A}x, y) - \Pi_{LP}(x, \mathscr{L}_{X_A}y) \\
&\overset{(5.17)}{=} \big(A[x, y] + [x, y]A\big) - [Ax + xA, y] - [x, Ay + yA] \\
&= -2[x, y]_A \\
&= -2\Pi_A(x, y),
\end{aligned}$$

for all $x, y \in \mathfrak{so}(n)$, proving the statement. $\triangle$

Problem 5.36 Write an expression in local coordinates of the Poisson tensor Π_A defined by the Poisson bracket (5.16). $\triangle$

The condition expressed in Formula (5.13) of Definition 5.25 is highly non-trivial as it is shown in the following:

Example 5.37 (Linear Poisson Pencils) As in the Example 2.9, let $\mathfrak{g}$ be a finite-dimensional Lie algebra, $A \in \mathrm{Hom}(\Lambda^2\mathfrak{g}, \mathbb{K})$ and let $\{\cdot, \cdot\}_A$ be the Poisson bracket defined in (2.13). Then, $(\{\cdot, \cdot\}_{LP}, \{\cdot, \cdot\}_A)$ is a compatible pair of Poisson brackets *if and only if* A is a two-cocycle of $\mathfrak{g}$ with values in the trivial $\mathfrak{g}$-module $\mathbb{K}$ (see Sect. 1.6 in Appendix C). In fact, if one writes $\Pi_{LP} = \frac{1}{2}\sum_{i,j}\sum_k c_{ij}^k x_k \xi_i \xi_j$ and $\Pi_A = \frac{1}{2}\sum_{m,l} A_{ml}\xi_m\xi_l$, where $x_1, \ldots, x_n$ is a system of linear coordinates on $\mathfrak{g}^*$ and $A_{ml} = A(x_m, x_l)$ for all m, l, then

$$[\![\Pi_{LP}, \Pi_A]\!] = \sum_s \frac{\partial \Pi_{LP}}{\partial \xi_s} \frac{\partial \Pi_A}{\partial x_s} + \frac{\partial \Pi_A}{\partial x_s} \frac{\partial \Pi_{LP}}{\partial x_s}$$

$$= \frac{1}{2}\sum_{l,m,i}\left(\sum_s A_{is} c_{lm}^s \xi_i \xi_l \xi_m\right)$$

$$= \frac{1}{2}\sum_{l,m,i}\left(\sum_s A_{is} c_{lm}^s + A_{ms} c_{il}^s + A_{ls} c_{mi}^s\right)\xi_i \xi_l \xi_m.$$

Then, the condition $[\![\Pi_{LP}, \Pi_A]\!] = 0$ turns out to be equivalent to

$$\sum_s A_{is} c_{lm}^s + A_{ms} c_{il}^s + A_{ls} c_{mi}^s = 0, \ \forall i, l, m,$$

which proves our statement. If A is a coboundary, i.e., $A(x, y) = \langle \alpha, [x, y] \rangle$ for all $x, y \in \mathfrak{g}$ and some $\alpha \in \mathfrak{g}^*$, then the bracket (2.13) becomes the frozen argument bracket, with argument α (see Example 2.14). In particular, the previous considerations implies that for all $\alpha \in \mathfrak{g}^*$, the frozen argument bracket with argument α is compatible with the linear Poisson bracket. Note that the Hamiltonian vector field of $x \in \mathfrak{g}$ with respect to the frozen bracket at α is the *constant* vector field $X_x(\beta) = \mathrm{ad}_x^\sharp(\alpha)$, for all $\beta \in \mathfrak{g}^*$. This implies that the symplectic foliation of $\{\cdot, \cdot\}_\alpha$ is obtained by parallel translation of the linear subspace generated by the vectors $\mathrm{ad}_x^\sharp(\alpha)$, for all $x \in \mathfrak{g}$, i.e., the parallel translation of the tangent space to the coadjoint orbit $\mathscr{O}_\alpha$ at the point α. For example, if $\mathfrak{g} = (\mathbb{R}^3, \times)$ (see Example C.11 in Appendix C), the symplectic foliation of the frozen Poisson structure at $\alpha \in \mathfrak{g}^*$, after the usual identification of $\mathfrak{g}^*$ with $\mathfrak{g}$, is the family of parallel planes orthogonal to x_α (Fig. 5.1).

It is worth noting that in this example, the leaves of the symplectic foliations of the two compatible Poisson structures $\{\cdot, \cdot\}_{LP}, \{\cdot, \cdot\}_\alpha$ are not contained one in the other. $\triangle$

Fig. 5.1 Compatible linear
and frozen Poisson structures

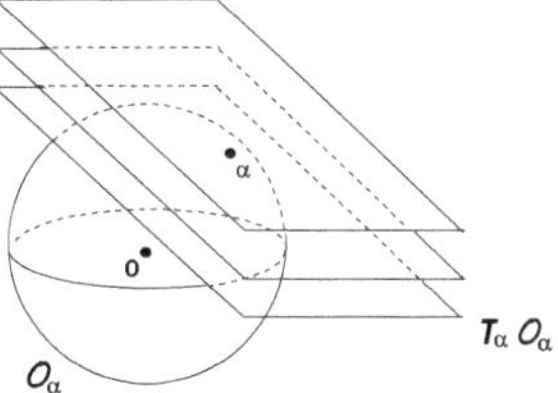

An interesting and important example of the bi-Hamiltonian structure is the
following:

Example 5.38 On $M = \mathbb{R}^3$

$$\Pi_0 = x_1 \frac{\partial}{\partial x_2} \wedge \frac{\partial}{\partial x_3} + x_2 \frac{\partial}{\partial x_3} \wedge \frac{\partial}{\partial x_1} + x_3 \frac{\partial}{\partial x_1} \wedge \frac{\partial}{\partial x_2} \quad \text{and}$$

$$\Pi_1 = \frac{x_1}{I_1} \frac{\partial}{\partial x_3} \wedge \frac{\partial}{\partial x_2} + \frac{x_2}{I_2} \frac{\partial}{\partial x_1} \wedge \frac{\partial}{\partial x_3} + \frac{x_3}{I_3} \frac{\partial}{\partial x_2} \wedge \frac{\partial}{\partial x_1},$$

where $I_i > 0$ for all $i = 1, \ldots, 3$ are two *linear* Poisson tensors, the first
corresponding to the Lie algebra structure defined by the cross-product (see
Example 2.52). The corresponding sharp maps, written in matrix form with respect
to the coordinates x_is, are

$$\Pi_0^\sharp = \begin{bmatrix} 0 & -x_3 & x_2 \\ x_3 & 0 & -x_1 \\ -x_2 & x_1 & 0 \end{bmatrix} \quad \text{and} \quad \Pi_1^\sharp = \begin{bmatrix} 0 & \frac{x_3}{I_3} & -\frac{x_2}{I_2} \\ -\frac{x_3}{I_3} & 0 & \frac{x_1}{I_1} \\ \frac{x_2}{I_2} & -\frac{x_1}{I_1} & 0 \end{bmatrix}.$$

If $H = \frac{1}{2} \sum_{i=1}^{3} \frac{x_i^2}{I_i}$, the integral curves of

$$X_H = \Pi_0^\sharp(dH) = \begin{bmatrix} 0 & -x_3 & x_2 \\ x_3 & 0 & -x_1 \\ -x_2 & x_1 & 0 \end{bmatrix} \begin{bmatrix} \frac{x_1}{I_1} \\ \frac{x_2}{I_2} \\ \frac{x_3}{I_3} \end{bmatrix} = \begin{bmatrix} x_2 x_3 \left(\frac{1}{I_2} - \frac{1}{I_3} \right) \\ x_1 x_3 \left(\frac{1}{I_1} - \frac{1}{I_3} \right) \\ x_1 x_2 \left(\frac{1}{I_1} - \frac{1}{I_2} \right) \end{bmatrix}, \tag{5.20}$$

are the solutions of the following system of ODE

$$\begin{cases} \dot{x}_1 = \dfrac{I_2 - I_3}{I_2 I_3} x_2 x_3 \\[2mm] \dot{x}_2 = \dfrac{I_3 - I_1}{I_1 I_3} x_1 x_3 \\[2mm] \dot{x}_3 = \dfrac{I_1 - I_2}{I_1 I_2} x_1 x_2 \end{cases}$$

and describe the movement of a rigid body in $\mathbb{R}^3$ freely rotating around its center of mass (see Sect. 4.1 in Chap. 6). It is worth to observe that if $K = \sum_{i=1}^{3} x_i^2$, then

$$\Pi_0^\sharp(dH) = \Pi_1^\sharp(dK),$$

i.e., the vector field (5.20) admits a *bi-Hamiltonian* representation (see Definition 5.42). We refer the reader to Chap. 6 for a more extensive and general discussion about the bi-Hamiltonian formulations of the rigid body dynamics. $\triangle$

Finally, we discuss why the presence of two bivectors should be relevant in the theory of Hamiltonian systems (see [165]).

3.1   Lenard-Magri Chains

Unless specified otherwise, in this subsection, Π_0 and Π_1 denote two *arbitrary* bivectors (i.e., we are not requiring them to be Poisson) and Π_0 and Π_1 are not necessarily compatible, meaning the condition $[\![\Pi_0, \Pi_1]\!] = 0$ is not enforced. While the theory of Lenard-Magri chains is usually cast in a bi-Hamiltonian context, the majority of the results presented here does not require this framework. Let $\{\cdot, \cdot\}_i$ be the bilinear skew-symmetric bracket associated with Π_i for $i = 0, 1$, which satisfies Leibniz identity (so it is a bi-derivation), but not Jacobi identity in general.

Proposition 5.39 *Let M be a manifold equipped with two bivectors Π_0 and Π_1 Suppose that $\{H_i\}_{i \geq 1} = H_1, H_2, H_3, \ldots, H_n, \ldots$ is a sequence of functions, possibly infinite, defined on M, and suppose that the following condition holds:*

$$\Pi_0^\sharp(dH_i) = \Pi_1^\sharp(dH_{i+1}), \tag{5.21}$$

for all $i \geq 1$. Then

$$\{H_n, H_m\}_0 = 0 = \{H_n, H_m\}_1, \tag{5.22}$$

for all $n, m \geq 1$. In other words, if (5.21) holds, the functions $\{H_i\}_{i \geq 1}$ are in involution with respect to both brackets $\{\cdot, \cdot\}_1$ and $\{\cdot, \cdot\}_0$ induced by the bivectors Π_0 and Π_1, respectively.

Proof Without loss of generality, suppose that $m > n$. Then, using (5.21), one has that

$$\{H_n, H_m\}_0 = \Pi_0(dH_n, dH_m) = \Pi_1(dH_{n+1}, dH_m) = -\Pi_1(dH_m, dH_{n+1}),$$

which, if $m = n + 1$, entails the left-hand side of (5.22). On the other hand, in this case, the right-hand side of (5.22) holds true since (assume $m = n + 1$):

$$\{H_n, H_m\}_1 = \{H_n, H_{n+1}\}_1 = -\{H_{n+1}, H_n\}_1 = -\Pi_1(dH_{n+1}, H_n)$$
$$= -\Pi_0(dH_n, dH_n) = 0.$$

On the other hand, if $m > n + 1$, one has

$$\{H_n, H_m\}_0 = \Pi_0(dH_n, dH_m) = \Pi_1(dH_{n+1}, dH_m) = -\Pi_1(dH_m, dH_{n+1})$$
$$= -\Pi_0(dH_{m-1}, dH_{n+1}) = \Pi_0(dH_{n+1}, dH_{m-1}).$$

Then, if $m = n + 2$, the first of the equalities in (5.22) holds, and the second follows easily too since

$$\{H_n, H_m\}_1 = \{H_n, H_{n+2}\}_1 = \Pi_1(H_n, H_{n+2}) = -\Pi_1(H_{n+2}, H_n)$$
$$= -\Pi_0(H_{n+1}, H_n) = \Pi_0(H_n, H_{n+1}) = \Pi_1(H_{n+1}, H_{n+1}) = 0.$$

It should be now clear how to proceed if $m > n + 2$. $\qquad\square$

Definition 5.40 (Lenard-Magri Chain) Given any two bivectors Π_0 and Π_1, a sequence of functions $\{H_i\}_{i \geq 1}$ satisfying (5.21) is called a *Lenard-Magri chain* with respect to Π_1, Π_0. If $\{H\}_{i \geq 1}$ is finite, say of length n, one requires that H_n is Casimir of Π_0. $\qquad\triangle$

Remark 5.41 Usually in the literature, a Lenard-Magri chain is associated with a pair of compatible Poisson structures Π_0 and Π_1. However, this is not necessary.

Note that if a Lenard-Magri chain $\{H_i\}_{i \geq 1}$ is such with respect to *two Poisson structures* Π_0 and Π_1, not necessarily compatible, one can associate with it a sequence of Hamiltonian vector fields $\{X_i\}_{i \geq 1}$, defined by

$$\Pi_0^\sharp(dH_i) = X_i = \Pi_1^\sharp(dH_{i+1}), \ \forall i \geq 1 \tag{5.23}$$

and that each X_i in (5.23) is Hamiltonian with respect to *both* the Poisson structures Π_0, Π_1. To underline the Hamiltonian nature of these vector fields, in case Π_1 and Π_0 are Poisson structures, we write X_i as $X_{H_i} = \Pi_0^\sharp(dH_i) = \{H_i, \cdot\}_0$ and $Y_{H_{i+1}} = \Pi_1^\sharp(dH_{i+1}) = \{H_{i+1}, \cdot\}_1$. Of course, in the case of a Lenard-Magri chain associated with two Poisson structures Π_1 and Π_0 not necessarily compatible, $X_{H_i} = Y_{H_{i+1}}$. $\qquad\triangle$

More generally one will say that

Definition 5.42 A vector field $X \in \mathfrak{X}(M)$ is *bi-Hamiltonian* with respect to the Poisson structures Π_0 and Π_1 (not necessarily compatible) if there exist $H, K \in C^\infty(M)$ such that

$$\Pi_0^\sharp(dH) = X = \Pi_1^\sharp(dK).$$

$\triangle$

Example 5.43 An example of a bi-Hamiltonian vector field is (5.20) in Example 5.38. In particular, the functions H, K provide a simple example of Lenard-Magri chain, with $H = H_1$ and $K = H_2$. Note that K is a Casimir of Π_0. In this specific example, the two bivectors Π_1 and Π_0 are Poisson and are compatible. $\triangle$

The following corollary is a direct consequence of Proposition 5.39 and of Remark 5.41. It shows that if (M, Π_0, Π_1) is a manifold equipped with two bivectors and $H \in C^\infty(M)$ belongs to a Lenard chain $\{H_i\}_{i \geq 1}$ with respect to Π_0, Π_1, the vector field $X = \Pi_0^\sharp(dH)$ is endowed with a rich supply of first integrals.

Corollary 5.44 *Let M be a manifold and suppose that a Lenard-Magri chain $\{H_i\}_{i \geq 1}$ is given with respect to two bivectors Π_0 and Π_1. For each fixed $i \geq 1$, denote with X_i the vector field $X_i = \Pi_0^\sharp(dH_i) \overset{(5.21)}{=} \Pi_1^\sharp(dH_{i+1})$. Then the following statements hold:*

(i) Each element H_j of the Lenard-Magri chain $\{H_j\}_{j \geq 1}$ is a first integral of X_i.
(ii) Furthermore, if Π_0 and Π_1 are two Poisson bi-vectors, not necessarily compatible, each vector field X_i is bi-Hamiltonian and $[X_i, X_j] = 0$ for all i, j.

Proof For the first point, $X_i(H_j) = \{H_i, H_j\}_0 = 0$. For the second, since Π_1 and Π_0 are Poisson, $X_i = X_{H_i} = Y_{H_{i+1}}$ (using the notation in Remark 5.41), so each of the X_i is Hamiltonian with respect to both Π_1 and Π_0. Furthermore, $[X_i, X_j] = [X_{H_i}, X_{H_j}] = X_{\{H_i, H_j\}_0} = 0$, where $[X_{H_i}, X_{H_j}] = X_{\{H_i, H_j\}_0}$ follows from the fact that Π_0 is Poisson. $\square$

How can we produce a Lenard-Magri chain on a given bi-Hamiltonian manifold (M, Π_0, Π_1)? The answer to this question is enclosed in the following:

Proposition 5.45 *Let $H \in C^\infty(M)[\lambda]$ be a Casimir of the Poisson pencil $\Pi = \Pi_0 - \lambda \Pi_1$. Then the coefficients of H form a Lenard-Magri chain.*

Proof $C^\infty(M)[\lambda]$ is the ring of polynomials with coefficients in $C^\infty(M)$. In other words, if $H \in C^\infty(M)[\lambda]$, then $H = \sum_{i=0}^n H_i \lambda^i$, where $H_i \in C^\infty(M)$, for all i. Then if H is Casimir of Π, one has that $\{H, f\} = 0$ for all $f \in C^\infty(M)$, i.e.,

$$0 = \{H, f\}_0 - \lambda\{H, f\}_1 = \sum_i \lambda^i \{H_i, f\}_0 - \sum_i \lambda^{i+1}\{H_i, f\}_1, \forall f \in C^\infty(M),$$

which implies that for all $f \in C^\infty(M)$,

$$\{H_0, f\}_0 = 0 \quad \{H_n, f\}_1 = 0 \quad \{H_{i+1}, f\}_0 = \{H_i, f\}_1, \ \forall 0 \le i \le n$$

where $H_{n+1} = 0$. In other words, if $H = \sum_{i=0}^n H_i \lambda^i \in C^\infty(M)[\lambda]$ is a Casimir of $\Pi = \Pi_0 - \lambda \Pi_1$, then H_0 is a Casimir of Π_0, H_n is a Casimir of Π_1, and $\{H_i\}_{i \ge 0}$ is a (finite) Lenard-Magri chain with respect to (Π_0, Π_1). $\qquad\square$

Note that one could generalize the previous proposition considering $H \in C^\infty(M)[[\lambda]]$, the ring of formal power series in λ with coefficients in $C^\infty(M)$.

Before concluding this section, we want to explore further how the existence of two compatible Poisson tensors affects a Lenard-Magri chain associated to them.

We start with the following:

Proposition 5.46 *Let Π be a Poisson tensor on a manifold M. Then for any $\gamma, \beta \in \Omega^1(M)$, one has*

$$\left(\mathscr{L}_{\Pi^\sharp(\gamma)}\Pi^\sharp\right)(\beta) = -\Pi^\sharp\left(i_{\Pi^\sharp(\beta)}d\gamma\right). \tag{5.24}$$

Proof The proof follows from the following computation:

$$\left(\mathscr{L}_{\Pi^\sharp(\gamma)}\Pi^\sharp\right)(\beta) = \mathscr{L}_{\Pi^\sharp(\gamma)}(\Pi^\sharp\beta) - \Pi^\sharp\left(\mathscr{L}_{\Pi^\sharp(\gamma)}\beta\right) =$$

$$\left[\Pi^\sharp(\gamma), \Pi^\sharp(\beta)\right] - \Pi^\sharp\left(\mathscr{L}_{\Pi^\sharp(\gamma)}\beta\right) =$$

$$\Pi^\sharp\{\gamma, \beta\}_\Pi - \Pi^\sharp\left(\mathscr{L}_{\Pi^\sharp(\gamma)}\beta\right) =$$

$$\Pi^\sharp\left(\mathscr{L}_{\Pi^\sharp(\gamma)}\beta - \mathscr{L}_{\Pi^\sharp(\beta)}\gamma + d\langle\gamma, \Pi^\sharp(\beta)\rangle\right) - \Pi^\sharp\left(\mathscr{L}_{\Pi^\sharp(\gamma)}\beta\right) =$$

$$\Pi^\sharp\left(d\langle\gamma, \Pi^\sharp(\beta)\rangle - \mathscr{L}_{\Pi^\sharp(\beta)}\gamma\right) =$$

$$\Pi^\sharp\left(d\langle\gamma, \Pi^\sharp(\beta)\rangle - i_{\Pi^\sharp(\beta)}d\gamma - di_{\Pi^\sharp(\beta)}\gamma\right) =$$

$$-\Pi^\sharp\left(i_{\Pi^\sharp(\beta)}d\gamma\right),$$

where on the third line we have used the fact that Π is Poisson; on the fourth line, we have used the definition of $\{\alpha, \beta\}_\Pi$; and on the sixth line, we used Cartan's formula for the Lie derivative of differential forms. $\qquad\square$

The following result shows how the existence of a genuine bi-Hamiltonian structure affects an associated Lenard-Magri chain of one-forms.

Proposition 5.47 *Let (M, Π_0, Π_1) be a bi-Hamiltonian manifold and let $\gamma_0, \gamma_1, \ldots, \gamma_n, \ldots$, a Lenard-Magri chain of one-forms on M, i.e., $\Pi_0^\sharp(\gamma_i) = X_i = Y_{i-1} = \Pi_1^\sharp(\gamma_{i-1})$, with say $\Pi_0^\sharp(\gamma_0) = 0$. For simplicity, assume that the symplectic foliation of Π_0 is regular with leaves that are embedded symplectic submanifolds. If $\gamma_0, \ldots, \gamma_n$ are closed on M, then $i_S^*\gamma_{n+1}$ is closed as an element of $\Omega^1(S)$, where $i_S : S \hookrightarrow M$ is the embedding of any symplectic leaf S of Π_0 in M.*

Proof Since $\Pi_0 + \Pi_1$ is Poisson, we can apply (5.24) to $\Pi_0 + \Pi_1$ with $\gamma = \gamma_n$ and obtain, for any $\beta \in \Omega^1(M)$:

$$\left(\mathscr{L}_{(\Pi_0^\sharp + \Pi_1^\sharp)(\gamma_n)} (\Pi_0 + \Pi_1)^\sharp \right)(\beta) = 0. \tag{5.25}$$

Using the fact that Π_0 and Π_1 are Poisson, the fact that γ_n is closed and again (5.24), one easily gets that (5.25) is equivalent to

$$\left(\mathscr{L}_{\Pi_0^\sharp(\gamma_n)} \Pi_1^\sharp \right)(\beta) + \left(\mathscr{L}_{\Pi_1^\sharp(\gamma_n)} \Pi_0^\sharp \right)(\beta) = 0. \tag{5.26}$$

Exploiting the Lenard-Magri property, i.e., $\Pi_0^\sharp(\gamma_n) = \Pi_1^\sharp(\gamma_{n-1})$, the fact that γ_{n-1} is closed, and also $\Pi_1^\sharp(\gamma_n) = \Pi_0^\sharp(\gamma_{n+1})$ (Lenard-Magri property), one obtains that (5.26) is equivalent to

$$\left(\mathscr{L}_{\Pi_0^\sharp(\gamma_{n+1})} \Pi_0^\sharp \right)(\beta) = 0.$$

Applying finally again (5.24) to this last expression, we get

$$\Pi_0^\sharp \left(i_{\Pi_0^\sharp(\beta)} d\gamma_{n+1} \right) = 0. \tag{5.27}$$

Therefore, $i_{\Pi_0^\sharp(\beta)} d\gamma_{n+1} \in \ker(\Pi_0^\sharp)$, but this kernel can be identified with the annihilator of the image of $\Pi_0^\sharp$ (see Problem 2.19). So this means that (5.27) is equivalent to

$$0 = \left\langle \Pi_0^\sharp(\alpha), i_{\Pi_0^\sharp(\beta)} d\gamma_{n+1} \right\rangle = d\gamma_{n+1}\left(\Pi_0^\sharp(\beta), \Pi_0^\sharp(\alpha) \right), \quad \forall \alpha,\ \beta \in \Omega^1(M).$$

What the last equation says is that $i_S^* \gamma_{n+1}$ is closed as an element of $\Omega^1(S)$, where S is any symplectic leaf of Π_0. $\qquad\square$

The previous proposition can be stated in more general terms, using the concept of foliated closed form, but we will not enter this here. Let us discuss some examples that highlight the issue pointed out by Proposition 5.47.

First, consider the Hopf fibration $p : \mathbb{S}^3 \to \mathbb{S}^2$, which is a smooth surjective submersion with fiber F diffeomorphic to $\mathbb{S}^1$. Since p is a submersion, its fibers are embedded submanifolds. On each fiber, there is the closed one-form $\alpha = d\theta$. However, α does not extend to a closed one-form γ on $\mathbb{S}^3$. Indeed, suppose it does, then $\gamma = df$ for a global function $f : \mathbb{S}^3 \to \mathbb{R}$, since the first de Rham cohomology group of $\mathbb{S}^3$ is trivial. On the other hand, $d\theta = \alpha = i_F^* \gamma = i_F^*(df) = d(i_F^* f)$, but there is no global function on $\mathbb{S}^1$ having differential $d\theta$, since $\int_{\mathbb{S}^1} d\theta = 2\pi$, while $\int_{\mathbb{S}^1} d(i_F^* f) = f(F(2\pi)) - f(F(0)) = 0$. Thus, α cannot be extended to a closed one-form on $\mathbb{S}^3$.

As a simpler example of this phenomenon concerning *two-forms*, take the following situation, which is also related to a linear Poisson structure on $\mathbb{R}^3$ we introduced before: Consider $\mathbb{R}^3$ with a foliation $\mathcal{F}$ given by concentric spheres centered at the origin, together with the singular leaf given by the origin itself. Equip each leaf S_r (sphere of radius r) with the closed two-form ω_r, which is the standard area form on S_r. Fix $r > 0$; we show that there is no closed extension α of ω_r to a neighborhood of S_r of the form $D_{a,b} := \cup_{s \in [a,b]} S_s$ with $0 < a < r < b$. Indeed, suppose such a closed α exists, and then by Stokes theorem

$$0 = \int_{D_{a,b}} d\alpha = \int_{S_b} \alpha_{|S_b} - \int_{S_a} \alpha_{|S_a} = \int_{S_b} \omega_b - \int_{S_a} \omega_a \neq 0,$$

since the surface areas of S_b and S_a are different.

Similarly, consider $\mathbb{R}^2$ with a foliation $\mathcal{F}$ given by concentric circles centered at the origin, together with the singular leaf given by the origin itself. Equip each leaf C_r (circle of radius r) with the closed one-form $\omega_r = rd\theta$, which is the standard length form on C_r. Fix $r > 0$, and then there is no closed extension α of ω_r to a neighborhood of C_r of the form $D_{a,b} := \cup_{\rho \in [a,b]} C_\rho$ with $0 < a < r < b$, as we argued above. However, if we restrict the problem to the circular sector $\mathcal{D} := \{(\rho, \theta) \in [a, b] \times (0, 2\pi)\}$, then one can easily check that the form $\alpha = d(\rho\theta)$, which is even exact on $\mathcal{D}$, restricts on $C_r \cap \mathcal{D}$ to $\omega_{r|\mathcal{D}}$.

Going back to Proposition 5.47, we remark that sometimes, in the literature, it is incorrectly stated that γ_{n+1} is closed; in particular, it is incorrectly stated that all γs are closed if γ_0 is in the kernel of $\Pi_0^\sharp$ and γ_1 is closed. On the other hand, if one is interested only in a local statement, like the one presented above for the foliation in concentric circles in the plane, restricted to a circular sector, then it is possible to obtain a stronger result. We leave this to the reader.

At this point, it should be clear why Lenard-Magri chains are relevant to the theory of completely integrable systems. Unfortunately, it is not an easy task to produce them in a general situation. In the bi-Hamiltonian setting, especially in the restricted class of *Poisson-Nijenhuis manifolds*, the goal is greatly simplified. Thus, we will restrict our attention to Poisson-Nijenhuis manifolds, which are equipped with a bi-Hamiltonian structure defined by a Poisson tensor together with a compatible Nijenhuis operator (see [141, 165, 167, 199]).

4 Poisson-Nijenhuis Manifolds

After introducing the general notion of a bi-Hamiltonian structure and discussing its relevance to detect the integrability in the theory of Hamiltonian systems, we are left with the problem of clarifying when this technology can be profitably applied to concrete problems. More precisely, given a Hamiltonian vector field X_H, one would like to understand how to produce a second Hamiltonian representation for it and how to find a Lenard-Magri chain, the elements of which are first integrals of X_H. The notion of *Poisson-Nijenhuis* manifold, introduced below, goes into this direction.

In Sect. 2, we discussed the conditions one should impose on a pair of two-forms, one of those symplectic, to generate on the underlying manifold a bi-Hamiltonian structure. Now, with a slight change of perspective, starting from a manifold M endowed with a Poisson tensor Π_0 and an endomorphism N of TM, we want to investigate under which assumptions $\Pi_1^\sharp = N\Pi_0^\sharp$ becomes a compatible Poisson structure with Π_0. The first, and more obvious, condition is expressed by

$$ {}^t\Pi_1{}^\sharp = -\Pi_1^\sharp, \tag{5.28} $$

meaning that Π_1 is a *skew-symmetric* element of $\mathrm{Hom}_{C^\infty(M)}\left(\Omega^1(M), \mathfrak{X}(M)\right)$. More explicitly, (5.28) means

$$ \langle \beta, \Pi_1^\sharp(\alpha) \rangle = -\langle \Pi_1^\sharp(\beta), \alpha \rangle, \ \forall \alpha, \beta \in \Omega^1(M). $$

Since $\Pi_1^\sharp = N\Pi_0^\sharp$, the previous identity becomes

$$ -\langle \Pi_1^\sharp(\beta), \alpha \rangle = \langle \beta, \Pi_1^\sharp(\alpha) \rangle = \langle \beta, N\Pi_0^\sharp(\alpha) \rangle = \langle {}^t N(\beta), \Pi_0^\sharp(\alpha) \rangle = -\langle \Pi_0^{\sharp\,t} N(\beta), \alpha \rangle, $$

for all $\alpha, \beta \in \Omega^1(M)$. In other words,

Proposition 5.48 *The tensor* Π_1, *defined by* $\Pi_1^\sharp = N\Pi_0^\sharp \in \mathrm{Hom}_{C^\infty(M)}$ $\left(\Omega^1(M), \mathfrak{X}(M)\right)$, *is a bivector field if and only if*

$$ N\Pi_0^\sharp = \Pi_0^{\sharp\,t} N. \tag{5.29} $$

The second condition is less intuitive and is given as a condition on the $\mathbb{K}$-bilinear form

$$ \mathcal{R}_{\Pi_0,N} : \Omega^1(M) \times \mathfrak{X}(M) \to \mathfrak{X}(M) $$

defined by

$$ \mathcal{R}_{\Pi_0,N}(\alpha, X) = (\mathscr{L}_{\Pi_0^\sharp(\alpha)} N)X - \Pi_0^\sharp\left(\mathscr{L}_X{}^t N(\alpha)\right) + \Pi_0^\sharp(\mathscr{L}_{NX}\alpha), \tag{5.30} $$

for all $\alpha \in \Omega^1(M)$ and $X \in \mathfrak{X}(M)$.

Problem 5.49 Show that the $\mathbb{K}$-bilinear form defined in (5.30) is $C^\infty(M)$-bilinear. $\triangle$

Problem 5.50 Given the $C^\infty(M)$-bilinear form defined in (5.30), let

$$ {}^t\mathcal{R}_{\Pi_0,N} : \Omega^1(M) \times \Omega^1(M) \to \Omega^1(M) $$

defined by

$$\left\langle {}^t\mathcal{R}_{\Pi_0,N}(\alpha,\beta), X\right\rangle = \left\langle \beta, \mathcal{R}_{\Pi_0,N}(\alpha, X)\right\rangle, \ \forall \alpha, \beta \in \Omega^1(M), \ X \in \mathfrak{X}(M).$$

(i) Show that

$$
{}^t\mathcal{R}_{\Pi_0,N}(\alpha,\beta) = \left(\mathscr{L}_{\Pi_0^\sharp(\alpha)}\,{}^tN\right)(\beta) - \left(\mathscr{L}_{\Pi_0^\sharp(\beta)}\,{}^tN\right)(\alpha) + d\left\langle \Pi_0^\sharp(\beta),\,{}^tN(\alpha)\right\rangle
$$
$$
- {}^tN\left(d\langle \Pi_0^\sharp(\beta), \alpha\rangle\right), \ \forall \alpha, \beta \in \Omega^1(M).
$$

(ii) Show that

$$
{}^t\mathcal{R}_{\Pi_0,N}(\alpha,\beta) = \left(\{{}^tN(\alpha),\beta\}_{\Pi_0} + \{\alpha,{}^tN(\beta)\}_{\Pi_0} - {}^tN\{\alpha,\beta\}_{\Pi_0}\right) - \{\alpha,\beta\}_{\Pi_1},
$$

$$\tag{5.31}$$

$\forall \alpha, \beta \in \Omega^1(M)$, where

(a) Π_1 is the bivector field defined by $N\Pi_0^\sharp$
(b) $\{\cdot,\cdot\}_A$ is the $\mathbb{K}$-bilinear form on $\Omega^1(M)$ defined, by $A \in \mathrm{Hom}_{C^\infty}$ $(\Omega^1(M), X(M))$, as in Formula (2.36)

$$\triangle$$

To explain the role of $\mathcal{R}_{\Pi_0,N}$ in our discussion, first we prove the following

Lemma 5.51 *If Π_0 and N are as above and satisfy (5.28), then*

$$ {}^t\mathcal{R}(df, dg) = \mathscr{L}_{X_f}(d_N g) - i_{X_g} dd_N f + d_N\langle df, X_g\rangle, \tag{5.32}$$

where $d_N f = {}^tN df$ (see Formula 5.129), $X_f = \Pi_0^\sharp(df)$, and $Y_f = \Pi_1^\sharp(df)$, for all $f \in C^\infty(M)$.

Proof This is a (more or less) straightforward computation.

$$
{}^t\mathcal{R}(df, dg) \overset{(5.31)}{=} \{{}^tN(df), dg\}_{\Pi_0} + \{df,{}^tN(dg)\}_{\Pi_0} - {}^tN\{df, dg\}_{\Pi_0} - \{df, dg\}_{\Pi_1}
$$

$$
\overset{(2.36)}{=} \mathscr{L}_{\Pi_0^\sharp({}^tNdf)} dg - \mathscr{L}_{X_g}({}^tNdf) - d\Pi_0({}^tNdf, dg)
$$

$$
\overset{(2.36)}{+} \mathscr{L}_{X_f}({}^tNdg) - \mathscr{L}_{\Pi_0^\sharp({}^tNdg)} df - d\Pi_0(df, {}^tNdg)
$$

$$
- {}^tN\{df, dg\}_{\Pi_0} - \{df, dg\}_{\Pi_1}
$$

$$
= \mathscr{L}_{X_f}({}^tNdg) - \mathscr{L}_{X_g}({}^tNdf) - {}^tN\{df, dg\}_{\Pi_0} - \{df, dg\}_{\Pi_1}.
$$

$$\tag{5.33}$$

On the other hand, using again (2.36)

$$\{df, dg\}_{\Pi_1} = \mathscr{L}_{Y_f} dg - \cancel{\mathscr{L}_{Y_g} df} - \cancel{d\Pi_1(df, dg)} = d\langle dg, Y_f\rangle = d\langle {}^t N dg, \Pi_0^\sharp(df)\rangle$$
$$= -d\langle {}^t N df, X_g\rangle, \tag{5.34}$$

and

$$\{df, dg\}_{\Pi_0} = \mathscr{L}_{X_f} dg - \cancel{\mathscr{L}_{X_g} df} - \cancel{d\Pi_0(df, dg)} = d\langle dg, X_f\rangle = d\{f, g\}_0. \tag{5.35}$$

Finally, since $\mathscr{L}_{X_g}({}^t N df) = i_{X_g} dd_N f + d i_{X_g} d_N f$, from (5.33), (5.34), and (5.35), one obtains

$${}^t\mathcal{R}(df, dg) = \mathscr{L}_{X_f}({}^t N dg) - i_{X_g} dd_N f - \cancel{d i_{X_g} d_N f} - d_N\{f, g\}_0 + \cancel{d\langle {}^t N df, X_g\rangle}$$
$$= \mathscr{L}_{X_f}({}^t N dg) - i_{X_g} dd_N f + d_N\langle df, X_g\rangle,$$

as stated. $\square$

Finally, the following result explains the reason to introduce (5.30).

Proposition 5.52 *If* $\mathcal{R}_{\Pi_0, N} = 0$, *the brackets defined by*

$$\{f, g\}_0 = \langle dg, \Pi_0^\sharp(df)\rangle \quad and \quad \{f, g\}_1 = \langle dg, \Pi_1^\sharp(df)\rangle, \ \forall f, g \in C^\infty(M),$$

are compatible (see Definition 5.14).

Proof For all $f, g, h \in C^\infty(M)$

$$\langle {}^t\mathcal{R}(df, dg), X_h\rangle \overset{(5.32)}{=} \langle \mathscr{L}_{X_f}(d_N g) - i_{X_g} dd_N f + d_N\langle df, X_g\rangle, X_h\rangle. \tag{5.36}$$

We analyze separately the three terms of right-hand side of the previous identity. For what concerns the first, one can write

$$\langle \mathscr{L}_{X_f}(d_N g), X_h\rangle = \mathscr{L}_{X_f}\langle d_N g, X_h\rangle - \langle d_N g, [X_f, X_h]\rangle$$
$$= \mathscr{L}_{X_f}\langle dg, N X_h\rangle - \langle dg, N[X_f, X_h]\rangle$$
$$= \mathscr{L}_{X_f}\langle dg, \Pi_1^\sharp(dh)\rangle - \langle dg, N X_{\{f, h\}_0}\rangle$$
$$= \mathscr{L}_{X_f}\langle dg, \Pi_1^\sharp(dh)\rangle - \langle dg, \Pi_1^\sharp(d\{f, h\}_0)\rangle$$
$$= \Pi_0(df, d\{h, g\}_1) + \Pi_1(dg, d\{f, h\}_0). \tag{5.37}$$

The second term yields

$$\langle i_{X_g} dd_N f, X_h \rangle = dd_N f(X_g, X_h) = X_g \langle d_N f, X_h \rangle - X_h \langle d_N f, X_g \rangle$$
$$- \langle d_N f, [X_g, X_h] \rangle$$
$$= X_g \langle df, \Pi_1^{\sharp}(dh) \rangle - X_h \langle df, \Pi_1^{\sharp}(dg) \rangle - \langle df, \Pi_1^{\sharp}(d\{g, h\}_0) \rangle$$
$$= \Pi_0(dg, d\{h, f\}_1) + \Pi_0(dh, d\{f, g\}_1) + \Pi_1(df, d\{g, h\}_0).$$
$$(5.38)$$

Finally, working out the last term

$$\langle d_N \langle df, X_g \rangle, X_h \rangle = \langle d \langle df, X_g \rangle, \Pi_1^{\sharp}(dh) \rangle = \langle d\{g, f\}_0, \Pi_1^{\sharp}(dh) \rangle$$
$$= \Pi_1(dh, d\{g, f\}_0). \qquad (5.39)$$

Inserting (5.37), (5.38), and (5.39) in (5.36), one concludes that

$$\langle {}^t\mathcal{R}(df, dg), X_h \rangle$$
$$= \{f, \{h, g\}_1\}_0 + \{g, \{f, h\}_0\}_1 + \{g, \{f, h\}_1\}_0 + \{h, \{g, f\}_1\}_0 + \{f, \{h, g\}_0\}$$
$$+ \{h, \{g, f\}_0\}_1,$$

for all $f, g, h \in C^{\infty}(M)$, which suffices to close the proof. $\qquad \square$

It is interesting to mention the following.

Corollary 5.53 *Under the same assumptions of the Proposition 5.52, if Π_0 is invertible, then the compatibility between $\{\cdot, \cdot\}_1$ and $\{\cdot, \cdot\}_0$ implies that $\mathcal{R}_{\Pi_0, N} = 0$.*

Proof In fact, one has that ${}^t\mathcal{R}_{\Pi_0, N} = 0$ if and only if ${}^t\mathcal{R}_{\Pi_0, N}(\alpha, \beta) = 0$ for all $\alpha, \beta \in \Omega^1(M)$. Since ${}^t\mathcal{R}_{\Pi_0, N}$ is tensor, it suffices to check this condition on every pair of exact one-forms, i.e., one should check that $\langle {}^t\mathcal{R}(df, dg), X \rangle = 0$ for all $f, g \in C^{\infty}(M)$ and for all $X \in \mathfrak{X}(M)$. But if Π_0 is invertible, $\mathfrak{X}(M)$ is (locally) generated by the Hamiltonian vector fields. $\qquad \square$

Remark 5.54 Note in the hypothesis of Proposition 5.52, there is no mention about the torsion of N. On the other hand, the properties of Π_0 being a Poisson tensor and the condition (5.28) are crucial for its proof. $\qquad \triangle$

The result enclosed in the above proposition justifies the following.

Definition 5.55 A Poisson bivector field Π_0 and an endomorphism N of TM are called *compatible* if

$$N\Pi_0^{\sharp} = \Pi_0^{\sharp t} N \quad \text{and} \quad \mathcal{R}_{\Pi_0, N} = 0. \qquad (5.40)$$

$$\triangle$$

Now we can state the following important result (see [141, 167]).

Theorem 5.56 (Magri-Morosi Theorem) *If Π_0 is a Poisson bivector field compatible with a Nijenhuis operator N, then*

$$[\![\Pi_1, \Pi_1]\!] = 0 \quad and \quad [\![\Pi_1, \Pi_0]\!] = 0,$$

where Π_1 is the bivector field defined by $\Pi_1^{\sharp} = N\Pi_0^{\sharp}$. In other words, under the above assumptions on N and Π_0, the pair (Π_0, Π_1) defines a bi-Hamiltonian structure on M.

Proof The theorem consists of two statements. The first one is the claim that under the above hypothesis, Π_1 is a Poisson tensor. To prove it, recall that the compatibility between Π_0 and N implies that Π_1 is skew-symmetric and that $\mathcal{R}_{\Pi_0,N} = 0$. The latter implies the compatibility between $\{\cdot, \cdot\}_1$ and $\{\cdot, \cdot\}_0$ (see Proposition 5.52), which, in turn, is equivalent to

$$[X_f, Y_g] + [Y_f, X_g] = Y_{\{f,g\}_0} + X_{\{f,g\}_1}, \ \forall f, g \in C^{\infty}(M)$$

(see Proposition 5.15). From the previous formula, it follows that

$$X_{\{f,g\}_1} = [X_f, NX_g] + [NX_f, X_g] - N[X_f, X_g] \overset{(5.131)}{=} [X_f, X_g]_N. \qquad (5.41)$$

Applying N to the leftmost and to the rightmost members of (5.41) above, one arrives to

$$Y_{\{f,g\}_1} = N[X_f, X_g]_N. \qquad (5.42)$$

If $T_N = 0$, then $N[X_f, X_g]_N = [Y_f, Y_g]$, for all $f, g \in C^{\infty}(M)$, which, plugged into (5.42), entails the following identity:

$$[Y_f, Y_g] = Y_{\{f,g\}_1},$$

which is equivalent to claim that $\{\cdot, \cdot\}_1$ satisfies the Jacobi identity. As far as the second part of theorem is concerned, recall that if Π_1 and Π_0 are two Poisson tensors, then the condition $[\![\Pi_1, \Pi_0]\!] = 0$ is equivalent to the compatibility of the corresponding Poisson brackets (see Propositions 5.24 and 5.28). $\qquad \square$

Remark 5.57 We would like to observe that Theorem 5.56 can be proven using the following two identities, which hold for all $\alpha, \beta \in \Omega^1(M)$,

$$[\![\Pi_1, \Pi_1]\!](\alpha, \beta) = N[\![\Pi_0, \Pi_0]\!]({}^t N(\alpha), \beta) + T_N\left(\Pi_0^{\sharp}(\alpha), \Pi_0^{\sharp}(\beta)\right)$$

$$- N\mathcal{R}_{\Pi_0,N}(\alpha, \Pi_0^{\sharp}(\beta)),$$

and

$$\llbracket \Pi_1, \Pi_0 \rrbracket(\alpha, \beta) = 2N\big(\llbracket \Pi_0, \Pi_0 \rrbracket(\alpha, \beta)\big) - \Pi_0^\sharp\big({}^t\mathcal{R}_{\Pi_0, N}(\alpha, \beta)\big) - \mathcal{R}_{\Pi_0, N}\big(\beta, \Pi_0^\sharp(\alpha)\big)$$
$$+ \mathcal{R}_{\Pi_0, N}\big(\alpha, \Pi_0^\sharp(\beta)\big).$$

For a proof, we refer the reader to [167]. $\triangle$

The tensor defined in (5.30) is known in the literature as the Magri-Morosi concomitant. Before moving on, we would like to present an alternative expression for this important tensor, which sometimes comes in handy.

Lemma 5.58 *For all $\alpha \in \Omega^1(M)$ and $X \in \mathfrak{X}(M)$*

$$\mathcal{R}_{\Pi, N}(\alpha, X) = N(\mathcal{L}_X \Pi^\sharp)(\alpha) - \mathcal{L}_X(N\Pi^\sharp)(\alpha) - (\mathcal{L}_{NX} \Pi^\sharp)(\alpha) + (\mathcal{L}_X \Pi^\sharp)^t N(\alpha).$$
$$(5.43)$$

Proof Starting from (5.30), one can compute

$$\mathcal{L}_{\Pi^\sharp(\alpha)}(N)X - \Pi^\sharp(\mathcal{L}_X({}^t N\alpha)) + \Pi^\sharp(\mathcal{L}_{NX}\alpha)$$
$$= \mathcal{L}_{\Pi^\sharp(\alpha)}(NX) - N(\mathcal{L}_{\Pi^\sharp(\alpha)}X) - \Pi^\sharp(\mathcal{L}_X({}^t N\alpha)) + \Pi^\sharp(\mathcal{L}_{NX}\alpha)$$
$$= -\mathcal{L}_{NX}(\Pi^\sharp(\alpha)) + N(\mathcal{L}_X(\Pi^\sharp(\alpha)) - \Pi^\sharp(\mathcal{L}_X({}^t N\alpha)) + \Pi^\sharp(\mathcal{L}_{NX}\alpha)$$
$$= -(\mathcal{L}_{NX}\Pi^\sharp)(\alpha) - \Pi^\sharp(\mathcal{L}_{NX}\alpha) + N(\mathcal{L}_X(\Pi^\sharp\alpha)) - \Pi^\sharp(\mathcal{L}_X({}^t N\alpha))$$
$$+ \Pi^\sharp(\mathcal{L}_{NX}\alpha)$$
$$= -(\mathcal{L}_{NX}\Pi^\sharp)(\alpha) + N(\mathcal{L}_X \Pi^\sharp)(\alpha) + N\Pi^\sharp(\mathcal{L}_X\alpha) - \Pi^\sharp(\mathcal{L}_X {}^t N)\alpha$$
$$- \Pi^{\sharp t}N(\mathcal{L}_X\alpha)$$
$$= -(\mathcal{L}_{NX}\Pi^\sharp)(\alpha) + N(\mathcal{L}_X \Pi^\sharp)(\alpha) - \Pi^\sharp(\mathcal{L}_X({}^t N\alpha)) + \Pi^{\sharp t}N(\mathcal{L}_X\alpha)$$
$$= -(\mathcal{L}_{NX}\Pi^\sharp)(\alpha) + N(\mathcal{L}_X \Pi^\sharp)(\alpha) - \mathcal{L}_X(\Pi^{\sharp t}N\alpha) + \Pi^{\sharp t}N(\mathcal{L}_X\alpha)$$
$$+ (\mathcal{L}_X \Pi^\sharp)^t N(\alpha)$$
$$= -(\mathcal{L}_{NX}\Pi^\sharp)(\alpha) + N(\mathcal{L}_X \Pi^\sharp)(\alpha) - \mathcal{L}_X(N\Pi^\sharp)(\alpha) + (\mathcal{L}_X \Pi^\sharp)^t N(\alpha).$$

$\square$

Finally, we can introduce the following important notion.

Definition 5.59 (Poisson-Nijenhuis Structures and Poisson-Nijenhuis Manifolds) A *Poisson-Nijenhuis* structure on M, PN-structure hereafter, is a pair (Π, N) of a Poisson structure *compatible* with a Nijenhuis tensor (see Definition 5.55). A manifold endowed with a PN-structure is called a Poisson-Nijenhuis manifold, PN-manifold hereafter. $\triangle$

Remark 5.60 It is clear that every PN-manifold is bi-Hamiltonian (see Definition 5.29). On the other hand, not every bi-Hamiltonian manifold is a PN-manifold. In fact, for such a bi-Hamiltonian manifold, the relation $\Pi_1^\sharp = N\Pi_0^\sharp$ forces each leaf of the symplectic foliation of Π_1 to be contained in a leaf of the symplectic foliation of Π_0. This condition is not fulfilled by the bi-Hamiltonian structure presented in Example 5.37, which, for this reason, does not come from a PN-structure. $\triangle$

Finally, we come to explain the reason of the relevance of the PN-structure in the theory of Hamiltonian systems. Recall that for a general bi-Hamiltonian structure, there is no guarantee about the existence of a Lenard-Magri chain. Here below we will see how such a deficiency is cured in the PN case.

Lemma 5.61 *Let* (M, Π, N) *be a* PN-*manifold and let*

$$kI_k = \mathrm{tr}\left(N^k\right), \forall k \geq 1. \tag{5.44}$$

Then

$$ {}^tN(dI_k) = dI_{k+1}, \forall k \geq 1. \tag{5.45}$$

Proof The proof follows from the cyclic property of the trace and from the following identity:

$$\mathscr{L}_{NX}({}^tN) = \mathscr{L}_X({}^tN)^t N, \ \forall X \in \mathfrak{X}(M). \tag{5.46}$$

More precisely, given $X \in \mathfrak{X}(M)$,

$$
\begin{aligned}
\langle dI_{k+1}, X\rangle &= \frac{1}{k+1}\mathscr{L}_X\left(\mathrm{tr}\, N^{k+1}\right) \\
&= \frac{1}{k+1}\mathscr{L}_X\left(\mathrm{tr}\,{}^t N^{k+1}\right) \ \text{since } \mathrm{tr}(A) = \mathrm{tr}({}^t A) \\
&= \mathrm{tr}\left({}^t N^{k-1}\mathscr{L}_X({}^tN)^t N\right) \ \text{the cyclic property of the trace} \\
&= \mathrm{tr}\left({}^t N^{k-1}\mathscr{L}_{NX}({}^tN)\right) \ \text{Formula (5.46)} \\
&= \frac{1}{k}\mathrm{tr}\left(\mathscr{L}_{NX}({}^tN^k)\right) \ \text{again the cyclic property} \\
&= \langle dI_k, NX\rangle \\
&= \langle {}^t N(dI_k), X\rangle.
\end{aligned}
$$

$\square$

Using the previous lemma, one can easily prove the following.

Proposition 5.62 *The functions $\{I_k\}_{k\geq 1}$ defined in (5.44) form a Lenard-Magri chain with respect to the bi-Hamiltonian structure (Π_0, Π_1), where $\Pi_1^\sharp = N\Pi_0^\sharp$.*

Proof Since ${}^t N(dI_k) = dI_{k+1}$ for all k, $\Pi_0^{\sharp\, t} N(dI_k) = \Pi_0^\sharp(dI_{k+1})$, implying that

$$\Pi_1^\sharp(dI_k) = \Pi_0^\sharp(dI_{k+1}), \ \forall k \geq 1,$$

since $\Pi_0^{\sharp\, t} N = N\Pi_0^\sharp$, proving the statement. $\qquad\qquad\square$

Then

Corollary 5.63 *The functions $\{I_k\}_{k\geq 1}$ are pairwise in involution with respect to both $\{\cdot, \cdot\}_0$ and $\{\cdot, \cdot\}_1$.*

Remark 5.64 Note that N is a recursion operator for each of the X_{I_i}s (see Definition 5.12). In particular, Lemma 5.61 specializes Proposition 5.13 to the framework of Poisson-Nijenhuis structures. $\qquad\qquad\triangle$

We close this part with the following observations.

Remark 5.65 (Iterated Brackets and Multi-Hamiltonian Structures) As it was explained above, PN-manifolds carry a bi-Hamiltonian structure. Although this observation would suffice to justify their introduction in the theory of integrable systems, it does not quite do justice to their relevance to this theory. Here below, aiming to give a more complete presentation, a few more properties of these structures will be presented. The main reference for what follows is [141]. First, we observe that every (non-negative) integer power of a Nijenhuis operator is still a Nijenhuis operator (see [152]). The proof of this statement can be obtained with a two-step induction. With the first one, it is shown that if $T_N = 0$, then

$$[NX, N^{k+1}Y] = N[X, N^{k+1}Y]+N^{k+1}[NX, Y]-N^{k+2}[X, Y], \ \forall X, Y \in \mathfrak{X}(M), \ k.$$

With the second step, one proves that

$$\left[N^l X, X^k Y\right] = N^l\left[X, N^k Y\right] + N^k\left[N^l X, Y\right] - N^{l+k}[X, Y], \ \forall X, Y \in \mathfrak{X}(M), \ k, l.$$

Choosing $n = k$ in the previous formula, one deduces that N^k is a Nijenhuis operator for all k. In this way, every Nijenhuis operator gives birth to a family of Lie brackets. More precisely, every $[\cdot, \cdot]_k : \mathfrak{X}(M) \times \mathfrak{X}(M) \to \mathfrak{X}(M)$ defined by

$$[X, Y]_k = \left[N^k X, Y\right] + \left[X, N^k Y\right] - N^k[X, Y], \ \forall X, Y \in \mathfrak{X}(M)$$

where N is a Nijenhuis operator and is a Lie bracket (see Sect. 6.3.3). As proven in [141, Section 1.3], this family of Lie brackets has several interesting properties, e.g., (1) they are compatible pair-wise, (2) the k-*torsion* of N^m is zero for all m, k, and (3) $N^m : (\mathfrak{X}(M), [\cdot, \cdot]_{k+m}) \to (\mathfrak{X}(M), [\cdot, \cdot]_k)$ is a morphism of Lie algebras for all m, k. In (2), what we called the k-torsion of N^m is $T_{N^m, k} : \mathfrak{X}(M) \times \mathfrak{X}(M) \to \mathfrak{X}(M)$

defined by

$$T_{N^m, k}(X, Y) = [N^m, N^m]_k - N^m([N^m X, Y] + [X, N^m Y] - N^m[X, Y]),$$

$$\forall X, Y \in \mathfrak{X}(M).$$

These observations become particularly relevant when applied to a PN-manifold. In fact, the compatibility conditions (5.40), together with the previous observations about the existence of the iterated brackets, imply the following.

Proposition 5.66 *If (M, Π_0, N) is a PN-manifold, then for all k, l:*

(i) (Π_k, N^l) *is a PN-structure.*
(ii) (Π_k, Π_l) *are compatible Poisson structures.*

where $\Pi_k = N^k \Pi_0$, for all $k \geq 0$.

The previous result can be phrased saying that every PN-manifold is endowed with a multi-Hamiltonian structure, i.e., with a family of pair-wise compatible Poisson structures. The proof of Proposition 5.66 can be found in [141, Section 5]. For a non-inductive proof of the same result, we refer the reader to [186]. $\triangle$

We will now present a few simple examples of bi-Hamiltonian systems, whose underlying bi-Hamiltonian description comes from a Poisson-Nijenhuis structure. We start with a generic framework that will be reconsidered and analyzed in more details at the end of this chapter. In particular, the first example is meant to explore more closely the relation between the integrability *à la* Liouville and the existence of a bi-Hamiltonian description for a given Hamiltonian system. This topic will be discussed in more generality at the end of this chapter (see Sect. 5). To this end, let (M, ω) be a $2n$-dimensional symplectic manifold and let $\pi : M \to B$ be a Lagrangian fibration with compact and connected fibers. Let $(J, \theta) = (J_1, \ldots, J_n, \theta_1, \cdots \theta_n)$ be a set of action-angle variables defined in a neighborhood of a given fiber of π (see Chap. 4). Suppose that $H \in C^\infty(M)$ defines a Hamiltonian vector field tangent to fibers of π so that (M, ω, H) is a completely integrable Hamiltonian system. Recall that under these assumptions, $X_H = \sum_{k=1}^n \frac{\partial H}{\partial J_k} \frac{\partial}{\partial \theta_k}$. In the next example, we give sufficient conditions so that (1) Π_0 can be completed to a Poisson-Nijenhuis structure in the chosen tubular neighborhood of the given fiber of π and (2) X_H is a bi-Hamiltonian vector field with respect to given Poisson-Nijenhuis structure. The first example is taken from [224] (see also [225]).

Example 5.67 Suppose that $M \simeq \mathbb{T}^n \times U$ is a tubular neighborhood of a given fiber of a Lagrangian fibration with compact and connected fibers supported on a symplectic manifold of dimension $2n$. Suppose further that on M is defined a set of action-angle coordinates (J, θ). In other words on M the sympletic form can be written as $\sum_{i=1}^{n} dJ_i \wedge d\theta_i$. If Π_0 is the Poisson bivector defined by ω, i.e., $\Pi_0 = \sum_{i=1}^{n} \frac{\partial}{\partial J_i} \wedge \frac{\partial}{\partial \theta_i}$ and if

$$N = \sum_{k=1}^{n} A_k(J_k) \left(dJ_k \otimes \frac{\partial}{\partial J_k} + d\theta_k \otimes \frac{\partial}{\partial \theta_k} \right)$$

where each A_k depends only on the corresponding J_k; then one has

$$^{t}\Pi_0^{\sharp} N = \Pi_0^{\sharp} N \quad \text{and} \quad \mathcal{R}_{\Pi_0, N} = 0, \tag{5.47}$$

which implies that (Π_0, N) is a Poisson-Nijenhuis structure (see Theorem 5.56). Note that the Poisson bivector field defined by the skew-symmetric endomorphism $^{t}N\Pi_0^{\sharp}$ is

$$\Pi_1 = \sum_{k=1}^{n} A_k(J_k) \frac{\partial}{\partial J_k} \wedge \frac{\partial}{\partial \theta_k}.$$

We prove now that if $H = \sum_{i=1}^{n} h_i(J_i)$, where $h_i \in C^{\infty}(U)$, then there exists $K \in C^{\infty}(M)$ such that $\Pi_0^{\sharp} dH = \Pi_1^{\sharp} dK$. In fact, if

$$\alpha_i = \frac{\partial h_i}{\partial J_i} A_i^{-1}, \ \forall i = 1, \ldots, n,$$

then $\alpha = \sum_{i=1}^{n} \alpha_i dJ_i$ is exact. This follows applying to α the relative Poincaré theorem (see Lemma 1.64). To this end, note that $d\alpha = 0$ and that the restriction of α to $\mathbb{T}^n$ is constant, i.e., $\alpha_{|\mathbb{T}^n} = \sum_{k=1}^{n} a_k dJ_k$, where $a_k = \alpha_{k|\mathbb{T}^n}$. Applying Formula (1.38) to the present case, one obtains

$$Q\,d\alpha + dQ\alpha = \rho_1^{*}\alpha - \rho_0^{*}\alpha = \alpha - \sum_{k=1}^{n} a_k dJ_k,$$

i.e.,

$$\alpha = dQ\alpha + \sum_{k=1}^{n} \alpha_i dJ_i = d\left(Q\alpha + \sum_{k=1}^{n} a_k J_k \right),$$

showing that α is exact in a suitable neighborhood of the chosen fiber. If $K := Q\alpha + \sum_{k=1}^{n} a_k J_k$,

$$\Pi_0^{\sharp}(dH) = \Pi_1^{\sharp}(\alpha) = \Pi_1^{\sharp}(dK).$$

$\triangle$

Problem 5.68 Prove (5.47). (Hint: it follows by a direct computation in the coordinates (J, θ).)   △

The previous example suggests that for the Hamiltonian H to have a bi-Hamiltonian representation coming from a PN structure, it is necessary for H to be not only dependent on the action variables alone but also to be separated in those variables. For this, see Sect. 5.

On a more concrete side, we present here below two elementary, but still not-trivial, examples of Poisson-Nijenhuis structure giving a bi-Hamiltonian representation of the harmonic oscillator and of the Kepler system (see [62] and [171, 242]).

Example 5.69 (Isotropic Harmonic Oscillator) Let $(M, \omega) = (\mathbb{R}^{2n}, \omega_0)$, $\omega_0 = \sum_{i=1}^{n} dp_i \wedge dq_i$, with Darboux coordinates (q, p) and let $H = \frac{1}{2} \sum_{i=1}^{n} (p_i^2 + q_i^2)$ be the Hamiltonian function of the n-dimensional isotropic harmonic oscillator. Recall that $h_i = \frac{p_i^2 + q_i^2}{2}$, for $i = 1, \ldots, n$, are (generically) independent first integrals of H pairwise in involution with respect to the Poisson bracket defined by ω_0 (see Examples 1.118 and 1.126). Then

$$\Pi_0 = \sum_{i=1}^{n} \frac{\partial}{\partial p_i} \wedge \frac{\partial}{\partial q_i} \quad \text{and} \quad \Pi_1 = \sum_{i=1}^{n} h_i \frac{\partial}{\partial p_i} \wedge \frac{\partial}{\partial q_i}$$

is a pair of compatible Poisson structures, defining a bi-Hamiltonian structure on $\mathbb{R}^{2n}$. The vector field

$$X = \sum_{i=1}^{n} \left(p_i \frac{\partial}{\partial q_i} - q_i \frac{\partial}{\partial p_i} \right) \tag{5.48}$$

is bi-Hamiltonian, more precisely $\Pi_0^\sharp(dH_1) = X = \Pi_1^\sharp(dH_0)$, where $H_0 = \sum_{i=1}^{n} \log h_i$ and $H_1 = H$. Moreover,

$$N := \Pi_1^\sharp \Pi_0^{\sharp,-1} = \sum_{i=1}^{n} h_i \left(dp_i \otimes \frac{\partial}{\partial p_i} + dq_i \otimes \frac{\partial}{\partial q_i} \right)$$

is a recursion operator for the bi-Hamiltonian vector field (5.48). Note that (Π_0, N) is a Poisson-Nijenhuis structure for the harmonic oscillator. In particular

$$\Pi_k = \sum_{i=1}^{n} h_i^k \frac{\partial}{\partial p_i} \wedge \frac{\partial}{\partial q_i}, \quad \forall k \geq 0. \tag{5.49}$$

       △

Example 5.70 (Kepler System) Recall the action-angle variables $(J_r, J_\theta, J_\varphi, \theta_r, \theta_\vartheta, \theta_\varphi)$ for the Kepler problem (see Sect. 2.2 in Chap. 4). Let us write 1 for r, 2 for ϑ, and 3 for φ. The Hamiltonian of this dynamical system, written in these coordinates, is $H = -\frac{1}{(J_1+J_2+J_3)^2}$, while the symplectic form, in a shorthand notation, becomes $\omega = \sum_{i=1}^{3} dJ_i \wedge d\theta_i$. The Hamiltonian vector field of H is easily computed to be

$$X = \frac{2}{(J_1 + J_2 + J_3)^3} \left(\frac{\partial}{\partial \theta_1} + \frac{\partial}{\partial \theta_2} + \frac{\partial}{\partial \theta_3} \right).$$

Let

$$S = \frac{1}{2} \begin{bmatrix} J_1 & J_2 & J_3 \\ J_2 - J_3 & J_1 + J_3 & J_3 \\ J_3 - J_2 & J_2 & J_1 + J_2 \end{bmatrix}, \tag{5.50}$$

and $\omega' = \sum_{i,j=1}^{3} S_{ij} dJ_i \wedge d\theta_j$. Then ω' is a symplectic form and $i_X \omega' = -dK$ where $K = -\frac{2}{J_1+J_2+J_3}$. In other words, the vector field X admits bi-Hamiltonian formulation, and H, K form a Lenard-Magri chain. Moreover the non-degeneracy of the symplectic form yields the existence of an (unique) endomorphism N such that $\omega'(X, Y) = \omega(NX, Y)$, for all vector fields X, Y. From the representation of ω and ω' in terms of the action-angle coordinates, one deduces that

$$N = \sum_{i,j=1}^{3} S_{ij} \left(dJ_i \otimes \frac{\partial}{\partial J_j} + d\theta_i \otimes \frac{\partial}{\partial \theta_j} \right),$$

where the S_{ij}s are defined in (5.50).   △

4.1 ΩN-Manifolds and Darboux-Nijenhuis Coordinates

In this subsection, we will consider a class of bi-Hamiltonian manifolds even more restricted than the PN ones, which we introduce in the following (see [81, 82]).

Definition 5.71 (ΩN-Manifold) An ΩN-manifold is a PN-manifold (M, N, Π_0), whose Poisson tensor Π_0 is invertible, i.e., it comes from a symplectic form ω_0 (see also Sect. 2). In analogy to the PN-case, the pair (ω_0, N) will be called an ΩN-structure.   △

We will now translate the compatibility conditions defining a PN-structure (see 5.40), into compatibility conditions for the pair (ω_0, N) to define an ΩN-structure. To this end, we start with the following.

Lemma 5.72 *Let (M, ω_0) be a symplectic manifold and let $N \in \mathrm{End}_{C^\infty(M)}(TM)$. The $C^\infty(M)$-bilinear form defined by*

$$\omega_1(X, Y) = \omega_0(NX, Y), \quad \forall X, Y \in \mathfrak{X}(M). \tag{5.51}$$

is skew-symmetric, i.e., ω_1 is a two-form, if and only if (5.29) holds, i.e., if and only if

$$\Pi_0^{\sharp t} N = N \Pi_0^{\sharp},$$

where Π_0 is the Poisson bivector field defined by ω_0 via Formula (2.28).

Proof First note that (5.29) is equivalent to

$$^t N \Pi_0^{\sharp-1} = \Pi_0^{\sharp-1} N. \tag{5.52}$$

Then

$$\omega_1(X, Y) = \omega_0(NX, Y) = \langle \omega_0^{\flat} NX, Y \rangle \overset{(2.28)}{=} -\langle \Pi_0^{\sharp-1} NX, Y \rangle$$

$$= \langle X, {}^t N \Pi_0^{\sharp-1} Y \rangle \overset{(5.52)}{=} \langle X, \Pi_0^{\sharp-1} NY \rangle$$

$$= -\langle \omega_0^{\flat} NY, X \rangle = -\omega_0(NY, X) = -\omega_1(Y, X).$$

On the other hand,

$$\omega_1(X, Y) = -\omega_1(Y, X), \ \forall X, Y \in \mathfrak{X}(M) \iff \omega_0(NX, Y)$$

$$= \omega_0(X, NY) \ \forall X, Y \in \mathfrak{X}(M). \tag{5.53}$$

Then

$$\langle \Pi_0^{\sharp-1} NX, Y \rangle = -\langle \omega_0^{\flat} NX, Y \rangle \overset{(5.53)}{=} -\omega_0(X, NY) = -\langle \omega_0^{\flat} X, NY \rangle$$

$$= \langle {}^t N \Pi_0^{\sharp-1} X, Y \rangle.$$

$$\square$$

Problem 5.73 Prove the equivalence in (5.53).   △

While the previous result translates the condition expressed in Formula (5.29) to a condition on the bilinear form ω_1 defined in (5.51), the following one gives a set of equivalent conditions for the brackets

$$\{f, g\}_1 = \omega_1(X_f, X_g) \quad \text{and} \quad \{f, g\}_0 = \omega_0(X_f, X_g) \tag{5.54}$$

to be compatible. The next proposition is taken from [57].

Proposition 5.74 *Let (M, ω_0) be a symplectic manifold and let $N \in \mathrm{End}_{C^\infty(M)}(TM)$ be such that ω_1 defined in (5.51) is a two-form. Then the following conditions are equivalent:*

(i) $d\omega_1 = 0$.
(ii) The Magri-Morosi concomitant $\mathcal{R}_{\Pi_0, N}$ (see 5.30) vanishes.
(iii) $i_{\mathcal{L}_{X_f} N}\omega_0 = -2dd_N f$ for all $f \in C^\infty(M)$,

where d_N and $i_N\omega$ are defined in Formula (5.129) and, respectively, (5.128).

Proof First note that if $X \in \mathfrak{X}(M)$ and $f \in C^\infty(M)$

$$
i_{\mathcal{R}_{\Pi_0, N}(df, X)}\omega_0 \overset{(5.30)}{=} i_{\mathcal{L}_{\Pi_0^\sharp(df)}(N)X}\omega_0 - i_{\Pi_0^\sharp\left(\mathcal{L}_{X^t N(df)}\right)}\omega_0 + i_{\Pi_0^\sharp\left(\mathcal{L}_{NX(df)}\right)}\omega_0
$$

$$
\overset{(2.28)}{=} i_{\mathcal{L}_{X_f}(N)X}\omega_0 + \mathcal{L}_{X^t}N(df) - \mathcal{L}_{NX}(df)
$$

$$
= i_{\mathcal{L}_{X_f}(N)X}\omega_0 + \cancel{di_{X^t}N(df)} + i_X d^t N(df) - \cancel{di_{NX}(df)}
$$

$$
\overset{(5.129)}{=} i_{\mathcal{L}_{X_f}(N)X}\omega_0 + i_X dd_N f, \tag{5.55}
$$

where in the penultimate equality we have used Cartan's formula $\mathcal{L}_x \alpha = d \circ i_X \alpha + i_X \circ d\alpha$. Now note that

$$
\omega_0\big((\mathcal{L}_{X_f} N)X, Y\big) = \omega_0\big(X, (\mathcal{L}_{X_f} N)Y\big), \quad \forall X, Y \in \mathfrak{X}(M). \tag{5.56}
$$

In fact, using $\mathcal{L}_{X_f}\omega_0 = 0$

$$
\omega_0\big((\mathcal{L}_{X_f} N)X, Y\big) = \omega_0(\mathcal{L}_{X_f}(NX), Y) - \omega_0(N\mathcal{L}_{X_f}X, Y)
$$
$$
= \mathcal{L}_{X_f}\big(\omega_0(NX, Y)\big) - \omega_0(NX, \mathcal{L}_{X_f}Y) - \omega_0(N\mathcal{L}_{X_f}X, Y)
$$
$$
= \mathcal{L}_{X_f}\big(\omega_0(X, NY)\big) - \omega_0(X, N\mathcal{L}_{X_f}Y) - \omega_0(\mathcal{L}_{X_f}X, NY)
$$

which yields

$$
\cancel{\omega_0(\mathcal{L}_{X_f}X, NY)} + \omega_0(X, (\mathcal{L}_{X_f}N)Y) + \cancel{\omega_0(X, N\mathcal{L}_{X_f}Y)} - \cancel{\omega_0(X, N\mathcal{L}_{X_f}Y)}
$$
$$
- \cancel{\omega_0(\mathcal{L}_{X_f}X, NY)}
$$

proving (5.56). Then

$$
\langle i_{(\mathcal{L}_{X_f}N)X}\omega_0, Y \rangle = \omega_0((\mathcal{L}_{X_f}N)X, Y) \overset{(5.128)}{=} \frac{1}{2}i_{\mathcal{L}_{X_f}N}\omega_0(X, Y), \quad \forall X, Y \in \mathfrak{X}(M),
$$

which together with (5.55) entails that

$$\mathcal{R}_{\Pi_0,N} = 0 \iff \frac{1}{2}i\,\mathscr{L}_{X_f}N\,\omega_0 + dd_N f = 0, \ \forall f \in C^\infty(M). \tag{5.57}$$

Now we compute

$$i_{X_f}d\omega_1 \overset{\text{Exercise 5.75}}{=} -\frac{1}{2}i_{X_f}d_N\omega_0 \overset{(5.129)}{=} -\frac{1}{2}i_{X_f}(i_N \circ d - d \circ i_N)\omega_0$$

$$=\frac{1}{2}i_{X_f}d(i_N\omega_0) = \frac{1}{2}\mathscr{L}_{X_f}(i_N\omega_0) - \frac{1}{2}di_{X_f}i_N\omega_0$$

$$=\frac{1}{2}i\,\mathscr{L}_{X_f}N\,\omega_0 - \frac{1}{2}di_{X_f}i_N\omega_0$$

$$=\frac{1}{2}i\,\mathscr{L}_{X_f}N\,\omega_0 + dd_N f, \tag{5.58}$$

where the last identity follows from the following computation:

$$\langle i_{X_f}i_N\omega_0, v\rangle = i_N\omega_0(X_f, v) = \omega_0(NX_f, v) + \omega_0(X_f, Nv)$$

$$\overset{(5.53)}{=} 2\omega_0(X_f, Nv) = -2\langle df, Nv\rangle \overset{(5.129)}{=} -2\langle d_N f, v\rangle.$$

Summarizing

$$\mathcal{R}_{\Pi_0,N} = 0 \overset{(5.57)}{\iff} \frac{1}{2}i\,\mathscr{L}_{X_f}N\,\omega_0 + dd_N f = 0 \overset{(5.58)}{\iff} i_{X_f}d\omega_1 = 0, \ \forall f \in C^\infty(M).$$

which proves the equivalences since $d\omega_1 = 0$ if and only if $i_{X_f}d\omega_1 = 0$ for all $f \in C^\infty(M)$. $\qquad\square$

Problem 5.75 Prove that under the assumptions above

$$d\omega_1 = -\frac{1}{2}d_N\omega_0.$$

(Hint: One can prove the previous identity by a direct computation using the definition of d_N.) $\qquad\triangle$

The previous result yields the following.

Corollary 5.76 *If any of the three equivalent conditions in Proposition 5.74 is fulfilled, then the brackets $\{\cdot, \cdot\}_1, \{\cdot, \cdot\}_0$ are compatible (see Definition 5.14).*

Proof See Proposition 5.15. $\qquad\square$

As the reader will have noticed, the previous results do not depend on the property of N to be torsionless. Adding this hypothesis, one arrives to the following.

Theorem 5.77 *If (M, ω_0) is a symplectic manifold and $N \in \mathrm{End}_{C^\infty(M)}(TM)$ are such that:*

(i) $\omega_0(NX, Y) = \omega_0(X, NY)$, for all $X, Y \in \mathfrak{X}(M)$
(ii) $d\omega_1 = 0$, where ω_1 is defined as in (5.51)
(iii) $T_N = 0$

then (M, ω_0, N) is an ΩN-manifold.

Proof Let Π_0 and $\{\cdot, \cdot\}_0$ denote, as usual, the Poisson tensor defined by ω_0 (see Formula (2.28) and, respectively, the corresponding Poisson bracket). If $\omega_0(NX, Y) = \omega_0(X, NY)$ for all X, Y, ω_1 defined in (5.53) is a two-form implying that $\Pi_0^{\sharp t} N = N \Pi_0^{\sharp}$ (see Lemma 5.72). If $d\omega_1 = 0$, then $\{\cdot, \cdot\}_1$, defined in (5.54), and $\{\cdot, \cdot\}_0$ are compatible (see Corollary 5.76). This condition is equivalent to $[\![\Pi_0, \Pi_1]\!] = 0$, where Π_1 is the bivector field associated with $\{\cdot, \cdot\}_1$. Finally, if Π_0 and Π_1 are compatible, then N is torsion-free and Π_1 is a Poisson bivector field (see Theorem 5.17). In other words, under the assumptions of the theorem, (M, Π_0, N) is a PN-manifold, and the invertibility of Π_0 closes the proof. $\qquad\square$

We close this part with the following observation.

Remark 5.78 In analogy to what we noticed in Remark 5.65, every ΩN-manifold (M, ω_0, N) carries a multi-Hamiltonian structure defined by the family $\{\Pi_k\}_{k \geq 0}$, where $\Pi_k^{\sharp} = N^k \Pi_0^{\sharp}$ for all $k \geq 0$ and Π_0 is the Poisson bivector field defined by ω_0 via (2.28). More precisely, in this enhanced case, one can prove easily that (1) for $k \geq 0$; ω_k defined by $\omega_k(X, Y) = \omega_0(N^k X, Y)$ for all $X, Y \in \mathfrak{X}(M)$ is skew-symmetric, i.e., it is a two-form; and that (2) $\omega_k(N^l X, Y) = \omega_k(X, N^l Y)$, for all l, k and $X, Y \in \mathfrak{X}(M)$. A more difficult task is to show that $d\omega_k = 0$ for all $k \geq 2$. Of course these properties follow at once from Proposition 5.66. $\qquad\triangle$

4.1.1 Darboux-Nijenhuis Coordinates

Now we start the proof of the existence of a *normal form* for the pair (ω_0, N). More precisely we will prove that under suitable hypothesis, it is possible to introduce local coordinates at every point of an ΩN-manifold, which reduces both ω_0 and N to a normal form. First prove the following.

Lemma 5.79 *For each $m \in M$, $\ker(N_m - \lambda(m)\,\mathrm{id}_m)$ is even-dimensional. In particular, the maximum number of distinct eigenvalues of N_m is obtained when each eigenvalues has algebraic multiplicity equal to two.*

Proof If $\Pi_1^{\sharp} = N\Pi_0^{\sharp}$, for any given $m \in M$, we have that

$$\dim \ker\left(N_m - \lambda(m)\,\mathrm{id}_m\right) = \dim \ker\left(\Pi_{1m}^{\sharp} - \lambda(m)\Pi_{0m}^{\sharp}\right)$$
$$= \dim M - \dim \mathrm{im}\left(\Pi_{1m}^{\sharp} - \lambda(m)\Pi_{0m}^{\sharp}\right),$$

which is an even number since M is symplectic, and $\left(\Pi^\sharp_{1\,m} - \lambda(m)\Pi^\sharp_{0m}\right)$ is a skew-symmetric linear map, which restricts to a non-degenerate linear map on $T^*_m M/\ker\left(\Pi^\sharp_{1\,m} - \lambda(m)\Pi^\sharp_{0m}\right) \simeq \mathrm{im}\left(\Pi^\sharp_{1\,m} - \lambda(m)\Pi^\sharp_{0m}\right)$. The second part of the statement follows now from the well-known fact that eigenvectors corresponding to distinct eigenvalues are independent. $\square$

Before moving on with our discussion, we note that every eigenvalue λ of $N \in \mathrm{End}_{C^\infty(M)}(TM)$ is a smooth function such that

$$\det\left(N_m - \lambda(m)\,\mathrm{id}_m\right) = 0,$$

for all m. Now we can introduce the following.

Definition 5.80 Let (M, ω_0, N) be a $2n$-dimensional ΩN-manifold. A point $m \in M$ is called *regular* if there exist $n = \frac{\dim M}{2}$ eigenvalues $\lambda_1, \ldots, \lambda_n$ of N_m such that:

(i) $\lambda_i(m) \neq \lambda_j(m)$, $\forall i \neq j$
(ii) $d\lambda_1 \wedge \cdots \wedge d\lambda_n(m) \neq 0$

A ΩN-manifold is called regular if all its points are regular. A recursion operator N is called regular if its eigenvalues satisfy the two previous conditions at all points of M. $\triangle$

Remark 5.81

1. The above definition of regular point (and regular ΩN-manifold is taken from [166] (see also [177]). We would like to alert the reader that what in these notes is called regular is sometimes called regular semi-simples (see, e.g., [81, 82]).
2. Note that since $m \rightsquigarrow \dim \mathrm{im}\,(N_n)$ is *lower semi-continuous*, if m is a regular point, there exists a neighborhood of m whose points are all regular.

$\triangle$

These observations yield the following.

Theorem 5.82 *If (M, ω_0, N) is a $2n$-dimensional regular ΩN-manifold, every $H \in C^\infty(M)$ such that $\{H, \lambda\}_0 = 0$ for all eigenvalues λ of N is the Hamiltonian of a completely integrable system defined on M. In particular, the eigenvalues of N form a complete set of first integrals for such an integrable system.*

Proof Since N is regular, the eigenvalues of N are distinct and independent on M. Moreover, Corollary 5.63 implies that they Poisson commute with each other on the dense open subset of M where the relations $I_k = I_k(\lambda)$, for $k = 1, \ldots, n$, can be inverted. $\square$

The previous theorem gives sufficient conditions for a Hamiltonian system to be completely integrable: it is sufficient to prove that it admits an ΩN-formulation with regular recursion operator. The inverse problem, i.e., the problem to decide under which assumptions a completely integrable system admits a bi-Hamiltonian formulation, will be discussed in Sect. 5.

Going back to our proposal of finding a normal form for a regular ΩN-manifold, one starts to observe that

Lemma 5.83 *Every regular point of an ΩN-manifold has a neighborhood where the eigenvalues of N can be completed to a set of Darboux coordinates for ω_0.*

Proof In fact, every regular point m is contained in a neighborhood where the eigenvalues of N $\lambda_1, \ldots, \lambda_n$ are independent. On such a neighborhood, the relations

$$kI^k = \sum_{l=1}^{n} \lambda_l^k, \; k \geq 1, \tag{5.59}$$

can be inverted, and the Corollary 5.63 implies

$$\{\lambda_i, \lambda_j\}_0 = 0, \; \forall i, j = 1, \ldots, n.$$

In other words, on suitable neighborhood U of m, $\lambda_1, \ldots, \lambda_n$ are n-independent functions, pairwise in involution with respect to $\{\cdot, \cdot\}_0$. Under these assumptions, by the Carathéodory-Jacobi-Lie Theorem (see Theorem 1.114), one can find a neighborhood $V \subset U$ of m and a set of independent functions $\mu_1, \ldots, \mu_n$, completing $\lambda_1, \ldots, \lambda_n$ to a system of local coordinates on V and such that

$$\{\mu_i, \mu_j\}_0 = 0 \quad \text{and} \quad \{\mu_i, \lambda_j\}_0 = \delta_{ij}, \forall i, j = 1, \ldots, n.$$

Note that the restriction of ω_0 to V can be written as

$$\omega_0 = \sum_{i=1}^{n} d\mu_i \wedge d\lambda_i. \tag{5.60}$$

$\square$

Remark 5.84 It is worth observing that the functions μ_is are defined uniquely only up to a *gauge* transformation defined by

$$\mu_i' = \mu_i + \frac{\partial S}{\partial \lambda_i}, \; \forall i = 1, \ldots, n,$$

where S is a smooth function defined in a neighborhood of m and depending only the variables λ_is. $\triangle$

A second step is contained in the

Lemma 5.85 *On the Darboux neighborhood of the regular point m defined in Lemma 5.83,*

$$N\frac{\partial}{\partial \lambda_i} = \lambda_i \frac{\partial}{\partial \lambda_i} + \sum_{k=1}^{n} a_{ik}\frac{\partial}{\partial \mu_k}, \quad \forall i = 1, \ldots, n$$

where the a_{ij}s are smooth functions, independent on the μ_is and such that $a_{ij} = -a_{ji}$ for all $i, j = 1, \ldots, n$.

Proof We start the proof with the following computation:

$$\lambda_1^{k-1}{}^t N(d\lambda_1) + \cdots \lambda_n^{k-1}{}^t N(d\lambda_n) = {}^t N(dI^k) \stackrel{(5.45)}{=} dI^{k+1} = \lambda_1^k d\lambda_1 + \cdots \lambda_n^k d\lambda_n,$$

from which it follows that $\lambda_i^{k-1}{}^t N(d\lambda_i) = \lambda_i^k d\lambda_i$, for all $i = 1, \ldots, n$. This, in turn, by the hypothesis on the λ_is, yields the following important identities:

$$ {}^t N(d\lambda_k) = \lambda_k d\lambda_k, \quad \forall k = 1, \ldots, n. \tag{5.61}$$

Then

$$\left\langle N\frac{\partial}{\partial \mu_i}, d\lambda_j \right\rangle = \left\langle \frac{\partial}{\partial \mu_i}, {}^t N(d\lambda_j) \right\rangle \stackrel{(5.61)}{=} \lambda_j \left\langle \frac{\partial}{\partial \mu_i}, d\lambda_j \right\rangle = 0, \tag{5.62}$$

and

$$\left\langle N\frac{\partial}{\partial \mu_i}, d\mu_j \right\rangle \stackrel{(5.60)}{=} -\left\langle N\frac{\partial}{\partial \mu_i}, i_{\frac{\partial}{\partial \lambda_j}}\omega_0 \right\rangle = -\omega_0 \left(\frac{\partial}{\partial \lambda_j}, N\frac{\partial}{\partial \mu_i} \right)$$

$$\stackrel{(5.53)}{=} -\omega_0 \left(N\frac{\partial}{\partial \lambda_j}, \frac{\partial}{\partial \mu_i} \right) \stackrel{(5.60)}{=} \left\langle d\lambda_i, N\frac{\partial}{\partial \lambda_j} \right\rangle = \lambda_i \delta_{ij},$$

which, together with (5.62), gives

$$N\frac{\partial}{\partial \mu_i} = \lambda_i \frac{\partial}{\partial \mu_i}, \quad \forall i = 1, \ldots, n. \tag{5.63}$$

On the other hand

$$\left\langle N\frac{\partial}{\partial \lambda_i}, d\lambda_j \right\rangle = \left\langle \frac{\partial}{\partial \lambda_i}, {}^t N(d\lambda_j) \right\rangle = \lambda_j \left\langle \frac{\partial}{\partial \lambda_i}, d\lambda_j \right\rangle = \lambda_j \delta_{ij},$$

and if

$$a_{ij} = \omega_0 \left(N\frac{\partial}{\partial \lambda_i}, \frac{\partial}{\partial \lambda_j} \right), \quad \forall i, j, \tag{5.64}$$

a simple computation shows that $a_{ij} = -a_{ji}$ for all i, j, and one can write

$$N \frac{\partial}{\partial \lambda_i} = \lambda_i \frac{\partial}{\partial \lambda_i} + \sum_{k=1}^{n} a_{ik} \frac{\partial}{\partial \mu_k}, \quad \forall i = 1, \ldots, n. \tag{5.65}$$

We are left to show that the functions defined in (5.64) do not depend on the μ_is. To this end, we compute

$$T_N \left(\frac{\partial}{\partial \mu_i}, \frac{\partial}{\partial \lambda_j} \right)$$

$$= \left[N \frac{\partial}{\partial \mu_i}, N \frac{\partial}{\partial \lambda_j} \right] - N \left(\left[N \frac{\partial}{\partial \mu_i}, \frac{\partial}{\partial \lambda_j} \right] + \left[\frac{\partial}{\partial \mu_i}, N \frac{\partial}{\partial \lambda_j} \right] \right)$$

$$\overset{(5.63) \text{ and } (5.65)}{=} \left[\lambda_i \frac{\partial}{\partial \mu_i}, \lambda_j \frac{\partial}{\partial \lambda_j} + \sum_{k=1}^{n} a_{jk} \frac{\partial}{\partial \mu_k} \right]$$

$$- N \left(\left[\lambda_i \frac{\partial}{\partial \mu_i}, \frac{\partial}{\partial \lambda_j} \right] + \left[\frac{\partial}{\partial \mu_i}, \lambda_j \frac{\partial}{\partial \lambda_j} + \sum_{k=1}^{n} a_{jk} \frac{\partial}{\partial \mu_k} \right] \right)$$

$$= -\lambda_j \frac{\partial}{\partial \mu_j} + \lambda_i \sum_{k=1}^{n} \frac{\partial a_{jk}}{\partial \mu_i} \frac{\partial}{\partial \mu_k} + \lambda_j \frac{\partial}{\partial \mu_j} - \sum_{k=1}^{n} \frac{\partial a_{jk}}{\partial \mu_i} N \frac{\partial}{\partial \mu_k}$$

$$= \sum_{k=1}^{n} \frac{\partial a_{jk}}{\partial \mu_i} (\lambda_i - \lambda_k) \frac{\partial}{\partial \mu_k}.$$

Since N is a Nijenhuis operator and since $\lambda_k \neq \lambda_i$ for all $i \neq k$, one concludes that for a given i and j,

$$\frac{\partial a_{jk}}{\partial \mu_i} = 0, \ \forall k \neq i.$$

For $k = i$, the statement follows from the skew-symmetry of the a_{ij}. $\qquad \square$

Before we come to the main result of this section, it is worth recording the following.

Corollary 5.86 *On the Darboux neighborhood of any regular point (see Lemma 5.83), the following identities hold:*

$$^{t}N(d\mu_i) = \lambda_i d\mu_i + \sum_{k=1}^{n} a_{ki} d\lambda_k, \ \forall i = 1, \ldots, n. \tag{5.66}$$

Proof The proof follows at once from (5.63) and (5.65). $\qquad \square$

Finally, we can state the following important result.

Theorem 5.87 (Darboux Theorem for ΩN-Structures) *Every regular point of a ΩN-manifold is contained in a Darboux neighborhood whose Darboux coordinates (μ, λ) reduce the Nijenhuis operator to the following normal form:*

$$N = \sum_{k=1}^{n} \lambda_k \left(d\lambda_k \otimes \frac{\partial}{\partial \lambda_k} + d\mu_k \otimes \frac{\partial}{\partial \mu_k} \right)$$

Proof As already observed while the λs are uniquely defined, the μs are defined only up an arbitrary function of the λ (see Remark 5.84). To make a good use of this freedom, one observes that the two-form $\eta = \sum_{i<j} a_{ij} d\lambda_i \wedge d\lambda_j$ is closed. In fact, computing ω_1 in the coordinates (μ, λ), one finds that $\eta = \omega_1 + \sum_{i=1}^{n} \lambda_i d\lambda_i \wedge d\mu_i$. The condition $d\eta = 0$ is equivalent to the following identities:

$$\frac{\partial a_{ij}}{\partial \lambda_k} + \frac{\partial a_{ki}}{\partial \lambda_j} + \frac{\partial a_{jk}}{\partial \lambda_i} = 0, \; \forall i, j, k. \tag{5.67}$$

Moreover, computing

$$T_N \left(\frac{\partial}{\partial \lambda_i}, \frac{\partial}{\partial \lambda_j} \right)$$

$$= \left[\lambda_i \frac{\partial}{\partial \lambda_i}, \lambda_j \frac{\partial}{\partial \lambda_j} \right] + \left[\lambda_i \frac{\partial}{\partial \lambda_i}, \sum_{k=1}^{n} a_{jk} \frac{\partial}{\partial \mu_k} \right] + \left[\sum_{k=1}^{n} a_{ik} \frac{\partial}{\partial \mu_k}, \lambda_j \frac{\partial}{\partial \lambda_j} \right]$$

$$+ \left[\sum_{k=1}^{n} a_{ik} \frac{\partial}{\partial \mu_k}, \sum_{k=1}^{n} a_{jk} \frac{\partial}{\partial \mu_k} \right]$$

$$- N \left(\left[\lambda_i \frac{\partial}{\partial \lambda_i}, \frac{\partial}{\partial \lambda_j} \right] + \left[\sum_{k=1}^{n} a_{ik} \frac{\partial}{\partial \mu_k}, \frac{\partial}{\partial \lambda_j} \right] + \left[\frac{\partial}{\partial \lambda_i}, \lambda_j \frac{\partial}{\partial \lambda_j} \right] \right.$$

$$\left. + \left[\frac{\partial}{\partial \lambda_i}, \sum_{k=1}^{n} a_{jk} \frac{\partial}{\partial \mu_k} \right] \right)$$

$$= \lambda_i \sum_{k=1}^{n} \frac{\partial a_{jk}}{\partial \lambda_i} \frac{\partial}{\partial \mu_k} - \lambda_j \sum_{k=1}^{n} \frac{\partial a_{ik}}{\partial \lambda_j} \frac{\partial}{\partial \mu_k} + \sum_{k=1}^{n} \lambda_k \frac{\partial a_{ik}}{\partial \lambda_j} \frac{\partial}{\partial \mu_k} - \sum_{k=1}^{n} \lambda_k \frac{\partial a_{jk}}{\partial \lambda_i} \frac{\partial}{\partial \mu_k},$$

which, together with the skew-symmetry of a_{ij}, yields the following identities:

$$\lambda_i \frac{\partial a_{jk}}{\partial \lambda_i} + \lambda_j \frac{\partial a_{ki}}{\partial \lambda_j} + \lambda_k \left(\frac{\partial a_{ik}}{\partial \lambda_j} + \frac{\partial a_{kj}}{\partial \lambda_i} \right) = 0, \; \forall i, j, k. \tag{5.68}$$

Using (5.67), (5.68) assumes the following form:

$$\lambda_i \frac{\partial a_{jk}}{\partial \lambda_i} + \lambda_j \frac{\partial a_{ki}}{\partial \lambda_j} + \lambda_k \frac{\partial a_{ij}}{\partial \lambda_k} = 0, \ \forall i, j, k. \tag{5.69}$$

Let S be an arbitrary function of the λs. Then from

$$\lambda_i' = \lambda_i \quad \text{and} \quad \mu_i' = \mu_i + \frac{\partial S}{\partial \lambda_i}, \ \forall i = 1, \dots, n.$$

one gets

$${}^t N(d\lambda_i') = \lambda_i' d\lambda_i' \quad \text{and} \quad {}^t N(d\mu_i') \overset{(5.66)}{=} \lambda_i d\mu_i + \sum_{k=1}^{n} a_{ki} d\lambda_k + \sum_{k=1}^{n} \frac{\partial^2 S}{\partial \lambda_k \partial \lambda_i} \lambda_k d\lambda_k.$$

Summing and subtracting $\lambda_i d(\frac{\partial S}{\partial \lambda_i})$, the second identity of the previous ones can be rewritten as

$${}^t N(d\mu_i') = \lambda_i d\mu_i' + \sum_{k=1}^{n} \left(a_{ki} + (\lambda_k - \lambda_i) \frac{\partial^2 S}{\partial \lambda_i \partial \lambda_k} \right) d\lambda_k \tag{5.70}$$

We will argue now that it is possible to choose S such that

$$a_{ki} = (\lambda_i - \lambda_k) \frac{\partial^2 S}{\partial \lambda_i \partial \lambda_k} = 0, \ \forall i, k = 1, \dots, n.$$

This will follow from $d\omega_1 = 0$ and from $T_N = 0$. More precisely, the *side-by-side* difference between the product of the (5.67) by λ_k and the (5.69) is

$$(\lambda_k - \lambda_j) \frac{\partial a_{ki}}{\partial \lambda_j} + (\lambda_k - \lambda_i) \frac{\partial a_{jk}}{\partial \lambda_i} = 0,$$

which, after dividing both sides by $(\lambda_k - \lambda_j)(\lambda_k - \lambda_i)$, yields

$$\frac{\partial}{\partial \lambda_j} \left(\frac{a_{ki}}{\lambda_k - \lambda_i} \right) = \frac{\partial}{\partial \lambda_i} \left(\frac{a_{kj}}{\lambda_k - \lambda_j} \right),$$

from which one deduces the existence of a smooth function f_k such that

$$\frac{a_{ki}}{\lambda_i - \lambda_k} = \frac{\partial f_k}{\partial \lambda_i}, \ \forall i = 1, \dots, n.$$

On the other hand, since

$$\frac{\partial f_k}{\partial \lambda_i} = \frac{a_{ki}}{\lambda_i - \lambda_k} = \frac{a_{ik}}{\lambda_k - \lambda_i} = \frac{\partial f_i}{\partial \lambda_k}, \quad \forall i, k,$$

one deduces the existence of a smooth function S such that

$$f_i = \frac{\partial S}{\partial \lambda_i}, \quad \forall i = 1, \ldots, n.$$

Plugging S in (5.70), one can summarize the above discussion saying that in a neighborhood of every regular point of a ΩN-manifold, one can find a set of Darboux coordinates (λ, μ) such that

$${}^t N(d\lambda_i) = \lambda_i d\lambda_i \quad \text{and} \quad {}^t N(d\mu_i) = \lambda_i d\mu_i, \quad \forall i = 1, \ldots, n,$$

where the λ_i's are the eigenvalues of N. Note that, in the previous formulas, we dropped the primes everywhere. To conclude the proof, it suffices to note that in this coordinate, one has

$$N\frac{\partial}{\partial \lambda_i} = \lambda_i \frac{\partial}{\partial \lambda_i} \quad \text{and} \quad N\frac{\partial}{\partial \mu_i} = \lambda_i \frac{\partial}{\partial \mu_i}, \quad \forall i = 1, \ldots, n.$$

$$\square$$

The previous theorem is an existence result for a distinguished sets of coordinates in a suitable neighborhood of every regular point of an ΩN-manifold. Because of their important role, it is worth giving the following.

Definition 5.88 (Darboux-Nijenhuis and Special Darboux-Nijenhuis Coordinates) The local coordinates (x, p) on an ΩN-manifold (M, ω_0, N) are called *Darboux-Nijenhuis, DN*-coordinates hereafter, if they are Darboux coordinates for the symplectic form ω_0 and if they reduce the Nijenhuis operator to a diagonal form, i.e., (x, p) are Darboux-Nijenhuis if:

(DN1) $\omega_0 = \sum_{i=1}^n dp_i \wedge dx_i$

(DN2) $N = \sum_{i=1}^n \lambda_i \left(dx_i \otimes \frac{\partial}{\partial x_i} + dp_i \otimes \frac{\partial}{\partial p_i} \right)$

The coordinates $(\lambda_1, \ldots, \lambda_n, \mu_1, \ldots, \mu_n)$ introduced above under the assumption of regularity will be called *special* Darboux-Nijenhuis coordinates. $\triangle$

Note that if (x, p) are *DN*-coordinates, then

$$\omega_1 = \sum_{i=1}^n \lambda_i dp_i \wedge dx_i$$

which implies $\{p_i, x_i\}_1 = \lambda_i \delta_{ij}$ for all $i, j = 1, \ldots, n$ all the other brackets being equal to zero.

Example 5.89 (DN-**Coordinates on Cotangent Bundles**) Every Nijenhuis operator $N \in \mathrm{End}(TQ)$ defines an ΩN-structure on T^*Q, i.e., the pair $(\Omega, \tilde{N})$, where Ω is the canonical symplectic structure and $\tilde{N}$ is the lift of N to T^*Q (see Example 5.21). In a suitable neighborhood of any regular point $m \in Q$ (see Definition 5.80), one can write

$$N = \sum_{i=1}^{n} \lambda_i d\lambda_i \otimes \frac{\partial}{\partial \lambda_i}, \tag{5.71}$$

where $\lambda_1, \ldots, \lambda_n$ are the eigenvalues of N. Let $(\lambda_1, \ldots, \lambda_n, \mu_1, \ldots, \mu_n)$ be the corresponding Darboux coordinates on T^*Q, and let $\Omega = \sum_{i=1}^{n} d\mu_i \wedge d\lambda_i$ be the canonical symplectic form. We show now that the (λ, μ)s are DN-coordinates for the pair $(\Omega, \tilde{N})$. First we compute $\tau_N^* \Theta$ (see Formula D.14)

$$\langle (\tau_N^* \Theta)_\alpha, v \rangle = \langle \Theta_{\tau_N(\alpha)}, (\tau_N)_{*,\alpha} v \rangle$$

$$= \langle \tau_N(\alpha), \pi_{*,\tau_N(\alpha)} (\tau_N)_{*,\alpha} v \rangle$$

$$= \langle \tau_N(\alpha), (\pi \circ \tau_n)_{*,\alpha} v \rangle$$

$$= \langle \tau_N(\alpha), \pi_{*,\alpha} v \rangle$$

$$= \langle \alpha, N(\pi_{*,\alpha} v) \rangle.$$

Writing $\alpha = \sum_{i=1}^{n} \mu_i(\alpha) d\lambda_i$, the previous computation yields

$$\left\langle (\tau_N^* \Theta)_\alpha, \frac{\partial}{\partial \lambda_i} \right\rangle = \left\langle \alpha, N(\pi_{*,\alpha} \frac{\partial}{\partial \lambda_i}) \right\rangle \stackrel{(5.71)}{=} \left\langle \alpha, \lambda_i \frac{\partial}{\partial \lambda_i} \right\rangle = \mu_i(\alpha) \lambda_i,$$

$$\left\langle (\tau_N^* \Theta)_\alpha, \frac{\partial}{\partial \mu_i} \right\rangle = 0,$$

i.e., $\tau_N^* \Theta = \sum_{i=1}^{n} \mu_i \lambda_i d\lambda_i$, which implies

$$\tau_N^* \Omega = \sum_{i=1}^{n} \lambda_i d\mu_i \wedge d\lambda_i.$$

This formula, together with the one defining the lift of N to T^*Q (see Sect. 3 in Appendix D), implies

$$\tilde{N}\left(\frac{\partial}{\partial \lambda_i} \right) = \lambda_i \frac{\partial}{\partial \lambda_i} \quad \text{and} \quad \tilde{N}\left(\frac{\partial}{\partial \mu_i} \right) = \lambda_i \frac{\partial}{\partial \mu_i},$$

which yields

$$\tilde{N} = \sum_{i=1}^{n} \lambda_i \left(d\lambda_i \otimes \frac{\partial}{\partial \lambda_i} + d\mu_i \otimes \frac{\partial}{\partial \mu_i} \right),$$

showing that, under the hypothesis above about N, the DN-coordinates for $(\Omega, \tilde{N})$ are just the Darboux coordinates on T^*Q defined by the eigenvalues of N. Observe that the *minimal* polynomial of $\tilde{N}$ *coincides* with the *characteristic* polynomial of N. $\triangle$

A more concrete example is presented below.

Example 5.90 In Example 5.69, it was observed that the n-dimensional isotropic harmonic oscillator has a bi-Hamiltonian representation coming from a Poisson-Nijenhuis structure (Π_0, N) on $\mathbb{R}^{2n}$ defined, in the physical coordinates (q, p), by

$$\Pi_0 = \sum_{i=1}^{n} \frac{\partial}{\partial p_i} \wedge \frac{\partial}{\partial q_i} \quad \text{and} \quad N = \sum_{i=1}^{n} h_i \left(dp_i \otimes \frac{\partial}{\partial p_i} + dq_i \otimes \frac{\partial}{\partial q_i} \right),$$

where $h_i(p_i, q_i) = \frac{p_i^2 + q_i^2}{2}$. Moreover, in Example 4.35, it was observed that the functions h_i can be used to define a set of action-angle variables (J, θ), where $J_i = \frac{2\pi}{\varpi_i} h_i$ and $\theta_i = \frac{\varpi_i}{2\pi} k_i$ (see Formula (4.37) in the above recalled example). Another simple computation shows that writing Π_0 and N in the variables (J, θ) so defined, one gets

$$\Pi_0 = \sum_{i=1}^{n} \frac{\partial}{\partial J_i} \wedge \frac{\partial}{\partial \theta_i} \quad \text{and} \quad N = \sum_{i=1}^{n} J_i \left(dJ_i \otimes \frac{\partial}{\partial J_i} + d\theta_i \otimes \frac{\partial}{\partial \theta_i} \right),$$

which implies that the action-angle variables are a set of DN-coordinates for the isotropic harmonic oscillator. In these coordinates, one has $\Pi_k = \sum_{i=1}^{n} J_i^k \frac{\partial}{\partial J_i} \wedge \frac{\partial}{\partial \theta_i}$, see Formula (5.49). $\triangle$

4.2 ΩN-Manifolds and Lax Pairs

We will see now how on a suitable neighborhood U of every regular point of an ΩN-manifold, it is possible to define a *hierarchy* of vector fields whose flows admit a Lax-type representation. The main references for this part are [166, 177]. For the definition of Lax representation and its relevance in the theory of integrable systems, we refer the reader to Sect. 2.2 of Chap. 6. The first step in this direction is to define another set of local coordinates on U, which, in spite of not being of Darboux for the symplectic form ω_0, have a remarkable commuting properties with respect to

$\{\cdot, \cdot\}_0$ and $\{\cdot, \cdot\}_1$. Let the I_ks be as in Formula (5.44) and let

$$J_k = \sum_{i=1}^{n} \mu_i \lambda_i^{k-1}. \tag{5.74}$$

Then

Proposition 5.91 *The (I, J)s form a system of local coordinates on U. Moreover,*

$$\{I_r, I_s\}_0 = 0, \quad \{J_r, I_s\}_0 = (r + s - 2)I_{r+s-2} \quad and \quad \{J_r, J_s\}_0 = (s - r)J_{r+s-2}, \tag{5.75}$$

for all r, s.

Proof To prove the first statement, it suffices to note that since

$$dI_k = \sum_{i=1}^{n} \lambda_i^{k-1} d\lambda_i \quad \text{and} \quad dJ_k = \sum_{i=1}^{n} \lambda_i^{k-1} d\mu_i + (k-1) \sum_{i=1}^{n} \mu_i \lambda_i^{k-2} d\lambda_i$$

the Jacobian matrix of the change of variables from the (λ, μ)s to the (I, J)s has the following form

$$\left[\begin{array}{c|c} V(\lambda) & \star \\ \hline 0 & V(\lambda) \end{array} \right], \tag{5.76}$$

where

$$V(\lambda) = \begin{bmatrix} 1 & \lambda_1 & \lambda_1^2 & \cdots & \lambda_1^{n-1} \\ 1 & \lambda_2 & \lambda_2^2 & \cdots & \lambda_2^{n-1} \\ \vdots & & \cdots & & \vdots \\ 1 & \lambda_n & \lambda_n^2 & \cdots & \lambda_n^{n-1} \end{bmatrix}. \tag{5.77}$$

Since, by assumption, the λs are distinct, the determinant of (5.76) is non-zero (note that $V(\lambda)$ is a Vandermonde matrix for the λs). For what concerns the second statement, it suffices to prove the two (sets of) identities involving the Js. Both follow from a direct computation. For example,

$$\{J_r, I_s\}_0 = \sum_{k=1}^{n} \left(\frac{\partial J_r}{\partial \mu_k} \frac{\partial I_s}{\partial \lambda_k} - \frac{\partial J_r}{\partial \lambda_k} \frac{\partial I_s}{\partial \mu_k} \right) \overset{(5.74)}{=} \sum_{k=1}^{n} \lambda_k^{r-1} \lambda_k^{s-1} = \sum_{k=1}^{n} \lambda_k^{r+s-2}$$

$$= (r + s - 2)I_{r+s-2}$$

$\square$

To proceed further, first we need to prove the following preliminary result. Let (L, M) be two $n \times n$ matrices such that L is diagonalizable with distinct eigenvalues.

Lemma 5.92 *Then*

$$\mathrm{tr}\left(ML^k\right) = 0, \ \forall k = 1, \ldots, n-1. \tag{5.79}$$

if and only if there exists another matrix B such that

$$M = [B, L].$$

Proof The hypothesis on L implies that there exists an invertible matrix g such that $gLg^{-1} = D$ is diagonal such that $d_i \neq d_j$ for all $i \neq j$. Then (5.79) is equivalent to

$$\mathrm{tr}\left(D^k gMg^{-1}\right) = 0, \ \forall k = 1, \ldots, n-1. \tag{5.80}$$

The identities (5.80) yield a linear system of n homogeneous equations whose n unknowns are the diagonal elements of gMg^{-1}. Since the matrix of the coefficients of (5.80) is the (transposed of the) Vandermonde matrix (5.77), it admits, as unique solution, the trivial one. In other words, a matrix M satisfies (5.79) if and only if there exists an invertible matrix g such that $(gMg^{-1})_{ii} = 0$ for all $i = 1, \ldots, n$. To complete the proof, it suffices to define

$$S_{ij} = \frac{(gMg)_{ij}}{\lambda_i - \lambda_j}, \ \forall i \neq j.$$

In this case, $[S, D]_{ij} = (SD)_{ij} - (DS)_{ij} = S_{ij}(\lambda_i - \lambda_j) = \left(gMg^{-1}\right)_{ij}$, which implies $M = g^{-1}[S, D]g = [g^{-1}Sg, L]$, proving the implication. The other is clear.
□

To every I_k, we will associate the corresponding Hamiltonian vector field X_{I_k}, i.e., the vector field such that $i_{X_{I_k}}\omega_0 = -dI_k$. Writing $\frac{df}{dt_k}$ to denote $-\mathscr{L}_{X_{I_k}}f$ (see also Remark 8.38), the first two identities in (5.75) can be written as

$$\frac{dI_r}{dt_s} = 0 \quad \text{and} \quad \frac{dJ_r}{dt_s} = (r+s-2)I_{r+s-2}, \ \forall r \tag{5.81}$$

If (L, X) are two $n \times n$ matrices such that

$$rI_r = \mathrm{tr}(L^r) \quad \text{and} \quad J_s = \mathrm{tr}(XL^{r-1}), \tag{5.82}$$

first set of the (5.81) yields

$$0 = \frac{dI_r}{dt_s} = \mathrm{tr}\left(\frac{dL}{dt_s}L^{r-1}\right), \ \forall r,$$

which implies the existence of B_s such that

$$\frac{dL}{dt_s} = [B_s, L]. \tag{5.83}$$

On the other hand, looking at the second set of (5.81), one can compute

$$\frac{dJ_r}{dt_s} = \frac{d\,\mathrm{tr}(XL^{r-1})}{dt_s} = \mathrm{tr}\left(\frac{dX}{dt_s}L^{r-1} + X\frac{dL^{r-1}}{dt_s}\right)$$

$$\overset{(5.83)}{=} \mathrm{tr}\left(\frac{dX}{dt_s}L^{r-1} + X\left[B_s, L^{r-1}\right]\right),$$

which, together with (the second set of identities in) (5.81), entails

$$0 = \mathrm{tr}\left(\frac{dX}{dt_s}L^{r-1} + X[B_s, L^{r-1}] - L^{r+s-2}\right)$$

$$= \mathrm{tr}\left(\frac{dX}{dt_s}L^{r-1} + [X, B_s]L^{r-1} - L^{r+s-2}\right)$$

$$= \mathrm{tr}\left[\left(\frac{dX}{dt_s} + [X, B_s] - L^{s-1}\right)L^{r-1}\right], \; \forall r.$$

Using again Lemma 5.92, we obtain the existence of a matrix C_s such that

$$\frac{dX}{dt_s} = [B_s, X] + [C_s, L] + L^{s-1}.$$

Summarizing the previous discussion, one could say that starting from a *regular* ΩN-manifold of dimension $2n$, one can introduce the *non-canonical* coordinates (I, J), which in turn define a family of vector fields X_s such that:

(i) They commute, i.e., $[X_r, X_s] = 0$, for all r, s.
(ii) They are integrable, and their flows admit the following representation:

$$\frac{dL}{dt_s} = [B_s, L] \quad \text{and} \quad \frac{dX}{dt_s} = [B_s, X] + [C_s, L] + L^{s-1} \tag{5.84}$$

where (L, X) are two $n \times n$ matrices satisfying (5.82) and (B_s, C_s) are two $n \times n$ matrices whose existence is guaranteed by Lemma 5.92.

Remark 5.93 In the theory of integrable systems, a family of commuting vector fields is commonly referred to as a *hierarchy*. The derivative along the field X_k is denoted by $\frac{d}{dt_k}$, and t_k is referred as the kth time of the hierarchy. $\triangle$

Definition 5.94 (Extended Lax Pair and Extended Lax Representation) The (5.84) is called the *extended Lax representation* of (5.81). The couples (L, B_s) and (X, C_s) entering in (5.84) are called the *Lax pairs* of (5.81). $\triangle$

It is worth observing that the previous discussion can be read backward. More precisely, starting from a dynamical system defined on M and admitting an extended Lax representation (5.84), if the eigenvalues of L and X are (functionally) independent, it is possible to construct on M an ΩN-structure. In fact, if $\lambda_1, \ldots, \lambda_n$ and $\mu_1, \ldots, \mu_n$ are the eigenvalues of L and, respectively, of X, one can define the functions $I_1, \ldots, I_n$ and $J_1, \ldots, J_n$ as described in Formula (5.59) and, respectively, in Formula (5.74). Using these functions, one can define the brackets $\{\cdot, \cdot\}_0$ and $\{\cdot, \cdot\}_1$ imposing, for the first, the conditions (5.75) and then extending the result to a bi-derivation of $C^\infty(M)$. As far as $\{\cdot, \cdot\}_1$ is concerned, one imposes

$$\{I_k, I_s\}_1 = 0, \quad \{J_s, I_k\}_1 = (s + k - 1)I_{k+s-1} \quad \text{and} \quad \{J_k, J_s\}_1 = (s - k)J_{k+s-1},$$

on the (I, J)s, see (5.54), and then again extend the result to a bi-derivation of $C^\infty(M)$. The bracket so defined correspond to the two-forms $\omega_0 = \sum_{i=1}^{n} d\mu_i \wedge d\lambda_i$ and $\omega_1 = \sum_{i=1}^{n} \lambda_i d\mu_i \wedge d\lambda_i$ described in Theorem 5.87. Note that the I_k form a maximal family of pairwise Poisson commuting (w.r.t. $\{\cdot, \cdot\}_0$) first integrals of the initial dynamical system. For this reason, such a system is Liouville integrable, and by definition, it admits a bi-Hamiltonian description. In other words, we showed the following.

Theorem 5.95 *To every regular ΩN-manifold, one can associate a dynamical system, which is completely integrable and admits an extended Lax representation. On the other hand, if a dynamical system admits an extended Lax representation such that the eigenvalues of the matrices L and X are independent, then this dynamical system admits a bi-Hamiltonian representation.*

An important application of this theorem will be given in Chap. 8, where it will be used to introduce a bi-Hamiltonian representation of the rational Calogero-Moser dynamical system.

5 From Integrability to Bi-Hamiltonian Structures

In the previous sections, a particular emphasis was posed on the relation between the existence of a bi-Hamiltonian representation and the complete integrability of a given Hamiltonian system. More precisely, narrowing down the general class of bi-Hamiltonian manifolds to regular ΩN manifolds, one obtains the complete integrability of every Hamiltonian, which depends only on the eigenvalues of the recursion operator underlying the ΩN-structure (see Theorem 5.82). Moreover, it was shown that the integrable systems associated with ΩN manifolds possess a particular Lax-type representation and that every integrable system endowed with such a representation is associated to a regular ΩN-manifold (see Theorem 5.95). In this section, we turn our attention to the opposite problem, i.e., we will be interested to understand to what extent the Liouville integrability of a Hamiltonian system entails the existence of a bi-Hamiltonian representation for the given system. This problem will be addressed from a *semi-local* viewpoint. More precisely, starting from a completely integrable system, we will look for conditions guaranteeing the existence of a bi-Hamiltonian structure on a suitable neighborhood of a given Liouville torus. In particular in what follows, we will assume the existence of action-angle coordinates on any such domain (see Chap. 4). The main references are [41, 42, 88, 89, 147, 171, 242]. The theoretical framework is as follows.

Let (M, ω) be a symplectic manifold of dimension $2n$ and let $\pi : M \to B$ be a Lagrangian fibration whose fibers are connected and compact. Let (J, θ) be a set of action-angle coordinates defined on a suitable tubular neighborhood $V \simeq \mathbb{R}^n \times \mathbb{T}^n$ of the fiber over a given point $b \in B$. In these coordinates, $\omega = \sum_{k=1}^{n} dJ_k \wedge d\theta_k$, and the corresponding Poisson bivector field is

$$\Pi = \sum_{k=1}^{n} \frac{\partial}{\partial J_k} \wedge \frac{\partial}{\partial \theta_k} \tag{5.85}$$

(see Example 2.3). Two cases will be analyzed. In the first one, we suppose that a chosen set of action-angle variables *separates* the Hamiltonian of the system. In the second, we suppose that the Hamiltonian is *non-degenerate*. Note that the two cases are not mutually exclusive.

(i) Let $H \in C^\infty(V)$, $H = H(J)$, which is separated in the action variables, i.e.,

$$H = \sum_{k=1}^{n} h_k(J_k), \; h_k \in C^\infty(V), \; \forall k = 1, \ldots, n.$$

Then $X_H = \sum_{k=1}^{n} \frac{\partial h_k}{\partial J_k} \frac{\partial}{\partial \theta_k}$, where Π is as in (5.85). For this reason, every $f_1, \ldots, f_n$ such that

$$df_1 \wedge \cdots \wedge df_n \neq 0 \quad \text{and} \quad f_i = f_i(J_i), \; \forall i = 1, \ldots, n$$

will fulfill condition (4.57), defining the admissible symplectic form

$$\omega_f = \sum_{i=1}^{n} df_i \wedge d\theta_i = \sum_{i=1}^{n} \frac{\partial f_i}{\partial J_i} dJ_i \wedge d\theta_i \tag{5.86}$$

(see 4.56). Comparing the framework of Sect. (3.4) of Chap. 4 with the present one, it is worth noting that to define (5.86), we made the following choices: $\alpha_i = d\theta_i$, $\nu_i = \frac{\partial h_i}{\partial J_i}$ and $A_{ij} = \delta_{ij} = C_{ij}$. Moreover, since $i_{X_H} \omega_f = -\sum_{i=1}^{n} \frac{\partial h_i}{\partial J_i} \frac{\partial f_i}{\partial J_i} dJ_i$, the Hamiltonian of X_H with respect to ω_f will be known, in general, only implicitly. If $\Pi_f^\sharp$ is the Poisson tensors defined by ω_f (see 2.28) and Π is as in (5.85), then the endomorphism N defined by $\Pi_f^\sharp = N\Pi^\sharp$ is a recursion operator for X_H of the form

$$N = \sum_{k=1}^{n} A_k(J_k)\left(dJ_k \otimes \frac{\partial}{\partial J_k} + d\theta_k \otimes \frac{\partial}{\partial \theta_k}\right),$$

where $A_k(J_k) = \left(\frac{\partial f_k}{\partial J_k}\right)^{-1}$, for all $k = 1,\ldots,n$. Note that the isotropic harmonic oscillator is one (important) example of a completely integrable Hamiltonian system fulfilling the hypothesis of this example. More precisely, its Hamiltonian is separated in the action-angle variables, and at the same time, its frequencies are not independent (in fact they are constant) (see Example 4.35).

(ii) If $H \in C^\infty(V)$, $H = H(J)$ is *non-degenerate*, i.e.,

$$\det\left(\frac{\partial H}{\partial J_i \partial J_j}\right) \neq 0,$$

the frequencies of the system $\nu_1,\ldots,\nu_n$ (see Definition 4.30) can be used as (local) coordinates on V. Under these assumptions

$$\omega_\nu = \sum_{k=1}^{n} d\nu_k \wedge d\theta_k$$

is an admissible symplectic form for X_H, whose Hamiltonian with respect to $\tilde{\omega}$ is readily discovered to be

$$H_\nu = \frac{1}{2}\sum_{k=1}^{n} \nu_k^2$$

(see 4.61). Under these assumptions, one can show, as in the previous case, that the endomorphism N defined by $\Pi_\nu^\sharp = N_\nu \Pi^\sharp$, assumes the form

$$N_\nu = \sum_{k=1}^{n} B_k(\nu_k)\left(d\nu_k \otimes \frac{\partial}{\partial \nu_k} + d\theta_k \otimes \frac{\partial}{\partial \theta_k}\right), \tag{5.87}$$

and it is a recursion operator for the vector field X_H. We leave the reader to compute the coefficients B_ks in the right-hand side of (5.87).

It is worth mentioning that under the non-degeneracy hypothesis of the Hamiltonian function, one can prove the following result (see [41]).

Proposition 5.96 *Y is a symmetry of the Hamiltonian vector field $X = \Pi^{\sharp}(dH)$ if and only if*

$$Y = \sum_{k=1}^{n} a_k(J)\frac{\partial}{\partial \theta_k}. \tag{5.88}$$

Proof In fact if $Y = \sum_{l=1}^{n}\left(a_l(J,\theta)\frac{\partial}{\partial J_l} + b_l(J,\theta)\frac{\partial}{\partial \theta_l}\right)$

$$0 = [X_H, Y] = \sum_{l=1}^{n}\left(\sum_{k=1}^{n}\frac{\partial H}{\partial J_k}\frac{\partial b_l}{\partial \theta_k} - a_k\frac{\partial^2 H}{\partial J_k \partial J_l}\right)\frac{\partial}{\partial \theta_l} + \sum_{l=1}^{n}\left(\sum_{k=1}^{n}\frac{\partial H}{\partial J_k}\frac{\partial a_l}{\partial \theta_k}\right)\frac{\partial}{\partial J_l}$$

implying that

$$X_H(b_l) = \sum_{k=1}^{n} a_k\frac{\partial^2 H}{\partial J_k \partial J_l} \quad \text{and} \quad X_H(a_l) = 0, \ \forall l = 1, \dots, n, \tag{5.89}$$

The non-degeneracy hypothesis implies that if $f \in C^{\infty}(V)$ is such that $X_H(f) = 0$, then $f = f(J)$ (see Remark 5.97). This observation, together with the second (set of identities) in (5.89), implies $a_l = a_l(J)$ for all $l = 1, \dots, n$. On the other hand, given l and s,

$$\frac{\partial}{\partial \theta_s}\left(X_H(b_l)\right) = X_H\left(\frac{\partial b_l}{\partial \theta_k}\right) = \sum_{k=1}^{n}\frac{\partial a_k}{\partial \theta_s}\frac{\partial^2 H}{\partial J_k \partial J_l} = 0, \ \forall s,$$

which implies there exists $F_{s,l} \in C^{\infty}(V)$, $F_{s,l} = F_{s,l}(J)$, such that $\frac{\partial b_l}{\partial \theta_s} = F_{s,l}(J)$, i.e.,

$$b_l = \sum_{k=1}^{n} F_{k,l}(J)\theta_k + G_l(J),$$

for some $G_l \in C^{\infty}(V)$, $G_l = G_l(J)$. Since the b_ls are univalent functions on $V \simeq \mathbb{R}^n \times \mathbb{T}^n$, one concludes that, for every l, $F_{k,l} = 0$ for all $k = 1, \dots, n$. In other words, $b_l = b_l(J)$ for all $l = 1, \dots, n$. This condition, together with the first (set of identities) in (5.89), implies

$$\sum_{k=1}^{n} a_k\frac{\partial^2 H}{\partial J_k \partial J_l} = 0, \ \forall l = 1, \dots, n,$$

which, together with the assumption of non-degeneracy of the Hamiltonian H, forces $b_l = 0$ for all $l = 1, \dots, n$. This in turn yields (5.88). $\qquad\square$

Remark 5.97 For the sake of completeness, we prove that under the hypothesis of non-degeneracy of the Hamiltonian function, i.e.,

$$\det\left(\frac{\partial^2 H}{\partial J_i \partial J_j}\right) \neq 0, \tag{5.90}$$

a first integral of X_H is a function depending only on the actions variables. The idea of the proof is the following (see [84]). Assuming that $f \in C^\infty(V)$ is a first integral of X_H, i.e., $\mathscr{L}_{X_H} f = \{f, H\} = 0$, and expanding f in Fourier series

$$f(J, \theta) = \sum_{\vec{k} \in \mathbb{Z}^n} F_{\vec{k}}(J) \exp\left(i\vec{k} \cdot \vec{\theta}\right), \tag{5.91}$$

where $\vec{\theta} = (\theta_1, \ldots, \theta_n)$, one can write

$$0 = \{f, H\} = \sum_{\vec{k} \in \mathbb{Z}^n} \{F_{\vec{k}}, H\} \exp\left(i\vec{k} \cdot \theta\right) + \sum_{\vec{k} \in \mathbb{Z}^n} F_{\vec{k}} \left\{\exp\left(i\vec{k} \cdot \vec{\theta}|\right), H\right\}$$

$$= i \sum_{\vec{k} \in \mathbb{Z}^n} F_{\vec{k}} \vec{k} \cdot \vec{v}, \tag{5.92}$$

where $\vec{v} = (v_1, \ldots, v_n)$ and the v_ls are the frequencies of the system (see Definition 4.30). In the sum (5.92) for every $\vec{k} \in \mathbb{Z}^n \setminus \{\vec{0}\}$ such that $F_{\vec{k}} \neq 0$, one has

$$\vec{k} \cdot \vec{v} = \sum_{l=1}^{n} k_l v_l = 0,$$

which implies

$$\sum_{l=1}^{n} k_l \frac{\partial v_l}{\partial J_s} = \sum_{l=1}^{n} k_l \frac{\partial^2 H}{\partial J_l \partial J_s} = 0, \ \forall s = 1, \ldots, n,$$

which contradicts (5.90). In other words, in (5.91), the only non-zero term is the one corresponding to $\vec{k} = \vec{0}$, i.e., $f = F_{\vec{0}}(J)$. $\triangle$

We show now that complete integrability does not entail the existence of a bi-Hamiltonian representation of a Hamiltonian system. In what follows, we will work under the following assumptions:

(IS) Every complete integrable system (M, ω, H) will be

(IS1) Associated with a Lagrangian fibration with compact and connected fibers.

(IS2) Non-degenerate, i.e., its Hamiltonian will satisfy the condition.

$$\det\left(\frac{\partial^2 H}{\partial J_i \partial J_j}\right) \neq 0$$

where the J_is are a set of actions coordinates associated with the corresponding Lagrangian fibration.

(BH) Every bi-Hamiltonian structure will be defined by regular ΩN-structure whose associated closed two-form ω_1 (see Formula 5.51) will be non-degenerate, i.e., symplectic.

Settled these assumptions, let $V \simeq \mathbb{R}^n \times \mathbb{T}^n$ be the tubular neighborhood of a fiber of a Lagrangian fibration defined by (M, ω_0, H) as above, let (V, ω_0, H) be the induced integrable system on V, and let $\pi : V \to B$ be the corresponding projection on an open subset of $\mathbb{R}^n$.

Theorem 5.98 *If ω_1 is a symplectic form on V defining with ω_0, a bi-Hamiltonian representation for (V, ω_0, H), then:*

(i) The fibers of π are Lagrangian with respect to ω_1.
(ii) The recursion operator N, defined by $\omega_1^\flat = \omega_0^\flat \circ N$, projects on B and defines $\tilde{N} \in \mathrm{End}_{C^\infty(B)}(TB)$ such that $T_{\tilde{N}} = 0$.
(iii) If $K, I_k \in C^\infty(B)$ are (the unique) smooth functions such that $\pi^* K = H$ and $\pi^* I_k = J_k$, then ${}^t\tilde{N}dK, {}^t\tilde{N}dI_1, \ldots, {}^t\tilde{N}dI_n$ are closed one forms on B.

Proof To prove (i), it suffices to show that $\omega_1(X, Y) = 0$ for every *vertical* vector fields X, Y, where vertical stands for tangent to the fibers of π. To this end, recall that if (J, θ) is a set of action-angle coordinates on V, $X_H = \sum_{i=1}^n \frac{\partial H}{\partial J_k}\frac{\partial}{\partial \theta_k}$ and that X is vertical if and only if Y is a symmetry of X_H, i.e., $[X_H, Y] = 0$ (see Proposition 5.96). From these observations,

$$\left[X_H, N\frac{\partial}{\partial \theta_k}\right] = (\mathscr{L}_{X_H}N)\frac{\partial}{\partial \theta_k} + N\left[X_H, \frac{\partial}{\partial \theta_k}\right] = 0,$$

where the first term cancels since N is a recursion operator for X_H, while the second cancelation follows since $\frac{\partial}{\partial \theta_k}$ is vertical. The previous computation implies that the vertical vector fields are invariant with respect to N or, said differently, that N restricts to an endomorphism of the vertical bundle of π. Finally, if X, Y are both vertical,

$$\omega_1(X, Y) = \omega_0(NX, Y) = 0,$$

proving what was requested. To prove (ii), one can use the hypothesis on N to introduce on V (local) special Darboux-Nijenhuis coordinates (see Definition 5.88). These coordinates, together with the property of the λ_is of being functions only of the actions (see Proposition 5.96), will suffice to prove that N descends to a Nijenhuis operator on B. More precisely, if (μ, λ) is a set of special DN-coordinates, then

$$
N \frac{\partial}{\partial \theta_k} \overset{\text{Proposition 5.96}}{=} \sum_{i=1}^{n} \lambda_i \left(\frac{\partial \lambda_i}{\partial \theta_k} \frac{\partial \theta_k}{\partial \lambda_i} + \frac{\partial \mu_i}{\partial \theta_k} \frac{\partial}{\partial \mu_i} \right) = \sum_{i=1}^{n} \lambda_i \frac{\partial \mu_i}{\partial \theta_k} \frac{\partial}{\partial \mu_i}
$$

(see (DN2) in Definition 5.88), which implies that the coordinate vector fields $\frac{\partial}{\partial \mu_i}$s need to be vertical, i.e., they cannot have non-zero components along the $\frac{\partial}{\partial J_i}$s. For these reasons, N descends to an endomorphism $\tilde{N} \in \mathrm{End}_{C^\infty(B)} TB$. Since the λ_is are independent and they depend only on the action coordinates, they induce coordinates $\chi_1, \ldots, \chi_n$ on B, where $\pi^* \chi_i = \lambda_i$ for all $i = 1, \ldots, n$. Note that

$$
\tilde{N} = \sum_{k=1}^{n} \chi_k \frac{\partial}{\partial \chi_k} \otimes d\chi_k. \tag{5.93}
$$

Finally, since $\pi : V \to B$ is Lagrangian with respect to ω_1, every vertical vector field whose coefficients do not depend on the fiber variables will be Hamiltonian—more precisely will be Hamiltonian with respect to the pullback of a closed one form on the base B. At the same time,

$$
\omega_1 \left(\frac{\partial}{\partial \theta_i}, X \right) = \omega_0 \left(N \frac{\partial}{\partial \theta_i}, X \right) = \omega_0 \left(\frac{\partial}{\partial \theta_i}, NX \right) = -\langle dJ_i, NX \rangle
$$

$$
= -\langle {}^t N(dJ_i), X \rangle, \ \forall X \in \mathfrak{X}(V),
$$

i.e.,

$$
i_{\frac{\partial}{\partial \theta_i}} \omega_1 = -{}^t N(dJ_i), \ \forall i = 1, \ldots, n,
$$

which implies that $d\big({}^t N(dJ_i)\big) = 0$, for all $i = 1, \ldots, n$. On the other hand,

$$
{}^t N dJ_i = {}^t N \pi^* dI_i = \pi^* \, {}^t \tilde{N}(dI_i), \tag{5.94}
$$

which implies that $d\big({}^t \tilde{N}(dI_i)\big) = 0$, for all $i = 1, \ldots, n$ (note that π is submersive). Since $X_H = \sum_{i=1}^{n} \frac{\partial H}{\partial J_k} \frac{\partial}{\partial \theta_k}$, $i_{X_H} \omega_1 = -{}^t N dH$, implying that $d\big({}^t N dH\big) = 0$. If $H = \pi^* K$, the same argument used in (5.94) implies that ${}^t \tilde{N} dK$ is closed. $\qquad \square$

Keeping the same hypothesis of the theorem above, we have the following.

Lemma 5.99 *If V supports a bi-Hamiltonian representation for H, then one can define on B a set of* splitting coordinates, *i.e., a set of coordinates $\chi_1, \ldots, \chi_n$ such that*

$$\frac{\partial^2 f}{\partial \chi_i \partial \chi_j} = 0,$$

for all $f \in C^\infty(B)$ such that $\,{}^t\tilde{N}df = 0$.

Proof If $f \in C^\infty(B)$ is such that $\,{}^t\tilde{N}df$ is closed, (5.93) implies

$$d\left(\sum_{i=1}^n \chi_i \frac{\partial f}{\partial \chi_i} d\chi_i\right) = \sum_{i,k=1}^n \chi_i \frac{\partial^2 f}{\partial \chi_i \partial \chi_k} d\chi_i \wedge d\chi_k,$$

i.e.,

$$(\chi_i - \chi_k)\frac{\partial^2 f}{\partial \chi_i \partial \chi_k} = 0, \ \forall i \neq k,$$

forcing $\frac{\partial^2 f}{\partial \chi_i \partial \chi_k} = 0$, for all $i \neq k$. $\qquad\square$

The previous results yield the following.

Corollary 5.100 *Under the assumptions and the notations introduced above, if the integrable system (V, ω_0, H) admits a bi-Hamiltonian representation and if $J_1, \ldots, J_n$ is a set of action variable on V, there exist smooth functions $h_1, \ldots, h_n$ and, for every $i = 1, \ldots, n$, $\ell_{1,i}, \ldots, \ell_{n,i}$ such that*

$$K(\chi) = h_1(\chi_1) + \cdots + h_n(\chi_n) \quad and \quad I_i(\chi) = \ell_{i,1}(\chi_1) + \cdots + \ell_{i,n}(\chi_n). \qquad (5.95)$$

A few comments are now in order.

Remark 5.101

(i) It is worth stressing that in the proofs of Theorem 5.98 and of Lemma 5.99, the use of the special Darboux-Nijenhuis coordinates is not necessary. We refer to the original papers [41] and [42] for a more general proof of the same results.
(ii) Among the general assumptions we imposed above, the one about the invertibility of ω_1 can be relaxed. For the proof of the existence of the splitting coordinates in this more general setting, we refer the reader to the paper [89] and to the unpublished PhD thesis [88].

$\triangle$

The previous results, when applied to the following example, taken from [41], imply that integrability alone does not entail the existence of a bi-Hamiltonian representation for the underlying dynamical system.

Example 5.102 (An Example of Brouzet) Let $M = \mathbb{R}^2 \times \mathbb{T}^2$ be endowed with the symplectic form $\omega = dJ_1 \wedge d\theta_1 + dJ_2 \wedge d\theta_2$, let $H : M \to \mathbb{R}$ be defined by $H = J_1^3 + J_2^3 + J_1 J_2$, and let $\pi : \mathbb{R}^2 \times \mathbb{T}^2 \to \mathbb{R}^2$ be the projection on the first factor. The Js and the θs are the action and, respectively, the angle variables. A simple computation shows that H is not degenerate on the open subset $M_0 = \{(J_1, J_2) \,|\, 36 J_1 J_2 \neq 1\}$. We leave to the reader to show its integrability. Note that it suffices to observe that $f = J_1$ satisfies the following two properties: 1. $df \wedge dH \neq 0$ and 2. $\{f, H\} = 0$, where $\{\cdot, \cdot\}$ is the Poisson bracket defined by ω. Keeping the notation introduced above, we let $K, I_1, I_2 \in C^\infty(\mathbb{R}^2)$ be such that $\pi^* K = H$ and $\pi^* I_i = J_i$ for $i = 1, 2$. We will now show that starting from the data above, it is not possible to define on $\mathbb{R}^2$ a set of splitting coordinates for K, I_1 and I_2, which, in view of Corollary 5.100, implies that M cannot support a bi-Hamiltonian structure obtained completing ω with a compatible (eventually degenerate) closed two-form ω_1 (see point (ii) in the above remark). We will argue by contradiction, supposing that splitting coordinates χ_1, χ_2 can be defined and that (5.95) are satisfied with $\ell_{i,j}$ and h_i, $i, j = 1, 2$. Suppose first that I_1 and I_2 depend only on one of the χs, for example,

$$I_1 = \ell_{1,1}(\chi_1) \quad \text{and} \quad I_2 = \ell_{2,2}(\chi_2).$$

In this case,

$$\frac{\partial^2 K}{\partial \chi_1 \partial \chi_2} = \frac{\partial I_1}{\partial \chi_1} \frac{\partial I_2}{\partial \chi_2} \frac{\partial^2 K}{\partial I_1 \partial I_2} \neq 0,$$

contradicting the hypothesis on the χs of being separation variables for K. In this way, one is left to suppose that (at least) one of the Is depends on both χ_1 and χ_2, for example, that $I_1 = \ell_{1,1}(\chi_1) + \ell_{1,2}(\chi_2)$, with $\ell'_{1,1} \neq 0 \neq \ell'_{1,2}$. These last conditions imply that $\tau_i = \ell_{1,i}(\chi_i)$ are coordinates on $\mathbb{R}^2$, such that

$$I_1 = \tau_1 + \tau_2, \quad I_2 = f_1(\tau_1) + f_2(\tau_2)$$

$$\text{and } K = (\tau_1 + \tau_2)^3 + (f_1 + f_2)^3 + (\tau_1 + \tau_2)(f_1 + f_2), \tag{5.96}$$

where f_1, f_2s are functions of τ_1 and, respectively, τ_2, obtained from the representation of I_2 in terms of the χ_is. A simple chain-rule computation shows that $\frac{\partial^2}{\partial \chi_1 \partial \chi_2} = \frac{\partial \tau_1}{\partial \chi_1} \frac{\partial \tau_2}{\partial \chi_2} \frac{\partial^2}{\partial \tau_1 \partial \tau_2}$, which implies $\frac{\partial^2 K}{\partial \chi_1 \partial \chi_2} = 0 \iff \frac{\partial^2 K}{\partial \tau_1 \partial \tau_2} = 0$. Using this observation, computing the mixed second derivative with respect to the τs of the last term in (5.96) yields

$$f_1' + f_2' + 6(\tau_1 + \tau_2) + 6 f_1' f_2'(f_1 + f_2) = 0, \tag{5.97}$$

where the $'$ denotes the derivative with respect to τ_1 or τ_2. Note that in (5.97), $f_1' f_2' \neq 0$. In fact, f_1' and f_2' cannot be both zero, since the τs are coordinates. On the other hand, if $f_1' = 0$, $f_2' + 6\tau_2 = -6\tau_1$, which again contradicts the hypothesis on the τs of being coordinates. Since (5.97) is symmetric in f_1 and f_2, the same argument shows that $f_2' \neq 0$. Applying again $\frac{\partial^2}{\partial \tau_1 \partial \tau_2}$ to (5.97) yields

$$f_1'' f_2''(f_1 + f_2) + f_1'^2 f_2'' + f_1'' f_2'^2 = 0. \tag{5.98}$$

Note that the previous identity entails that $f_1'' f_2'' \neq 0$. In fact, using relation (5.98) and what we already have inferred about f_1' and f_2', one gets that $f_1'' f_2'' = 0$ if and only if $f_1'' = 0$ and $f_2'' = 0$. This would imply $f_1(\tau_1) = a_1 \tau_1 + b_1$ and $f_2(\tau_2) = a_2 \tau_2 + b_2$, where the a_is and the b_is are constants. Inserting these expressions into (5.97) entails

$$a_1 + a_2 + 6(\tau_1 + \tau_2) + 6a_1a_2(a_1\tau_1 + b_1 + a_2\tau_2 + b_2) = a_1 + a_2 + 6a_1a_2(b_1 + b_2)$$
$$+ 6(a_1^2 a_2 + 1)\tau_1 + 6(a_1 a_2^2 + 1)\tau_2 = 0$$

identically in τ_1, τ_2, which implies

$$a_1 + a_2 + 6(\tau_1 + \tau_2) = 0,$$
$$6(a_1^2 a_2 + 1) = 0,$$
$$6(a_1 a_2^2 + 1) = 0,$$

i.e.,

$$6(a_1^2 a_2 + 1) = 6(a_1 a_2^2 + 1) \Rightarrow a_1^2 a_2 = a_1 a_2^2 \overset{a_1 a_2 \neq 0}{\Longrightarrow} a_1 = a_2.$$

Using this result, one computes

$$dI_1 \wedge dI_2 \overset{(5.96)}{=} d(\tau_1 + \tau_2) \wedge d(a_1\tau_1 + b_1 + a_2\tau_2 + b_2)$$
$$= a_2 d\tau_1 \wedge d\tau_2 + a_1 d\tau_2 \wedge d\tau_1$$
$$= (a_2 - a_1) d\tau_1 \wedge d\tau_2$$
$$= 0,$$

contradicting the hypothesis on I_1, I_2 of being coordinates. In this way, (5.98) can be separated dividing its left and the right-hand side by $f_1'' f_2''$, i.e.,

$$\left(f_1 + \frac{f_1'^2}{f_1''}\right) = C_1 \tag{5.99}$$

$$\left(f_2 + \frac{f_2'^2}{f_2''} \right) = C_2, \tag{5.100}$$

for some C_1, C_2 constant. Since f_1 and f_2 were defined uniquely up to a constant, one can choose $C_1 = 0 = C_2$, which implies that both f_1 and f_2 are solution of the differential equation

$$\zeta + \frac{\zeta'^2}{\zeta''} = 0. \tag{5.101}$$

The identity $\int \zeta \zeta'' dt = \zeta \zeta' - \int \zeta'^2 dt$, together with (5.101), implies $\zeta \zeta' = 0$, yielding $\frac{1}{2} \frac{d\zeta^2}{dt} = a$, i.e.,

$$\zeta = \pm\sqrt{2at + b}, \tag{5.102}$$

where a, b are constants. Using (5.102) in (5.99) and (5.100), with the condition $C_1 = 0 = C_2$, one concludes

$$f_1 = \pm\sqrt{A_1\tau_1 + B_1} \quad \text{and} \quad f_2 = \pm\sqrt{A_2\tau_2 + B_2},$$

which is incompatible with the (5.97), by direct computation. This implies that there are no splitting coordinates for the completely integrable system described above, i.e., that such a system does not admit a bi-Hamiltonian representation. $\triangle$

With the previous example at hand, it is clear that the class of bi-Hamiltonian systems does not coincide with the class of those that are completely integrable. It is however important to make at least one more comment about this statement. Among the working hypothesis that led us to this conclusion, there are the following two:

(a) The Hamiltonian of (M, ω, H) is non-degenerate.
(b) The sought bi-Hamiltonian representation of (M, ω, H) is obtained completing the existing Hamiltonian description, i.e., it is obtained by finding a closed two-form ω_1 compatible with ω.

If one is willing to drop the condition (b) above, the Hamiltonian system described in Example 5.102 admits many bi-Hamiltonian description, as it is exemplified here below.

Example 5.103 (An Example of Landi, Marmo, and Vilasi) The Hamiltonian vector field of the Hamiltonian system of the Example 5.102 is $X_H = (3J_1^2 + J_2)\frac{\partial}{\partial\theta_1} + (3J_2^2 + J_1)\frac{\partial}{\partial\theta_2}$. The functions $v_1 = 3J_1^2 + J_2$ and $v_2 = 3J_2^2 + J_1$ can be used as coordinates where $dv_1 \wedge dv_2 \neq 0$, i.e., where the Hamiltonian function H

is non-degenerate. On this open set, for all f_1, $f_2 \in C^\infty(M)$ such that $f_1 = f_1(v_1)$ and $f_2 = f_2(v_2)$

$$\omega_0 = \sum_{i=1}^{2} dv_i \wedge d\theta_i \quad \text{and} \quad \omega_1 = \sum_{i=1}^{2} f_i(v_1) dv_i \wedge d\theta_i \tag{5.103}$$

are compatible two-forms, admissible with respect to X_H (see the examples at the beginning of the section). By construction, ω_0 is invertible, and $N = \omega_1^\flat \circ \omega_0^{\flat^{-1}}$ is the recursion operator for the underlying ΩN-structure, which, if written in the coordinates (v, θ)s, assumes the following form:

$$N = \sum_{i=1}^{2} \frac{1}{f_i} \left(dv_i \otimes \frac{\partial}{\partial v_i} + d\theta_i \otimes \frac{\partial}{\partial \theta_i} \right).$$

N is defined out of the zero-locus of the f_is, where ω_1 is symplectic. Note that this is not in contradiction with the conclusions after Example 5.102. In fact, as observed in [147], the symplectic form $\omega = \sum_{i=1}^{2} dJ_i \wedge d\theta_i$ does not belong to the family of closed two-forms defined in (5.103) by the choice of the functions f_is. $\triangle$

To close this circle of ideas, we introduce the following notion (see [88] and [89]). Let $\mathbb{A}_{\mathbb{K}}^{n+1}$ be the $(n + 1)$-dimensional affine space, with coordinates $(\xi_1, \dots, \xi_{n+1})$.

Definition 5.104 A hypersurface in $\mathbb{A}_{\mathbb{K}}^{n+1}$ is called of *translation* type, or a *translation hypersurface*, if it admits a parametrization of the type

$$\xi_i = a_{i,1}(t_1) + \cdots + a_{i,n}(t_n), \ i = 1, \dots, n+1, \tag{5.104}$$

where the $a_{i,j}$, $i = 1, \dots, n+1$, and $j = 1, \dots, n$, are smooth functions and $(t_1, \dots, t_n)$ are parameters. $\triangle$

Example 5.105 When $n = 2$, a simple example of translation (hyper)surface is obtained translating point-wise two curves $\gamma_1 = (t_1, 0, f_1(t_1))$ and $\gamma_2 = (0, t_2, f_2(t_2))$ along each other. Indeed such a surface admits the parametrization $\xi_1 = t_1, \xi_2 = t_2, \xi_3 = f_1(t_1) + f_2(t_2)$, therefore it is of the form (5.104). $\triangle$

Recall that if (M, ω, H) is a completely integrable system, the action variables define an affine structure on the base of the corresponding Lagrangian fibration (see Sect. 3.3 in Chap. 4). Using these notions, Corollary 5.100 can be rephrased saying that

Proposition 5.106 *Under the assumptions spelled above, a necessary condition for the existence of a bi-Hamiltonian representation for the integrable system (M, ω, H) is that the graph of K is a translation hypersurface, where we recall that K is the function defined in Theorem 5.98.*

Proof For fact that the graph of K is an hypersurface in $\mathbb{R}^{n+1}$, which we endow with the structure of affine space induced by the action variables (see Sect. 3.3 in Chap. 4). More precisely, its affine coordinates are $\xi_1 = I_i$ for $i = 1, \ldots, n$ and $\xi_{n+1} = K$. The splitting coordinates $\chi_1, \ldots, \chi_n$ defined in Lemma 5.99 provide the required parametrization (5.104) (see Formulas 5.95). $\qquad\square$

It is important to notice, one more time, that the structure of translation hypersurface is determined by the affine structure defined by the action coordinates. We will say that the graph of a non-degenerate, complete integrable Hamiltonian is of translation type with respect to the affine structure induced by the action variables, if the corresponding K is of this type. The previous proposition is nicely complemented by the following.

Proposition 5.107 *If (M, ω, H) is a completely integrable system whose Hamiltonian is non-degenerate and if its graph is a translation hypersurface with respect to the action variables, then X_H admits a bi-Hamiltonian representation obtained completing ω with a compatible closed two-form.*

Proof By hypothesis, one can find coordinates $\chi_1, \ldots, \chi_n$ on $\mathbb{R}^n$ and $h_1, \ldots, h_n \in C^\infty(\mathbb{R}^n)$, such that $K = h_1(\chi_1) + \cdots + h_n(\chi_n)$. Note that, eventually shrinking their domains, the $y_i = h_i(\chi_i)$s can be used as local coordinates. In fact, the $x_i = \pi^* \chi_i$s are independent functions of (only) the action variables. The non-degeneracy condition $\det \frac{\partial^2 H}{\partial J_i \partial J_j} \neq 0$ implies $\det \frac{\partial^2 H}{\partial x_i \partial x_j} \neq 0$. In this way, it is possible to conclude that $\det \frac{\partial^2 K}{\partial \chi_i \partial \chi_j} \neq 0$, which, together with the splitting property, forces $\frac{\partial^2 h_i}{\partial \chi_i^2} \neq 0$ for all $i = 1, \ldots, n$. Then the $\xi_i = \frac{\partial h_i}{\partial \chi_i}$s can be used as local coordinates and, for this reason,

$$dh_1 \wedge \cdots \wedge dh_n = \xi_1 \ldots \xi_n d\chi_1 \wedge \cdots \wedge d\chi_n \neq 0,$$

which implies that the $y_i = h_i$s can be used as local coordinates as desired. We can write $K = y_1 + \cdots + y_n$. Let $z_1, \ldots, z_n$ defined by $z_i = \pi^* y_i$, for all i. Since $H = \pi^* K$, one has $H = z_1 + \cdots + z_n$, and if $\varphi_1, \ldots, \varphi_n$ are the variables conjugate to the zs, the symplectic form $\omega = \sum_{i=1}^n dz_i \wedge d\varphi_i$ is compatible with $\omega_1 = \sum_{i=1}^n z_i dz_i \wedge d\varphi_i$, and $N := \omega_0^{\flat^{-1}} \circ \omega_1^{\flat}$ is a recursion operator for $X_H = \sum_{i=1}^n \frac{\partial}{\partial \varphi_i}$. A simple check shows that $\mathscr{L}_{X_H} N = 0$, and a computation shows that X_H is bi-Hamiltonian with respect to ω_0 and ω_1 with Hamiltonians $H = \sum_{i=1}^n z_i$ and $H_1 = \frac{1}{2} \sum_{i=1}^n z_i^2$. $\qquad\square$

From this section, it is clear that the class of bi-Hamiltonian systems is a proper subclass of integrable systems in the sense of the Arnold-Liouville-Mineur theorem. Actually more is true. Under suitable assumptions on the nature of the bi-Hamiltonian pair one is dealing with, it has been proved recently by H. Boualem and R. Brouzet that, generically, integrable systems in the sense of the Arnold-Liouville-Mineur theorem cannot be bi-Hamiltonian (in particular, the set of bi-Hamiltonian systems is meager in a suitable topology in the class of all integrable systems). We refer to the very interesting [37] for complete details.

6 Concluding Remarks and Further Topics

We conclude this chapter commenting on a few topics worthwhile to mention.

6.1 *Darboux-Nijenhuis Coordinates* à la *Falqui-Pedroni*

Under the hypothesis of regularity, it was shown that on every ΩN-manifold, it is possible to define the so-called DN-coordinates, i.e., Darboux coordinates diagonalizing the Nijenhuis operator. These coordinates were constructed in two successive steps: the first one, algebraic in flavor, consists in diagonalizing N; the second one, more analytical, consists in completing, by quadrature, the set of the eigenvalues of N to a set of Darboux coordinates. In [82] and [83] in substitution to this second step, an alternative recipe was proposed, more algebraic in nature, which is based on the following important notion

Definition 5.108 Let $\Phi(\lambda)$ be a smooth function on M depending on a parameter λ. Then Φ is called a *Nijenhuis function generator*, Nfg hereafter, if there exists a one-form α_Φ on M such that

$$^{t}Nd\Phi(\lambda) = \lambda d\Phi(\lambda) + \Delta_N(\lambda)\alpha_\Phi, \qquad (5.105)$$

where $\Delta_N(\lambda)$ is the minimal polynomial of N. $\triangle$

Remark 5.109 In the above definition, the dependence of Φ on λ is supposed to be smooth. Examples of this kind of functions are the characteristic and the minimal polynomial of N, which are polynomials in λ whose coefficients are smooth functions on M. The function Φ can be thought as a function on (two sets of) variables, i.e., $\Phi = \Phi(x, \lambda)$. Note that also the one-form α_Φ depends, in general, on λ. $\triangle$

Example 5.110 Let $\Delta_N(\lambda) = \lambda^n - \sum_{k=1}^{n} p_k \lambda^{n-k}$ be the minimal polynomial of N, where the p_k are smooth functions on M. Using the Newton formulas relating the p_is to the traces of the powers of N, one can write

$$^{t}Ndp_i = dp_{i+1} + p_i dp_1, \ 1 \le i \le n \quad \text{and} \quad p_{n+1} = 0 \qquad (5.106)$$

Then

$$
{}^t N d \Delta_N(\lambda) \;=\; {}^t N \left(d \left(\lambda^n - \sum_{k=1}^{n} p_k \lambda^{n-k} \right) \right) = - \sum_{k=1}^{n} {}^t N (dp_k) \lambda^{n-k}
$$

$$
\overset{(5.106)}{=} \; - \sum_{k=1}^{n} (dp_{k+1} + p_k dp_1) \lambda^{n-k}
$$

$$
= \; \lambda d \Delta_N(\lambda) + \Delta_N(\lambda) dp_1,
$$

i.e., the minimal polynomial of N is a Nfg. $\triangle$

So how can one use the notion of Nfg to complete the eigenvalues of N to a set of DN-coordinates? The idea is the following. First, one observes that, given $\lambda_1, \ldots, \lambda_n$, the sought coordinates $\mu_1, \ldots, \mu_n$ need to satisfy two (sets of) requirements:

(FG1) ${}^t N d\mu_i = \lambda_i d\mu_i$, for all i
(FG2) $\{\mu_i, \lambda_j\}_0 = \delta_{ij}$ and $\{\mu_i, \mu_j\}_0 = 0$ for all i, j

where $\{\cdot, \cdot\}_0$ is the Poisson bracket defined by the symplectic structure ω_0. Furthermore, we have the following.

Lemma 5.111 *If $\mu_1, \ldots, \mu_n$ are functions such that (FG1) holds for all i and $Y :=
-\Pi_0^\sharp(dI_1)$, then they satisfy (FG2) if and only if*

$$
\mathscr{L}_Y(\mu_i) = 1, \forall i = 1, \ldots, n.
$$

Proof One can compute

$$
\begin{aligned}
\lambda_i \{\mu_i, \lambda_j\}_0 &= \lambda_i \langle \Pi_0^\sharp(d\mu_i), d\lambda_j \rangle \\
&= -\lambda_i \langle d\mu_i, \Pi_0^\sharp(d\lambda_j) \rangle \\
&= -\langle {}^t N(d\mu_i), \Pi_0^\sharp(d\lambda_j) \rangle \\
&= -\langle d\mu_i, N\Pi_0^\sharp(d\lambda_j) \rangle \\
&= -\langle d\mu_i, \Pi_0^\sharp({}^t N d\lambda_j) \rangle \\
&= \lambda_j \langle \Pi_0^\sharp(d\mu_i), d\lambda_j \rangle \\
&= \lambda_j \{\mu_i, \lambda_j\}_0
\end{aligned}
$$

which yields $\{\mu_i, \lambda_j\}_0 = 0$ for all $i \neq j$. On the other hand,

$$
\{\mu_i, \lambda_i\}_0 = 1 \iff \Pi_0^\sharp(d\lambda_i) = -\frac{\partial}{\partial \mu_i}
$$

The condition $\{\mu_i, \mu_j\}_0 = 0$ for all i, j follows, analogously, computing $\lambda_i \{\mu_i, \mu_j\}_0$. $\qquad\square$

The evaluation of $\Phi(\lambda)$ at $f \in C^\infty(M)$ yields the element $\Phi(f) \in C^\infty(M)$, whose differential can be computed as

$$d(\mathrm{ev}_f(\Phi)) = d\Phi|_{\lambda=f} + \left.\frac{\partial \Phi}{\partial \lambda}\right|_{\lambda=f} df, \qquad (5.107)$$

where $\mathrm{ev}_f(\Phi) = \Phi(f)$. Then one can prove the following.

Lemma 5.112 *If Φ is a Nfg, then*

$$^t N d\Phi_i = \lambda_i d\Phi_i, \ \forall i = 1, \dots, n,$$

where $\Phi_i = \mathrm{ev}_{\lambda_i}(\Phi)$ for all i.

Proof The proof follows at once from (5.105), (5.107), and $\Delta_N(\lambda_i) = 0$ for all $i = 1, \dots, n$. $\qquad\square$

Then one arrives to the following crucial.

Proposition 5.113 *If $Y := -\Pi_0^\sharp(dI_1)$ and Φ is a Nfg such that*

$$\mathscr{L}_Y \Phi_i = 1, \ \forall i = 1, \dots, n \qquad (5.108)$$

then $\lambda_1, \dots, \lambda_n, \Phi_1, \dots, \Phi_n$ is a set of DN-coordinates.

Proof It follows immediately from the Lemmas 5.111 and 5.112. $\qquad\square$

We are now left with the problem to provide (at least) sufficient *algebraic* conditions guaranteeing the existence of a Nfg satisfying (5.108). To this end one first notices that the Nfg from a $\mathbb{K}$-algebra, i.e., any linear combination of Nfgs, is a Nfg, and the product of Nfgs is a Nfg. Moreover

Lemma 5.114 *The vector space of Nfgs is invariant with respect to $Y = -\Pi_0^\sharp(dI_1)$, i.e., if Φ is a Nfg, $\mathscr{L}_Y \Phi$ is Nfg as well.*

Proof In fact, since $I_1 = \sum_{i=1}^n \lambda_i$, then Y is an infinitesimal symmetry of N, i.e., $\mathscr{L}_Y N = 0$ and

$$^t N d(\mathscr{L}_Y \Phi) = {}^t N \mathscr{L}_Y(d\Phi)$$
$$= \mathscr{L}_Y({}^t N d\Phi) - (\mathscr{L}_Y {}^t N)d\Phi$$
$$= \mathscr{L}_Y(\lambda d\Phi + \Delta_N(\lambda)\alpha_\Phi)$$
$$= \lambda d\left(\mathscr{L}_Y(\Phi)\right) + \mathscr{L}_Y(\Delta_N(\lambda))\alpha_\Phi + \Delta_N(\lambda)\mathscr{L}_Y\alpha_\Phi,$$

where, in the last line, the cancellation follows again from the property of Y being a symmetry of N. $\square$

Let $C_Y^\infty(M)$ be the ring of smooth functions invariant with respect to Y and let Φ be a Nfg such that

$$Y^n(\Phi) = \sum_{k=0}^{n-1} a_k Y^k(\Phi), \qquad (5.109)$$

for some $n \geq 1$ and $a_k \in C_Y^\infty(M)$. Then

Proposition 5.115 *Equation* (5.108) *can be solved algebraically.*

Proof The condition (5.109) amounts to say that Φ generates a finitely generated $C_Y^\infty(M)$-module, which is invariant with respect to the action of the vector field Y. The proof follows analyzing separately the following two cases: either it has at least one non-zero eigenvalue or the restriction of Y to such a module is nilpotent. We refer the reader to [82] for further details. $\square$

We close this subsection with the following remark, where we would like to call the reader's attention to an important application of the theory of bi-Hamiltonian manifolds.

Remark 5.116 (Bi-Hamiltonian Structures and Separation of Variables) The Hamilton-Jacobi theory, together with the method of separation of variables, is central topics in classical mechanics (see the remark in (PR3)). This theory, born in the mid of the nineteenth century in the classical work of Jacobi (see [125]) and inscribed soon after in the realm of Riemannian geometry by Stäckel, Levi-Civita, and Eisenhart, is still a very active area of research (see, e.g., the monographs [129], [216], and the references therein). In this theory, a central role is played by the so-called separation relations, i.e., relations of the form $\phi_i(p_i, q_i, H_1, \ldots, H_n) = 0$ where the H_is are (independent) Hamiltonians of the system, the (p, q) are separating coordinates on the underlying manifold, and the ϕ_i are smooth function satisfying a suitable non-degeneracy condition. The existence of the separation relations entails that the (time-independent) Hamilton-Jacobi equations of the Hamiltonians can be solved by a common separated integral (see [82]). In the classical theory recalled above, the background manifold is the cotangent bundle of a Riemannian manifold, the separating coordinates are orthogonal fibered coordinates, and the Hamiltonians are smooth functions quadratic in the momenta. The recent introduction of the so-called *algebraic completely integrable* systems (a.c.i. systems hereafter; see, e.g., [6]), moved the playground of the separation of variables from the cotangent bundles to (finite dimensional) coadjoint orbits of, in general, infinite-dimensional Lie algebras. For this class of examples, the separating coordinates and the Hamiltonians are provided by the so-called spectral curve

$$\det(\mu \,\mathrm{id} - L(\lambda)) = 0, \qquad (5.110)$$

where $L = L(\lambda)$ is the Lax matrix with *spectral parameter*. More precisely, the Hamiltonians are the spectral invariants of (5.110), and the separation coordinates $(\mu_1, \ldots, \mu_n, \lambda_1, \ldots, \lambda_n)$ are provided by suitable points (μ_i, λ_i) of the spectral curve (see, e.g., [3, 4, 222]). In [82, 83], it was advocated a different approach to separation of variables, based on the theory of bi-Hamiltonian structures. In this proposal, the chosen background manifolds are endowed with a ΩN-structure, the separating coordinates are Darboux-Nijenhuis coordinates of the underlying ΩN-structure, and the Hamiltonians are separated functions on the chosen Darboux-Nijenhuis coordinates. For more information about this beautiful topic and for the application of these results to study the separability of the *Gel'fand-Zakharevich systems*, which is an important class of bi-Hamiltonian systems we did not touch in these lectures, we refer the reader to the above-cited references [82, 83]. $\triangle$

6.2 Master Symmetries and Bi-Hamiltonian Structures

In this subsection, we sketch an introduction to the theory of master and conformal symmetries and how these relate to the bi-Hamiltonian geometry. The concept of master symmetry was introduced in the framework of infinite-dimensional dynamical systems of solitonic type (see, e.g., [96]) but found interesting applications to finite-dimensional Hamiltonian dynamical systems (see, e.g., [201, 202]). We start introducing a very general definition. Recall that to every $X \in \mathfrak{X}(M)$ is associated with the Lie algebra of its (infinitesimal) symmetries, $\mathscr{L}(X) = \{Y \mid [Y, X] = 0\}$ (see Sect. 2 in this chapter). Now let χ be a vector field on M such that

$$[X, [X, \chi]] = 0. \tag{5.111}$$

Then both $\chi_1 = [\chi, X]$ and $\chi_2 = [\chi, [\chi, X]] \in \mathscr{L}(X)$. Note that in general, applying χ to χ_2 does not yield a symmetry of X; in fact,

$$[X, [\chi, \chi_2]] = [[X, \chi], \chi_2] + \underbrace{[\chi, [X, \chi_2]]}_{} = [[X, \chi], \chi_2] \overset{\text{in general}}{\neq} 0.$$

On the other hand, if $\mathscr{L}(X)$ were *abelian*, then $[X, [\chi, \chi_2]] = 0$, and $\chi_n = \mathrm{ad}_\chi^n(X) \in \mathscr{L}(X)$ for all $n \geq 1$. More in general,

Lemma 5.117 *If $\mathscr{L}(X)$ is Abelian, then it is invariant with respect to the linear operator* ad_χ.

In other words, a vector field χ satisfying (5.111) can be used to generate new symmetries of X from old ones. Obviously, this generation process will be trivial if $\chi \in \mathscr{L}(X)$. These ideas lead to the following notion.

Definition 5.118 A *master symmetry* of X is a vector field χ such that

$$[X, [X, \chi]] = 0 \quad \text{and} \quad [X, \chi] \neq 0.$$

$\triangle$

Example 5.119 A regular ΩN-manifold comes equipped with an atlas of DN-coordinates (I, J) (see Proposition 5.91). Formula (5.75) entails that each of the X_{J_k}s is a (locally defined) master symmetry of each of the X_{I_k}s, where $X_f = \{f, \cdot\}_0$ for all $f \in C^\infty(M)$. $\triangle$

The idea to generate symmetries of a given vector field using a master symmetry should be compared with what was briefly discussed at the beginning of this chapter, where it was used a recursion operator to generate new symmetries from old ones (see Sect. 2 in this chapter). In this context, master symmetries turn out to be useful when a recursion operator is not immediately available, which is the case of a vector field X having a bi-Hamiltonian representation not associated to a Poisson-Nijenhuis structure.

We will now sketch the connection between master symmetries and bi-Hamiltonian structure. To this end, we first introduce the notions of a conformal symmetry of a tensor field.

Definition 5.120 A *conformal symmetry* of a tensor fields T on M is an element $Z \in \mathfrak{X}(M)$ such that

$$\mathscr{L}_Z T = aT, \tag{5.112}$$

for some $a \in \mathbb{K}$. The a in the previous formula will be dubbed the *weight* of the conformal symmetry. $\triangle$

Note that a conformal symmetry Z corresponding to $a = 0$ is a symmetry of the tensor field. Moreover, we would like to warn the reader that the name weight for the factor a in (5.112) is not standard.

The first worthwhile observation is enclosed in the following proposition. Let N be a recursion operator for $X \in \mathfrak{X}(M)$ (see Definition 5.12). Suppose that $\mathscr{L}_Z X = aX$, $\mathscr{L}_Z N = bN$, for some $a, b \in \mathbb{K}$, and define $X_k := N^k X$ and $Z_k := N^k Z$, for all $k \geq 0$, with $X_0 = X$ and $Z_0 = Z$. Then

Proposition 5.121

$$\mathscr{L}_{Z_k} N = bN^k, \tag{5.113}$$

$$[Z_r, X_s] = (a + sb)X_{r+s}, \tag{5.114}$$

$$[Z_r, Z_s] = b(s - r)Z_{r+s}. \tag{5.115}$$

for all $r, s \geq 0$.

Proof This result is an amplification of the Proposition 5.13. More precisely, (5.113) follows (5.2) in Exercise 5.10. Let us prove (5.114).

$$
\begin{aligned}
[Z_r, X_s] &= \mathcal{L}_{Z_r} X_s = \mathcal{L}_{Z_r}(N^s X) \\
&= N^{s-k} \sum_{k=1}^{s} (\mathcal{L}_{Z_r} N) N^{k-1} X + N^s [Z_r, X] \\
&\overset{(5.2)}{=} N^{s+r-k} \sum_{k=1}^{s} b N^k X + N^s [Z_r, X] \\
&= sb N^{s+r} X + N^s [Z_r, X] \\
&= sb N^{s+r} X - N^s \mathcal{L}_X(N^r Z) \\
&= sb N^{s+r} X + N^{s+r} \mathcal{L}_Z X \\
&= sb N^{s+r} X + a N^{s+r} X \\
&= (a + sb) X_{s+r}.
\end{aligned}
$$

Since the proof of (5.115) goes similarly, it will be left as an exercise. $\qquad\square$

Remark 5.122 It is interesting to note that if in the previous proposition N is *invertible*, the vector field Z_n will satisfy (5.113), (5.114), and (5.115) for all $n \in \mathbb{Z}$. In particular, under this hypothesis, if $b = 1$, then Z_ns with the bracket given in (5.115) form the so-called Virasoro algebra without central extension. $\qquad\triangle$

Going back to the master symmetries, from the previous proposition, one easily deduces the following.

Corollary 5.123 *For all n, Z_n is a master symmetry of X.*

Proof In fact $\left[[Z_n, X], X\right] \overset{(5.114)}{=} a[X_n, X] \overset{(5.2)}{=} 0.$ $\qquad\square$

We come now to a relation between master symmetries and bi-Hamiltonian structures. Suppose that (M, N, Π_0) is an ΩN-manifold. For all $r \geq 0$, let Π_r be the Poisson bivector field defined by the skew-symmetric element $N^r \Pi_0^\sharp \in \mathrm{Hom}_{C^\infty(M)}(T^*M, TM)$, for all $r \geq 0$.

Theorem 5.124 *If $Z \in \mathfrak{X}(M)$ and $a, b \in \mathbb{K}$ are such that $\mathcal{L}_Z \Pi_0 = a\Pi_0$ and $\mathcal{L}_Z \Pi_1 = b\Pi_1$, then*

$$
\mathcal{L}_{Z_s} \Pi_r = (b + (r - s - 1)(b - a))\Pi_{r+s}, \quad \forall r, s. \tag{5.116}
$$

Moreover, if $X \in \mathfrak{X}(M)$ is such that $\Pi_1^\sharp(dH_0) = X = \Pi_0^\sharp(dH_1)$ for some $H_0, H_1 \in C^\infty(M)$ and $c \in \mathbb{K}$ is such that $\mathscr{L}_Z H_0 = cH_0$, then

$$d\langle dH_r, Z_s \rangle = (c + (r + s)(b - a))dH_{r+s}, \quad \forall r, s, \tag{5.117}$$

where $dH_r = {}^t N^r (dH_0)$ for all r.

Proof First recall that N is a recursion operator for X (see Remark 5.64). Then observe that

$$\mathscr{L}_Z \Pi_0^\sharp = a\Pi_0^\sharp, \tag{5.118}$$

$$\mathscr{L}_Z \Pi_1^\sharp = b\Pi_1^\sharp, \tag{5.119}$$

$$\mathscr{L}_Z \Pi_0^{\sharp,-1} = -a\Pi_0^{\sharp,-1}.$$

From the previous identities, one deduces

$$\mathscr{L}_Z N = (b - a)N. \tag{5.120}$$

After these preliminary remarks, one can compute

$$\begin{aligned}
\mathscr{L}_{Z_s} \Pi_r^\sharp &= \mathscr{L}_{Z_s}\left(N^r \Pi_0^\sharp\right) \\
&= N^{r-k} \sum_{k=1}^{r} (\mathscr{L}_{Z_s} N) N^{k-1} \Pi_0^\sharp + N^r \mathscr{L}_{Z_s} \Pi_0^\sharp \\
&\stackrel{(5.113)}{=} r(b - a)N^{r+s} \Pi_0^\sharp + N^r \left(\mathscr{L}_{Z_s} \Pi_0^\sharp\right).
\end{aligned} \tag{5.121}$$

To evaluate $\mathscr{L}_{Z_s} \Pi_0^\sharp$, one can proceed by induction on s. For $s = 1$,

$$\begin{aligned}
\mathscr{L}_{Z_1} \Pi_0^\sharp \quad &= \quad \mathscr{L}_{NZ} \Pi_0^\sharp \\
&\stackrel{(5.43)}{=} \quad N\left(\mathscr{L}_Z \Pi_0^\sharp\right) - \mathscr{L}_Z\left(N\Pi_0^\sharp\right) + \left(\mathscr{L}_Z \Pi_0^\sharp\right)^t N \\
&\stackrel{(5.118)\ \text{and}\ (5.119)}{=} \quad (2a - b)\Pi_1^\sharp.
\end{aligned}$$

Then, if

$$\mathscr{L}_{Z_{s-1}} \Pi_0^\sharp = (a - (s - 1)(b - a))\Pi_{s-1}^\sharp \tag{5.122}$$

$$\begin{aligned}
L_{Z_s} \Pi_0^\sharp \quad &= \quad \mathscr{L}_{NZ_{s-1}} \Pi_0^\sharp \\
&\stackrel{(5.43)}{=} \quad N\left(\mathscr{L}_{Z_{s-1}} \Pi_0^\sharp\right) - \mathscr{L}_{Z_{s-1}}\left(N\Pi_0^\sharp\right) + \left(\mathscr{L}_{Z_{s-1}} \Pi_0^\sharp\right)^t N
\end{aligned}$$

$$
\begin{aligned}
&= && N\left(\mathscr{L}_{Z_{s-1}}\Pi_0^\sharp\right) - \left(\mathscr{L}_{Z_{s-1}}N\right)\Pi_0^\sharp - N\left(\mathscr{L}_{Z_{s-1}}\Pi_0^\sharp\right) \\
&+ && \left(\mathscr{L}_{Z_{s-1}}\Pi_0^\sharp\right){}^t N \\
&\overset{(5.2)}{=} && -\left(N^{s-1}\mathscr{L}_Z N\right)\Pi_0^\sharp + \left(L_{Z_{s-1}}\Pi_0^\sharp\right){}^t N \\
&\overset{(5.120) \text{ and } (5.122)}{=} && -(b-a)N^s\Pi_0^\sharp + \left(a-(s-1)(b-a)\right)\Pi_{s-1}^\sharp {}^t N \\
&\overset{(5.29)}{=} && \left(a-(s-1)(b-a)\right)N\Pi_{s-1}^\sharp - (b-a)\Pi_s^\sharp \\
&= && (a-s(b-a))\Pi_s^\sharp.
\end{aligned}
$$

Inserting this result in (5.121), one gets

$$
\mathscr{L}_{Z_s}\Pi_r^\sharp = r(b-a)N^{r+s}\Pi_0^\sharp + (a-s(b-a))N^r\Pi_s^\sharp = (a+(r-s)(b-a))\Pi_{r+s}^\sharp,
$$

which proves (5.116). As far as (5.117) is concerned, one has

$$
\begin{aligned}
d\langle dH_r, Z_s\rangle &= d\langle {}^t N^r(dH_0), N^s Z\rangle \\
&= d\langle {}^t N^{r+s}(dH_0), Z\rangle \\
&= \mathscr{L}_Z\left({}^t N^{r+s}(dH_0)\right) \\
&= \mathscr{L}_Z\left({}^t N^{r+s}\right)dH_0 + {}^t N^{r+s}(\mathscr{L}_Z dH_0) \\
&= {}^t\left(\mathscr{L}_Z N^{r+s}\right)dH_0 + {}^t N^{r+s}(d\mathscr{L}_Z H_0) \\
&= (r+s)(b-a)\,{}^t N^{r+s}(dH_0) + c\,{}^t N^{r+s}(dH_0) \\
&= (c+(r+s)(b-a))dH_{r+s}.
\end{aligned}
$$

$\square$

Now we go back to the framework of Example 5. In particular, we assume that (M,ω) is a $2n$-dimensional symplectic manifold endowed with a Lagrangian fibration $\pi : M \to B$ with compact and connected fibers. Moreover, we let $V \subset M$ be a tubular neighborhood of a fiber of π where a set of action-angle coordinates (J,θ) is defined. Under the above assumptions, one can show that a master symmetry of X_H, where $H \in C^\infty(V)$, $H = H(J)$ is of the form

$$
Z = \sum_{l=1}^{n}\left(a_l(J)\frac{\partial}{\partial J_l} + b_l(J)\frac{\partial}{\partial \theta_l}\right), \tag{5.123}
$$

with $a_l, b_l \in C^\infty(V)$ depending only on the action-variables. In fact, if Z is a master symmetry, then $[X_H, Z]$ is a symmetry of X_H and, since H is non-degenerate

$$[X_H, Z] = \sum_{k=1}^{n} a_k(J) \frac{\partial}{\partial \theta_k}, \tag{5.124}$$

where the a_ls are suitable smooth functions θ-independent (see Example 5). If

$$Z = \sum_{l=1}^{n} \left(c_l(J, \theta) \frac{\partial}{\partial J_l} + d_l(J, \theta) \frac{\partial}{\partial \theta_l} \right),$$

comparing $[X_H, Z]$ with the right-hand side of (5.124), an argument analogue to the one used in Example 5 yields the representation (5.123). Moreover, let us introduce the following notion.

Definition 5.125 A master symmetry of degree r of $X \in \mathfrak{X}(M)$ is a vector field Z such that

$$\mathrm{ad}_X^r(Z) \neq 0 \quad \text{and} \quad \mathrm{ad}_X^{r+1}(Z) = 0. \tag{5.125}$$

$$\triangle$$

For example, a master symmetry of degree zero of X is a symmetry of X, while a master symmetry of degree one is a master symmetry in the sense of our original Definition 5.118. Having said that, under the assumption of compactness of the fibers of the Lagrangian fibration, one can prove that if Z is a master symmetry of X_H, its degree is less or equal to one. The proof of this statement goes along the lines of the one characterizing the master symmetries of degree zero, using compactness of the fibers of the Lagrangian fibration and non-degeneracy of the Hamiltonian function. More precisely, if Z were a master symmetry of degree two, then $[X_H, [X_H, Z]]$ should be a symmetry of X_H, and for this reason, it could be written as

$$\left[X_H, [X_H, Z] \right] = \sum_{l=1}^{n} \left(a_l(J) \frac{\partial}{\partial J_k} + b_k(J) \frac{\partial}{\partial \theta_k} \right). \tag{5.126}$$

On the other hand, writing $Z = \sum_{k=1}^{n} \left(c_k(J, \theta) \frac{\partial}{\partial J_k} + d_k(J, \theta) \frac{\partial}{\partial \theta_k} \right)$, computing $[X_H[X_H, Z]]$, and comparing the result with the right-hand side of (5.126), one deduces that the coefficients c_k, d_k do not depend on the angle variables, implying that Z is master symmetry of degree one, which contradicts the hypothesis on Z of being of degree two.

Remark 5.126 One can use a similar argument to the one sketched above to rule out the possibility of existence of master symmetries of degree greater than two.

Note that, to this end, both the hypothesis of compactness of the fibers and of non-degeneracy of the Hamiltonian are used in the same way they were used to prove that a symmetry of X_H has the representation given in (5.88). Finally, dropping the hypothesis of compactness, one can show that the Hamiltonian vector field X_H can have master symmetries of arbitrary degree. $\triangle$

Now we look back at Examples 5.69 and 5.70 from the point of view of the master symmetries.

Example 5.127 (Master Symmetries vs the Kepler System and the Harmonic Oscillator) We start with Example 5.69 (see also Example 5.90). Using the same notations of those examples, one can check that

$$Z = -\frac{1}{4} \sum_{k=1}^{n} h_k \left(p_k \frac{\partial}{\partial p_k} + q_k \frac{\partial}{\partial q_k} \right),$$

satisfies

$$[X, [X, Z]] = 0 \quad \text{and} \quad [X, Z] \neq 0,$$

i.e., Z is a master symmetry (of degree one) of Hamiltonian vector field (5.48). Another computation proves that

$$\mathscr{L}_Z \Pi_0 = \Pi_1,$$

showing that the bi-Hamiltonian structure of the harmonic oscillator defined in Example 5.69 is *exact* in the sense of Definition 5.33.

Let us consider now the Kepler system (see [224, 225]). Example 5.70 provides a bi-Hamiltonian representation for this system. We are now going to give another representation that is obtained as a by-product of the theory discussed above. Adopting the same notation used in that example, recall that in the action-angle variables, the Hamiltonian of the system is $H = -\frac{1}{(J_1+J_2+J_3)^2}$, while the canonical Poisson bivector field is $\Pi_0 = \sum_{k=1}^{3} \frac{\partial}{\partial J_k} \wedge \frac{\partial}{\partial \theta_k}$. The Hamiltonian vector is $\Pi_0^\sharp(dH) = \frac{1}{(J_1+J_2+J_3)^3} \left(\frac{\partial}{\partial \theta_1} + \frac{\partial}{\partial \theta_2} + \frac{\partial}{\partial \theta_3} \right)$. We now introduce a new set of coordinates (I, ϕ) via the following change of variables:

$$\begin{cases} I_1 = J_3 \\ I_2 = J_3 + J_2 \\ I_3 = J_3 + J_2 + J_1, \end{cases} \quad \text{and} \quad \begin{cases} \phi_1 = \theta_3 - \theta_2 \\ \phi_2 = \theta_2 - \theta_1 \\ \phi_3 = \theta_1. \end{cases}$$

A computation shows that this change of variables is indeed symplectic, i.e., in the new variables, the canonical Poisson structure looks like $\Pi_0 = \sum_{k=1}^{3} \frac{\partial}{\partial I_k} \wedge \frac{\partial}{\partial \phi_k}$. The variables (I, ϕ) so defined are the *Delauney elements* (see, e.g., [84]). If represented in these coordinates, the Hamiltonian vector field becomes $X_H = \frac{2}{I_3^3} \frac{\partial}{\partial \phi_3}$. Note in particular that the Hamiltonian function in the new coordinates is separated in the variables Is, i.e., $H = -\frac{1}{I_3^2}$. Defining

$$N = \sum_{k=1}^{3} I_k \left(dI_k \otimes \frac{\partial}{\partial I_k} + d\phi_k \otimes \frac{\partial}{\partial \phi_k} \right),$$

a simple computation shows that $T_N = 0$, and a slightly more complex one shows that $\mathcal{R}_{\Pi_0,N} = 0$, implying that (Π_0, N) is a Poisson-Nijenhuis structure with $\Pi_1 = \sum_{k=1}^{3} I_k \frac{\partial}{\partial I_k} \wedge \frac{\partial}{\partial \phi_k}$. If $K = -\frac{2}{3I_3^3}$, then

$$\Pi_0^\sharp(dH) = \Pi_1^\sharp(dK),$$

i.e., the Poisson-Nijenhuis structure described above yields a bi-Hamiltonian representation of the Kepler system. Note that this representation is not the one presented in Example 5.70. But there is more to be said about this structure. First, if

$$W = -\frac{1}{2} \sum_{k=1}^{3} I_k^2 \left(\frac{\partial}{\partial I_k} + \frac{\partial}{\partial \phi_k} \right),$$

then

$$\mathscr{L}_W \Pi_0 = \Pi_1,$$

showing that, also in this case, the bi-Hamiltonian structure defined by (Π_0, N) is exact in the sense of Definition 5.33. Moreover, another direct computation shows that

$$\left[X_H, [X_H, W] \right] = 0 \quad \text{but} \quad [X_H, W] \neq 0,$$

i.e., W is a master symmetry (of degree one) of X_H. Finally, if

$$Z = -\frac{1}{2} \sum_{k=1}^{3} I_k \left(\frac{\partial}{\partial I_k} + \frac{\partial}{\partial \phi_k} \right)$$

one can show that

$$\mathscr{L}_Z X_H = \frac{3}{2} X_H, \quad \mathscr{L}_Z \Pi_0 = \frac{1}{2}\Pi_0 \quad \text{and} \quad \mathscr{L}_Z \Pi_1 = 0,$$

implying that Z is a conformal symmetry of X_H, Π_0, and Π_1 of weight $\frac{3}{2}$, $\frac{1}{2}$, and, respectively, 0. In particular, Z is a symmetry of Π_1. Since $\Pi_1^\sharp = N\Pi_0^\sharp$, Formulas

(5.118), (5.119), and (5.120) imply that $\mathscr{L}_Z N = -\frac{1}{2}N$ and writing $Z_k = N^k Z$, $X_k = N^k X_H$, from Proposition 5.121 follows

$$[Z_r, X_s] = \left(\frac{3}{2} - \frac{s}{2}\right) X_{r+s} \quad \text{and} \quad [Z_r, Z_s] = \frac{(r-s)}{2} Z_{r+s}, \ \forall r, s \geq 0,$$

where $Z_0 = Z$ and $X_0 = X_H$. Moreover, Corollary 5.123 tells us that the Z_ks are all master symmetries for X_H, and finally, Theorem 5.124 says that

$$\mathscr{L}_{Z_s} \Pi_r = \frac{(r+1-s)}{2} \Pi_{r+s}, \ \forall r, s.$$

Note that restricting ourselves to the open set where N is invertible, the previous formulas can be extended to all integer values of k. $\triangle$

6.2.1 Final Considerations

We close this part with a few more technical comments. Until now, the notion of master symmetry was applied to a single vector field, but one can generalize this notion as follows. Let $\mathscr{A} \subset \mathfrak{G}$ be an Abelian Lie sub-algebra of the Lie algebra $\mathfrak{G}$. Then

Definition 5.128 $Z \in \mathfrak{G}$ is a master symmetry of $\mathscr{A}$ of degree r if for all $X_{i_1}, \ldots, X_{i_{r+1}} \in \mathscr{A}$

$$\mathrm{ad}_{X_{i_1}} \circ \cdots \circ \mathrm{ad}_{X_{r+1}}(Y) = 0. \tag{5.127}$$

The master symmetries of $\mathscr{A}$ of degree r form a vector space that will be denoted by $\mathcal{M}^r_{\mathscr{A}}$. $\triangle$

Note that $\mathscr{A} \subset \mathcal{M}^0_{\mathscr{A}}$. Moreover, $\mathcal{M}^r_{\mathscr{A}} \subset \mathcal{M}^{r+1}_{\mathscr{A}}$ for all r so that if one defines $\mathcal{M}_{\mathscr{A}}$ as the subset of all $Y \in \mathfrak{G}$ satisfying (5.127) for some $r \geq 0$, then $\mathcal{M}_{\mathscr{A}}$ is a filtered vector space whose filtration is defined by the $\{\mathcal{M}^r_{\mathscr{A}}\}_{r \geq 0}$. For all $r \geq -1$, let $\mathcal{M}^r_{\mathscr{A}}[1] := \mathcal{M}^{r+1}_{\mathscr{A}}$ and let $\mathcal{M}_{\mathscr{A}}[1]$ be the corresponding filtered vector space. Then

Lemma 5.129 $\mathcal{M}_{\mathscr{A}}[1]$ *is a filtered Lie algebra.*

Proof All we need to show is that if $Z \in \mathcal{M}^i_{\mathscr{A}}[1]$ and $W \in \mathcal{M}^j_{\mathscr{A}}[1]$, then $[Z, W] \in \mathcal{M}^{i+j}_{\mathscr{A}}[1]$, for all i, j. This is deduced from a direct computation that uses the Jacobi identity. Denote by I and J two index sets with $i + 1$ and, respectively, $j + 1$ elements. Let $X_{I \sqcup J}$ be the result of the composition of $I \sqcup J$ elements of $\mathscr{A}(M)$. Then

$$\mathrm{ad}_{X_{I \sqcup J}}\left([Z, W]\right) = \left[X_{I \sqcup J}, [Z, W]\right] = \sum_{I_1 \sqcup I_2 = I \sqcup J} \left[[X_{I_1}, Z], [X_{I_2}, W]\right].$$

In this sum, both the summands where $\sharp I_1 > i + 1$, $\sharp I_1 < i + 1$, are zero since in this case $[X_{I_1}, Z] = 0$ and, respectively, $[X_{I_2}, W] = 0$. On the other hand, if $\sharp I_1 = i + 1$ and $\sharp I_2 = j + 1$, then $[X_{I_1}, Z], [X_{I_2}, W] \in M_{\mathscr{A}}^{-1}[1]$, which implies $[[X_{I_1}, Z], [X_{I_2}, W]] = 0$ since $M_{\mathscr{A}}^{-1}[1]$ is Abelian. $\qquad\square$

To make a closer connection with the previous definition of master symmetry, one needs to consider the graded algebra associated with $M_{\mathscr{A}}[1]$, i.e., one needs to consider

$$\mathcal{M}_{\mathscr{A}}[1] = \oplus_{r \geq -1} \frac{\mathcal{M}^r[1]}{\mathcal{M}^{r-1}[1]}, \quad \mathcal{M}^{-2}[1] = \{0\},$$

endowed with the Lie bracket induced by the one defined in $M_{\mathscr{A}}[1]$. Trading the filtered Lie algebra with its associated graded Lie algebra, one recovers, in this general framework, the first condition in (5.125) (see Definition 5.125).

The previous considerations are very general, and for this reason, they could be useful, at least in principle, when applied to any Lie algebra. On the other hand, their relevance to the main topic of these notes stems from the analysis of the *higher-order* symmetries of integrable systems.

We recall that a completely integrable system on a $2n$-dimensional symplectic manifold (M, ω) is a maximal commutative Poisson subalgebra $\mathscr{A} \subset \mathfrak{G} = (C^\infty(M), \{\cdot, \cdot\})$, generated by any set of n functionally independent elements $f_1, \ldots, f_n \in C^\infty(M)$ such that $\{f_i, f_j\} = 0$, for all i, j. From this follows, in particular, that $g \in C^\infty(M)$ is such that $\{g, f_i\} = 0$ for all $i = 1, \ldots, n$ if and only if $g = F(f_1, \ldots, f_n)$ for some $F \in C^\infty(\mathbb{R}^n)$. After these reminders, one immediately sees that:

1. f is a master symmetry of $\mathscr{A}$ of degree zero if and only if $f \in \mathscr{A}$.
2. If f is a master symmetry of degree zero and g is a master symmetry of degree r, then $\{f, g\}$ is a master symmetry of degree $r - 1$.

In particular,

3. If f is a master symmetry of degree one, $\{f, g\}$ is a master symmetry of degree zero for all $f \in \mathscr{A}$.

From this last property, it follows that the master symmetries of degree one are the *angle-variables* of the integrable system described by $\mathscr{A}$. In fact, if $H \in \mathscr{A}$ and f is a *master symmetry* of H of degree one, since $\{H, f\} \in \mathscr{A}$, one can write, $\{H, f\} = \frac{df}{dt} = \omega(f_1, \ldots, f_n)$, where $f_1, \ldots, f_n$ is any set of generators for $\mathscr{A}$, where the time-derivative denotes, as usual, the derivative of its argument with respect to the flow generated by the Hamiltonian H. This explains why the existence of master symmetries (of degree one) is an important issue in classical mechanics. Being the (analogue of the) angle variables, they allow to linearize the equations describing the evolution of the dynamical system under investigation. Note that this point of view is fruitful both in the case when the Lagrangian fibration has compact fibers and non-compact fibers. In the first case, their usefulness stems from the fact that, although the angle variables exists from the general considerations discussed in Chap. 4, they are in general very difficult to find explicitly. On the other hand, when the fibers are not compact, even the existence of the angle variables, in *stricto sensu*, is a problematic issue.

6.3 *P N-Structures, Lie Algebroids, and Bidifferential Calculus*

In this subsection, we will see how the compatibility conditions between N and Π_0 on a PN-manifolds can be translated into a compatibility condition between the Cartan differential and another differential naturally defined by d and N. We will also comment about how PN-structure can be analyzed from the viewpoint of the theory of the Lie algebroids (see Sect. 4.3 in Chap. 2), and we conclude with a short introduction to the so called bidifferential calculus.

6.3.1 Lie Bialgebroids

We start with the following observation. Let $N \in \mathrm{End}_{C^\infty(M)}(TM)$ and, for all $k \geq 1$, let $i_N : \Omega^k(M) \to \Omega^k(M)$ defined by

$$(i_N\alpha)(X_1, \ldots, X_k) = \sum_{i=1}^{k} \alpha(X_1, \ldots, NX_i, \ldots, X_k). \tag{5.128}$$

One can easily prove that imposing $i_N f = 0$ for all $f \in C^\infty(M)$ and extending (5.128) by linearity to all $\Omega^\bullet(M)$, it remains defined a derivation of degree zero still called i_N. Recall that being a derivation of degree zero means that i_N is $\mathbb{K}$-linear and $i_N(\alpha \wedge \beta) = (i_N\alpha) \wedge \beta + \alpha \wedge (i_N\beta)$ for all $\alpha \in \Omega^k(M)$ and $\beta \in \Omega^n(M)$. Given i_N and d, one can define $d_N : \Omega^\bullet(M) \to \Omega^\bullet(M)$ as

$$d_N = i_N \circ d - d \circ i_N. \tag{5.129}$$

A simple computation, using (5.128) and the properties of d, shows that $d_N(\alpha \wedge \beta) = (d_N\alpha) \wedge \beta + (-1)^k \alpha \wedge d_N\beta$ for all $\alpha \in \Omega^k(M)$ and $\beta \in \Omega^\bullet(M)$. In other words, d_N is a derivation of degree one of $\Omega^\bullet(M)$. Note that d_N anti-commutes with d, i.e.,

$$d_N \circ d = -d \circ i_N \circ d = -(d \circ i_N \circ d - d \circ d \circ d_N) = -d \circ d_N.$$

We will give now a different representation for d_N. To this end, let $D_N : \Omega^\bullet(M) \to \Omega^{\bullet+1}(M)$ be defined on $\alpha \in \Omega^k(M)$ by the formula

$$(D_N\alpha)(X_1, \dots, X_{k+1}) \tag{5.130}$$

$$= \sum_{i=1}^{k+1} (NX_i)\Big(\alpha(X_1, \dots, \hat{X}_i, \dots, X_{k+1})\Big)$$

$$+ \sum_{i<j} (-1)^{i+j}\alpha\Big([X_i, X_j]_N, X_1, \dots, \hat{X}_i, \dots, \hat{X}_j, \dots, X_{k+1}\Big),$$

where

$$[X, Y]_N = [NX, Y] + [X, NY] - N[X, Y], \ \forall X, Y \in \mathfrak{X}(M). \tag{5.131}$$

Note that (5.129) defines a derivation of degree one of $\Omega^\bullet(M)$, which anti-commutes with d. For this reason, $D_N = d_N$ *if and only if* they coincide on the zero-forms, because every $\alpha \in \Omega^k(M)$ is (locally) a linear combination of forms of the type $\xi_1 \wedge \cdots \xi_k$, with $\xi_i \in \Omega^1(M)$ with $\xi_i = f_i dg_i$ for some (locally defined) smooth functions f_i, g_i. Let us compute

$$D_N f(X) \overset{(5.130)}{=} NX(f) = \langle df, NX \rangle = \langle {}^tN df, X \rangle,$$

on the other hand

$$\langle d_N f, X \rangle \overset{(5.129)}{=} \langle i_N(df) - \cancel{d i_N(f)}, X \rangle = \langle df, NX \rangle = \langle {}^tN df, X \rangle,$$

which shows that $d_N f = D_N f$, and for what is observed above, it yields that $d_N = D_N$. We can now prove that

Theorem 5.130 d_N *is a differential, i.e.,* $d_N^2 = 0$, *if and only if* $T_N = 0$.

Proof To conclude, using the same argument used to prove $d_N = D_N$, it is sufficient to prove that $d_N^2 f = 0$ for all $f \in C^\infty(M)$. To this end, we compute

$$(d_N^2 f)(X, Y) = d_N(d_N f)(X, Y)$$

$$\overset{(5.130)}{=} NX\langle d_N f, Y \rangle - NY\langle d_N f, X \rangle - \langle d_N f, [X, Y]_N \rangle$$

$$= NX\langle {}^tN df, Y \rangle - NY\langle {}^tN df, X \rangle - \langle df, N[X, Y]_N \rangle$$

$$\overset{(5.130)}{=} NX(NY(f)) - NY(NY(f)) - N[X, Y]_N(f)$$

$$= ([NX, NY] - N[X, Y]_N)(f),$$

which implies

$$\left(d_N^2 f\right)(X, Y) = 0 \iff T_N(X, Y)(f) = 0,$$

proving that $d_N^2 f = 0$ for all $f \in C^\infty(M)$ if and only if N is torsion-free. $\qquad\square$

We summarize the previous observation in the following.

Proposition 5.131 *If N is a Nijenhuis operator on M, then $(\Omega^\bullet(M), d_N)$, where d_N, defined in (5.129), is a complex.*

The previous proposition can be applied to every PN-manifold. We change now our viewpoint, and we look at the previous topic from the perspective of the theory of Lie algebroids. Recall that the tangent bundle of every manifold is endowed with a Lie algebroid structure whose anchor is the identity id $\in$ End$_{C^\infty(M)}(TM)$ and whose bracket is the standard Lie bracket on $\mathfrak{X}(M)$ (see Sect. 4.3 in Chap. 2 for a very short introduction and [162] for a very complete discussion of this topic). Once an endomorphism N of the tangent bundle is chosen, it is possible to define a new bracket on $\mathfrak{X}(M)$ as in (5.131). By direct inspection, the bracket so defined is skew-symmetric. The computations collected below (see Sect. 6.3.3) show that $T_N = 0$ is a sufficient condition for this bracket to satisfy the Jacobi identity. Moreover, since $T_N = 0$ is the same that $N[\cdot, \cdot]_N = [N\cdot, N\cdot]$, one can conclude the following.

Proposition 5.132 *If $N \in$ End$_{C^\infty(M)}(TM)$ is torsion-free, then $TM_N = (TM, N, [\cdot, \cdot]_N)$ is a Lie algebroid.*

Proof The only thing remains to check is

$$[X, fY]_N = f[X, Y]_N + NX(f)Y, \ \forall X, Y \in \mathfrak{X}(M), \ f \in C^\infty(M),$$

see (2.68). We leave it as an exercise for the reader. $\qquad\square$

Moreover, comparing (2.69) with (5.130), one can conclude that $(\Omega^\bullet(M), d_N)$ is the complex associated with the Lie algebroid TM_N. To connect this observation, the one in Sect. 4.3 of Chap. 2, and the conditions defining a PN-structure (see 5.40), we need to add a few more general observations. First, if both a vector bundle E and its dual vector bundle E^* carry a structure of Lie algebroid $(E, a, [\cdot, \cdot]_E)$ and, respectively, $(E^*, a_*, [\cdot, \cdot]_{E^*})$, then one can define two complexes (Γ_E, d_E) and (Γ_{E^*}, d_{E^*}), where $\Gamma_E = \Gamma(\Lambda^\bullet E^*)$ and, respectively, $\Gamma_{E^*} = \Gamma(\Lambda^\bullet E)$ and whose differentials are defined as in Formula (2.69) (see Sect. 4.3 in Chap. 2). A simple but important instance of this situation is enclosed in the following.

Example 5.133 Let (M, Π) be a Poisson manifold. Then TM_{id} and T^*M_Π are two Lie algebroids, the first defined on TM and the second on the dual vector bundle T^*M (see Examples 2.79 and 2.81). To the first Lie algebroid, it is associated with the de Rham complex of M, $(\Omega^\bullet(B), d)$; to the second one, it is associated with the complex $(\Gamma(\Lambda^\bullet TM), d_\Pi)$. If N is a torsion-free endomorphism of TM, then TM_N is also a Lie algebroid whose complex, as remarked above, is $(\Omega^\bullet(M), d_N)$. $\qquad\triangle$

Now comes a crucial point, which will focus on the case of the Lie algebroid $(E, a, [\cdot, \cdot]_E)$, being the dual case completely analogous. The Lie bracket $[\cdot, \cdot]_E$ admits a unique extension, still denoted with the same symbol, to $\Gamma(\Lambda^\bullet E)$, fulfilling the following conditions (see [137]):

(G1) $[\xi, \eta]_E = -(-1)^{nk}[\eta, \xi]_E$, for all $\xi \in \Gamma(\Lambda^n E)$, $\eta \in \Gamma(\Lambda^k E)$.

(G2) $[s, f]_E = a(s)f$, for all $s \in \Gamma(E)$ and $f \in C^\infty(M)$.

(G3) For every $\zeta \in \Gamma(\Lambda^{l+1} E)$, $[\zeta, \cdot]_E : \Gamma(\Lambda^\bullet E) \to \Gamma(\Lambda^\bullet E)$ is a derivation of degree l of $(\Gamma(\Lambda^\bullet E)), \wedge)$ i.e.,

$$[\zeta, \xi \wedge \eta]_E = [\zeta, \xi]_E \wedge \eta + (-1)^{ln} \xi \wedge [\zeta, \eta]_E, \ \forall \xi \in \Gamma\left(\Lambda^n E\right), \ \eta \in \Gamma\left(\Lambda^k E\right).$$

Note that the bracket so obtained is a *graded Lie bracket*, i.e., it satisfies

$$(-1)^{(k_1-1)(k_3-1)}[\![\eta_1, [\![\eta_2, \eta_3]\!]]\!] + (-1)^{(k_2-1)(k_1-1)}[\![\eta_2, [\![\eta_3, \eta_1]\!]]\!]$$
$$+ (-1)^{(k_3-1)(k_2-1)}[\![\eta_3, [\![\eta_1, \eta_2]\!]]\!] = 0,$$

for $\eta_i \in \Gamma(\Lambda^{k_i} E)$, $i = 1, 2, 3$.

Remark 5.134 Observe that this construction is the analogue of the one sketched in Chap. 2 (see Theorem 2.14), where it was remarked that on every manifold M, the Lie bracket between vector fields admits a unique extension to a bracket on multivector fields satisfying the set of conditions in the theorem recalled above. The bracket there defined was called a Schouten bracket, and for this reason, the one defined on $\Gamma(\Lambda^\bullet E)$, for a general Lie algebroid $(E, a, [\cdot, \cdot]_E)$, is sometimes called a *generalized Schouten bracket*. The datum of a graded Lie algebra endowed with a generalized Schouten bracket is called a Gesternhaber algebra. $\triangle$

Finally, following [138], one can prove the following:

Lemma 5.135 d_{E^*} *is a derivation of* $(\Gamma(\Lambda^\bullet E), [\cdot, \cdot]_E)$ *if and only if* d_E *is a derivation of* $(\Gamma(\Lambda^\bullet E^*), [\cdot, \cdot]_{E^*})$.

Starting from these preliminary observations and remarks, we can now introduce the following important notion:

Definition 5.136 A Lie bialgebroid is a pair of Lie algebroids $(E, a, [\cdot, \cdot]_E)$ and $(E^*, a_*, [\cdot, \cdot]_{E^*})$ in duality with each other such that $d_{E^*} : \Gamma(\Lambda^\bullet E) \to \Gamma(\Lambda^\bullet E)$ satisfies

$$d_{E^*}[\xi, \eta]_E = [d_{E^*}\xi, \eta] + (-1)^{n-1}[\xi, d_{E^*}\eta]_E,$$

for all $\xi \in \Gamma(\Lambda^n E)$ and $\eta \in \Gamma(\Lambda^k E)$. The Lie bialgebroid formed by a pair $(E, a, [\cdot, \cdot]_E)$ and $(E^*, a_*, [\cdot, \cdot]_{E^*})$ will be denoted by (E, E^*). $\triangle$

Example 5.137 If (B, Π) is a Poisson manifold, then $(TB_{\mathrm{id}}, T^*B_{\Pi})$ is a Lie bialgebroid. In fact $d_{\Pi} : \Gamma(\Lambda^{\bullet}TB) \to \Gamma(\Lambda^{\bullet}TB)$ is a derivation of the (generalized) Schouten bracket $[\cdot, \cdot]_{TM} = [\![\cdot, \cdot]\!]$, as follows from the (graded) Jacobi identity recalling that $d_{\Pi} = [\![\Pi, \cdot]\!]$. $\triangle$

Finally, we can state the sought characterization of the PN-structures in terms of Lie bialgebroids (see [138]).

Theorem 5.138 *Let (M, Π) be a Poisson manifold and let N be a torsion-free endomorphism of TM. Then the following conditions are equivalent:*

(i) Π and N are compatible (see 5.40).
(ii) d_{Π} is a derivation of $(\Gamma(\Lambda^{\bullet}TM), [\cdot, \cdot]_{TM_N})$.
(iii) d_N is a derivation of $(\Omega^{\bullet}(M), \{\cdot, \cdot\}_{\Pi})$.
(iv) (TM_N, T^*M_{Π}) is a Lie bialgebroid.

6.3.2 Bidifferential Calculi

We would like to close this part describing briefly the so-called bidifferential calculi, an approach to PN-structures whose starting point is the existence of a pair of compatible differentials on $\Omega^{\bullet}(M)$. More precisely, a *simple* bidifferential calculus on M consists of a pair of degree one anti-commuting differentials (d_1, d_2) of $(\Omega^{\bullet}(M), \wedge)$ such that, say, d_1 satisfies a Poincaré Lemma–type theorem. This means that:

1. $d_i : \Omega^{\bullet}(M) \to \Omega^{\bullet+1}(M)$ is $\mathbb{K}$-linear and $d(\alpha \wedge \beta) = d\alpha \wedge \beta + (-1)^k \alpha \wedge d\beta$ for all $\alpha \in \Omega^k(M)$ and $\beta \in \Omega^{\bullet}(M)$
2. $d_i^2 = 0$ for all $i = 1, 2$
3. $d_1 \circ d_2 + d_2 \circ d_1 = 0$
4. Every d_1-closed form is (locally) d_1-exact.

Given a bidifferential calculus on M, if

$$f \in C^{\infty}(M) \quad \text{and} \quad d_1 d_2 f = 0 = d_2 d_1 f,$$

then there exists $\{f_k\}_{k \geq 0}$ such that

$$f_0 = f \quad \text{and} \quad d_1 f_{k+1} = d_2 f_k, \ \forall k.$$

The f_ks will be, in general, only locally defined. Their existence follows observing that since $d_1 d_2 f_0 = 0$, then $d_2 f_0 = d_1 f_1$ for some f_1. But $0 = d_2(d_2 f_1) = d_2 d_1 f_1 = -d_1 d_2 f_1$, which implies that there exists f_2 such that $d_2 f_1 = d_1 f_2$. It should be clear now how to continue the process that yields the sequence of the f_ks. An important example of bidifferential calculus is obtained choosing $d_1 = d$, the Cartan differential. In this case, a fundamental observation in [94] tells that a derivation D of $(\Omega^{\bullet}(M), \wedge)$ of degree one anti-commutes with d if and only if

$D = d_N$, for some $N \in \mathrm{End}_{C^\infty(M)}(TM)$ (see 5.129). In this case, as it was shown above, $D^2 = 0$ if and only if $T_N = 0$. The connection between integrable systems and bidifferential calculi is enclosed in the following.

Proposition 5.139 *Let* (d, d_N) *be a bidifferential calculus on a simply connected manifold* M, *and suppose that* $\Pi_0 \in \Gamma(\Lambda^2 TM)$ *is such that*

$$\Pi_0\big({}^t N(\alpha), \beta\big) = \Pi_0\big(\alpha, {}^t N(\beta)\big), \quad \forall \alpha, \beta \in \Omega^1(M). \tag{5.132}$$

Let $f \in C^\infty(M)$ *such that* $d d_N f = 0$. *Then the functions* $\{f_k\}_{k \geq 0}$, *where*

$$f_0 = f \quad and \quad df_{k+1} = d_N f_k, \quad \forall k \geq 0, \tag{5.133}$$

are pair-wise in involution with respect to the bracket defined by Π_0 *via the formula* $\{g, h\}_0 = \Pi_0(dg, dh)$ *for all* $g, h \in C^\infty(M)$.

Proof Suppose $n > m$.

$$\Pi_0\big(df_n, df_m\big) \overset{(5.133)}{=} \Pi_0\big(d_N f_{n-1}, f_m\big) \overset{(5.132)}{=} \Pi_0\big(df_{n-1}, d_N f_m\big)$$
$$\overset{(5.133)}{=} \Pi_0\big(df_{n-1}, df_{m+1}\big),$$

which proves the statement as seen in the proof of Proposition 5.39. $\square$

Obviously, the previous proposition remains true a fortiori if Π_0 is a Poisson tensor. In other words, a bidifferential calculus compatible with a Poisson tensor in the sense of (5.132) induces a family of pairwise commuting first integrals of the *Hamiltonian* f_0. Note, however, that the existence of the f_ks is, in general, obstructed. To make a closer connection with the theory of the PN-manifolds, we suppose that the bivector field in the previous proposition is Poisson and that it comes from a symplectic structure ω_0. In this case

Corollary 5.140 *If the two-form* ω_1 *defined by* $\omega_1(X, Y) = \omega_0(NX, Y)$ *for all* $X, Y \in \mathfrak{X}(M)$ *is closed, then* (M, ω_0, N) *is a* ΩN-*manifold.*

Proof The first part of the statement follows at once from (5.132), since that condition is equivalent to $\Pi_0^{\sharp t} = N \Pi_0^{\sharp}$. The second part follows from Theorem 5.77.

$\square$

Remark 5.141 We briefly compare the notion of bidifferential calculi with that of a bidifferential complex induced by two compatible Poisson structures Π_1 and Π_0. Associated with the tensors Π_1 and Π_0, there are differentials $d_{\Pi_i} : \Gamma(\Lambda^\bullet TM) \to \Gamma(\Lambda^{\bullet+1} TM)$, where $i = 1, 0$ (see Sect. 4.3 of Chap. 2). These differentials satisfy $d_{\Pi_1}^2 = d_{\Pi_0}^2 = 0$ (since Π_1 and Π_0 are Poisson) and $d_{\Pi_1} \circ d_{\Pi_0} + d_{\Pi_0} \circ d_{\Pi_1} = 0$ (since Π_1 and Π_0 are compatible). However, it is not true in general that a d_{Π_i}-closed multivector is locally d_{Π_i} exact, since the corresponding Poisson cohomology is in general highly non-trivial, even locally. Even if Π_i has a globally constant rank, the corresponding Poisson cohomology is a complicate object (see the insightful [253] and [239]).   △

The notion of bidifferential calculus was introduced in [68] and further investigated in [56] and [57]. Proposition 5.139 and the Corollary 5.140 are borrowed from [57] and, respectively, [56]. For further developments on the use of bidifferential calculus in the theory of infinite dimensional integrable systems, see [159] and [17].

6.3.3 Some Computations

Here below, we collect a few (long) computations functional to the discussion presented in the last section. Let N be an endomorphism of the tangent bundle of M. Then

Proposition 5.142 *If $T_N = 0$, then the bracket $[\cdot, \cdot]_N : \mathfrak{X}(M) \times \mathfrak{X}(M) \to \mathfrak{X}(M)$ defined by*

$$[X, Y]_N = [NX, Y] + [X, NY] - N[X, Y]$$

satisfies the Jacobi identity.

Proof The proof follows from a direct computation. First observe that

$$\left[[X, Y]_N, Z\right]_N + \left[[Z, X]_N, Y\right]_N + \left[[Y, Z]_N, X\right]_N$$

$$= \left[N\Big([NX, Y] + [X, NY] - N[X, Y]\Big), Z\right]$$

$$+ \left[\underbrace{[NX, Y]}_{1} + \underbrace{[X, NY]}_{2} - N[X, Y], NZ\right]$$

$$- N\left[[NX, Y] + [X, NY] - N[X, Y], Z\right]$$

$$+ \left[N\Big([NZ, X] + [Z, NX] - N[Z, X]\Big), Y\right]$$

$$+ \left[\underbrace{[NZ, X]}_{2} + \underbrace{[Z, NX]}_{3} - N[Z, X], NY\right]$$

$$- N\left[[NZ, X] + [Z, NX] - N[Z, X], Y\right]$$

$$+ \left[N\Big([NY, Z] + [Y, NZ] - N[Y, Z]\Big), X\right]$$

$$+ \left[\underbrace{[NY, Z]}_{3} + \underbrace{[Y, NZ]}_{1} - N[Y, Z], NX\right]$$

$$- N\left[[NY, Z] + [Y, NZ] - N[Y, Z], X\right].$$

Then collecting the terms labelled by 1, 2, and 3, one obtains $-[[NZ, NX], Y]$, $-[[NY, NZ], X]$, and, respectively, $-[[NX, NY], Z]$. Then one can write

$$\left[[X, Y]_N, Z\right]_N + \left[[Z, X]_N, Y\right]_N + \left[[Y, Z]_N, X\right]_N$$

$$= \left[N\big([NX, Y] + [X, NY] - N[X, Y]\big) - [NX, NY], Z\right]$$

$$+ \left[N\big([NZ, X] + [Z, NX] - N[Z, X]\big) - [NZ, NX], Y\right]$$

$$+ \left[N\big([NY, Z] + [Y, NZ] - N[Y, Z]\big) - [NY, NZ], X\right]$$

$$- \underbrace{\left[N[X, Y], NZ\right]}_{a} - \underbrace{\left[N[Z, X], NY\right]}_{b} - \underbrace{\left[N[Y, Z], NX\right]}_{c}$$

$$- N\left[[NX, Y] + [X, NY] - N[X, Y], Z\right]$$

$$- N\left[[NZ, X] + [Z, NX] - N[Z, X], Y\right]$$

$$- N\left[[NY, Z] + [Y, NZ] - N[Y, Z], X\right].$$

Since in the formula above

$$a = T_N([X, Y], Z) + N\big([N[X, Y], Z] + [[X, Y], NZ] - N[[X, Y], Z]\big),$$

$$b = T_N([Z, X], Y) + N\big([N[Z, X], Y] + [[Z, X], NY] - N[[Z, X], Y]\big),$$

and, respectively,

$$c = T_N([Y, Z], X) + N\big([N[Y, Z], X] + [[Y, Z], NX] - N[[Y, Z], X]\big),$$

applying the Jacobi identity, one has

$$- a - b - c = -T_N([X, Y], Z) - N\big([N[X, Y], Z] + [[X, Y], NZ]\big)$$
$$+ \text{cyclic permutations in } X, Y, Z.$$

In other words,

$$\left[[X, Y]_N, Z\right]_N + \left[[Z, X]_N, Y\right]_N + \left[[Y, Z]_N, X\right]_N$$

$$= -2[T_N(X, Y), Z] - 2\left[T_N(Z, X), Y\right] - 2\left[T_N(Y, Z), X\right]$$

$$- \cancel{N[N[X, Y], Z]} + N\left[[X, Y], NZ\right]$$

$$- \cancel{N[N[Z, X], Y]} + N\left[[Z, X], NY\right]$$

$$- \cancel{N[N[Y, Z], X]} + N\left[[Y, Z], NX\right]$$

$$- N\big[[NX, Y] + [X, NY], Z\big] + N[N[X,Y],Z]$$
$$- N\big[[NZ, X] + [Z, NX], Y\big] + N[N[Z,X],Y]$$
$$- N\big[[NY, Z] + [Y, NZ], X\big] + N[N[Y,Z],X]$$
$$= -2\big[T_N(X, Y), Z\big] - 2\big[T_N(Z, X), Y\big] - 2\big[T_N(Y, Z), X\big]$$
$$+ N\big[[X, Y], NZ\big] - N\big[[NZ, X], Y\big] - N\big[[Y, NZ], X\big]$$
$$+ N\big[[Z, X], NY\big] - N\big[[X, NY], Z\big] - N\big[[NY, Z], X\big]$$
$$+ N\big[[Y, Z], NX\big] - N[Z, NX], Y\big] - N\big[[NX, Y], Z\big]$$

using the Jacobi identity and the skew-symmetry of T_N

$$= 2\big[T_N(Y, X), Z\big] + 2\big[T_N(X, Z), Y\big] + 2\big[T_N(Z, Y), X\big],$$

proving the statement. $\qquad\square$

In other words, if N is a Nijenhuis tensor, then $[\cdot, \cdot]_N$ is a Lie bracket on the vector space $\mathfrak{X}(M)$.

7 Bibliographical Notes

The main references for this long chapter are the surveys articles [80, 164, 165, 167], where the definitions and the main properties of bi-Hamiltonian and Poisson-Nijenhuis manifolds can be found. Other two important and nice references to be consulted are [141, 168]. In the latter, the reader will find a wealth of information about Poisson-Nijenhuis structures from a more algebraic point of view. We would like to warn the reader that Poisson-Nijenhuis and, more in general, bi-Hamiltonian structures played and still play an important role in the theory of infinite-dimensional Hamiltonian systems, where the geometric description leaves the place to a more algebraic one. For this reason, the reader should not be surprised to find in the above references many more applications to these infinite-dimensional systems than to the ones discussed in these notes. Section 4.2 and the construction of the Darboux-Nijenhuis coordinates in Sect. 4.1.1 are borrowed from the unpublished PhD dissertation [177] (see also [164–166]). The relation between integrability *à la* Liouville and the existence of a bi-Hamiltonian representation for a given Hamiltonian system, which is the content of Sect. 5, was investigated in depth in [39–41, 89] (see also the unpublished PhD dissertation [88, 167]). Examples 5.102 and 5.103 are taken from [41] and, respectively, [147]. The notion of translation hypersurface (see Definition 5.104) was introduced and used to detect the existence of a Poisson-Nijenhuis structure in the neighborhood of an invariant torus in [39, 40, 88, 89] (see also [42]). The content of Propositions 5.106 and 5.107 was borrowed from these references. Section 6.1 is based on the fundamental works [82, 83], where the relation between Poisson-Nijenhuis structures and separation of variables for the Hamilton-Jacobi equation was analyzed in depth (see Remark 5.116). As mentioned in Sect. 6.2, the notion of master symmetry was introduced in [96] in the framework of infinite-dimensional integrable systems, and it found important and interesting applications to the finite-dimensional case in [62, 87, 200–202] to cite a few references. While our presentation is based on [201, 202], Example 5.127 is borrowed from [224, 225]. Finally, Sect. 6.3, where the Poisson-Nijenhuis structures are described in terms of Lie algebroids, is meant to be a companion to Sect. 4.3 and has, as a main reference, [138]. The part about the bidifferential calculus, which is in our opinion very close in spirit to the Lie (bi)algebroid description of the PN-structure, was inspired by [56, 57] (see also [68] where the bidifferential calculus approach to integrable systems was introduced for the first time). Among the books where the reader can find an introduction to the theory of the bi-Hamiltonian structures, we would like to mention [228, 242]. Another nice reference, although more bent toward the infinite-dimensional case, is [32]. Finally, we would like to draw the reader's attention to the following fundamental references, [100–102], where a general program to classify bi-Hamiltonian structures was proposed.

Chapter 6
Rigid Bodies

The main aim of this chapter is to give a detailed presentation of the bi-Hamiltonian structure of a very classical integrable system, the n-dimensional-free rigid body with a fixed point. More precisely, it will be shown that the Hamiltonian description of this dynamical system can be enhanced to a bi-Hamiltonian one, which, however, does not come from a PN-structure. This presentation will occupy Sect. 3, where, after the description of this dynamical system, we provide a bi-Hamiltonian representation of the equations of motion. Then we introduce two (infinite) sequences of conserved quantities, the so-called *Mishchenko* and *Manakov* integrals, which form the Lenard-Magri chain for the bi-Hamiltonian structure previously introduced. The first two sections of this chapter are devoted to more general topics. More precisely, in the first one, we discuss the geometry of the cotangent bundle of a Lie group, with emphasis on the symplectic aspects, while the second one is dedicated to the so-called Euler equations, which are Hamiltonian equations defined on coadjoint orbits. The relevance of this part for the main topic of the present chapter stems from the fact that the dynamics of the free rigid body, which takes place on the cotangent bundle of $\mathrm{SO}_n(\mathbb{R})$, can be *reconstructed* from the solutions of the Euler equations on $\mathfrak{so}_n(\mathbb{R})$.

1 The Geometry of T^*G

In this section, we discuss in some details the geometry of the (co)tangent bundle of a Lie group G. First we will introduce a convenient geometrical setting to describe both the right and left G-actions, and then we will describe the moment maps associated with the (co)tangent lifts of these actions.

© The Author(s), under exclusive license to Springer Nature Switzerland AG 2026

A. Arsie, I. Mencattini, *Geometry of Integrable Systems*, Latin American Mathematics Series – UFSCar subseries,

https://doi.org/10.1007/978-3-031-96282-0_6

1.1 Left and Right Actions

Let G be a Lie group and let TG be its tangent bundle. The maps $t_r : TG \to G \times \mathfrak{g}$ and $t_l : TG \to G \times \mathfrak{g}$, defined by

$$t_r : \xi \rightsquigarrow \left(g, (R_{g^{-1}})_{*,g}\xi\right) \text{ and } t_l : \xi \rightsquigarrow \left(g, (L_{g^{-1}})_{*,g}\xi\right), \ \forall \xi \in T_g G \text{ and } \forall g \in G,$$

(6.1)

are two diffeomorphisms called *right* and, respectively, *left trivialization* of TG. Similarly, one can define right and left trivialization of T^*G. Denoting them with the same symbols $t_r, t_l : T^*G \to G \times \mathfrak{g}^*$, one has

$$t_r : \alpha \rightsquigarrow \left(g, (R_g)^*_{,e}\alpha\right), \text{ and } t_l : \alpha \rightsquigarrow \left(g, (L_g)^*_{,e}\alpha\right) \ \forall \alpha \in T^*_g G \text{ and } g \in G. \quad (6.2)$$

Note that $t_r \circ t_l^{-1} : G \times \mathfrak{g} \to G \times \mathfrak{g}$ is the diffeomorphism defined by

$$(g, x) \rightsquigarrow (g, \mathrm{Ad}_g x),$$

while $t_r \circ t_l^{-1} : G \times \mathfrak{g}^* \to G \times \mathfrak{g}^*$ becomes

$$(g, \alpha) \rightsquigarrow \left(g, \mathrm{Ad}^{\sharp}_{g^{-1}} \alpha\right).$$

Left and right trivializations of tangent and cotangent bundles of a Lie group give two *preferred reference systems*. In particular, borrowing the terminology usually used in fluid mechanics, we have that:

(i) The points of T^*G (TG) will be said to be in the *Lagrangian* or *material* representation.
(ii) The ones represented using the *left trivialization* of T^*G (TG) will be said to be in the *body* or *convective* representation.
(iii) The ones represented using the *right trivialization* of T^*G (TG) will said to be in the *spatial* or *Eulerian* representation.

Remark 6.1 Observe that the right and left trivializations of the cotangent and tangent bundles of G are obtained using a basis of non-vanishing global sections of these two vector bundles. More precisely, the left (right) trivialization of TG is obtained using a basis of left and right invariant vector fields, while the analogue trivializations of T^*G are obtained using a basis of left and right invariant one-forms. $\triangle$

1.2   The Cotangent Lift Again and the Corresponding Moment Maps

We will now give the explicit descriptions of the lift to tangent and cotangent bundle of both right and left actions of G on itself. We will focus our efforts on the case of cotangent bundles, leaving the details of the other one to the reader as an exercise. We refer the reader to the Appendix D for more information about the (co)tangent lifts of group actions. We start observing that right and left translations can be lifted to diffeomorphisms of the tangent bundle of G. More precisely, given $R_g, L_g :$ $G \to G$, the corresponding cotangent lifts are

$$\widetilde{R}_g : \alpha \rightsquigarrow \left(R_{g^{-1}}\right)^*_{,hg} \alpha \ \text{ and } \ \widetilde{L}_g : \alpha \rightsquigarrow \left(L_{g^{-1}}\right)^*_{,gh} \alpha, \ \forall \alpha \in T^*_g G \ \text{ and } g, h \in G.$$

Now let us trivialize T^*G via *left translations* (see Formula 6.1), and let us describe the cotangent lifts of left and right translations in this trivialization.

Proposition 6.2 *If $l_g : G \times \mathfrak{g}^* \to G \times \mathfrak{g}^*$ and $r_g : G \times \mathfrak{g}^* \to G \times \mathfrak{g}^*$ are the G-actions induced on the left trivialized cotangent bundle of G by the cotangent lift of the left and, respectively, right translations, then*

$$l_g : (h, \alpha) \rightsquigarrow (gh, \alpha) \tag{6.3}$$

$$r_g : (h, \alpha) \rightsquigarrow \left(hg, \mathrm{Ad}^{\sharp}_{g^{-1}} \alpha\right). \tag{6.4}$$

Proof Formula (6.3) follows from the following diagram (here $\alpha \in T^*_h G$):

$$
\begin{array}{ccc}
\alpha & \xrightarrow{\ t_l\ } & (h, (L_h)^*_{,e}\alpha) \\[2pt]
\widetilde{L}_g \Big\downarrow & & \Big\downarrow l_g \\[2pt]
\left(gh, (L_{g^{-1}})^*_{,gh}\alpha\right) & \xrightarrow{\ t_l\ } & \left(gh, (L_{gh})^*_{,e}(L_{g^{-1}})^*_{,gh}\alpha\right)
\end{array}
$$

while Formula (6.4) follows from the following one:

$$
\begin{array}{ccc}
\alpha & \xrightarrow{\ t_l\ } & (h, (L_h)^*_{,e}\alpha) \\[2pt]
\widetilde{R}_g \Big\downarrow & & \Big\downarrow r_g \\[2pt]
\left(hg, (R_{g^{-1}})^*_{,hg}\alpha\right) & \xrightarrow{\ t_l\ } & \left(hg, (L_{hg})^*_{,e}(R_{g^{-1}})^*_{,hg}\alpha\right)
\end{array}
$$

$\square$

Problem 6.3 Write the explicit formulas for the cotangent lift of right and left translations on the *right trivialized* cotangent bundle of G. $\triangle$

Problem 6.4 Let G be a Lie group and let TG be its tangent bundle:

(i) Give an explicit description of the tangent lifts of the *right* translations on the *left and right trivialized* tangent bundle.
(ii) Give an explicit description of the tangent lifts of the *left* translations on the *left and right trivialized* tangent bundle.

$\triangle$

Now we will describe the moment maps $\mu^l, \mu^r : T^*G \to \mathfrak{g}^*$ associated with the cotangent lifts of left and right translations. We start reminding that if G is a Lie group acting on a manifold N, the cotangent lift of the G-action is strongly Hamiltonian with moment map $\mu : T^*N \to \mathfrak{g}^*$, defined by

$$\langle \mu(\alpha), x \rangle = F_x(\alpha) = \langle \alpha, X_x(\pi(\alpha)) \rangle, \ \forall \alpha \in T^*N, \ x \in \mathfrak{g}$$

(see Formula 3.2). In the formula above, $\pi : T^*N \to N$ is the canonical projection, and X_x is the fundamental vector field defined by the G-action and associated with $x \in \mathfrak{g}$.

For every $x \in \mathfrak{g}$, let X_x^l and X_x^r be the fundamental vector fields associated with $x \in \mathfrak{g}$ and defined on G by the *left* and, respectively, *right translations*. More precisely

$$X_x^l(g) = \left.\frac{d}{dt}\right|_{t=0} \exp(-tx)g = \left.\frac{d}{dt}\right|_{t=0} R_g(\exp(-tx)) = -X_x^R(g), \qquad (6.5)$$

while

$$X_x^r(g) = \left.\frac{d}{dt}\right|_{t=0} g\exp(tx) = \left.\frac{d}{dt}\right|_{t=0} L_g(\exp(tx)) = X_x^L(g), \qquad (6.6)$$

where with X_x^R and X_x^L, we denoted the *right* and *left invariant* vector fields defined by the element $x \in \mathfrak{g}$ (see Appendix C). Then:

Proposition 6.5 *The moment maps for the cotangent lifts of the left and right translations are, respectively,*

$$\mu^r : T^*G : \xrightarrow{\ \alpha \rightsquigarrow (L_g)^*_{,e}\alpha\ } \mathfrak{g}^*,$$

$$\mu^l : T^*G : \xrightarrow{\ \alpha \rightsquigarrow -(R_g)^*_{,e}\alpha\ } \mathfrak{g}^*,$$

for all $\alpha \in T^*_g G$ and all $g \in G$.

Finally, we give explicit formulas for the moment maps for the cotangent lifts of the *right* and *left translations* on the *left trivialized* cotangent bundle (see Proposition 6.2). To this end, let us trivialize the *tangent* bundle TG using the left translations

$$TG \xrightarrow{v \rightsquigarrow (g,(L_{g^{-1}})_{*,g}v)} G \times \mathfrak{g}.$$

Applying this diffeomorphism to the fundamental vectors fields X^l_x and X^r_x, we get their expressions (still denoted with the same symbols) in the *body representation*:

$$X^l_x : g \rightsquigarrow \left(g, -(L_{g^{-1}})_{*,g} X^R_x(g)\right) = \left(g, -\operatorname{Ad}_{g^{-1}} x\right)$$

and, respectively,

$$X^r_x : g \rightsquigarrow \left(g, (L_{g^{-1}})_{*,g} X^L_x(g)\right) = (g, x).$$

Trivializing now T^*G using the left translations, we arrive to the formulas

$$\mu^l : G \times \mathfrak{g}^* : \xrightarrow{(g,\alpha) \rightsquigarrow -\operatorname{Ad}^\sharp_g \alpha} \mathfrak{g}^*$$

and, respectively,

$$\mu^r : G \times \mathfrak{g}^* : \xrightarrow{(g,\alpha) \rightsquigarrow \alpha} \mathfrak{g}^*,$$

which express the moment maps in the body representation, i.e., the moment maps defined by the actions (6.3) and, respectively, (6.4).

The previous discussion is summarized in the next table (Fig. 6.1).

Problem 6.6 Prove that the entries in the first line of the table above are the values of the fundamental vector fields and of the moment maps described in the Eulerian representation. $\triangle$

Remark 6.7 Observe that composing the *left* and *right* trivializations maps (6.2) with the projection $G \times \mathfrak{g}^* \to \mathfrak{g}^*$, $(g, \alpha) \rightsquigarrow \alpha$, we obtain the following two (smooth) maps $\pi^l, \pi^r : T^*G \to \mathfrak{g}^*$, respectively:

$$\pi^l : \alpha \rightsquigarrow (L_g)^*_{,e}\alpha \quad \text{and} \quad \pi^r : \alpha \rightsquigarrow (R_g)^*_{,e}\alpha$$

Fig. 6.1 Eulerian, Lagrangian, and body representation

	$X^r_x(g)$	$X^l_x(g)$	$\mu^r(\alpha)$	$\mu^l(\alpha)$
Eulerian	$(g, \operatorname{Ad}_g x)$	$(g, -x)$	$\operatorname{Ad}^\sharp_{g^{-1}} \alpha$	$-\alpha$
Lagrangian	$X^r_x(g)$	$X^l_x(g)$	$(L_g)^*_{,e}(\alpha)$	$-(R_g)^*_{,e}(\alpha)$
Body	(g, x)	$(g, -\operatorname{Ad}_{g^{-1}} x)$	α	$-\operatorname{Ad}^\sharp_g \alpha$

Fig. 6.2 Left G-action vs right G-action

By direct inspection, we see that $\mu^r = \pi^l$ and $\mu^l = -\pi^r$, which bring us to the following (almost) dual picture (Fig. 6.2).

$\triangle$

2 Hamiltonian Dynamics on T^*G and the Euler Equations

We start this section giving an explicit formula for the Liouville one-form Θ and for the canonical symplectic form $\Omega = d\Theta$ on the left trivialized cotangent bundle of G (see also Example 1.20). Let G be a Lie group and let T^*G be its cotangent bundle endowed with its canonical symplectic form Ω. Let us trivialize T^*G using left translations, $t_l^{-1} : G \times \mathfrak{g}^* \simeq T^*G, \alpha \rightsquigarrow (L_{g^{-1}})^*_{,g}\alpha$ (see 6.2). First, we describe how the Liouville one-form Θ and the canonical symplectic form $\Omega = d\Theta$ look like after pullback on $G \times \mathfrak{g}^*$ using left trivialization. Let $(g, \alpha) \in G \times \mathfrak{g}^*$ and $(v, \beta), (u, \gamma) \in T_{(g,\alpha)}(G \times \mathfrak{g}^*) \simeq T_g G \oplus \mathfrak{g}^*$ and let $\vartheta = (t_l^{-1})^*\Theta$ and $\omega = (t_l^{-1})^*\Omega$. Then:

Proposition 6.8

$$\langle \vartheta_{(g,\alpha)}, (v, \beta) \rangle = \langle \alpha, x \rangle \tag{6.7}$$

and

$$\omega_{(g,\alpha)}\big((v, \beta), (u, \gamma)\big) = \langle \beta, y \rangle - \langle \gamma, x \rangle - \langle \alpha, [x, y] \rangle \tag{6.8}$$

where $x, y \in \mathfrak{g}$ are such that $v = (L_g)_{*,e}x$ and $u = (L_g)_{*,e}y$.

Proof Recall that

$$t_l^{-1}(g, \alpha) = (L_{g^{-1}})^*_{,g}(\alpha), \ \forall (g, \alpha) \in G \times \mathfrak{g}^*, \tag{6.9}$$

and

$$\tag{6.10}$$

Formula (6.7) follows from the following computation:

$$\langle \vartheta_{(g,\alpha)}, (v,\beta)\rangle = \left\langle \left((t_l^{-1})^*\Theta\right)_{(g,\alpha)}, (v,\beta)\right\rangle = \left\langle \Theta_{t_l^{-1}(g,\alpha)}, (t_l^{-1})_{*,(g,\alpha)}(v,\beta)\right\rangle$$

$$\overset{(6.9)}{=} \left\langle \Theta_{(L_{g^{-1}})^*_{,g}(\alpha)}, (t_l^{-1})_{*,(g,\alpha)}(v,\beta)\right\rangle$$

$$= \left\langle (L_{g^{-1}})^*_{,g}(\alpha), \pi_{*,(L_{g^{-1}})^*_{,g}(\alpha)}(t_l^{-1})_{*,(g,\alpha)}(v,\beta)\right\rangle$$

$$\overset{(6.10)}{=} \left\langle (L_{g^{-1}})^*_{,g}(\alpha), p_{*.(g,\alpha)}(v,\beta)\right\rangle$$

$$= \left\langle (L_{g^{-1}})^*_{,g}(\alpha), v\right\rangle$$

$$= \left\langle \alpha, (L_{g^{-1}})_{*,g}v\right\rangle,$$

for all $(v,\beta) \in T_{(g,\alpha)}(G \times \mathfrak{g}^*)$. Let $(v,\beta) \in T_{(g,\alpha)}(G \times \mathfrak{g}^*) \simeq T_g G \oplus \mathfrak{g}^*$ and let:

(i) X_x^L be the (unique) left invariant vector field on G such that $X_x^L(g) = v$, i.e., $X_x^L(g) = \frac{d}{dt}\big|_{t=0} L_g(\exp(tx))$ (see 6.6).

(ii) $\mathcal{Y}_\beta$ be the constant vector field on $\mathfrak{g}^*$ such that $\mathcal{Y}_\beta(\alpha) = \beta$ for all $\alpha \in \mathfrak{g}^*$. Note that $\mathcal{Y}_\beta(\alpha) = \frac{d}{dt}\big|_{t=0} (\alpha + t\beta)$.

Now if $(v,\beta), (u,\gamma) \in T_{(g,\alpha)}(G \times \mathfrak{g}^*)$ and $X_y^L(g) = u$

$$d\vartheta_{(g,\alpha)}\big((v,\beta),(u,\gamma)\big) = d\vartheta_{(g,\alpha)}\Big(\big(X_x^L(g), \mathcal{Y}_\beta(\alpha)\big), \big(X_y^L(g), \mathcal{Y}_\gamma(\alpha)\big)\Big)$$

$$= \big(X_x^L(g), \mathcal{Y}_\beta(\alpha)\big)\langle \vartheta, \big(X_y^L, \mathcal{Y}_\gamma\big)\rangle - \big(X_y^L(g), \mathcal{Y}_\gamma(\alpha)\big)\langle \vartheta, \big(X_x^L, \mathcal{Y}_\beta\big)\rangle \qquad (6.11)$$

$$- \langle \vartheta, [\big(X_x^L, \mathcal{Y}_\beta\big), \big(X_y^L, \mathcal{Y}_\gamma\big)]\rangle(\alpha, v). \qquad (6.12)$$

For the reader convenience, we analyze (6.11) and (6.12) separately, starting from the latter. In this term, one has

$$\Big[\big(X_x^L, \mathcal{Y}_\beta\big), \big(X_y^L, \mathcal{Y}_\gamma\big)\Big] = \Big(\big[X_x^L, X_y^L\big], [\mathcal{Y}_\beta, \mathcal{Y}_\gamma]\Big) = \Big(X_{[x,y]}^L, 0\Big),$$

since $\mathcal{Y}_\beta$ and $\mathcal{Y}_\gamma$ are constant, while X_x^L and X_y^L are left invariant. Looking at the first summand in (6.11), one has

$$\big(X_x^L(g), \mathcal{Y}_\beta(\alpha)\big)\Big\langle \vartheta, \big(X_y^L, \mathcal{Y}_\gamma\big)\Big\rangle = \frac{d}{dt}\Big|_{t=0} \Big\langle \vartheta, \big(X_y^L, \mathcal{Y}_\gamma\big)\Big\rangle (L_g(\exp(tx)), \alpha + t\beta)$$

$$= \frac{d}{dt}\Big|_{t=0} \Big\langle \vartheta_{\big(L_g(\exp(tx)),\alpha+t\beta\big)}, \big(X_y^L(L_g(\exp(tx)),$$

$$\mathcal{Y}_\gamma(\alpha + t\beta)\big)\Big\rangle$$

$$\overset{(6.7)}{=} \frac{d}{dt}\Big|_{t=0} \Big\langle \alpha + t\beta, \big(L_{(g\exp(tx))^{-1}}\big)_{*,g\exp(tx)}$$

$$X_y^L(g\exp(tx))\Big\rangle$$

$$= \frac{d}{dt}\Big|_{t=0} \langle \alpha + t\beta, y\rangle,$$

where the equality follows since X_y^L is left invariant. In other words

$$\big(X_x^L(g), \mathcal{X}_\beta(\alpha)\big)\langle \vartheta, \big(X_y^L, \mathcal{X}_\gamma\big)\rangle = \langle \beta, y\rangle.$$

This concludes the proof, since the other summand in (6.11) can be computed in the same way. $\qquad\square$

Remark 6.9 Clearly, $t_l : (T^*G, \Omega) \to (G \times \mathfrak{g}^*, \omega)$ is a symplectomorphism. $\qquad\triangle$

Let $h : T^*G \to \mathbb{R}$ be a smooth function and let $X_h \in \mathfrak{X}(T^*G)$ be the corresponding Hamiltonian vector field. Let $H \in C^\infty(G \times \mathfrak{g}^*)$ be defined by $H := h \circ t_l^{-1}$. In the next proposition, we will present a formula that gives an explicit description of the vector field $X_H \in \mathfrak{X}(G \times \mathfrak{g}^*)$.

Remark 6.10 Note that since t_l is a simplectomorphism $X_H = t_l X_h$ (see Proposition 1.57, and recall that $H := h \circ t_l^{-1}$). $\qquad\triangle$

Let $f : G \times \mathfrak{g}^* \to \mathbb{R}$ be a smooth function, and let $g \in G$, $v \in T_gG$, and $\alpha, \beta \in \mathfrak{g}^*$. Then:

$$f_{*,(g,\alpha)}(v, \beta) = (f \circ i_g)_{*,\alpha}\beta + (f \circ i_\alpha)_{*,g}v,$$

where $i_g : \mathfrak{g}^* \to G \times \mathfrak{g}^*, \alpha \rightsquigarrow (g, \alpha)$ and $i_\alpha : G \to G \times \mathfrak{g}^*, g \rightsquigarrow (g, \alpha)$, for all $\alpha \in \mathfrak{g}^*$ and $g \in G$. Note that:

$$(f \circ i_\alpha)_{*,g} : T_gG \to \mathbb{R},$$

while:

$$(f \circ i_g)_{*,\alpha} : \mathfrak{g}^* \to \mathbb{R}$$

are linear maps, and for this reason, we can identify the former with one element of T_g^*G and the latter with one element of $\mathfrak{g}$, respectively.

Proposition 6.11 *Let $\alpha, \beta \in \mathfrak{g}^*$, $v \in T_gG$, and $x \in \mathfrak{g}$ such that $v = (L_g)_{*,e}x$. Then, given $H : G \times \mathfrak{g}^* \to \mathbb{R}$, the corresponding Hamiltonian vector field X_H can be written as*

$$X_H(g, \alpha) = \Big((L_g)_{*,e}(H \circ i_g)_{*,\alpha}, -(L_g)_{*,e}^*(H \circ i_\alpha)_{*,g} - \mathrm{ad}_{(H\circ i_g)_{*,\alpha}}^\sharp \alpha\Big) \qquad (6.13)$$

Proof First we write $X_H(g, \alpha)$ as a pair $(X^\alpha(g), \Xi^g(\alpha))$, where X^α is a vector field on G smoothly depending on α, while Ξ^g is a vector field on $\mathfrak{g}^*$ smoothly depending on g. Then we use the definition of a Hamiltonian vector field to write

$$\langle dH_{(g,\alpha)}, (v, \beta) \rangle = -\langle (i_{X_H}\omega)_{(g,\alpha)}, (v, \beta) \rangle = \omega_{(g,\alpha)}\big((v, \beta), (X^\alpha(g), \Xi^g(\alpha))\big).$$

Let now x, $y^{(g,\alpha)} \in \mathfrak{g}$ such that $(L_g)_{*,e}x = v$ and $(L_g)_{*,e}y^{(g,\alpha)} = X^\alpha(g)$. Note that since X^α is smoothly dependent on α, also the corresponding vector in $\mathfrak{g}$ will be so. Now, applying Formula (6.8), we obtain

$$\langle \beta, (H \circ i_g)_{*,\alpha} \rangle + \big\langle (L_g)^*_{,e}(H \circ i_\alpha)_{*,g}, x \big\rangle = \big\langle \beta, y^{(g,\alpha)} \big\rangle - \big\langle \Xi^g(\alpha), x \big\rangle$$

$$- \Big\langle \alpha, \big[x, y^{(g,\alpha)} \big] \Big\rangle$$

$$= \langle \beta, y^{(g,\alpha)} \rangle - \Big\langle \Xi^g(\alpha) + \mathrm{ad}^\sharp_{y^{(g,\alpha)}} \alpha, x \Big\rangle,$$

which yields

$$y^{(g,\alpha)} = (H \circ i_g)_{*,\alpha} \tag{6.14}$$

$$\Xi^g(\alpha) = -(L_g)^*_{,e}(H \circ i_\alpha)_{*,g} - \mathrm{ad}^\sharp_{y^{(g,\alpha)}} \alpha \tag{6.15}$$

The proof of the proposition follows now from Formulas (6.14) and (6.15). $\square$

In this way, one arrives to the following.

Corollary 6.12 *The Hamiltonian equations defined by the vector field (6.13) are*

$$\begin{cases} \dot{g}(t) &= (H \circ i_g)_{*,\alpha} \\ \dot{\alpha}(t) &= -(L_g)^*_{,e}(H \circ i_\alpha)_{*,g} - \mathrm{ad}^\sharp_{(H \circ i_g)_{*,\alpha}} \alpha \end{cases} \tag{6.16}$$

where the time dependence of the terms of the right-hand sides is understood.

We conclude this section with the observations enclosed in the following.

Problem 6.13 Let $(g, \alpha) \in G \times \mathfrak{g}^*$, $H \in C^\infty(\mathfrak{g}^*)$, and $\tilde{H} = H \circ \mu^l$. Define $\xi_{g,\alpha} = \mathrm{Ad}_{g^{-1}} H_{*,-\mathrm{Ad}^\sharp_g \alpha}$. Show that:

1. $(\tilde{H} \circ i_g)_{*,\alpha} = -\xi_{g,\alpha}$ and $(L_g)^*_{,e}(\tilde{H} \circ i_\alpha)_{*,g} = \mathrm{ad}^\sharp_{\xi_{g,\alpha}} \alpha$.
2. $X_{\tilde{H}}(g, \alpha) = \big(-(L_g)_{*,e}\xi_{g,\alpha}, 0\big)$.

(Hint: As far as the first part of point 1 is concerned, it suffices to show that

$$(\tilde{H} \circ i_g)_{*,\alpha}(\beta) = \frac{d}{dt}\bigg|_{t=0} H(-\mathrm{Ad}^\sharp_g(\alpha + t\beta)) = -(H)_{*,-\mathrm{Ad}^\sharp_g \alpha}(\mathrm{Ad}^\sharp_g \beta)$$

and apply the definition of $\mathrm{Ad}^\sharp$. To get the second part of point 1, it suffices to prove that, given any $x \in \mathfrak{g}$

$$(L_g)^*_{,e}\left(\widetilde{H} \circ i_\alpha\right)_{*,s}(x) = \left.\frac{d}{dt}\right|_{t=0} H\left(-\mathrm{Ad}^\sharp_{g\,\exp(tx)}\,\alpha\right) = -(H)_{*,-\mathrm{Ad}^\sharp_g \alpha}\left(\mathrm{Ad}^\sharp_g \mathrm{ad}^\sharp_x \alpha\right)$$

and conclude that

$$(L_g)^*_{,e}\left(\widetilde{H} \circ i_\alpha\right)_{*,s}(x) = \langle \mathrm{ad}^\sharp_{\xi_{g,\alpha}}(\alpha), x \rangle.$$

Finally, to solve part 2, it suffices to use the results of part 1 to compute the second entry in Formula (6.13) for the Hamiltonian vector field $X_{\widetilde{H}}$.) $\triangle$

2.1 Euler Equations

If $\widetilde{H} : G \times \mathfrak{g}^* \to \mathbb{R}$ is a *left invariant* function, i.e., if $(l^*_g \widetilde{H})(h, \alpha) = \widetilde{H}(h, \alpha)$, $\forall g, h \in G$, $\alpha \in \mathfrak{g}^*$, then $(\widetilde{H} \circ i_\alpha)_{*,g} = 0$, $\forall g \in G$, $\alpha \in \mathfrak{g}^*$, and Formula (6.16) reduces to

$$\begin{cases} \dot{g}(t) &= (L_g)_{*,e}(H \circ i_g)_{*,\alpha} \\ \dot{\alpha}(t) &= -\,\mathrm{ad}^\sharp_{(H \circ i_g)_{*,\alpha}}\,\alpha. \end{cases}$$

Now it is worth observing that

Lemma 6.14 *The left invariant functions on $G \times \mathfrak{g}^*$ are in one-to-one correspondence with the functions on $\mathfrak{g}^*$.*

Proof In fact, since $\mu^r : G \times \mathfrak{g}^* \to \mathfrak{g}^*$ is surjective, $\mu^{r,*} : C^\infty(\mathfrak{g}^*) \to C^\infty(G \times \mathfrak{g}^*)$ is injective, and $\mu^{r,*}(f)$ is left invariant. On the other hand, every $F \in C^\infty(G \times \mathfrak{g}^*)$ left invariant is determined uniquely by its restriction to $\{e\} \times \mathfrak{g}^*$, which can be identified with a (unique) smooth function on $\mathfrak{g}^*$. $\square$

Let $H \in C^\infty(\mathfrak{g}^*)$ and $\widetilde{H} = H \circ \mu^r$ the corresponding left invariant function. Then

$$\left(\widetilde{H} \circ i_g\right)(\alpha) = H(\alpha) \quad \text{and} \quad \left(\widetilde{H} \circ i_g\right)_{*,\alpha} = H_{*,\alpha} = dH_\alpha$$

for all $\alpha \in \mathfrak{g}^*$. Since $dH_\alpha \in \mathfrak{g}$ for all $\alpha \in \mathfrak{g}^*$

$$\begin{cases} \dot{g}(t) &= (L_g)_{*,e} dH_\alpha \\ \dot{\alpha}(t) &= -\,\mathrm{ad}^\sharp_{dH_\alpha}\,\alpha. \end{cases} \tag{6.17}$$

Remark 6.15 One can complement the previous result observing that since the left invariant functions on $G \times \mathfrak{g}^*$ are in one-to-one correspondence with the left invariant functions on T^*G, the latter are in one-to-one correspondence with $C^\infty(\mathfrak{g}^*)$.   △

Let us look more closely at the second formula in (6.17). If $H \in C^\infty(\mathfrak{g}^*)$, one can observe that for all $\alpha \in \mathfrak{g}^*$, $dH_\alpha = H_{*,\alpha} \in \mathfrak{g}$, i.e., dH is a smooth map between $\mathfrak{g}^*$ and $\mathfrak{g}$. This observation, together with (2.30), implies that the Hamiltonian vector field $X_H \in \mathfrak{X}(\mathfrak{g}^*)$ is defined by

$$X_H(\alpha) = -\operatorname{ad}^\sharp_{dH_\alpha}(\alpha) \tag{6.18}$$

for all $\alpha \in \mathfrak{g}^*$.

Definition 6.16 (Euler Equations) The equations

$$\dot{\alpha}(t) = -\operatorname{ad}^\sharp_{dH_{\alpha(t)}}\alpha(t) \tag{6.19}$$

describing the integral curves of vector field (6.18) are called the *Euler equations*.

△

The previous remarks suggest the following strategy to solve the Hamiltonian equations defined by any G-invariant $h \in C^\infty(T^*G)$:

(i) After choosing the initial conditions $\alpha_0 = \alpha(0)$, one solves (6.19) for the function $H \in C^\infty(\mathfrak{g}^*)$ corresponding to H (see Remark 6.15).
(ii) Using this solution, one solves (the first equation in) (6.17) to find the curve $g = g(t)$ such that

$$(L_{g(t)^{-1}})_{*,g(t)}\dot{g}(t) = dH_{\alpha(t)}, \ \forall t.$$

(iii) Finally, one arrives to the sought solution applying isomorphism $t_l^{-1} : G \times \mathfrak{g}^* \to T^*G$ to the pair $(g(t), \alpha(t))$ previously obtained.

Remark 6.17 In the previous three-step procedure, it is summarized the so-called *Reconstruction of the Dynamics* theorem (see [175] and also Remark 3.88). The function $H \in C^\infty(\mathfrak{g}^*)$ corresponding to the G-invariant Hamiltonian $h \in C^\infty(T^*G)$ is sometimes called a collective Hamiltonian (see [113, 114]).   △

2.2 Lax Representation of the Equations of Motion

In this subsection, we aim to introduce the so-called Lax representation of the equation of motion. As it will be clarified hereafter, the existence of such a representation yields a plethora of first integrals for the underlying dynamical system, and for this reason, it is a clue to its integrability. We start discussing the following case. Let $L_0 \in \mathfrak{gl}_n(\mathbb{K})$ and let $g = g(t)$ be a (smooth) curve in $\mathrm{GL}_n(\mathbb{K})$ with $g(0) = I$. Define $L(t) = g^{-1} L_0 g$.

Lemma 6.18 *Then:*

 (i) *The characteristic polynomial $p_\lambda(t)$ of $L(t)$ does not depend on t.*
(ii) *If $B(t) = g^{-1}\dot{g}$, then $L(t)$ satisfies the following differential equation:*

$$\frac{dL}{dt} = [L, B]. \tag{6.20}$$

Proof Since $L(t) = g^{-1} L_0 g$, $p_\lambda(t) = \det(L(t) - \lambda I) = \det(L_0 - \lambda I)$, which proves (i). On the other hand:

$$\frac{dL}{dt} = \frac{d(g^{-1} L_0 g)}{dt} = \frac{dg^{-1}}{dt} L_0 g + g^{-1} L_0 \frac{dg}{dt}$$

$$= -g^{-1}\frac{dg}{dt}g^{-1} L_0 g + g^{-1} L_0 g g^{-1} \frac{dg}{dt} = L(t)B(t) - B(t)L(t)$$

where $B(t) = g^{-1}\dot{g}$. $\qquad\qquad\square$

Definition 6.19 Equation (6.20) is called *a Lax equation* for the time-dependent matrix $L = L(t)$ and (L, B) is called *a Lax* pair. $\qquad\triangle$

Remark 6.20 A couple of observations are now in order.

 (i) Lemma 6.18 shows that the characteristic polynomial of a matrix L, whose time evolution is described by a Lax equation, is constant during the evolution. This implies that the coefficients of this polynomial are *constant* during the evolution of the matrix. These coefficients are symmetric functions of the eigenvalues of the matrix, i.e., $p_L(\lambda) = \lambda^n + \sum_{k=1}^{n-1}(-1)^k e_k(\lambda_1, \ldots, \lambda_n)\lambda^{n-k}$, where $e_k(\lambda_1, \ldots, \lambda_n) = \sum_{l_1 < \cdots < l_k} \lambda_{l_1} \cdots \lambda_{l_k}$, for all $k = 1, \ldots, n$. Since these symmetric functions are generically independent, one can conclude that the eigenvalues of the matrix L are (generically) first integrals. Because of this property, it is commonly said that the Lax equations (6.20) define a *isospectral deformation* of the matrix L.
(ii) The Lax representation of the equations of motion given in (6.20) is clearly *not* unique. In fact, the equations represented by a Lax-pair (L, B) can be represented as well by a Lax-pair of the form $(L, B + M)$, where M is any matrix which commuting with L.

$\triangle$

Viceversa

Proposition 6.21 (Lax) *Suppose that the time-dependent matrix $L = L(t)$ satisfies the Lax equation (6.20). If $g = g(t) \in \mathrm{GL}_n(\mathbb{K})$ is the solution of $\dfrac{dg}{dt} = gB$ with initial condition $g(0) = I$, then $g(t)L(t)g(t)^{-1} = L(0)$.*

Remark 6.22 We can summarize the previous discussion saying that if a matrix satisfies Eq. (6.20), then its characteristic polynomial does not depend on t. In particular, if the matrix L_0 is generic (i.e., diagonalizable with distinct eigenvalues), then its eigenvalues are constant. $\triangle$

Corollary 6.23 *The functions $F_k = \mathrm{tr}(L^k)$ are constant along the curve $t \rightsquigarrow L(t)$.*

Proof The F_k are the coefficients of the characteristic polynomial, which, by the proposition, does not depend on t. Another proof is by direct calculation:

$$\frac{dF_k}{dt} = \mathrm{tr}\left(\frac{dL^k}{dt}\right) = k\ \mathrm{tr}\left(\frac{dL}{dt}L^{k-1}\right) = k\ \mathrm{tr}\left([L,B]L^{k-1}\right) = 0$$

by the cyclic property of the trace and the fact that $\mathrm{tr}(AB) = \mathrm{tr}(BA)$ for any pair of square matrices. $\square$

Remark 6.24 There are two natural generalizations of the previous remarks:

(i) On one hand, one can generalize the previous considerations to any Lie algebra $\mathfrak{g}$. In fact $B = g^{-1}\dot{g}$ is the (left) *Maurer-Cartan* form of $\mathrm{GL}_n(\mathbb{K})$ (see Definition C.104 in Appendix C). Starting from this observation, one can define the Lax equation for a *time-dependent* element $\eta \in \mathfrak{g}$ as

$$\frac{d\eta}{dt} = \mathrm{ad}_{\eta(t)}\left(L_{g^{-1}(t),*}\dot{g}(t)\right) = \left[\eta(t), L_{g^{-1}(t),*}\dot{g}(t)\right],$$

where $g = g(t)$ is a smooth curve passing through $e \in G$ and $\dot{g}(t)$ is the tangent vector to this curve at the time t.

(ii) On the other hand, one can extend the isospectral deformations of matrices to suitable classes of *linear operators* acting on infinite-dimensional vector spaces. This was indeed the original framework where such deformations were introduced by Peter Lax in his studies of the KdV equation (see, e.g., [151]).

$\triangle$

We now come to our main example. In fact, under a mild hypothesis on the Lie algebra $\mathfrak{g}$, it can be shown that the Euler equations (6.19) admits a Lax representation. To this end, let $X_H(\cdot) = -\mathrm{ad}^{\sharp}_{dH}(\cdot)$ be the Hamiltonian vector field associated with $H \in C^\infty(\mathfrak{g}^*)$, whose corresponding Euler equations are (6.19). Recall that every Hamiltonian vector field on $\mathfrak{g}^*$ is tangent to the coadjoint orbits. More precisely, the tangent space to every point of a coadjoint orbit is the linear span of the Hamiltonian vector fields evaluated at that point (see Example 2.65).

Suppose now that $\mathfrak{g}$ is a quadratic Lie algebra, and let $B : \mathfrak{g} \otimes \mathfrak{g} \to \mathbb{K}$ be the corresponding bilinear form (see Definition C.17 in Appendix C). Note that B defines a diffeomorphism $\check{B} : \mathfrak{g} \to \mathfrak{g}^*$, which identifies *adjoint* with *coadjoint* orbits. Under this correspondence, to each Hamiltonian vector field on $\mathfrak{g}^*$, there is an associated unique vector field on $\mathfrak{g}$ tangent to the adjoint orbits. In what follows, an explicit form for this correspondence will be presented.

Let $x_\alpha = \check{B}^{-1}(\alpha)$, where $\check{B}^{-1} : \mathfrak{g}^* \to \mathfrak{g}$ is the linear isomorphism defined by B (see Remark 2.7 in Appendix C).

Proposition 6.25 *The system of ODEs on $\mathfrak{g}$ corresponding to (6.19) via $\check{B}^{-1}$ has the following Lax-type representation:*

$$\dot{x}_\alpha = [x_\alpha, dH_\alpha]. \tag{6.21}$$

Proof Applying the (the time-independent) $\check{B}^{-1}$ to both sides of (6.19), one obtains

$$\dot{x}_\alpha = -x_{\mathrm{ad}^\sharp_{dH_\alpha}(\alpha)}.$$

Using the isomorphism $\check{B} : \mathfrak{g} \to \mathfrak{g}^*$, we can write

$$B\left(x_{\mathrm{ad}^\sharp_{dH_\alpha}(\alpha)}, y\right) = \langle \mathrm{ad}^\sharp_{dH_\alpha}(\alpha), y \rangle = -\langle \alpha, [dH_\alpha, y] \rangle = -B(x_\alpha, [dH_\alpha, y])$$

$$= B([dH_\alpha, x_\alpha], y)$$

for all $y \in \mathfrak{g}$, where in the last equality, we used the ad-invariance of B. Comparing the last term with the first one and using the non-degeneracy of B, one concludes that

$$x_{\mathrm{ad}^\sharp_{dH_\alpha}(\alpha)} = [dH_\alpha, x_\alpha],$$

which implies

$$\dot{x}_\alpha = -x_{\mathrm{ad}^\sharp_{dH_\alpha}(\alpha)} = [x_\alpha, dH_\alpha].$$

$\square$

In particular, for $K \in C^\infty(\mathfrak{g})$, the isospectral equations (6.21) becomes

$$\dot{x} = [x, \nabla_x K], \ \forall x \in \mathfrak{g} \tag{6.22}$$

(see Formula 2.10).

We remark that equations in Lax form have been extensively used also for numerical linear algebra problems, where they provide analogue algorithms to solve problems in this area. We provide some ideas related to this in Sect. 5.1, also commenting on the origin of the Toda system.

After discussing all these general issues, we come now to the main character of this chapter: the dynamics of the rigid body. The next section will be devoted to the introduction of this classical dynamical system, to its integrability, and to the description of possible bi-Hamiltonian formulations of its equations of motion. Our focus will be the *free* (n-dimensional) case though; in the final part of the chapter, we will comment on the heavy (i.e., subject to gravity) one as well.

3 The n-Dimensional Free Rigid Body

Following [211] (see also [9, 12]), we will study the *free* rotation of an n-*dimensional rigid* body about a *fixed* point O. We start clarifying the meaning of the italicized terms in the previous sentence. First, an n-dimensional body is a compact set $\mathscr{B} \subset \mathbb{R}^n$, endowed with its Euclidean inner product and with the corresponding distance, $d : \mathbb{R}^n \times \mathbb{R}^n \to \mathbb{R}$. $\mathscr{B}$ is called rigid if for all $P, Q \in \mathscr{B}$, $d(P, Q)$ is constant during the motion. This condition is commonly dubbed the *rigidity constraint*. The fixed point about which the motion (rotation) takes place will be assumed to be the origin of $\mathbb{R}^n$. Finally, the rotation is free because no (active) forces will act on $\mathscr{B}$. Under these assumptions, using the same notations as in [211], if $f(t, P)$ denotes the position at the time t of a point of $\mathscr{B}$ whose position at the time $t = 0$ was P, the rigidity constraint implies $f(t, P) = B(t)P$, where $B(t) \in O(n)$ for all t. Since the motion is smooth (as a function of t) and $f(0, P) = P$ for every $P \in \mathscr{B}$, one can conclude that $B = B(t)$ describes a *smooth* curve in $\mathrm{SO}_n(\mathbb{R})$. Assuming the mass density of $\mathscr{B}$ constant (equal to 1) its *kinetic energy* is $\mathscr{K}(t) = \frac{1}{2} \int_{\mathscr{B}} \|\dot{f}(t, P)\|^2 dP$, where $\|\cdot\|$ is the Euclidean norm in $\mathbb{R}^n$. Since

$$\dot{f}(t, P) = \frac{d}{ds}\bigg|_{s=t} B(s)P = \dot{B}(t)P = \dot{B}(t)B^{-1}(t)f(t, P),$$

one observes that, defining $\Omega_s(t) = \dot{B}(t)B(t)^{-1}$,

$$\|\dot{f}(t, P)\|^2 = \|B(t)^{-1}\Omega_s(t)B(t)P\|^2.$$

Using a more Lie group theoretical notation, one can observe that

$$\Omega_s(t) = (R_{B(t)^{-1}})_{*,B(t)}\dot{B}(t)$$

or, equivalently,

$$\dot{B}(t) = (R_{B(t)})_{*,\mathrm{id}}\Omega_s(t),$$

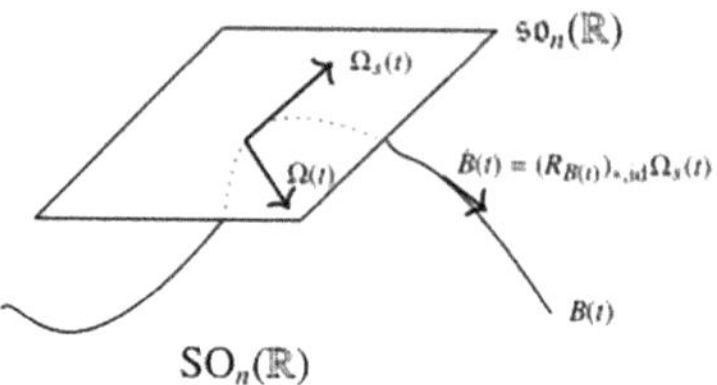

Fig. 6.3 Angular velocities in the body and space coordinates

i.e., $\dot{B}$ is the *right invariant* vector field on $SO_n(\mathbb{R})$ defined by $\Omega_s \in \mathfrak{so}_n(\mathbb{R})$. Moreover

$$B^{-1}(t)\dot{B}(t) = B^{-1}(t)\Omega_s(t)B(t) = \mathrm{Ad}_{B^{-1}(t)}\,\Omega_s(t),$$

see Fig. 6.3.

Thus, defining

$$\Omega(t) = \left(L_{B(t)^{-1}}\right)_{*,B(t)}\dot{B}(t)$$

one can write $\|\dot{f}(t, P)\|^2 = \|\Omega(t)P\|^2$, i.e.,

$$\mathscr{K}(t) = \frac{1}{2}\int_{\mathscr{B}} \|\Omega(t)P\|^2 dP. \tag{6.23}$$

Definition 6.26 Ω_s and Ω are called the angular velocities in the *space* and, respectively, *body* coordinate system (see Sect. 1). $\triangle$

Remark 6.27 $\dot{B}$, Ω_s, and Ω are the velocity vectors in the Lagrangian, Eulerian, and, respectively, Body representation (see Sect. 1). In particular, Ω_s and Ω represent the velocity vector $\dot{B}$ with respect to a *fixed spatial* reference system and, respectively, to a *fixed body* reference system (see the table at the end of Sect. 1.1, Eqs. 6.5 and 6.6). $\triangle$

Let $\langle \cdot, \cdot \rangle$ be the standard Euclidean inner product and let dP be the standard Lebesgue measure on $\mathbb{R}^n$. If $\mathscr{C} \subset \mathbb{R}^n$ is a compact subset not properly contained in any hyperplane, i.e., *not-flat*, then $(\cdot, \cdot) : \mathfrak{so}_n(\mathbb{R}) \times \mathfrak{so}_n(\mathbb{R}) \to \mathbb{R}$, defined by

$$(X, Y) = \int_{\mathscr{C}} \langle XP, YP \rangle dP, \tag{6.24}$$

is a non-degenerate, symmetric, bilinear form on $\mathfrak{so}_n(\mathbb{R})$, where $P \in \mathbb{R}^n$. In fact, the function $P \rightsquigarrow \langle XP, YP \rangle$ is continuous, and if there is $X \in \mathfrak{so}_n(\mathbb{R})$ such that $\int_{\mathscr{C}}\langle XP, YP \rangle dP = 0$, for all $Y \in \mathfrak{so}_n(\mathbb{R})$, then in particular, it is zero for $Y = X$. In this case however, $\int_{\mathscr{C}}\langle XP, XP \rangle dP = 0$ is possible if and only if $XP = 0$ for all $P \in \mathscr{C}$. Since the linear span of $\mathscr{C}$ is $\mathbb{R}^n$ because $\mathscr{C}$ is not flat, this yields that X is the zero matrix. On the other hand, symmetry and bilinearity are clear. Note that

(6.23), written in terms of (6.24), becomes

$$\mathscr{K}(t) = \frac{1}{2}(\Omega(t), \Omega(t)),$$

where the integral representing the right-hand side is computed over $\mathscr{B}$. The rest of this subsection will be devoted to prove the following result:

Theorem 6.28 *Let $\mathscr{B} \in \mathbb{R}^n$ be a not-flat rigid body. Then there exists an orthonormal basis of $\mathbb{R}^n$, uniquely determined by the mass distribution of $\mathscr{B}$ and a linear map $\Psi_\Upsilon : \mathfrak{so}_n(\mathbb{R}) \to \mathfrak{so}_n(\mathbb{R})$, defined by*

$$\Psi_\Upsilon(X) = \Upsilon X + X\Upsilon, \ \forall X \in \mathfrak{so}_n(\mathbb{R}), \tag{6.25}$$

where

$$\Upsilon = diag(\lambda_1, \ldots, \lambda_n) \quad with \quad \lambda_i + \lambda_j > 0, \ \forall i \neq j, \tag{6.26}$$

such that the equations of the free motion of $\mathscr{B}$ around the origin of $\mathbb{R}^n$ assume the form

$$\dot{M} = [M, \Omega], \tag{6.27}$$

where

$$M = \Psi_\Upsilon(\Omega). \tag{6.28}$$

The equations (6.27) are Hamiltonian on each adjoint orbit of $\mathfrak{so}_n(\mathbb{R})$ with Hamiltonian function

$$H(M) = -\frac{1}{4} \operatorname{tr}(M\Omega). \tag{6.29}$$

Before delving into the proof of the statement above, we note that the condition on Υ expressed in Theorem 6.28 implies that the linear map Ψ_Υ is *non-degenerate* and *symmetric* with respect to the trace-form. In fact

$$\Psi_\Upsilon(X)_{ij} \overset{(6.25)}{=} (\Upsilon X)_{ij} + (X\Upsilon)_{ij} \overset{(6.26)}{=} (\lambda_i + \lambda_j)X_{ij}, \ \forall i, j = 1, \ldots, n. \tag{6.30}$$

Moreover

$$\operatorname{tr}(\Psi_\Upsilon(A), B) \overset{(6.25)}{=} \operatorname{tr}(A\Psi_\Upsilon(B)), \ \forall A, B \in \mathfrak{so}_n(\mathbb{R}). \tag{6.31}$$

Finally, we can state the following.

Definition 6.29 The Euler vector field $\mathscr{E} \in \mathfrak{so}_n(\mathbb{R})$ is the vector field defined by the right-hand side of (6.27), i.e.,

$$\mathscr{E}(M) = \left[M, \Psi_{\Upsilon}^{-1}(M) \right], \quad \forall M \in \mathfrak{so}_n(\mathbb{R}). \tag{6.32}$$

$$\triangle$$

Proof (of the Theorem 6.28.) Using the symmetric bilinear form (6.24) and the trace form defined by

$$(\!(X, Y)\!) = -\frac{1}{2}\, \mathrm{tr}(XY),$$

one can introduce $\mathscr{I} : \mathfrak{so}_n(\mathbb{R}) \to \mathfrak{so}_n(\mathbb{R})$ as the unique linear operator such that

$$(X, Y) = (\!(\mathscr{I}X, Y)\!), \tag{6.33}$$

for all X, Y. To find an explicit formula for $\mathscr{I}$, observe that (6.33) implies

$$\frac{1}{2}\, \mathrm{tr}((\mathscr{I}X)Y) = \int_{\mathscr{B}} \langle XYP, P \rangle dP. \tag{6.34}$$

Choosing $Y = e_{ij} - e_{ji}$ and using the skew-symmetry of $\mathscr{I}X$, a simple computation shows that

$$\frac{1}{2}\, \mathrm{tr}\left((\mathscr{I}X)(e_{ij} - e_{ji})\right) = (\mathscr{I}X)_{ji}.$$

On the other hand, looking at the right-hand side of (6.34), one computes

$$\left(X(e_{ij} - e_{ji})P\right)_{kl} = (Xe_{ij}P)_{kl} - (Xe_{ji}P)_{kl} = \sum_{l=1}^{n} \left(X_{li}P_j - X_{lj}P_i\right) e_l,$$

where $e_1, \ldots, e_n$ is the canonical basis in $\mathbb{R}^n$, which implies

$$\langle X\left(e_{ij} - e_{ji}\right)P, P \rangle = \sum_{l=1}^{n} \left(X_{li}P_j - X_{lj}P_i\right) P_l.$$

If, for all i, j, $\Lambda_{ij} = \int_{\mathscr{B}} P_i P_j dP$, the previous computation entails that

$$\int_{\mathscr{B}} \langle X(e_{ij} - e_{ji})P, P \rangle dP = \sum_{l=1}^{n} \Lambda_{jl}X_{li} - \Lambda_{il}X_{lj} = \sum_{l=1}^{n} \Lambda_{jl}X_{li} + \Lambda_{il}X_{jl}.$$

In other words

$$(\mathscr{I}X)_{ji} = \sum_{l=1}^{n} \Lambda_{jl} X_{li} + \Lambda_{il} X_{jl} = (\Lambda X + X\Lambda)_{ji}, \ \forall i, j = 1, \ldots, n,$$

i.e.,

$$\mathscr{I}(X) = \Lambda X + X\Lambda, \ \forall X \in \mathfrak{so}_n(\mathbb{R}).$$

Since Λ is a $n \times n$ symmetric matrix, there exists $g \in \mathrm{SO}_n(\mathbb{R})$ such that $\Upsilon :=$ $g\Lambda g^{-1} = \mathrm{diag}(\lambda_1, \ldots, \lambda_n)$. If $\Psi_\Upsilon : \mathfrak{so}_n(\mathbb{R}) \to \mathfrak{so}_n(\mathbb{R})$ is defined by $\Psi_\Upsilon(X) = g\mathscr{I}(g^{-1} X g)g^{-1}, \ \forall X \in \mathfrak{so}_n(\mathbb{R})$, then

$$\Psi_\Upsilon(X) = \Upsilon X + X\Upsilon.$$

Moreover

$$(\!(\Psi_\Upsilon(X), Y)\!) = (X, Y), \ \forall X, Y \in \mathfrak{so}_n(\mathbb{R}). \tag{6.35}$$

In fact

$$
\begin{aligned}
(\!(\Psi_\Upsilon(X), Y)\!) &= -\frac{1}{2} \mathrm{tr}\left(\mathscr{I}(g^{-1} X g)g^{-1} Y g\right) \\
&= (\!(\mathscr{I}(g^{-1} X g), g^{-1} Y g)\!) \\
&= (g^{-1} X g, g^{-1} Y g) \\
&= \int_{\mathscr{B}} \langle g^{-1} X g(P), g^{-1} Y g(P)\rangle dP \\
&= \int_{\mathscr{B}} \langle X g(P), Y g(P)\rangle dP \\
&= \int_{\mathscr{B}} \langle X P, Y P\rangle dP \\
&= (X, Y).
\end{aligned}
$$

Note that since $(\!(\Psi_\Upsilon(X), X)\!) = (X, X) \geq 0$ for all $X \neq 0$, Ψ_Υ is a positive operator. Since

$$\Psi_\Upsilon(e_{ij} - e_{ji}) = (\lambda_i + \lambda_j)(e_{ij} - e_{ji})$$

for all i, j, one concludes that $\lambda_i + \lambda_j > 0$ for all $i \neq j$. It is worth noting that the $\lambda_i + \lambda_j$, for $i \neq j$ are classically known as the *principal inertia* moments of $\mathscr{B}$ (see Sect. 4.1). Now, let

$$H(M) := \frac{1}{2}\left(M, \Psi_\Upsilon^{-1}(\Omega)\right) \stackrel{(6.35)}{=} \frac{1}{2}((M, \Omega)) \stackrel{(6.29)}{=} -\frac{1}{4}\operatorname{tr}(M\Omega),$$

for all $M \in \mathfrak{so}_n(\mathbb{R})$. Then H is a (smooth) $\mathrm{SO}_n(\mathbb{R})$-invariant function on $\mathfrak{so}_n(\mathbb{R})$, whose gradient is readily computed

$$
\begin{aligned}
\nabla_M H(A) &= \frac{d}{ds}\bigg|_{t=0} H(M + tA) \\
&= -\frac{1}{4}\frac{d}{ds}\bigg|_{t=0} \operatorname{tr}\left((M + tA)\Psi_\Upsilon^{-1}(M + tA)\right) \\
&= -\frac{1}{4}\operatorname{tr}\left(A\Psi^{-1}_{\Upsilon(M)+M\Psi_\Upsilon^{-1}}(A)\right) \\
&= -\frac{1}{2}\operatorname{tr}\left(\Psi_\Upsilon^{-1}(M)A\right),
\end{aligned}
$$

i.e.,

$$\nabla_M H = \Psi_\Upsilon^{-1}(M) = \Omega. \tag{6.36}$$

One can conclude applying (6.22). $\square$

Remark 6.30 The relation of the previous result with the classical case of the (three-dimensional) free rigid-body theory will be briefly discussed in Sect. 4.1. $\triangle$

After setting the Hamiltonian framework for the evolution equations of the n-dimensional free rigid body, we will now address the problem of their integrability.

3.1 Integrability

We start noticing that (6.27) describes a Hamiltonian dynamical system evolving on an adjoint orbit $\mathcal{O} \subset \mathfrak{so}_n(\mathbb{R})$, whose symplectic form is the one induced by the linear Poisson structure of $\mathfrak{so}_n(\mathbb{R})$ (see Example 2.8 and Formula (6.22)). For this reason, to prove the integrability *à la Liouville* of (6.27), it suffices to show the existence of $f_1, \ldots, f_n \in C^\infty(\mathcal{O})$, $n = \frac{1}{2}\dim\mathcal{O}$, generically independent and in involutions with respect to the Poisson bracket defined by the symplectic form. As a first step in this direction, we prove the following.

Lemma 6.31 *The* generic *adjoint orbit of* $\mathfrak{so}_n(\mathbb{R})$ *has dimension* $\frac{n(n-1)}{2} - \lfloor\frac{n}{2}\rfloor$, *where* $\lfloor\frac{n}{2}\rfloor$ *is the greatest integer, less or equal to* $\frac{n}{2}$.

Proof Given $A \in \mathfrak{so}_n(\mathbb{R})$, the dimension of the orbit $\mathcal{O}_A$ under the adjoint action of $\mathrm{SO}_n(\mathbb{R})$ is equal to $\dim \mathfrak{so}_n(\mathbb{R}) - \dim(G_A)$, where G_A is the stabilizer of A. The orbit is called generic if A is a regular element, i.e., if G_A is a maximal torus, whose dimension is $\lfloor \frac{n}{2} \rfloor$. Of course, $\dim \mathfrak{so}_n(\mathbb{R}) = \frac{n(n-1)}{2}$. $\qquad\square$

Remark 6.32 A maximal torus in $\mathrm{SO}_n(\mathbb{R})$ is a subgroup isomorphic to the direct product of n-copies of $\mathrm{SO}_2(\mathbb{R})$. If $n = 2m$, in the standard representation defined by $n \times n$-matrices, its elements are of the form

$$
\begin{bmatrix}
T_1 & 0 & \dots & 0 \\
0 & T_2 & \dots & \vdots \\
\vdots & \vdots & \ddots & \\
0 & \dots & \dots & T_n
\end{bmatrix},
$$

while if $n = 2m + 1$, they look like

$$
\begin{bmatrix}
T_1 & 0 & \dots & \dots & 0 \\
0 & T_2 & \dots & \dots & \vdots \\
\vdots & \vdots & \ddots & \dots & \vdots \\
0 & \dots & \dots & T_n & 0 \\
0 & 0 & \dots & 0 & 1
\end{bmatrix},
$$

where all the T_is belong to $\mathrm{SO}_2(\mathbb{R})$. $\qquad\triangle$

Example 6.33 For $n = 3$, the generic adjoint orbit is (diffeomorphic) to a two-dimensional sphere, while (the unique) non-generic one is zero-dimensional. In this case, every maximal torus is isomorphic to $\mathrm{SO}_2(\mathbb{R})$. $\qquad\triangle$

The second step consists in the following important observation.

Proposition 6.34 (Manakov, [170]) *Equation* (6.27) *is equivalent to*

$$
\frac{d(M + \Upsilon^2 \xi)}{dt} = [M + \Upsilon^2 \xi, \Omega + \Upsilon \xi], \tag{6.37}
$$

where ξ is called the spectral parameter.

Proof For the proof, it suffices to compute the commutator on the right-hand side of (6.37) using (6.28) and observe that the left-hand side is $\dot{M}$. $\qquad\square$

The previous result entails that to prove the *Liouville* integrability of (6.27), it suffices to show that the functions

$$
f_{k,\xi}(M) = \frac{1}{k} \operatorname{tr} \left(M + \Upsilon^2 \xi \right)^k, \quad k = 2, \dots, n \tag{6.38}
$$

provide

(I1) $\frac{1}{2}\left(\frac{n(n-1)}{2} - \left[\frac{n}{2}\right]\right)$ functions on each adjoint orbit $\mathcal{O}$ (see Remark 6.36 below)

(I2) In involution with respect the Poisson bracket induced on $\mathcal{O}$ by the linear Poisson structure of $\mathfrak{so}_n(\mathbb{R})$,

(I3) In involution with the Hamiltonian (6.29)

Example 6.35 The $f_{k,\xi}$s up to $k = 3$ look like the following:

1. $f_{1,\xi}(M) = \mathrm{tr}(M) + \xi\,\mathrm{tr}\left(\Upsilon^2\right) = \xi\,\mathrm{tr}\left(\Upsilon^2\right)$.

2. $f_{2,\xi}(M) = \frac{\mathrm{tr}\left(M^2\right)}{2} + \xi\,\mathrm{tr}\left(M\Upsilon^2\right) + \frac{\xi^2\,\mathrm{tr}\left(\Upsilon^4\right)}{2} = \frac{\mathrm{tr}\left(M^2\right)}{2} + \frac{\xi^2\,\mathrm{tr}(\Upsilon^4)}{2}$.

3. $f_{3,\xi}(M) = \frac{\mathrm{tr}\left(M^3\right)}{3} + \xi\,\mathrm{tr}\left(M^2\Upsilon^2\right) + \xi^2\,\mathrm{tr}\left(M\Upsilon^4\right) + \frac{\xi^3\,\mathrm{tr}\left(\Upsilon^6\right)}{3} = \xi\,\mathrm{tr}\left(M^2\Upsilon^2\right) + \frac{\xi^3\,\mathrm{tr}\left(\Upsilon^6\right)}{3}$.

where it was used the cyclic property of the trace and the fact that the trace of a skew-symmetric matrix is zero. $\triangle$

Remark 6.36 To prove the integrability à la Liouville of (6.27), one property should be added to the previous list: the (generic) functional independence of the integrals stemming from the $f_{k,\xi}$s. This point will not be addressed in the present notes since it is more technical than the other ones and its discussion would take us far away from the main stream of the present notes. Nevertheless, we would like to point the reader's attention to the work [35], where this issue is studied in depth. $\triangle$

First, it is worth noting that the functions (6.38) are invariant by conjugation with elements of $\mathrm{SO}_n(\mathbb{R})$, and for this reason, they descend to smooth functions on the adjoint orbits of $\mathfrak{so}_n(\mathbb{R})$. Then, given k, expanding (6.38) in power of ξ as

$$f_{k,\xi}(M) = \sum_{l=0}^{k} h_{k,l}(M)\xi^{k-l}, \tag{6.39}$$

one sees that $h_{k,l}(M) = 0$ for all l such that $k - l$ is an odd (positive) integer. This implies that, for each k, $f_{k,\xi}$ defines $\lfloor\frac{k-1}{2}\rfloor$ non-trivial Hamiltonians on each adjoint orbit. In fact, $h_{k,k}(M) = \frac{\mathrm{tr}(M^k)}{k}$, which, if not identically zero for parity reasons, is a Casimir function of Π_{LP}, the linear Poisson structure, since $\nabla_M h_{k,k} = -2M^{k-1}$, for all $k \geq 1$. Counting the coefficients of (6.39) in degree between 1 and $k - 1$ and recalling that for $k - l$ odd the corresponding coefficient is identically zero, one proves the statement. To conclude the proof of (I1) above, it is sufficient to compute the sum

$$\sum_{k=2}^{n} \left\lfloor\frac{k-1}{2}\right\rfloor = \frac{1}{2}\left(\frac{n(n-1)}{2} - \left\lfloor\frac{n}{2}\right\rfloor\right).$$

Definition 6.37 Following [187], we will call $h_{k,l}$ in the expansion of $f_{k,\xi}$ a *Manakov* integral of order (k, l). $\triangle$

The proof of the involutivity of the Manakov integrals with respect to the linear Poisson structure of $\mathfrak{so}_n(\mathbb{R})$ (see (I2)) will follow after completing such a Poisson structure to a bi-Hamiltonian representation of (6.27). With such a bi-Hamiltonian structure at hand, we will be able to prove that the Manakov integrals satisfy the Lenard-Magri relations, see Proposition 5.39.

3.2 Bi-Hamiltonian Structure

We start our discussion recalling that if A is a real, symmetric, $n \times n$ matrix, then $[\cdot, \cdot]_A : \mathfrak{so}_n(\mathbb{R}) \otimes \mathfrak{so}_n(\mathbb{R}) \to \mathfrak{so}_n(\mathbb{R})$ defined by $[x, y]_A = xAy - yAx$, $\forall x, y \in \mathfrak{so}_n(\mathbb{R})$ is a Lie bracket, whose corresponding Lie-Poisson structure can be written as

$$\{f, g\}_A(x) = B\left(x, [\nabla_x f, \nabla_x g]_A\right)$$

$$= -\frac{1}{2}\operatorname{tr}\left(x[\nabla_x f, \nabla_x g]_A\right), \ \forall x \in \mathfrak{so}_n(\mathbb{R}), \ f, g \in C^\infty(\mathfrak{so}_n(\mathbb{R}))$$

(see Examples 5.35 and 2.8).

As promised, we can now present the sought bi-Hamiltonian representation of (6.27).

Proposition 6.38 *The Euler vector field* (6.32) *is bi-Hamiltonian with respect to the pair* $(\Pi_{LP}, \Pi_{\Upsilon^2})$ *where* Π_{LP} *is the linear Poisson structure, while* Π_{Υ^2} *is the* frozen *one with constant* Υ^2.

Proof The sharp maps of Π_{LP} and Π_{Υ^2} (see 2.24) are, respectively,

$$\Pi^\sharp_{LP,M}(\nabla_M H) = [M, \nabla_M H] \quad \text{and} \quad \Pi^\sharp_{\Upsilon^2,M}(\nabla_M H) = M\nabla_M H\Upsilon^2 - \Upsilon^2\nabla_M HM, \tag{6.40}$$

$\forall M \in \mathfrak{so}_n(\mathbb{R})$, $H \in C^\infty(\mathfrak{so}_n(\mathbb{R}))$ (see Formula (2.10); below, the reader is asked to prove (6.40)). Choosing H as in (6.29), (6.31) implies that $\nabla_M H = \Omega$ (see 6.36), which plugged in the first of the (6.40) yields

$$\Pi^\sharp_M(\nabla_M H) = [M, \nabla_M H] = \mathscr{E}(M).$$

We claim that if $K \in C^\infty(\mathfrak{so}_n(\mathbb{R}))$ is defined by

$$K(M) = \frac{1}{4}\operatorname{tr}\left(\Upsilon^{-1}M\Upsilon^{-1}\Psi_\Upsilon^{-1}(M)\right), \tag{6.41}$$

then $\mathcal{E}(M) = \Pi^{\sharp}_{\Upsilon^2,M}(\nabla_M K)$. In fact

$$
\begin{aligned}
\nabla_M K(A) \;&=\; \frac{d}{ds}\Big|_{t=0} K(M+tA)\\[2mm]
&\overset{(6.41)}{=} \frac{1}{4}\,\mathrm{tr}\left(\Upsilon^{-1}M\Upsilon^{-1}\Psi_{\Upsilon}^{-1}(A) + \Upsilon^{-1}\Psi_{\Upsilon}^{-1}(M)\Upsilon^{-1}A\right)\\[2mm]
&\overset{(6.31)}{=} \frac{1}{4}\,\mathrm{tr}\left(\Upsilon^{-1}\Psi_{\Upsilon}^{-1}(M)\Upsilon^{-1}A + \Psi_{\Upsilon}^{-1}(\Upsilon^{-1}M\Upsilon^{-1})A\right)\\[2mm]
&\overset{(6.30)}{=} \frac{1}{2}\,\mathrm{tr}\left(\Upsilon^{-1}\Psi_{\Upsilon}^{-1}(M)\Upsilon^{-1}A\right),
\end{aligned}
$$

which entails that

$$
\nabla_M K = -\Upsilon^{-1}\Psi_{\Upsilon}^{-1}(M)\Upsilon^{-1}, \quad \forall M \in \mathfrak{so}_n(\mathbb{R}).
$$

Plugging this result in the second expression in (6.40), one gets

$$
\begin{aligned}
\Pi^{\sharp}_{\Upsilon^2,M}(\nabla_M K) \;&=\; \Upsilon\Psi_{\Upsilon}^{-1}(M)\Upsilon^{-1}M - M\Upsilon^{-1}\Psi_{\Upsilon}^{-1}(M)\Upsilon\\[2mm]
&\overset{(6.28)}{=} \Upsilon\Omega\Upsilon^{-1}M - M\Upsilon^{-1}\Omega\Upsilon\\[2mm]
&\overset{(6.28)}{=} \Upsilon\Omega^2 - \Omega^2\Upsilon.
\end{aligned}
$$

Finally, using (6.28) one more time, one arrives to

$$
\mathcal{E}(M) = \Pi^{\sharp}_{\Upsilon^2,M}(\nabla_M K),
$$

which concludes the proof. $\square$

Problem 6.39 Prove Formula (6.40). (Hint: see Example 2.8.)

$\triangle$

We conclude showing that the Manakov integrals (see Definition 6.37) satisfy the Lenard-Magri relations with respect to the Poisson pair $(\Pi_{LP}, \Pi_{\Upsilon^2})$. To this end, first we observe that for all $k \geq 1$

$$
\nabla_M f_{\xi,k} = (-M + \Upsilon^2\xi)^{k-1} - (M + \Upsilon^2\xi)^{k-1}. \tag{6.42}
$$

Problem 6.40 Prove (6.42). (Hint: it is useful to observe that $\mathrm{tr}(GM) = \frac{1}{2}\,\mathrm{tr}(G - G^t)M$, for all $G \in \mathfrak{gl}(n)$ and all $M \in \mathfrak{so}_n(\mathbb{R})$.)

$\triangle$

Then we note that Formula (6.42) implies the following.

Proposition 6.41 *For all $k \geq 1$ and all $M \in \mathfrak{so}_n(\mathbb{R})$*

$$\xi \Pi^{\sharp}_{\Upsilon^2, M}(\nabla_M f_{\xi,k}) = \Pi^{\sharp}_{LP, M}(\nabla_M f_{\xi,k+1}). \tag{6.43}$$

Proof Using (6.42) together with (6.40), one sees that proving this statement amounts to show that

$$\xi M(-M + \Upsilon^2\xi)^{k-1}\Upsilon^2 - \xi M(M + \Upsilon^2\xi)^{k-1}\Upsilon^2 - \xi\Upsilon^2(-M + \Upsilon^2\xi)^{k-1}M$$

$$+ \xi\Upsilon^2(M + \Upsilon^2\xi)^{k-1}M = [M, (-M + \Upsilon^2\xi)^k - (M + \Upsilon^2\xi)^k].$$

To prove this identity, it suffices to observe that its right-hand side can be written as follows:

$$M(-M + \Upsilon^2\xi)^{k-1}(-M + \Upsilon^2\xi) - M(M + \Upsilon^2\xi)^{k-1}(M + \Upsilon^2)$$

$$- (-M + \Upsilon^2\xi)(-M + \Upsilon^2\xi)^{k-1}M + (M + \Upsilon^2\xi)(M + \Upsilon^2\xi)^{k-1}M$$

which, expanded further, gives

$$\xi M(-M + \Upsilon^2\xi)^{k-1}\Upsilon^2 - \xi M(M + \Upsilon^2\xi)^{k-1}\Upsilon^2 - \xi\Upsilon^2(-M + \Upsilon^2\xi)^{k-1}$$

$$M + \xi\Upsilon^2(M + \Upsilon^2\xi)^{k-1}M,$$

which is what we wanted to prove. $\qquad\square$

After all these preliminary observations, it will be easy to prove the following.

Theorem 6.42 *The Manakov integrals satisfy the Lenard-Magri relations, i.e., for all $k \geq 1$ and $M \in \mathfrak{so}_n(\mathbb{R})$,*

$$\Pi^{\sharp}_{\Upsilon^2, M}(\nabla_M h_{k,l}) = \Pi^{\sharp}_{LP, M}(\nabla_M h_{k+1,l}), \; \forall l. \tag{6.44}$$

Moreover, $h_{k,k}$ is a Casimir of Π_{LP} (see also the comment below Formula (6.39)).

Proof It suffices to apply (6.43) where $f_{k,\xi}$ is as in (6.39). Note that $\Pi^{\sharp}_{LP,M}(\nabla_M h_{k,k}) = 0$ for all k, proving the last statement. $\qquad\square$

In particular, the previous result has the following.

Corollary 6.43 *The Manakov integrals are in involution with respect to both $\{\cdot, \cdot\}_{LP}$ and $\{\cdot, \cdot\}_{\Upsilon^2}$.*

Proof Despite the fact that this statement follows from the general theory of the bi-Hamiltonian structures (see Sect. 3.1 in Chap. 5), we will present its proof below. Since the Lenard-Magri identities (6.44) imply that

$$\{h_{k,l}, h_{s,r}\}_{\Upsilon^2} = \{h_{k+1,l}, h_{s,r}\}_{LP}, \tag{6.45}$$

for all k, l, s, r, to conclude, it suffices to prove that the $h_{k,l}$s are in involution with respect to $\{\cdot, \cdot\}_{LP}$. Since $\{h_{1,l}, h_{r,s}\}_{kk} = 0$ and $\{h_{k,k}, h_{r,s}\}_{LP} = 0$ for all r, s, let $k \geq 2$ and $l < k$. In this case,

$$\{h_{k,l}, h_{r,s}\}_{LP} \overset{(6.45)}{=} -\{h_{r,s}, h_{k-1,l}\}_{\Upsilon^2} \overset{(6.45)}{=} -\{h_{r+1,s}, h_{k-1,l}\}_{LP}.$$

If $l = k - 1$, then $\{h_{k,l}, h_{r,s}\}_{LP} = 0$. On the contrary, one can iterate the previous computation until this condition will be satisfied. $\square$

Now we aim to prove that the Manakov integrals are in involution with (6.29), with respect to the Poisson structure induced by Π_{LP} on $\mathscr{O}$ (see point (I3)). We show that this result can be obtained as a consequence of a more general one, which, being of independent interest in the theory of the rigid body, will be presented below.

3.3   Mishchenko Integrals

Up to this point, it was shown that (6.27) admits a bi-Hamiltonian representation, with respect Π_{LP} and Π_{Υ^2} and that the Manokov integrals are in involution with respect to both these Poisson structures, i.e., that they form a bi-Hamiltonian hierarchy. These observations are not sufficient to conclude that (6.27) is completely integrable in the sense of Liouville. In fact, it is easily checked that (6.29) does not belong to the Manakov hierarchy, and for this reason, nobody can guarantee that it Poisson commutes with the Manakov integrals. To save the day, we will introduce another hierarchy of functions on $\mathfrak{so}_n(\mathbb{R})$, the so-called (extended) Mishchenko hierarchy, having, among many others, the following three properties:

  (i) It contains (6.29).
 (ii) Its elements Poisson commute with respect to both Π_{LP} and Π_{Υ^2}.
(iii) Its elements commute with the Manakov integrals.

We start with the following.

Definition 6.44 The Mishchenko integrals of order $r \in \mathbb{Z}$ are the functions $m_r \in C^\infty(\mathfrak{so}_n(\mathbb{R}))$ defined by

$$m_r(M) = \begin{cases} -\frac{1}{4} \operatorname{tr} \left(\sum_{s=0}^{r-1} \Upsilon^s M \Upsilon^{r-s-1} \Omega \right) & \text{if} \quad r \geq 1, \\ 0 & \text{if} \quad r = 0, \\ \frac{1}{4} \operatorname{tr} \left(\sum_{s=r}^{-1} \Upsilon^s M \Upsilon^{r-s-1} \Omega \right) & \text{if} \quad r \leq -1. \end{cases}$$

where, as usual, $\Omega = \Psi_{\Upsilon}^{-1}(M)$. $\triangle$

Note that m_1 is (6.29), while m_{-1} is (6.41). Moreover,

Lemma 6.45 *The Mishchenko integrals of even and positive order are, up to a constant factor, Manakov integrals.*

Problem 6.46 Prove the statement of this lemma.   △

To move to the second point above, one needs to compute the gradient of m_r at the generic point M. The result of this computation is the content of the following.

Lemma 6.47

$$\nabla_M m_r = \begin{cases} \sum_{s=0}^{r-1} \Upsilon^{r-1-s} \Psi_\Upsilon^{-1}(M) \Upsilon^s & \text{if } r \geq 1, \\ 0 & \text{if } r = 0, \\ -\sum_{s=r}^{-1} \Upsilon^{s-r-1} \Psi_\Upsilon^{-1}(M) \Upsilon^s & \text{if } r \leq -1. \end{cases} \tag{6.46}$$

Proof The proof consists in the computation of $\nabla_M m_r(A) = \frac{d}{ds}\big|_{t=0} m_r(M + tA)$, and we will not present it here since it goes along the same lines of the proof Proposition 6.38. Nevertheless, it is worth noting that the proof uses the following identity:

$$\Psi_\Upsilon^{-1}(\Upsilon^k M \Upsilon^n) = \Upsilon^k \Psi_\Upsilon^{-1}(M) \Upsilon^n, \ \forall n, k \in \mathbb{Z}.$$

□

Then one can prove the following.

Proposition 6.48 *The Mishchenko integrals satisfy the following recursion relations:*

$$\Pi_{\Upsilon,M}^\sharp(\nabla_M m_r) = \Pi_{LP,M}^\sharp(\nabla_M m_{r+2}), \ \forall M \in \mathfrak{so}_n(\mathbb{R}), \ r \in \mathbb{Z}. \tag{6.47}$$

Proof Following [187], we divide the proof in five cases: $r \leq -3, r = -2, r = -1$, $r = 0$, and $r \geq 1$. The case $r = -1$ was proven in Proposition 6.38. If $r = -2$, we need to prove $\Pi_{\Upsilon,M}^\sharp(\nabla_M m_{-2}) = 0$, for all M. Using (6.46), one sees that

$$\nabla_M m_{-1} = -\Upsilon^{-2} \Psi_\Upsilon^{-1}(M) \Upsilon^{-1} - \Upsilon^{-1} \Psi_\Upsilon^{-1}(M) \Upsilon^{-1}$$

$$= -\Upsilon^{-2}(\Psi_\Upsilon^{-1}(M) \Upsilon + \Upsilon \Psi_\Upsilon^{-1}(M)) \Upsilon^{-2}$$

$$= -\Upsilon^{-2} M \Upsilon^{-2},$$

which, inserted in the second formula of (6.40), gives what is required. The case $r = 0$ is treated similarly to the previous one. More precisely, one sees that for $r = 0$, one needs to show that $\Pi_{LP,M}^\sharp(\nabla_M m_2) = 0$ for all M. This follows from the first of (6.40) together with the fact that $(\nabla_M m_2) = M$. The last two cases, $r \geq 1$ and $r \leq -3$, can be analyzed in similar way and will be left to the reader as an exercise.   □

Problem 6.49 Complete the proof of the previous proposition showing that the identity (6.47) holds also for $r \geq 1$ and $r \leq -3$.   △

Let $\{\cdot, \cdot\}_{LP}$ and $\{\cdot, \cdot\}_{\Upsilon^2}$ be the Poisson brackets defined by Π_{LP} and, respectively, Π_{Υ^2}. Then, as consequence Proposition 6.48, one can prove the following.

Corollary 6.50 *For all $n, k \in \mathbb{Z}$,*

$$\{m_n, m_k\}_{LP} = 0 \quad and \quad \{m_n, m_k\}_{\Upsilon^2} = 0.$$

Proof Modulo a minor difference, the proof goes as the one of the Lenard-Magri recursion relations in Proposition 5.39. More precisely, if m and k have the same parity, one argues as in Proposition 5.39, while if they have different parities, together with the argument in Proposition 5.39, one uses $m_0 = 0$.   □

Problem 6.51 Fill in the details of the proof of the previous corollary.   △

Going back to the Manakov integrals, now one can prove the following.

Proposition 6.52 *The Manakov integrals are in involution with the Mishchenko integrals with respect to the Poisson brackets $\{\cdot, \cdot\}_{LP}$ and $\{\cdot, \cdot\}_{\Upsilon^2}$, i.e.,*

$$\{h_{k,l}, m_r\}_{LP} = 0 \quad and \quad \{h_{k,l}, m_r\}_{\Upsilon^2} = 0, \tag{6.48}$$

for all $r \in \mathbb{Z}$, $k \geq 1$, and $0 \leq l \leq k$.

Proof If r is even and positive, there is nothing to prove (see Lemma 6.45). Since (6.44) and (6.47) imply, respectively,

$$\{h_{k,l}, m_r\}_{\Upsilon^2} = \{h_{k+1,l}, m_r\}_{LP} \quad and \quad \{m_r, h_{k,l}\}_{\Upsilon^2} = \{m_{r+2}, h_{k,l}\}_{LP} \tag{6.49}$$

for all k, l, r as above, to conclude, it suffices to prove (6.48) for $\{\cdot, \cdot\}_{LP}$. Note that for $l = 0$ and $l = k$, there is nothing to prove, since $h_{k,0}$ is constant, while $h_{k,k}$ is a Casimir of Π_{LP}. Assuming $k > 1$ and $l \leq k - 1$, skew-symmetry of the Poisson brackets and (6.49) imply

$$\{h_{k,l}, m_r\}_{LP} = \{h_{k-1,l}, m_{r+2}\}_{LP}, \tag{6.50}$$

which, since $h_{k-1,k-1}$ is a Casimir of Π_{LP}, concludes the proof if $l = k - 1$. Analogously, one can iterate this process as many times it is necessary, so that the first index of h becomes equal to l.   □

Problem 6.53 Prove (6.50).   △

Finally, one can state the following.

Corollary 6.54 *The Manakov integrals are first integrals for (6.29).*

Proof As it was already noticed, (6.29) is the Mishchenko integral m_2, which, because of Proposition 6.52, Poisson commutes, with respect to both $\{\cdot, \cdot\}_{LP}$ and $\{\cdot, \cdot\}_\Upsilon$, with $h_{k,l}$ for all k, l. $\qquad\qquad\qquad\qquad\qquad\square$

4 Concluding Remarks and Further Topics

In these last subsections, we collect a few more information about the case of the three-dimensional rigid body.

4.1 The Three-Dimensional Case

The case $n = 3$ is, for obvious reasons, particularly important (see, e.g., [10]). In this case, the isomorphism (C.1) (see Example C.11 in Appendix C) identifies the (Lie algebra) of the configuration space of the rigid body with the three-dimensional Euclidean space endowed with the Lie algebra structure defined by the cross-product. Under this isomorphism, the adjoint orbits of $\mathfrak{so}_3(\mathbb{R})$ correspond to two-dimensional spheres in $\mathbb{R}^3$ centered at the origin (see Example C.75), and (6.28) corresponds to the linear map $\Psi_I : \mathbb{R}^3 \to \mathbb{R}^3$ defined by

$$\Psi_I(v) = I v, \ \forall v \in \mathbb{R}^3,$$

where $I = \mathrm{diag}(I_1, I_2, I_3)$ with $I_1 = \lambda_2 + \lambda_3$, $I_2 = \lambda_1 + \lambda_3$, $I_3 = \lambda_1 + \lambda_2$ and the right-hand side of the previous formula denotes the standard multiplication of matrix by a (column) vector. The linear map Ψ_I and the I_is so defined are classically known as the inertia tensor and, respectively, the *principal moments of inertia* of $\mathscr{B}$. In particular, if $\omega = (\omega_1, \omega_2, \omega_3) \in \mathbb{R}^3$ is the vector corresponding to $\Omega \in \mathfrak{so}_3(\mathbb{R})$ via (C.1), then $m = I\omega$, i.e., $m_i = I_i\omega_i$, for all $i = 1, 2, 3$, where $M = (m_1, m_2, m_3)$. In this way, (6.29) becomes, in the *moment-space* coordinates

$$H(m) = \frac{1}{2}\langle m, I^{-1}m \rangle = \frac{1}{2}\sum_{i=1}^{3} \frac{m_i^2}{I_i}, \tag{6.51}$$

or, equivalently, in the *angular-velocities* coordinates

$$H(\omega) = \frac{1}{2}\langle I\omega, \omega \rangle = \frac{1}{2}\sum_{i=1}^{3} I_i\omega_i^2.$$

In the moment-space coordinates, the Euler vector field is $\mathscr{E}(m) = m \times I^{-1}m$, and the corresponding Lax equations become

$$\begin{cases} \dot{m}_1 = \dfrac{I_2 - I_3}{I_2 I_3} m_2 m_3 \\[2mm] \dot{m}_2 = \dfrac{I_3 - I_1}{I_1 I_3} m_1 m_3 \\[2mm] \dot{m}_3 = \dfrac{I_1 - I_2}{I_1 I_2} m_1 m_2. \end{cases} \tag{6.52}$$

The same equations, in the angular-velocity representation, become

$$\begin{cases} I_1 \dot{\omega}_1 = (I_2 - I_3)\omega_2 \omega_3 \\ I_2 \dot{\omega}_2 = (I_3 - I_1)\omega_1 \omega_3 \\ I_3 \dot{\omega}_3 = (I_1 - I_2)\omega_1 \omega_2. \end{cases} \tag{6.53}$$

Definition 6.55 (Euler Rigid Body)   Equations (6.52) and (6.53) are called the *Euler rigid-body equations*. They describe the motion of a (three-dimensional) rigid body rotating about its center of mass, in absence of external forces.   △

In the three-dimensional case, the proof of the complete integrability of (6.27) is much simpler, and it follows simply observing that the Euler vector field representing (6.27) is tangent to the (co)adjoint orbits of $SO_3(\mathbb{R})$, which are (diffeormorphic) to spheres centered at the origin of $\mathbb{R}^3$. Note that the spheres are the level sets of the *momentum integral*, $m_1^2 + m_2^2 + m_3^2$, while the integral curve of $\mathscr{E}$ is obtained as the intersection between these surfaces with ellipsoids representing the level sets of the Hamiltonian function (6.51) (Fig. 6.4).

After these observations, we make a few comments about the explicit solutions of (6.53). There are three cases of increasing complexity one should consider. The first, and the easiest one, is when $I_1 = I_2 = I_3$. In this case, the $\dot{\omega}_i = 0$ for all $i = 1, 2, 3$, whose solutions is $\omega_i(t) = \omega_i(0)$, for all i. The intermediate case is when two of the principal moments are equal and different from the third one. For

Fig. 6.4 Integral curves of $\mathscr{E}$

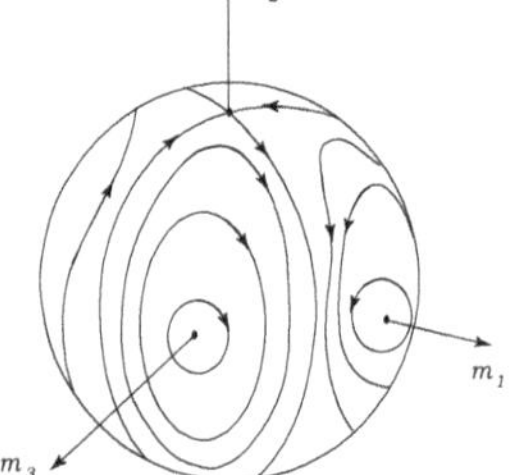

example, $I_1 = I_2 = \lambda$, and $I_3 = \mu \neq \lambda$. In this case, (6.53) reduces to

$$\begin{cases} \lambda\dot{\omega}_1 = (\lambda - \mu)\omega_2\omega_3 \\ \lambda\dot{\omega}_2 = (\mu - \lambda)\omega_1\omega_3 \\ \dot{\omega}_3 = 0. \end{cases} \Rightarrow \begin{cases} \dot{\omega}_1 = \kappa\omega_2 \\ \dot{\omega}_2 = -\kappa\omega_1 \end{cases}$$

where $\kappa = \frac{\omega_3(0)(\lambda-\mu)}{\lambda}$. This linear system is readily integrated, and its solutions are

$$\begin{cases} \omega_1(t) = \omega_{1,0}\cos(\kappa t) + \omega_{2,0}\sin(\kappa t) \\ \omega_2(t) = -\omega_{1,0}\sin(\kappa t) + \omega_{2,0}\cos(\kappa t), \end{cases}$$

where $\omega_{i,0} = \omega_i(0)$, for $i = 1, 2$; in other words, under the above assumptions, ω_3 is constant, while ω_1 and ω_2 *move around* the axis determined by ω_3 with an angular velocity equal to κ. This motion is called *precession*. The third and more complicated case needs the introduction of so-called Jacobi elliptic functions (see Sect. 4.3). In this case, supposing $I_3 > I_2 > I_1$, a solution of (6.53) is obtained from (6.57), writing

$$\omega_1 = C_1\mathrm{cn}(vt, k), \quad \omega_2 = C_2\mathrm{sn}(vt, k) \quad \text{and} \quad \omega_3 = C_3\mathrm{dn}(vt, k),$$

where C_1, C_2, C_3, v, k are constants (k is the modulus of the elliptic functions), which will be determined in terms of the physical constants I_1, I_2, I_3, H, m^2.

4.2 Heavy Tops

In this subsection, complementing Example 1.124, we will briefly comment on the integrability of the so-called spinning tops, i.e., a (three-dimensional) rigid body with a fixed point moving in a gravitational field. Since there are several and very detailed presentations of this topic available in the literature, below we will just summarize, somehow superficially, the content of [123] (see also [114], to which we refer the reader for the missing details). The Lie algebra behind the geometric mechanics of the heavy tops is $\mathfrak{e}(3)$, the three-dimensional Euclidean Lie algebra (see Example 2.71). Recall that $\mathfrak{e}(3) = \mathfrak{so}_3(\mathbb{R}) \ltimes \mathbb{R}^3$, which we identify with $\mathbb{R}^3 \ltimes \mathbb{R}^3$ (see Example C.11). After choosing a basis (in both copies) of $\mathbb{R}^3$, one disposes of coordinates that, according with Example 2.71, will be denoted as $(m_1, m_2, m_3, f_1, f_2, f_3)$. Written in these coordinates, the Hamiltonian function of a spinning top assumes the following form:

$$H(m, f) = \frac{1}{2}\sum_{i=1}^{3}\frac{m_i^2}{I_i} + \sum_{i=1}^{3}x_i f_i, \tag{6.54}$$

where $I_i > 0$ for all i and (x_1, x_2, x_3) is a vector in $\mathbb{R}^3$. Written in a more invariant way, the previous formula becomes

$$H(m, f) = \frac{1}{2}\langle m, I^{-1}m \rangle + \langle f, x \rangle,$$

where $\langle \cdot, \cdot \rangle$ denotes the Euclidean inner product of $\mathbb{R}^3$ and I is the diagonal matrix (I_1, I_2, I_3). These formulas should be compared with (6.51). Using (2.64), one can write the Hamiltonian vector field corresponding to (6.54)

$$\Pi^\sharp(dH) = \begin{bmatrix} 0 & -m_3 & m_2 & 0 & -f_3 & f_2 \\ m_3 & 0 & -m_1 & f_3 & 0 & -f_1 \\ -m_2 & m_1 & 0 & -f_2 & f_1 & 0 \\ 0 & -f_3 & f_2 & 0 & 0 & 0 \\ f_3 & 0 & -f_1 & 0 & 0 & 0 \\ -f_2 & f_1 & 0 & 0 & 0 & 0 \end{bmatrix} \begin{bmatrix} \frac{m_1}{I_1} \\ \frac{m_2}{I_2} \\ \frac{m_3}{I_3} \\ x_1 \\ x_2 \\ x_3 \end{bmatrix}$$

$$= \begin{bmatrix} \left(\frac{I_2-I_3}{I_2 I_3}\right)m_2 m_3 + f_2 x_3 - f_3 x_2 \\ \left(\frac{I_3-I_1}{I_1 I_3}\right)m_1 m_3 + f_3 x_1 - f_1 x_3 \\ \left(\frac{I_1-I_2}{I_1 I_2}\right)m_1 m_2 + f_1 x_2 - f_2 x_1 \\ \frac{m_3 f_2}{I_3} - \frac{m_2 f_3}{I_2} \\ \frac{m_1 f_3}{I_1} - \frac{m_3 f_1}{I_3} \\ \frac{m_2 f_1}{I_2} - \frac{m_1 f_2}{I_1} \end{bmatrix}, \tag{6.55}$$

in other words, written in the coordinates $(m_1, m_2, m_3, f_1, f_2, f_3)$, the Hamiltonian equations of (6.54) with respect to the linear Poisson structure of $\mathfrak{e}(3)$ are

$$\left\{ \begin{aligned} \dot{m}_1 &= \frac{I_2 - I_3}{I_2 I_3} m_2 m_3 + f_2 x_3 - f_3 x_2 \\ \dot{m}_2 &= \frac{I_3 - I_1}{I_1 I_3} m_1 m_3 + f_3 x_1 - f_1 x_3 \\ \dot{m}_3 &= \frac{I_1 - I_2}{I_1 I_2} m_1 m_2 + f_1 x_2 - f_2 x_1 \\ \dot{f}_1 &= \frac{m_3 f_2}{I_3} - \frac{m_2 f_3}{I_2} \\ \dot{f}_2 &= \frac{m_1 f_3}{I_1} - \frac{m_3 f_1}{I_3} \\ \dot{f}_3 &= \frac{m_2 f_1}{I_2} - \frac{m_1 f_2}{I_1}, \end{aligned} \right.$$

which can be written as

$$\left\{ \begin{aligned} \dot{m} &= m \times I^{-1}m + f \times x \\ \dot{f} &= f \times I^{-1}m, \end{aligned} \right. \tag{6.56}$$

where f and m are (columns) vectors in $\mathbb{R}^3$, whose components are the f_is and, respectively, the m_is, while I is, again, a diagonal 3×3 matrix, whose diagonal elements are the I_is. It is probably worthwhile to make a few comments about the physical meaning of the ingredients present in the system (6.56). First of all, all the quantities involved are measured in a *body frame*, a reference system at rest relatively to the body. The vector x is a constant vector measuring the distance between the origin of the chosen body reference system and the center of mass of the latter. This is a constant vector (relative to the chosen system of coordinates). The vector f measures the strength of the gravitational field. With respect to the non-inertial system of coordinates, this is (in general) not a constant vector. Finally, m is angular momentum with respect to the body frame, the same as it appears in (6.51) and $I = \mathrm{diag}(I_1, I_2, I_3)$, with $I_i > 0$ as the inertia tensor. The latter is defined by the mass distribution of the body, and it defines, as in the free case, a non-degenerate quadratic form that identifies the space of the momenta with the one of the angular velocities, $\omega = I^{-1}m$. As far as the integrability of (6.56) is concerned, note that these equations define a dynamical system on the (co)adjoint orbits of $\mathfrak{e}(3)$, of which the generic ones are four-dimensional, given as the zero set of the two quadratic Casimirs $C_1 = \sum_{i=1}^{3} f_i^2$ and $C_2 = \sum_{i=1}^{3} f_i m_i$ (see Example 2.71). This implies that the integrability of (6.56) will follow from the existence of a first integral of (6.55) independent of H. As it was already mentioned in Example 1.124, this additional integral is known in only three cases, which we spell out (again) below.

4.2.1 Euler-Poinsot Case

In this case, the center of mass of the rigid body coincides with the origin of the non-inertial system of coordinates. In this way, the system behaves like a free rigid body, whose integrability was already discussed above. As already noticed at the end of Sect. 4.1, the second first integral is provided by the square of the angular momentum, $m_1^2 + m_2^2 + m_3^2$. Note that in this case, since $x = 0$, the first equation in (6.56) misses the term $f \times x$.

4.2.2 Lagrange Case

This is traditionally known as the symmetric top. In this case, there exists a principal reference frame at rest with the body whose principal moments are $I = (\lambda, \lambda, \mu)$. Moreover, the axis of revolution lies along the line connecting the origin of the reference frame to center of mass of the system. Choosing $x_1 = 0 = x_2$ then $K = m_3$ is a second independent integral. In fact,

$$\{H, m_3\} = \langle dm_3, X_H \rangle \overset{(6.55)}{=} 0,$$

proving the integrability of the Lagrange top.

4.2.3 Kowalevski Case

In this case, there exists a principal frame at rest with the body, such that the inertia tensor is $I = \mathrm{diag}(2\lambda, 2\lambda, \lambda)$, and the center of mass has coordinates $(\kappa, 0, 0)$. In this case, $K = (m_1^2 - m_2^2 + 4\lambda\kappa f_1)^2 + (2m_1 m_2 + 4\lambda\kappa f_2)^2$ turns out to be a first integral for the system as a direct computation, using again (6.55), shows that $\{H, K\} = 0$, proving the integrability of the system.

Remark 6.56 In fluid dynamics, the (6.56) are called the *Kirchhoff* equations. They describe the motion of a rigid body immersed in an ideal, incompressible fluid whose motion is potential and is at rest at infinity. As the Euler equation of the free rigid body is the finite-dimensional analogue of the Euler equations for the ideal fluid dynamics, the Kirchhoff equations are the finite-dimensional analogue of the so-called magneto-hydrodynamics equations (see [12] for more information about these, and many more, beautiful topics). △

4.3 Jacobi Elliptic Functions

In this short subsection, we sketch an introduction to the Jacobi elliptic functions. Here we follow the nice presentation [182]. Let $0 < k < 1$. The *Jacobi elliptic functions*

$$\mathrm{sn}(t, k), \mathrm{cn}(t, k), \mathrm{dn}(t, k) \tag{6.57}$$

are the solutions of the following system of non-linear ODE

$$\left\{ \begin{array}{l} \dot{x} = yz \\ \dot{y} = -xz \\ \dot{z} = -k^2 xy, \end{array} \right\}. \tag{6.58}$$

with initial condition $(0, 1, 1)$ for $t = 0$. A simple calculation shows that the functions $x^2 + y^2$ and $k^2 x^2 + z^2$ are *first integrals* of the system (6.58) and, consequently, that

$$\mathrm{sn}^2(t, k) + \mathrm{cn}^2(t, k) = 1 \quad \text{and} \quad k^2 \mathrm{sn}^2(t, k) + \mathrm{dn}^2(t, k) = 1, \; \forall t \in \mathbb{R}.$$

From these identities, it follows that

$$|\mathrm{sn}(t, k)| \le 1, \quad |\mathrm{cn}(t, k)| \le 1 \quad \text{and} \quad \kappa = \sqrt{1 - k^2} \le \mathrm{dn}(t, k) \le 1.$$

Furthermore, it can be shown that the Jacobi elliptic functions are real analytic functions of t and k. The parameters k and κ are called the *modulus* and, respectively, the *complementary modulus*. As k approaches 0 from the *right*,

$sn(t, k) \to \sin(t)$, $cn(t, k) \to \cos(t)$, and $dn(t, k) \to 1$, while as it approaches 1 from the *left*, $sn(t, k) \to \tanh(t)$, $cn(t, k) \to \cos(t)$, and $dn(t, k) \to \operatorname{sech}(t)$, where $\tanh(t) = \frac{e^t - e^{-t}}{e^t + e^{-t}}$ and $\operatorname{sech}(t) = \frac{2}{e^t + e^{-t}}$ are the hyperbolic *tangent* and, respectively, *secant*. In other words, the Jacobi elliptic functions, as k changes between 0 and 1, interpolate between the trigonometric and the hyperbolic world.

Finally, calling $x(t) = sn(t, k)$, from the (6.58), one can deduce that $\frac{dt}{dx} = \frac{1}{\sqrt{(1 - x^2)(1 - k^2 x^2)}} > 0$, for all $x \in (-1, 1)$. Then

$$t(x) = \int_0^x \frac{dy}{\sqrt{(1 - y^2)(1 - k^2 y^2)}}$$

is the smooth inverse of the function $sn(x, t)$ on the interval $(-1, 1)$. Since t is finite at $x = 1$ and $x = -1$ and $t(\pm 1) = \pm K$, $t(x)$ is a continuous function on the closed interval $[-1, 1]$.

Note that for $k = 0$, the previous integral is

$$\int_0^x \frac{dy}{\sqrt{1 - y^2}} = \arcsin x,$$

i.e., the inverse function of $\sin x$.

5 Bibliographical Notes

This chapter was devoted to the free rigid-body dynamics with emphasis on some of its bi-Hamiltonian aspects. The analysis of the rigid-body motion is a very classical topic, going back to the Euler's *Theoria motus corporum solidorum seu rigidorum*, published in 1760 (see [70]). For a general introduction to the foundations of mechanics, where plenty of information about the dynamics of the rigid body can also be found, we suggest the beautiful presentation of Trusdell [237]. For what concerns the first and the second sections of our exposition, we refer to [1, 158, 175] where the reader will find more information about the use of the moment map for the reconstruction of the dynamics. Other useful references for this topic are [113, 114, 123]. For more information, bending toward the theory of non-linear evolution equation, about the Lax representation, we suggest the review paper [151]. Our description of the dynamics of the n-dimensional rigid body was borrowed by [211] (see also [1, 175]). On the other hand, for a very nice and readable introduction to the relation between the rigid-body dynamics and many aspects of fluid mechanics, we strongly suggest the monograph [12] (see also [9, 10]). Finally, coming to our presentation of the bi-Hamiltonian description of the rigid body, our main reference was [187]. For more information about this topic, we also suggest to consult the original papers [35, 170, 184]. Finally, for a more algebro-geometric presentation of the integrability of the rigid body, we suggest [19], while for a more differential geometric perspective, we refer to [59].

Chapter 7
The Toda System

This chapter aims to be an introduction to the so-called Toda system, or Toda *lattice*, a dynamical system consisting of particles moving under a nearest-neighbor potential of exponential type.

The Toda system is named after the physicist Morikazu Toda, who introduced it in 1967 in a seminal paper where he proved that the chains of particles moving under nearest-neighbor potential of exponential-type exhibit solution of soliton type, even though already in 1954, Heinz Rutishauser constructed a continuous analogue of the quotient-difference algorithm in numerical linear algebra that contains the finite non-periodic Toda flow as a special case, essentially already written in the Flaschka's coordinates. (For more on this and the beautiful relations with algorithms in numerical linear algebra, see Sect. 5.1 of this chapter.)

Since then, this dynamical system underwent many generalization, and nowadays, it is difficult to underestimate its importance because of the central role it plays in many areas of pure and applied mathematics, from representation theory to numerical analysis and mathematical physics. The Toda system has two main incarnations, the *open* and the *closed* or *periodic* one. The difference between the two hinges on the boundary conditions prescribed on them, and it can be visualized, by saying that in the open case, the particles lie on a line, while in the closed one, they are disposed on a circle. In these notes, we will be concerned only with the open case. After introducing this system in a more formal way, we will show that it admits a Lax representation, and then we will prove its complete integrability. The final section will be devoted to a few results, making a first connection between this system and the theory of Lie groups and Lie algebras.

A. Arsie, I. Mencattini, *Geometry of Integrable Systems*, Latin American Mathematics Series – UFSCar subseries,
https://doi.org/10.1007/978-3-031-96282-0_7

1   The Open Toda System

As it was previously mentioned, the open Toda lattice is a dynamical system consisting of n particles moving on a line under the potential

$$U(x) = \sum \exp(x_i - x_{i+1}). \tag{7.1}$$

In the above formula, x_i denotes the position of the i-th particle, and one supposes that the initial conditions satisfy $x_1 < x_2 < \cdots < x_n$. In this idealized model, the particles tend to move away from each other. In fact we can think of the Toda lattice as a mathematical model of a system of n particles connected by *springs*, pushing with a force proportional to $\exp(-x) - 1$, where x denotes the distance between two neighboring particles. The open Toda lattice is characterized by the following boundary conditions. We suppose that the system consists of $n + 2$ particles, numbered from 0 to $n + 1$, with the particles labelled by 0 and $n + 1$ placed at $-\infty$ and at $+\infty$, respectively. Under these assumptions, the potential function (7.1) becomes

$$U(x) = \overline{\exp(x_0 - x_1)} + \sum_{i=1}^{n-1} \exp(x_i - x_{i+1}) + \overline{\exp(x_n - x_{n+1})},$$

where the cancellations follow from the chosen boundary conditions. The corresponding Hamiltonian function assumes the following form:

$$H(p, x) = \frac{1}{2} \sum_{i=1}^{n} p_i^2 + \sum_{i=1}^{n-1} \exp(x_i - x_{i+1}) \tag{7.2}$$

yielding the following Hamiltonian equations:

$$\begin{cases} \dot{x}_k = p_k \\ \dot{p}_k = \exp(x_{k-1} - x_k) - \exp(x_k - x_{k+1}) \end{cases} \tag{7.3}$$

for $k = 1, \ldots, n$. For $k = 1$, these equations read

$$\begin{cases} \dot{x}_1 = p_1 \\ \dot{p}_1 = -\exp(x_1 - x_2), \end{cases}$$

while for $k = n$

$$\begin{cases} \dot{x}_n = p_n \\ \dot{p}_n = \exp(x_{n-1} - x_n), \end{cases}$$

due to the boundary conditions introduced above. In this way, the Hamiltonian vector field corresponding to (7.2) is

$$
X_H = \sum_{k=1}^{n} p_k \frac{\partial}{\partial x_k} + \sum_{k=1}^{n} \left(\exp(x_{k-1} - x_k) - \exp(x_k - x_{k+1}) \right) \frac{\partial}{\partial p_k}, \tag{7.4}
$$

where the boundary terms of the second sum need to be treated according to the previous observations. The (7.3) are equivalent to the following system of second-order equations:

$$
\ddot{x}_j = \exp(x_{j-1} - x_j) - \exp(x_j - x_{j+1}), \quad j = 1, \ldots, n \tag{7.5}
$$

whose solution describes the time evolution of the system on its configuration space. Once more, note that the boundary conditions $x_0 = -\infty$ and $x_{n+1} = +\infty$ yield

$$
\ddot{x}_1 = -\exp(x_1 - x_2) \quad \text{and} \quad \ddot{x}_n = \exp(x_{n-1} - x_n),
$$

for $j = 1$ and, respectively, $j = n$. One of the main goals of the present chapter is to show that the (7.3) is a completely integrable system. More precisely, we shall prove the following result [189–191]:

Theorem 7.1 (Moser)

(i) *The system (7.3) possesses n first integrals $F_1, \ldots, F_n$, which are polynomial functions in the variables*

$$
\left(p_1, \ldots, p_n, \exp(x_1 - x_2), \ldots, \exp(x_{n-1} - x_n) \right).
$$

(ii) *The functions $F_1, \ldots, F_n$ are generically independent and in involution with respect to the Poisson bracket defined by the standard symplectic form on $\mathbb{R}^{2n}$.*
(iii) *The solution of (7.5) can be expressed as* rational functions of $\left(\exp(-\lambda_1 t), \ldots, \exp(-\lambda_n t) \right)$, *where $\lambda_1, \ldots, \lambda_n$ are distinct real numbers depending on the initial conditions.*

Before reading the following remark, we suggest to review the content of Sect. 4 in Chap. 3.

Remark 7.2 (First Reduction of the System) The Hamiltonian (7.2) is *translational invariant*, i.e.,

$$
H(p_1, \ldots, p_n, x_1 + t, \ldots, x_n + t) = H(p_1, \ldots, p_n, x_1, \ldots, x_n),
$$

for $t \in \mathbb{R}$. The infinitesimal generator of this one-parameter group of diffeomorphisms is $X = \sum_{i=1}^{n} \frac{\partial}{\partial x_i}$, and the corresponding Hamiltonian function is $P = \sum_{i=1}^{n} p_i$, i.e., the total momentum. We can use this first integral to reduce the Toda system to a submanifold of the *big phase-space*, which is $\mathbb{R}^{2n}$ with its canonical symplectic structure, applying the Marsden-Weinstein-Meyer reduction. This goes as follows: let us work in a barycentric system of coordinates, so that $P = \sum_{i=1}^{n} p_i = 0$. The vector field $X = \sum_{i=1}^{n} \frac{\partial}{\partial x_i}$ defines a coisotropic distribution on the manifold $P = 0$ (the integral curves of this vector field are lines defined by the equations $\dot{x}_i = 1$ for all $i = 1, \ldots, n$). Modding out this distribution, we finally get the reduced phase-space. This is diffeomorphic to $\mathbb{R}^{2n-2}$, and it is endowed with the (global) coordinates $(p_1, \ldots, p_{n-1}, \xi_1, \ldots, \xi_{n-1})$, where $\xi_i = x_i - x_{i+1}$, for all $i = 1, \ldots, n - 1$. The reduced symplectic form in these coordinates assumes the canonical form $\widetilde{\omega} = \sum_{i=1}^{n-1} dp_i \wedge d\xi_i$, while the reduced Hamiltonian assumes the form

$$\widetilde{H}(p, \xi) = \frac{1}{2} \sum_{i=1}^{n-1} p_i^2 + \sum_{i=1}^{n-1} \exp \xi_i - \frac{1}{2} \left(\sum_{i=1}^{n-1} p_i \right)^2 .$$

△

After setting the general framework, we will introduce an important representation of Eq. (7.5), which will play a crucial role in the proof of their complete integrability.

2   Flaschka Transform and the Integrability of the Toda System

In this section, we show that Eq. (7.5) can be written in the Lax form. This result is due to Hermann Flaschka (see [93]), who realized that Eq. (7.5) can be simplified by the change of variables

$$\begin{cases} a_j = \frac{1}{2} \exp\left(\frac{x_j - x_{j+1}}{2}\right), & j = 1, \ldots n - 1 \\ b_j = -\frac{1}{2} p_j, & j = 1, \ldots n, \end{cases} \tag{7.6}$$

where it is assumed that $a_0 = 0 = a_n$. Note that these are $2n - 2$ independent variables: in fact the conservation of the total momentum implies $\sum_{j=1}^{n} b_j = \text{const}$. We shall assume that $\sum_{j=1}^{n} p_j = 0$, i.e., we shall work in a barycentric reference system. Now, one can introduce a class of matrices that will play a fundamental role in what follows.

Definition 7.3 An $n \times n$ matrix L with real entries will be called a *Jacobi* matrix if:

(J1) It is symmetric.
(J2) It is traceless.
(J3) It is *tridiagonal*, i.e., $L_{ij} = 0$ if $j \notin \{i - 1, i, i + 1\}$ for all $i = 1, \ldots, n$;
(J4) all its off-diagonal entries are non-negative.

The set of all $n \times n$ Jacobi matrices will be denoted with Jac_n. In other words, $L \in \mathrm{Jac}_n$ if and only if

$$L = \begin{pmatrix} b_1 & a_1 & 0 & \cdots & & 0 \\ a_1 & b_2 & a_2 & \cdots & & \vdots \\ 0 & a_2 & \ddots & \ddots & & 0 \\ \vdots & \vdots & \ddots & \ddots & & a_{n-1} \\ 0 & 0 & \cdots & a_{n-1} & & b_n \end{pmatrix} \tag{7.7}$$

with $a_i > 0$ and $\sum_{i=1}^{n} b_i = 0$. $\qquad\triangle$

Remark 7.4 It is worth mentioning that by a Jacobi matrix often is meant a real matrix, satisfying all the requirements in Definition 7.3 but (J2). In what follows, unless differently specified, for a Jacobi matrix, we will always mean a *traceless* Jacobi matrix. $\qquad\triangle$

The *Flaschka coordinates* $(a_1, \ldots, a_{n-1}, b_1, \ldots, b_n)$ introduced in (7.6) yield the following very important result [93].

Theorem 7.5 (Flaschka) *The Eq. (7.5) can be written in Lax form*

$$\frac{dL}{dt} = [B, L], \tag{7.8}$$

with $L \in \mathrm{Jac}_n$, and

$$B = \begin{pmatrix} 0 & a_1 & 0 & \cdots & & 0 \\ -a_1 & 0 & a_2 & \cdots & & \vdots \\ 0 & -a_2 & 0 & \ddots & & 0 \\ \vdots & \vdots & \ddots & \ddots & & a_{n-1} \\ 0 & 0 & \cdots & -a_{n-1} & & 0 \end{pmatrix} \tag{7.9}$$

Proof Rewriting the equations in (7.5) in the variables $(a_1, \ldots, a_{n-1}, b_1, \ldots, b_n)$, one obtains

$$\begin{cases} \dot{a}_j = a_j \left(b_{j+1} - b_j \right), \ j = 1, \ldots, n-1 \\ \dot{b}_j = 2 \left(a_j^2 - a_{j-1}^2 \right), \ j = 1, \ldots, n, \end{cases} \tag{7.10}$$

with boundary conditions $a_0 = 0 = a_n$. One can check that equations in (7.10) are equivalent to $\frac{dL}{dt} = [B, L]$, where the matrices L and B are given in (7.7) and (7.9), respectively. $\qquad\square$

Remark 7.6   Note that (7.8) differs from (6.20) by a sign. This difference obviously does not interfere with the properties of the Lax representation presented in Sect. 2.2 of Chap. 6, and it is due only to our choice to keep the same conventions adopted in the original references (see, for instance, [93, 189, 190]). $\triangle$

Example 7.7   For the three-particle Toda lattice, (7.10) assumes the following form:

$$\dot{b}_1 = 2a_1^2, \quad \dot{b}_2 = 2\left(a_2^2 - a_1^2\right) \quad \text{and} \quad \dot{b}_3 = -2a_2^2$$

$$\dot{a}_1 = a_1(b_2 - b_1) \quad \text{and} \quad \dot{a}_2 = a_2(b_3 - b_2),$$

where we used the boundary conditions $a_0 = 0$ and $a_3 = 0$. In this case,

$$L = \begin{pmatrix} b_1 & a_1 & 0 \\ a_1 & b_2 & a_2 \\ 0 & a_2 & b_3 \end{pmatrix} \quad \text{and} \quad B = \begin{pmatrix} 0 & a_1 & 0 \\ -a_1 & 0 & a_2 \\ 0 & -a_2 & 0 \end{pmatrix},$$

which yields

$$[B, L] = \begin{pmatrix} 2a_1^2 & a_1(b_2 - b_1) & 0 \\ a_1(b_2 - b_1) & 2(a_2^2 - a_1^2) & a_2(b_3 - b_2) \\ 0 & a_2(b_3 - b_2) & -2a_2^2 \end{pmatrix},$$

which exemplifies the statement of Theorem 7.5. $\triangle$

Remark 7.8

(i) Note that $B = L_+ - L_-$, where $L_\pm$ are the *strictly upper* (+) and, respectively, *strictly lower* (-) triangular part of L.
(ii) In view of the Lax representation of the equations in (7.10), one can think of the Toda system as a dynamical system whose flow takes place in the set Jac_n. Later in this chapter, we shall give a Lie theoretic interpretation of this set of matrices.

$\triangle$

In this section, we aim to present a complete proof of Theorem 7.1. To this end, first observe that the existence of a Lax representation for (7.5) entails the existence of a certain number of first integrals (see Sect. 2.2). Here we just remind the reader that once the equation of motion has been expressed in Lax form $\frac{dL}{dt} = [B, L]$,, every function of the eigenvalues of L is constant along the flow of the dynamical system. Among the many possible choices, a prominent role is played by the symmetric functions of the eigenvalues of L, defined by

$$F_k = \frac{\mathrm{tr}\left(L^k\right)}{k} = \frac{1}{k} \sum_{i=1}^{n} \lambda_i^k, \ k = 1, \ldots, n. \tag{7.11}$$

Problem 7.9 Let L be as in (7.7).

(i) Show that F_1 and F_2 are, up to multiplicative constants, the total momentum and, respectively, the Hamiltonian of the Toda system.
(ii) Compute F_3 as function of $(p_1, \ldots, p_n, x_1, \ldots, x_n)$.

$\triangle$

2.1 Open Toda System in Hessenberg Form

Theorem 7.5 presents the so-called symmetric form of the open Toda system in the Flaschka variables. Below we introduce a different, though equivalent, Lax representation of (7.10), which is commonly known as the *Hessenberg form* of the Toda system. To this end, given L as in (7.7), let $D = \mathrm{diag}(d_1, \ldots, d_n)$ where $d_1 = 1$ and $d_i = a_1^{-1} \cdots a_{i-1}^{-1}$ for all $i = 2, \ldots n$. A simple computation shows that conjugating $L \in \mathrm{Jac}_n$ with the corresponding D yields the following *Hessenberg matrix*:

$$
Y = DLD^{-1} = \begin{pmatrix}
b_1 & a_1^2 & 0 & \cdots & 0 \\
1 & b_2 & a_2^2 & \cdots & \vdots \\
0 & 1 & \ddots & \ddots & 0 \\
\vdots & \vdots & \ddots & \ddots & a_{n-1}^2 \\
0 & 0 & \cdots & 1 & b_n
\end{pmatrix}
$$

(see [135, 136] and also the monograph [67] for general information about the properties of this class of matrices). Given Y as above, if

$$
F = \begin{pmatrix}
0 & a_1^2 & 0 & \cdots & 0 \\
0 & 0 & a_2^2 & \cdots & \vdots \\
0 & 0 & \ddots & \ddots & 0 \\
\vdots & \vdots & \ddots & \ddots & a_{n-1}^2 \\
0 & 0 & \cdots & 0 & 0
\end{pmatrix}
$$

a direct computation shows that the Lax-type equation

$$
\frac{dY}{dt} = 2[F, Y] \tag{7.12}
$$

is equivalent to (7.10). Let us check this claim again in the case of the three-particle Toda lattice, when

$$Y = \begin{pmatrix} b_1 & a_1^2 & 0 \\ 1 & b_2 & a_2^2 \\ 0 & 1 & b_3 \end{pmatrix} \quad \text{and} \quad F = \begin{pmatrix} 0 & a_1^2 & 0 \\ 0 & 0 & a_2^2 \\ 0 & 0 & 0 \end{pmatrix}.$$

In this case,

$$\frac{dY}{dt} = \begin{pmatrix} \dot{b}_1 & 2\dot{a}_1 a_1 & 0 \\ 0 & \dot{b}_2 & 2\dot{a}_2 a_2 \\ 0 & 0 & \dot{b}_3 \end{pmatrix} \quad \text{and} \quad [F, Y] = \begin{pmatrix} a_1^2 & a_1^2(b_2 - b_1) & 0 \\ 0 & a_2^2 - a_1^2 & a_2^2(b_3 - b_2) \\ 0 & 0 & -a_2^2 \end{pmatrix},$$

which shows for the three-particle Toda lattice, (7.12) is equivalent to (7.10) (see Example 7.7). Instead, for the integrability in the sense of Liouville of the Toda flow on generic Hessenberg coadjoint orbits and for the case of classical simple real split Lie algebras, see instead the insightful [155].

2.2   Complete Integrability of the Open Toda Lattice

Now we discuss the complete integrability of the open Toda lattice. We have already proven that this system admits a Lax representation. This implies that the functions $F_1, \ldots, F_n$, where the F_i are defined in Eq. (7.11), are first integrals of the Toda lattice. Since the Hamiltonian is $H_2 = \frac{1}{2} \operatorname{tr}(L^2)$ (see Problem 7.9), to complete the proof of the complete integrability, it suffices to show that the functions $F_1, \ldots, F_n$ are generically independent and that they Poisson commute with respect to the canonical Poisson bracket defined on $\mathbb{R}^{2n}$.

Proposition 7.10   *The functions $F_1, \ldots, F_n$ are independent.*

Proof   A straightforward calculation shows that for every $k = 1, \ldots, n$, the function F_k can be written as

$$F_k = \frac{1}{k} \sum_{i=1}^{n} p_i^k + f(p, x),$$

where $f(p, x)$ is a polynomial in the p_is of degree strictly less than k. Thus, the Jacobian matrix of the functions $F_1, \ldots, F_n$ (i.e., the matrix whose entries are the components of the differentials of the functions F_k) reads

$$
\begin{pmatrix}
1 & \cdots & 1 & \frac{\partial f_1}{\partial x_1} & \cdots & \frac{\partial f_1}{x_n} \\
p_1 & \cdots & p_n & \frac{\partial f_2}{\partial x_1} & \cdots & \frac{\partial f_2}{\partial x_n} \\
\vdots & & \vdots & \vdots & & \vdots \\
p_1^{n-1} & \cdots & p_n^{n-1} & \frac{\partial f_n}{\partial x_1} & \cdots & \frac{\partial f_n}{\partial x_n}
\end{pmatrix}
$$

This matrix has (generically) rank n since the left $n \times n$ principal minor is a Vandermonde determinant. $\qquad\square$

Next we show that the functions $F_1, \ldots, F_n$ are in involution. Let us start with the following observation. Let L be as in (7.7).

Lemma 7.11 *The spectrum of L is real and simple.*

Proof Since L is symmetric, it is diagonalizable with real eigenvalues. Let λ be an eigenvalue of L and u a corresponding eigenvector. Since the entries a_i of L are positive, the equation $Lu = \lambda u$ yields a set of $n - 1$ linear equations, which, for each k, express uniquely the entry u_k of u in terms of the entry u_1, of the eigenvalue λ, and of the entries of the matrix L. The set of these equations is

$$
\begin{cases}
a_1 u_2 + b_1 u_1 = \lambda u_1 \\
a_{k-1} u_{k-1} + b_k u_k + a_k b_{k+1} = \lambda u_k, \quad k = 2, \ldots, n-1 \\
a_{n-1} u_{n-1} + b_n u_n = \lambda u_n.
\end{cases}
\tag{7.13}
$$

These equations entail that *any* eigenvector associated with the eigenvalue λ is a scalar multiple of a vector whose entries depend on λ and on the entries of the matrix L. $\qquad\square$

Phrased differently, Lemma 7.11 states that L has n distinct, real eigenvalues $\lambda_1 < \lambda_2 < \cdots < \lambda_n$. A first consequence of this result is that the F_ks are smooth functions of $(a_1, \ldots, a_{n-1}, b_1, \ldots, b_n)$ and, because of Formula (7.6), they are smooth functions of the variables $(p_1, \ldots, p_n, x_1, \ldots, x_n)$ as well. Suppose now that λ is an eigenvalue of L and that u is a corresponding eigenvector, i.e., $Lu = \lambda u$. Then

Lemma 7.12 *The partial derivatives of λ with respect to the p_is and, respectively, to the x_is are given by*

$$
\frac{\partial \lambda}{\partial p_k} = -\frac{u_k^2}{2} \quad \text{and} \quad \frac{\partial \lambda}{\partial x_k} = a_k u_{k+1} u_k - a_{k-1} u_k u_{k-1},
\tag{7.14}
$$

for all $k = 1, \ldots, n$. In particular, for $k = 1$

$$
\frac{\partial \lambda}{\partial x_1} = a_1 u_1 u_2
$$

and for $k = n$

$$\frac{\partial \lambda}{\partial x_n} = -a_{n-1} u_{n-1} u_n.$$

Proof With no loss of generality, we can assume that u is normalized with respect to the (standard) Euclidean inner product $(\cdot \,|\, \cdot)$ so that $(u \,|\, u) = 1$. Then $\lambda = (Lu \,|\, u)$. Taking now the derivative with respect to p_k of both sides of this identity yields

$$\frac{\partial \lambda}{\partial p_k} = \left(\frac{\partial L}{\partial p_k} u \,\Big|\, u \right) + \left(L \frac{\partial u}{\partial p_k} \,\Big|\, u \right) + \left(Lu \,\Big|\, \frac{\partial u}{\partial p_k} \right),$$

where the sum of the last two summands is equal to zero. In fact, since L is a symmetric and $(u \,|\, u) = 1$

$$\left(L \frac{\partial u}{\partial p_k} \,\Big|\, u \right) + \left(Lu \,\Big|\, \frac{\partial u}{\partial p_k} \right) = 2 \left(Lu \,\Big|\, \frac{\partial u}{\partial p_k} \right) = 2\lambda \left(u \,\Big|\, \frac{\partial u}{\partial p_k} \right) = 0.$$

Then the previous computation, together with (7.7) and (7.6), yields

$$\frac{\partial \lambda}{\partial p_k} = \left(\frac{\partial L}{\partial p_k} u \,\Big|\, u \right) = -\frac{u_k^2}{2}, \quad \forall k = 1, \ldots, n,$$

proving the first set of formulas in (7.14). For the second one, proceeding in the same way, one first proves that

$$\frac{\partial \lambda}{\partial x_k} = \left(\frac{\partial L}{\partial x_k} u \,\Big|\, u \right), \quad \forall k = 1, \ldots, n,$$

and then, a direct computation, that uses again (7.7) and (7.6), entails that

$$\frac{\partial \lambda}{\partial x_k} = a_k u_{k+1} u_k - a_{k-1} u_k u_{k-1}, \quad \forall k = 1, \ldots, n.$$

This formula, together with the boundary conditions $a_0 = 0 = a_n$, implies that for $k = 1$

$$\frac{\partial \lambda}{\partial x_1} = a_1 u_1 u_2,$$

while for $k = n$,

$$\frac{\partial \lambda}{\partial x_n} = -a_{n-1} u_{n-1} u_n,$$

as required. $\qquad\square$

Let u and v be two eigenvectors of L, corresponding two distinct eigenvalues, say λ and, respectively, μ. Under these assumptions, one can prove that

$$\{\lambda, \mu\} = \sum_{k=1}^{n} \left(\frac{\partial \lambda}{\partial p_k} \frac{\partial \mu}{\partial x_k} - \frac{\partial \lambda}{\partial x_k} \frac{\partial \mu}{\partial p_k} \right) = 0.$$

We present below all the steps necessary to prove this identity, starting with the following computation:

$$\{\lambda, \mu\} \stackrel{(7.14)}{=} \frac{1}{2} \sum_{k=1}^{n} v_k^2 (a_k u_{k+1} u_k - a_{k-1} u_k u_{k-1}) - \frac{1}{2} \sum_{k=1}^{n} u_k^2 (a_k v_{k+1} v_k - a_{k-1} v_k v_{k-1}) =$$

$$= \frac{1}{2} \sum_{k=1}^{n} v_k u_k a_k (v_k u_{k+1} - u_k v_{k+1}) + \frac{1}{2} \sum_{k=1}^{n} v_k u_k a_{k-1} (u_k v_{k-1} - v_k u_{k-1}),$$

i.e.,

$$\{\lambda, \mu\} = \frac{1}{2} \sum_{k}^{n} v_k u_k (R_k + R_{k-1}), \tag{7.15}$$

where $R_j = a_j (u_{j+1} v_j - u_j v_{j+1})$ for $j = 1, \ldots, n$. The next one is a statement enclosed in the following lemma, whose proof follows almost immediately from the definition of the matrix L.

Lemma 7.13 *If λ, μ and u, v as above, then*

$$v_j u_j = \frac{1}{\lambda - \mu} (R_{j-1} - R_j). \tag{7.16}$$

Proof We prove the formula for $j = 2, \ldots, n - 1$, leaving the remaining cases to the reader as exercises. In the chosen interval, the j-th component of $Lu = \lambda u$ and of $Lv = \mu v$ is given by

$$a_{j-1} u_{j-1} + a_j u_{j+1} + b_j u_j = \lambda u_j \quad \text{and} \quad a_{j-1} v_{j-1} + a_j v_{j+1} + b_j v_j = \mu v_j$$

(see Formula 7.13). Multiplying the first equation by v_j and the second by u_j and subtracting term by term, one arrives to (7.16). $\qquad\square$

After substituting the expression for $v_k u_k$ (see 7.16) into (7.15), one gets

$$\{\lambda, \mu\} = \frac{1}{2(\lambda - \mu)} \sum_{k=1}^{n} \left(R_{k-1}^2 - R_k^2 \right) = 0,$$

where the last equality follows from the boundary conditions $a_n = 0 = a_0$. Summarizing

Proposition 7.14 *The eigenvalues of the matrix L commute with respect to the standard Poisson bracket of $\mathbb{R}^{2n}$. In particular, the functions $F_1, \ldots, F_n$ (see Formula 7.11) are in involution with respect to such a Poisson structure.*

This result, together with the proof of the (generic) independence of $F_1, \ldots, F_n$ and with the fact that the Hamiltonian of the open-Toda lattice is F_2, shows that the open Toda lattice is a complete integrable system.

Remark 7.15 In Sect. 3, we present the so-called Adler-Kostant-Symes Theorem, or *AKS Theorem*, which yields a more conceptual proof of the complete integrability of the Toda system. The AKS Theorem, in turn, can be put in the (even) more general framework of the so-called R-*matrix theory*. In this context, the proof of the integrability of the Toda lattice is a consequence of its Lie theoretical interpretation, which will be outlined at the end of this chapter.   △

After settling the proof of the complete integrability of the open Toda system, in the next subsection, we aim to present an explicit set of solutions of this remarkable dynamical system.

2.3 Explicit Representation of the Solutions

In this subsection, we will present explicit formulas for the solutions of (7.5) or, equivalently, of (7.10). To this end, we start noticing that the eigenvalues $\lambda_1, \ldots, \lambda_n$ of L are (functionally) independent functions of the physical variables $(p_1, \ldots, p_n, x_1, \ldots, x_n)$.

Problem 7.16 Using the result of Proposition 7.10, show that the eigenvalues of the Lax matrix L are independent. (Hint: this follows from the fact that the λs are constants of motion and satisfy the condition $\lambda_1 < \cdots < \lambda_n$, and also note that the determinant of the Jacobian of the map

$$\mathcal{F} : \mathbb{R}^n \to \mathbb{R}^n$$

$$(\lambda_1, \ldots, \lambda_n) \mapsto (F_1, \ldots, F_n)$$

is a Vandermonde determinant.)   △

Since $\lambda_1, \ldots, \lambda_n$ are independent constants of motion, they can be taken as *action-variables* of the Toda system. Following Moser, we shall now give a recipe to find a set of *angle-variables* conjugate to the λs, and using these variables, we shall be able to describe the solutions of the system (7.5) in a closed form. An important preliminary result, of independent interest, is enclosed in the theorem below, where it is shown that the space of Jacobi matrices can be put in a one-to-one correspondence with a space of suitable rational functions. This, as most of the results enclosed in this chapter, is due to Moser (see [189, 190]).

Our presentation starts with the definition of the relevant space of rational functions. Let

$$\mathrm{Rat}_n = \left\{ f(z) \mid f(z) = \sum_{i=1}^{n} \frac{x_i^2}{z - \lambda_i}, \ \lambda_1 < \cdots < \lambda_n, \ x_i > 0 \ \text{and} \ \sum_{i=1}^{n} x_i^2 = 1 \right\}$$

(7.17)

Theorem 7.17 (Moser) *There is a one-to-one correspondence between the set of Jacobi matrices* Jac_n *and the set* Rat_n *as defined in (7.17).*

Proof Associating a function $f(z) \in \mathrm{Rat}_n$ with a Jacobi matrix L (see Definition 7.3) is quite easy. Let L be a $n \times n$ Jacobi matrix. We have already proven that L has a simple spectrum. Let us suppose that its eigenvalues satisfy the condition $\lambda_1 < \cdots < \lambda_n$ and let $(v_1, \ldots, v_n)$ be an orthonormal basis of eigenvectors (which exists because L is symmetric), i.e., $L v_i = \lambda_i v_i$ for all i and $(v_i, v_j) = \delta_{ij}$, where $(\cdot, \cdot)$ is the standard Euclidean inner product in $\mathbb{R}^n$. Furthermore, let $e_n = (0, \ldots, 1)$ and let

$$f(z) = \left((z - L)^{-1} e_n, e_n \right).$$

We claim that this function belongs to Rat_n. To prove this claim, it suffices to write the vector e_n in terms of the eigenvectors $v_1, \ldots, v_n$, i.e.,

$$e_n = \sum_{j=1}^{n} (e_n, v_j) v_j$$

and observe that

$$(z - L)^{-1} e_n = \sum_{j=1}^{n} (z - \lambda_j)^{-1} (v_j, e_n) v_j , \qquad (7.18)$$

(see Problem 7.18). If x_j denotes the absolute value of the component of e_n along v_j, i.e., $x_j = |(e_n, v_j)|$, then

$$f(z) = \left((z - L)^{-1} e_n, e_n \right) = \sum_{j=1}^{n} \frac{x_j^2}{z - \lambda_j}.$$

To close the first part of the proof, it remains to show that $\sum_{j=1}^{n} x_j^2 = 1$ (see Problem 7.18). Now we observe that $f(z) = \left((z - L)^{-1} e_n, e_n \right)$ admits a representation in terms of a finite continued fraction, i.e.,

$$f(z) = \cfrac{1}{z - b_n - \cfrac{a_{n-1}^2}{z - b_{n-1} - \cfrac{a_{n-2}^2}{\ddots \cfrac{}{z - b_2 - \frac{a_1^2}{z-b_1}}}}} \tag{7.19}$$

where the (a_i, b_i) coincide with those that define the matrix L (note that the a_is of the matrix L are positive so that they are uniquely defined by their squares).

To show the existence of the representation (7.19), let $V = (V_1, \ldots, V_n)$ be the vector in $\mathbb{R}^n$ whose components are

$$\begin{cases} V_k = (-1)^{k+n} \frac{\Delta_{k-1}}{\Delta_n}, \ k = 1, \ldots, n - 1, \\ V_n = \frac{\Delta_{n-1}}{\Delta_n}, \end{cases}$$

where $\Delta_i = \det(z - L_k)$ and L_k is matrix obtained deleting from L all but the first k rows and k columns. Then

$$(z - L)V = e_n \tag{7.20}$$

and

$$f(z) = \big((z - L)^{-1} e_n, e_n\big) = V_n = \frac{\Delta_{n-1}}{\Delta_n} \tag{7.21}$$

(see Problem 7.18). On the other hand, the Δ_ks satisfy the recursion relation

$$\begin{cases} \Delta_k = (z - b_k)\Delta_{k-1} - a_{k-1}^2 \Delta_{k-2}, \ k = 3, \ldots, n, \\ \Delta_1 = (z - b_1) \\ \Delta_2 = (z - b_2)\Delta_1 - a_1^2 \end{cases} \tag{7.22}$$

which can be easily obtained by expanding the determinant of the matrix L with respect to the last row. Formulas (7.22) imply

$$\frac{\Delta_{n-1}}{\Delta_n} = \frac{1}{z - b_n - a_{n-1}^2 \frac{\Delta_{n-2}}{\Delta_{n-1}}}.$$

From this and from Formulas (7.21) and (7.22), one can deduce the representation (7.19) of $f(z)$ in terms of finite continued fraction.

We come now to the hard part of the proof. We need to show that to every rational function

$$f(z) = \sum_{i=1}^{n} \frac{x_i^2}{z - \lambda_i}$$

with $\lambda_1 < \cdots < \lambda_n$, $x_i > 0$ for all $i = 1, \ldots, n$, and with $\sum_{i=1}^{n} x_i^2 = 1$, it is associated with a unique Jacobi matrix L as in Formula (7.7). To this end, we shall show that every rational function $f(z)$ with the properties described above has a unique representation as a finite continued fraction as in Formula (7.19). Suppose it is given a rational function $f(z) = \sum_{i=1}^{n} \frac{x_i^2}{z - \lambda_i}$ whose poles and residues satisfy the hypothesis of the proposition. Then

$$f(z) = \frac{P_{n-1}(z)}{Q_n(z)} = \frac{\sum_{i=1}^{n} x_i^2 \prod_{j \neq i}(z - \lambda_j)}{\prod_{i=1}^{n}(z - \lambda_i)}$$

as $\sum_{i=1}^{n} x_i^2 = 1$, $P_{n-1}(z)$ is a monic polynomial. After easy but tedious computations, one can write

$$f(z) = \frac{1}{z - A - B f_1(z)} \tag{7.23}$$

where

$$A = \sum_{i=1}^{n} \lambda_i^2 x_i^2 - \left(\sum_{i=1}^{n} \lambda_i x_i^2 \right)^2 > 0 \text{ and } B = \sum_{i=1}^{n} \lambda_i x_i^2 \tag{7.24}$$

and $f_1(z) = \frac{P_{n-2}(z)}{P_{n-1}(z)}$. Here $P_{n-2}(z)$ is a monic polynomial of degree $n - 2$ whose coefficients are rational functions of the λs and of the xs. Formula (7.23) and Problem 7.18 imply that

$$f_1(z) = \frac{1}{B}\left(z - A - \frac{1}{f(z)} \right)$$

is a rational function with $n - 1$ real poles $\mu_1 < \cdots < \mu_{n-1}$, so that it can be represented in terms of partial fractions as

$$f_1(z) = \sum_{i=1}^{n-1} \frac{X_i}{z - \mu_i}.$$

We want to show that the poles and residues of the function $f_1(z)$ have the same properties of the poles and residues of the function $f(z)$. First, the expression in partial fractions of the function $f(z)$ yields

$$X_i = \operatorname{res}_{z=\mu_i} \frac{1}{B}\left(z - A - \frac{1}{f(z)}\right)$$

$$= \lim_{z \to \mu_i} \frac{(z - \mu_i)}{B}\left(z - A - \frac{1}{f(z)}\right) = -\frac{1}{Bf'(\mu_i)} > 0.$$

On the other hand, since $f_1(z) = \frac{P_{n-2}(z)}{P_{n-1}(z)}$ with $P_{n-2}(z)$ a monic polynomial, we conclude that $\sum_{i=1}^{n-1} X_i = 1$. Thus $f_1(z)$ is a rational function whose poles and residues satisfy the same properties as those of $f(z)$. Finally, proceeding with the function $f_1(z)$ as we did with the function $f(z)$, and iterating this process a finite number of times, we arrive to the desired representation of the rational function $f(z)$ in terms of a finite continued fraction. Now, to define the Jacobi matrix associated with the rational function $f(z)$, it suffices to define $a_{n-1}^2 = A > 0$, $b_n = B$, and so on. This concludes the proof. $\qquad\square$

Problem 7.18

 (i) Prove Formula (7.18).

 (ii) Prove that the sum of the squares of the residues of the function $f(z)$ is 1.

 (iii) Prove Formulas (7.20) and (7.21).

 (iv) Prove that a rational function $f(z) = \sum_{i=1}^{n} \frac{x_i^2}{z - \lambda_i}$, such that:

    (a) $\lambda_1 < \cdots < \lambda_n$

    (b) $x_i > 0$ with $\sum_{i=1}^{n} x_i^2 = 1$

    has $n - 1$ zeroes separating the poles.

 (v) Let $f(z)$ be a rational function as in Problem 7.18. Prove that the representation of $f(z)$ given in Formula (7.23) holds true with the coefficients A, B given in Formula (7.24) when $n = 2$ and when $n = 3$.

Hints: For (i), it suffices to apply the spectral theorem to the linear operator $(z - L)^{-1}$. Point (ii) follows from the fact that e_n is a unit vector in $\mathbb{R}^n$. $\qquad\triangle$

Define the following subsets of $\mathbb{R}^{2n}$:

$$\Lambda = \left\{ (\lambda_1, \ldots, \lambda_n, x_1, \ldots, x_n)\mid \lambda_1 < \cdots < \lambda_n, \ \sum_{i=1}^{n} \lambda_i = 0, \right.$$

$$\left. \sum_{i=1}^{n} x_i^2 = 1, \ x_i > 0, \ \forall i = 1, \ldots, n \right\}$$

and

$$D = \left\{ (a_1, \ldots, a_{n-1}, b_1, \ldots, b_n) \mid \sum_{i=1}^{n} b_i = 0, \ a_i > 0 \ \forall \, i = 1, \ldots, n \right\}.$$

There are one-to-one correspondences $\mathrm{Rat}_n \longleftrightarrow \Lambda$ and $\mathrm{Jac}_n \longleftrightarrow D$. Using these identifications, Theorem 7.17 can be rephrased as follows.

Corollary 7.19 *There is a one-to-one correspondence*

$$\phi : \Lambda \to D.$$

Proposition 7.17 and this corollary tell us that the set of Jacobi matrices Jac_n is parametrized by the rational functions belonging to the set Rat_n. On this set one has coordinates $(\lambda_1, \ldots, \lambda_n, x_1, \ldots, x_n)$. These coordinates are a crucial ingredient of the following.

Proposition 7.20 (Moser) *The system (7.10) is equivalent to*

$$\begin{cases} \dot{\lambda}_j = 0 \\ \dot{x}_j = - \left(\lambda_j - \sum_{k=1}^{n} \lambda_k x_k^2 \right) x_j \ \ j = 1, \ldots, n. \end{cases} \tag{7.25}$$

Proof Since we have already proved that the λ_is are first integrals, we are left to show that

$$\dot{x}_j = - \left(\lambda_j - \sum_{k=1}^{n} \lambda_k x_k^2 \right) x_j \ \ j = 1, \ldots n. \tag{7.26}$$

Let $v_k(0)$ be an eigenvector for L with eigenvalue λ_k at $t = 0$, i.e., $L(0)v_k(0) = \lambda_k v_k(0)$. Since the evolution of L may be written as $L(t) = U(t)L(0)U(t)^{-1}$, where $U = U(t)$ is an invertible matrix depending on t (see Sect. 2.2 in Chap. 6) and λ_k is t-independent, one has

$$L(t)U(t)v_k(0) = \lambda_k U(t)v_k(0),$$

whence one deduces that, for every t,

$$v_k(t) := U(t)v_k(0),$$

is an eigenvector of $L(t)$ whose eigenvalue is λ_k. Furthermore,

$$\frac{dv_k(t)}{dt} = \dot{U}(t)v_k(0) = B(t)U(t)v_k(0) = Bv_k(t), \tag{7.27}$$

where B is the matrix defined in (7.9) (see Lemma 6.18). Thinking of the elements of $\mathbb{R}^n$ as column-vectors, if $e_n \in \mathbb{R}^n$ is the nth vector of the canonical basis and $(\cdot \mid \cdot)$ is the standard Euclidean inner product in $\mathbb{R}^n$, then

$$\frac{d(v_k(t) \mid e_n)}{dt} = \left(\frac{dv_k(t)}{dt} \mid e_n \right) \overset{(7.27)}{=} (B(t)v_k(t) \mid e_n) \overset{(7.9)}{=} -(v_k(t) \mid B(t)e_n)$$

$$\overset{(7.9)}{=} -a_{n-1}(t)(v_k(t) \mid e_{n-1}). \tag{7.28}$$

On the other hand, since $L(t)v_k(t) = \lambda_k v_k(t)$, and L is a symmetric matrix, one has

$$(v_k(t) \mid L(t)e_n) = (L(t)v_k(t) \mid e_n) = \lambda_k(v_k(t) \mid e_n),$$

which, in virtue of (7.7), is equivalent to

$$a_{n-1}(t)(v_k(t) \mid e_{n-1}) + b_n(t)(v_k(t) \mid e_n) = \lambda_k(v_k(t) \mid e_n). \tag{7.29}$$

Combining (7.28) with (7.29), and choosing v_k such that $x_k(t) := (v_k(t) \mid e_n) > 0$, dropping the time-dependence, one can write

$$\frac{dx_k}{dt} = -a_{n-1}(v_k \mid e_{n-1}) = (b_n - \lambda_k)x_k. \tag{7.30}$$

Since $\sum_{i=1}^n x_i^2 = 1$, we have that $\sum_{i=1}^n \dot{x}_i x_i = 0$, which implies

$$b_n = \sum_{i=1}^n \lambda_i x_i^2 . \tag{7.31}$$

Formula (7.26) follows now from Eqs. (7.31) and (7.30). $\qquad\qquad\qquad\qquad\square$

Proposition 7.20 yields an explicit formula for the solution of the system (7.10). More precisely

Corollary 7.21 *The system* (7.25) *has the solution*

$$\begin{cases} \lambda_j(t) = \lambda_j(0) \\ x_j^2(t) = \dfrac{x_j(0) \exp(-2\lambda_j t)}{\sum_{k=1}^n x_k^2(0) \exp(-2\lambda_k t)}, \end{cases}$$

for all $j = 1, \ldots, n$.

Proof It suffices to show that, for all $j = 1, \ldots, n$,

$$x_j^2(t) = \frac{x_j(0) \exp(-2\lambda_j t)}{\sum_{k=1}^n x_k^2(0) \exp(-2\lambda_k t)}.$$

To this end, for every $j = 1, \ldots, n$, let y_j such that

$$x_j = \frac{y_j}{\left(\sum_{k=1}^{n} y_k^2\right)^{\frac{1}{2}}}. \tag{7.32}$$

Applying this change of variables

$$\dot{x}_j = -\left(\lambda_j - \sum_{k=1}^{n} \lambda_k x_k^2\right) x_j, \quad j = 1, \ldots, n, \tag{7.33}$$

becomes the following linear system in the y_is (see Problem 7.22)

$$\dot{y}_j = -\lambda_j y_j, \quad j = 1, \ldots, n, \tag{7.34}$$

which is easily integrated, proving the statement. $\qquad\square$

Problem 7.22 Using the change of variables defined in (7.32), prove that the system (7.33) becomes the system (7.34). Hint: the proof follows from the formula

$$\dot{x}_j = \frac{\dot{y}_j \sum_{l=1}^{n} y_l^2 - y_j \sum_{l=1}^{n} \dot{y}_l y_l}{\left(\sum_{l=1}^{n} y_l^2\right)^{\frac{3}{2}}}, \quad j = 1, \ldots, n.$$

$$\triangle$$

3 The Toda System and the Adler-Kostant-Symes Construction

In this section, we will describe a result commonly known as the *Adler-Kostant-Symes Theorem*, *AKS Theorem* from now on, which yields a construction of Hamiltonian systems on (co)adjoint orbits of reductive Lie algebras (see Example C.13 in Appendix C) endowed with a rich supply of first integrals coming from the Ad-invariant functions defined on the underlying Lie algebra. Using this method, we will show how to recover the open Toda system in both its symmetric and its Hessenberg form (see Sect. 2.1). At the end of the section, we will comment about the so-called R-matrix formalism, which is a powerful generalization of the Adler-Kostant-Symes method.

3.1 AKS Theorem

We start setting up the stage for the main result of this section. First, we prove the following.

Lemma 7.23 *Let* $(\mathfrak{g}, B)$ *be a quadratic Lie algebra (see Definition C.17). If* $f \in C^\infty(\mathfrak{g})$ *is Ad-invariant, then*

$$\mathrm{ad}_x \, \nabla_x f = 0, \ \forall x \in \mathfrak{g}. \tag{7.35}$$

Proof Since f is Ad-invariant $\left.\frac{d}{dt}\right|_{t=0} f(\mathrm{Ad}_{\exp(ty)} x) = 0$ for all $x, y \in \mathfrak{g}$, which implies $\mathrm{ad}_x^\sharp (df_x) = 0$ for all x. Applying to both sides of this identity $\check{B}^{-1}$, the G-invariance of B yields

$$0 = \check{B}^{-1}\big(\mathrm{ad}_x^\sharp(df_x)\big) = \mathrm{ad}_x \, \check{B}^{-1}(df_x) = \mathrm{ad}_x \, \nabla_x f$$

(see Formula 2.10). $\square$

A few observations are in order:

 (i) If G is connected, f is Ad-invariant if and only if (7.35) holds for all $x \in \mathfrak{g}$.
 (ii) The Ad-invariant functions on $\mathfrak{g}$ correspond, via the isomorphism $\check{B}$, to the Casimirs of the linear Poisson structure of $\mathfrak{g}^*$. Said differently, the Ad-invariant functions on $\mathfrak{g}$ *are* the Casimirs of the linear Poisson structure (2.11).
 (iii) In the proof of the previous lemma, we claimed, implicitly, that the ad-invariance of B is equivalent to the fact that B intertwines the adjoint with the coadjoint actions of $\mathfrak{g}$ on $\mathfrak{g}$ and, respectively, on $\mathfrak{g}^*$.

Let $(\mathfrak{g}, B)$ be as above, and suppose that $\mathfrak{g}$ is endowed with a vector space decomposition $\mathfrak{g} = \mathfrak{a} \oplus \mathfrak{b}$, where both $\mathfrak{a}$ and $\mathfrak{b}$ are Lie subalgebras of $\mathfrak{g}$. This decomposition comes together with the projections

$$\pi_\mathfrak{a} : \mathfrak{g} \to \mathfrak{a} \quad \text{and} \quad \pi_\mathfrak{b} : \mathfrak{g} \to \mathfrak{b}.$$

Recall that B defines an isomorphism $\check{B} : \mathfrak{g} \to \mathfrak{g}^*$, and suppose it restricts to an isomorphism between

$$\mathfrak{a}^\perp = \{x \in \mathfrak{g} \,|\, B(x, y) = 0, \ \forall y \in \mathfrak{a}\}$$

and $\mathfrak{b}^*$. Under these assumptions,

Lemma 7.24 $\check{B}$ *restricts to an isomorphism between* $\mathfrak{b}^\perp$ *and* $\mathfrak{a}^*$ *as well.*

Proof The non-degeneracy of B implies that $\check{B} : \mathfrak{a}^\perp \to \mathfrak{b}^*$ is an injective linear map. To conclude, it suffices to notice that $\mathfrak{a}^\perp$ and $\mathfrak{b}^*$ have the same dimension. $\square$

This lemma implies that

Lemma 7.25 B *defines an isomorphism between* $\mathfrak{g}$ *and* $\mathfrak{b}^\perp \oplus \mathfrak{a}^\perp$.

Proof First notice that $\mathfrak{g}^* = (\mathfrak{a} \oplus \mathfrak{b})^* = \mathfrak{a}^* \oplus \mathfrak{b}^*$. Then the previous lemma implies that $\mathfrak{g}^* \simeq \mathfrak{b}^\perp \oplus \mathfrak{a}^\perp$. To conclude, it suffices to recall that $\check{B}$ is an isomorphism between $\mathfrak{g}$ and $\mathfrak{g}^*$. $\square$

Remark 7.26 The vector space decomposition $\mathfrak{g} = \mathfrak{b}^\perp \oplus \mathfrak{a}^\perp$ comes together with the two projections $\pi_{\mathfrak{a}^\perp}$ and $\pi_{\mathfrak{b}^\perp}$, the first from $\mathfrak{g}$ to $\mathfrak{a}^\perp$ along $\mathfrak{b}^\perp$ and the second from $\mathfrak{g}$ to $\mathfrak{b}^\perp$ along $\mathfrak{a}^\perp$. Note that $\pi_{\mathfrak{a}} \oplus \pi_{\mathfrak{b}} = \mathrm{id}_{\mathfrak{g}} = \pi_{\mathfrak{a}^\perp} \oplus \pi_{\mathfrak{b}^\perp}$, but it is *not* true that $\pi_{\mathfrak{a}} \oplus \pi_{\mathfrak{a}^\perp} = \mathfrak{g}$ $\triangle$

The following example plays an important role in the theory of the Toda system.

Example 7.27 Let $\mathfrak{g} = \mathfrak{sl}_n(\mathbb{R})$ be endowed with the $\mathfrak{g}$-invariant, non-degenerate bilinear form $B : \mathfrak{g} \times \mathfrak{g} \to \mathbb{R}$ defined by $B(x, y) = \mathrm{tr}(xy^t)$, where y^t is the transposed of y. The Lie algebra $\mathfrak{g}$ has the following vector space decomposition $\mathfrak{g} = \mathfrak{a} \oplus \mathfrak{b}$, where

$$\mathfrak{a} = \{x \in \mathfrak{g} \mid x_{ij} = 0, \ \forall j > i\} \quad \text{and} \quad \mathfrak{b} = \mathfrak{so}_n(\mathbb{R}),$$

i.e., $\mathfrak{b}$ is the Lie algebra of the skew-symmetric real $n \times n$ matrices, and $\mathfrak{a}$ is the Lie algebra of the real, *traceless $n \times n$ lower-triangular* matrices. One can easily prove that $\mathfrak{b}^\perp \overset{\check{B}}{=} \mathrm{Symm}_0(n)$, the vector space of the $n \times n$ real *traceless symmetric* matrices, while $\mathfrak{a}^\perp$ is the vector space of the real, $n \times n$ *strictly upper-triangular* matrices. To describe the projections defined by the decompositions $\mathfrak{g} = \mathfrak{a} \oplus \mathfrak{b}$ and $\mathfrak{g} = \mathfrak{b}^\perp \oplus \mathfrak{a}^\perp$, note that the Lie algebra $\mathfrak{g}$ decomposes as

$$\mathfrak{g} = \mathfrak{g}_+ \oplus \mathfrak{g}_0 \oplus \mathfrak{g}_-, \tag{7.36}$$

where $\mathfrak{g}_\pm$ are the vector spaces of the $n \times n$, real strictly upper $(+)$ and lower $(-)$ triangular matrices, while $\mathfrak{g}_0$ is the vector space of the $n \times n$ traceless diagonal ones. Using (7.36), one can define

$$\pi_{\mathfrak{a}} x = x_- + x_0 + x_+^t \tag{7.37}$$

$$\pi_{\mathfrak{b}} x = x_+ - x_+^t \tag{7.38}$$

$$\pi_{\mathfrak{a}^\perp} x = x_+ - x_-^t$$

$$\pi_{\mathfrak{b}^\perp} x = x_- + x_0 + x_-^t,$$

for all $x \in \mathfrak{g}$. Note that $\pi_{\mathfrak{a}} x + \pi_{\mathfrak{a}^\perp} x \neq x$ (see Remark 7.26). $\triangle$

Going back to the general setting, using the isomorphism $\mathfrak{b}^\perp \overset{\check{B}}{\simeq} \mathfrak{a}^*$, one can endow $\mathfrak{b}^\perp$ with a Poisson bracket, whose explicit construction goes as follows. First, let $\tilde{f} \in C^\infty(\mathfrak{g})$ be an extension of a given $f \in C^\infty(\mathfrak{b}^\perp)$, and note that for every $x \in \mathfrak{b}^\perp$, $\pi_{\mathfrak{a}} \nabla_x \tilde{f}$ depends only on f and not on its extension to $\mathfrak{g}$. One way to see it is to observe that $\nabla_x \tilde{f}$ can be decomposed in two direct summands one in $\mathfrak{b}^\perp$ and the other in $\mathfrak{b}$. Applying $\pi_{\mathfrak{a}}$, which is the projection to $\mathfrak{a}$ along $\mathfrak{b}$, only the former will survive, proving the statement (note that $x \in \mathfrak{b}^\perp$). After all these preliminary comments, the Poisson bracket on $f, g \in C^\infty(\mathfrak{b}^\perp)$ is defined as follows:

$$\{f, g\}_{\mathfrak{b}^\perp}(x) = B(x, [\pi_{\mathfrak{a}} \nabla_x \tilde{f}, \pi_{\mathfrak{a}} \nabla_x \tilde{g}]), \ \forall x \in \mathfrak{b}^\perp, \tag{7.39}$$

where $\tilde{f}, \tilde{g} \in C^\infty(\mathfrak{g})$ are any two extensions of f and, respectively, g. Before stating the main result of this section, it is worth making still one general comment. Let $j : \mathfrak{b}^\perp \to \mathfrak{g}^*$ be defined by $j = i \circ \check{B}$, where $i : \mathfrak{a}^* \to \mathfrak{g}^*$ is the canonical immersion. Then

Lemma 7.28 *The Hamiltonian vector field with respect to (7.39) with Hamiltonian $f \in C^\infty(\mathfrak{b}^\perp)$ is given by the formula*

$$X_f(x) = \pi_{\mathfrak{b}^\perp}\left[x, \pi_\mathfrak{a}\nabla_x\tilde{f}\right], \ \forall x \in \mathfrak{b}^\perp, \tag{7.40}$$

where $\pi_{\mathfrak{b}^\perp} : \mathfrak{g} \to \mathfrak{b}^\perp$ was defined in Remark 7.26.

Proof In fact, if $f, g \in C^\infty(\mathfrak{b}^\perp)$,

$$\begin{aligned}
\{f, g\}_{\mathfrak{b}^\perp}(x) &\stackrel{(7.39)}{=} B\left(x, \left[\pi_\mathfrak{a}\nabla_x\tilde{f}, \pi_\mathfrak{a}\nabla_x\tilde{g}\right]\right) \\
&= B\left(\left[x, \pi_\mathfrak{a}\nabla_x\tilde{f}\right], \pi_\mathfrak{a}\nabla_x\tilde{g}\right) \\
&= B\left(\pi_{\mathfrak{b}^\perp}\left[x, \pi_\mathfrak{a}\nabla_x\tilde{f}\right], \pi_\mathfrak{a}\nabla_x\tilde{g}\right) \\
&= B\left(\pi_{\mathfrak{b}^\perp}\left[x, \pi_\mathfrak{a}\nabla_x\tilde{f}\right], \nabla_x\tilde{g}\right) \\
&= \langle dg_x, \pi_{\mathfrak{b}^\perp}\left[x, \pi_\mathfrak{a}\nabla_x\tilde{f}\right]\rangle,
\end{aligned}$$

which proves the statement since $\{f, g\}_{\mathfrak{b}^\perp}(x) = \langle dg_x, X_f(x)\rangle$, for all $x \in \mathfrak{b}^\perp$. $\square$

We can now state and prove our main result. Under the assumptions made above

Theorem 7.29 (Adler-Kostant-Symes Theorem) *If $\tilde{f}, \tilde{g} \in C^\infty(\mathfrak{g}^*)$ are Ad-invariant and if $f = j^*\tilde{f}$, $g = j^*\tilde{g}$, then:*

(i) $\dot{x} = [\pi_\mathfrak{b}(\nabla_x f), x]$, for all $x \in \mathfrak{b}^\perp$
(ii) $\{f, g\}_{\mathfrak{b}^\perp} = 0$

Proof To prove (i), we need to compute the Hamiltonian vector field with respect to (7.39) whose Hamiltonian is f, i.e.,

$$\begin{aligned}
X_f(x) &\stackrel{(7.40)}{=} \pi_{\mathfrak{b}^\perp}\left[x, \pi_\mathfrak{a}\nabla_x\tilde{f}\right] \\
&= \pi_{\mathfrak{b}^\perp}\left[x, \nabla_x\tilde{f} - \pi_\mathfrak{b}\nabla_x\tilde{f}\right] \\
&\stackrel{(7.35)}{=} \pi_{\mathfrak{b}^\perp}\left[\pi_\mathfrak{b}\nabla_x\tilde{f}, x\right] \\
&= \left[\pi_\mathfrak{b}\nabla_x\tilde{f}, x\right],
\end{aligned}$$

where the last equality follows from $[\mathfrak{b}, \mathfrak{b}^\perp] \subset \mathfrak{b}^\perp$. To prove part (ii), it suffices to compute

$$\{f, g\}_{\mathfrak{b}^\perp}(x) \overset{(7.39)}{=} B\left(x, \left[\pi_\mathfrak{a}\nabla_x \tilde{f}, \pi_\mathfrak{a}\nabla_x \tilde{g}\right]\right)$$

$$= B\left(x, \left[\nabla_x \tilde{f} - \pi_\mathfrak{b}\nabla_x \tilde{f}, \nabla_x \tilde{g} - \pi_\mathfrak{b}\nabla_x \tilde{g}\right]\right)$$

$$\overset{(7.35)}{=} B\left(x, \left[\pi_\mathfrak{b}\nabla_x \tilde{f}, \pi_\mathfrak{b}\nabla_x \tilde{g}\right]\right)$$

$$= 0,$$

for all $x \in \mathfrak{b}^\perp$ since $\mathfrak{b}$ is a Lie subalgebra of $\mathfrak{g}$. $\qquad\square$

As remarked above, the linear Poisson structure on $\mathfrak{a}^*$ can be transferred to $\mathfrak{b}^\perp$ using the isomorphism $\mathfrak{b}^\perp \overset{\check{B}}{\simeq} \mathfrak{a}^*$ (see Formula 7.39). Let $\mathfrak{A}$ be a Lie group whose Lie algebra is $\mathfrak{a}$. Without loss of generality, $\mathfrak{A}$ can be chosen connected and simply connected. Said that, in the next proposition, an explicit formula for the induced $\mathfrak{A}$-action on $\mathfrak{b}^\perp$ will be given.

Proposition 7.30 *Under the identification* $\mathfrak{b}^\perp \overset{\check{B}}{\simeq} \mathfrak{a}^*$, $\mathfrak{b}^\perp$ *inherits a* $\mathfrak{A}$-*action defined by the following formula:*

$$g.x = \pi_{\mathfrak{b}^\perp} \operatorname{Ad}_g x,$$

$\forall x \in \mathfrak{b}^\perp$ *and* $g \in \mathfrak{A}$.

Proof The sought formula will be obtained, requiring that for all $g \in \mathfrak{A}$,

$$g.x = \check{B}^{-1} \circ \operatorname{Ad}_g^\sharp \circ \check{B}x, \tag{7.41}$$

for all $x \in \mathfrak{b}^\perp$. On the other hand, for all $x \in \mathfrak{b}^\perp$, $y \in \mathfrak{a}$, and $g \in \mathfrak{A}$

$$\left\langle \operatorname{Ad}_g^\sharp \check{B}x, y \right\rangle = B(\operatorname{Ad}_g x, y)$$

$$= B(\pi_{\mathfrak{a}^\perp} \operatorname{Ad}_g x + \pi_{\mathfrak{b}^\perp} \operatorname{Ad}_g x, y)$$

$$= B(\pi_{\mathfrak{b}^\perp} \operatorname{Ad}_g x, y)$$

$$= \left\langle \check{B}(\pi_{\mathfrak{b}^\perp} \operatorname{Ad}_g x), y \right\rangle,$$

implying that

$$\pi_{\mathfrak{b}^\perp} \operatorname{Ad}_g x = \check{B}^{-1} \circ \operatorname{Ad}_g^\sharp \circ \check{B} x, \ \forall x \in \mathfrak{b}^\perp, \ g \in \mathfrak{A},$$

which, compared with (7.41), entails the thesis. $\qquad\square$

In the next example, we will apply the previous result to recover the Lax representation of the Toda system in its symmetric form (see Formula 7.8).

Example 7.31 Below we adopt the assumptions and notations of Example 7.27. Following [5] and [230], we start noticing that the orbit of

$$
z = \begin{pmatrix} 0 & 1 & 0 & \cdots & 0 \\ 1 & 0 & 1 & \cdots & \vdots \\ 0 & 1 & \ddots & \ddots & 0 \\ \vdots & \vdots & \ddots & \ddots & 1 \\ 0 & 0 & \cdots & 1 & 0 \end{pmatrix} \in \mathfrak{b}^{\perp},
$$

under the $\mathfrak{A}$-action defined in (7.41) coincides with the set of the Jacobi matrices, i.e., $\mathfrak{A}.z = \mathrm{Jac}_n$ (see Definition 7.3). Since the bilinear form B identifies the Ad-invariant functions on $\mathfrak{g}$ with the $\mathrm{Ad}^\sharp$-invariant functions on $\mathfrak{g}^*$, and since in our example $B = \mathrm{tr}$, the functions $f_k \in C^\infty(\mathfrak{g})$ defined by $f_k(x) = \frac{\mathrm{tr} x^k}{k}$ will define Hamiltonian systems on $\mathfrak{b}^\perp$ of the type defined in (i) of Theorem 7.29. To find out the explicit formulas, we need to compute $\nabla_x f_k$. Let $y \in \mathfrak{g}$

$$
\begin{aligned}
B(\nabla_x f_k, y) = \langle df_{k,x}, y \rangle &= \frac{d}{dt}\bigg|_{t=0} f_k(x + ty) \\
&= \frac{1}{k}\frac{d}{dt}\bigg|_{t=0} \mathrm{tr}\left((x + ty')^k\right) \\
&= \mathrm{tr}\left(x^{k-1}y\right) \\
&= B\left(x^{k-1}, y\right),
\end{aligned}
$$

since x^{k-1} is symmetric. This computation implies that

$$
\nabla_x f_k = x^{k-1}. \tag{7.42}
$$

For every $x \in \mathfrak{b}^\perp$, (i) in Theorem 7.29 reads $\dot{x} = [\pi_\mathfrak{b}\nabla_x f_2, x] \overset{(7.42)}{=} [\pi_\mathfrak{b}x, x]$, and if $x = L \in \mathrm{Jac}_n$, one recovers (7.8) as required. $\triangle$

3.2 R-Matrix Formalism

We will now give a quick introduction to the so-called R-matrix formalism, which, as it was already mentioned above, represents a powerful generalization of the AKS method. Because of its relevance, nowadays, there is a huge amount of literature treating the results presented below. For more information, we refer the reader to the original paper [218] and to the presentation given in [213] and [139]. Let $(\mathfrak{g}, [\cdot, \cdot])$ be a (finite) dimensional Lie algebra and let $R \in \mathrm{End}(\mathfrak{g})$ such that $[\cdot, \cdot]_R : \mathfrak{g} \otimes \mathfrak{g} \to \mathfrak{g}$ defined by

$$[x, y]_R = \frac{1}{2} \left([R(x), y] + [x, R(y)] \right), \quad x, y \in \mathfrak{g}, \tag{7.43}$$

is a Lie bracket. Under these assumptions, 7.43 induces the Lie-Poisson bracket $\{\cdot, \cdot\}_R$ on $\mathfrak{g}^*$ defined by

$$\{f, g\}_R(\alpha) := \langle \alpha, [df_\alpha, dg_\alpha]_R \rangle, \ \forall f, g \in C^\infty(\mathfrak{g}^*), \ \alpha \in \mathfrak{g}^*$$

(see Example 2.8). We will explain now how, starting from this framework, one can define Hamiltonian system with a large supply of first integrals. To this end, for every $f \in C^\infty(\mathfrak{g}^*)$, let $X_{f,R}$ denote its Hamiltonian vector field *with respect to the Poisson bracket* $\{\cdot, \cdot\}_R$, i.e., $X_{f,R}(g) = \{f, g\}_R$ for all $g \in C^\infty(\mathfrak{g}^*)$ or, equivalently, $X_{f,R} = \Pi_R^\sharp(df)$, where Π_R is the Poisson bivector defined by $\{\cdot, \cdot\}_R$ (see Sect. 1.2). Note that the Lie brackets $[\cdot, \cdot]$ and $[\cdot, \cdot]_R$ yield two *different* coadjoint actions on $\mathfrak{g}^*$

$$\left\langle \mathrm{ad}_x^\sharp(\alpha), y \right\rangle = -\langle \alpha, [x, y] \rangle \quad \text{and} \quad \left\langle \mathrm{ad}_x^{\sharp, R}(\alpha), y \right\rangle = -\frac{1}{2}\langle \alpha, [R(x), y]$$

$$+[x, R(y)] \rangle, \ \forall x, y \in \mathfrak{g}, \ \alpha \in \mathfrak{g}^*.$$

The definition of $\mathrm{ad}^{\sharp, R}$ implies

$$\{f, g\}_R(\alpha) = \frac{1}{2}\left\langle \mathrm{ad}_{dg_\alpha}^\sharp(\alpha), R(df_\alpha) \right\rangle - \langle \mathrm{ad}_{df_\alpha}^\sharp(\alpha), R(dg_\alpha) \rangle), \ \forall f, g \in C^\infty(\mathfrak{g}^*), \alpha \in \mathfrak{g}^*. \tag{7.44}$$

In particular

Proposition 7.32 *If f is a Casimir of the linear Poisson structure defined by $[\cdot, \cdot]$, then*

$$X_{f,R}(\alpha) = -\frac{1}{2} \, \mathrm{ad}_{R(df_\alpha)}^\sharp(\alpha), \forall \alpha \in \mathfrak{g}^*. \tag{7.45}$$

Proof The proof follows directly from (7.44), noticing that under the assumption on f, $X_f(\alpha) = \Pi_\alpha^\sharp(df_\alpha, \cdot) = 0$, for all $\alpha \in \mathfrak{g}^*$. □

Together with the previous result, it is worth noticing that (7.44) implies that

Corollary 7.33 *If f and g are two Casimirs of the Lie-Poisson bracket defined by $[\cdot, \cdot]$, then $\{f, g\}_R = 0$.*

Note, however, that generally, the Casimir functions of the *original* linear Poisson structure do not lie in the center of the Poisson algebra $(C^\infty(\mathfrak{g}^*), \{\cdot, \cdot\}_R)$, and for this reason, they can be used as Hamiltonians functions for non-trivial Hamiltonian vector fields with respect to the *new* Poisson structure defined by R. Corollary 7.33 implies that each of these Hamiltonian vector fields comes equipped with plenty of first integrals, i.e., the Casimirs of the original linear Poisson structure. Before moving on with our presentation, it is worth making a comment about the assumption we made about R, i.e., that (7.43) is a Lie bracket. The non-trivial part of this hypothesis is that the bracket $[\cdot, \cdot]_R$ satisfies the Jacobi identity, which, if spelled out, can be phrased as follows:

Proposition 7.34 ([213]) *The bracket (7.43) satisfies the Jacobi identity if and only if*

$$[F_R(x, y), z] + [F_R(z, x), y] + [F_R(y, z), x] = 0, \ \forall x, y, z \in \mathfrak{g},$$

where

$$F_R(x, y) = [Rx, Ry] - R([Rx, y] + [x, Ry]), \ \forall x, y \in \mathfrak{g}.$$

The proof of this result is by a direct computation, and it will be left to the reader as an exercise. On the other hand, it is worth noting that this result provides sufficient conditions for (7.43) is a Lie bracket. More precisely

Corollary 7.35 *If there exists $c \in \mathbb{R}$ such that*

$$F_R(x, y) = -c^2[x, y], \ \forall x.y \in \mathfrak{g}, \tag{7.46}$$

then (7.43) is a Lie bracket.

In spite of the fact that the proof of this result is almost evident, it opens the door to a systematic way to define new Lie algebra structures starting from an old one and a solution of (7.46). Following the terminology of [213], (7.46) for $c = 0$ is called *Classical Yang-Baxter Equation* (CYBE), while when $c \neq 0$, it is called *modified Classical Yang-Baxter Equation* (mCYBE) . The solutions of the CYBE are called Classical R-matrices, while the solutions of the mCYBE are called modified R-matrices.

Going back to our main concern, we present a formula that expresses (7.45) and the corresponding system of ODE, in terms of a gradient like system, in the same flavor of (2.10). To this end, we assume that $(\mathfrak{g}, B)$ is a quadratic Lie algebra. Under this assumption, (7.44) induces the following linear Poisson bracket on $\mathfrak{g}$

$$\{f, g\}_R(x) = B\left(x, [\nabla_x f.\nabla_x g]_R\right), \ \forall x \in \mathfrak{g}, \ f, g \in C^\infty(\mathfrak{g})$$

(see 2.11). Using (7.43), this formula can be rewritten as

$$\{f, g\}_R(x) = \frac{1}{2} B\left(x, [R(\nabla_x f), \nabla_x g] + [\nabla_x f, R(\nabla_x g)]\right). \tag{7.47}$$

If $f \in C^\infty(\mathfrak{g})$ is an Ad-invariant function, then Lemma 7.23 yields

Proposition 7.36

$$X_{R,f}(x) = \frac{1}{2}[x, R(\nabla_x f)], \ \forall x \in \mathfrak{g}.$$

Proof Since f is Ad-invariant

$$B(x, [\nabla_x f, R(\nabla_x g)]) \overset{(7.35)}{=} 0,$$

and (7.47) reduces to

$$\{f, g\}_R(x) = \frac{1}{2} B(x, [R(\nabla_x f), \nabla_x g]),$$

which, by the $\mathfrak{g}$-invariance of B, can be written as

$$\{f, g\}_R(x) = \frac{1}{2} B([x, R(\nabla_x f)], \nabla_x g).$$

In other words,

$$\langle dg_x, X_{R,f}(x)\rangle = \langle dg_x, \Pi_R^\sharp(df_x)\rangle = \frac{1}{2} B([x, R(\nabla_x f)], \nabla_x g)$$

$$= \frac{1}{2}\langle dg_x, [x, R(\nabla_x f)]\rangle,$$

which is what we wanted to prove. $\qquad\qquad\square$

The previous argument implies that for each Ad-invariant f, the system of ODE defined by the $X_{R,f}$ assumes the following Lax-type form

$$\dot{x} = \frac{1}{2}[x, R(\nabla_x f)] \tag{7.48}$$

(see Proposition 6.25).

In the next example, we will see how to deduce the AKS Theorem from the R-matrix formalism.

Example 7.37 Suppose that $(\mathfrak{g}, B)$ is a quadratic Lie algebra endowed with a vector space decomposition $\mathfrak{g} = \mathfrak{a} \oplus \mathfrak{b}$ where both $\mathfrak{a}$ and $\mathfrak{b}$ are Lie subalgebras of $\mathfrak{g}$. First, we prove the following.

Proposition 7.38 *If $R : \mathfrak{g} \to \mathfrak{g}$ is the linear operator on $\mathfrak{g}$ defined by*

$$R := \pi_\mathfrak{a} - \pi_\mathfrak{b},$$

then (7.43) is a Lie bracket.

Proof The skew-symmetry follows by direct inspection. To prove that (7.43) satisfies the Jacobi identity, one observes that

$$[x, y]_R = \frac{1}{2}\left([\pi_\mathfrak{a} x - \pi_\mathfrak{b} x, y] + [x, \pi_\mathfrak{a} y - \pi_\mathfrak{b} y]\right)$$

$$= \frac{1}{2}\left([\pi_\mathfrak{a} x - \pi_\mathfrak{b} x, \pi_\mathfrak{a} y + \pi_\mathfrak{b} y] + [\pi_\mathfrak{a} x + \pi_\mathfrak{b} x, \pi_\mathfrak{a} y - \pi_\mathfrak{b} y]\right)$$

$$= [\pi_\mathfrak{a} x, \pi_\mathfrak{a} y] - [\pi_\mathfrak{b} x, \pi_\mathfrak{b} y].$$

Then the formula

$$[[x, y]_R, z]_R = [\pi_\mathfrak{a}([x, y]_R, \pi_\mathfrak{a} z] - [\pi_\mathfrak{b}([x, y]_R, \pi_\mathfrak{b} z]$$

$$= [[\pi_\mathfrak{a} x, \pi_\mathfrak{a} y], \pi_\mathfrak{a} z] - [[\pi_\mathfrak{b} x, \pi_\mathfrak{b} y], \pi_\mathfrak{b} z],$$

together with the hypothesis that both $\mathfrak{a}$ and $\mathfrak{b}$ are Lie subalgebras of $\mathfrak{g}$, entails the proof of the claim. $\square$

Choosing R as in the previous proposition, for every Ad-invariant f, one can write

$$\dot{x} \stackrel{(7.48)}{=} \frac{1}{2}[x, R(\nabla_x f)] = \frac{1}{2}[x, \pi_\mathfrak{a} \nabla_x f - \pi_\mathfrak{b} \nabla_x f]$$

$$\stackrel{(7.35)}{=} \frac{1}{2}[x, \nabla_x f] - [x, \pi_\mathfrak{b} \nabla_x f]$$

$$= [\pi_\mathfrak{b} \nabla_x f, x],$$

recovering the description of the system of ODE given in Theorem 7.29. $\triangle$

3.2.1   The Toda Orbit from the Hessenberg Viewpoint

Recall that the Jac_n is conjugate to a set of Hessenberg matrices and that (7.10) admit a non-symmetric Lax representation (see Sect. 2.1). In the final part of this chapter, the Hessenberg description of Jac_n will be studied from a more Lie theoretical viewpoint. The following results are borrowed from [142, 143]. We start setting up the notations will be used hereafter. In particular, we set $\mathfrak{g} = \mathfrak{sl}_n(\mathbb{R})$ and $G = \mathrm{SL}_n(\mathbb{R})$, i.e., G denotes the Lie group of the $n \times n$ real invertible matrices whose determinant is equal to 1; note that $\mathfrak{g} = \mathrm{Lie}\, G$. Let $\mathfrak{b}_-, \mathfrak{n}_-$ and $\mathfrak{b}_+$ be the Lie subalgebras of $\mathfrak{g}$ defined by

$$\mathfrak{b}_- = \{x \in \mathfrak{g}, \ x_{ij} = 0 \text{ if } i < j\}, \mathfrak{n}_- = \{x \in \mathfrak{g}, \ x_{ij} = 0 \text{ if } i \le j\} \text{ and}$$

$$\mathfrak{b}_+ = \{x \in \mathfrak{g}, \ x_{ij} = 0 \text{ if } i > j\}.$$

Furthermore, let $B_- = \{g \in G, \ g_{ij} = 0 \text{ if } i < j\}$ be the Lie group whose Lie algebra is $\mathfrak{b}_-$. As far as terminology is concerned, $\mathfrak{b}_{\pm}$ are examples of *solvable* Lie algebras, called *Borel* subalgebras (more precisely, they are maximal solvable subalgebras); B_- is a *Borel* subgroup of G, and $\mathfrak{n}_-$ is a *nilpotent* subalgebra of $\mathfrak{g}$ (see Definition C.14). Note that $\mathfrak{g} = \mathfrak{b}_+ \ltimes \mathfrak{n}_-$, i.e., $\mathfrak{g}$ is the *semi-direct product* of $\mathfrak{n}_+$ with $\mathfrak{b}_+$ (see Example C.65). In particular,

$$\mathfrak{g} = \mathfrak{b}_+ \oplus \mathfrak{n}_-. \tag{7.49}$$

Finally, let $B : \mathfrak{g} \times \mathfrak{g} \to \mathbb{R}$ be the bilinear form defined by

$$B(x, y) = \mathrm{tr}(xy).$$

Having pointed out this, we start observing that

Lemma 7.39 *The isomorphism* $\mathfrak{g} \overset{\check{B}}{\simeq} \mathfrak{g}^*$ *restricts to the following isomorphisms:*

$$\mathfrak{b}_+ \overset{\check{B}}{\simeq} \mathfrak{b}_-^* \quad \text{and} \quad \mathfrak{n}_- \overset{\check{B}}{\simeq} \mathfrak{b}_-^\circ,$$

where $\mathfrak{b}_-^\circ$ *is the annihilator of* $\mathfrak{b}_-$ *in* $\mathfrak{g}^*$.

Proof The proof follows by a direct computation. In particular, if E_{ij} is the $n \times n$ elementary matrix with 1 in the (i, j)-entry and 0 elsewhere, it suffices to compute $\mathrm{tr}(x E_{ij})$ for $i \ne j$ and $\mathrm{tr}\big(x(E_{ii} - E_{i+1\,i+1})\big)$ for $i = 1, \ldots, n - 1$. In the first case, we get $\mathrm{tr}(x E_{ij}) = x_{ji} = 0$, for all $i > j$. The other case is left to the reader as an exercise. $\qquad\square$

The decomposition (7.49) yields the following projections: $\pi_+ : \mathfrak{g} \to \mathfrak{b}_+$ and $\pi_- : \mathfrak{g} \to \mathfrak{n}_-$, where $\mathrm{id}_\mathfrak{g} = \pi_+ + \pi_-$. In this way, we are back to the setting of the AKS method. In particular, Proposition 7.30 can be rephrased as follows.

Proposition 7.40 $\mathfrak{b}_+$ *is a* B_-*-space whose* B_-*-action is defined by*

$$g.b = \pi_+ \mathrm{Ad}_g \, b$$

for all $g \in B_-$ *and* $b \in \mathfrak{b}_+$.

Modulo a simple notational translation, the proof is the same as the one of Proposition 7.30, and for this reason, it will not be repeated here. We would like to stress once more that this result entails that the isomorphism $\mathfrak{b}_+ \overset{\check{B}}{\simeq} \mathfrak{b}_-^*$:

(i) Intertwines the B_--actions, i.e.,

$$
\begin{array}{ccc}
\mathfrak{b}_+ & \overset{\check{B}}{\longrightarrow} & \mathfrak{b}_-^* \\
{\scriptstyle \pi_+ \, \mathrm{Ad}_g} \downarrow & & \downarrow {\scriptstyle \mathrm{Ad}_g^\sharp} \\
\mathfrak{b}_+ & \overset{\check{B}}{\longrightarrow} & \mathfrak{b}_-^*
\end{array}
$$

commutes for all $g \in B_-$,

(ii) Defines a one-to-one correspondence between the set of the B_--orbits in $\mathfrak{b}_+$ and the set of the coadjoint orbits in $\mathfrak{b}_-^*$

Now let $\mathcal{T}_n$ be the set of all $n \times n$ real matrices X such that

$$
X = \begin{pmatrix}
\beta_1 & \alpha_1 & 0 & \cdots & & 0 \\
0 & \beta_2 & \alpha_2 & \cdots & & \vdots \\
0 & 0 & \ddots & \ddots & & 0 \\
\vdots & \vdots & & \ddots & \ddots & \alpha_{n-1} \\
0 & 0 & \cdots & & 0 & \beta_n
\end{pmatrix}
\tag{7.50}
$$

with $\sum_{i=1}^n \beta_i = 0$ and $\alpha_i > 0$ for all $i = i, \ldots, n-1$. As a consequence of what is discussed in Sect. 2.1, $\mathcal{T}_n \overset{1:1}{\simeq} \mathrm{Jac}_n$. Moreover

Theorem 7.41 $\mathcal{T}_n$ *is in a one-to-one correspondence with the orbit* $B_-.e \subset \mathfrak{b}_+$, *where*

$$
e = \begin{pmatrix}
0 & 1 & 0 & \cdots & 0 \\
0 & 0 & 1 & \cdots & \vdots \\
0 & 0 & \ddots & \ddots & 0 \\
\vdots & \vdots & \ddots & \ddots & 1 \\
0 & 0 & \cdots & 0 & 0
\end{pmatrix},
$$

As a consequence, the vector space Jac_n *can be identified with a coadjoint orbit of the group* B_-. *More precisely, if*

$$
f = \begin{pmatrix}
0 & 0 & 0 & \cdots & 0 \\
1 & 0 & 0 & \cdots & \vdots \\
0 & 1 & \ddots & \ddots & 0 \\
\vdots & \vdots & \ddots & \ddots & 0 \\
0 & 0 & \cdots & 1 & 0
\end{pmatrix},
$$

then Jac_n *is in a one-to-one correspondence with the translate of the coadjoint orbit* $B_-.e$ *by the matrix* f, *i.e.,*

$$
\mathrm{Jac}_n \simeq f + B_-.e.
$$

We are now in the position to write an explicit formula for the symplectic structure defined on Jac_n. To this end, it suffices to compute the symplectic form defined on the coadjoint orbit $B_-.e$ by the Lie-Poisson structure on $\mathfrak{b}_-^*$. Theorem 7.41 claims that every matrix belonging to $B_-.e$ is as in (7.50) with the stated conditions on its elements.

Let $(y_1, \ldots, y_{n-1}, x_1, \ldots, x_{n-1})$ be the coordinates on this set of matrices defined by

$$
\begin{cases}
x_i(X) = \alpha_i \\
y_i(X) = \sum_{j=1}^{i} \beta_j,
\end{cases}
$$

for all $i = 1, \ldots, n-1$. Here, $x_i(X) = \mathrm{tr}(E_{i+1,i}X)$ and $y_i(X) = \sum_{j=1}^{i} \mathrm{tr}(E_{jj}X)$ for all $i = 1, \ldots, n-1$. In this way, the Lie-Poisson structure of $\mathfrak{b}_-^*$, restricted to the orbit $B_-.e$, is defined as follows:

$$
\{x_i, x_j\} = 0 = \{y_i, y_j\} \quad \text{and} \quad \{x_i, y_j\} = \delta_{ij}x_j, \ \forall i, j = 1, \ldots, n-1.
$$

In fact, by definition,

$$
\{x_i, x_j\}_{B_-.e} = \mathrm{tr}\left(\left[E_{i+1\,i}, E_{j+1\,j}\right]B_-.e\right) = 0,
$$

$$
\{y_i, y_j\}_{B_-.e} = \mathrm{tr}\left(\left[\sum_{l=1}^{i} E_{ll}, \sum_{k=1}^{j} E_{kk}\right]B_-.e\right) = 0,
$$

and

$$\{x_i, y_j\}_{B_-.e} = \mathrm{tr}\left(\left[E_{i+1\,i}, \sum_{k=1}^{j} E_{kk}\right] B_-.e\right) = \delta_{ij} E_{j+1\,j}$$

for all $i, j = 1, \ldots, n - 1$. In other words, we proved the following.

Proposition 7.42 *Written in the coordinates* $(y_1, \ldots, y_{n-1}, x_1, \ldots, x_{n-1})$, *the symplectic form* ω *on the orbit* $B_-.e$ *assumes the following expression:*

$$\omega = \sum_{j=1}^{n-1} dy_j \wedge \frac{dx_j}{x_j} \tag{7.51}$$

Via the correspondence defined in Theorem 7.41, Jac_n *is a symplectic manifold whose symplectic form is defined in* (7.51).

Proof The first part follows from the discussion above. The last part follows observing that the $(y_1, \ldots, y_{n-1}, x_1, \ldots, x_{n-1})$ are coordinates also on Jac_n. $\square$

3.2.2 Solution by Factorization

Recall that in the application of the AKS method to the Toda system, $\mathfrak{g} = \mathfrak{sl}_n(\mathbb{R})$, $\mathfrak{b} = \mathfrak{so}_n(\mathbb{R})$, and $\mathfrak{a}$ was the Lie algebra of the real, $n \times n$ traceless, lower-triangular matrices (see Example 7.27). We will now show how this decomposition, when read at the level of the corresponding Lie groups, entails the solution of (7.8). What follows is borrowed, modulo minor differences, from [230, 231] (see also [194]). The construction presented hereafter is based on the so-called QL-factorization (or *QL-decomposition*), which states that every invertible matrix g admits a unique decomposition as a product $g = kl$, where k is orthogonal and l is an invertible lower-triangular matrix.

Remark 7.43 It is worth noting that in the literature, there is a different, though equivalent under our assumptions, decomposition known as *QR-decomposition*, which consists in factorizing g as $g = kr$, where g is orthogonal and r is *upper-triangular* (see, e.g., [181]). One can prove the existence of the QL-factorization starting from the QR-one by simply observing that if J is the $n \times n$ matrix obtained from the identity inverting the order of the columns, i.e., $J = [e_n, e_{n-1}, \ldots, e_2, e_1]$, where e_i is the i-th column of the identity matrix and the QR factorization of JgJ yields the QL-factorization of g (see [230, 231]). $\triangle$

Now, let $L_0 \in \mathrm{Jac}_n$ and let $k(t), l(t)$ be the factors of the QL decomposition of $\exp(tL_0) \in \mathrm{SL}_n(\mathbb{R})$ for all $t \in \mathbb{R}$. Then

Proposition 7.44

$$L(t) = k(t)^{-1} L_0 k(t) \tag{7.52}$$

is a solution of (7.8).

Proof Computing the time derivative of the left- and right-hand side of (7.52) yields

$$\dot{L}(t) = -k(t)^{-1}\dot{k}(t)k(t)^{-1}L_0 k(t) + k(t)^{-1}L_0\dot{k}(t)$$

$$= -k(t)^{-1}\dot{k}(t)k(t)^{-1}L_0 k(t) + k(t)^{-1}L_0 k(t)k(t)^{-1}\dot{k}(t)$$

$$= k(t)^{-1}\dot{k}(t)L(t) - L(t)k(t)^{-1}\dot{k}(t),$$

i.e.,

$$\dot{L}(t) = [k(t)^{-1}\dot{k}(t), L(t)].$$

One easily proves that $k(t)^{-1}\dot{k}$ is skew-symmetric. Furthermore, since $\exp(tL_0) = k(t)l(t)$, one has

$$k(t) = \exp(tL_0)l(t)^{-1}$$

$$k(t)^{-1} = l(t)\exp(-tL_0)$$

$$\dot{k}(t) = \exp(tL_0)L_0 l(t)^{-1} - \exp(tL_0)l(t)^{-1}\dot{l}(t)l(t)^{-1}.$$

All this implies

$$k(t)^{-1}\dot{k}(t) = l(t)L_0 l(t)^{-1} - \dot{l}(t)l(t)^{-1},$$

i.e.,

$$l(t)L_0 l(t)^{-1} = k(t)^{-1}\dot{k}(t) + \dot{l}(t)l(t)^{-1}, \ \forall t \in \mathbb{R}.$$

This identity, together with $\exp(tL_0) = k(t)l(t)$ and $L(t) = k(t)^{-1}L_0 k(t)$, entails

$$L(t) = k(t)^{-1}\dot{k}(t) + \dot{l}(t)l(t)^{-1}, \forall t \in \mathbb{R}. \tag{7.53}$$

Note that while $k(t)^{-1}\dot{k}(t)$ belongs to $\mathfrak{so}_n(\mathbb{R})$ for all t, $\dot{l}(t)l(t)^{-1}$ is a lower-triangular matrix. To conclude the proof of the proposition, it suffices to show that for all t

$$k(t)^{-1}\dot{k}(t) = \pi_\kappa L(t) \quad \text{and} \quad \dot{l}(t)l(t)^{-1} = \pi_\mathfrak{a} L(t) \, \forall t \in \mathbb{R},$$

where $\pi_\mathfrak{b}, \pi_\mathfrak{a}$ are defined in (7.38) and, respectively, in (7.37). To this end,

$$\pi_\mathfrak{b} L(t) \overset{(7.53)}{=} \pi_\mathfrak{b}\left(k(t)^{-1}\dot{k}(t)\right) + \pi_\mathfrak{b}\left(\dot{l}(t)l(t)^{-1}\right)$$

$$\overset{(7.38)}{=} \left(k(t)^{-1}\dot{k}(t)\right)_+ - \left(k(t)^{-1}\dot{k}(t)\right)_+^t + \left(\dot{l}(t)l(t)^{-1}\right)_+ - \left(\dot{l}(t)l(t)^{-1}\right)_+^t$$

$$= k(t)^{-1}\dot{k}(t),$$

where the cancellations take place since $\dot{l}(t)l(t)^{-1}$ is lower-triangular part and A_+ is the projection on the (strictly) upper-triangular part of A. On the other hand, the last equality follows from the fact that $k(t)^{-1}\dot{k}(t)$ is skew-symmetric. Moreover

$$\pi_\mathfrak{a} L(t) \overset{(7.53)}{=} \pi_\mathfrak{a}\left(k(t)^{-1}\dot{k}(t)\right) + \pi_\mathfrak{a}\left(\dot{l}(t)l(t)^{-1}\right),$$

where

$$\pi_\mathfrak{a}\left(k(t)^{-1}\dot{k}(t)\right) \overset{(7.37)}{=} \left(k(t)^{-1}\dot{k}(t)\right)_- + \left(k(t)^{-1}\dot{k}(t)\right)_0 + \left(k(t)^{-1}\dot{k}(t)\right)_+^t = 0$$

$$\pi_\mathfrak{a}\left(k(t)^{-1}\dot{k}(t)\right) \overset{(7.37)}{=} \left(\dot{l}(t)l(t)^{-1}\right)_- + \left(\dot{l}(t)l(t)^{t-1}\right)_0 + \left(\dot{l}(t)l(t)^{-1}\right)_+^t = \dot{l}(t)l(t)^{-1},$$

concluding the proof of the proposition. $\square$

As remarked in [231], it is worth noting that since the QL-factorization of $\exp(tL_0)$ can be performed for all t, the solutions obtained above will be defined for all t as well. In spite of this remarkable property, the solutions of (7.8) obtained via the previous method are, in general, not very explicit since they presuppose the knowledge of the QL-factorization of the time-exponential. In the case of the two-particle Toda system, the previous construction yields what follows (see [111]).

Example 7.45 If $L_0 \in \mathrm{Jac}_2$ is defined by

$$L_0 = \begin{pmatrix} b_1(0) & a(0) \\ a(0) & b_2(0) \end{pmatrix} = \begin{pmatrix} 0 & a \\ a & 0 \end{pmatrix},$$

then

$$\exp(tL_0) = \begin{pmatrix} \cosh(ta) & \sinh(at) \\ \sinh(ta) & \cosh(at) \end{pmatrix}.$$

The QL-factors of this matrix are

$$k(t) = \frac{1}{\sqrt{f_a(t)}}\begin{pmatrix} \cosh(ta) & \sinh(at) \\ -\sinh(at) & \cosh(ta) \end{pmatrix} \quad \text{and} \quad l(t) = \frac{1}{\sqrt{f_a(t)}}\begin{pmatrix} 1 & 0 \\ g_a(t) & f_a(t) \end{pmatrix}$$

where $f_a(t) = \cosh^2(at) + \sinh^2(at)$ and $g_a(t) = 2\sinh(at)\cosh(at)$. Then

$$L(t) = k(t)^{-1} L_0 k(t) = \frac{a}{\sqrt{f_a(t)}} \begin{pmatrix} -g_a(t) & 1 \\ 1 & g_a(t), \end{pmatrix}$$

entailing

$$a(t) = \frac{a}{\cosh^2(at) + \sinh^2(at)} \quad \text{and} \quad b(t) = -2a \frac{\sinh(at)\cosh(at)}{\sinh^2(at) + \cosh^2(at)}.$$

$\triangle$

After this long discussion about the solutions of the open Toda system, we go back to a topic of a more geometrical flavor. More precisely in the next subsection, we will introduce *a* bi-Hamiltonian formulation of the open Toda system. We refer the reader to Chap. 5 for the basic notions about bi-Hamiltonian geometry.

4 A Bi-Hamiltonian Representation for the Toda System

In this section, we prove that there exists a Poisson bivector field Π_1, which, together with the canonical one of $\mathbb{R}^{2n}$, defines a bi-Hamiltonian representation for the vector field (7.4). Recall that the phase-space of the Toda system is $\mathbb{R}^{2n}$ endowed with its canonical symplectic form $\omega_0 = \sum_{i=1}^{n} dp_i \wedge dx_i$. The corresponding Poisson bivector field is, following our conventions, $\Pi_0 = \sum_{i=1}^{n} \frac{\partial}{\partial p_i} \wedge \frac{\partial}{\partial x_i}$ (see Formula (2.28)). The first result we want to present is enclosed in the following.

Theorem 7.46 *The Toda system described by the Hamiltonian* (7.2) *is bi-Hamiltonian with respect to the PN-structure given by* Π_0 *and the recursion operator N below:*

$$\Pi_0 = \sum_{i=1}^{n} \frac{\partial}{\partial p_i} \wedge \frac{\partial}{\partial x_i} \quad and$$

$$N = \sum_{i=1}^{n} p_i \left(dx_i \otimes \frac{\partial}{\partial x_i} + dp_i \otimes \frac{\partial}{\partial p_i} \right) + \sum_{i<j} \left(dp_j \otimes \frac{\partial}{\partial x_i} - dp_i \otimes \frac{\partial}{\partial x_j} \right)$$

$$+ \sum_{i=1}^{n-1} \exp(x_i - x_{i+1}) \left(dx_i \otimes \frac{\partial}{\partial p_{i+1}} - dx_{i+1} \otimes \frac{\partial}{\partial p_i} \right). \tag{7.54}$$

The corresponding compatible pair of Poisson tensors is given by

$$\Pi_0 \quad and \quad \Pi_1 = \sum_{i<j} \frac{\partial}{\partial x_i} \wedge \frac{\partial}{\partial x_j} + \sum_{i=1}^{n} p_i \frac{\partial}{\partial p_i} \wedge \frac{\partial}{\partial x_i} + \sum_{i=1}^{n-1} \exp(x_i - x_{i+1}) \frac{\partial}{\partial p_i} \wedge \frac{\partial}{\partial p_{i+1}}.$$

$$(7.55)$$

Proof For the proof that Π_0 and N define a PN-structure (see [63]). From this, it follows that Π_0 and Π_1 are a compatible Poisson pair, where $\Pi_1^\sharp = N\Pi_0^\sharp$. Note that for this computation, the basis of one-forms on $\mathbb{R}^{2n}$ is ordered as $dx_1, \ldots, dx_n, dp_1, \ldots, dp_n$. Once this is established, it is enough to prove that there exists $K \in C^\infty(\mathbb{R}^{2n})$ such that $\Pi_0^\sharp(dH) = \Pi_1^\sharp(dK)$, where H is the Hamiltonian (7.2). A simple computation shows that $K = \sum_{i=1}^{n} p_i$ satisfies this condition, concluding the proof. It is worth noting that K is total linear momentum of the Toda system. $\qquad\square$

Example 7.47 (The Three-Particle Toda System) In this case, the recursion operator (7.54) assumes the following form:

$$N = \begin{bmatrix} p_1 & 0 & 0 & 0 & 1 & 1 \\ 0 & p_2 & 0 & -1 & 0 & 1 \\ 0 & 0 & p_3 & -1 & -1 & 0 \\ 0 & -e^{(x_1-x_2)} & 0 & p_1 & 0 & 0 \\ e^{(x_1-x_2)} & 0 & -e^{(x_2-x_3)} & 0 & p_2 & 0 \\ 0 & e^{(x_2-x_3)} & 0 & 0 & 0 & p_3 \end{bmatrix}.$$

$$\triangle$$

It is worth observing that the results of Sect. 6.2 can be applied to the present case. More precisely, one can prove the following.

Proposition 7.48 *The vector field*

$$Z = \sum_{i=1}^{n} (n+1-2i) \frac{\partial}{\partial x_i} + \sum_{i=1}^{n} p_i \frac{\partial}{\partial p_i} \tag{7.56}$$

is a conformal symmetry *of Π_0, Π_1, and K (see Definition 5.120). More precisely*

$$\mathscr{L}_Z \Pi_0 = -\Pi_0 \quad \mathscr{L}_Z \Pi_1 = 0 \quad and \quad \mathscr{L}_Z K = K.$$

Proof The first and the last identity can be proved with a direct computation. For example,

$$\mathscr{L}_Z K = \sum_{i,j=1}^{n} \mathscr{L}_{p_i \frac{\partial}{\partial p_i}} p_j = \sum_{i=1}^{n} p_i = K,$$

and

$$
\mathcal{L}_Z \Pi_0 = \sum_{i=1}^{n} \left[Z, \frac{\partial}{\partial p_i} \right] \wedge \frac{\partial}{\partial x_i} + \sum_{i=1}^{n} \frac{\partial}{\partial p_i} \wedge \left[Z, \frac{\partial}{\partial x_i} \right]
$$

$$
= \sum_{i=1}^{n} \sum_{j=1}^{n} \left[p_j \frac{\partial}{\partial p_j}, \frac{\partial}{\partial p_i} \right] \wedge \frac{\partial}{\partial x_i}
$$

$$
- \sum_{i=1}^{n} \frac{\partial}{\partial p_i} \wedge \frac{\partial}{\partial x_i}.
$$

To prove the remaining identity, it suffices to note that Z is Hamiltonian with respect to Π_1 with Hamiltonian function $f = -\sum_{i=1}^{n} x_i$. To prove this statement, one can argue by induction on n. It is simple to check that it holds for n, say, equal to 3. Supposing that $\Pi_1^{\sharp}(df) = Z$ for $n - 1$, i.e.,

$$
- \sum_{i=1}^{n-1} \Pi_1^{\sharp}(dx_i) = \sum_{i=1}^{n-1} (n - 2i) \frac{\partial}{\partial x_i} + \sum_{i=1}^{n-1} p_i \frac{\partial}{\partial p_i},
$$

we are left to show that it holds for n. To this end, one notes that

$$
\Pi_1^{\sharp}(df) = - \sum_{i=1}^{n-1} \Pi_1^{\sharp}(dx_i) - \Pi_1^{\sharp}(dx_n) = \sum_{i=1}^{n-1} (n - 2i) \frac{\partial}{\partial x_i} + \sum_{i=1}^{n-1} p_i \frac{\partial}{\partial p_i}
$$

$$
- (n - 1) \frac{\partial}{\partial x_n} + \sum_{i=1}^{n-1} \frac{\partial}{\partial x_i} + p_n \frac{\partial}{\partial p_n}.
$$

The sum of the first with the fourth summand yields $\sum_{i=1}^{n-1} (n + 1 - 2i) \frac{\partial}{\partial x_i}$, while the sum of the second with the last summand yields $\sum_{i=1}^{n} p_i \frac{\partial}{\partial p_i}$. Then

$$
\Pi_1^{\sharp}(df) = \sum_{i=1}^{n-1} (n + 1 - 2i) \frac{\partial}{\partial x_i} - (n - 1) \frac{\partial}{\partial x_n} + \sum_{i=1}^{n} p_i \frac{\partial}{\partial p_i} = \sum_{i=1}^{n} (n + 1 - 2i) \frac{\partial}{\partial x_i}
$$

$$
+ \sum_{i=1}^{n} p_i \frac{\partial}{\partial p_i},
$$

proving the statement. $\qquad\square$

The results enclosed in the previous proposition, together with Formula 5.120, yield

$$\mathscr{L}_Z N = N.$$

Moreover, if $X = \Pi_1^{\sharp}(dK)$, then

$$\mathscr{L}_Z X = \mathscr{L}(\Pi_1^{\sharp}(dK)) = (\cancel{\mathscr{L}_Z \Pi_1^{\sharp}})(dK) + \Pi_1^{\sharp} d(\mathscr{L}_Z K) = \Pi_1^{\sharp}(dK) = X.$$

In other words

Lemma 7.49 *The vector field in* (7.56) *is a conformal symmetry of conformal weight equal to* 1 *for both X and N.*

The previous lemma implies the following.

Proposition 7.50 *If* $X_k = N^k X$ *and* $Z_k = N^k Z$ *for all* $k \geq 0$, *then*

$$\mathscr{L}_{Z_k} N = N^k, \quad [Z_r, X_s] = (s+1)X_{r+s} \quad and \quad [Z_r, Z_s] = (s-r)Z_{r+s},$$

$$(7.57)$$

for all $r, s \geq 0$. *Moreover,*

$$\mathscr{L}_{Z_s} \Pi_r = (r - s - 1)\Pi_{r+s}, \quad and \quad d\langle dH_r, Z_s \rangle = (r + s + 1)dH_{r+s}, \; \forall r, s,$$

$$(7.58)$$

where $dH_r = {}^t N^r (dH_0)$ *for all* r.

Proof The proof of (7.57) follows from Proposition 5.121, taking $a = 1 = b$ and noticing that N is a recursion operator for X. Equation (7.58) follows from Theorem 5.124 with $a = -1$ and $b = 0$. $\qquad\square$

4.1 A Bi-Hamiltonian Structure in Flaschka Coordinates

The Toda equations (7.10) written in Flaschka coordinates also admit a bi-Hamiltonian representation, which will be presented here below. To this end, we start observing that (7.6) can be interpreted in a more geometrical terms as the components of a map $\mathscr{F}$ from $\mathbb{R}^{2n}$ onto $\mathbb{R}^{2n-1}$, which a simple computation shows to be a *submersion*. Let $Z = \sum_{k=1}^{n} \frac{\partial}{\partial x_i}$ and let

$$\gamma_m^Z(t) = \begin{cases} x_i(t) = & x_i(0) + t \\ p_i(t) = & p_i(0), \; \forall i = 1, \ldots, n, \end{cases}$$

$$(7.59)$$

be its integral curve going through $m_0 = (p(0), x(0)) \in \mathbb{R}^{2n}$ when $t = 0$. Formula (7.59) implies that if $m_0 \in \mathscr{F}^{-1}(a, b)$, then $\gamma_{m_0}^Z \in \mathscr{F}^{-1}(a, b)$ for all $t \in \mathbb{R}$, and since the fibers of $\mathscr{F}$ are one-dimensional submanifolds of $\mathbb{R}^{2n}$, it follows that they coincide with the integral curves of the vector field Z. In other words, the formulas enclosed in (7.6) define a fiber bundle $\mathscr{F} : \mathbb{R}^{2n} \to \mathbb{R}^{2n-1}$, whose fibers are the integral curves of vector field Z. Note that these integral curves can, in turn, be identified with the orbits of the action of Lie group $(\mathbb{R}, +)$, obtained integrating the (infinitesimal) action defined by Z.

Furthermore, note that $\mathscr{L}_Z \Pi_0 = 0$ and $\mathscr{L}_Z \Pi_1 = 0$, where Π_0 and Π_1 are as in (7.55), as it can be shown by a direct computation or simply noticing that both Poisson tensors are *translation-invariant*. Under these assumptions, Theorem 2.74 applied to Π_0 and Π_1 yields the existence of two Poisson tensors $\tilde{\Pi}_0$ and $\tilde{\Pi}_1$ on $\mathbb{R}^{2n-1}$ (see in particular Example 2.75). More precisely, a direct computation using (7.6) and (7.55) yields the following result.

Proposition 7.51 *The reduced Poisson tensors $\tilde{\Pi}_0$ and $\tilde{\Pi}_1$ assume the following form:*

$$\tilde{\Pi}_0 = \frac{1}{4} \sum_{k=1}^{n-1} a_k \frac{\partial}{\partial a_k} \wedge \left(\frac{\partial}{\partial b_k} - \frac{\partial}{\partial b_{k+1}} \right) \tag{7.60}$$

$$\tilde{\Pi}_1 = \frac{1}{4} \sum_{k=1}^{n-1} a_k a_{k+1} \frac{\partial}{\partial a_k} \wedge \frac{\partial}{\partial a_{k+1}} + \sum_{k=1}^{n-1} a_k^2 \frac{\partial}{\partial b_k} \wedge \frac{\partial}{\partial b_{k+1}}$$

$$+ \frac{1}{2} \sum_{k=1}^{n-1} a_k b_k \frac{\partial}{\partial b_k} \wedge \frac{\partial}{\partial a_k} + \frac{1}{2} \sum_{k=1}^{n-1} a_k b_{k+1} \frac{\partial}{\partial a_k} \wedge \frac{\partial}{\partial b_{k+1}}, \tag{7.61}$$

where $a_n = 0$.

Note that the Hamiltonians $H = \frac{1}{2} \sum_{k=1}^{n} p_k^2 + \sum_{k=1}^{n-1} \exp(x_k - x_{k+1})$ and $K = \sum_{i=1}^{n} p_k$, if written in the Flaschka coordinates (7.6), assume the following form:

$$\tilde{H} = 4 \sum_{i=1}^{n-1} a_i^2 + 2 \sum_{i=1}^{n} b_i^2 \quad \text{and} \quad \tilde{K} = -2 \sum_{i=1}^{2} b_i.$$

This representation and another simple computation entail that

$$\tilde{\Pi}_0(d\tilde{H}) = \sum_{i=1}^{n-1} a_i (b_{i+1} - b_i) \frac{\partial}{\partial a_i} + 2 \sum_{i=1}^{n-1} a_i^2 \left(\frac{\partial}{\partial b_i} - \frac{\partial}{\partial b_{i+1}} \right)$$

$$= \sum_{i=1}^{n-1} a_i (b_{i+1} - b_i) \frac{\partial}{\partial a_i} + 2 \sum_{i=1}^{n} a_i \left(a_i^2 - a_{i-1}^2 \right) \frac{\partial}{\partial b_i}, \tag{7.62}$$

with the boundary conditions $a_0 = 0 = a_n$, showing that $\tilde{\Pi}^\sharp(d\tilde{H})$ is the vector field whose corresponding system of ODEs is (7.10). A similar computation yields

$$\tilde{\Pi}_1(d\tilde{K}) = \sum_{i=1}^{n-1} a_i(b_{i+1} - b_i)\frac{\partial}{\partial a_i} + 2\sum_{i=1}^{n} a_i\left(a_i^2 - a_{i-1}^2\right)\frac{\partial}{\partial b_i},$$

still with $a_0 = 0 = a_n$, which implies that the Toda vector field (7.4), in the Flaschka coordinates, admits a Hamiltonian representation with respect to both $\tilde{\Pi}_0$ and $\tilde{\Pi}_1$, whose Hamiltonian functions are $\tilde{H}$ and, respectively, $\tilde{K}$, i.e.,

$$\tilde{\Pi}_0(d\tilde{H}) = \tilde{\Pi}_1(d\tilde{K}).$$

It is worth observing that the reduction from (7.4) to (7.62) is performed in the framework of Theorem 2.78. Comparing the statement of the latter result with the present case, one sees that $M = P = \mathbb{R}^{2n}$ and E is the vector bundle whose fiber, at each point, is the one-dimensional vector space generated by the value of Z at the chosen point. In this case, the condition on M of being conserved by $X_H = X_K$ is trivially verified. To close this circle of ideas, we are left to show that the reduced Poisson structures are compatible. To this end, observe that if $\{\cdot, \cdot\}_{0,\sim}$ and $\{\cdot, \cdot\}_{1,\sim}$ are the Poisson brackets defined by $\tilde{\Pi}_0$ and, respectively, $\tilde{\Pi}_1$ and $F_1, F_2, F_3 \in \mathbb{R}^{2n-1}$, then

$$\left\{\{F_1, F_2\}_{0,\sim}, F_3\right\}_{1,\sim} \circ \mathscr{F} + \left\{\{F_1, F_2\}_{1,\sim}, F_3\right\}_{0,\sim} \circ \mathscr{F} + \circlearrowleft (F_1, F_2, F_3) \circ \mathscr{F}$$

$$= \left\{\{F_1 \circ \mathscr{F}, F_2 \circ \mathscr{F}\}_{0,\sim}, F_3 \circ \mathscr{F}\right\}_{1,\sim} + \left\{\{F_1 \circ \mathscr{F}, F_2 \circ \mathscr{F}\}_{1,\sim}, F_3 \circ \mathscr{F}\right\}_{0,\sim}$$

$$+ \circlearrowleft (F_1 \circ \mathscr{F}, F_2 \circ \mathscr{F}, F_3 \circ \mathscr{F})$$

$$= 0,$$

since $\{\cdot, \cdot\}_0$ and $\{\cdot, \cdot\}_1$ are the Poisson brackets of Π_0 and Π_1, which are compatible Poisson tensors (see Theorem 7.46). This observation, together with the property of $\mathscr{F}$ of being submersive, entails that

$$\left\{\{F_1, F_2\}_{0,\sim}, F_3\right\}_{1,\sim} + \left\{\{F_1, F_2\}_{1,\sim}, F_3\right\}_{0,\sim} + \circlearrowleft (F_1, F_2, F_3) = 0,$$

for all $F_1, F_2, F_3 \in \mathbb{R}^{2n-1}$, which implies that $\tilde{\Pi}_0$ and $\tilde{\Pi}_1$ are compatible Poisson tensors (see Chap. 5). In other words, the bi-Hamiltonian structure of the open Toda system (see Theorem 7.46), defined on $\mathbb{R}^{2n}$, reduces, via the Flaschka map $\mathscr{F}$, to a bi-Hamiltonian structure on $\mathbb{R}^{2n-1}$, which entail a bi-Hamiltonian representation of the Toda vector field in the Flaschka coordinates. What is enclosed in the previous discussion is a particular case of an extension to bi-Hamiltonian setting of the

Poisson reduction theorem of Marsden and Ratiu (see Sect. 4.2 of Chap. 2). In more general terms, this result can be stated as follows (see [199]).

Theorem 7.52 *Let (P, Π_0, Π_1) be a bi-Hamiltonian manifold, (M, i) a submanifold of P, where $i : M \to P$ is the corresponding immersion map, and let E be a subvector bundle of $T_M P$, the restriction of $T P$ to M. Suppose that:*

(i) *For all $f_1, f_2 \in C^\infty(P)$ such that $df_1|_E = 0$, $df_2|_E = 0$, $d\{f_1, f_2\}_i|_E = 0$, for $i = 0, 1$, where $\{\cdot, \cdot\}_i$ is the Poisson brackets defined by Π_i, $i = 0, 1$.*

(ii) *$\Pi_i^\sharp(E^\circ) \subset TM + E$, where $E^\circ \subset T^*P$ is the annihilator of E.*

(iii) *The distribution defined by $E \cap TM$ is completely integrable and that the corresponding space of leaves $M/\sim$ is a smooth manifold. Moreover suppose that the canonical projection $\pi : M \to M/\sim$ is a smooth submersion.*

Then on $M/\sim$, there exists a unique pair of compatible Poisson structures $(\tilde{\Pi}_0, \tilde{\Pi}_1)$ such that for all $F_1, F_2 \in C^\infty(M/\sim)$,

$$\{f_1, f_2\}_i \circ i = \{F_1, F_2\}_{i,\sim} \circ \pi, \ i = 0, 1,$$

for all $f_1, f_2 \in C^\infty(P)$ smooth extensions of $F_1 \circ \pi$ and, respectively, of $F_2 \circ \pi$, such that $d(F_1 \circ \pi)|_E = 0 = d(F_2 \circ \pi)|_E$.

A few remarks are now in order.

Remark 7.53 The previous result is a straightforward extension of Theorem 2.74 of Chap. 2 to the bi-Hamiltonian framework. More precisely, Theorem 7.52 follows at once from the result of Marsden and Ratiu *modulo* the proof of the compatibility of the reduced Poisson structures, which follows from π being a submersion. Finally, it could be worth observing that the reduced bi-Hamiltonian structure of the Toda system is obtained from Theorem 7.52 choosing $P = \mathbb{R}^{2n}$, Π_0, and Π_1 as in Theorem 7.46, $M = P$, and E as the vector bundle whose fiber at point is the vector space generated by the value of Z at that point. The corresponding reduced space $M/\sim$ turns out to be $\mathbb{R}^{2n-1}$, and the projection map π is the Flaschka map $\mathscr{F}$. $\triangle$

Looking back to the bi-Hamiltonian reduction of the Toda system, it is worth noting that since the reduced space is odd dimensional, both $\tilde{\Pi}_0$ and $\tilde{\Pi}_1$ are singular, i.e., none of them comes from a symplectic structure. For this reason, there is no obvious recursion operator associated with the bi-Hamiltonian structure defined by $(\tilde{\Pi}_0, \tilde{\Pi}_1)$. In fact, it is possible to prove that this reduced bi-Hamiltonian structure is *not* a PN-structure, i.e., it does not admit an associated recursion operator. This was first observed in [87].

To explain why this is the case, we need to introduce first a few notions. Recall that, by definition, a vector field X on the total space of a fibration $\pi : M \to B$ is *projectable* if $\pi_{*,p}(X(p))$ is independent on $p \in \pi^{-1}(b)$. In particular, if X is projectable, then there exists a unique $\tilde{X} \in \mathfrak{X}(B)$, defined as $\tilde{X}(b) = \pi_{*,p}(X(p))$, for any $p \in \pi^{-1}(b)$. Clearly, $\tilde{X}$ is π-related to X,, and every vector field on B comes from a projectable vector field on the total space M.

One has the following:

Proposition 7.54 *Let $\pi : M \to B$ be a smooth surjective submersion. If X is vector field on M that is projectable, and if Y is any π-vertical vector field (i.e., Y is tangent everywhere to the fibers of π or equivalently $\pi_* Y = 0$), then $\mathcal{L}_Y X = [Y, X]$ is also a vertical vector field.*

Proof First, observe that if X_1 and X_2 are two vector fields on M that are projectable, with associated vector fields $\tilde{X}_1$ and $\tilde{X}_2$ on B, then $[X_1, X_2] = \mathcal{L}_{X_1} X_2$ is projectable and indeed $\pi_*[X_1, X_2] = [\tilde{X}_1, \tilde{X}_2]$, (see Problem 7.55). Then observe that a vertical vector field Y is by definition projectable to B with projection the identically vanishing vector field. Thus, $\mathcal{L}_Y X = [Y, X]$ is projectable, and $\pi_*[Y, X] = [0, \tilde{X}] = 0$; therefore, $\mathcal{L}_Y X \in \ker(\pi_*)$, and therefore, it is vertical. $\square$

It is possible to prove that the condition expressed in Proposition 7.54 for projectability is not only necessary but also sufficient. However, we don't need this here.

Problem 7.55 *Let $\pi : M \to B$ be a smooth surjective submersion and let X_1, X_2 projectable vector fields on M. Prove that $[X_1, X_2]$ is projectable with $\pi_*([X_1, X_2]) = [\tilde{X}_1, \tilde{X}_2]$ where $\tilde{X}_i$ is the projection of X_i to B for $i = 1, 2$.* $\triangle$

Now we introduce the following:

Proposition 7.56 *Let Π_0 and Π_1 be bivectors on a manifold M, let $\pi : M \to B$ be a smooth submersion to a manifold B, and let $\tilde{\Pi}_i = \pi_* \Pi_i$ and $i = 0, 1$ be the corresponding bivectors on B. Suppose there exists an endomorphism $N : TM \to TM$ such that $\Pi_1^\sharp = N \circ \Pi_0^\sharp$. If $\tilde{N}$ is an endomorphism of TB such that $\tilde{\Pi}_1^\sharp = \tilde{N} \circ \tilde{\Pi}_0^\sharp$, then necessarily,*

$$\tilde{N}(\pi_* \Pi_0^\sharp(\pi^*\alpha)) = \pi_*(N(\Pi_0^\sharp(\pi^*\alpha))), \quad \forall \alpha \in \Omega^1(B). \tag{7.63}$$

Proof Since $\tilde{\Pi}_i$ are obtained from Π_i using π_*, one has that for all $\alpha \in \Omega^1(B)$, $\tilde{\Pi}_0^\sharp(\alpha) = \pi_* \Pi_0^\sharp(\pi^*\alpha)$ and

$$\tilde{\Pi}_1^\sharp(\alpha) = \pi_* \Pi_1^\sharp(\pi^*\alpha). \tag{7.64}$$

In particular, the last two relations tell us that the vector fields $X_i = \Pi_i^\sharp(\pi^*\alpha)$ on M are projectable to B.

Using that $\tilde{\Pi}_1^\sharp = \tilde{N} \circ \tilde{\Pi}_0^\sharp$ and $\Pi_1^\sharp = N \circ \Pi_0^\sharp$ in (7.64), one gets $\tilde{N}(\tilde{\Pi}_0^\sharp(\alpha)) = \pi_*(N(\Pi_0^\sharp(\pi^*\alpha)))$. Since $\tilde{\Pi}_0^\sharp(\alpha) = \pi_* \Pi_0^\sharp(\pi^*\alpha)$, substituting in the previous expression, one obtains (7.63). $\square$

Combining what we have seen so far, one gets the following.

Corollary 7.57 *The projectable vector fields $\mathfrak{X}_{proj}(M)$ form a Lie subalgebra of the Lie algebra $(\mathfrak{X}(M), [\cdot, \cdot])$, and the vertical vector fields $\mathfrak{X}_v(M)$ are a Lie ideal in $\mathfrak{X}_{proj}(M)$. Furthermore, π_* induces a Lie algebra isomorphism from $\mathfrak{X}_{proj}(M)/\mathfrak{X}_v(M)$ to $\mathfrak{X}(B)$.*

Proof This follows at once from Proposition 7.54, its proof, and Exercise 7.55. $\quad\square$

Corollary 7.58 *Let Π_i, $\tilde{\Pi}_i$, $i = 0, 1$, N, and $\tilde{N}$ be like in Proposition 7.56. For $\tilde{N}$ to exist, it is necessary and sufficient that N sends projectable vector fields to projectable vector fields and vertical vector fields to vertical vector fields.*

Proof This follows at once from Proposition 7.56 and Corollary 7.57. $\quad\square$

Now if we apply N to a vertical vector field $V := f \sum_{i=1}^{n} \frac{\partial}{\partial x^i}$ where f is an arbitrary smooth function of all the variables in $\mathbb{R}^{2n}$, we get

$$N(V) = f \sum_{i=1}^{n} p_i \frac{\partial}{\partial x^i} + \sum_{i=1}^{n-1} f \exp\left(x^i - x^{i+1}\right) \frac{\partial}{\partial p_{i+1}} - \sum_{i=1}^{n-1} f \exp\left(x^i - x^{i+1}\right) \frac{\partial}{\partial p_i},$$

which is definitely not vertical since any vertical vector field of the form $g \sum_{i=1}^{n} \frac{\partial}{\partial x^i}$ for some smooth g.

Thus, using Corollary 7.58, we conclude that $\tilde{N}$ cannot exist and there is no PN structure on the reduced space of the Toda system.

Remark 7.59 As a final observation, we note that although there is no recursion operator associated with the bi-Hamiltonian structure defined in (7.60) and (7.61), the latter encloses higher Poisson structures due to the reducibility of the master symmetries of the unreduced system (see [87] and the references therein). $\quad\triangle$

5 Concluding Remarks and Further Topics

In this final section, we will point to a few topics related to what is discussed in the main body of this chapter and that we think could be worth being further explored. The presentation will be mostly informal.

5.1 Applications of Equations in Lax Form and Rutishauser's System

Here we discuss briefly applications of Lax equations in the area of numerical linear algebra, optimization, and analogue computations. For much more on this, we refer to the monograph [115] and the review article [51].

Many algorithms in linear algebra can be viewed as discrete dynamical systems, so it is natural to look for continuous dynamical systems that either interpolate a specific discrete dynamical system or produce the same output as time grows unbounded. For instance, some variants of the Toda flow are continuous analogues of the QR algorithm [231] and of other algorithms related to numerical linear algebra, and they can be cast in Lax form. It turns out that this relation is older than commonly known. Already in the 1950s, Rutishauser constructed a continuous analogue of the quotient-difference algorithm that contains the finite non-periodic Toda flow as a special case, essentially already written in the Flaschka's coordinates (see [214]). The flow found by Rutishauser already in 1954 is given by

$$\dot{Q}_i = E_i - E_{i-1}, \quad i = 1, \ldots, n, \tag{7.65}$$

$$\dot{E}_i = E_i(Q_{i+1} - Q_i), \quad i = 1, \ldots, n - 1, \quad E_0 = E_n = 0. \tag{7.66}$$

Equations (7.65) and (7.66) look like Flaschka's form of the finite non-periodic Toda equations. Indeed, the change of variables $Q_i = 2a_i$ and $E_i = 4b_i^2$ transforms the Rutishauser's system into the Toda equations. Equations (7.65) and (7.66) are actually more general, since the Toda flow corresponds to the subcase $E_i > 0$, $i = 1, \ldots, n - 1$.

A nice feature of the Rutishauser's approach is that it constructs the continuous flow via a limiting process directly from the discrete eigenvalue algorithm, so it clarifies the direct connection between the Toda flow and numerical linear algebra, and it can also be applied to other discrete algorithms (see [245]).

Similarly, many problems in optimization involving specific structures give rise to equations in Lax form. For instance, the set of symmetric matrices with a fixed spectrum or the set of matrices with fixed singular values is equipped with the structure of a homogenous space. Various problems on these homogeneous spaces (for instance, select among the symmetric matrices with a fixed spectrum one with a specific structure, like being diagonal) can be formulated as an optimization problem, the solution of which is obtained via a suitable gradient flow. It turns out that, with the appropriate choice of the metric, these gradient flows can be written in Lax form. More precisely, these can be written in what is called a double-bracket form (a special Lax form):

$$\dot{A} = [A, [A, N]], \tag{7.67}$$

where A and N are symmetric matrices. This equation is often known as Brockett's equation since it was discovered by R. Brockett in [38]. There the author discusses how to map the data associated with a linear programming problem into $A(0)$ and N so that if $A(t)$ is the corresponding solution of (7.67), then $\lim_{t \to +\infty} A(t)$ provides a solution of the associated linear programming problem. Equation (7.67) provides a system that can deal with a variety of combinatorial optimization problems, like sorting a list of numbers with a specific ordering, in continuous time. For this, it is enough to choose N as a diagonal matrix with distinct entries having the desired ordering and give as input a symmetric matrix $A(0)$ whose eigenvalues are the initial list of numbers (to be precise, one needs to perturb a bit $A(0)$ since a symmetric matrix is an equilibrium of the Brockett's vector field, but it is an unstable one if the ordering of the diagonal entries of $A(0)$ is not the same as the ordering given by N). Then, in this case, $\lim_{t \to +\infty} A(t)$ is a diagonal matrix with the initial list of number sorted according to the order given by the diagonal elements of N. This is a remarkable instance of an analogue algorithm, a setup that is nowadays massively explored in the area of quantum computing.

Furthermore, many completely integrable Hamiltonian systems that show a gradient-like behavior on common level surfaces of their integrals admit a Lax representation in double bracket form (see [33, 86, 193]).

Of course, it is not true in general that equations in Lax form are necessarily integrable, not even if they are in double-bracket form. The following equation:

$$\dot{A} = \left[[N, A + A^T], A \right] + \nu \left[[A^T, A], A \right], \tag{7.68}$$

where N is a diagonal constant $n \times n$ matrix and A is a real $n \times n$ matrix, A^T denotes the transpose of A, and ν is a positive constant introduced in [73] and further studied and generalized in [74] and in [14]. Equation (7.68) clearly reduces to Brockett's equation in the case A is symmetric, but it is more general since it can diagonalize a non-symmetric matrix. In particular, it was proved in [14] that if $A(0)$ is a real matrix with a real spectrum, which is possibly degenerate, and if $A(t)$ is the corresponding solution of (7.68), then $\lim_{t \to +\infty} A(t)$ exists, and moreover, it is a diagonal form of $A(0)$ if $A(0)$ is diagonalizable, or it is the diagonal part of a Jordan normal form for $A(0)$ if $A(0)$ is not diagonalizable.

Concluding, we mention few other systems of Lax type that isospectrally deform an initial matrix to a target form. The system

$$\dot{A} = \left[\pi([A^T, A]), A \right],$$

where A is an $n \times n$ real matrix and π is the projection to upper triangular matrices deforms Hessenberg matrices to Jacobi matrices isospectrally [15]. More precisely, if $A(0)$ is upper Hessenberg matrix with simple spectrum and subdiagonal elements different from zero, then $\lim_{t \to +\infty} A(t)$ exists, and it is a tridiagonal symmetric matrix with the same sign pattern in the codiagonal elements as $A(0)$. A generalization of this system to complex Hessenberg matrices was obtained in [18].

As we mentioned, complete integrability of a system in Lax form is not guaranteed, and even if the system is integrable, it is not so easy to construct explicit solutions. A case where this is possible is the following. Let $\pi_+ : \mathfrak{g} \to \mathfrak{g}$ be any R-matrix on a Lie algebra $\mathfrak{g}$. Then the explicit solution for the equation in Lax form given by

$$\dot{A} = [A, \pi_+ A], \quad A(0) = A_0,$$

can be obtained in a kind of exponential form (see [75] for details).

For much more on the interaction between equations in Lax form, integrable systems, numerical linear algebra algorithms, and matrix factorizations, see the beautiful [66].

5.2 Toda and Root Systems

The notion of *Generalized Toda System* (GTS from now on) was introduced in [34], and then it was analyzed in depth in [142, 143] (see also [208] and the references therein). The data for the construction of a GTS is an n-dimensional (real) quadratic vector space (V, B) and a set of vectors $\{u_1, \ldots, u_k\} \subset V$. From these data, one can define $H : V \times V \to \mathbb{R}$ by

$$H(p, x) = \frac{1}{2} B(p, p) + \sum_{i=1}^{k} \exp(B(x, u_i)), \, \forall (p, x) \in V \times V. \tag{7.69}$$

If $\{e_1, \ldots, e_n\} \subset V$ is an B-orthonormal basis and if

$$u_l = e_l - e_{l+1}, \, l = 1, \ldots, n - 1, \tag{7.70}$$

then

$$H(p, x) = \frac{1}{2} \sum_{i=1}^{n} p_i^2 + \sum_{i=1}^{n-1} \exp(x_i - x_{i+1}), \tag{7.71}$$

i.e., one recovers (7.2), the Hamiltonian of the open Toda system. Now, one can wonder under what assumptions on the data $(V, B, u_1, \ldots, u_k)$ the corresponding (7.69) will have some (or all) of the marvelous properties of (7.71). A (partial answer) to this question goes as follows. If $\Delta = \{u_1, \ldots, u_k\} \subset V$ is a *simple root system* of a simple Lie algebra, then (7.69) is completely integrable, and the corresponding equations of motion admit a Lax representation. Note that (7.71) defines an Hamiltonian system on $V \times V \overset{B}{\simeq} T^*V$ with respect to its canonical symplectic form. Here we will not give the details of the proof of this statement, which can be found in [34] (see also [208]). We just recall that the a simple root system is a particular system of generators of a *root system*, which, roughly speaking, is a finite subset of a finite dimensional, real quadratic vector space, which is invariant with respect to the group generated by the reflections through the hyperplanes defined by its elements. Root systems are classified and turn out to be in one-to-one correspondence with the isomorphism classes of *simple Lie algebras* (see [219] for a nice and detailed exposition of these results). For example, the root system generated by (7.70) is the one associated with the Lie algebra $\mathfrak{sl}_n(\mathbb{R})$. We close this subsection observing that the results contained in Sect. 4 of this chapter have been extended to the case of the GTSs associated with the root systems of type A_n, B_n, C_n and D_n (see [61]).

5.3 R-Matrices and Factorization

As it was mentioned in Sect. 3.2, the AKS method can be generalized introducing the notion of a R-matrix. Here we would like to add a few comments to what is discussed above. The main references for what follows are [212, 213, 218]. See also [154] for a much more general point of view and more recent developments. Let $\mathfrak{g}$ be a Lie algebra, let R be a solution of the mCYBE with $c^2 = 1$ (see 7.46), and let $R_\pm : \mathfrak{g} \to \mathfrak{g}$ be

$$R_\pm = \frac{1}{2}(R \pm \mathrm{id}_\mathfrak{g}).$$

Denoting by $\mathfrak{g}_R$ the Lie algebra whose underlying vector space is $\mathfrak{g}$ and Lie bracket is $[\cdot, \cdot]_R$, one can prove that

Lemma 7.60 $R_\pm : \mathfrak{g}_R \to \mathfrak{g}$ *are two Lie algebra homomorphisms so that their images* $\mathfrak{g}_\pm$ *are Lie subalgebras of* $\mathfrak{g}$.

Moreover, let $i_R : \mathfrak{g} \to \mathfrak{g} \times \mathfrak{g}$ be defined by $i_R = (R_+, R_-) \circ \Delta$ where $\Delta : \mathfrak{g} \to \mathfrak{g} \times \mathfrak{g}$ is the diagonal morphism, i.e., $\Delta x = (x, x)$. Under these assumptions

Lemma 7.61 *Every* $x \in \mathfrak{g}$ *is decomposed as* $x = x_+ - x_-$, *where* $(x_+, x_-) = i_R(x)$ *for all* $x \in \mathfrak{g}$.

In other words, a solution of the mCYBE entails a splitting of the underlying Lie algebra $\mathfrak{g}$ in the direct sum of two Lie algebras, obtained as the images of the two Lie algebra morphisms $R_\pm$. Note that the choice of $c = \pm 1$ is irrelevant. In particular, if $c \neq 0$, up to a suitable rescaling, one can reduce (7.46) to this case. Observe however that the Lie algebra decomposition just described cannot be obtained starting from a solution of the CYBE, i.e., (7.46) when $c = 0$. The following example is meant to illustrate the content of the previous lemma.

Example 7.62 Suppose that $\mathfrak{g}$ is a Lie algebra that admits a vector space decomposition $\mathfrak{g} = \mathfrak{a} \oplus \mathfrak{b}$, where $\mathfrak{a}$ and $\mathfrak{b}$ are Lie subalgebras of $\mathfrak{g}$. If $\pi_\mathfrak{a} : \mathfrak{g} \to \mathfrak{a}$ and $\pi_\mathfrak{b} : \mathfrak{g} \to \mathfrak{b}$ are the corresponding projections, one can show that $R = \pi_\mathfrak{a} - \pi_\mathfrak{b}$ satisfies (7.46) with $c^2 = 1$. In fact, since $[x, y]_R = [\pi_\mathfrak{a} x, \pi_\mathfrak{a} y] - [\pi_\mathfrak{b} x, \pi_\mathfrak{b} y]$ (see Proposition 7.38),

$$R\Big([Rx, y] + [x, Ry]\Big) - [Rx, Ry] = [\pi_\mathfrak{a} x, \pi_\mathfrak{a} y] + [\pi_\mathfrak{b} x, \pi_\mathfrak{b} y] + \overline{[\pi_\mathfrak{b} x, \pi_\mathfrak{a} y]}$$

$$+ \overline{[\pi_\mathfrak{a} x, \pi_\mathfrak{b} y]} = [x, y],$$

for all $x, y \in \mathfrak{g}$. Moreover, since $\mathrm{id}_{\mathfrak{g}} = \pi_{\mathfrak{a}} + \pi_{\mathfrak{b}}$, $R_+ = \pi_{\mathfrak{a}}$ and $R_- = -\pi_{\mathfrak{b}}$. In this way

$$R_+\left([x, y]_R\right) = \pi_{\mathfrak{a}}\left([\pi_{\mathfrak{a}}x, \pi_{\mathfrak{a}}y] - [\pi_{\mathfrak{b}}x, \pi_{\mathfrak{b}}y]\right) = [\pi_{\mathfrak{a}}x, \pi_{\mathfrak{a}}y] = [R_+x, R_+y],$$

and $R_-([x, y]_R) = [R_-x, R_-y]$, for all $x, y \in \mathfrak{g}$, in agreement with the statement in Lemma 7.60. As far as the statement in Lemma 7.61 is concerned, in this case, there is nothing to prove. $\triangle$

Let G be a Lie group whose Lie algebra is $\mathfrak{g}$ and let $G_\pm \subset G$ be two Lie subgroups of G whose Lie algebras are $\mathfrak{g}_\pm$.

Theorem 7.63 *Every $g \in G$, contained in a suitable neighborhood of the identity $e \in G$, admits a factorization as $g = h_1 h_2^{-1}$, for $h_1 \in G_1$ and $h_2 \in G_2$.*

The factorization of G as product of G_1 and G_2 holds, in general, only locally in a neighborhood of the identity of G; however, this result has a very nice consequence when applied to dynamical systems. More precisely, let R be a solution of the mCYBE and let f be a Casimir of the linear Poisson structure on $\mathfrak{g}^*$ defined by the Lie bracket of $\mathfrak{g}$. Then

Theorem 7.64 *For any $\alpha \in \mathfrak{g}^*$, if $g_\pm(t)$ are two (smooth) curves in $G_\pm$ factorizing the curve $\exp(td f_\alpha)$ in a suitable neighborhood of e, i.e., if*

$$\exp(td f_\alpha) = g_+(t)g_-^{-1}(t),$$

for every $|t| < \epsilon$, then the integral curve $\alpha(t)$ of (7.45), solution of $\dot{\alpha} = -\frac{1}{2}\,\mathrm{ad}^\sharp_{R(d f_\alpha)}(\alpha)$ with initial condition $\alpha(0) = \alpha$, is

$$\alpha(t) = \mathrm{Ad}^\sharp_{g_+^{-1}(t)}\,\alpha = \mathrm{Ad}^\sharp_{g_-^{-1}(t)}\,\alpha.$$

This result extends to the framework of the R-matrix theory, the discussion in Sect. 3.2.2 at the end of Sect. 3.

6 Bibliographical Notes

As we mentioned at the beginning of this chapter, the Toda system was introduced at the end of the 1960s by Morikazu Toda, in his deep analysis of the dynamics of the anharmonic lattices (see [235]). The proof of the integrability of the periodic Toda system goes back to Henon and Flaschka (see [116] and [93]), while the Lax representation of this system was discovered independently by Flaschka and Manakov in [92] and [169]. A deep analysis of the non-periodic case goes back to fundamental work of Moser (see [189, 190]), to which we refer the reader for more information about the content of Sects. 1 and 2 in this chapter. The analysis of this dynamical system from the viewpoint of Lie theory was initiated in [34] and further developed in [139] (see also [142]), to which we refer for more information about the content of Sect. 3 of this chapter. See also the beautiful and very readable presentations in [111, 208]. Other useful references for this sections are [5, 230–232] and [43] from where we borrowed the discussion about the AKS Theorem. The bi-Hamiltonian representation of the open Toda system, as presented in Sect. 4, is due to Das and Okubo (see [63]), while its relations with the master symmetries was studied in depth in [60, 87]. The extension of the Poisson reduction to the bi-Hamiltonian setting can be found in [164, 168] (see also [206] and the references therein). Another nice reference treating the problem of the bi-Hamiltonian reduction in the PN case is [241]. Finally, for more details about the content of Sect. 5.2, we refer again to [208] (see also [61]), while for more information about the content of Sect. 5.3, we refer to [139] and to the references therein. The role of the Toda system and its relevance to other areas of mathematics are discussed in depth in the very nice and comprehensive review papers [136, 236]. A far-reaching generalization of the integrability of the Toda flow on a generic orbit is treated in depth in the insightful [65].

Chapter 8
Calogero-Moser Systems

This chapter is meant to be an introduction to the so-called *rational Calogero-Moser* system, CM-system hereafter, a dynamical system of n-point particles disposed online and interacting via a potential proportional to the inverse square of the distance of its constituents. This system was introduced by Francesco Calogero at the beginning of the 1970s. Its complete integrability was proved by Jürgen Moser in 1975, using analytical methods borrowed from scattering theory. Soon after the fundamental work of Moser, in 1976, Mikhail Olshanetsky and Askold Perelomov published another proof of the integrability of this dynamical system more geometric in spirit, and in 1978, David Kazhdan, Bertram Kostant, and Shlomo Sternberg gave yet another proof of the complete integrability of the CM-system using Hamiltonian reduction. In our presentation, after introducing the system and its Lax representation, we will follow the ideas of Kazhdan, Kostant, and Sternberg to prove its integrability. It is worth mentioning that the rational CM-system is just one of the three possible incarnations of this important dynamical system, the other two being the *trigonometric* and the *elliptic* one. In the former case, the particles interact under a potential proportional to $\frac{1}{\sin^2 x}$, while in the latter, the potential is of the type $\frac{1}{\wp(x)}$. Finally, we would like to stress that in recent years, the theory of the CM-system(s) has become a crossroad of several areas of mathematics and theoretical physics, such as representation theory, symplectic geometry, algebraic geometry, combinatorics, and supersymmetric Yang-Mills theory, just to cite a few. For much more on this, see the insightful [78].

1   Description of the System

The rational CM-system is a dynamical system describing the motion of n identical particles that move on the real line subject to a repulsive potential, proportional to the inverse square of the distance of the particles, and an elastic attractive potential. Normalizing the mass of each particle to one, the Hamiltonian of the CM-system is the function

$$H(p, x) = \frac{1}{2} \sum_{i=1}^{n} p_i^2 + b \sum_{1 \leq i < j \leq n} \frac{1}{(x_i - x_j)^2} + \frac{a^2}{2} \sum_{j=1}^{n} x_j^2, \tag{8.1}$$

and the corresponding Hamiltonian vector field is

$$X_H = \sum_{i=1}^{n} p_i \frac{\partial}{\partial x_i} - \sum_{i=1}^{n} \sum_{k \neq i} \frac{2b}{(x_k - x_i)^3} \frac{\partial}{\partial p_i} - a^2 \sum_{i=1}^{n} x_i \frac{\partial}{\partial p_i}. \tag{8.2}$$

From the Hamilton equations

$$\begin{cases} \dot{x}_i = p_i \\ \dot{p}_i = -\sum_{k \neq i} \frac{2b}{(x_k - x_i)^3} - a^2 x_i \end{cases} \tag{8.3}$$

we can read off the equations of motion

$$\ddot{x}_i = -\sum_{k \neq i} \frac{2b}{(x_k - x_i)^3} - a^2 x_i, \quad i = 1, \ldots, n. \tag{8.4}$$

Collisions are forbidden, so that the configuration space of the system is an open subset of $\mathbb{R}^n$:

$$\mathrm{Conf} = \{(x_1, \ldots, x_n) \in \mathbb{R}^n \mid x_1 < \cdots < x_n\},$$

so that the phase-space of the CM-system can be identified with an open subset of $T^*\mathbb{R}^n$, endowed with its canonical symplectic form.

Problem 8.1 The goal of this exercise is to prove an interesting observation of Perelomov: the CM-system we have introduced so far can always be reduced to the case $a = 0$ and $b = 1$. Define $X_i(T)$ via the formula $x_i(t) = c \cos(at) X_i(T)$, where $T(t) = \tan(at)$ for all i. Show that

$$\ddot{x}_i + a^2 x_i = \frac{ca^2}{(\cos(at))^3} \frac{d^2}{dT^2} X_k(T),$$

where $\ddot{x}_i$ means second derivative with respect to t. Show that if $X_i(T)$ $i = 1, \ldots, n$ satisfy Eq. (8.4) with $a = 0$ and $b = 1$, then $x_i(t)$ and $i = 1, \ldots, n$ satisfy the equation

$$\ddot{x}_i = -\sum_{i \neq k} \frac{2c^4 a^2}{(x_k - x_i)^3} - a^2 x_i.$$

Therefore, comparing with (8.4) with $b = c^4 a^2$, the transformation takes the original system into one with new $a = 0$ and new $b = 1$. (In principle, this may lead to complex c, but this is not a problem since all solutions have complex continuation; see [191], pages 157–158.) $\triangle$

Due to Exercise 8.1, it is not restrictive to consider the CM-system with $a = 0$ and $b = 1$, which is what we are doing in the rest of the chapter. In this case, the corresponding potential is given by:

$$V(x) = \sum_{1 \leq i < j \leq n} \frac{1}{(x_i - x_j)^2}. \tag{8.5}$$

and the equation of motions are:

$$\ddot{x}_i = -\sum_{k \neq i} \frac{2}{(x_k - x_i)^3}, \quad i = 1, \ldots, n. \tag{8.6}$$

Our main goal in this chapter is to prove that the Hamiltonian system described by (8.1) is completely integrable. To achieve this result, we will show that Eq. (8.6) can be recast in a Lax form. To this end, we start introducing the following class of matrices, which will play a prominent role in what follows.

Definition 8.2 A $n \times n$ Hermitian matrix L whose entries are

$$L_{kl} = \delta_{kl} p_k + \iota \frac{(1 - \delta_{kl})}{(x_k - x_l)}, \quad \text{where } \iota = \sqrt{-1}, \tag{8.7}$$

is called a *Calogero-Moser matrix*, CM-matrix from now on. Note that the entries of the matrix in Eq. (8.7) are rational functions of $(p_1, \ldots, p_n, x_1, \ldots, x_n)$. $\triangle$

Now we state the main result of this section.

Theorem 8.3

(i) *The rational CM-system is a completely integrable system.*
(ii) *A set of solutions of the Eq. (8.6) is provided by the distinct eigenvalues of the matrix*

$$S_{\text{CM}} = diag(x_1(0), \ldots, x_n(0)) + t L(x(0), p(0)).$$

The proof of this theorem will be presented in the next subsections, where it will be shown that (8.6) admit a Lax representation.

1.1 Lax Representation of the CM Equations

It was a remarkable observation by Moser that the CM-system *admits* a Lax representation where the first matrix of the relevant Lax pair (L, M) is a CM-matrix (see (8.7)), while the second one is defined by

$$M_{kl} = \iota \delta_{kl} \sum_{s=1}^{n} \frac{(1 - \delta_{ks})}{(x_k - x_s)^2} - \iota \frac{(1 - \delta_{kl})}{(x_k - x_l)^2} \tag{8.8}$$

where δ_{kl} is the Kronecker delta. Note that L is *Hermitian*, while M is *skew-Hermitian*.

Example 8.4 For a three-particle CM-system, one has:

$$L = \begin{pmatrix} p_1 & \frac{\iota}{x_1 - x_2} & \frac{\iota}{x_1 - x_3} \\ \frac{\iota}{x_2 - x_1} & p_2 & \frac{\iota}{x_2 - x_3} \\ \frac{\iota}{x_3 - x_1} & \frac{\iota}{x_3 - x_2} & p_3 \end{pmatrix} \quad \text{and}$$

$$M = \begin{pmatrix} \frac{\iota}{(x_1 - x_2)^2} + \frac{\iota}{(x_1 - x_3)^2} & \frac{-\iota}{(x_1 - x_2)^2} & \frac{-\iota}{(x_1 - x_3)^2} \\ \frac{-\iota}{(x_1 - x_2)^2} & \frac{\iota}{(x_2 - x_1)^2} + \frac{\iota}{(x_2 - x_3)^2} & \frac{-\iota}{(x_2 - x_3)^2} \\ \frac{-\iota}{(x_1 - x_3)^2} & \frac{-\iota}{(x_2 - x_3)^2} & \frac{\iota}{(x_3 - x_1)^2} + \frac{\iota}{(x_3 - x_2)^2} \end{pmatrix}.$$

$$\triangle$$

We will now explain in what sense the CM-system admits Lax representation with a Lax pair (L, M). To simplify the computation, let

$$\begin{cases} z_{kl} = \frac{1}{x_k - x_l}, & \text{if } k \neq l \\ z_{kl} = 0 \text{ otherwise.} \end{cases} \tag{8.9}$$

We start our discussion considering the simplest possible example, the two-particles CM-system.

Example 8.5 In this case, using the notation just introduced

$$L = \begin{pmatrix} p_1 & \iota z_{12} \\ \iota z_{21} & p_2 \end{pmatrix} \quad \text{and} \quad M = \begin{pmatrix} \iota z_{12}^2 & -\iota z_{12}^2 \\ -\iota z_{12}^2 & \iota z_{12}^2 \end{pmatrix},$$

entailing

$$[L, M] = \begin{pmatrix} 2z_{12}^3 & \imath(p_2 - p_1)z_{12}^2 \\ \imath(p_1 - p_2)z_{12}^2 & -2z_{12}^3 \end{pmatrix}. \tag{8.10}$$

On the other hand,

$$\dot{z}_{kl} = \frac{\dot{x}_l - \dot{x}_k}{(x_k - x_l)^2} = (p_l - p_k)z_{kl}^2$$

yields

$$\dot{L} = \begin{pmatrix} \dot{p}_1 & \imath(p_2 - p_1)z_{12}^2 \\ \imath(p_1 - p_2)z_{12}^2 & \dot{p}_2 \end{pmatrix},$$

showing that $\frac{dL}{dt} = [L, M]$ if (8.6) holds. $\triangle$

In spite of the result enclosed in the previous example, the matrix equation $\frac{dL}{dt} = [L, M]$ with L and B defined as above is not equivalent to the system (8.3). In fact, while the latter describes a dynamical system with $n - 1$ degrees of freedom, being H translational-invariant, the Lax equation depends a priori on all z_{ij} with $i < j$, as one can check already in the case of the three-particle CM-system (see Example 8.4). Obviously, this dependence is highly redundant because of the following relations:

$$z_{ij}^{-1} + z_{jk}^{-1} = z_{ik}^{-1}, \ \forall i \neq j \neq k \quad \text{and} \quad z_{ij} + z_{ji} = 0, \ \forall i, j. \tag{8.11}$$

In any case, if H is as in (8.1), then

Theorem 8.6 *The system*

$$\begin{cases} \dot{z}_{kl} = (p_l - p_k)z_{kl}^2, \\ \dot{p}_s = -\sum_{k \neq i} \frac{2}{(x_k - x_s)^3} \end{cases} \tag{8.12}$$

with $p_s = \dot{x}_s$ for all s is equivalent to

$$\frac{dL}{dt} = [L, M], \tag{8.13}$$

with (L, M) defined in (8.7) and, respectively, (8.8).

Proof Following Moser, let

$$L = Y + \imath Z_1 \quad \text{and} \quad M = \imath D_2 - \imath Z_2, \tag{8.14}$$

where $Y = \mathrm{diag}(p_1, \ldots, p_n)$ and

$$(Z_\alpha)_{ij} = z_{ij}^\alpha, \ \forall i \neq j \quad \text{and} \quad (Z_\alpha)_{ii} = 0, \ \alpha = 1, 2, \tag{8.15}$$

$$(D_\alpha)_{ij} = 0, \ \forall i \neq j \quad \text{and} \quad (D_\alpha)_{ii} = \sum_{k \neq i} z_{ik}^\alpha, \ \alpha = 1, 2, 3. \tag{8.16}$$

Then the right-hand side of (8.13) can be written as

$$[L, M] = \imath [Y, D_2] - [Z_1, D_2] - \imath [Y, Z_2] + [Z_1, Z_2].$$

To proceed with the proof, it is necessary to compute the summands of the left-hand side of the previous identity. We start with

$$[Z_1, Z_2]_{kl} = \sum_{s=1}^{n} (Z_1)_{ks}(Z_2)_{sl} - (Z_2)_{ks}(Z_1)_{sl} \overset{(8.15)}{=} \sum_{s=1}^{n} (z_{ks} z_{sl}^2 - z_{ks}^2 z_{sl}),$$

and

$$[D_2, Z_1]_{kl} = \sum_{s=1}^{n} (D_2)_{ks}(Z_1)_{sl} - (Z_1)_{ks}(D_2)_{sl} = (D_2)_{kk}(Z_1)_{kl} - (Z_1)_{kl}(D_2)_{ll}$$

$$\overset{(8.16)}{=} \left(\sum_{s=1}^{n} z_{ks}^2 - \sum_{s=1}^{n} z_{ls}^2 \right) z_{kl},$$

which yield $[Z_1, Z_2] + [D_2, Z_1] = \sum_{s=1}^{n} Q_{kl,s}$, where

$$Q_{kl,s} = (z_{ks} - z_{kl}) z_{ls}^2 + (z_{kl} - z_{sl}) z_{sk}^2. \tag{8.17}$$

The left-hand side of (8.17) can be reordered as follows:

$$z_{ks} z_{ls}^2 - z_{sl} z_{sk}^2 - (z_{ls}^2 - z_{ks}^2) z_{kl} = z_{ks} z_{sl}(z_{sl} - z_{ks}) - (z_{ls}^2 - z_{ks}^2) z_{kl}$$

$$= (z_{sl} - z_{ks})(z_{ks} z_{sl} - z_{kl}(z_{sl} + z_{ks}))$$

and note that if $k \neq s \neq l$,

$$P_{kl,s} = z_{ks} z_{sl} - z_{kl}(z_{sl} + z_{ks}) = z_{ks} z_{sl} z_{kl}(z_{kl}^{-1} - z_{ks}^{-1} - z_{sl}^{-1}) \overset{(8.11)}{=} 0.$$

On the other hand, if $k \neq l$,

$$P_{kl,s} = z_{ks} z_{sl} - z_{kl}(z_{sl} + z_{ks}) = -z_{kl}^2, \ \text{if } s = k;$$

$$P_{kl,s} = z_{ks} z_{sl} - z_{kl}(z_{sl} + z_{ks}) = -z_{kl}^2, \ \text{if } s = l$$

(see (8.11)). These computations entail

$$
Q_{kl,s} = (z_{sl} - z_{ks})P_{kl,s} = \begin{cases} 0 \text{ if } k \neq l \neq s, \\ (z_{kl} - z_{ks})(-z_{kl}^2) = -z_{kl}^3 \text{ if } k \neq l \text{ and } s = k \\ (z_{kl} - z_{ks})(-z_{kl}^2) = z_{kl}^3 \text{ if } k \neq l \text{ and } s = l. \end{cases}
$$

In other words, given k, l such that $k \neq l$,

$$
\sum_{s=1}^{n} Q_{kl,s} = \sum_{s \neq k,\, s \neq l} \cancel{Q_{kl,s}} + \cancel{Q_{kl,k}} + \cancel{Q_{kl,l}} = 0.
$$

On the other hand, if $k = l$,

$$
[Z_1, Z_2]_{kk} - [Z_1, D_2]_{kk} = \sum_{s=1}^{n}(Z_1)_{ks}(Z_2)_{sk} - (Z_2)_{ks}(Z_1)_{sk} - \sum_{s=1}^{n}(Z_1)_{ks}(D_2)_{sk}
$$

$$
- (D_2)_{ks}(Z_1)_{sk}
$$

$$
= 2\sum_{s=1}^{n} z_{ks}^3 - \cancel{(Z_1)_{kk}(D_2)_k} + \cancel{(D_2)_k(Z_1)_{kk}}
$$

$$
= 2D_3,
$$

which yields

$$
[L, M] = [Y, M] + 2D_3.
$$

Computing the entries of the first summand of the right-hand side of the previous identity, one finds

$$
[Y, M]_{kl} \overset{(8.14)}{=} \imath([Y, D_2]_{kl} - [Y, Z_2]_{kl}) = \imath \sum_{s=1}^{n}(Y_{ks}(D_2)_{sl} - (D_2)_{ks}Y_{sl})
$$

$$
- \imath \sum_{s=1}^{n}(Y_{ks}(Z_2)_{sl} - (Z_2)_{ks}Y_{sl})
$$

$$
= \imath(p_k(D_2)_l - (D_2)_k p_l)\delta_{kl} - \imath(p_k - p_l)z_{kl}^2,
$$

which yields

$$
[L, M]_{kl} = [Y, M]_{kl} + 2(D_3)_{kl} = \begin{cases} \imath(p_l - p_k)z_{kl}^2 \text{ if } k \neq l, \\ 2(D_2)_{kk} = 2\sum_{s=1}^{n} z_{ks}^3 \text{ if } k = l, \end{cases}
$$

which proves (8.12). $\square$

Remark 8.7 Both the theorem above and its proof are due to Moser, see [190]. $\triangle$

On the other hand, it is worth noting that if X_H is as in (8.2), one can show

$$X_H(L) = [L, M],$$

see Example 8.9, which entails

Corollary 8.8 *The functions* $F_k = \frac{1}{k}\,\mathrm{tr}(L^k)$ *and* $k = 1, \ldots, n$ *are first integrals for* (8.6).

Example 8.9 For the two-particle CM-system, one has

$$X_H(L) = \begin{pmatrix} X_H(p_1) & \imath X_H(z_{12}) \\ \imath X_H(z_{21}) & X_H(p_2) \end{pmatrix} \overset{(8.2)}{=} \begin{pmatrix} 2z_{12}^3 & \imath(p_2 - p_1)z_{12}^2 \\ \imath(p_1 - p_2)z_{12}^2 & 2z_{21}^3 \end{pmatrix} \overset{(8.10)}{=} [L, M].$$

Note that the action of X_H on the matrix valued function L is component-wise. The general case does not present more difficulties, and it will be left as an exercise for the reader. $\triangle$

Example 8.10 The function $F_1 = \mathrm{tr}(L) = \sum_{i=1}^{n} p_i$, i.e., the linear momentum of the CM-system. $\triangle$

Problem 8.11 Compute F_2 and F_3. $\triangle$

As already noticed, since L is Hermitian and M is skew-Hermitian, (8.13) can be thought of as a dynamical system defined by the vector field $X_M = [\cdot, M]$ on the vector space of Hermitian matrices $\mathfrak{herm}(n) = \{B \in \mathfrak{gl}_n(\mathbb{C})\mid B = \overline{B}^t\}$. The consequences of this observation will be explored in the next subsection. We close this subsection introducing an interesting extension of the Lax representation of the CM-system.

1.1.1 Extended Lax Representation

The Lax equation $\dot{L} = [L, M]$ described above can be enhanced with a second matrix equation, which will play an important role in the description of a bi-Hamiltonian representation of the CM-system. This second matrix equation describes the evolution of $X = \mathrm{diag}(x_1, \ldots, x_n)$ along the flow defined by the Hamiltonian vector field of the CM-system. More precisely

Proposition 8.12 *The time-evolution of* X *with respect to the Hamiltonian vector field* (8.2) *can be written in terms of the matrices* L, M *as follows:*

$$X_H(X) = L + [X, M]. \tag{8.18}$$

Proof Since X_H acts on X entry by entry (see Example 8.9), $X_H(X) = \text{diag}(p_1, \ldots, p_n)$. On the other hand,

$$[X, M]_{kl} = \sum_{s=1}^{n}(X_{ks}M_{sl} - M_{ks}X_{sl}) = M_{kl}(x_k - x_l)$$

$$\overset{(8.8)}{=} \iota\delta_{kl}(x_k - x_l)\sum_{s=1}^{n}\frac{(1 - \delta_{ks})}{(x_k - x_s)^2} - \iota\frac{(1 - \delta_{kl})}{(x_k - x_l)}$$

$$= -\frac{\iota}{x_k - x_l}, \text{ if } k \neq l, \ 0 \text{ otherwise,}$$

which proves (8.18). $\qquad\square$

Finally, one can introduce the following.

Definition 8.13 If L, M and X are as above

$$\dot{L} = [L, M] \quad \text{and} \quad \dot{X} = L + [X, M] \tag{8.19}$$

are called the *extended Lax representation* of (8.3), where $\dot{A} = X_H(A)$ for all matrix A. $\qquad\triangle$

The existence of the extended Lax representation for the CM-system was observed in [24] and by Perelomov (see [208] and the references therein). As already remarked, everything else in this section is due to Moser. In particular, the definition of the matrices L, M (see (8.7) and (8.8)) and the proof of Theorem 8.6 are borrowed from [190]. In Sect. 4, it will be shown that (8.19) can be *extended* to the so-called full *CM-hierarchy*.

2 Complete Integrability

This subsection will be devoted to the presentation of a proof of the complete integrability of the rational CM- system. As we have already hinted in the introduction to this chapter, the integrability of the CM-system was at first proved by Moser using arguments borrowed from scattering theory. Instead of following his reasoning, we will overview the approach of [130] based on the Hamiltonian reduction. Before diving into the details of their argument, we shall sketch the main idea behind it. The starting point of our presentation is the following *picture* suggested by the existence of the Lax representation (8.13). We will think of $\mathfrak{herm}(n)$ as the *configuration* space of a dynamical system of the Lax-type $\frac{dB}{dt} = [B, \xi]$ where $\xi \in \mathfrak{u}(n)$. The functions $F_k = \frac{1}{k}\text{tr}(B^k)$ are first integrals of this system, and if we pull them back to $M = T^*\mathfrak{herm}(n) \simeq \mathfrak{herm}(n) \times \mathfrak{herm}(n)$ by the canonical projection, they define a set of functions $\{f_k\}_{k \geq 1}$ in involution with respect to the Poisson structure there defined by the canonical symplectic form. In particular, the function f_2 defines on M a linear *flow* that restricts to the level sets of the moment map μ associated with the cotangent lift of the $U(n)$-action $\varphi : U(n) \times \mathfrak{herm}(n) \to \mathfrak{herm}(n)$, $\varphi(g, B) = gBg^{-1}$. One of the main insights of [130] is the choice of $\alpha \in \mathfrak{u}(n)$ such that:

(i) The canonical projection $\pi_\alpha : \mu^{-1}(\alpha) \to \mu^{-1}(\alpha)/G_\alpha$ has a *section* whose points are in a one-to-one correspondence with the pairs (D, L) where D is diagonal and L is a CM-matrix.

(ii) The reduced phase-space $\mu^{-1}(\alpha)/G_\alpha$ is symplectomorphic to $T^*\mathbb{R}^n$ endowed with its canonical symplectic form.

From the points (i) and (ii) above, it follows that the restriction of the f_ks to $\mu^{-1}(\alpha)$ is a generically independent set of functions, and they descend to the reduced phase-space to generate a maximal commutative subalgebra of the Poisson algebra of functions on the phase-space of the CM-system, proving its complete integrability.

In what remains of the present subsection, we will give a more detailed account of the steps sketched above. We begin with the following.

Lemma 8.14 *The cotangent bundle* $M = T^*\mathfrak{herm}(n)$ *can be identified with* $\mathfrak{herm}(n) \times \mathfrak{herm}(n)$. *Moreover, this isomorphism identifies* $\mathfrak{herm}(n) \times \mathfrak{herm}(n)$ *with* $\mathbb{R}^{2n^2}$ *as symplectic vector spaces, both endowed with their standard symplectic structures.*

Proof Using the trace form

$$(A, B) = \mathrm{tr}(AB^\dagger) = \mathrm{tr}(AB), \tag{8.20}$$

one can identify $\mathfrak{herm}(n)$ with its dual space so that $T^*\mathfrak{herm}(n)$ is identified with $\mathfrak{herm}(n) \times \mathfrak{herm}(n)$, which is a real vector space of dimension equal to $2n^2$. Under the previous identification, the canonical symplectic form of $T^*\mathfrak{herm}(n)$ can be written as

$$\Omega = \mathrm{tr}(dP \wedge dX) \tag{8.21}$$

where (P, X) is a shorthand notation for the canonical symplectic coordinates on M. To make the notation more explicit, one can expand the right-hand side of (8.21) in terms of the entries of the matrices (P, X), to obtain

$$\Omega = \sum_{i,j=1}^{n} dP_{ij} \wedge dX_{ji},$$

which yields the identification of $\mathfrak{herm}(n) \times \mathfrak{herm}(n)$ endowed with the symplectic form (8.21) with $\mathbb{R}^{2n^2}$ endowed with its standard symplectic form. In particular, note that, given i, j, P_{ij}, X_{ji} are a pair of *canonical* and *conjugate* variables on M with respect to $\Omega = \mathrm{tr}(dP \wedge dX)$. $\qquad\square$

Let $\pi : M \to \mathfrak{herm}(n)$ be defined by $\pi(A, B) = B$ for all $(A, B) \in M$, $F_k = \frac{1}{k}\mathrm{tr}B^k$, and $f_k = \pi^* F_k$, $k = 1, \ldots, n$. Then

Lemma 8.15 *The functions f_k are in involution with respect to the Poisson structure defined by Ω.*

Problem 8.16 Prove Lemma 8.15. [Hint: the functions f_k are really functions only of B, but the symplectic form is...]. $\qquad\triangle$

Moreover, observe that for each $\xi \in \mathfrak{u}(n)$, the cotangent lift $\tilde{X}_\xi$ of the vector field $X_\xi := [\cdot, \xi]$ is defined by

$$\tilde{X}_\xi(A, B) = ([A, \xi], [B, \xi]) \tag{8.22}$$

for all $(A, B) \in M$ (see Appendix D). To explain this formula, note that $U(n)$ acts on M by simultaneous conjugation

$$\tilde{\varphi}(g, (A, B)) = (gAg^{-1}, gBg^{-1}) \tag{8.23}$$

and the vector field $\tilde{X}_\xi$ defined in Eq. (8.22) is the fundamental vector field corresponding to $\xi \in \mathfrak{u}(n)$.

Problem 8.17 Prove (8.22).

Hint: As far as (i) concerns, recall that the fundamental vector field associated with a left G-action is defined in Formula (C.5), which, if applied in the present context, becomes

$$\tilde{X}_\xi((A, B)) = \left.\frac{d}{dt}\right|_{t=0} \tilde{\varphi}_{(A,B)}(\exp(-t\xi)) = \left.\frac{d}{dt}\right|_{t=0} (\exp(-t\xi)A\exp(t\xi),$$

$$\exp(-t\xi)B\exp(t\xi)).$$

$\qquad\triangle$

Lemma 8.18 *The action (8.23) is strongly Hamiltonian, and a corresponding moment map is defined by*

$$\mu(A, B) = [A, B], \tag{8.24}$$

for all $(A, B) \in M$.

Proof The first part of the statement follows from the observation that (8.23) is the cotangent lift of $\varphi : U(n) \times \mathfrak{herm}(n) \to \mathfrak{herm}(n)$ defined by

$$\varphi(g, A) = gAg^{-1}. \tag{8.25}$$

To prove that a corresponding moment map is as in (8.24), one notes that the fundamental vector field associated with $\xi \in \mathfrak{u}(n)$ by (8.25) is

$$X_\xi(A) = [A, \xi], \ \forall A \in \mathfrak{herm}(n).$$

Then

$$F_\xi(\alpha) = \langle \alpha, X_\xi(A) \rangle, \ \forall \alpha \in T_A^*(\mathfrak{herm}(n))$$

(see (D.1)). Identifying $T_A^*\mathfrak{herm}(n) \simeq \mathfrak{herm}(n)^*$ with $\mathfrak{herm}(n)$ via the trace form (8.20), one has

$$F_\xi(\alpha) = \langle \alpha, X_\xi(A) \rangle = \mathrm{tr}(A_\alpha[A, \xi]) = \mathrm{tr}([A_\alpha, A]\xi) = -\mathrm{tr}([A_\alpha, A]\xi^\dagger)$$

$$= \mathrm{tr}([A, A_\alpha]\xi^\dagger)$$

which, together with (3.2) and the abovementioned identification of $T_A \mathfrak{herm}(n)^* \simeq \mathfrak{herm}(n)$, yields

$$\mu(A, A_\alpha) = [A, A_\alpha],$$

entailing (8.24). $\qquad\qquad\qquad\qquad\qquad\qquad\qquad\qquad\qquad\qquad\qquad\qquad\quad \Box$

Remark 8.19 Note that in the proof of the previous lemma, A_α is the (unique) element in $\mathfrak{herm}(n)$ such that $\langle \alpha, B \rangle = \mathrm{tr}(A_\alpha B)$ for all $B \in \mathfrak{herm}(n)$. It is also worth noticing that the non-degenerate bilinear form (8.20) is the restriction to $\mathfrak{herm}(n)$ of the trace form on $\mathfrak{gl}_n(\mathbb{C})$ defined by the same formula. The restriction of this form to $\mathfrak{u}(n)$ is $(A, B) = -\mathrm{tr}(AB)$ for all $A, B \in \mathfrak{u}(n)$. $\qquad\qquad \triangle$

After these preliminary remarks, we can start to spell out the ingredients of the proof of the complete integrability of the CM-system. Since we want to use the Hamiltonian reduction, an important point of the proof will be the choice of an appropriate element $\alpha \in \mathfrak{u}(n)$, a regular value for the moment map μ, from which it will be possible to recover the phase-space of the CM-system as the reduced symplectic space $\frac{\mu^{-1}(\alpha)}{G_\alpha}$. A crucial step of the solution proposed by the authors of [130] is enclosed in the following.

Theorem 8.20 (Kazhdan-Kostant-Sternberg, [130]) *There exists $\alpha \in \mathfrak{u}(n)$ such that every G_α-orbit in $\mu^{-1}(\alpha)$ contains a unique pair of matrices (D, L), where D is defined by*

$$D_{ij} = \delta_{ij} x_i,$$

and L is a CM-matrix.

Proof Let $v \in \mathbb{C}^n$ be a *column* vector and $v^\dagger$ its *transposed conjugate*. Denote with $v \otimes v^\dagger$ the *rank-one* linear transformation on $\mathbb{C}^n$ defined by $v \otimes v^\dagger(u) = (\bar{v}, u)v =$

$(\sum_{j=1}^{n} \bar{v}_j u_j)v$, where v_j, u_j are components of v and, respectively, u. We start our proof showing that if

$$\alpha = \imath(-\operatorname{id} + v \otimes v^\dagger), \quad v^\dagger = (1, \ldots, 1) \tag{8.26}$$

and $(A, B) \in \mu^{-1}(\alpha)$, then A can be diagonalized by some element $g \in G_\alpha$, the isotropy group of α in $U(n)$. In fact, since A is an Hermitian matrix, it is possible to find a unitary matrix g such that $D = gAg^{-1}$, with D diagonal. Let $E = gBg^{-1}$ and $w = gv$. Since $(A, B) \in \mu^{-1}(\alpha)$

$$[D, E] = \imath(-\operatorname{id} + w \otimes w^\dagger).$$

Since D is diagonal, the right-hand side of the previous identity is an $n \times n$ matrix with zeros on the main diagonal. This condition yields the following constraints on w:

$$w_k \overline{w_k} = |w_k|^2 = 1, \ \forall k = 1, \ldots, n,$$

i.e., $w_k = e^{\imath\theta_k}$ for $k = 1, \ldots, n$. Multiplying g on the left by $h = \operatorname{diag}(e^{-\imath\theta_1}, \ldots, e^{-\imath\theta_n})$, we will obtain a new unitary matrix g', which diagonalizes A and, at the same time, fixes v entering the definition of α (see Formula (8.26)). Summarizing, if $(A, B) \in \mu^{-1}(\alpha)$, it is possible to diagonalize A using an element of the isotropy group of α. Note that the matrix representing the element α of (8.26) is

$$\alpha = \begin{pmatrix} 0 & \imath & \imath & \imath & \cdots \\ \imath & 0 & \imath & \imath & \cdots \\ \imath & \imath & 0 & \imath & \cdots \\ \imath & \imath & \imath & \ddots & \cdots \end{pmatrix} \tag{8.27}$$

To conclude the proof, observe that if $(x_1, \ldots, x_n)$ are the eigenvalues of A, that is, the diagonal entries of D, and if $L = g'Bg'^{-1}$, since (D, L) satisfy the equation

$$[D, L] = \alpha,$$

one has

$$[D, L]_{ij} = (x_i - x_j)L_{ij} = \alpha_{ij} = \imath,$$

for all $i \neq j$. This computation entails that:

1. $x_i \neq x_j$ for all $i \neq j$
2. $L_{ij} = \dfrac{\imath}{x_i - x_j}$
3. The diagonal entries $(p_1, \ldots, p_n)$ of L are arbitrary

$\square$

A few observations are in order.

Remark 8.21

(i) The matrix D obtained in the previous theorem is unique up to the permutation of the diagonal entries.

(ii) The tuple $(x_1, \ldots, x_n, p_1, \ldots, p_n)$ can be used as a set of local coordinates on the quotient $\mu^{-1}(\alpha)/G_\alpha$.

(iii) The theorem above can be rephrased, saying that, under the stated assumptions, there exists a section $S_{KKS} : \mu^{-1}(\alpha)/G_\alpha \to \mu^{-1}(\alpha)$ (where KKS stands for Kazhdan-Kostant-Sternberg) of the smooth submersion $p : \mu^{-1}(\alpha) \to \mu^{-1}(\alpha)/G_\alpha$. The points of this section are pairs of matrices (D, L), where D is diagonal and L is a CM-matrix.

(iv) Although we are working with the reductive Lie group $U(n)$, everything, in the previous theorem, takes place in the simple Lie group $SU(n)$. In fact, the image of the moment map $\mu : M \to \mathfrak{u}(n)$ defined by $\mu(A, B) = [A, B]$ for all A, B is contained in the Lie algebra $\mathfrak{su}(n)$, and so we could have used from the very beginning the corresponding Lie group. However, it seems more natural to work with the group $U(n)$ since the Moser matrices belong to the full space of the Hermitian matrices. More comments on this issue can be found in Sect. 3.

$$\triangle$$

Problem 8.22 If α is as in (8.27), describe the isotropy group $G_\alpha \subset U(n)$, and find its dimension. $\triangle$

Coming back to our main concern, i.e., the proof of the complete integrability of the CM-system, in the next result, we prove that the reduced phase-space described in Theorem 8.20 is the *right one*, i.e., it is (symplectomorphic to) the cotangent bundle of $\mathbb{R}^n$.

Theorem 8.23 *The natural map*

$$\chi : \mu^{-1}(\alpha)/G_\alpha \to \mathbb{R}^n/S_n \times \mathbb{R}^n \simeq T^*\mathbb{R}^n \tag{8.28}$$

$$(D, L) \mapsto (x_1, \ldots, x_n, p_1, \ldots, p_n)$$

is a symplectomorphism once we endow $T^*\mathbb{R}^n$ with its canonical symplectic structure and $\mu^{-1}(\alpha)/G_\alpha$ with the symplectic form induced via the Marsden-Weinstein-Meyer reduction. In the formula above, S_n denotes the group of permutations of n-elements.

Proof From the remark above, it follows that there is a one-to-one correspondence between the points of the quotient space $\mu^{-1}(\alpha)/G_\alpha$ and $2n$-tuples of real numbers, where the first n numbers $(x_1, \ldots, x_n)$ are unordered and pairwise different. This proves the existence of the one-to-one map χ defined in Eq. (8.28). We are left to prove that such a map is a symplectomorphism. This will followed by computing the restriction of the symplectic form $\Omega = \mathrm{tr}(dP \wedge dX)$ to the level set $\mu^{-1}(\alpha)$ and observing that the resulting form is invariant by the simultaneous conjugation with any element in $U(n)$. This last property follows from the properties of the trace. On the other hand, since in every G_α-orbit in $\mu^{-1}(\alpha)$ there is, up to *permutations* of the entries of D, only one pair of the form (D, L), it suffices to compute $\mathrm{tr}(dL \wedge dD)$ to show that the map χ has the property stated in the theorem. This will be left to the reader. □

Remark 8.24 The outcome of Theorems 8.20 and 8.23 is that the Marsden-Weinstein-Meyer reduced phase-space $\mu^{-1}(\alpha)/G_\alpha$, where μ and α are as in (8.27) and, respectively, as in (8.24), can be identified with $T^*\mathbb{R}^n$ endowed with its canonical symplectic structure $\omega_0 = \sum_{i=1}^n dp_i \wedge dx_i$. It is worth to stress that the canonical coordinates (x, p) on the symplectic quotient are the *eigenvalues* of an Hermitian matrix A and, respectively, the *diagonal entries* of an Hermitian matrix B, where (A, B) is just an arbitrary pair of matrices in $\mu^{-1}(\alpha)$. In the proof of Theorem 8.20, the arbitrariness of the choice made at the very beginning, of diagonalizing A and not B, should catch the reader attention. Making this choice would entail a completely equivalent description of the quotient. More precisely, one can show that there exists $g \in G_\alpha$ such that $X = gBg^{-1} = \mathrm{diag}(\lambda_1, \ldots, \lambda_n)$, and using the same argument used in the proof of Theorem 8.20, one shows that $Z = gAg^{-1}$ is the Hermitian matrix defined by

$$Z_{ij} = \frac{\iota}{\lambda_i - \lambda_j} \quad \text{and} \quad Z_{ii} = \mu_i, \ \forall i, j = 1, \ldots, n,$$

where the μ_is are arbitrary real numbers. The set of coordinates (μ, λ) are symplectic, i.e., the reduced phase-space carries the symplectic form $\sum_{i=1}^n d\lambda_i \wedge d\mu_i$. In particular, the map $(p, x) \rightsquigarrow (-\lambda, \mu)$ is a symplectomorphism between the two reduced phase-spaces.   △

Using Theorem 8.23, we shall be able to prove the complete integrability of the CM-system.

To this end, first, we observe that

Proposition 8.25 *The Hamiltonian flow of the function $f_2 = \pi^*(F_2)$ on the total space M is linear, and it restricts to $\mu^{-1}(\alpha)$.*

Proof Recalling (8.21), the Hamiltonian equations defined by $H \in C^\infty(M)$ are

$$\begin{cases} \dot{X}_{ji} = \frac{\partial H}{\partial P_{ij}} \\ \dot{P}_{ij} = -\frac{\partial H}{\partial X_{ji}} \end{cases}$$

for all $i, j = 1, \ldots, n$. Choosing $H = f_2 = \frac{\operatorname{tr} P^2}{2} = \frac{1}{2} \sum_{k,l=1}^{n} P_{kl} P_{lk}$, we have

$$\frac{\partial H}{\partial P_{ij}} = \frac{1}{2} \sum_{k,l=1}^{n} \delta_{ki} \delta_{lj} P_{lk} + \frac{1}{2} \sum_{k,l=1}^{n} \delta_{li} \delta_{jk} P_{kl} = \frac{1}{2} P_{ji} + \frac{1}{2} P_{ji} = P_{ji}$$

$$\frac{\partial H}{\partial X_{ji}} = 0,$$

which yields

$$\begin{cases} P_{ij}(t) = P_{ij}(0) \\ X_{ij}(t) = X_{ij}(0) + t P_{ij}(0), \end{cases}$$

for all $i, j = 1, \ldots, n$. In other words, the flow defined by the Hamiltonian f_2 is linear,

$$(A, B) \rightsquigarrow (A + tB, B). \tag{8.29}$$

To prove the last statement, it suffices to observe that if $(A_0, B_0) = (A(0), B(0)) \in \mu^{-1}(\alpha)$, then

$$[A(t), B(t)] \stackrel{(8.29)}{=} [A_0 + tB_0, B_0] = [A_0, B_0] = \alpha, \ \forall t$$

(see also Remark 8.26). $\qquad\qquad\qquad\qquad\qquad\qquad\qquad\qquad\qquad\qquad\qquad\quad \square$

A few comments are now in order.

Remark 8.26

(i) In the discussion above, we kept the notation (P, X) to denote *momentum* and *position* coordinates on M, while in our construction of the reduced phase-space, the *positions* and the *momenta* are enclosed in the matrices A and, respectively, B. Recall in fact that positions and momenta in the reduced phase-space $\mu^{-1}(\alpha)/G_\alpha$ are the eigenvalues of A and, respectively, the diagonal entries of $L = gBg^{-1}$, for some $g \in G_\alpha$. On the other hand, observe that we used $\pi : M \to \mathfrak{herm}(n)$ defined by $\pi(A, B) = B$, which, implicitly, subtends the idea that the Bs play the role of *positions* on M, while the As play the one of momenta. For this reason, moving from the *geometrical* picture provided by $\pi : M \to \mathfrak{herm}(n)$ to the *dynamical* one defined by $\pi_\alpha : \mu^{-1}(\alpha) \to \mu^{-1}(\alpha)/G_\alpha$, we assist to a sort of switching between the positions and momenta.

(ii) Note that since the f_ks are invariant for conjugation, their restrictions to $\mu^{-1}(\alpha)$ coincide with the restrictions of the same functions to the section S_{KKS} (see point (iii) of Remark 8.21). For this reason, we will write these restrictions as $f_k = \frac{\mathrm{tr}L^k}{k}$.

$\triangle$

Finally we can prove the following.

Proposition 8.27 *The CM-system is completely integrable.*

Proof We need to show that the functions $(f_1, \ldots, f_n)$ descend to the reduced space $\mu^{-1}(\alpha)/G_\alpha$ remaining generically independent and in involution. The last part of the statement follows from Lemma 8.15 and from the general theory of the Hamiltonian reduction: a set of functions that are in involution on the *big* phase-space (M, ω) will still be in involution when restricted to the level set of the moment map $\mu : M \to \mathfrak{g}^*$. Then, since $\pi : \mu^{-1}(\alpha) \to \mu^{-1}(\alpha)/G_\alpha$ is a Poisson map, if those functions are G_α-invariant, they will descend to a set of functions that are again in involution on the reduced phase-space $\mu^{-1}(\alpha)/G_\alpha$. Applying Lemma 8.15, we conclude that $(f_1, \ldots, f_n)$ descend to the quotient to generate a commutative Poisson subalgebra of the algebra of functions on $\mu^{-1}(\alpha)/G_\alpha$. On the other hand, using the inverse of the map χ defined in Theorem 8.23, we can pull back the functions $(f_1, \ldots, f_n)$ on $T^*\mathbb{R}^n$, where they can be written as

$$(\chi^{-1})^* f_k = \sum_{i=1}^{n} p_i^k + g_k(p, x),$$

for $k = 1, \ldots, n$, where the functions $g_k(p, x)$ are polynomial functions in the p_is of degree strictly less than k. To prove the functional independence of the functions $(\xi^{-1})^* f_k,$, it suffices to note that the differential of the map

$$\left((\chi^{-1})^* f_1, \ldots, (\chi^{-1})^* f_n \right) : T^*\mathbb{R}^n \to \mathbb{R}^n$$

has generically rank n, as it has an $n \times n$ minor of Vandermonde type. $\qquad \square$

Remark 8.28 It is worth to stress that $(\chi^{-1})^* f_1 = \sum_{i=1}^{n} p_i$ and $(\chi^{-1})^* f_2 = \frac{1}{2} \sum_{i=1}^{n} 2p_i^2 + \sum_{i<j} \frac{1}{(x_i - x_j)^2}$, i.e., $(\chi^{-1})^* f_1$ and $(\chi^{-1})^* f_2$, are, respectively, the linear momentum and the Hamiltonian of the CM-system. In particular, the CM-flow is obtained as Hamiltonian reduction of the linear flow defined on $\mu^{-1}(\alpha)$ by the restriction of f_2 to this level set. $\qquad \triangle$

2.1 Explicit Solutions

In this subsection, we will give an explicit formula for the solutions of (8.6). In particular, it will be shown that

Proposition 8.29 *The eigenvalues of the matrix*

$$S_t = \mathrm{diag}(x_1(0), \ldots, x_n(0)) + tL(p(0), x(0))$$

form a complete set of solutions of the system (8.6).

Proof Restrict the Hamiltonian flow defined by f_2 to the level set $\mu^{-1}(\alpha)$ (see the remark above). As a consequence of Theorem 8.20, one can find a one-parameter family of matrices $\{V_t\}_{t \in \mathbb{R}}$, with $V_t \in G_\alpha$ for all t, such that

$$(V_t D_t V_t^{-1}, V_t L_t V_t^{-1}) = (V_0(D_0 + tL_0)V_0^{-1}, V_0 L_0 V_0^{-1}),$$

where $D_t = \mathrm{diag}(x_1(t), \ldots, x_n(t))$ and $L_t = L(p(t), x(t))$. Defining $U_t = V_t^{-1} V_0$, one has

$$(U_t^{-1} D_t U_t, U_t^{-1} L_t U_t) = (D_0 + tL_0, L_0), \tag{8.30}$$

for all t. This identity entails that the functions (of (p, x)) defined by $\frac{1}{k}\mathrm{tr}(L_0^k) = \frac{1}{k}\mathrm{tr}(L_t^k)$, for $k = 1, \ldots, n$, do not depend on t, so that they are first integrals. Moreover, from the equality of the first entries, it follows that the $x_k = x_k(t)$ are the eigenvalues of the matrix $D_0 + tL_0$ as stated in the proposition. $\qquad\square$

3 Hamiltonian Reduction

In Theorem 8.20, it was proved that the points of the reduced phase-space $\mu^{-1}(\alpha)/G_\alpha$ are in a one-to-one correspondence with the pairs (D, L), where D is a diagonal matrix and L is a CM-matrix. Using this result, we concluded that the dimension of the reduced phase-space is $2n$ and observed that a suitable system of local coordinates is given by the $2n$-tuple $(x_1, \ldots, x_n, p_1, \ldots, p_n)$ (see Remark 8.21). In this subsection, we will comment more about a few Lie theoretic aspects of the reduction process. In particular, we show that working with the simple Lie group $\mathrm{SU}(n)$ yields the same results obtained in the previous section. To this end, we start spelling out the steps of the computation of the dimension of the reduced phase-space $\mu^{-1}(\alpha)/G_\alpha$ under the assumption that $G = \mathrm{U}(n)$. The dimension of the stabilizer $G_\alpha \subset \mathrm{U}(n)$ is given in the following.

Lemma 8.30 *Given $\alpha = \iota(-\,\mathrm{id} + v \otimes v^*)$, $G_\alpha \subset \mathrm{U}(n)$ is isomorphic to the group* $\mathrm{U}(1) \times \mathrm{U}(n-1)$. *Then its (real) dimension is $n^2 - 2n + 2$.*

Let $Z(\mathrm{U}(n)) = \mathrm{U}(1)$ and $Z(G_\alpha)$ be the centers of the groups $\mathrm{U}(n)$ and, respectively, G_α. The previous lemma entails that $Z(\mathrm{U}(n)) = Z(G_\alpha)$. Hence

Corollary 8.31 *The dimension of the quotient space $\mu^{-1}(\alpha)/G_\alpha$ is to 2n.*

Proof Here we need to observe that since the moment map takes values in $\mathfrak{su}(n)$, dimension of which is $n^2 - 1$, the level set $\mu^{-1}(\alpha)$ is a submanifold of dimension $2n^2 - (n^2 - 1) = n^2 + 1$. Then, when we pass to the quotient $\mu^{-1}(\alpha)/G_\alpha$, only $n^2 - 2n + 1$ dimensions out of $n^2 - 2n + 2$ will contribute to the dimension of the quotient, as $Z(G_\alpha)$ acts trivially on $\mu^{-1}(\alpha)$. From these observations, the proof of the statement follows. $\qquad\square$

Let us now discuss the case of the group $\mathrm{SU}(n)$. In this case, the moment map, which is given by the same formula as for the case of $G = \mathrm{U}(n)$, takes values in the Lie algebra $\mathfrak{su}(n)$. The dimension of the level set $\mu^{-1}(\alpha)$ is the same as in the $\mathrm{U}(n)$ case, i.e., $n^2 + 1$. On the other hand, the stabilizer of the element α is

$$G'_\alpha = \{g \in \mathrm{U}(1) \times \mathrm{U}(n-1) \mid \det(g) = 1\}.$$

The last condition is imposed by the hypothesis that $G'_\alpha \subset \mathrm{SU}(n)$. Then the dimension of the stabilizer is

$$\dim \ G'_\alpha = (n-1)^2.$$

Note that, also in this case, the center of G'_α is nontrivial, and it is equal to

$$Z(G'_\alpha) = \{g \in \mathrm{U}(1) \times Z(\mathrm{U}(n-1)) \mid \det(g) = 1\}.$$

By this characterization, it follows that $Z(G'_\alpha)$ does not act trivially on the level set $\mu^{-1}(\alpha)$. Moreover, it is possible to show that

Lemma 8.32 *The stabilizer of every $(A, B) \in \mu^{-1}(\alpha)$ in $\mathrm{SU}(n)$ is a discrete subgroup.*

From this consideration, it follows the following.

Corollary 8.33 *The dimension of the quotient $\mu^{-1}(\alpha)/G'_\alpha$ is 2n.*

Proof Again, we apply the formula

$$\dim \ \mu^\alpha/G'_\alpha = \dim \ \mu^{-1}(\alpha) - \dim \ \mathrm{SU}(n) - \dim \ G'_\alpha$$

$$= 2n^2 - (n^2 - 1) - (n-1)^2 = 2n.$$

$$\square$$

4   Extended Lax Hierarchy, Bi-Hamiltonian Representation, and Master Symmetries

The bi-Hamiltonian representation of the Hamiltonian vector field (8.2) follows at once from Theorem 5.95 and from the existence of the extended Lax representation for the CM-system (see 1.1.1). To make more precise this statement, we show how (8.19) can be extended to the full *CM-hierarchy*, which we introduce below. Recall that in Proposition 8.36, it was shown that the explicit solutions of (8.6) can be obtained after the *reduction* of the linear flow defined in (8.30). In the proof of this proposition, it was remarked that such a flow is obtained from the invariant function f_2. Moreover, in general, for every $k \geq 1$, the restriction of the invariant function f_k to the level set $\mu^{-1}(\alpha)$ defines

$$\varphi_t^k(D_0, L_0) = (D_0 + t L_0^{k-1}, L_0), \ \forall t \in \mathbb{R}. \tag{8.31}$$

Definition 8.34   The family $\{\varphi_t^k\}_{k \in \mathbb{N}, t \in \mathbb{R}}$ so defined is called the *CM-hierarchy*. $\triangle$

Note that $\varphi_t^k \varphi_s^l = \varphi_s^l \varphi_t^k$ for all k, l and for all t, s, i.e., the flows of the CM-hierarchy commute with each other. Moreover, it is worth observing that (8.31) is compatible with the constraint defined by the moment map (8.24). In particular, if $\mu(D_0, L_0) = \alpha$, then $\mu(D_0 + t L_0^{k-1}, L_0) = \alpha$ for all t (see also Proposition 8.25 and Remark 8.28). Following [208] (see also [24]), we show that the flows of the CM-hierarchy have an infinitesimal counterpart, which yields a hierarchy of extended Lax representations (see 8.19). The general statement goes as follows. Under the hypothesis and notation adopted in Proposition 8.25, consider the Hamiltonian flows defined by the $f_k = \pi^* F_k$, i.e., $(A, B) \xrightarrow{f_k} (A + t B^{k-1}, B)$. Let $\{g_t\}_{t \in \mathbb{R}}$ be a family of unitary matrices such that $A + t B^{k-1} = g_t D_t g_t^{-1}$ where D_t is diagonal. Computing the *time*-derivative of both sides of this identity, omitting the time-dependence to simplify the notation, yields

$$\begin{aligned}
B^{k-1} &= \dot{g} D g^{-1} + g \dot{D} g^{-1} - g D g^{-1} \dot{g} g^{-1} \\
&= \dot{g} g^{-1} g D g^{-1} + g \dot{D} g^{-1} - g D g^{-1} \dot{g} g^{-1} \\
&= [\dot{g} g^{-1}, g D g^{-1}] + g \dot{D} g^{-1} \\
&= [g g^{-1} \dot{g} g^{-1}, g D g^{-1}] + g \dot{D} g^{-1} \\
&= g([g^{-1} \dot{g}, D] + \dot{D}) g^{-1},
\end{aligned}$$

i.e.,

$$\dot{D} = [D, g^{-1} \dot{g}] + g^{-1} B^{k-1} g.$$

The *time-derivative* of $L := g^{-1}Bg$ is $\dot{L} = -g^{-1}\dot{g}g^{-1}B + g^{-1}B\dot{g}$, which can be written as

$$\dot{L} = [L, g^{-1}\dot{g}]$$

In other words

Theorem 8.35 *The time-evolution of the pair* $(D, L) \in S_{kks}$ *under the flow defined by* f_k *admits the following extended Lax representation:*

$$\dot{D} = [D, M_k] + L^{k-1} \quad and \quad \dot{L} = [L, M_k], \tag{8.32}$$

where $M_k = g^{-1}\dot{g}$.

A couple of comments are in order:

(i) The starting point of the previous arguments was the existence of a *linear flow* in M, obtained by solving the Hamilton equations defined by the Hamiltonian function f_k. This flow should be labeled by t_k, and all the time-derivative computed above should be taken with respect to such a parameter. In a more geometrical language, such time-derivative is the Lie derivative with respect to X_{f_k}.

(ii) In accordance with the previous point, the parameter t in the family $\{g_t\}_{t\in\mathbb{R}}$ should be better denoted by t_k to recall that the time evolution is defined by the kth-Hamiltonian f_k.

Applying the previous theorem to the members of the CM-hierarchy (8.31), one arrives to the following.

Proposition 8.36 *If* (D_0, L_0) *is a pair of matrices as in* (8.31) *and if* $(D_0 + tL_0^{k-1}, L_0)$ *is the k-th flow of the CM-hierarchy, then there exists a family of matrices* $\{g_t\}_{t\in\mathbb{R}}$, $g_t \in G_\alpha$ *for all t, such that* $D_t = g_t^{-1}(L_0 + tL_0^{k-1})g_t$ *is diagonal. If* $L_t = g_t^{-1}L_0 g_t$, *then the time evolution of* (D_t, L_t) *admits the extended Lax representation* (8.32).

Proof The proof of the statement is identical to the one of Theorem 8.35; the only difference is that g_t can be chosen in G_α for all t. This follows from Theorem 8.20 and from the observation that $D_0 + tL_0^{k-1}$ belongs to the level set $\mu^{-1}(\alpha)$ for all $t \in \mathbb{R}$. $\qquad\square$

In this way, one concludes that the equations in (8.6) admit an extended Lax representation, and applying Theorem 5.95, it is possible to conclude the following.

Theorem 8.37 *The Hamiltonian vector field* (8.2) *admits a bi-Hamiltonian representation.*

Proof The proof is essentially already contained in the discussion, which led to the statement of Theorem 5.95. More precisely, if one restricts to the locus S_{KKS} (see Remark 8.21), where the set of the eigenvalues of L together with the diagonal entries of D form a set of local coordinates, the functions $I_k = \frac{1}{k}\mathrm{tr}(L^k)$ and $J_k = \mathrm{tr}(DL^{k-1})$, for $k = 1,\ldots n$ form a system of local coordinates (see Proposition 5.91). Under these assumptions, the brackets $\{\cdot,\cdot\}_0$ and $\{\cdot,\cdot\}_1$ defined on these coordinates by

$$\{I_k, I_l\}_0 = 0, \quad \{I_k, J_l\}_0 = (k+l-2)I_{k+l-2} \quad \text{and} \quad \{J_k, J_l\}_0 = (k-l)J_{k+l-2} \tag{8.33}$$

$$\{I_k, I_l\}_1 = 0, \quad \{I_k, J_l\}_1 = (k+l-1)I_{k+l-1} \quad \text{and} \quad \{J_k, J_l\}_1 = (k-l)J_{l+k-1}, \tag{8.34}$$

are compatible and define a bi-Hamiltonian structure on the underlying manifold. To close the proof, it suffices to note that $\{I_2, f\}_0 = \{I_1, f\}_1$ for all smooth function f. Indeed, this follows from the previous formulas noticing that

$$\{I_2, I_k\}_0 = 0 = \{I_2, I_k\}_1, \ \forall k \quad \text{and} \quad \{I_2, J_k\}_0 = \{I_1, J_k\}_1, \ \forall k,$$

implying that the vector field (8.2) is bi-Hamiltonian. It is worth noting that the bracket $\{\cdot,\cdot\}_0$ is defined starting from the extended Lax representation. In fact, writing $\frac{d}{dt_k}$ to denote the Lie derivative along the Hamiltonian vector field X_{f_k}, one has

$$\frac{d}{dt_k}I_l = 0 \quad \text{and} \quad \frac{d}{dt_k}J_l = (l+k-2)I_{l+k-2} \tag{8.35}$$

Indeed

$$\frac{dI_l}{dt_k} = \frac{1}{k}\frac{d(\mathrm{tr}L^l)}{dt_k} = \mathrm{tr}\left(\frac{dL}{dt_k}L^{l-1}\right) \stackrel{(8.32)}{=} \mathrm{tr}([L, B_k]L^{k-1}) = 0,$$

while

$$\frac{dJ_l}{dt_k} = \frac{d}{dt_k}\mathrm{tr}(DL^{l-1}) = \mathrm{tr}\left(\frac{dD}{dt_k}L^{l-1} + D\frac{dL^{l-1}}{dt_k}\right)$$

$$\stackrel{(8.32)}{=} \mathrm{tr}\left([D, B_k]L^{l-1} + L^{l+k-2} + D\frac{dL^{l-1}}{dt_k}\right)$$

$$\stackrel{(8.32)}{=} \mathrm{tr}\left([B_k, L^{l-1}]D + L^{l+k-2} + D\frac{dL^{l-1}}{dt_k}\right)$$

$$= \mathrm{tr}(L^{l+k-2}) = (l+k-2)I_{l+k-2},$$

as required. $\square$

Remark 8.38 Formulas (8.33) and (8.34) differ from the ones introduced in Sect. 4.2 of Chap. 5 by a sign of the brackets where both the Is and the Js appear. This difference is irrelevant for what concerns the results presented in this chapter, and it is a consequence of the choice to identify the Lie derivative of X_{f_k} with $\frac{d}{dt_k}$ and not with $-\frac{d}{dt_k}$. $\triangle$

It is worth to note that the functions (I, J)s introduced above provide a very simple description of the CM-system. More precisely, (8.35) implies that these coordinates provide a *linearization* of the Hamiltonian system (8.3). In particular, under the hypothesis on (D, L), while the I_ks form a set of *action variables* for (8.2), the J_ks should be thought as a set of *angle coordinates* for this system. Using the terminology introduced in Sect. 6.2 of Chap. 5, the I_ks form a set of master symmetry of degree zero, while the J_ks are a set of master symmetries of degree one. More in general, one can prove that for every n analytical functions $g_1, \ldots, g_n$ and every $r_1, \ldots, r_n \in \mathbb{N}$,

$$\mathrm{tr}(g_1(L)Q^{r_1} \cdots g_n(L)Q^{r_n}),$$

is a master symmetry of (8.1) of degree $r = \sum_{i=1}^{n} r_i$, see [24] and [202].

Example 8.39 (Two-Particles CM-System) In this simple case

$$(D, L) = \left(\begin{pmatrix} x_1 & 0 \\ 0 & x_2 \end{pmatrix}, \begin{pmatrix} p_1 & \frac{\iota}{x_1-x_2} \\ \frac{\iota}{x_2-x_1} & p_2 \end{pmatrix} \right)$$

entailing

$$I_1 = p_1 + p_2 \quad \text{and} \quad I_2 = \frac{1}{2}(p_1^2 + p_2^2) + \frac{1}{(x_1 - x_2)^2}$$

$$J_1 = x_1 + x_2 \quad \text{and} \quad J_2 = x_1 p_1 + x_2 p_2.$$

It is interesting to note that if $\{\cdot, \cdot\}$ is the canonical bracket on $\mathbb{R}^4$ with coordinates (p_i, x_i), i.e., $\{x_1, x_2\} = 0 = \{p_1, p_2\}$ and $\{p_i, x_j\} = \delta_{ij}$, then

$$\{I_1, I_2\} = 0 = \{I_1, I_2\}_0, \quad \{I_1, J_1\} = 2 = \{I_1, J_1\}_0 \quad \text{and} \quad \{I_1, J_2\} = I_1 = \{I_1, J_2\}_0$$

$$\{I_2, J_1\} = I_1 = \{I_2, J_1\}_0 \quad \text{and} \quad \{I_2, J_2\} = 2I_2 = \{I_2, J_2\}_0,$$

implying that the bracket $\{\cdot, \cdot\}_0$ is the canonical Poisson bracket. $\triangle$

Remark 8.40 It is worth to note that the bi-Hamiltonian structure of [166] was obtained in [23] using a method of Hamiltonian reduction. $\triangle$

5 Concluding Remarks and Further Topics

In this last section, we will discuss a few further topics related to the CM-system. The first one concerns the construction of a special set of DN-coordinates whose existence is tightly connected to the existence of the bi-Hamiltonian representation of the CM-system discussed in the previous section. Then we will turn our attention to some relations of the CM-system with the so-called Kadomtsev-Petviashvili equation, a non-linear, dispersive, wave (type) equation admitting solitonic solutions. Finally, in the last subsection, we will discuss briefly another class of Calogero-Moser systems, the so-called trigonometric ones.

5.1 Spectral Coordinates

In this subsection, we will present a special set of Darboux coordinates called *canonical spectral coordinates*, csc from now on, for the CM-system that turned out to be DN-coordinates for the bi-Hamiltonian structure introduced in the previous section (see Sects. 4.1 and 6.1). The csc were introduced by Sklyanin in 2009 and then analyzed from the viewpoint of bi-Hamiltonian geometry in [81] and from the perspective of symplectic reduction in [109], to which we refer the reader for further details. We start our presentation making a comment about the name of the set of coordinates to be introduced. If a dynamical system admit a Lax representation with spectral parameter, i.e., if the operator L depends both on the physical variables, i.e., momenta and positions, and on an additional set of parameters (very often just one), one can introduce the so-called spectral curve, i.e., the set on points of $(\lambda, z) \in \mathbb{C}^2$ cut by the equation $\det(\lambda - L(z)) = 0$. In this algebro-geometric approach to the analysis of the integrable systems, the relevant Hamiltonian, or more in general, multi-Hamiltonian description of the dynamical system is obtained as a by-product of its algebro-geometric description. In particular, the Darboux coordinates for the abovementioned Hamiltonian structures can be obtained as the coordinates of the poles of a suitable (normalized) eigenvector of the Lax operator (see, e.g., [3, 145, 222]). Because of their origin, the coordinates so obtained are often dubbed *spectral*. In spite of the Lax matrix (8.7) *not* depending on a spectral parameter, in analogy to what is done in this enhanced setting, Sklyanin conjectured that $(\tilde{\lambda}_i, \tilde{\mu}_i)$ defined by

$$\Delta(\tilde{\lambda}_i) = 0 \quad \text{and} \quad \tilde{\mu}_i = \left.\frac{\mathscr{E}(\lambda)}{\Delta'(\lambda)}\right|_{\lambda=\tilde{\lambda}_i}, \quad \forall i = 1, \ldots, n, \tag{8.36}$$

should form a set of Darboux coordinates for the CM-systems, where $\Delta(\lambda) = \det(\lambda - L)$, $\Delta'(\lambda) = \frac{d\Delta}{d\lambda}$, and $\mathscr{E}(\lambda) = \operatorname{tr}(u^t X \operatorname{adj}(\lambda - L)u)$ where $u^t = (1, \ldots, 1)$, $X = \operatorname{diag}(x_1, \ldots, x_n)$, and $\operatorname{adj}(A)$ denote the *adjugate* of the matrix A, i.e., the transpose of the cofactor matrix of A. The proof of the Sklyanin conjecture in the bi-Hamiltonian setting starts observing that

$$\mathcal{G}(\lambda) = \mathrm{tr}(X \, \mathrm{adj}(\lambda - L))$$

is a Nijenhuis function generator (see Definition 5.108). If Π_0 is the canonical Poisson tensor, $Y = \Pi_0^{\sharp}(d\,\mathrm{tr}L) = \sum_{i=1}^{n} \frac{\partial}{\partial x_i}$, since $\mathrm{tr}L = \sum_{i=1}^{n} p_i$, and one can prove that

$$Y(\mathcal{G}(L)) = \mathrm{tr}(\mathrm{adj}(\lambda - L) = \Delta'(\lambda) \tag{8.37}$$

(see Formula (34) in [81]). This formula and the observation that $\mathrm{adj}(\lambda - L)$ depend only on the differences $x_i - x_j$ entails $Y(Y(\mathcal{G})) = 0$. These observations yield

$$Y\left(\frac{\mathcal{G}(\lambda)}{Y(\mathcal{G}(\lambda))}\right) = 1,$$

which implies that the

$$\mu_i = \left.\frac{\mathcal{G}(\lambda)}{Y(\mathcal{G}(\lambda))}\right|_{\lambda=\lambda_i}, \quad \forall i = 1, \ldots, n, \tag{8.38}$$

and the λ_i form a set of DN-coordinates for the bi-Hamiltonian structure defined in (8.33) and (8.34) (see Proposition 5.113). We now spell out the previous construction in the simplest possible case

Example 8.41 (Twoparticle CM-System) In this case

$$(\lambda - L) = \begin{pmatrix} \lambda - p_1 & \frac{\iota}{x_2 - x_1} \\ \frac{\iota}{x_1 - x_2} & \lambda - p_2 \end{pmatrix}$$

(see (8.7)) implies that

$$\Delta(\lambda) = \lambda^2 - (p_1 + p_2)\lambda + p_1 p_2 - \frac{1}{(x_1 - x_2)^2}, \tag{8.39}$$

and

$$\mathrm{adj}(\lambda - L) = \begin{pmatrix} \lambda - p_2 & \frac{\iota}{x_1 - x_2} \\ \frac{\iota}{x_2 - x_1} & \lambda - p_1 \end{pmatrix}. \tag{8.40}$$

Formula (8.39) leads to

$$\lambda_1 = \frac{p_1 + p_2 + \sqrt{(p_1 - p_2)^2 - 4z_{12}z_{21}}}{2} \quad \text{and}$$

$$\lambda_2 = \frac{p_1 + p_2 - \sqrt{(p_1 - p_2)^2 - 4z_{12}z_{21}}}{2},$$

where z_{ij} is as in (8.9), while (8.40) implies

$$\mathscr{G}(\lambda) = x_1(\lambda - p_2) + x_2(\lambda - p_1). \tag{8.41}$$

Note that

$$Y(\mathscr{G}(\lambda)) \overset{(8.41)}{=} \left(\frac{\partial}{\partial x_1} + \frac{\partial}{\partial x_2} \right) (x_1(\lambda - p_2) + x_2(\lambda - p_1)) = 2\lambda - p_1 - p_2$$

$$\overset{(8.39)}{=} \Delta'(\lambda),$$

in agreement with (8.37). Finally, using (8.9) in (8.38), one obtains

$$\mu_1 = \frac{1}{2} \left[x_1 + x_2 + \frac{(x_1 - x_2)(p_1 - p_2)}{\sqrt{(p_1 - p_2)^2 - (4z_{12}z_{21})}} \right] \quad \text{and}$$

$$\mu_2 = \frac{1}{2} \left[x_1 + x_2 - \frac{(x_1 - x_2)(p_1 - p_2)}{\sqrt{(p_1 - p_2)^2 - (4z_{12}z_{21})}} \right]. \tag{8.42}$$

$$\triangle$$

To move on with our discussion about the Sklyanin conjecture, it is necessary to recall that in the proof of the existence of the DN-coordinates of a (regular) ΩN-manifold, the conjugate variables of the eigenvalues of Nijenhuis operator were defined only up to a *gauge transformation* defined by

$$\tilde{\lambda}_i = \lambda_i \quad \text{and} \quad \tilde{\mu}_i = \mu_i + f_i(\lambda_1, \ldots, \lambda_i), \ \forall i = 1, \ldots, n, \tag{8.43}$$

where the f_is are subjected to the integrability conditions

$$\frac{\partial f_i}{\partial \lambda_j} = \frac{\partial f_j}{\partial \lambda_i}, \ \forall i, j = 1, \ldots, n,$$

(see Remark 5.84). Although both (λ_i, μ_i) and $(\lambda_i, \tilde{\mu}_i)$ are conjugate variables, if (λ_i, μ_i) are DN-coordinates, unless the f_i are separated in the λs, $(\lambda_i, \tilde{\mu}_i)$ are not DN anymore. More precisely, if $\{\cdot, \cdot\}_0$ and $\{\cdot, \cdot\}_1$ are a pair of compatible Poisson brackets defining a ΩN-structure on the underlying manifold, then:

(i) $(\lambda_i, \tilde{\mu}_i)$ are canonical coordinates for $\{\cdot, \cdot\}_0$
(ii) They reduce $^t N$ to the following block-diagonal form

$$^t N = \begin{pmatrix} \Lambda & \beta \\ 0 & \Lambda \end{pmatrix}$$

where $\Lambda = \mathrm{diag}(\lambda_1, \ldots, \lambda_n)$ and $\beta_{ji} = \lambda_j \frac{\partial f_i}{\partial \lambda_j}$, $\forall i, j$, where the f_i are as in (8.43)

(iii) $\{\tilde{\mu}_i, \lambda_j\}_1 = \delta_{ij}\lambda_j$ and $\{\tilde{\mu}_i, \tilde{\mu}_j\}_1 = B_{ij}$, where $B_{ij} = \beta_{ij} - \beta_{ji}$, for all $i, j = 1, \ldots, n$

These observations justify the following.

Definition 8.42 If (μ_i, λ_i) are DN-coordinates, the $(\tilde{\mu}_i, \tilde{\lambda}_i)$ defined in (8.43) are called *Magnetic Darboux Nijenhuis coordinates, MDN-coordinates* from now on. $\triangle$

We stress one more time that in spite of the name, (8.43) are, in general, Darboux coordinates for one Poisson bracket, the $\{\cdot, \cdot\}_0$ in our presentation, but they are *not* DN-coordinates. After all these observations and remarks, we can finally state the following.

Proposition 8.43 *The coordinates defined in* (8.36) *form a set of MDN-coordinates for the bi-Hamiltonian structure defined in* (8.33) *and* (8.34).

To prove this proposition, it suffices to show that

$$\tilde{\mu}_i - \mu_i \overset{\substack{(8.36) \text{ and } (8.38)}}{=} \left.\frac{\mathscr{E}(\lambda)}{\Delta'(\lambda)}\right|_{\lambda=\lambda_i} - \left.\frac{\mathscr{G}(\lambda)}{Y(\mathscr{G}(\lambda))}\right|_{\lambda=\lambda_i} = f_i(\lambda_1, \ldots, \lambda_n), \ \forall i = 1, \ldots, n,$$

For the details of the proof of this proposition, we refer the reader to [81]. Here below we consider again the case of the two-particle CM-system.

Example 8.44 (Two-Particles CM-System) Since we have already computed the μ_is, we are left to compute the $\tilde{\mu}_i$s (see (8.36)). To this end, first note that $\mathrm{tr}(u^t X A u) = x^t A u$ where $u^t = (1, \ldots, 1)$ $X = \mathrm{diag}(x_1, \ldots, x_n)$, $x^t = (x_1, \ldots, x_n)$, and A is any $n \times n$ matrix. From this observation, it follows

$$\frac{\mathscr{E}(\lambda)}{\Delta'(\lambda)} = \frac{(\lambda - p_2)x_1 + \iota x_1 z_{12} + (\lambda - p_1)x_2 + \iota x_2 z_{21}}{2\lambda - p_1 - p_2},$$

which, evaluated at λ_1 and λ_2, gives

$$\tilde{\mu}_1 = \frac{(x_1 + x_2)}{2} + \frac{(p_1 - p_2)(x_1 - x_2)}{2\sqrt{\xi}} + \frac{\iota}{\sqrt{\xi}},$$

$$\tilde{\mu}_2 = \frac{(x_1 + x_2)}{2} - \frac{(p_1 - p_2)(x_1 - x_2)}{2\sqrt{\xi}} - \frac{\iota}{\sqrt{\xi}},$$

where $\xi = (p_1 - p_2)^2 - 4z_{12}z_{21}$. Computing the difference with the corresponding μ_i (see (8.42)) yields

$$f_1(\lambda_1, \lambda_2) = \frac{\iota}{\lambda_1 - \lambda_2} \quad \text{and} \quad f_2(\lambda_1, \lambda_2) = \frac{\iota}{\lambda_2 - \lambda_1}.$$

$\triangle$

The previous formulas lead us to the last part of this subsection. In fact, as it can be checked by a simple computation, the f_is so defined can be obtained by the following formula:

$$f_i = \frac{\iota}{2} \frac{\Delta''(\lambda)}{\Delta'(\lambda)}\bigg|_{\lambda=\lambda_i}, \quad i = 1, 2,$$

which can be seen as the shadow of the following general relation

$$\mathscr{E}(\lambda) = \mathscr{G}(\lambda) + \frac{\iota}{2}\Delta''(\lambda),$$

which was conjectured in [81] and nicely proven in [109], to which we refer the reader for further details.

5.2 Calogero-Moser System and the Kadomtsev-Petviashvili Equation

In this subsection, we aim to introduce the so-called *CM-KP correspondence*, a very interesting relation between the CM-system and a class of non-linear PDEs known under the name of Kadomtsev-Petviashvili hierarchy, KP-hierarchy from now on, of which the Kadomtsev-Petviashvili equation, KP-equation from now on, is just one its members. The latter is a non-linear, dispersive wave equation in $2+1$-dimension, i.e., with two spatial and one temporal variables, (x, y) and t, and putting equal to one all the numerical coefficients is written as

$$u_{yy} = (u_t - uu_x + u_{xxx})_x \tag{8.44}$$

where $u = u(t, x, y)$, while u_x denotes the partial derivative of u with respect to x and so on. The class of solutions of (8.44), which are rational in x and tend to zero for $x \to \infty$, are of particular relevance for us. It can be proven (see [144]) that these functions are of the following form:

$$u(x, y, t) = \sum_{i=1}^{n} \frac{1}{(x - x_i(y, t))^2}, \tag{8.45}$$

where, again, all the numerical coefficients were put equal to one. It was shown in [144] (see also [208]) that under the flow defined by (8.44), the poles of (8.45) behaves as a completely integrable Hamiltonian system. Note that while the x_is describe the positions of the pole/particles, y, t are the times of the systems. More precisely

Theorem 8.45 (Krichever) *The motion of the poles of (8.45) are described by the following equations:*

$$\frac{\partial x_j}{\partial y} = \frac{\partial H_2}{\partial p_j} \quad and \quad \frac{\partial p_j}{\partial y} = -\frac{\partial H_2}{\partial x_j}$$

$$\frac{\partial x_j}{\partial t} = \frac{\partial H_3}{\partial p_j} \quad and \quad \frac{\partial p_j}{\partial y} = -\frac{\partial H_3}{\partial x_j}, \ \forall j = 1, \ldots, n,$$

where H_2 and H_3 are the second and the third Hamiltonian of the CM-system, i.e., $H_2 = \frac{1}{2}\mathrm{tr}L^2$ and, respectively, $H_3 = \frac{1}{3}\mathrm{tr}L^3$, where L is as in (8.7).

In [220], this beautiful result was extended to the so-called KP-hierarchy, an infinite family of (formal) Lax-type equations of the form

$$\frac{\partial L}{\partial x_k} = [B_k, L], \ k \geq 1, \tag{8.46}$$

where $L = \partial + \sum_{j=1}^{\infty} f_j \partial^{-j}$ is a formal pseudo-differential operator of order one and $B_k = (L^k)_+$ is the differential part of the kth power of of L. The family (8.46) can be thought as an infinite set of (non-linear) evolution equations in the coefficients f_js with respect to the infinite times x_js, the arguments of the f_js. It can be proved that in spite of (8.46) depending on infinitely many functions, its solutions can be obtained using only one function called the τ-*function*. We will not develop in any details this beautiful theory, for which we refer the reader to the following references [215, 217] (see also [203] and the more recent [134] and the references therein). On the other hand, with the few observations we made above, we can at least state the following.

Theorem 8.46 (Shiota) *Let $\tau = \tau(\mathbf{t})$, where $\mathbf{t} = (x, t_2, t_3, \ldots)$, be a monic polynomial in x whose coefficients depend on $\mathbf{t}' = (t_2, t_3, \ldots)$. Then τ is a τ-function of the KP-hierarchy if and only if the motion of the zeros of τ is governed by the CM-hierarchy (see Definition 8.34).*

In other words, still following [220], $\tau = \prod_{i=1}^{n}(x - x_i(\mathbf{t}'))$ is a τ function for (8.46) if and only if

$$\frac{\partial}{\partial t_n}\begin{pmatrix} x_i \\ \xi_i \end{pmatrix} = (-1)^n \begin{pmatrix} \frac{\partial H_n}{\partial \xi_i} \\ -\frac{\partial H_n}{\partial x_i} \end{pmatrix}, \ i = 2, 3, \ldots$$

where $\xi_i = \frac{1}{2}\frac{\partial x_i}{\partial t_2}$, $H_n = \mathrm{tr}L^n$ where L is a *Moser type matrix*, (8.7), more precisely, $L = L(\mathbf{t}')$ such that

$$L_{ij} = \frac{1}{x_i - x_j}, \ i \neq j \quad and \quad L_{ii} = \xi_i, \ \forall i. \tag{8.47}$$

It is worth to note that the polynomial τ functions in the sense of the theorem above have the following determinantal representation:

$$\tau(\mathbf{t}) = \det\left(D_0 + \sum_{k \geq 1} kt_k(-L_0)^{k-1}\right),$$

where $D_0 = \mathrm{diag}(x_1(0), \ldots, x_n(0))$ and L_0 is the matrix (8.47) at $\mathbf{t}' = 0$. These results were framed in a more geometrical setting in the fundamental reference [252]. We cannot summarize in a few lines the results contained in this mathematical masterpiece. Here we will limit ourselves to the following comments. In the first place, in the *classical* presentation the of the CM-system, the one we proposed in this chapter, collisions among the particles are forbidden. This is so because the particles are *real*, i.e., they occupy points of the real line, and the potential is *repulsive* (see (8.5)). To encompass the possibility of collisions between particles, in [252], Wilson proposed to *complexify* the picture of [130]. More precisely, he considered

$$\overline{\mathscr{C}_n} = \{(X, Z) \in \mathfrak{gl}_n(\mathbb{C}) \mid \mathrm{rk}([X, Z] + \mathrm{id}) = 1\},$$

endowed with the natural $\mathrm{PGL}_n(\mathbb{C})$-action, where $\mathrm{PGL}_n(\mathbb{C}) = \mathrm{GL}_n(\mathbb{C})/\mathbb{C}^*$, by simultaneous conjugation, and he proved that if $\mathscr{C}_n' \subset \mathscr{C}_n = \overline{\mathscr{C}_n}/\mathrm{PGL}_n(\mathbb{C})$ is the set of equivalence classes represented by the pairs (X, Z) where X is diagonalizable, then it can be identified with the phase-space of the complex n-particles CM-system whose dynamics, as in the real case, can be obtained after the reduction of suitable linear flows defined on the *big phase-space* $\overline{\mathscr{C}_n}$. In this enhanced picture, $\mathscr{C}_n$ can be seen as a *completed phase-space* for the CM-system, the sought space where the collisions between the particles come to be. A construction of $\mathscr{C}_n$, also presented in [252], closer to the construction of the phase-space of the CM-system presented in [130], goes as follows. Let $V_n = \mathfrak{gl}_n(\mathbb{C}) \oplus \mathfrak{gl}_n(\mathbb{C}) \oplus \mathbb{C}^n \oplus \mathbb{C}^n$ be endowed with the following: $\mathrm{GL}_n(\mathbb{C})$-action

$$g.(X, Z, v, w) = (gXg^{-1}, gZg^{-1}, gv, wg^{-1}), \ \forall g \in \mathrm{GL}_n(\mathbb{C}), \ (X, Z, v, w) \in V_n,$$
$$(8.48)$$

where v is considered a *column-vector* and w a *row-vector*, and let $\tilde{\mathscr{C}}_n \subset V_n$ be defined as the set of tuples in V_n such that

$$[X, Z] + vw = -\mathrm{id}. \tag{8.49}$$

Since $\tilde{\mathscr{C}}_n$ is $\mathrm{GL}_n(\mathbb{C})$-invariant, one can consider $\tilde{\mathscr{C}}_n/\mathrm{GL}_n(\mathbb{C})$ and observe that, set-theoretically, this space is in a one-to-one correspondence with the $\mathscr{C}_n$ introduced above. This identification stems from the fact that two tuples $(X, Z, v, w), (X, Z, v', w') \in \tilde{\mathscr{C}}_n$ if and only if there exists $a \in \mathbb{C}^*$ such that

$v' = av$, $w' = a^{-1}w$. On the other hand, (X, Z, v, w) and (X, Z, v', w') determine the same point in $\overline{\mathscr{C}}_n$. Now it can be proven that [252]

Theorem 8.47 (Wilson) *$\mathscr{C}_n$ is a smooth connected affine variety of (complex) dimension $2n$. Moreover if in (X, Z, v, w), say Z is diagonalizable, then its eigenvalues are distinct and its $\mathrm{GL}_n(\mathbb{C})$-orbit contains a representative such that:*

(i) $X = diag(x_1, \ldots, x_n)$.
(ii) All the entries of the vectors v and w are equal to 1.
(iii) Z is a CM-matrix, i.e.,

$$Z_{ii} = p_i \quad and \quad Z_{ij} = \frac{1}{x_i - x_j}, \ if \ i \neq j,$$

where the x_is are the eigenvalues of X. Such a pair (X, Z) is unique up to the permutations of the pairs (p_i, x_i).

In this way, one recovers in this enhanced case the classical result of [130] (see also Theorem 8.20). Note that the (complex) classical phase-space of the CM-system, described in the second part of the previous theorem, sits inside $\mathscr{C}_n$ as dense subspace, and it represents the *collisionless* locus of the CM-system. Finally, still borrowing the notation of [252], as in the classical case the CM-system on $\mathscr{C}_n$ is obtained by the reduction of the $\mathrm{GL}_n(\mathbb{C})$-invariant flows $(X, Z, v, w) \overset{t}{\rightsquigarrow} (X + kt(-Z)^{k-1}, Z, v, w)$ on $\tilde{\mathscr{C}}_n$, defined by the Hamiltonians $H_k = \mathrm{tr}Z^k$. We close this summary with two observations.

5.2.1 Hamiltonian Reduction

As the attentive reader will have noticed, the construction of $\mathscr{C}_n$ recalled above is achieved via Hamiltonian reduction. More precisely, after identifying V_n with the cotangent bundle $\mathfrak{gl}_n(\mathbb{C}) \oplus \mathbb{C}^n$ and after observing that (8.48) is nothing else than the lifting to $T^*(\mathfrak{gl}_n(\mathbb{C}) \oplus \mathbb{C}^n)$ of the $\mathrm{GL}_n(\mathbb{C})$-action $g.(X, w) = (gXg^{-1}, wg^{-1})$, where w is a row-vector, one concludes that (8.48) is strongly Hamiltonian with moment map given by

$$\mu(X, Z, v, w) = [X, Z] + vw$$

(see (8.49)). Note that we are working in the holomorphic setting and that we are exactly in the same framework of Sect. 5.2.1 of Chap. 3. Comparing these results with the ones in Sect. 5.2.1, one concludes that the completed phase-space of the CM-system $\mathscr{C}_n$ is diffeomorphic, as hyperKähler manifold, to D_n, the Hilbert scheme of n points on $\mathbb{C}^2$. One is obtained from the other one via a diffeomorphism induced by a suitable rotation of the hyperKähler structure.

5.2.2   Adelic Grassmannian

We close this subsection with a short summary of the CM-KP correspondence as
it is presented in [252]. To this end, it is worth recalling that the solutions of the
equations of the KP-hierarchy can be obtained from a unique function, of infinite
many variables, called the τ-function (see [215, 217]) and that the geometrical
incarnation of these functions is the so-called *Sato-Segal-Wilson Grassmannnian*,
an infinite dimensional analogue of the Grassmannian of the k-dimesional planes in
a n-dimensional vector space. From this viewpoint, the rational solutions decaying
at infinity of Theorem 8.46 are generated by τ-functions corresponding to a
Grassmannian strictly smaller than the Sato-Segal-Wilson one, introduced for the
first time in [251], there dubbed *adelic Grassmannian*, and then studied in depth
in [252]. In particular, in this reference, it is shown that the Adelic Grassmannian
Gr^{ad} is a *cellular space* and that the union of the cells of dimension equal to n is
in a one-to-one correspondence with $\mathscr{C}_n$, i.e., it was proven the existence of a set
theoretic bijection

$$\bigcup_{n \geq 0} \mathscr{C}_n \xrightarrow{\beta} \mathrm{Gr}^{ad} .$$

From this point of view, the CM-KP correspondence amounts to the statement that
the map β intertwines the flows of the CM-system with the ones of the KP-hierarchy.

5.3   *The Trigonometric CM-System*

The n-particles trigonometric CM-system, also called n-particles Calogero-
Sutherland system, is a dynamical system of n particles moving on a circle and
interacting with a potential $V(x) \sim \frac{1}{\sin^2 x}$, where x denotes the distance between
the particles. Setting all the physical constant equal to 1, the Hamiltonian of this
system is

$$H(p, x) = \frac{1}{2} \sum_{k=1}^{n} p_k^2 + \sum_{i \neq j} \frac{1}{\sin^2(x_i - x_j)} . \tag{8.50}$$

To prove the integrability of this system, which was introduced in [229], we will
follow [130]. More precisely, we will show that (8.50) can be recovered, via
symplectic reduction of $M = T^* \mathrm{U}(n)$, from a simpler system. In complete analogy
with the case of the rational CM-system, we will define a moment map on M using
a cotangent lift of a suitable G-action on the group manifold $\mathrm{U}(n)$. Let us start

defining G as the direct product $U(n) \times U(n)$, and let us consider the following G-action on $U(n)$

$$c_{(A_1, A_2)}(B) = A_1 B A_2^{-1} \tag{8.51}$$

for all A_1, A_2 and $B \in U(n)$. We identify $T^* U(n)$ with $U(n) \times \mathfrak{u}(n)$ using the *left trivialization* (see Sect. 1 in Chap. 6) and the non-degenerate trace form $(\beta_1, \beta_2) = \mathrm{tr}(\beta_1 \beta_2)$. Using this identification, one can prove the following.

Proposition 8.48 *The cotangent lift of the action defined in* (8.51) *is given by*

$$C_{(A_1, A_2)}(B, \beta) = \left(A_1 B A_2^{-1}, A_2 \beta A_2^{-1} \right),$$

for every A_1, A_2, $B \in U(n)$ and $\beta \in \mathfrak{u}(n)$. Moreover, the corresponding moment map $\mu : T^* U(n) \simeq U(n) \times \mathfrak{u}(n) \to \mathfrak{u}(n)$ is given by

$$\mu(B, \beta) = B \beta B^{-1} - \beta,$$

for all $B \in U(n)$ and $\beta \in \mathfrak{u}(n)$.

To proceed further with the process of Hamiltonian reduction, one needs to choose a suitable $\alpha \in \mathfrak{u}(n)$. In complete analogy with the rational case, let

$$\alpha = \begin{pmatrix} 0 & \iota & \iota & \iota & \cdots \\ \iota & 0 & \iota & \iota & \cdots \\ \iota & \iota & 0 & \iota & \cdots \\ & & & \ddots & \\ \iota & \iota & \iota & & \cdots \end{pmatrix},$$

see (8.27). If $G_\alpha \subset U(n)$ denotes the stabilizer α, one can prove that

Proposition 8.49

(i) *For every $(B, \beta) \in T^* U(n) \simeq U(n) \times \mathfrak{u}(n)$ such that $\mu(B, \beta) = \alpha$, one can diagonalize B using $A' \in G_\alpha$.*
(ii) *Every G_α orbit in $\mu^{-1}(\alpha)$ has a representative (D, L_{trg}), where*

$$D_{lk} = e^{\iota \theta_k} \delta_{lk} \quad and \quad (L_{\mathrm{trg}})_{lk} = p_k \delta_{lk} + \frac{\iota(1 - \delta_{lk})}{(e^{\iota(\theta_l - \theta_k)} - 1)},$$

for all $l, k = 1, \dots, n$.

The matrix L_{trg} is called a trigonometric *CM-matrix or a* Calogero-Sutherland *matrix.*

Proof The proof of this proposition goes along the same lines of the proof of Theorem 8.20. As far as (i) concerns, one first diagonalizes B using some unitary matrix A. If $D = ABA^{-1}$ and $\gamma = A\beta A^{-1}$, then

$$D\gamma D^{-1} - \gamma = A\alpha A^{-1} = \iota(-\operatorname{id} + (Av \otimes (Av)^{\dagger})).$$

Using the same notations and the same arguments used in the proof of Theorem 8.20, one can argue that if $w = Av$, then $|w_k|^2 = 1$ for all $k = 1, \ldots, n$, which yields $w_k = e^{\iota \eta_k}$ for all $k = 1, \ldots, n$. If we multiply the matrix A on the left by $D' = \operatorname{diag}(e^{-\iota \eta_1}, \ldots, e^{-\iota \eta_n})$, we get a matrix A', which diagonalizes B and fixes α. Note that $D = ABA^{-1} = \operatorname{diag}(e^{\iota \theta_1}, \ldots, e^{\iota \theta_n})$, since D is a unitary diagonal matrix. The proof of part (ii) goes as follows. Let

$$(A'BA'^{-1}, A'\beta A'^{-1}) = (D, L_{\mathrm{trg}})$$

and note that

$$DL_{\mathrm{trg}}D^{-1} - L_{\mathrm{trg}} = \alpha,$$

where the left-hand side of the previous identity is a skew-Hermitian $n \times n$ matrix with zero entries on the main diagonal. The other entries can be found by a direct computation

$$(DL_{\mathrm{trg}}D^{-1} - L_{\mathrm{trg}})_{kl} = \iota,$$

which implies

$$(L_{\mathrm{trg}})_{kl}e^{\iota(\theta_l - \theta_k)} - (L_{\mathrm{trg}})_{kl} = \iota$$

for all $k \neq l$ with $k, l = 1, \ldots, n$. Now the rest of the statement follows. $\square$

Remark 8.50 From the proof of point (ii) of the previous proposition, one easily deduces that in the entries of the diagonal matrix D, $\theta_i \neq \theta_j$ for all $i \neq j$. This observation implies that the quotient space $\mu^{-1}(\alpha)/G_\alpha$ can be parametrized by the $2n$-tuple of real numbers $(p_1, \ldots, p_n, \theta_1, \ldots, \theta_n)$ with $\theta_i \neq \theta_j$ if $i \neq j$ and $\theta_i \in [0, 2\pi)$, with no condition on the p_is. $\triangle$

Problem 8.51 Prove that $\operatorname{tr}(L_{\mathrm{trg}}^2)$ equals, up to a numerical factor, the Hamiltonian of the Calogero-Sutherland system. $\triangle$

Finally, it is still necessary to double check that the reduction provides the *right* symplectic structure. To this end, first note that the phase-space of the n-particles trigonometric CM-system can be identified with the cotangent bundle of the Cartesian product of n-copies of S^1, which, in turn, can be identified with the cartesian product of n-copies of $S^1 \times \mathbb{R}$ where $\mathbb{R} \simeq \operatorname{Lie} S^1$. The corresponding canonical symplectic form can be written as $\omega = \sum_{k=1}^{n} dp_k \wedge d\theta_k$, where θ_k is the (local) angular coordinate on S_k^1 and p_k is the corresponding fibered coordinate. Recall now that the identification $T^* U(n) \simeq U(n) \times \mathfrak{u}(n)$ was obtained using the left trivialization and the bilinear form $(\beta_1, \beta_2) = \mathrm{tr}(\beta_1\beta_2)$ for all $\mathfrak{u}(n)$, which, together with Formula (6.8), yields

$$\omega_{(D,L)}\big((\dot{D}_1, L_1), (\dot{D}_2, L_2)\big) = \mathrm{tr}(L_1 d_2) - \mathrm{tr}(L_2 d_1) - \mathrm{tr}(L[d_1, d_2])$$

$$= \mathrm{tr}(L_1 d_2) - \mathrm{tr}(L_2 d_1),$$

where the D_is are tangent vectors to $U(n)$ at D and the d_is are then translated to $T_{\mathrm{id}} U(n)$ of the D_is, which we identify with two diagonal matrices whose entries are the *positions* θ_ks. Since the d_is are diagonal $[d_1, d_2] = 0$. From this formula, it follows that the reduced symplectic form is the canonical symplectic form of the cotangent bundle of the configuration space of the trigonometric CM-system.

6 Bibliographical Notes

The classical (as opposed to the quantum) CM-system was introduced in [46], where the author also conjectured its complete integrability. As already mentioned in the introduction to this chapter, this conjecture was proven by Moser in [189], where he proposed the Lax representation of the equations of motion. In the fundamental work [130], the complete integrability of both the rational and the trigonometric CM-system via Hamiltonian reduction of a linear system was shown. Our presentation is based on the references [189] and [130]. The extended Lax representation of the equations of motion of the rational CM-system is due to Olshanetsky and Perelomov (see [208] and the references therein), while the bi-Hamiltonian representation of the system can be found in [166] (see also [165] and the unpublished PhD thesis [177]). The spectral coordinates for the rational CM-system, discussed in Sect. 5.1, were obtained in [81] (see also [109]). As already recalled above, in [23], the bi-Hamiltonian structure of [166] was recovered via a process of Hamiltonian reduction. The history of the relation between the hierarchies of non-linear evolution equations and the finite dimensional Hamiltonian systems is very interesting and quite complicate. One of the first instances where a correspondence between the solutions of these two classes of systems was observed is contained in the fundamental work [8], where the case of the CM-system and the KdV equation was studied in depth. This can be considered, at least at some extent, the starting point of the material presented in Sect. 5.2. As we have already mentioned in the introduction to this chapter, in the recent years, the theory of the CM-systems has been at the crossroad of many different areas of pure mathematics. Here we would only turn the attention of the reader to a few references, which point to some of these new and very exciting topics. The first is [78], where many relations between the (classical and quantum) CM-systems and some topics of Lie theory are beautifully presented. Then we would like to mention [26] and [27], where the authors make a deep analysis of the KP-CM correspondence from the viewpoint of the theory of the D-modules. Finally, it is worthwhile to point the reader's attention to [156] and [157], where a class of integrable spin Calogero-Moser systems is studied in depth.

Appendix A
Elements of Symplectic Linear Algebra

In this appendix, we collect a few definitions and results of symplectic linear algebra. All the vector spaces will be real and finite dimensional.

1 Symplectic Vector Spaces

A bilinear form on vector space V is a map

$$\omega : V \times V \longrightarrow \mathbb{R},$$

with the following properties:

(i) $\omega(w, \alpha v + \beta u) = \alpha \omega(w, v) + \beta \omega(w, u)$ for all $u, v, w \in V$ and $\alpha, \beta \in \mathbb{R}$ (linearity in the second entry).
(ii) $\omega(\alpha v + \beta u, w) = \alpha \omega(v, w) + \beta \omega(u, w)$ for all $u, v, w \in V$ and $\alpha, \beta \in \mathbb{R}$ (linear in the first entry).

It is called *non-degenerate* if $\omega(v, u) = 0$ for all $u \in V$ implies $v = 0$ and skew-symmetric if $\omega(u, v) = -\omega(v, u)$ for all $u, v \in V$.

Definition A.1 A bilinear form ω is called symplectic if it is non-degenerate and skew-symmetric. A pair (V, ω) is called a symplectic vector space. $\triangle$

Remark A.2 To every bilinear form ω on V, it is associated with the linear map $A : V \to V$, defined by $A(v) = \omega(v, \cdot)$. The entries of the matrix describing A in the basis $(e_1, \ldots, e_n)$ are $A_{jk} = \langle A(e_j), e_k^* \rangle = \omega(e_j, e_k)$, where $(e_1^*, \ldots, e_n^*)$ is the dual basis of $(e_1, \ldots, e_n)$. $\triangle$

© The Author(s), under exclusive license to Springer Nature Switzerland AG 2026
A. Arsie, I. Mencattini, *Geometry of Integrable Systems*, Latin American
Mathematics Series – UFSCar subseries,
https://doi.org/10.1007/978-3-031-96282-0

Lemma A.3 *Let (V, ω) be a symplectic vector space and let A be the associated linear map. Then:*

(i) Any matrix representing A is skew-symmetric.
(ii) A is an isomorphism.

In particular

Lemma A.4 *A vector space V carries a symplectic form if and only if its dimension is even.*

Example A.5 Let W be any vector space and let $W^* = \mathrm{Hom}_{\mathbb{R}}(V, \mathbb{R})$. Define $V = W \oplus W^*$ and $b : V \times V \longrightarrow \mathbb{R}$ as follows:

$$b((w_1, \phi_1), (w_2, \phi_2)) = \phi_2(w_1) - \phi_1(w_2), \ \forall \, (w_1, \phi_1), \ (w_2, \phi_2) \in V.$$

Then b defines a symplectic form on V. To prove this statement, it suffices to show that b is non-degenerate, since bilinearity and skew-symmetry are clear from the definition. Then, suppose that there exists $(w, \phi) \in V$ such that $b((w, \phi), (v, \psi)) = 0$ for all $(v, \psi) \in V$. In this case $\psi(w) - \phi(v) = 0$ for very (v, ψ). To conclude it is enough to consider the case $(v, \psi) = (0, \psi)$ and, respectively $= (v, 0)$. $\triangle$

Before moving to the next result, we introduce the following definition. Let U be any subset of V.

Definition A.6 The *ω-orthogonal complement*, or *symplectic orthogonal complement*, to U is

$$U^\omega = \{v \in V \mid \omega(v, u) = 0, \ \forall u \in U\}.$$

$\triangle$

From this follows at once that:

 (i) U^ω is a vector subspace of V
 (ii) If $U_1 \subset U_2 \subset V$, then $U_2^\omega \subset U_1^\omega$
(iii) If U is a vector subspace of V, then $(U^\omega)^\omega = U$.

The following result states that every linear symplectic form can be written in a canonical form or, equivalently, that in every symplectic vector space, there exist distinguished bases with respect to which the symplectic form assumes a particularly simple expression. This result is the skew-analog of the Gram-Schmidt process to diagonalize non-degenerate symmetric bilinear forms. Moreover, it is the algebraic avatar of the *Darboux Theorem*, which provides the existence of local *canonical* coordinate systems on every symplectic manifold (see Sect. 3.3 of Chap. 1).

Theorem A.7 (Linear Darboux Theorem) *Let (V, ω) be a symplectic vector space. Then there is a basis $(e_1, \ldots, e_n, f_1, \ldots, f_n)$ in V such that $\omega(e_i, e_j) = \omega(f_i, f_j) = 0$ and $\omega(e_i, f_j) = \delta_{ij}$.*

Proof Let $v \in V$, $v \neq 0$. Since ω is non-degenerate there exists $w \in V$ such that $\omega(v, w) \neq 0$. After eventually normalizing, one can assume that $\omega(v, w) = 1$. Let us call $v = e_1$ and $u = f^1$. If dim $V = 2$, we are done. On the contrary, let $V' = \text{span} \langle e_1, f_1 \rangle^{\omega}$ be the ω-orthogonal complement of the vector space generated by (e_1, f_1). Let $v' \in V'$ be any non-zero vector. Using again the property of ω of being non-degenerate, one can find another vector, say $w' \in V$, such that $\omega(v', w') \neq 0$. Since $v' \in V'$, also w' belong to V'. Up to normalization, the vector v' will be e_2 and w' will be f_2. As in the previous step, let us consider, V'', the ω-orthogonal complement to the vector space generated by (e_1, e_2, f_1, f_2). If $V'' = \{0\}$ we are done, if not we will find two vectors v'' and w'' in V'' such that $\omega(v'', w'') \neq 0$. If dim $V = 2n$, this process will produce the desired basis after n-steps. The basis so defined has by construction the desired property. $\square$

Definition A.8 (Symplectic Basis)   The bases defined in Theorem A.7 are called *symplectic*. $\triangle$

Let (V, ω) be a symplectic vector space and let $(e_1, \ldots, e_n, f_1, \ldots, f_n)$ be a symplectic basis. For each $i = 1, \ldots, n$, let $x_i : V \to \mathbb{R}$ and $p_i : V \to \mathbb{R}$ the functions defined by $x_i(v) = \langle e_i^*, v \rangle$ and $p_i(v) = \langle f_i^*, v \rangle$, for all $v \in V$. Since the functions so defined are linear, $dx_i = x_i = e^{i*}$ and $dp_i = p_i = f_i^*$, for all $i = 1, \ldots, n$. The functions $(x_1, \ldots, x_n, p_1, \ldots, p_n)$ form a set of *linear symplectic coordinates* for (V, ω). Using these coordinates, one can write

$$\omega = \sum_{i=1}^{n} dp_i \wedge dx_i.$$

In particular, Darboux Theorem A.7 can be rephrased saying that, if (V, ω) is a symplectic vector space, then there exists a set of coordinates on V such that $\omega = \sum_{i=1}^{n} dp_i \wedge dx_i$. The non-linear generalization of this result is discussed in Sect. 3.3 in Chap. 1.

1.1   Subspaces of a Symplectic Vector Space

Let (V, ω) be a symplectic vector space. By using the symplectic form ω, one can single out some distinguished classes of subspaces of V.

Definition A.9   A vector subspace U of V is said to be:

  (i)   *Isotropic* if $U \subset U^{\omega}$
  (ii)   *Coisotropic* if $U^{\omega} \subset U$
  (iii)   *Lagrangian* if $U = U^{\omega}$
  (iv)   *Symplectic* if $U \cap U^{\omega} = \{0\}$

$\triangle$

Proposition A.10

(i) *If $U \subset V$ is a symplectic vector subspace of V, then the restriction of the symplectic form ω to U is a symplectic form.*

(ii) *A subspace L of V is Lagrangian if and only if it is isotropic and coisotropic.*

(iii) *If $\dim V = 2n$, then the dimension of every* isotropic *subspace of V is less than or equal to n, while the dimension of every* coisotropic *subspace of V is greater than or equal to n. The dimension of every* Lagrangian *subspace of V is equal to n.*

Proof

(i) The restriction of the symplectic form ω to U is a bilinear and skew-symmetric form. If U is symplectic, such a restriction is also non-degenerate. In fact, the existence of $v \in U$ such that $\omega(v, u) = 0$ for all $u \in U$ would imply that $v \in U^{\omega}$, so that $v \in U \cap U^{\omega}$. But this intersection contains only the zero-vector.

(ii) It follows from the definition of Lagrangian subspace.

(iii) Suppose that U is an isotropic subspace of V whose dimension is greater than n. Since U is isotropic, every vector in U is ω-orthogonal to every other vector in U. Fix a basis for U, say $(u_1, \ldots, u_m)$. Since ω is non-degenerate, for each element u_i of the basis, there is an element v_i in V such that $\omega(u_i, v_i) \neq 0$. None of these elements v_i can belong to U (since U is isotropic); on the other hand, the elements $(v_1, \ldots, v_m)$ are linearly independent since ω is non-degenerate. But this is impossible as $m > n$. The statements about the coisotropic and Lagrangian subspaces follow in a similar way.

$\square$

Corollary A.11 *If $U \subset V$ is isotropic (coisotropic), then U^{ω} is coisotropic (isotropic).*

Example A.12 Every line through the origin of V is an *isotropic* vector subspace of V. Every hyperplane through the origin of V is a *coisotropic* subspace of V. If $\dim V = 2$, then every line through its origin is a *Lagrangian* subspace. $\triangle$

Proposition A.13 *Let (V, ω) be a symplectic vector space. Every Lagrangian vector subspace $L \subset V$ has a Lagrangian complement, i.e., there exists another Lagrangian subspace L' of V such that $V = L \oplus L'$.*

Proof The proof follows from the linear Darboux Theorem (see Theorem A.7). In fact, let us fix a basis of L. Since L is Lagrangian and ω is non-degenerate, one can find a set of linearly independent vectors that belong to $V \setminus L$. The proof of the proposition follows now by dimensional reason, since $\dim L = \frac{1}{2} \dim V$. $\square$

1.2   Symplectic Linear Group

Let (V, ω) be a symplectic vector space. Then

Definition A.14 (Symplectic Group) The symplectic (linear) group of (V, ω) is

$$\mathrm{Sp}(V) = \{f \in \mathrm{GL}(V) \mid \omega\big(f(v), f(u)\big) = \omega(v, u), \ \forall u, v \in V\}$$

$$\triangle$$

Remark A.15 An easy check shows that $\mathrm{Sp}(V)$ is an abstract subgroup of $\mathrm{GL}(V)$. Since the condition defining $\mathrm{Sp}(V)$ is *closed*, it follows from Theorem C.2 in Appendix C that $\mathrm{Sp}(V)$ is a Lie subgroup of $\mathrm{GL}(V)$. $\hfill \triangle$

The Lie algebra of $\mathrm{Sp}(V)$ is a Lie subalgebra of $\mathfrak{gl}(V)$, and it will be denoted as $\mathfrak{sp}(V)$. To characterize the elements $A \in \mathfrak{gl}(V)$ that belong to $\mathfrak{sp}(V)$, one can proceed as follows. First note that $f \in \mathrm{Sp}(V)$ if and only if $\omega\big(f(v), u\big) = \omega\big(v, f^{-1}(u)\big)$, for all $u, v \in V$. Then, one can say that $A \in \mathfrak{gl}(V)$ belongs to $\mathfrak{sp}(V)$ if and only if

$$\omega\big(\exp(tA)u, v\big) = \omega(u, \exp(-tA)), \ \forall t \in \mathbb{R}. \tag{A.1}$$

Computing now the time derivative at $t = 0$ of both sides of (A.1), one gets that $A \in \mathfrak{gl}(V)$ belongs to $\mathfrak{sp}(V)$ *if and only if*:

$$\omega(Au, v) + \omega(u, Av) = 0, \ \forall u, v \in V.$$

In other words, we proved that

$$\mathfrak{sp}(V) = \{A \in \mathfrak{gl}(V) \mid \omega(Au, v) = \omega(u, Av), \ \forall u, v \in V\}.$$

Note that the choice of a symplectic basis $(e_1, \ldots, e_n, f_1, \ldots, f_n)$ let us identify V with $\mathbb{R}^{2n}$, and the symplectic form ω with the $2n \times 2n$ matrix

$$J = \begin{pmatrix} 0 & I \\ -I & 0 \end{pmatrix},$$

where I and 0 are the identity and, respectively, the zero $n \times n$ matrices. Moreover, given $v, w \in V$, one can write $v = \sum_{i=1}^{n} v_i e_i + \widetilde{v}_i f_i$ and $w = \sum_{i=1}^{n} w_i e_i + \widetilde{w}_i f_i$, so that

$$\omega(v, w) = \sum_{i=1}^{n} v_i \widetilde{w}_i - \widetilde{v}_i w_i = w^T J v; \tag{A.2}$$

where, in the previous formula, v and w are columns-vectors.

Remark A.16 It is worth making a remark about Formula (A.2). The meaning of the first term in that formula should be clear from the computation above. The scalar $\omega(v, w)$ is obtained in a two-step process: first by *contracting* ω with v and then by *contracting* the linear form $i_v\omega$ previously obtained to get $\omega(v, w) = \langle i_v\omega, w\rangle$. This two-step process, if performed using a symplectic basis, is equivalent to multiplying (from left to right) the row vector w^T with the matrix J and then with the vector v.

$\triangle$

Let $\mathrm{Sp}_n(\mathbb{R})$ be the subgroup of $\mathrm{GL}_{2n}(\mathbb{R})$ whose elements are the matrices satisfying the condition $A^T J A = J$. Then:

Proposition A.17 $\mathrm{Sp}(V) \simeq \mathrm{Sp}_n(\mathbb{R})$.

Proof Once a symplectic basis of V has been chosen, every element $f \in \mathrm{Sp}(V)$ is represented by a matrix A_f. From the identifications above, one obtains that

$$v^T J w = \omega(v, w) = \omega(fv, fw) = (A_f v)^T J(A_f w), \quad \forall v, w \in V.$$

Then:

$$v^T J w = v^T\big(A_f^T J A_f\big)w \Rightarrow J = A_f^T J A_f$$

because of the non-degeneracy of the symplectic form. The statement follows from the fact the map $f \rightsquigarrow A_f$ is a bijection. $\qquad\square$

Theorem A.18 *For each n, $\mathrm{Sp}_n(\mathbb{R})$ is a Lie group whose dimension is $2n^2 + n$.*

Proof The first part of the statement follows from the fact that the relations of the elements of $\mathrm{Sp}_n(\mathbb{R})$ define a closed subset in $\mathrm{GL}_{2n}(\mathbb{R})$ and that a closed subgroup of a Lie group has a natural structure of Lie group (see, e.g., [244]). For the second part, it suffices to compute the dimension of the Lie algebra $\mathfrak{sp}_n(\mathbb{R})$ of $\mathrm{Sp}_n(\mathbb{R})$. This is the set of $2n \times 2n$ matrices A such that

$$A^T J + J A = 0. \tag{A.3}$$

This identity is obtained noticing that $A \in \mathfrak{sp}_n(\mathbb{R})$ if and only if

$$\exp(tA)^T J \exp(tA) = J \quad \text{for all} \quad t \in \mathbb{R},$$

taking the derivative with respect to t, and evaluating it for $t = 0$. Now if we write

$$A = \begin{pmatrix} a & b \\ c & d \end{pmatrix} \tag{A.4}$$

with a, b, c and d $n \times n$ blocks, the condition (A.3) reads

$$\begin{cases} b = b^T \\ c = c^T \\ a = -d^T. \end{cases}$$

Each of the first two identity gives $\frac{n(n-1)}{2}$ conditions, since the matrices B and C are symmetric, while the last gives n^2 conditions. So the dimension of the Lie algebra $\mathfrak{sp}_n(\mathbb{R})$ is equal to $(2n)^2 - n(n-1) - n^2 = 2n^2 + n$. $\qquad\square$

If (V, ω) is a symplectic vector space, one can define on the direct sum $\mathcal{V} = V \oplus V$ a symplectic form by letting $B = \omega \oplus (-\omega)$, or, more explicitly,

$$B\big((v_1, w_1), (v_2, w_2)\big) = \omega(v_1, v_2) - \omega(w_1, w_2).$$

Non-degeneracy and skew-symmetry follow easily from the definition. Then one arrives at the following characterization of the elements of $\mathrm{Sp}(V)$ in terms of *Lagrangian* subspaces of $(\mathcal{V}, B)$.

Proposition A.19 $f \in \mathrm{GL}(V)$ *is symplectic* if and only if *the graph of f*

$$\Gamma_f = \{(u, v) \in \mathcal{V} | v = f(u)\},$$

is a Lagrangian vector subspace of $\mathcal{V}$ (see also Proposition 1.91).

Proof If $f : V \longrightarrow V$ is a linear map, one has

$$B\big((v, f(v)), (w, f(w))\big) = \omega(v, w) - \omega(f(v), f(w)).$$

The right-hand side of the previous equality is equal to zero for each pair of vectors $v, w \in V$ if and only if $f \in \mathrm{Sp}(V)$. $\qquad\square$

Proposition A.20 *The symplectic linear group acts transitively on the set of complementary pairs of Lagrangian subspaces.*

Proof If $L, L' \subset V$ are Lagrangian and $V = L \oplus L'$, by choosing bases of L and L', we get a symplectic basis for V, and *vice versa*, a symplectic basis defines such a Lagrangian pair. Then the claim follows from the fact that the symplectic group acts transitively on the space of symplectic bases. $\qquad\square$

Appendix B
Elements of Differential Geometry

Most of these notes use tools from (elementary) differential geometry. This appendix collects a list of those notions and some of the main results that play a role in the theory of integrable systems. We refer the reader to the cited literature for more information. In particular, among the several texts that present the basic concepts of differential geometry, the following are closer to our needs: [90, 133, 183, 227, 244].

1 Distributions and Generalized Distributions

Definition B.1 (Distributions)   Let M be a manifold of dimension k.

(i) A *distribution* $\mathcal{D}$ of dimension d with $0 \leq d \leq k$ on M is choice of a d-dimensional vector subspace of $T_m M$, for all $m \in M$. This is also sometimes called the *rank* of $\mathcal{D}$; compare below with the definition of generalized distributions.
(ii) A distribution of dimension d is called *smooth* if each $m \in M$ has a neighborhood U and local vector fields $X_1, \ldots, X_d \in \mathfrak{X}(U)$ such that $\mathcal{D}(m') \subset T_{m'} M$ is the span$\langle X_1(m'), \ldots, X_d(m') \rangle$ for all $m' \in U$.
(iii) A vector field $X \in \mathfrak{X}(M)$ is said to *belong to* or to *lie in* $\mathcal{D}$ if $X(m) \in \mathcal{D}(m)$ for all $m \in M$.
(iv) A distribution $\mathcal{D}$ is called *involutive* if the vector space of all vector fields belonging to $\mathcal{D}$ is a Lie subalgebra of $\mathfrak{X}(M)$.

$\triangle$

Example B.2   The choice of a vector in $T_m M$, for all $m \in M$, defines a one-dimensional distribution. If such a distribution is smooth, then it is automatically involutive because of dimensional reason. $\triangle$

© The Author(s), under exclusive license to Springer Nature Switzerland AG 2026
A. Arsie, I. Mencattini, *Geometry of Integrable Systems*, Latin American
Mathematics Series – UFSCar subseries,
https://doi.org/10.1007/978-3-031-96282-0

Definition B.3 (Integral Submanifold) An immersed submanifold (N, i) of M is called an *integral submanifold* of the distribution $\mathscr{D}$ if

$$i_{*,n}(T_n N) = \mathscr{D}(i(n)), \text{ for all } n \in N.$$

$$\triangle$$

Moreover

Definition B.4 (Completely Integrable Distribution) A distribution will be called *integrable* or *completely integrable*, if for each point $m \in M$ there exists an integral submanifold N of $\mathscr{D}$ passing for m (i.e., $m \in N$). $\triangle$

The Frobenius theorem links the *complete integrability* with the *involutiveness* of a smooth distribution. Below is its statement.

Theorem B.5 (Frobenius Theorem) *A smooth distribution $\mathscr{D}$ defined on M is integrable if and only if it is involutive. In particular, if $\mathscr{D}$ is a d-dimensional smooth and involutive distribution, then for every $m \in M$ passes a unique maximal, d-dimensional, connected integral submanifold of $\mathscr{D}$, and every connected integral submanifold of $\mathscr{D}$ is contained in a maximal one.*

There is a dual version of the Frobenius theorem that turns out to be very useful. Before stating it, one needs to introduce a few more notions. Let $\mathscr{D} : m \rightsquigarrow \mathscr{D}(m)$ be a smooth distribution of dimension p defined on M. A k-form β is said to *annihilate* $\mathscr{D}$ if

$$\beta_m(v_1, \ldots, v_k) = 0, \ \forall v_1, \ldots, v_k \in \mathscr{D}(m) \text{ and } m \in M.$$

Moreover, $\xi = \sum_i \xi_i \in \Omega^\bullet(M)$ is said to *annihilate* $\mathscr{D}$ if all its *homogeneous components* ξ_i annihilate $\mathscr{D}$. Having said that, given $\mathscr{D}$, one can define:

$$\mathcal{I}(\mathscr{D}) = \{\xi \in \Omega^\bullet(M) | \xi \text{ annihilates } \mathscr{D}\}.$$

Then:

(i) $\mathcal{I}(\mathscr{D})$ is an *ideal* in $\Omega^\bullet(M)$.
(ii) If M has dimension n and $\mathscr{D}$ has dimension k, then $\mathcal{I}(\mathscr{D})$ is *locally generated* by $n - k$ independent one-forms.

Remark B.6 The point (ii) in the previous statement means that for each point $m \in M$ exists a neighborhood U containing it and $\alpha_1, \ldots, \alpha_{n-k} \in \Omega^1_M(U)$ linearly independent such that, if $\xi \in \mathcal{I}(\mathscr{D})$, then $\xi|_U$ belongs to the ideal generated by $\alpha_1, \ldots, \alpha_{n-k}$. Furthermore, if $\xi \in \Omega^\bullet(M)$ is such that, for each $m \in M$, its restriction to a suitable neighborhood U containing m belongs to the ideal generated by $\alpha_1, \ldots, \alpha_{n-k} \in \Omega^1_M(U)$, which generate $\mathcal{I}(\mathscr{D})$ on U, then $\xi \in \mathcal{I}(\mathscr{D})$. $\triangle$

In other words, to each smooth distribution, $\mathscr{D}$ is associated with an ideal $\mathcal{I}(\mathscr{D})$. On the other hand, the converse holds true.

Proposition B.7 *If M is a n-dimensional manifold and I is an ideal in $\Omega^\bullet(M)$ locally generated by $n - k$ one-forms, then there exists a unique smooth distribution of dimension k such that $I = I(\mathscr{D})$.*

Definition B.8 (Differential Ideal) An ideal I of $\Omega^\bullet(M)$ is called *differential* if $dI \subset I$. △

We can finally state the following important result.

Theorem B.9 *A smooth distribution $\mathscr{D}$ is involutive if and only if the corresponding $I(\mathscr{D})$ is a differential ideal.*

The Frobenius theorem tells us that the existence of a smooth and involutive distribution on a manifold M is equivalent to the existence of a *foliation*, i.e., to a partition of M into disjoint and connected subsets $\{\mathcal{M}_\alpha\}_{\alpha \in A}$, such that for each $\alpha \in A$, $(\mathcal{M}_\alpha, i_\alpha : \mathcal{M}_\alpha \to M)$ is an immersed submanifold of M (which is a maximal integral submanifold of $\mathscr{D}$). It is important to note that the manifolds so obtained have all the same dimension. Moreover, observe that for each $m \in M$, $\mathscr{D}(m) \subset T_m M$ is the tangent space to the integral submanifold of $\mathscr{D}$ passing through the point m.

For more information about foliations, see the monograph [47] and the references [227, 244] for the proof of the Frobenius theorem.

What follows contains a generalization of the concepts of distribution and of foliation. Also in this case, we will not give the proofs of the results involved, but rather we will point the reader to the relevant bibliographical references.

We start generalizing the concept of a distribution. Let M be a manifold.

Definition B.10 (Generalized Distribution) A *generalized distribution* $\mathscr{D}$ on M is a subset of TM such that $\mathscr{D}(m) := \mathscr{D} \cap T_m M$ is a vector subspace of $T_m M$ for all $m \in M$. △

Given a generalized distribution on M:

 (i) The dimension of $\mathscr{D}(m)$ is called the *rank* of $\mathscr{D}$ at the point m.
 (ii) A *smooth section* of $\mathscr{D}$ on an open subset $U \subset M$ is a vector field X such that $X(m) \in \mathscr{D}(m)$ for all $m \in U$.
(iii) An immersed and connected submanifold (N, i) of M is called an *integral manifold* of $\mathscr{D}$ if $i_{*,n}(T_n N) \subset T_{i(n)}M$, where $i : N \to M$ is the canonical injection. The integral submanifold is said of *maximal dimension* at $m \in M$ if $i_{*,n}(T_n N) = T_{i(n)}M$.
(iv) A generalized distribution $\mathscr{D}$ is called *smooth* if for all $m \in M$ and for all $v \in \mathscr{D}(m)$, there exists a smooth section of $\mathscr{D}$ such that $X(m) = v$.

Note that the above definitions are not significantly different from the analogue ones given to state the Frobenius theorem. The key difference is that, in the generalized case, the rank in general is *not* (locally) constant. In particular, a generalized distribution of constant rank is a distribution in the previous sense. Let us point out that in the literature, what we call generalized distribution is often called distribution, while distributions in our sense are often called *regular* distributions.

We come now to the relevant definition of involutivity for the case of generalized distributions.

Definition B.11 Let $\mathscr{D}$ be a generalized distribution. Then:

(i) $\mathscr{D}$ is called *involutive* if it is *invariant* with respect to the *local flows* generated by its smooth sections.

(ii) $\mathscr{D}$ is called *completely integrable* if for every point $m \in M$ there exists an integral manifold of $\mathscr{D}$ of maximal dimension containing m.

$\triangle$

Remark B.12 Note if $\mathscr{D}$ is a (regular) smooth distribution, it is involutive in the sense of point (iv) of Definition B.1 if and only if it is involutive in the sense of Definition B.11. But in the case of generalized distributions, the two concepts are not equivalent. $\triangle$

One comes now to the result that generalizes the Frobenius theorem to the case of generalized distributions.

Theorem B.13 (Stefan-Sussmann Theorem)   *A generalized distribution is completely integrable if and only if it is involutive.*

In other words, to every involutive generalized distribution, one can associate its *generalized foliation*, $\Psi = \{\mathcal{M}_\alpha\}_{\alpha \in A}$ of M, whose leaves $\mathcal{M}_\alpha$ are the maximal integral submanifolds of $\mathscr{D}$. The difference between the present case and the Frobenius one, discussed before, is that the dimension of the leaves is *not* constant. Let $\Psi = \{\mathcal{M}_\alpha\}_{\alpha \in A}$ be a generalized foliation on M. Then:

Definition B.14 A leaf $\mathcal{M}_\alpha \in \Psi$ is called *regular* if it has an open neighborhood V that intersects leaves of only the same dimension as $\mathcal{M}_\alpha$. A leaf will be called *singular* if it is not regular, i.e., if such an open neighborhood is not available. In the same vein, a point of M will be called *regular* if it belongs to a regular leaf, *singular* if it belongs to a singular one. $\triangle$

Remark B.15 The theory of generalized distributions and the results about their complete integrability plays an important role in Poisson geometry. In fact, the symplectic foliation, defined on every Poisson manifold, is a generalized one. More precisely, it is the generalized foliation associated with the generalized distribution generated by the local Hamiltonian vector fields. Note that such a distribution is, by its definition, completely integrable in the sense of Theorem B.13.

To convince the reader that the symplectic foliation of a Poisson manifold is, in general, a generalized one, it suffices to consider the case of the linear Poisson structures: the leaves of such a foliation cannot have constant rank since the origin of $\mathfrak{g}^*$ is always a zero-dimensional singular leaf in the sense of Definition B.14. $\triangle$

1.1   Time-Dependent Vector Fields and Homotopy Operators

First, we consider both an important generalization of the concept of vector field and of the corresponding one-parameter group of diffeomorphisms, and we comment about their relation with the so-called isotopies. In what follows, every map will be *smooth*.

Definition B.16 (Time-Dependent Vector Field) A *time-dependent vector field* on a manifold M is a smooth map $X : \mathbb{R} \times M \to TM$, such that for all $m \in M$, $X(t, m) \in T_m M$, for all $t \in \mathbb{R}$. $\hspace{2cm} \triangle$

In other words, in spite of the fact that a time-dependent vector field is *not* a single vector field, it can be thought as a family of vector fields $\{X_t\}_{t\in\mathbb{R}}$ smoothly depending on a parameter $t \in \mathbb{R}$.

To each time-dependent vector field X, one can associate a vector field $\widehat{X} \in \mathfrak{X}(\mathbb{R} \times M)$, called, sometimes, its *suspension*. By definition, $\widehat{X} = X + \frac{\partial}{\partial t}$. The one-parameter group of diffeomorphisms $\{\widehat{\varphi}_t\}_{t\in\mathbb{R}}$ defined by $\widehat{X}$ is given by

$$\widehat{\varphi}_t(s, m) = \big(t + s, g_t(s, m)\big), \ \forall m \in M, \ t, s \in \mathbb{R}, \tag{B.1}$$

where $g_t : \mathbb{R} \times M \to M$ is a smooth function, smoothly depending on the parameter $t \in \mathbb{R}$. Since $\widehat{\varphi}_t$ defines a one-parameter group of diffeomorphisms, $\widehat{\varphi}_t \circ \widehat{\varphi}_s = \widehat{\varphi}_{s+t}$, for all $t, s \in \mathbb{R}$ such that both sides of this equality are defined. From this observation, one can easily deduce that

$$g_t\big(u + s, g_u(s, m)\big) = g_{t+u}(s, m), \tag{B.2}$$

for all $m \in M$ and all $s, t, u \in \mathbb{R}$, such that both sides of this equality are defined. Formula (B.1) implies that for each $t, s \in \mathbb{R}$, $g_t(s, \cdot)$ is a diffeomorphism. In fact, given $t, s \in \mathbb{R}$, since $\widehat{\varphi}_{-t}\big(\widehat{\varphi}_t(s, m)\big) = (s, m)$ for all $m \in M$, one sees that $g_{-t}\big(s + t, g_t(s, m)\big) = m$, for all $m \in M$, proving the statement.

More precisely, since $\widehat{\varphi}_{-t}\big(\widehat{\varphi}_t(s, m)\big) = (s, m)$, for all $t, s \in \mathbb{R}$ and $m \in M$, one sees that, for every given $t, s \in \mathbb{R}$, $g_{-t}\big(s + t, g_t(s, m)\big) = m$, for all $m \in M$.

Given a time-dependent vector field X and given the associated family of diffeomorphisms $\{g_t\}_{t\in\mathbb{R}}$ as above, let $\{\mathscr{F}_{t,s}\}_{t,s\in\mathbb{R}}$ be the family of diffeomorphisms of M defined by

$$\mathscr{F}_{t+s,s}(m) = g_t(s, m), \ \forall s, t \in \mathbb{R}, \ m \in M.$$

Then the property expressed in Formula (B.2), if read on the maps $\mathscr{F}$, becomes

$$\mathscr{F}_{t+s+u,s}(m) = \mathscr{F}_{t+s+u,u+s}\big(\mathscr{F}_{u+s,s}(m)\big), \forall s, t, u \in \mathbb{R}, \ m \in M.$$

In other words, to a time-dependent vector field X is associated a family of diffeomorphisms $\{\mathscr{F}\}_{t,s \in \mathbb{R}}$ of M such that:

$$\mathscr{F}_{s,u} \circ \mathscr{F}_{u,t} = \mathscr{F}_{s,t}, \quad \forall t, s, u \in \mathbb{R}.$$

The diffeomorphisms $\mathscr{F}_{t,s}$ are called *evolution operators*, and they are defined, requiring that $t \rightsquigarrow \mathscr{F}_{t,s}(m)$ be the integral curve of the time-dependent vector field X starting at the point m at time $t = s$. In other words

$$\frac{d}{dt}\mathscr{F}_{t,s}(m) = X\big(t, \mathscr{F}_{t,s}(m)\big)$$

with the condition $\mathscr{F}_{t,t}(m) = m$.

Proposition B.17 *If M is compact and X_t is a time-dependent vector field defined on M, then there is a smooth isotopy $\{\phi_t\}_{t \in I}$ (see Definition B.19) such that:*

$$\frac{d\phi_s}{ds}\bigg|_{s=t}(m) = X_t(\phi_t(m)), \ \forall\, m \in M \ and\, \forall t \in I.$$

Remark B.18 Note that every vector field, i.e., every time-independent vector field, X can be suspended via the formula $\widehat{X} = X + \frac{\partial}{\partial t}$. Assuming X is complete, the one-parameter group of diffeomorphisms defined by $\widehat{X}$ is $\hat{\varphi}_t(s, m) = (s + t, g_t(m))$, where $\{g_t\}_{t \in \mathbb{R}} = \{\varphi_t^X\}_{t \in \mathbb{R}}$ is the one-parameter group of diffeomorphisms defined by X. Note that the hypothesis of completeness of X is not really necessary, and the previous statement can be easily adjusted to this case. In the time-independent case, the evolution operator $\mathscr{F}_{t,s}$ is still defined in the same way, but in this case, it depends only on the difference $t - s$, and not on the initial time.   △

Let now M and N be two manifolds and let $I = [0, 1]$.

Definition B.19 (Homotopy and Isotopy) Two maps $\psi_0, \psi_1 : M \to N$ are called *homotopic* if there exists a continuous map $\psi : I \times M \to N$ such that $\psi_t := \psi(t, \cdot) : M \to N$ is a map and $\psi(0, \cdot) = \psi_0$ and $\psi(1, \cdot) = \psi_1$. In this case, the map ψ is a called a homotopy between ψ_0 and ψ_1. If ψ_0 and ψ_1 are two diffeomorphisms and there exists a homotopy $\psi : I \times M \to N$ between them, such that ψ_t is a diffeomorphism for all $t \in I$, ψ_0 and ψ_1 are called *isotopic*, and ψ is called an *isotopy*   △

Let us take $M = N$ in the previous definition. Now given an isotopy $\{\psi_t\}_{t \in \mathbb{R}}$ of M, we can define a time-dependent vector field $\{X_t\}_{t \in \mathbb{R}}$ via the formula

$$X_t\big(\psi_t(m)\big) := \frac{d}{ds}\bigg|_{s=t}\psi_s(m), \ \forall t \in \mathbb{R}, \ m \in M. \tag{B.3}$$

On the other hand, from the previous discussion, it follows that under suitable assumption of completeness, a family of time-dependent diffeomorphisms of the manifold M can be associated with a time-dependent vector field X. With a little more effort, one can prove that under suitable hypothesis (e.g., M compact; see [117]), if $\{X_t\}_{t\in\mathbb{R}}$ is a time-dependent vector field on M, then there exists an isotopy $\{\psi_t\}_{t\in\mathbb{R}}$ such that (B.3) holds true.

We want to prove the following important result.

Proposition B.20 (Homotopy Lemma)   *If ψ_1, ψ_0 are two homotopic maps between M and N, then there exists $\mathcal{H} \in \mathrm{Hom}_{\mathbb{R}}\left(\Omega^k(N), \Omega^{k-1}(M)\right)$ such that*

$$\psi_1^* - \psi_0^* = d \circ \mathcal{H} + \mathcal{H} \circ d. \tag{B.4}$$

Proof We first prove the statement for $M = \mathbb{R}^m$ and $N = \mathbb{R}^n$. Define

$$i_t : \mathbb{R}^m \xrightarrow[x \rightsquigarrow (t,x)]{} I \times \mathbb{R}^m$$

so that $\psi \circ i_t = \psi_t : \mathbb{R}^m \to \mathbb{R}^n$, for all $t \in [0, 1]$. Then, given $\omega \in \Omega^k(\mathbb{R}^n)$

$$\psi_1^*\omega - \psi_0^*\omega = (\psi \circ i_1)^*\omega - (\psi \circ i_0)^*\omega = \int_0^1 \frac{d}{dt}(\psi \circ i_t)^*\omega\, dt \tag{B.5}$$

which by construction is a k-form on $\mathbb{R}^m$. On the other hand, let $\Psi^h : I \times \mathbb{R}^m \to I \times \mathbb{R}^m$, the map defined by

$$\Psi^h(t, x) = (t + h, x), \ \forall x \in \mathbb{R}^m \text{ and } t, h \in I, \text{ such that } t + h \in I.$$

Then

$$\left.\frac{d}{dh}\right|_{h=0} \Psi^h(t, x) = \frac{\partial}{\partial t} := (1, 0, \dots, 0) \in \mathbb{R} \times \mathbb{R}^m,$$

and, since $i_{t+h} = \Psi^h \circ i_t : \mathbb{R}^m \to I \times \mathbb{R}^m$,

$$\frac{d}{dt}i_t^* = \left.\frac{d}{dh}\right|_{h=0} i_{t+h}^* = i_t^* \circ \left.\frac{d}{dh}\right|_{h=0} \left(\Psi^h\right)^* = i_t^* \circ \mathcal{L}_{\frac{\partial}{\partial t}},$$

by the definition of the Lie derivative. In particular, one deduces that $\frac{d}{dt}i^*$ is a $\mathbb{R}$-linear map between $\Omega^k(I \times \mathbb{R}^m)$ and $\Omega^k(\mathbb{R}^m)$. Using this result, one computes

$$\frac{d}{dt}\left(\psi \circ i_t\right)^* = \frac{d}{dt}\left(i_t^* \circ \psi^*\right) = \left(\frac{d}{dt} \circ i_t\right)^* \circ \psi^*$$

$$= i_t^* \circ \mathcal{L}_{\frac{\partial}{\partial t}} \circ \psi^*$$

which using Cartan's formula $\mathcal{L}_X\alpha = d \circ i_X\alpha + i_X \circ d\alpha$

$$= i_t^* \circ \left(d \circ i_{\frac{\partial}{\partial t}} + i_{\frac{\partial}{\partial t}} \circ d \right) \circ \psi^*$$

$$= d \circ \left(i_t^* \circ i_{\frac{\partial}{\partial t}} \circ \psi^* \right) + \left(i_t^* \circ i_{\frac{\partial}{\partial t}} \circ \psi^* \right) \circ d.$$

Comparing the result of this computation with Formula (B.5), one gets the proof of the proposition, when $M = \mathbb{R}^m$ and $N = \mathbb{R}^n$, defining

$$\mathcal{H} = \int_0^1 \left(i_t^* \circ i_{\frac{\partial}{\partial t}} \circ \psi^* \right) dt. \tag{B.6}$$

The proof of the proposition for N and M general manifolds follows from the previous proof using a partition of unity argument (see, e.g., [107]). $\square$

Remark B.21 (Poincaré Lemma)   Note that if $M = \mathbb{R}^m = N$ and $\psi_1, \psi_0 : \mathbb{R}^m \to \mathbb{R}^m$ are defined by $\psi_1(x) = x$ and, respectively, $\psi_0(x) = (0, \ldots, 0)$, for all $x \in \mathbb{R}^m$, then $\psi : I \times \mathbb{R}^m \to \mathbb{R}^m$, defined by $\psi(t, x) = tx$ for all $x \in \mathbb{R}^m$, is a homotopy between ψ_1 and ψ_0. A simple computation shows that $\psi_0^* \omega = 0$ for all $\omega \in \Omega^k(\mathbb{R}^m)$ and for all k. Then, if ω is closed, since $\psi_1^* \omega = \omega$, Formula (B.4) implies that

$$d(\mathcal{H}(\omega)) = \omega,$$

giving a proof of the Poincaré Lemma. $\triangle$

Appendix C
Lie Groups, Lie Algebras and Fiber Bundles

In this appendix, we collect some of the basic notions about Lie groups, Lie algebras, and fiber bundles, which have been used in these notes. For more information about these topics, the reader could consult, for example, the monographs [72, 209, 227] and [244].

1 Lie Groups and Lie Algebras

We start with the following definition.

Definition C.1 (Lie Groups, Homomorphisms and Lie Subgroups) A *Lie group* G is:

(i) A smooth manifold endowed with a structure of an abstract group compatible with the differentiable structure. This means that $m : G \times G \to G$ and $i : G \to G$ are two smooth maps (here m is the *multiplication* map, and i is the *inversion* map).

(ii) A map $\Phi : G \to H$ between two Lie groups is called a *homomorphism* of Lie groups if is smooth, and it is a homomorphism of the underlying abstract groups.

(iii) A pair (H, ψ) is called a *Lie subgroup* of G if:

    (a) (H, Ψ) is a submanifold of G (i.e., $\Psi : H \to G$ is an *injective immersion*)

    (b) H is a Lie group

    (c) $\Psi : H \to G$ is a group homomorphism

$\triangle$

© The Author(s), under exclusive license to Springer Nature Switzerland AG 2026
A. Arsie, I. Mencattini, *Geometry of Integrable Systems*, Latin American
Mathematics Series – UFSCar subseries,
https://doi.org/10.1007/978-3-031-96282-0

Here we state, without proof, the following important result:

Theorem C.2 (Closed Lie Subgroups) *A Lie subgroup (H, Ψ) is a closed subgroup of G if and only if the map Ψ is an embedding, i.e., if and only if Ψ is an onto homeomorphism. In particular, closed subgroups of a Lie groups are Lie subgroups (see [244, Theorem 3.21]).*

Example C.3 In order to clarify the previous definitions and the previous theorem, let G be the Lie group $S^1 \times S^1$ where S^1 can be thought as the circle of complex numbers of modulus one with group operation given by multiplication. Then $S^1 \times \{0\}$ and $\{0\} \times S^1$ are Lie subgroups of G that are also closed Lie subgroups and are also submanifolds. The same is true if one considers for any co-prime pair of integers (k_1, k_2):

$$H_{k_1, k_2} := \left\{ \left(e^{ik_1 t}, e^{ik_2 t} \right), t \in \mathbb{R} \right\}.$$

However, there are also many Lie subgroups of G that are not submanifolds (and are not closed Lie subgroups). In fact, for any irrational number α,

$$H_\alpha := \left\{ \left(e^{it}, e^{i\alpha t} \right), t \in \mathbb{R} \right\},$$

is a Lie subgroup of G, which is not closed and which is not a submanifold of G. In particular, compact Lie groups may have non-compact subgroups. $\triangle$

Here, we also record the following important result:

Theorem C.4 (Basic Structure) *Let G be a Lie group. Let G^0 be the connected component of G containing the identity.*

(i) Then G^0 is a normal subgroup of G, the index of which is equal to the number of connected components of G.
(ii) There exists $U \subset G^0$, which is an open neighborhood of the identity of G, such that $G^0 = \bigcup_{i=1}^{\infty} U$ (meaning that every element of G^0 can be written as a product of finitely many elements of U).

For a proof of this result, see, for example, the book [72].

The *infinitesimal analogue* of a Lie group is a Lie algebra.

Definition C.5 (Lie Algebra)   Let $\mathfrak{g}$ be a vector space.

(i) A *Lie bracket* $[\cdot, \cdot]$ over $\mathfrak{g}$ is a *bilinear* map $[\cdot, \cdot] : \mathfrak{g} \times \mathfrak{g} \to \mathfrak{g}$ such that:

    (a) $[x, y] = -[y, x]$ (skew–symmetry)
    (b) $[[x, y], z] + [[z, x], y] + [[y, z], x] = 0$ (Jacobi identity)

    for every $x, y, z \in \mathfrak{g}$. A vector space endowed with a Lie bracket is called a *Lie algebra*.

(ii) A linear map $\psi : \mathfrak{g} \to \mathfrak{g}'$ between two Lie algebra is called a *homomorphism* of Lie algebras if $\psi([x, y]) = [\psi(x), \psi(y)]'$ for all $x, y \in \mathfrak{g}$.

(iii) A vector subspace $\mathfrak{h}$ of a Lie algebra $\mathfrak{g}$ is called a *Lie subalgebra* if $[\mathfrak{h}, \mathfrak{h}] \subset \mathfrak{h}$. It is called a *Lie ideal* if $[\mathfrak{g}, \mathfrak{h}] \subset \mathfrak{h}$.

Note that every ideal of $\mathfrak{g}$ is a Lie subalgebra of $\mathfrak{g}$. Also remember that a Lie algebra is called *Abelian* or *commutative* if $[\mathfrak{g}, \mathfrak{g}] = \{0\}$. $\triangle$

Example C.6 (Center of $\mathfrak{g}$) Let $\mathfrak{g}$ be a Lie algebra. Then $\mathscr{Z}_{\mathfrak{g}} = \{x \in \mathfrak{g}|\, [x, \mathfrak{g}] = 0\}$ is an ideal of $\mathfrak{g}$, called the *center* of $\mathfrak{g}$. $\triangle$

Example C.7 ($\mathrm{GL}_n(\mathbb{K})$ & $\mathfrak{gl}_n(\mathbb{K})$) Let $\mathrm{GL}_n(\mathbb{K})$ be the set of all $n \times n$ *invertible* matrices endowed with the induced Euclidean topology (we think $\mathrm{GL}_n(\mathbb{K}) \subset \mathrm{Mat}_n(\mathbb{K}) \simeq \mathbb{K}^{n^2}$ with the Euclidean topology). The *usual row by columns* matrix multiplication and the operation of taking the *inverse* of a given matrix are *smooth maps*, since their outcomes can be written component wise as *rational functions* (with denominators always different from zero) on the entries of the matrices. In this way $\mathrm{GL}_n(\mathbb{K})$ becomes a Lie group if endowed with these two operations. On the other hand, since the multiplication row by columns is *associative*, this product induces a bracket, which for all pair of matrices (A, B) is defined as $[A, B] = AB - BA$ for all $A, B \in \mathrm{Mat}_n(\mathbb{K})$, which turns this vector space into a Lie algebra, which will be denoted as $\mathfrak{gl}_n(\mathbb{K})$. $\triangle$

Example C.8 Let $\mathscr{A}$ be an *associative algebra* (this means that $\mathscr{A}$ is a vector space with a product $\cdot$ such that $a \cdot (b \cdot c) = (a \cdot b) \cdot c$ for all $a, b, c \in \mathscr{A}$). Then $(a, b) \rightsquigarrow [a, b]_{\mathscr{A}} = ab - ba$ is a Lie bracket. The Lie algebra $(A, [\cdot, \cdot]_A)$ will be denoted with the symbol $\mathscr{A}_{\mathrm{Lie}}$. Note that if $\mathscr{A}$ and $\mathscr{B}$ are two associative algebras, any morphism $\phi : \mathscr{A} \to \mathscr{B}$ (this means that $\phi(ab) = \phi(a)\phi(b)$ for all $a, b \in \mathscr{A}$) is a also a morphism between the corresponding Lie algebras. This will be denoted with ϕ_L.$\triangle$

Example C.9 (Direct Product) Let $\mathfrak{a}$ and $\mathfrak{b}$ be two Lie algebras and let $\mathfrak{g} = \mathfrak{a} \oplus \mathfrak{b}$ be endowed with the bracket

$$\big[(x_1, y_1), (x_2, y_2)\big]_{\mathfrak{g}} = \big([x_1, x_2]_{\mathfrak{a}}, [y_1, y_2]_{\mathfrak{b}}\big), \ \forall x_1, x_2 \in \mathfrak{a} \text{ and } y_1, y_2 \in \mathfrak{b}.$$

Then $\big(\mathfrak{g}, [\cdot, \cdot]_{\mathfrak{g}}\big)$ is a Lie algebra called the direct product of $\mathfrak{a}$, $\mathfrak{b}$. The maps $i_{\mathfrak{a}} : \mathfrak{a} \to \mathfrak{g}, x \rightsquigarrow (x, 0), i_{\mathfrak{b}} : \mathfrak{b} \to \mathfrak{g}$, and $y \rightsquigarrow (0, y)$ are morphisms of Lie algebras. Note that $\mathfrak{a}$, $\mathfrak{b}$ centralize each other in $\mathfrak{g}$. $\triangle$

Example C.10 (Some Classical Lie Groups and Their Lie Algebras) We give the definitions of the matrix Lie groups and matrix Lie algebras that play some role in this work.

The Lie groups:

(i) $\mathrm{SL}_n(\mathbb{K}) = \big\{g \in \mathrm{Gl}_n(\mathbb{K})|\, \det(g) = 1\big\}$, *Special linear group*

(ii) $\mathrm{O}_n(\mathbb{K}) = \big\{g \in \mathrm{GL}_n(\mathbb{K})|\, gg^t = \mathrm{id}\,\big\}$, *Orthogonal group*

(iii) $\mathrm{SO}_n(\mathbb{K}) = \big\{g \in \mathrm{O}_n(\mathbb{K})|\, \det(g) = 1\big\}$, *Special orthogonal group*

(iv) $\mathrm{U}(n) = \left\{ g \in \mathrm{GL}_n(\mathbb{C}) \mid gg^{\dagger} = \mathrm{id} \right\}$, *Unitary group*
 (v) $\mathrm{SU}(n) = \left\{ g \in \mathrm{U}(n) \mid \det(g) = \mathrm{id} \right\}$, *Special unitary group*

Their Lie algebras:

  (i) $\mathfrak{sl}_n(\mathbb{K}) = \{A \in \mathfrak{gl}_n(\mathbb{K}) \mid \mathrm{tr}(A) = 0\}$, *Special linear Lie algebra*
 (ii) $\mathfrak{o}_n(\mathbb{K}) = \mathfrak{so}_n(\mathbb{K}) = \{A \in \mathfrak{gl}_n(\mathbb{K}) \mid A + A^t = 0\}$, *orthogonal Lie algebra*
(iii) $\mathfrak{u}_n = \{A \in \mathfrak{gl}_n(\mathbb{C}) \mid A + A^{\dagger} = 0\}$, *Unitary Lie algebra*
(iv) $\mathfrak{su}_n = \{A \in \mathfrak{u}_n \mid \mathrm{tr}(A) = 0\}$, *Special unitary Lie algebra*

$$\triangle$$

Example C.11 ($\mathfrak{so}_3(\mathbb{R})$ **vs** $\mathbb{R}^3$) The Lie algebra of the Lie group $\mathrm{SO}_3(\mathbb{R})$ is $\mathfrak{so}_3(\mathbb{R})$, the vector space of 3×3 skew-symmetric matrices, whose Lie bracket is $[A, B] = A \cdot B - B \cdot A$. Looking at dimensions, $\mathfrak{so}_3(\mathbb{R}) \simeq \mathbb{R}^3$ as a vector space. On the other hand, $\mathbb{R}^3$ becomes a Lie algebra if we define a bracket between two vectors $v = (v_1, v_2, v_3)$ e $u = (u_1, u_2, u_3)$ as $[v, u] = v \times u$ where the $\times$ is the *usual cross product* in $\mathbb{R}^3$. It is a nice exercise to show that the map $i : \mathbb{R}^3 \to \mathfrak{so}_3(\mathbb{R})$ defined by

$$i(x, y, z) = A_{(x,y,z)} = \begin{pmatrix} 0 & -z & y \\ z & 0 & -x \\ -y & x & 0 \end{pmatrix} \tag{C.1}$$

defines a Lie algebras isomorphism between $(\mathbb{R}^3, \times)$ and $(\mathfrak{so}_3(\mathbb{R}), [\cdot, \cdot])$. Under this isomorphism, the product rows by columns between the A_v and the $u \in \mathbb{R}^3$ become $A_v u = v \times u$. Endow now $\mathbb{R}^3$ with the standard Euclidean scalar product, $(v|u) = \sum_{i=1}^{3} v_i u_i$, and $\mathfrak{so}_3(\mathbb{R})$ with the non-degenerate bilinear form

$$(A|B) = -\frac{1}{2}\,\mathrm{tr}(AB) \tag{C.2}$$

then $v \cdot u = (A_v | B_u)$ for all $v, u \in \mathbb{R}^3$. $\triangle$

Remark C.12 Note that the previous example could be put in a slightly more general setting as follows. Let V be a n-dimensional vector space and let $(\cdot | \cdot)$ be a definite-positive inner product, i.e., a symmetric and bilinear form such that $(v \mid v) > 0$ for all $v \neq 0$. An orthonormal basis $E = (e_1, \ldots, e_n)$ of $\left(V, (\cdot | \cdot)\right)$ defines a *volume-form* $\eta \in \Lambda^n V$, $\eta = e_1 \wedge \cdots \wedge e_n$, which is independent of E if one chooses any other orthonormal basis obtained applying to E an element of $\mathrm{SO}(V)$, the orthogonal group associated with $(V, (\cdot | \cdot))$ whose elements have *determinant* equal to one. Let now $e_{i_1} \wedge \cdots \wedge e_{i_d} \in \Lambda^d V$ and let $(j_1, \ldots, j_{n-d})$ be a sequence of integers such that $(i_1, \ldots, i_d, j_1, \ldots, j_{n-d})$ is an *even permutation* of $(1, \ldots, n)$, and let us define $\star : \Lambda^d V \to \Lambda^{n-d} V$ as the map obtained extending by linearity the map defined on the generators of $\Lambda^d V$ by $\star(e_{i_1} \wedge \cdots \wedge e_{i_d}) = e_{j_1} \wedge \cdots \wedge e_{j_{n-d}}$. This map is called the Hodge $\star$-product. One can show that this linear map is well defined, i.e., it is independent of the choice made to define it. Moreover, $\star : \Lambda^d V \to \Lambda^{n-d} V$ is such that $\star \circ \star = (-1)^{d(n-d)}\,\mathrm{id}_{\Lambda^d V}$. In particular, if n is *odd*, $\star$ is an *isomorphism* for each d, and it coincides with its own inverse (see, e.g., [31]).

If V is three-dimensional, then the Hodge $\star$-product defines a bilinear and skew-symmetric bracket $[\cdot, \cdot] : V \times V \to V$, defined via $[v, u] = \star(v \wedge u)$. A simple computation shows that this bracket satisfies the Jacobi identity, and for this reason, it defines a structure of a Lie algebra on V. Note that this Lie bracket, by construction, in compatible with the inner product $(\cdot | \cdot)$, i.e., $\big(V, [\cdot, \cdot], (\cdot | \cdot)\big)$ is a *quadratic* Lie algebra.

Note that if $V = \mathbb{R}^3$ and $(\cdot | \cdot)$ is the *standard* Euclidean product, $*(v \wedge u) = v \times u$, for all $u, v \in V$, where $\times$ is the standard *cross* product on $\mathbb{R}^3$. $\triangle$

Example C.13 (Simple, Semi-simple, and Reductive Lie Algebras) A Lie algebra is called:

(i) *Simple* if it does not have any non-trivial ideals (i.e., ideals besides the zero ideal and the Lie algebra itself)
(ii) *Semi-simple* if it does not have any non-trivial Abelian ideals
(iii) *Reductive* if $\mathfrak{g} = \mathscr{Z}_{\mathfrak{g}} \oplus [\mathfrak{g}, \mathfrak{g}]$, where $\mathscr{Z}_{\mathfrak{g}}$ is the *center* of $\mathfrak{g}$ (see Definition C.6).

$\triangle$

We also add this further information.

Definition C.14 A finite-dimensional Lie algebra $\mathfrak{g}$ is called *solvable* if given

$$D^n(\mathfrak{g}) = \big[D^{n-1}(\mathfrak{g}), D^{n-1}(\mathfrak{g})\big] \text{ for } n \geq 2, \text{ and } D^1(\mathfrak{g}) = \mathfrak{g}$$

there exists n such that $D^n(\mathfrak{g}) = \{0\}$. An ideal $\mathfrak{a} \subset \mathfrak{g}$ is solvable if it is solvable as a Lie algebra. Since the sum of solvable ideals is a solvable ideal, every Lie algebra $\mathfrak{g}$ has a *maximal* solvable ideal, $\mathfrak{r}(\mathfrak{g})$, called the *radical* or the *solvable radical* of $\mathfrak{g}$. $\triangle$

Finally we mention the following result:

Theorem C.15 (Levi Decomposition) *Let $\mathfrak{g}$ be a finite-dimensional Lie algebra. Then:*

$$\mathfrak{g} = \mathfrak{s} \ltimes \mathfrak{r}(\mathfrak{g}),$$

i.e., $\mathfrak{g}$ is the semi-direct product *(see Example C.65) of $\mathfrak{r}$ with $\mathfrak{s}$, where $\mathfrak{s}$ is a semi-simple Lie subalgebra of $\mathfrak{g}$ called a* Levi factor *of $\mathfrak{g}$, which is unique up to isomorphism.*

Remark C.16 Note that while it is not true that every Lie group is a matrix Lie group (see, e.g., [209, pp. 395–398]), it is true that every (finite-dimensional) Lie algebra is a Lie algebra of matrices. This fact is a consequence of the so-called Ado theorem (see [209, pp. 415–417]). $\triangle$

Here we also record the following definition:

Definition C.17 (Quadratic Vector Spaces and Quadratic Lie Algebras) A vector space V is called *quadratic* if it carries a bilinear form

$$B : V \otimes V \to \mathbb{K},$$

(i) *Non-degenerate*, i.e., $B(x, y) = 0$ for all $y \in V$ implies $x = 0$.
(ii) *Symmetric*, i.e., $B(x, y) = B(y, x)$ for all $x, y \in V$.

We will say that $\mathfrak{g}$ is a quadratic Lie algebra if $\mathfrak{g}$ is a quadratic vector space whose bilinear form is $\mathfrak{g}$-invariant, i.e., $(\mathrm{ad}_x y \mid z) + (y \mid \mathrm{ad}_x z) = 0$ for all $x, y, z \in \mathfrak{g}$. Moreover, a vector subspace $W \subset V$ will be called *isotropic* if $(v \mid u) = 0$ for all $u, v \in W$. $\triangle$

Remark C.18 Note that if a vector space V is endowed with a non-degenerate bilinear form $B : V \otimes V \to \mathbb{K}$, then V can be identified with its dual via the linear map $\check{B} : V \to B^*$

$$\check{B} : v \rightsquigarrow \alpha_v$$

where $\langle \alpha_v, u \rangle = B(v, u)$ for all $u \in V$. $\triangle$

Since B is non-degenerate, $\check{B}$ is invertible, and its inverse $(\check{B})^{-1} : V^* \to V$ takes each $\alpha \rightsquigarrow v_\alpha$, where $B(v_\alpha, u) = \langle \alpha, u \rangle$ for all $u \in V$.

Remark C.19 In this work, the bilinear form B will always be positive definite. $\triangle$

Example C.20 (Simple Lie Algebras) Let $\mathfrak{g}$ be a Lie algebra and $x \in \mathfrak{g}$. Define $f_x : \mathfrak{g} \to \mathfrak{g}$ via $f_x(z) = [x, z]$ for all $z \in \mathfrak{g}$. Then, for any $x, y \in \mathfrak{g}$, we can define:

$$B(x, y) = \mathrm{tr}\left(f_x \circ f_y\right)$$

This is called the *Killing form*. It is bilinear and ad-invariant (this follows from the Jacobi identity). It is in general degenerate since its kernel coincides with the center of $\mathfrak{g}$. A result due to Cartan tells us that $\mathfrak{g}$ is simple if and only if its Killing form is not degenerate (see, e.g., [209]) $\triangle$

In the next subsection, it is shown how to associate to every Lie group G a Lie algebra $\mathrm{Lie}(G) = \mathfrak{g}$, which can be considered as the *infinitesimal* analogue of G.

1.1 Lie Algebras of Lie Groups and the Exponential Map

For every $g \in G$, one can define the *left* and the *right translations* by

$$L_g : G \xrightarrow{h \rightsquigarrow gh} G \ \text{ and } \ R_g : G \xrightarrow{h \rightsquigarrow hg} G$$

The following properties can easily be checked:

(i) L_g and R_g are a diffeomorphism of G.
(ii) $R_g \circ L_h = L_h \circ R_g$.
(iii) The map $g \rightsquigarrow L_g$ and the map $g \rightsquigarrow R_g$ are a *homomorphism* and an *anti-homomorphism* (meaning that $R_{gh} = R_h \circ R_g$), respectively, of (abstract) groups.

The next step is to study how the vector fields behave under the action of the *left translations* (the case of the right ones is completely analogous).

Since for all $g \in G$ L_g is a diffeomorphism, given $X \in \mathfrak{X}(G)$, one has:

$$(L_g X)(h) = (L_g)_{*,L_{g^{-1}}(h)}\big(X(L_{g^{-1}}(h))\big) = (L_g)_{*,g^{-1}h}\big(X(g^{-1}h)\big), \ \forall g, h \in G$$

(see Formula 1.28). In particular:

Definition C.21 If G is a Lie group:

(i) $X \in \mathfrak{X}(G)$ will be called *left invariant* if

$$(L_g X)(h) = X(h) \ \ \forall h \in G.$$

(ii) The vector space of all left invariant vector fields will be denoted with $\mathfrak{X}^L(G)$.

(See Formula 1.29.)   △

Remark C.22 In a completely analogous way, one can define the notion of a right invariant vector field. The space of all the right invariant vector fields will be denoted with $\mathfrak{X}^R(G)$.   △

The following property will be crucial in what follows:

Lemma C.23 *The vector spaces $\mathfrak{X}^L(G)$ and $\mathfrak{X}^R(G)$ are Lie subalgebras of $\mathfrak{X}(G)$.*

Since L_g is a diffeomorphism:

$$(L_g)_{*,e} : T_e G \to T_g G$$

is a linear isomorphism so that, for any $x \in T_e G$, as g moves on G,

$$X_x^L(g) = (L_g)_{*,e}(x)$$

is a map from G to the its tangent bundle such that $\pi\big(X_x^L(g)\big) = g$ (here π is the canonical projection from TG to G).

We can now state the following important result:

Lemma C.24 *For every $x \in T_e G$:*

(i) *The map $g \rightsquigarrow X_x^L(g)$ is a left invariant vector field.*

(ii) The linear map:

$$l : T_eG \xrightarrow{\ x \rightsquigarrow X_x^L\ } \mathfrak{X}^L(G)$$

is a linear isomorphism.

(iii) The inverse of the map l is the restriction to $\mathfrak{X}^L(G)$ of the evaluation map

$$ev_e : \mathfrak{X}(G) \xrightarrow{\ X \rightsquigarrow X(e)\ } T_eG$$

We can now define the Lie algebra structure on the tangent space at the identity of G.

Definition C.25 Let G be a Lie group. The tangent space T_eG at the identity of G can be endowed with a structure of Lie algebra, defined by the unique Lie bracket $[\cdot, \cdot]^L$ such that

$$[x, y]^L = \left[X_x^L, X_y^L\right](e) \tag{C.3}$$

for all $x, y \in T_eG$. We will denote with $\mathfrak{g}^L$ the Lie algebra $\left(T_eG, [\cdot, \cdot]^L\right)$ so defined.
$\triangle$

Remark C.26 (Left vs Right Lie Bracket) By the same token, we could define equally well a structure of Lie algebra on T_eG using the Lie algebra of the *right invariant vector fields*. Here we just note the two Lie algebras $\mathfrak{g}^L$ and $\mathfrak{g}^R$ turn out to be *anti-isomorphic*. Such an anti-isomorphism is provided by the *differential* at the identity $e \in G$ of the map $i : G \to G$.
$\triangle$

Convention C.27 *In these notes, unless specified otherwise, the Lie algebra of the Lie group G will always be the Lie algebra $\mathfrak{g}^L$. For this reason, in what follows, we will simply write $\mathfrak{g}$ instead of $\mathfrak{g}^L$.*

Definition C.28 A Lie group G is called simple, semi-simple, or reductive if its Lie algebra $\mathfrak{g}$ is simple, semi-simple, or reductive (see Example C.13).
$\triangle$

Example C.29 If $G = \mathrm{GL}_n(\mathbb{K})$, then $\mathfrak{g} = \mathfrak{gl}_n(\mathbb{K})$.
$\triangle$

Example C.30 (Direct Product 2) Let A, B be two Lie groups. The Cartesian product $G = A \times B$, endowed with the product $\cdot : \big((a_1, b_1), (a_2, b_2)\big) \rightsquigarrow (a_1 a_2, b_1 b_2)$, is a Lie group whose unit element is (e_A, e_B) and $(a, b)^{-1} = (a^{-1}, b^{-1})$ for all $a \in A$, $b \in B$. The two canonical maps $i_A : A \to G, a \rightsquigarrow (a, e_B)$ and $i_B : B \to G, b \rightsquigarrow (e_A, b)$, are morphisms of Lie groups. Note that:

(i) A, B centralize each other in G
(ii) The Lie algebra of G is isomorphic to the direct product of the Lie algebras of A and B (see Example C.9).

$\triangle$

In this way, we defined a correspondence between Lie groups and Lie algebras $G \rightsquigarrow \mathfrak{g}$. There is much more to be said about such a correspondence. In particular, we have the following.

Theorem C.31 *Let H, G be two Lie groups and $\mathfrak{h}, \mathfrak{g}$ the corresponding Lie algebras.*

(i) Given a Lie group homomorphism $\phi : H \to G$, the map $\phi_{*,e} : \mathfrak{h} \to \mathfrak{g}$ is a Lie algebra homomorphism.

(ii) If H is simply connected and $\psi : \mathfrak{h} \to \mathfrak{g}$ is a homomorphism of Lie algebras, then there exists a unique Lie group homomorphism $\Psi : H \to G$ such that $(\Psi)_{*,e_H} = \psi$.

While the first part of the previous theorem is a somehow a simple statement to check, the second part is a deep result of Lie theory (see, e.g., [244, Theorem 3.27]).

The left (right) invariant vector fields are complete. Let $\gamma_e^x : \mathbb{R} \to G$ be the integral curve of the left invariant vector field X_x (here $x \in \mathfrak{g}$ as above). Then it can be shown that

Lemma C.32 *The map $t \rightsquigarrow \gamma_e^x(t)$ is a Lie group homomorphism (we think of $\mathbb{R}$ with its additive structure). Moreover, every continuous group homomorphism between the real line and G is of this form.*

Definition C.33 The integral curve of a left invariant vector field going through the identity $e \in G$ is called a *one-parameter group homomorphism* of the Lie group G.

△

Let γ_e^x be the one-parameter group corresponding to the left invariant vector field X_x and let:

$$\exp(x) = \gamma_e^x(1) \tag{C.4}$$

Definition C.34 (Exponential Map)   The map defined in Formula (C.4) is called *exponential map*.

△

Example C.35   If $G = \mathrm{GL}_n(\mathbb{K})$, $g = \mathfrak{gl}_n(\mathbb{K})$, the map defined in Formula (C.4) is the *usual exponential map of matrices*.

△

The following proposition contains some important properties of the exponential map.

Proposition C.36 *The exponential map defined in Formula (C.4):*

(i) Is smooth

(ii) Is $(\exp)_{*,0} = \mathrm{id}_{\mathfrak{g}}$. In particular, there exist $U \subset \mathfrak{g}$ and $V \subset G$ open neighborhoods of $0 \in \mathfrak{g}$ and $e \in G$ such that $\exp|_U : U \to V$ is a diffeomorphism.

(iii) *Let G and H be two Lie groups and $\mathfrak{g}$ and $\mathfrak{h}$ be the corresponding Lie algebras. Then if $\phi : G \to H$ is a morphism of Lie groups, then:*

$$\exp \circ \phi_{*,e} = \phi \circ \exp$$

1.2   G-Actions and $\mathfrak{g}$-Actions

Let us suppose that N is a smooth manifold and that G is a Lie group.

Definition C.37 We say that G acts on N on the *left* if there exists a smooth map

$$\varphi : G \times N \xrightarrow{\;(g,n)\rightsquigarrow\varphi(g,n)=:gn\;} N$$

such that the following two conditions are verified:

(i) $\varphi(e, n) = n, \;\; \forall n \in N$.
(ii) $\varphi\big(g, \varphi(h, n)\big) = \varphi(hg, n), \;\; \forall g, h \in G$ and $n \in N$.

$\triangle$

Remark C.38 (Right Actions) In a similar way, we can define also a *right G-action* on N as a map $\psi(n, g) = ng$ for all $n \in N$ and $g \in G$. $\triangle$

If G acts on N via a left(right) G-action, we will say that N is a *left(right) G-manifold*.

Example C.39 Every Lie group is naturally a left (right) G-manifold with respect to the left (right) translations. $\triangle$

Let us focus on the case of the *left G*-actions. Suppose N be a left G-manifold.

Lemma C.40 *Then for all $g \in G$ and $n \in N$:*

(i) *The map $\varphi_g : N \to N$, $\varphi_g(n) = \varphi(g, n)$ is a diffeomorphism.*
(ii) *The map $\varphi_n : G \to N$, $\varphi_n(g) = \varphi(g, n)$ is smooth.*
(iii) *The map $g \rightsquigarrow \varphi_g$ (see point (i)) is a homomorphism of groups between G and $\mathscr{D}(N)$, the group of diffeomorphisms of N.*

1.3   Infinitesimal Action

The infinitesimal notion of G-manifold is enclosed in the following:

Definition C.41 ($\mathfrak{g}$-Manifolds and $\mathfrak{g}$-Actions)   Let $\mathfrak{g}$ be a Lie algebra. A $\mathfrak{g}$-manifold is a smooth manifold N endowed with a $\mathfrak{g}$-action, i.e., with a *homomorphism* of Lie algebras $\mathfrak{g} \rightsquigarrow \mathfrak{X}(N)$. $\triangle$

Proposition C.42 (Infinitesimal Left Actions) *Every G-manifold carries a natural structure of $\mathfrak{g}$-manifold. If N is a G-manifold, the image in $\mathfrak{X}(N)$ of $x \in \mathfrak{g}$ will be denoted with X_x and will be called the* fundamental vector field *defined by the element x (in general, we will not explicitly mention the action).*

Proof Take $n \in N$ and consider the curve $t \rightsquigarrow \varphi_n(\exp(-tx))$ and define

$$X_x(n) = \left.\frac{d}{dt}\right|_{t=0} \varphi_n(\exp(-tx)). \tag{C.5}$$

Then the map $X_x : N \rightsquigarrow TN$ so defined satisfies $\pi \circ X_x = \mathrm{id}_N$, where $\pi : TN \to N$ is the canonical projection. Since this map is *smooth* (see, e.g., [244]), it defines a vector field on N. Now we will show that for all $x, y \in \mathfrak{g}$

$$X_{[x,y]} = \left[X_x, X_y\right] \tag{C.6}$$

which entails that the (linear) map $x \rightsquigarrow X_x$ defines an *infinitesimal action* of G on N. To prove (C.6), it suffices to compute $(\mathcal{L}_{X_x} X_y)(m)$, for $m \in N$. Using Formula (1.31), this turns out to be equivalent to compute

$$\left.\frac{d}{dt}\right|_{t=0} (\varphi_{\exp(tx)})_{*,\exp(-tx)m} X_y(\exp(-tx)m), \tag{C.7}$$

where X_x is defined in Formula (C.5). We suggest the reader to compare the previous formula with (1.31). The two differ for the signs in front of the t variable. This difference is a consequence of the definition of X_x (see Formula C.5). To proceed with the computation, first recall the G-action $\varphi : G \times N \to N$ defines two families of smooth maps, $\{\varphi_g\}_{g \in G}$, where $\varphi_g : N \to N$ for all $g \in G$, and $\{\varphi_n\}_{n \in N}$, where $\varphi_n : G \to N$ (see Lemma C.40). Moreover, observe that Formula (C.5) is equivalent to:

$$X_x(n) = -(\varphi_n)_{*,e}(x), \ \forall n \in N \text{ and } x \in \mathfrak{g}$$

and note that

$$\varphi_g \circ \varphi_{gn} = c_g \circ \varphi_n, \ \forall n \in N.$$

where c_g is defined in Formula (C.15). Computing the differential at the identity of the maps in the previous equation, one gets $(\varphi_g)_{*,n} \circ (\varphi_n)_{*,e}(y) = (\varphi_{gn})_{*,e} \circ (c_g)_{*,e}(y)$, for all $y \in \mathfrak{g}$, which is equivalent to:

$$(\varphi_g)_{*,n}(X_y(n)) = X_{\mathrm{Ad}_g y}(gn), \ \forall g \in G, \ n \in N, \text{ and } y \in \mathfrak{g}$$

(see Formula C.16). Taking $g = \exp(tx)$ and $n = \exp(-tx)m$ in the previous formula, and substituting it into (C.7), one arrives to:

$$(\mathscr{L}_{X_x} X_y)(m) = \left.\frac{d}{dt}\right|_{t=0} X_{\mathrm{Ad}_{\exp(tx)y}}(m) = X_{[x,y]}(m),$$

for all $x, y \in \mathfrak{g}$ and $m \in N$. $\qquad\qquad\qquad\qquad\qquad\qquad\qquad\qquad\qquad\qquad\qquad\square$

It is also interesting to note that if $x \in \mathfrak{g}$ and if X_x is the corresponding fundamental vector field, then:

Proposition C.43

$$\varphi_g X_x = X_{\mathrm{Ad}_{g^{-1}} x}, \ \forall g \in G. \tag{C.8}$$

Proof

$$\begin{aligned}
(\varphi_g X_x)(n) &= \left(\varphi_g^{-1}\right)_{*,\varphi_g(n)}\big(X(\varphi_g(n))\big)\\[2mm]
&= \left(\varphi_g^{-1}\right)_{*,gn} \left.\frac{d}{dt}\right|_{t=0} \exp(-tx)gn \ \text{ where } gn = \varphi_g(n)\\[2mm]
&= \left.\frac{d}{dt}\right|_{t=0} \big(g^{-1}\exp(-tx)gn\big)\\[2mm]
&= \left.\frac{d}{dt}\right|_{t=0} \exp(-t\,\mathrm{Ad}_{g^{-1}} x)n\\[2mm]
&= X_{\mathrm{Ad}_{g^{-1}} x}(n),
\end{aligned}$$

for all $n \in N$ and $g \in G$. $\qquad\qquad\qquad\qquad\qquad\qquad\qquad\qquad\qquad\qquad\qquad\square$

Remark C.44 Note that:

$$\begin{aligned}
(\mathscr{L}_{X_x} X_y)(m) &= \left.\frac{d}{dt}\right|_{t=0} (\varphi_{\exp(tx)})_{*,\exp(-tx)m} X_y(\exp(-tx)m)\\[2mm]
&\overset{(1.28)}{=} \left.\frac{d}{dt}\right|_{t=0} (\varphi_{\exp(-tx)} X_y)(m),\\[2mm]
&\overset{(\mathrm{C.8})}{=} \left.\frac{d}{dt}\right|_{t=0} X_{\mathrm{Ad}_{\exp(tx)} y}(m),
\end{aligned}$$

which provides a different proof of Formula (C.6). $\qquad\qquad\qquad\qquad\qquad\qquad\triangle$

Remark C.45 (Infinitesimal Right Actions) If φ were a *right* action instead of a *left* one (as we assumed above), then Formula (C.5) defining X_x should be modified as follows:

$$X_x(n) = \left.\frac{d}{dt}\right|_{t=0} \varphi_n(\exp(tx)).$$

$\triangle$

Let us recall some (general) definitions:

Definition C.46 (Isotropy Group) Let N be a G-manifold and $n \in N$. Then

$$G_n = \{g \in G \mid \varphi_g(n) = n\}$$

is a (abstract) subgroup of G called the *isotropy group* or the *stabilizer* of n. $\triangle$

Proposition C.47 *For all $n \in N$, G_n is a closed subgroup of G; in particular, it is a Lie subgroup G.*

Proof For each $n \in N$, $G_n = \varphi_n^{-1}(n)$, which proves the first part of the statement. The second one follows from the first and from Theorem C.2. $\square$

Definition C.48 For each $n \in N$, we will denote with $\mathfrak{g}_n$ the Lie algebra of the Lie group G_n, and we will call it the *isotropy Lie algebra* of the point n. $\triangle$

Definition C.49 A G-action on N is called:

(i) *Free* if $G_n = \{e\}$ for all $n \in N$
(ii) *Locally free* if $\mathfrak{g}_n = \{0\}$ for all $n \in N$ or, equivalently, if G_n is a *discrete* subgroup of G for all $n \in N$
(iii) *Transitive* if for every pair of points $n, n' \in N$, there exists $g \in G$ such that $\varphi(g, n) = n'$
(iv) *Proper* if for all K_1 and K_2 compact subsets of N, the subset of all $g \in G$ such that $\varphi(g, K_1) \cap K_2 \neq \emptyset$ is compact in G

$\triangle$

1.4   Homogeneous G Spaces

In this subsection, we will briefly discuss an important class of G-manifold, called *G-homogeneous spaces* or G-homogeneous manifolds.

Definition C.50 (G-Homogeneous Manifold) A manifold N is called *G-homogeneous* if it carries a *transitive G-action*. $\triangle$

Remark C.51 When G is connected, any G-homogeneous manifold N will be also connected since such a manifold will be the image of the map $\varphi_n : G \to N$, for any given $n \in N$. Moreover, there is a *set bijection* between N and the set of all the classes G/G_n (again, n here is an arbitrary point of N), which induces a *manifold* structure on such a set. Moreover, the tangent space at the point $n \in N$ can be identified with the vector space $\mathfrak{g}/\mathfrak{g}_n$, where $\mathfrak{g}$ is the Lie algebra of G and $\mathfrak{g}_n$ is the Lie algebra of G_n. Note furthermore that N, seen as set of all the classes G/G_n, is naturally a *right G-space* and *left G_n-space* via the actions.   △

Among the homogeneous G-manifolds, the ones for which G is a vector space play an important role in these notes. For this reason, we survey below the main results about the structure of this class of manifolds. Let M be a manifold and V a vector space and let:

$$\varphi : V \times M \to M$$

be a smooth action. Suppose now that for every $m \in M$, the map $(\varphi_m)_{*,0} : V \to T_m M$ is an isomorphism. Then, since

$$(\varphi_m)_{*,v} = (\varphi_v)_{*,m} \circ (\varphi_m)_{*,0},$$

it follows that, for all $m \in M$ and $v \in V$,

$$(\varphi_m)_{*,v} : T_v V \to T_{\varphi_m(v)} M$$

is an isomorphism, i.e., that the action φ of V on M is *locally free* (see (ii) in Definition C.49). Since $(\varphi_m)_{*,0}$ is invertible, there exist $U \subset V$ and $U' \subset M$ open subsets such that $0 \in U$ $m \in U'$ and such that $\varphi_{m|_U} : U \to U'$ is a diffeomorphism. Since $\varphi_m(U) \subset \varphi_m(V) = \mathscr{O}_m$, it follows that for every point of $\mathscr{O}_m$, there exists an open neighborhood of this point totally contained in it, implying that $\mathscr{O}_m$ is open. Since the same argument applies to $\mathscr{O}_{m'}$, for every $m' \in M$, it follows that $\mathscr{O}_m$ is also closed. Finally, since M is connected, the previous argument implies that:

Lemma C.52 *Under the above assumptions, the V-action φ on M is transitive. In particular, $M \simeq V/\Lambda_m$ where Λ_m is the stabilizer of m in V.*

Remark C.53 This isomorphism is obtained via the map $\varphi_m : V \to M$ and therefore *depends* on the choice of the point m. On the other hand, since for each $m, m' \in M$, Λ_m is *conjugated* to $\Lambda_{m'}$ and since V is *Abelian*, it follows that $\Lambda_m = \Lambda_{m'}$. For this reason, in what follows, we will simply write $M \simeq V/\Lambda$.   △

To this remark, we add the following simple, but important, observation.

Lemma C.54 *The subgroup Λ is discrete, i.e., for all $v \in \Lambda$, there exists an open neighborhood $U \subset V$ of v such that $U \cap \Lambda = \{v\}$.*

Proof In fact, since $\Lambda = \Lambda_m = \ker\left((\varphi_m)_{*,0} : V \to T_m M\right)$ is trivial, the Lie algebra of the stabilizer of (every) $m \in M$ has dimension equal to zero.   □

Remark C.55 Observe that saying that $(\varphi_m)_{*,0}$ is invertible for every $m \in M$ is equivalent to say that $T_m M$ is spanned by $X_{v_1}(m), \ldots, X_{v_n}(m)$, where X_v is the fundamental vector field defined by the element $v \in V$ and $v_1, \ldots, v_n$ is a basis of the vector space V. $\triangle$

Then, to characterize the V-actions φ on M with $(\varphi_m)_{*,0}$ invertible for every $m \in M$, it suffices to characterize the *discrete subgroups* of V. Such a characterization is provided in the following lemma.

Lemma C.56 (Après Vladimir I. Arnold) *Every discrete subgroup of V is isomorphic to $\Lambda = \mathbb{Z}v_1 \oplus \cdots \oplus \mathbb{Z}v_k$ with $1 \leq k \leq dim(V)$ and where $v_1, \ldots, v_k$ are k-linearly independent vectors.*

Proof Here we follow very closely the proof given by Arnold in his *Mathematical Methods of Classical Mechanics*. The statement is (trivially) true if $\Lambda = \{0\}$. If not, let $v_0 \neq 0$ belonging to Λ and let us define the line l_{v_0} (the line going through $0 \in V$ generated by v_0). After choosing a (Euclidean) inner product, we can consider the set of all $v \in l_{v_0}$ such that $\|v\| \leq \|v_0\|$. Since Λ is discrete, the intersection between this closed segment and Λ will contain only a finite number of points. Among those, choose one having minimal distance from the origin. Let us call v_1 the chosen point. Then, the intersection $l_{v_0} \cap \Lambda$ will be the set of the form $\mathbb{Z}v_1$. (Why? Well, think of the distance from the origin of the vector $v - mv_1$ if there exists a vector $v = kv_1$, with $k \in (m, m+1)$ where $m \in \mathbb{Z} \ldots$). If $\Lambda \subset l_{v_1}$, we are done. If not, let $u \in \Lambda \backslash \mathbb{Z}v_1$ and use the inner product to define the orthogonal projection λv_1 of u along the line l_{v_1} (here $\lambda \in [m, m+1]$ for some $m \in \mathbb{Z}$). Call δ the distance of the vector u from the line l_{v_1}, and consider the right circular cylinder whose axis lies along the line l_{v_1}, whose altitude is $h = \|\lambda v_1\|$, and whose radius is equal to δ. Since Λ is discrete, its intersection with this cylinder will be a finite set. Now, among those points, choose the one having minimal distance from the line l_{v_1}, and call it v_2. Then observe that the distance of any $u \in \Lambda \backslash l_{v_1}$ to the line l_{v_1} is greater than or equal to the distance of v_2 to the same line. In fact, if $u \in \Lambda \backslash l_{v_1}$, $u - mv_1 \in \Lambda \backslash l_{v_1}$, and these two points have the same distance from the line l_{v_1}. On the other hand, if we denote with $u' = \mu v_1$ the orthogonal projection of u along the line l_{v_1}, then $\mu \in [n, n + 1]$ for some $n \in \mathbb{Z}$, which implies that $u' - nv_1 = (\mu - n)v_1$ with $\mu - n \in [0, 1]$. But $u' - nv_1$ is the orthogonal projection of the vector $u - nv_1$ on the line l_{v_1}. This vector belongs to Λ, and it is contained in the cylinder described above (since the two vectors u and $u - nv_1$ have the same distance from the line l_{v_1}, which is less than the distance of the vector v_2), but this contradicts the hypothesis on v_2. Consider now $\mathbb{Z}v_1 \oplus \mathbb{Z}v_2$. Then, $\Lambda \cap (\mathbb{R}v_1 \oplus \mathbb{R}v_2) = \mathbb{Z}v_1 \oplus \mathbb{Z}v_2$. In fact, suppose we can find a vector $v \in \Lambda \cap (\mathbb{R}v_1 \oplus \mathbb{R}v_2)$, which cannot be written as a linear combination with integer coefficients of the vectors v_1 and v_2. Then $v = \lambda_1 v_1 + \lambda_2 v_2$, with $\lambda_i \in \mathbb{R}\backslash\mathbb{Z}$ for $i = 1, 2$. Since $n_i \leq \lambda_i \leq n_{i+1}$ for some $n_i \in \mathbb{Z}$ $(i = 1, 2)$, then the point $v - n_1 v_1 - n_2 v_2$ will belong to Λ, and it will be closer to the line l_{v_1} than the point v_2 (check this please), bringing us to a contradiction. In this way we define a plane, generated by two elements of Λ. Then if we cannot find any point in Λ, which does not belong to $\mathbb{Z}v_1 \oplus \mathbb{Z}v_2$, we are done. Otherwise, we can find such a point and call it w. Then we can find, among the vectors of Λ, which do not belong to the subspace generated by v_1 and v_2, a vector having minimal distance from $\mathbb{R}v_1 \oplus \mathbb{R}v_2$ (why?). Calling this vector v_3, we end up with the subgroup $\mathbb{Z}v_1 \oplus \mathbb{Z}v_2 \oplus \mathbb{Z}v_3$. If this does not exhaust Λ, then we can proceed further, in a way that should be clear now. Note that the vectors $v_1, \ldots, v_k$ obtained following the procedure sketched above are linearly independent so that the same procedure will stop, at most, after $n = \dim(V)$ steps. $\qquad\qquad\square$

Remark C.57 The previous argument shows that every discrete subgroup of $\mathbb{R}^n$ admits a subset of $\mathbb{R}$-linearly independent vectors. In particular, any discrete subgroup of $\mathbb{R}^n$ of maximal rank admits a set of generators, which form a basis of $\mathbb{R}^n$, i.e., any discrete subgroup of $\mathbb{R}^n$ is a *lattice*. $\qquad\qquad\triangle$

Then we arrive to the following important result (see Lemma C.52 and Remark C.53).

Proposition C.58 *If M is a n-dimensional manifold endowed with a transitive and locally free action of an n-dimensional vector space, then $M \simeq \mathbb{R}^{n-k} \times \mathbb{T}^k$ where $\mathbb{T}^k$ is a k-dimensional torus and k is the rank of discrete subgroup Λ. In particular, if M is compact, then it is diffeomorphic to a torus of the same dimension.*

So how can we get a transitive action of a finite-dimensional vector space on a manifold M? This is a possible way. Let $n = \dim(M)$ and let $X_1, \ldots, X_n$ be n-vector fields globally defined on M, such that:

(i) $[X_i, X_j] = 0$ for all $i, j = 1, \ldots, n$.
(ii) $X_1, \ldots, X_n$ are *complete* and *linearly independent*.

The completeness is equivalent to the condition that for each i, the flow $\varphi^i(t, m) = \varphi^{X_i}(t, m) =: \mathbb{R} \times M \to M$ is defined for all $t \in \mathbb{R}$. On the other hand, the condition $[X_i, X_j] = 0$ for all i, j implies that the corresponding flows satisfy $\varphi^i \circ \varphi^j = \varphi^j \circ \varphi^i$. In this way, for each $(t_1, \ldots, t_n) \in \mathbb{R}^n$ and $m \in M$, we can define:

$$\varphi : \mathbb{R}^n \times M : \xrightarrow{\ \ ((t_1,\ldots,t_n),m) \rightsquigarrow \varphi^1_{t_1} \circ \cdots \circ \varphi^n_{t_n}(m)\ \ } M. \qquad (C.9)$$

A simple computation shows that the previous formula defines an action of $\mathbb{R}^n$ on M. In particular, note that: $\varphi^1_{t_1} \circ \cdots \circ \varphi^n_{t_n}(m) = \varphi^{\sigma(1)}_{t_{\sigma(1)}} \circ \cdots \circ \varphi^{\sigma(n)}_{t_{\sigma(n)}}(m)$ for every $\sigma \in \Sigma_n$, permutation of $1, \ldots, n$. Finally, since the vector fields $X_1, \ldots, X_n$ are linearly independent, using what we have discussed above, we can conclude that the action defined by Formula (C.9) is transitive. Summarizing:

Proposition C.59 *If M is a n-dimensional manifold and $X_1, \ldots, X_n$ are complete, commuting, and linearly independent vector fields, then M is diffeomorphic to the quotient of (a n-dimensional) vector space V by a discrete subgroup $\Lambda \subset V$. Such a diffeomorphism is not canonical, since it depends on the choice of a point $m \in M$.*

A *G*-manifold in general is not homogeneous, but it always *breaks up* into homogeneous pieces. In fact, given $n \in N$, let:

$$\mathscr{O}_n = \left\{ n' \in N \mid n' = \varphi_n(g), \; \exists\, g \in g \right\}.$$

Then:

Definition C.60 $\mathscr{O}_n$ is called the *G-orbit* of the point n. $\triangle$

Here we collect some elementary properties of the *G*-orbits.

Lemma C.61 *Every G-orbit is:*

(i) A connected (since G is connected by hypothesis), smooth, and immersed submanifold of N
(ii) Locally closed in N

Moreover, if $\mathscr{O}$ and $\mathscr{O}'$ are two orbits, then or they coincide, or they are disjoint.

1.5 Linear Actions

When $N = V$ is a vector space (which we assume finite dimensional) among the examples of *G*-actions, we find the *linear ones*. To give a linear action of *G* on *V* is equivalent to give a Lie group homomorphism $\varphi : G \to \mathrm{GL}(V)$ (here $GL(V)$ is the group of invertible endomorphisms of *V*).

Definition C.62 A linear action of *G* on *V* is (also) called a *representation* of *G* on the vector space *V*. In this case, we will say that *V* is a *G*-module. $\triangle$

Note that every linear *G*-action on *V* corresponds to a Lie algebra homomorphism $\varphi_{*,e} : g \to \mathrm{End}(V)$, intertwined by the exponential map (see item (iii) in Proposition C.36). More precisely, every representation $\rho : G \to \mathrm{GL}(V)$ corresponds to a representation of *G* on *V*, defined by: $\rho_{*,e} : g \to \mathrm{End}(V)$,

$$x \rightsquigarrow \frac{d}{dt}\bigg|_{t=0} \rho(\exp(tx))v := r(x)v \tag{C.10}$$

for all $x \in g$ and all $v \in V$. The linear map ρ satisfies $\rho([x, y]) = \rho(x) \circ \rho(y) - \rho(y) \circ \rho(x)$, for all $x, y \in g$, i.e., ρ is a *morphism* of Lie algebras (see Proposition C.42). We can finally introduce the following concept.

Definition C.63 Let g be a Lie algebra and let *V* be a vector space. *V* is called a (left) g-*module* if it carries a linear (left) g-action. This is equivalent to the existence of homomorphism of Lie algebras $\phi : g \to \mathrm{End}(V)$. The pair (ϕ, V) will be called a *representation* of g. $\triangle$

Remark C.64 Every representation (ρ, V) of the Lie group G defines a representation (ρ^*, V^*) via the formula

$$\langle \rho_g^* \xi, v \rangle = \langle \xi, \rho_{g^{-1}} v \rangle, \ \forall \ \in \xi \in V^* \text{ and } v \in V \text{ and } g \in G. \tag{C.11}$$

Similarly, every representation (r, V) of the Lie algebra $\mathfrak{g}$ induces a representation $(r^\sharp, V^*)$ via the following formula:

$$\left\langle r^\sharp(x)\xi, v \right\rangle = -\langle \xi, r(x)v \rangle, \ \forall \xi \in V^*, \ v \in V, \text{ and } x \in \mathfrak{g}. \tag{C.12}$$

Note that if (r, V) is the infinitesimal representation of $\mathfrak{g}$ associated with (ρ, V) of G (see Formula C.10), then:

$$r^\sharp(x)\xi = \left. \frac{d}{dt} \right|_{t=0} \rho^\sharp(\exp(tx))\xi, \ \forall \xi \in V^* \text{ and } x \in \mathfrak{g}. \tag{C.13}$$

$$\triangle$$

Example C.65 (Semi-direct Product)   If (V, ρ) is a representation of the group G,

$$(g_1, v_1)(g_2, v_2) = (g_1 g_2, v_1 + \rho_{g_1}(v_2)) \tag{C.14}$$

is a group law on $G \times V$. More precisely, it is an associative binary operations such that $(e, 0)(g, v) = (g, v) = (e, 0)(g, v)$, $(g^{-1}, -\rho_{g^{-1}}(v)(g, v) = (e, 0) = (g, v)(g^{-1}, -\rho_{g^{-1}}(v)$ for all $(g, v) \in G \times V$. Then $G \times V$ with the multiplication defined in (C.14) is called the semi-direct product of G with V, and it is denoted with $G \ltimes_\rho V$. If G is a Lie group, then $G \ltimes V$ is also a Lie group, equipping V and $G \times V$ with their obvious smooth manifold structures. Analogously, if (V, r) is a representation of a Lie algebra $\mathfrak{g}$, then

$$\left[(x_1, v_1), (x_2, v_2) \right] = \left([x_1, x_2], r_{x_1}(v_2) - r_{x_2}(v_1) \right)$$

is a Lie bracket on $\mathfrak{g} \times V$, which, in this way, becomes a Lie algebra called the semi-direct product of $\mathfrak{g}$ with V and indicated with $\mathfrak{g} \ltimes_r V$. Note that the Lie algebra of $G \ltimes_\rho V$ is $\mathfrak{g} \ltimes_r V$, where (V, r) is the infinitesimal representation of $\mathfrak{g}$ associated with (V, ρ). $\triangle$

Let us consider in more detail the following important case. Suppose G is a Lie group and $\mathfrak{g}$ the corresponding Lie algebra. For all $g \in G$, we can define the map

$$c_g = R_{g^{-1}} \circ L_g = L_g \circ R_{g^{-1}}. \tag{C.15}$$

This map is a *diffeomorphism* (since it is composition of two diffeomorphisms) and *fixes* the identity e, i.e., $c_g(e) = e$ (this is true for all $g \in G$). Then $(c_g)_{*,e} \in \mathrm{GL}(\mathfrak{g})$

for all $g \in G$. In this way, we have defined a map from G to $\mathrm{GL}(\mathfrak{g})$ such that $g \rightsquigarrow (c_g)_{*,e}$. If we call:

$$(c_g)_{*,e} = \mathrm{Ad}_g \tag{C.16}$$

then we have:

Lemma C.66 (Adjoint Representation and Adjoint Action) *The map* Ad : $G \to \mathrm{GL}(\mathfrak{g})$ *is a Lie group homomorphism. Such a map is called the adjoint representation of G, while the induced map* $\mathrm{Ad}_{(\cdot)}(\cdot) : G \times \mathfrak{g} \to \mathfrak{g}$ *is called the adjoint action of G. Note that it is a left action.*

The differential at the identity e of Ad is the linear map $\mathrm{Ad}_{*,e} : \mathfrak{g} \to \mathrm{End}(\mathfrak{g})$, which is a Lie algebra homomorphism (see Theorem C.31).

Definition C.67   The map $\mathrm{Ad}_{*,e} : \mathfrak{g} \to \mathrm{End}(\mathfrak{g})$ is called the *adjoint representation* of $\mathfrak{g}$ on $\mathfrak{g}$, and it will be denoted with ad, i.e.,

$$\mathrm{ad} = \mathrm{Ad}_{*,e} .$$

$\triangle$

In particular:

Proposition C.68   *Given $x \in \mathfrak{g}$* $\mathrm{ad}_x(y) = [x, y]$ *for all $y \in \mathfrak{g}$.*

Remark C.69   The induced map $\mathrm{ad}_{(\cdot)}(\cdot) : \mathfrak{g} \times \mathfrak{g} \to \mathfrak{g}$ is called the *adjoint action* (note that this is the *infinitesimal action* associated with the adjoint action of G) $\triangle$

The adjoint representations of G and $\mathfrak{g}$ induce a representation of G and, respectively, $\mathfrak{g}$ on $\mathfrak{g}^*$ (see Formulas C.11 and C.12). More precisely, let us define $\mathrm{Ad}^\sharp : G \to \mathrm{GL}(\mathfrak{g}^*)$ and $\mathrm{ad}^\sharp : \mathfrak{g} \to \mathrm{End}(\mathfrak{g}^*)$ as follows:

$$\langle \mathrm{Ad}_g^\sharp \, \alpha, x \rangle = \langle \alpha, \mathrm{Ad}_{g^{-1}} x \rangle \text{ and } \langle \mathrm{ad}_x^\sharp \, \alpha, y \rangle = -\langle \alpha, \mathrm{ad}_x y \rangle,$$

for all $\alpha \in \mathfrak{g}^*$ and all $x, y \in \mathfrak{g}$.

Definition C.70 (Coadjoint Representation)   The map $\mathrm{Ad}^\sharp$ is called the *coadjoint representation* of G, while $\mathrm{ad}^\sharp$ is called the coadjoint representation of $\mathfrak{g}$. $\triangle$

Remark C.71   (i) It is a simple computation to show that:

$$\mathrm{ad}_x^\sharp \, \alpha = \left. \frac{d}{dt} \right|_{t=0} \mathrm{Ad}_{\exp(tx)}^\sharp \, \alpha,$$

for all $x \in \mathfrak{g}$ and $\alpha \in \mathfrak{g}^*$.

(ii) Note that the induced map $\mathrm{Ad}_{(\cdot)}^\sharp(\cdot) : G \times \mathfrak{g}^* \to \mathfrak{g}^*$ is a (linear) *left action*, which is called the *coadjoint action* of G. The corresponding infinitesimal $\mathfrak{g}$-

action is defined by the map $\mathrm{ad}^{\sharp}_{(\cdot)}(\cdot) : \mathfrak{g} \times \mathfrak{g}^* \to \mathfrak{g}^*$, which is called the *coadjoint action*.

$\triangle$

Example C.72 To the adjoint and the coadjoint representations are associated Lie groups $G \ltimes_{\mathrm{ad}} \mathfrak{g}$ and $G \ltimes_{\mathrm{ad}^\sharp} \mathfrak{g}^*$, the semi-direct products of G with $\mathfrak{g}$ and $\mathfrak{g}^*$ defined by the correspondent representations. $\triangle$

Example C.73 ($\mathrm{SO}_3(\mathbb{R})$**-Orbits 1**) Let $G = \mathrm{SO}_3(\mathbb{R})$ act on $V = \mathbb{R}^3$ on the left via the usual multiplication rows-by-columns. For every v, the $\mathrm{SO}_3(\mathbb{R})$-orbit of v is a sphere centered in the origin of $\mathbb{R}^3$ with radius equal to $\|v\|$, the length of the vector v (if $v = 0$, the v-orbit coincides with the vector itself). If $v \neq 0$, the stabilizer $G_v \subset \mathrm{SO}_3(\mathbb{R})$ of v is isomorphic to the $\mathrm{SO}_2(\mathbb{R})$ and coincide with all the rotation whose axis in $\mathbb{R}^3$ is spanned by the same vector v. On the other hand, if $v = 0$, then the stabilizer $G_0 \subset \mathrm{SO}_3(\mathbb{R})$ coincides with $\mathrm{SO}_3(\mathbb{R})$. $\triangle$

Definition C.74 The G-orbits for the adjoint and coadjoint action of G are called *adjoint* and, respectively, *coadjoint* orbits. $\triangle$

Example C.75 ($\mathrm{SO}_3(\mathbb{R})$**-Orbits 2**) Let us consider $\mathrm{SO}_3(\mathbb{R})$ and let us see how the coadjoint orbits look like in this case (see Example C.11 in Appendix C). In that example, we learned that there is an isomorphism of Lie algebras between $(\mathfrak{so}_3(\mathbb{R}), [\cdot, \cdot])$ and $(\mathbb{R}^3, \times)$ (see Formula C.1). Moreover, we proved that given $v, u \in \mathbb{R}^3$, then $A_v u = v \times u$, and endowing $\mathbb{R}^3$ with the usual Euclidean product and $\mathfrak{so}_3(\mathbb{R})$ with the trace form $(A | B) = -\frac{1}{2} \mathrm{tr}(AB)$, then the map $v \rightsquigarrow A_v$ is an isometry. We prove now that $A_{gv} = \mathrm{Ad}_g A_v$ for all $g \in \mathrm{SO}_3(\mathbb{R})$ and $v \in \mathbb{R}^3$. Here $\mathrm{Ad}_g A_v = g A_v g^{-1}$ is the adjoint action of G on its Lie algebra $\mathfrak{so}_3(\mathbb{R})$. In fact

$$\left(g A_v g^{-1} u \right) \cdot w = \left(A_v g^{-1} u \right) \cdot \left(g^{-1} w \right) = \left(v \times (g^{-1} u) \right) \cdot g^{-1} w$$
$$= \left(g^{-1} u \times g^{-1} w \right) \cdot v = \left(g^{-1}(u \times w) \right) \cdot v$$

Then

$$\left(g A_v g^{-1} u \right) \cdot w = \left(g^{-1}(u \times w) \right) \cdot v = \left((gv) \times u \right) \cdot w = A_{gv} u \cdot w,$$

for every u, w in $\mathbb{R}^3$ and $g \in \mathrm{SO}_3(\mathbb{R})$. From this result, it follows that the $\mathrm{SO}_3(\mathbb{R})$-adjoint orbits are two-dimensional spheres centered in the origin of $\mathbb{R}^3$.

Using the isomorphisms $\mathbb{R}^3 \simeq \mathfrak{so}_3(\mathbb{R})$ (defined by Formula (C.1)), $\mathbb{R}^3 \simeq (\mathbb{R}^3)^*$ (induced by the Euclidean scalar product), and $\mathfrak{so}_3(\mathbb{R}) \simeq \mathfrak{so}_3(\mathbb{R})^*$ (defined via the bilinear form (C.2)), it follows that the orbits of the coadjoint action of $\mathrm{SO}_3(\mathbb{R})$ on $\mathfrak{so}_3(\mathbb{R})^*$ can be identified with the adjoint orbits, i.e., they are spheres centered in the origin of $\mathbb{R}^3$. Note that identifying $\alpha \in \mathfrak{so}_3(\mathbb{R})$, $\alpha \neq 0$, with a non-zero vector in $\mathbb{R}^3$, the group stabilizer in $\mathrm{SO}_3(\mathbb{R})$ of this point turns out to be isomorphic to $S^1 \simeq \mathrm{SO}_2(\mathbb{R})$. In this way, we get that $\mathscr{O}_\alpha \simeq \mathrm{SO}_3(\mathbb{R})/\mathrm{SO}_2(\mathbb{R})$, which is the *usual* description of S^2 as a homogeneous space. $\triangle$

Example C.76 (Coadjoint Action of Semi-direct Products) Let G be a *matrix* Lie group, $\mathfrak{g}$ its Lie algebra, and (V, ρ) a representation of G. Let (V, r) be the corresponding infinitesimal representation of its Lie algebra $\mathfrak{g}$. Under the assumption made on G, $G \ltimes_\rho V$ and $\mathfrak{g} \ltimes_r V$ embed into $\mathrm{GL}_{n+1}(\mathbb{K})$ as follows:

$$(g, v) \rightsquigarrow \begin{pmatrix} g & v \\ 0 & 1 \end{pmatrix} \quad \text{and} \quad (x, v) \rightsquigarrow \begin{pmatrix} x & v \\ 0 & 0 \end{pmatrix}. \tag{C.17}$$

In these representations, g and x occupy n $n \times n$-block, v is a column n-vector, and the 0 in the first column and second row represents a n-vector whose entries are all zeros. We will now give an explicit formula for the coadjoint action of $G \ltimes_\rho V$ on the dual of its Lie algebra, which will be identified with $\mathfrak{g}^* \oplus V^*$ (see [114] and [120]). To this end, first, let $\langle \cdot, \cdot \rangle : \mathfrak{g}^* \oplus V^* \times \mathfrak{g} \ltimes_r V \to \mathbb{K}$ be defined by

$$\langle (\alpha, p), (x, v) \rangle = \langle \alpha, x \rangle + \langle p, v \rangle,$$

where the two brackets in the right-hand side of the previous formula denote the pairings between $\mathfrak{g}$ and $\mathfrak{g}^*$ and, respectively, V and V^*. Using the representations in (C.17) and recalling that for matrix Lie group $\mathrm{Ad}_g\, x = gxg^{-1}$, for all $g \in G$ and $x \in \mathfrak{g}$, one arrives to the following formula:

$$\left\langle \mathrm{Ad}^\sharp_{(g,v)}(\alpha, p), (x, u) \right\rangle = \left\langle (\alpha, p), \mathrm{Ad}_{(g,v)^{-1}}(x, u) \right\rangle$$

$$= \left\langle (\alpha, p), \left(\mathrm{Ad}_{g^{-1}} x, \rho_{g^{-1}}(r_x(v)) + \rho_{g^{-1}}(u) \right) \right\rangle.$$

The sought formula is obtained from the rightmost term of the previous formula, i.e.,

$$\left\langle \mathrm{Ad}^\sharp_{(g,v)}(\alpha, p), (x, u) \right\rangle = \left\langle \left(\mathrm{Ad}^\sharp_g \alpha + (\rho^\sharp_g p) \diamond v, \rho^\sharp_g(p) \right), (x, u) \right\rangle, \quad \forall (x, u) \in \mathfrak{g} \ltimes_r V,$$

where for all $(p, v) \in V \oplus V^*$, where $p \diamond v$ is the element in $\mathfrak{g}^*$ defined by

$$\langle p \diamond v, x \rangle = \langle p, r_x(v) \rangle, \quad \forall x \in \mathfrak{g}.$$

In other words,

$$\mathrm{Ad}^\sharp_{(g,v)}(\alpha, p) = \left(\mathrm{Ad}^\sharp_g \alpha + (\rho^\sharp_g p) \diamond v, \rho^\sharp_g(p) \right),$$

$\forall (\alpha, p) \in \mathfrak{g}^* \oplus V^*$, $(g, v) \in G \ltimes_\rho V$. $\qquad\qquad\qquad\qquad\qquad\qquad\triangle$

In the next subsection, we present the notions about Lie group and Lie algebra cohomology we used in this work. We start with the *infinitesimal aspects*, i.e., with the Lie algebra cohomology, then we will go to the Lie group cohomology, and finally, we will make some remarks about the relations between the two.

1.6 Lie Algebra Cohomology

In what follows, unless explicitly stated, all G and $\mathfrak{g}$-actions will be assumed to be left actions.

Let V be a $\mathfrak{g}$-module and let (V, ϕ) be the corresponding representation of $\mathfrak{g}$ (see C.63). For all $n \geq 0$, let us consider the vector space of the *k-cochains* $\mathscr{C}^n(\mathfrak{g}, V) = \mathrm{Hom}_{\mathbb{K}}(\Lambda^n \mathfrak{g}, V)$ and let $\mathscr{C}^\bullet(\mathfrak{g}, V) = \bigoplus_{n \geq 0} \mathscr{C}^n(\mathfrak{g}, V)$. For all $x \in \mathfrak{g}$ and $v \in V$, let us write $x.v$ instead of $\phi(x)(v)$ and define the map $\delta : \mathscr{C}^n(\mathfrak{g}, V) \to \mathscr{C}^{n+1}(\mathfrak{g}, V)$ via:

$$(\delta f)(x_1, \cdots, x_{n+1}) = \sum_{i=1}^{n+1} (-1)^{i-1} x_i.f\left(x_1, \ldots, \hat{x}_i, \ldots, x_{n+1}\right)$$

$$+ \sum_{1 \leq i < j \leq n+1} (-1)^{i+j} f\left([x_i, x_j], x_1, \ldots, \hat{x}_i, \ldots, \hat{x}_j, \ldots, x_{n+1}\right)$$

Lemma C.77 *The map δ squares to zero, i.e., $\delta^2 = 0$.*

Definition C.78 The pair $(\mathscr{C}^\bullet(\mathfrak{g}, V), \delta)$ is called the *complex of cochains of $\mathfrak{g}$ with values in V*, commonly known as the *Cartan-Eilenberg* complex of $\mathfrak{g}$ with values in V. Define the vector spaces:

(i) $\mathscr{Z}^k(\mathfrak{g}, V) = \ker(\delta : \mathscr{C}^k(\mathfrak{g}, V) \to \mathscr{C}^{k+1}(\mathfrak{g}, V))$ *group of k-cocycles* of $\mathfrak{g}$ with values in V.

(ii) $\mathscr{B}^k(\mathfrak{g}, V) = \mathrm{im}(\delta : \mathscr{C}^{k-1}(\mathfrak{g}, V) \to \mathscr{C}^k(\mathfrak{g}, V))$ *group of k–coboundaries* of $\mathfrak{g}$ with values in V.

(iii) $H^k(\mathfrak{g}, V) = \dfrac{\mathscr{Z}^k(\mathfrak{g}, V)}{\mathscr{B}^k(\mathfrak{g}, V)}$ *k–cohomology group* of $\mathfrak{g}$ with values in V.

Example C.79 (Low-Degree Cases) In low degree, we find:

(i) $\mathscr{C}^0(\mathfrak{g}, V) \simeq V$, and if f is a zero-cochain, $(\delta f)(x) = x.f$ so that $\mathscr{Z}^0(\mathfrak{g}, V) = \mathrm{Inv}(V)$, i.e., the space of all elements in V on which $\mathfrak{g}$ acts *trivially*. On the other hand, since $\mathscr{B}^0(\mathfrak{g}, V) = \{0\}$, we conclude that $H^0(\mathfrak{g}, V) = \mathrm{Inv}(V)$.

(ii) If $f \in \mathscr{C}^1(\mathfrak{g}, V)$, then $(\delta f)(x, y) = x.f(y) - y.f(x) - f([x, y])$ for all $x, y \in \mathfrak{g}$. Then: $f \in \mathscr{C}^1(\mathfrak{g}, V)$ if and only if $x.f(y) - y.f(x) = f([x, y])$ for all $x, y \in \mathfrak{g}$.

(iii) Given $f \in \mathscr{C}^2(\mathfrak{g}, V)$:

$$(\delta f)(x, y, z) = x.f(y, z) - y.f(x, z) + z.f(x, y)$$

$$- f([x, y], z) + f([x, z], y) - f([y, z], x),$$

forall $x, y, z \in \mathfrak{g}$.

△

Example C.80 An important class of $\mathfrak{g}$-modules is offered by the *homogeneous* vector subspaces of the tensor algebra of $\mathfrak{g}$. On those subspaces, $x \in \mathfrak{g}$ acts as derivation:

$$\mathrm{ad}_x(x_1 \otimes \cdots \otimes x_n) = \sum_{i=1}^{n} x_1 \otimes \cdots \otimes \mathrm{ad}_x(x_i) \otimes \cdots \otimes x_n$$

i.e., extending the *adjoint representation* to a homomorphism of Lie algebras between $\mathfrak{g}$ and the Lie algebra of derivations of (each of the homogeneous components of) $T_{\mathfrak{g}}$. The cases that play some role in these notes are:

(i) $V = \mathbb{K}$. In this case, the adjoint action is the *trivial action*, i.e., $x.\alpha = \mathrm{ad}_x(\alpha) = 0$ for all $x \in \mathfrak{g}$ and $\alpha \in \mathbb{K}$, so that Formula (C.18) reduces to

$$(\delta f)(x_1, \cdots, x_{n+1})$$

$$= \sum_{1 \le i < j \le n+1}^{n+1} (-1)^{i+j} f\left([x_i, x_j], x_1, \ldots, \hat{x}_i, \ldots, \hat{x}_j, \ldots, x_{n+1}\right). \qquad (C.18)$$

If $\mathfrak{g}$ is Abelian, then $\delta = 0$ and $H^k(\mathfrak{g}, \mathbb{K}) = \Lambda^k \mathfrak{g}$ for all k. On the other hand, the general case of the low degrees goes as follows:

(a) $H^0(\mathfrak{g}, \mathbb{K}) = \mathbb{K}$ (see item (i) of Example C.79).

(b) $H^1(\mathfrak{g}, \mathbb{K}) = \dfrac{\mathfrak{g}}{[\mathfrak{g}, \mathfrak{g}]}$ (see item (ii) of Example C.79).

(c) $H^2(\mathfrak{g}, \mathbb{K})$. In this case $f \in \mathscr{Z}^2(\mathfrak{g}, \mathbb{K})$ if and only if

$$(\delta f)(x, y, z) = f([y, x], z) + f([x, z], y) + f([z, y], x)$$

$$= 0, \ \forall x, y, z \in \mathfrak{g} \qquad (C.19)$$

(see point (iii) of Example C.79).

The relevance of the group $H^2(\mathfrak{g}, \mathbb{K})$ comes from the following observation: There a is one-to-one correspondence between $H^2(\mathfrak{g}, \mathbb{K})$ and the *classes of equivalence* of *Lie algebra extensions* of the type:

$$0 \longrightarrow \mathbb{K} \longrightarrow \widetilde{\mathfrak{g}} \longrightarrow \mathfrak{g} \longrightarrow 0 \qquad (C.20)$$

As a vector space, $\widetilde{\mathfrak{g}} \simeq \mathfrak{g} \oplus \mathbb{K}$, while the Lie bracket is given as:

$$[x \oplus a, y \oplus b] = [x, y] \oplus f(x, y) \qquad (C.21)$$

where $f \in \mathscr{Z}^2(\mathfrak{g}, K)$. The condition $\delta f = 0$ implies that (C.21) satisfies the Jacobi identity. Extensions of the type (C.20) are called *central*. This observation can be generalized as follows. Let $\mathfrak{g}$ be a Lie algebra and $\mathfrak{h}$ another

Lie algebra that is also a $\mathfrak{g}$-module (see Definition C.63). To the direct sum of vector spaces $\mathfrak{g} \oplus \mathfrak{h}$, one can give a straightforward structure of Lie algebra by letting $[(x_1, y_1), (x_2, y_2)] = ([x_1, x_2], [y_1, y_2])$. But suppose one is looking for nontrivial Lie algebra structures, i.e., *nontrivial extensions of* $\mathfrak{g}$ *by* $\mathfrak{h}$, by exploiting the $\mathfrak{g}$-module structure of $\mathfrak{h}$. Recall that the extensions of $\mathfrak{g}$ by $\mathfrak{h}$ are described, and actually classified, by the group $H^2(\mathfrak{g}, \mathfrak{h})$. More precisely let $c \in \mathscr{Z}^2(\mathfrak{g}, \mathfrak{h})$, i.e., an element in $\mathscr{C}^2(\mathfrak{g}, \mathfrak{h})$ that satisfies the (cocycle) condition $(\delta c)(x_1, x_2, x_3) = 0$, i.e.,

$$x_1.c(x_2, x_3) - c([x_1, x_2], x_3) + \text{ciclic permutations in } x_1, x_2, x_3 = 0,$$

for all $x_1, x_2, x_3 \in \mathfrak{g}$ (see Formula C.18). One can check that by letting

$$[(x_1, y_1), (x_2, y_2)] = \big([x_1, x_2], x_2.y_1 - x_1.y_2 + [y_1, y_2] + c(x_1, x_2)\big),$$

one defines a Lie algebra structure in $\widetilde{g} = \mathfrak{g} \oplus \mathfrak{h}$. This is called an *extension of* $\mathfrak{g}$ *by* $\mathfrak{h}$. The reader can check that

$$0 \to \mathfrak{h} \to \widetilde{g} \to \mathfrak{g} \to 0$$

is an exact sequence of Lie algebras.

(ii) $V = \mathfrak{g}^n = \underbrace{\mathfrak{g} \otimes \cdots \otimes \mathfrak{g}}_{n-\text{times}}$. V is a $\mathfrak{g}$-module with respect to the *adjoint action*.

More precisely, the $\mathfrak{g}$-action is defined by:

$$\mathrm{ad}_x(x_1 \otimes \cdots \otimes \cdots x_n) = \sum_{i=1}^{n} x_1 \otimes \cdots \otimes \mathrm{ad}_x x_i \cdots \otimes x_n.$$

The case $n = 2$ is particularly relevant for us, and we observe that the decomposition $\mathfrak{g} \otimes \mathfrak{g} \simeq \mathfrak{g} \odot \mathfrak{g} \oplus \Lambda^2\mathfrak{g}$ in *symmetric* and *skew-symmetric* parts is compatible with the $\mathfrak{g}$-module structure. (Here $\odot$ is the symmetrized tensor product, i.e., $x \odot y = \frac{1}{2}x \otimes y + \frac{1}{2}y \otimes x$.)

$\triangle$

Theorem C.81 (Whitehead's Lemmas) *If $\mathfrak{g}$ is a semi-simple Lie algebra and M is a $\mathfrak{g}$-module, then $H^i(\mathfrak{g}, M) = 0$ for $i = 1, 2$.*

Remark C.82 This result is commonly known *the first Whitehead's lemma* when $i = 1$ and *the second Whitehead's lemma* when $i = 2$. $\triangle$

1.7 *Universal Enveloping Algebras*

We refer the reader to Example C.8 for all the relevant definitions about associative Lie algebras. Let $T_{\mathfrak{g}} = (T^{\bullet}\mathfrak{g}, \otimes)$ be the tensor algebra of $\mathfrak{g}$. This is a *graded, associative*, but *non-commutative* algebra, whose homogeneous subspace $\mathfrak{g}^n = \underbrace{\mathfrak{g} \otimes \cdots \otimes \mathfrak{g}}_{n-\text{times}}$ of degree n is generated (as vector space) by the monomial of the type $x_{i_1} \otimes \cdots \otimes x_{i_n}$. Consider the *two-sides ideal*

$$I = T_{\mathfrak{g}}\big(x \otimes y - y \otimes x - [x, y]\big)T_{\mathfrak{g}}.$$

Definition C.83 (Universal Enveloping Algebra) The *universal enveloping algebra* $\mathscr{U}(\mathfrak{g})$ of $\mathfrak{g}$ is the algebra is $\dfrac{T_{\mathfrak{g}}}{I}$ whose product is induced on the quotient vector space by $\otimes$, i.e., if $\overline{X}, \overline{Y} \in \mathscr{U}(\mathfrak{g})$ are the classes of the monomials $X \in \mathfrak{g}^k$ and $Y \in \mathfrak{g}^l$, then $\overline{X} \cdot \overline{Y}$ is the class of the monomial $X \otimes Y \in \mathfrak{g}^{k+l}$.

Remark C.84 Some observations are now in order:

(i) We observe that $\mathscr{U}(\mathfrak{g})$ is a *unital* and *associative algebra*. Since the ideal I is non-homogeneous, $\mathscr{U}(\mathfrak{g})$ is *not* a graded algebra. Nevertheless, it is a *filtered algebra*, endowed with the filtration $\mathbb{K} = \mathscr{U}_0(\mathfrak{g}) \subset \mathscr{U}_1(\mathfrak{g}) \subset \cdots \subset \mathscr{U}_n(\mathfrak{g}) \subset \cdots$ where $\mathscr{U}_n(\mathfrak{g})$ is the subspace of $\mathscr{U}(\mathfrak{g})$ generated by *monomial of length at most n*, i.e., monomial like $x_{i_1} \cdot x_{i_n}$ with $x_{i_1}, \ldots, x_{i_n} \in \mathfrak{g}$. Note that $\mathscr{U}_i(\mathfrak{g}) \cdot \mathscr{U}_j(\mathfrak{g}) \subset \mathscr{U}_{i+j}(\mathfrak{g})$ for all $i, j \geq 0$ and that $\mathscr{U}(\mathfrak{g}) = \cup_{k \geq 0}\mathscr{U}_k(\mathfrak{g})$, and observe that $\mathscr{U}_1(\mathfrak{g}) \simeq \mathfrak{g}$ so that there is a *natural* homomorphism of Lie algebra $i : \mathfrak{g} \to \mathscr{U}(\mathfrak{g})_{\text{Lie}}$.

(ii) The adjective universal is due to the fact that $\mathscr{U}(\mathfrak{g})$ has the following property: suppose that $\mathscr{A}$ is an associative algebra and that $j : \mathfrak{g} \to \mathscr{A}_{\text{Lie}}$ is a morphism of Lie algebras. Then there is a *unique* morphism of unital and associative algebras $\phi : \mathscr{U}(\mathfrak{g}) \to \mathscr{A}$, which makes the following diagram into a commutative diagram of Lie algebras:

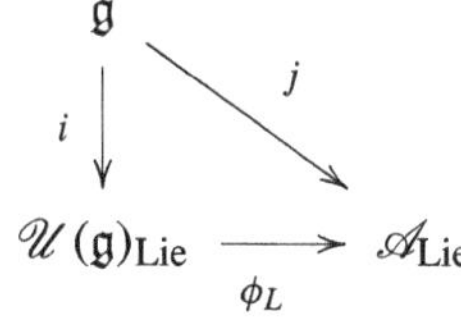

$\triangle$

The *graded algebra* associated with $\mathscr{U}(\mathfrak{g})$ is:

$$\mathrm{Gr}(\mathscr{U}(\mathfrak{g})) = \bigoplus_{k \geq 0} \frac{\mathscr{U}_k(\mathfrak{g})}{\mathscr{U}_{k-1}(\mathfrak{g})}, \quad \mathscr{U}_{-1}(\mathfrak{g}) = \{0\}.$$

Here we note that:

(i) $\mathfrak{g} \simeq \dfrac{\mathscr{U}_1(\mathfrak{g})}{\mathscr{U}_0(\mathfrak{g})}$ so that there exists a linear map $i : \mathfrak{g} \to \mathrm{Gr}(\mathscr{U}(\mathfrak{g}))$.

(ii) $\mathrm{Gr}(\mathscr{U}(\mathfrak{g}))$ endowed with its obvious multiplication is a *commutative algebra*. In fact, if for every $k \geq 0$, $x_{i_1} \cdots x_{i_k} - x_{\sigma(i_1)} \cdots x_{\sigma(i_k)} \in \mathscr{U}_{k-1}(\mathfrak{g})$, for all $\sigma \in \Sigma_k$, the permutation group on k elements. This is clear when σ is a transposition. The statement in the case of a general permutation follows now from the fact that every permutation is the product of transpositions.

On the other hand, the *symmetric algebra* $S_{\mathfrak{g}}$ of $\mathfrak{g}$ is $\dfrac{T_{\mathfrak{g}}}{J}$, where J is the two-side ideal $T_{\mathfrak{g}}\left(x \otimes y - y \otimes x\right)T_{\mathfrak{g}}$. $S_{\mathfrak{g}}$ is a *commutative algebra* such that:

(i) There exists a natural *injective* linear map $j : \mathfrak{g} \to S_{\mathfrak{g}}$.

(ii) $S_{\mathfrak{g}}$ has the following *universal property*. Given a commutative algebra $\mathscr{C}$ and a linear map $f : \mathfrak{g} \to \mathscr{C}$, there is a unique map of commutative algebras $\phi : S_{\mathfrak{g}} \to \mathscr{C}$ that closes the following to a commutative diagram:

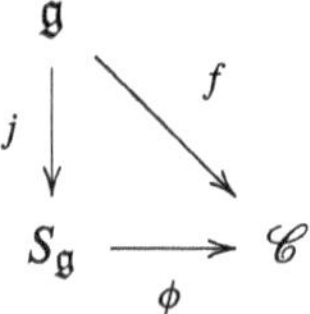

Remark C.85 (Exterior Algebra of $\mathfrak{g}$)   Together with the symmetric algebra, we can introduce the *exterior algebra* $\Lambda^\bullet \mathfrak{g}$ of $\mathfrak{g}$. This can be defined as the associative algebra obtained by taking the quotient of the tensor algebra $T_{\mathfrak{g}}$ by the two-side ideal $J = T_{\mathfrak{g}}(x \otimes y + y \otimes x)T_{\mathfrak{g}}$. The exterior algebra $\Lambda^\bullet \mathfrak{g}$ is finite dimensional and graded, $\Lambda^\bullet \mathfrak{g} \simeq \oplus_{i=1}^{\dim \mathfrak{g}} \Lambda^i \mathfrak{g}$, where $\Lambda^i \mathfrak{g}$ is a vector space of dimension $\binom{\dim \mathfrak{g}}{i}$ generated by the monomials of the type $\xi = x_{k_1} \wedge \cdots \wedge x_{k_i}$. Note that each of the ξ is totally skew-symmetric, meaning that $x_{\sigma(k_1)} \wedge \cdots x_{\sigma(k_i)} = (-1)^{|\sigma|} x_{k_1} \wedge \cdots \wedge x_{k_i}$, where $\sigma \in \Sigma_i$ and $|\sigma|$ is its signature. Note that also $\Lambda^\bullet \mathfrak{g}$ can be characterized by a universal property. $\triangle$

After these preliminary remarks, we can state the following important result. Let $\mathscr{C} = \mathrm{Gr}(\mathscr{U}(\mathfrak{g}))$ and let $f = i : \mathfrak{g} \to \mathrm{Gr}(\mathscr{U}(\mathfrak{g}))$. Then:

Theorem C.86 (Poincaré-Birkhoff-Witt) *The corresponding map* $\phi : S_{\mathfrak{g}} \to \mathrm{Gr}(\mathscr{U}(\mathfrak{g}))$ *in the above diagram is an isomorphism of commutative algebras.*

An equivalent statement is the following one:

Theorem C.87 *Choose a basis* $x_1, \ldots, x_n$ *of* $\mathfrak{g}$. *Then the ordered monomials* $x_1^{\alpha_1} \cdots x_n^{\alpha_n}$ *with* $(\alpha_1, \ldots \alpha_n) \in \mathbb{Z}_+^n$ *is a basis of* $\mathscr{U}(\mathfrak{g})$.

Observe that when $\mathfrak{g}$ is Abelian, $\mathscr{U}(\mathfrak{g}) \simeq S_{\mathfrak{g}}$, while for general $\mathfrak{g}$, the Poincaré-Birkhoff-Witt theorem tells us that we still have an isomorphism $\mathscr{U}(\mathfrak{g}) \simeq S_{\mathfrak{g}}$ but *only* at the level of vector spaces.

The universal enveloping algebra is an example of a *quasi-commutative algebra*, which is an associative, unital filtered algebra $\mathscr{A}$, whose associated graded algebra $\mathrm{Gr}(\mathscr{A})$ is commutative. We state the following interesting result.

Proposition C.88 (Quasi-commutative vs Poisson Algebras) *Let $\mathscr{A}$ be a quasi-commutative algebra. Then* $\mathrm{Gr}(\mathscr{A})$ *is a Poisson algebra.*

Proof To define the Poisson bracket on $\mathrm{Gr}(\mathscr{A})$, it suffices to define it on the homogeneous components of the associated algebra. To this end, let:

$$\{\cdot,\cdot\} : \frac{\mathscr{A}_i}{\mathscr{A}_{i-1}} \times \frac{\mathscr{A}_j}{\mathscr{A}_{j-1}} \to \frac{\mathscr{A}_{i+j-1}}{\mathscr{A}_{i+j-2}}, \quad (\overline{x},\overline{y}) \rightsquigarrow (xy - yx) \bmod \mathscr{A}_{i+j-2} \qquad \text{(C.22)}$$

where $x \in \mathscr{A}_i$ and $y \in \mathscr{A}_j$ are two lifts of $\overline{x}$ and, respectively, $\overline{y}$. The proof of the proposition follows at once after showing that such a bracket is well defined, in particular that given x, y as above $xy - yx \in \mathscr{A}_{i+j-1}$ and that the result does not depend on the choice of the two lifts. We leave to the reader to check the Leibniz identity. Given that the proof that the above bracket is Poisson follows from the fact that $\mathscr{A}$ is a associative algebra. $\qquad\square$

In particular, given a Lie algebra $\mathfrak{g}$, the graded algebra associated with $\mathscr{U}(\mathfrak{g})$ is a Poisson algebra. Thanks to the Poincaré-Birkhoff-Witt theorem, such a Poisson structure can be transferred to the symmetric algebra $S_{\mathfrak{g}}$. We conclude this discussion with the following proposition:

Proposition C.89 *The Poisson bracket induced on $S_{\mathfrak{g}}$ by the one defined on* $\mathrm{Gr}(\mathscr{U}(\mathfrak{g}))$ *coincides with the linear Poisson structure.*

Proof It enough to check this statement on the restriction of (C.22) on the components of degree equal to one:

$$\{\cdot,\cdot\} : \frac{\mathscr{U}_1(\mathfrak{g})}{\mathscr{U}_0(\mathfrak{g})} \times \frac{\mathscr{U}_1(\mathfrak{g})}{\mathscr{U}_0(\mathfrak{g})} \to \frac{\mathscr{U}_1(\mathfrak{g})}{\mathscr{U}_0(\mathfrak{g})}, \quad (\overline{x},\overline{y}) \rightsquigarrow (xy - yx) \bmod \mathscr{U}_0(\mathfrak{g})$$

which shows that $\{\overline{x},\overline{y}\} = \overline{[x,y]}$, which proves what is requested (remember that $\dfrac{\mathscr{U}_1(\mathfrak{g})}{\mathscr{U}_0(\mathfrak{g})} \simeq \mathfrak{g}$ and that $\mathscr{U}_0(\mathfrak{g}) \simeq \mathbb{K}$). $\qquad\square$

1.8   Lie Group Cohomology

We introduce now the relevant (for our goals) Lie group cohomology. This will be done in complete analogy with the case of the Lie algebra cohomology we introduced above. In particular, given a Lie group G we will say that the vector space V is a G-module if V carries a G action, i.e., if there is a map $\Phi : G \to \mathrm{GL}(V)$ that is a Lie group homomorphism (see C.62). Given $g \in G$ and $v \in V$, we will write $g.v$ to denote $\Phi(g)(v)$. For each $k \geq 0$, we define $\mathscr{C}^k(G^k, V)$ as the space of all smooth maps $f : G^k = \underbrace{G \times \cdots \times G}_{k--\text{times}} \to V$. Such a space will be called the space of the *smooth k-cochains* of G with values in V.

We defined now the map $\delta : \mathscr{C}^k(G, V) \to \mathscr{C}^{k+1}(G, V)$ via

$$(\delta f)(g_1, \ldots, g_{k+1}) = g_1.f(g_2, \ldots, g_{k+1}) + \sum_{i=1}^{k}(-1)^i f(g_1, \ldots, g_i g_{i+1}, \ldots, g_{k+1})$$

$$+ (-1)^{k+1} f(g_1, \ldots, g_k)$$

Lemma C.90 *The map δ squares to zero, i.e., $\delta^2 = 0$.*

Definition C.91 The pair $(\mathscr{C}^\bullet(G, V), \delta)$ is called the *complex of the smooth cochains of G with values in V*. Define the vector spaces:

(i) $\mathscr{Z}^k(G, V) = \ker(\delta : \mathscr{C}^k(G, M) \to \mathscr{C}^{k+1}(G, V))$ *space of k-cocycles* of G with values in V.
(ii) $\mathscr{B}^k(G, V) = \mathrm{im}(\delta : \mathscr{C}^{k-1}(G, V) \to \mathscr{C}^k(G, V))$ *space of k-coboundaries* of G with values in V.
(iii) $H^k(G, V) = \dfrac{\mathscr{Z}^k(G, V)}{\mathscr{B}^k(G, V)}$ *k-cohomology group* of G with values in V.

Remark C.92 (Inhomogeneous vs Homogeneous Co-chains)
It is worth mentioning that there are other complexes that can be used to compute Lie group cohomology. One of those complexes is defined as follows. Fix a G-module V as above, and for all $k \geq 0$, define $\widetilde{\mathscr{C}}^k(G, V)$ as the vector space of all smooth V-valued maps $\widetilde{f} : G^k = \underbrace{G \times \cdots \times G}_{k-\text{times}} \to V$ such that $\widetilde{f}(g g_0, \ldots, g g_k) = g.\widetilde{f}(g_0, \ldots, g_k)$ for all $g, g_0, \ldots, g_k \in G$. Define now $\widetilde{\delta} : \widetilde{\mathscr{C}}^k(G, V) \to \widetilde{\mathscr{C}}^{k+1}(G, V)$

$$(\widetilde{\delta}\widetilde{f})(g_0, \ldots, g_{k+1}) = \sum_{i=1}^{k+1}(-1)^i \widetilde{f}(g_0, \ldots, \hat{g}_i, \ldots, g_{k+1})$$

which squares to zero, making $(\widetilde{\mathscr{C}}^\bullet(G, V), \widetilde{\delta})$ into a complex. This is called the complex of the *homogeneous cochains* of G with values in V,, as opposed to the one introduced before, which is the complex of the *inhomogeneous cochains* of G with values in V. There is an isomorphism between $\varphi : \widetilde{\mathscr{C}}^\bullet(G, V) \to \mathscr{C}^\bullet(G, V)$ defined as follows:

(i) $(\varphi \widetilde{f}) = f(e)$ for $k = 0$
(ii) $(\varphi \widetilde{f})(g_1, \ldots, g_k) = \widetilde{f}(e, g_1, g_1 g_2, \ldots, g_1 g_2 \cdots g_{k-1} g_k)$ for $k > 0$

whose inverse is defined as:

(i) $\left(\varphi^{-1} f\right)(g_0) = g_0 . \tilde{f}$ for $k = 0$

(ii) $\left(\varphi^{-1} f\right)(g_0, \ldots, g_k) = g_0 . f\left(g_0^{-1} g_1, \ldots, g_0^{-1} g_k\right)$ for $k > 1$.

It can be shown that φ intertwines the differentials δ and $\tilde{\delta}$ so that it defines an isomorphism between the cohomology groups of the two chain complexes (the map φ defined here is what is called a *quasi-isomorphism* between $(\mathscr{C}^{\bullet}(G, V), \delta)$ and $(\widetilde{\mathscr{C}}^{\bullet}(G, V), \tilde{\delta}))$. $\triangle$

Example C.93 ($T_{\mathfrak{g}}$ **as a** G**-Module**) An important example of G-module is the Lie algebra $\mathfrak{g}$ acted on by G via the adjoint action. Such an action can be extended to all the tensor algebra of $\mathfrak{g}$ as follows. For all $n = 0$, define $\mathrm{Ad}_g\, a = a$ for all $a \in \mathbb{K}$ and $g \in G$, while for $n > 0$, define $\mathrm{Ad}_g(x_1 \otimes \cdots \otimes x_n) = \mathrm{Ad}_g\, x_1 \otimes \cdots \otimes \mathrm{Ad}_g\, x_n$ for all $x_1 \otimes \cdots \otimes x_n \in \mathfrak{g}^n$ and all $g \in G$. Extend now this to an element of $\mathrm{End}(\mathfrak{g}^n)$, which becomes a G-module. These definitions define on a full tensor algebra a structure of a G-module. Note that this action extends to the exterior algebra $\Lambda_{\mathfrak{g}}$ and to the symmetric algebra $S_{\mathfrak{g}}$ and more in general to all the vector subspaces of $T_{\mathfrak{g}}$. $\triangle$

Example C.94 (Low Degrees and One-Cocycles) For $k = 0$, Formula (C.23) becomes $(\delta f)(g) = -f(g)$, while for $k = 1$, it looks like $(\delta f)(g_1, g_2) = g_1 . f(g_2) - f(g_1 g_2) + f(g_1)$. From this last equation, we see that $f \in \mathscr{C}^1(G, M)$ is a *one-cocycle* if and only if

$$f(g_1 g_2) = f(g_1) + g_1 . f(g_2), \ \forall\, g_1, g_2 \in G. \tag{C.23}$$

$\triangle$

2   Fiber Bundles and Connections

We present some very basic notions about fiber bundles and connections. For more details, see the monograph [133].

2.1   Locally Trivial Fiber Bundles

Definition C.95 (Trivial and Locally Trivial Fibrations) Suppose that N, F, B are smooth manifold and that $\pi : N \to B$ is a smooth and surjective map.

(i) Then (N, B, F, π) is called a *trivial fibration* or *trivial fiber bundle* if there exists a diffeomorphism $\phi : N \to B \times F$ such that the following diagram

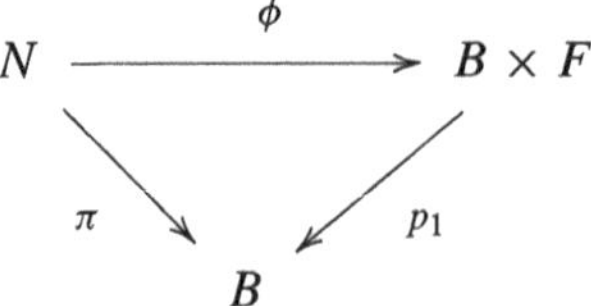

is commutative. Here the map p_1 is the *projection onto* the first factor. In this case, we will say that the trivial fibration is *realized* by the diffeomorphism ϕ.

(ii) The quadruple (N, B, F, π) will be called a *locally trivial fibration* or *locally trivial fiber bundle* if there exists an open covering $\{U_i\}_{i \in I}$ and a family of diffeomorphisms $\{\phi_i\}_{i \in I}$ such that for each $i \in I$,

$$\left(\pi^{-1}(U_i), U_i, F, \pi|_{U_i}\right)$$

is a trivial fibration realized by the diffeomorphism ϕ_i, i.e., such that the following is a commutative diagram for all $i \in I$

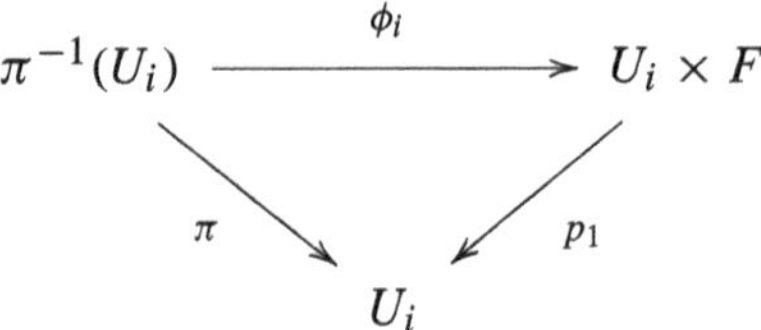

Note that $\phi_i(n) = (\pi(n), g_i(n))$ for some smooth map $g_i : \pi^{-1}(U_i) \to F$.

(iii) In both cases, we will say that manifolds N, B, and F are *total space*, the *base*, and the *typical fiber* of the fibration, respectively. The map π will be called the *canonical projection* of the fibration.

The collection $\{U_i\}_{i \in I}$ is called a *trivializing open covering*, and each U_i is called a *trivializing open subset*.
For all $b \in U$, $\pi^{-1}(b) = F_b \simeq F$ is called the *fiber over the point b*. They are all diffeomorphic (but non-canonically) to the typical fiber.

Remark C.96 Note that the same definition could be given in the topological category.   △

Remark C.97 In these notes, we will mainly consider fibrations that are locally trivial. For this reason, unless differently stated, *fibration* (fiber bundle) will be synonymous of *locally trivial fibration* (locally trivial fiber bundle).   △

Given a fibration and a trivializing open covering as above, to each trivializing open subset U, we can associate the set of *local sections*, i.e., smooth maps, $s : U \to \pi^{-1}(U)$ such that $\pi \circ s = \mathrm{id}_U$. We will denote the space of these maps with $\Gamma_U(N)$ or with $\Gamma(U, N)$. Note that at this level of generality, this space has no algebraic structure. It is possible to show that there is a one-to-one correspondence between $\Gamma_U(N)$ and the space of smooth map between the U and F so that we will often think of a local-section $s \in \Gamma_U(N)$ as the corresponding map $s : U \to F$. In particular, assume $s \in \Gamma_U(N)$. Such a map, together with the diffeomorphism ϕ_U, defines a map $\sigma_U : U \to U \times F$ given by $\sigma(b) = \big(\pi(s(b)), \phi_U(s(b))\big) = \big(b, \phi_U(s(b))\big)$, which explains the previous remark about the correspondence between $\Gamma_U(N)$ and the space of smooth maps between U and F. Suppose now that U, V are two trivializing open sets for the fibration $\pi : N \to B$ with non-empty intersection. Suppose that $\phi_U : \pi^{-1}(U) \to U \times F$ and $\phi_V : \pi^{-1}(V) \to V \times F$ are the corresponding diffeomorphisms; then we define $\lambda_{UV} : U \cap V \to \mathscr{D}(F)$ such that if $s_U \in \Gamma_U(N)$ and $s_V \in \Gamma_V(N)$, then $s_U = \lambda_{UV} s_V$, i.e., $s_U(b) = \lambda_{UV}(b) s_V(b)$ for all $b \in U \cap V$. In other words, the values of two different sections over the same trivializing open subset differ by a *translation* on the fiber, which depends on the point $b \in U \cap V$. Finally, consider now three trivializing open sets U, V, W such that $U \cap V \cap W \neq \emptyset$. Then it is possible to show that (i) $\lambda_{UV} \circ \lambda_{VW} \circ \lambda_{UV} = \mathrm{id}_F$. On the other hand, for any pair of trivializing open sets U, V with non-empty intersection, (ii) $\lambda_{UV} \circ \lambda_{VU} = \mathrm{id}_F$. Summarizing, given a fibration and a trivializing open covering as above, we have defined a family of maps $\{\lambda_{UV}\}$ defined on the double intersection and with values into the group $\mathscr{D}(F)$ having abovementioned properties (i) and (ii). On the other hand, given B and F manifolds, an open covering $\{U_i\}_{i \in I}$ of B and a family of maps each of those defined on the double intersections, taking values in $\mathscr{D}(F)$ and sharing properties (i) and (ii), one can construct a fibration having base B typical fiber F and total space $N = \bigsqcup_{b \in B} F_b$.

Example C.98 (Vector Bundles)   The tangent and the cotangent bundle of every manifold B are examples of bundles in the sense of definition above, whose typical fiber is a vector space, and for this reason are called *vector bundles*. While a bundle admits always local sections, in general, it does not admit any *global section*, i.e., a smooth map $\sigma : B \to N$ such that $\pi \circ \sigma = \mathrm{id}_B$. On the other hand, every vector bundle (E, B, π, V) admits always a global section, which is given by assigning to every point $b \in B$ the zero vector of the fiber $V_b, 0_b \in V_b$. This section is called the *zero-section* Z of the vector bundle, and its image is diffeomorphic to the base B, an explicit diffeomorphism being provided by the restriction to Z of the projection π. $\triangle$

We are now going to focus on a particular class of fiber bundles called *principal bundles*.

2.2   Principal Bundles

Let G be a Lie group and let N be a manifold carrying a G-action, $\varphi : G \times M \to M$, such that:

(i) The action φ is *free*.

(ii) The set of the G-orbits B can be endowed with structure of a smooth manifold, with respect to which the canonical projection $\pi : M \to B$ is a smooth.

(iii) There exists an open covering $\{U_i\}_{i \in I}$ of B and, for every $i \in I$, a G-*equivariant diffeomorphism* ϕ_i such that the following diagram

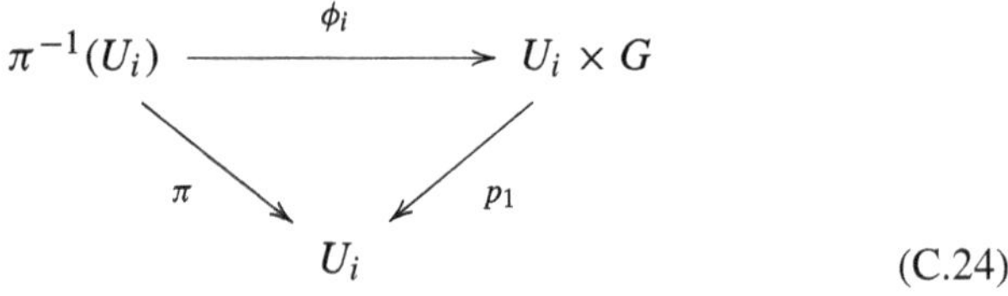

$$\tag{C.24}$$

commutes for all $i \in I$. In the diagram, $p_1 : (m, g) \rightsquigarrow m$ is the *projection onto the first factor*, while the property of G-equivariance of the map ϕ_i means that:

(a) $\phi_i(m) = (\pi(m), f_i(m))$ where

(b) $f_i : \pi^{-1}(U_i) \to G$ is *smooth* and $f_i(\varphi(g, m)) = g f_i(m)$ for all $m \in \pi^{-1}(U_i)$ and $g \in G$.

Remark C.99 From (ii) and (iii), one conclude that (M, B, G, π) is a fiber bundle.

△

Definition C.100 (Principal Bundle) The quadruple (M, B, G, π) is called a *principal bundle*. The Lie group G is called the *structure group* of the principal bundle.

Remark C.101 We note the following two differences between the case of the principal bundle and the more general fiber bundles discussed above.

(i) For principal bundles, the diffeomorphisms ϕ_i are G-equivariant.

(ii) Moreover, the typical fiber is the Lie group G. In this way, for every $b \in B$, $\pi^{-1}(b)$ is diffeomorphic, though, non-canonically to G. Note also that each fiber coincides with a G-orbit since we assume that the G-action is free.

△

Observe that to a principal bundle (M, B, G, π) and to a representation (V, ρ) of G, one can associate a vector bundle (E, B, V, p) via the following construction. On the set $M \times V$, one defines $(m_1, v_1) \simeq (m_2, v_2)$ if and only if there exists $g \in G$ such that $m_2 = \varphi_g(m_1)$ and $v_2 = \rho(g^{-1})v_1$. Then $\simeq$ is an equivalence relation, and the set of equivalence classes inherits from M and V the structure of a differentiable manifold we denote by $M \times_\rho V$. The (well-defined) natural map $p : M \times_\rho V \to B$, defined by $p([m, v]) = \pi(m)$, turns out to be a smooth submersion with respect to the smooth structures of $M \times_\rho V$ and B. Furthermore, every open covering trivializing the principal bundle will be also a trivializing open covering for $(M \times_\rho V, B, V, p)$. The vector bundle so defined is called the *associated* vector bundle to (M, B, G, π) and to (V, ρ). An important example is the so-called adjoint bundle, $M \times_{\mathrm{Ad}} \mathfrak{g}$, the associated vector bundle defined by the adjoint representation $(\mathfrak{g}, \mathrm{Ad})$.

Definition C.102 (Connection One-Form) A connection one-form $\mathcal{A}$ on (M, B, G, π) is an element of $\Omega^1_M(\mathfrak{g})$, i.e., a one-form on M with values in $\mathfrak{g}$, such that

$$\langle \mathcal{A}_m, X_x(m) \rangle = x, \ \forall m \in M, \ x \in \mathfrak{g}; \tag{C.25}$$

and

$$\varphi_g^* \mathcal{A} = \mathrm{Ad}_g \circ \mathcal{A}, \ \forall g \in G \tag{C.26}$$

where (C.26) stands for

$$\langle (\varphi_g^* \mathcal{A})_m, v \rangle = \mathrm{Ad}_g \langle \mathcal{A}_m, v \rangle,$$

for all $g \in G$, $m \in M$, and $v \in T_m M$.

The choice of a connection one-form $\mathcal{A}$ is *equivalent* to the definition of $\mathcal{D}^{\mathcal{A}}$, a *G-invariant distribution* on M whose dimension is equal to dim B. In fact, given $\mathcal{A}$, for each $m \in M$, let

$$\mathcal{D}^{\mathcal{A}}(m) = \ker(\mathcal{A}_m : T_m M \to \mathfrak{g}).$$

Since $\mathcal{A}$ satisfies (C.25),

$$\dim(\mathcal{D}^{\mathcal{A}}(m)) = \dim(M) - \dim(\mathfrak{g}) = \dim B.$$

On the other hand, (C.26) entails

$$(\varphi_g)_{*,m} \mathcal{D}^{\mathcal{A}}(m) \subset \mathcal{D}_{\mathcal{A}}(\varphi_g(m)),$$

and since φ_g is a diffeomorphism, the G-invariance of $\mathcal{D}^{\mathcal{A}}$ follows. The distribution so defined is called *horizontal*. Clearly, for every $m \in M$,

$$T_m M = \mathrm{Vert}_m \oplus \mathcal{D}^{\mathcal{A}}(m),$$

and $\pi_{*,m}$ restricts to an isomorphism between $\mathcal{D}^{\mathcal{A}}(m)$ and $T_{\pi(m)} B$, defining a splitting of the exact sequence of vector spaces

$$0 \longrightarrow \mathrm{Vert}_m \longrightarrow T_m M \overset{\pi_{*,m}}{\longrightarrow} T_{\pi(m)} B \longrightarrow 0,$$

How far $\mathcal{D}^{\mathcal{A}}$ is from being integrable is measured by $C^{\mathcal{A}} \in \Omega^2_M(\mathfrak{g})$, the *curvature* of $\mathcal{A}$. More precisely, for each $m \in M$, let $h_m : T_m M \to T_m M$ be the *projection map* to the horizontal subspace $\mathcal{D}^{\mathcal{A}}(m)$, and define

$$C^{\mathcal{A}}_m(u, v) := d\mathcal{A}_m(h_m(u), h_m(v)), \ \forall u, v \in T_m M.$$

Then $C^{\mathcal{A}} \in \Omega^2_M(\mathfrak{g})$ and

$$\varphi_g^* C^{\mathcal{A}} = \mathrm{Ad}_g \circ C^{\mathcal{A}}, \tag{C.27}$$

where

$$\mathrm{Ad}_g \circ C^{\mathcal{A}}_m(u, v) = \mathrm{Ad}_g \left(C^{\mathcal{A}}_m(u, v) \right),$$

for all $m \in M$ and $u, v \in T_m M$.

Theorem C.103 *For each $m \in M$ and $u, v \in T_m M$*

$$d\mathcal{A}_m(u, v) = -[\mathcal{A}_m(u), \mathcal{A}_m(v)]_{\mathfrak{g}} + C^{\mathcal{A}}_m(u, v) \tag{C.28}$$

and $\mathcal{D}^{\mathcal{A}}$ is Frobenius integrable if and only if $C^{\mathcal{A}}_m = 0$, for all $m \in M$. In (C.28), $[\cdot, \cdot]_{\mathfrak{g}}$ denotes the Lie bracket of $\mathfrak{g}$.

Proof The proof of (C.28) follows from the analysis of the cases (i) $u, v \in \mathrm{Vert}_m$, (ii) $u, v \in \mathcal{D}^{\mathcal{A}}(m)$, and (iii) $u \in \mathcal{D}^{\mathcal{A}}(m)$, $v \in \mathrm{Vert}_m$. In the cases (i) and (ii), (C.28) follows by a direct inspection. On the other hand, if (iii) is verified, let $X \in \mathfrak{X}(M)$ be any horizontal vector field extending u, i.e., X is such that $X(m) = u$ and $X(m') \in \mathcal{D}^{\mathcal{A}}(m')$ for all m' in a suitable neighborhood of m and let $x \in \mathfrak{g}$ such that $X_x(m) = v$. Then

$$d\mathcal{A}_m(u, v) = X(m)\langle \mathcal{A}, X_x \rangle - X_x(m)\langle \mathcal{A}, X \rangle - \mathcal{A}([X, X_x])(m) = -\mathcal{A}([X, X_x])(m).$$

In fact, $X(m)\langle \mathcal{A}, X_x \rangle = 0$ since $\langle \mathcal{A}_m, X_x(m) \rangle = x$ for each $m \in M$, i.e., $\langle \mathcal{A}, X_x \rangle$ is constant, and $\langle \mathcal{A}, X \rangle(m) = \langle \mathcal{A}_m, X(m) \rangle = 0$ since $X(m) \in \mathcal{D}^{\mathcal{A}}(m)$. On the other hand

$$[X_x, X](m) = \mathscr{L}_{X_x} X(m) = \frac{d}{dt}\bigg|_{t=0} (\varphi_{\exp(-tx)})_{*, \varphi_{\exp(tx)}(m)} X(\varphi_{\exp(tx)}(m)) \in \mathcal{D}^{\mathcal{A}}(m),$$

because the distribution $\mathcal{D}^{\mathcal{A}}$ is G-invariant. This implies that $\mathcal{A}([X, X_x])(m) = 0$, i.e., $d\mathcal{A}_m(u, v) = 0$. On the other hand, $C^{\mathcal{A}}_m(u, v) = 0$ being $v \in \mathrm{Vert}_m$. This proves (C.28). To conclude the proof, it suffices to prove that if X, Y are horizontal, i.e., $X(m), Y(m) \in \mathcal{D}^{\mathcal{A}}(m)$, for all $m \in M$, then $[X, Y]$ is horizontal if and only if $C^{\mathcal{A}}_m(X(m), Y(m)) = 0$ for all $m \in M$. To this end, note that (C.28) implies that

$$d\mathcal{A}_m(X(m), Y(m)) = -C^{\mathcal{A}}_m(X(m), Y(m)), \ \forall m \in M,$$

and at the same time

$$d\mathcal{A}_m(X(m), Y(m)) = -\mathcal{A}([X, Y])(m), \ \forall m \in M,$$

i.e.,

$$\mathcal{A}([X, Y])(m) = C_m^{\mathcal{A}}(X(m), Y(m)), \ \forall m \in M.$$

Since $C^{\mathcal{A}}$ is horizontal, i.e., it is zero every time it hits a vertical vector, and since $\mathcal{D}^{\mathcal{A}}$ is smooth, this formula implies that $C_m^{\mathcal{A}} = 0$ for all $m \in M$, if and only if $\mathcal{A}([X, Y])(m) = 0$ for all X, Y horizontal and $m \in M$, which is equivalent to what we wanted to show. $\square$

Example C.104 (Maurer-Cartan Forms)
   The (left) Maurer-Cartan form θ^L is the one-form θ on G with values in $\mathfrak{g}$ defined by:

$$\theta_g^L(v) = (L_{g^{-1}})_{*,g} v$$

for all $v \in T_g G$ and $g \in G$. The form θ^L is the *unique* left invariant $\mathfrak{g}$-valued form such that $\theta_e^L(x) = x$ for all $x \in \mathfrak{g}$ and such that

$$R_h^* \theta^L = \mathrm{Ad}_{h^{-1}} \theta^L, \ \forall h \in G. \tag{C.29}$$

Furthermore, it satisfies the following identity:

$$d\theta^L + \left[\theta^L, \theta^L\right] = 0. \tag{C.30}$$

The unicity follows from the left invariance, which, in turn, follows from a direct computation, which we leave as an exercise for the reader (it is necessary to show that $(L_h)_{*,g}^* \theta_{hg}^L(v) = \theta_g^L v$ for all $v \in T_g G$ and for all $g, h \in G$). Spelling out (C.30), one gets

$$\left(d\theta^L\right)_g(u, v) = -\left[\theta_g^L(u), \theta_g^L(v)\right], \ \forall u, v \in T_g G, \ g \in G.$$

Then to prove (C.30), it suffices to compute $d\theta^L\left(X_x^L, X_y^L\right)(g)$, where X_x^L, X_y^L are the (unique) left-invariant vector fields such that $X_x^L(g) = u$ and $X_y^L(g) = v$. We have:

$$d\theta^L\left(X_x^L, X_y^L\right)(g)$$

$$= X_x^L(g)\theta^L\left(X_y^L\right) - X_y^L(g)\theta^L\left(X_x^L\right) - \theta^L\left([X_x^L, X_y^L]\right)(g)$$

$$= -\theta^L\left(X_{[x,y]}^L\right)(g) \tag{C.31}$$

$$= -[x, y] \tag{C.32}$$

$$= -\left[\theta_g^L(u), \theta_g^L(v)\right]. \tag{C.33}$$

Equality (C.31) follows from the definitions of θ^L and of X_x^L. More precisely, since $X_x^L(g) = (L_g)_{*,g}(x)$, we have that $\theta_g^L(X_x^L(g)) = x$ for all $g \in G$, implying that $g \rightsquigarrow \theta^L(X_x^L)(g)$ and $g \rightsquigarrow \theta^L(X_y^L)(g)$ are constant functions on G. Equality (C.32) follows from (our convention about) the definition of the Lie bracket on $\mathfrak{g}$ (see Formula C.3). Finally, (C.33) follows from $X_x^L(g) = u$, $X_y^L(g) = v$ and from the definition of left invariant vector fields. To prove (C.29), one computes:

$$\begin{aligned}
(R_h)^*_{,g}\theta_{gh}^L(v) &= \theta_{gh}^L\big((R_h)_{*,g}v\big) \\
&= (L_{(gh)^{-1}})_{*,gh}(R_h)_{*,g}v \\
&= (L_{h^{-1}} \circ L_{g^{-1}})_{gh} \circ (R_h)_{*,g}v \\
&= (L_{h^{-1}} \circ R_h)_{*,e} \circ (L_{g^{-1}})_{*,h}v \\
&= \mathrm{Ad}_{h^{-1}}(L_{g^{-1}})v,
\end{aligned}$$

for all $v \in T_g G$ and $g, h \in G$. Note that (C.29) tells us that θ is a $\mathfrak{g}$-valued G-invariant one-form defined on G, where G carries the structure of a *right G-manifold*, defined by the right translations, $R_g : h \rightsquigarrow hg$ for all $g, h \in G$, while $\mathfrak{g}$ is endowed with its canonical structure of a G-module defined by the adjoint action. In a completely analogous way, one can introduce the *right Maurer-Cartan form* θ^R via the formula

$$\theta_g^R(v) = (R_{g^{-1}})_{*,g}(v), \ \forall v \in T_g G, \ g \in G.$$

Trading *left* for *right*, one can prove that all the statements true for θ^L remain true for θ^R. In particular, θ^R is the unique, right invariant $\mathfrak{g}$-valued one-form on G, which restricts to the identity on $T_e G \simeq \mathfrak{g}$ and that satisfies

$$d\theta^R - \left[\theta^R, \theta^R\right] = 0. \tag{C.34}$$

Note that both right and left Maurer-Cartan forms on G are indeed connections form on G seen as a principal bundle over a *point*. In this way, Formulas (C.30) and (C.34) are simply the (obvious) statements that the corresponding horizontal distributions are integrable. $\triangle$

Let $\{U_i\}_{i \in I}$ be a trivializing open covering of (M, B, G, π). For every i, let $s_i : U_i \to \pi^{-1}(U_i)$ be the local section defined by $s_i = \phi_i^{-1} \circ \sigma_i$, where $\sigma_i : U_i \to U_i \times G$ is the section of $p_1 : U_i \times G \to U_i$, defined by $\sigma_i(b) = (b, e)$, for all $b \in U_i$ (see Diagram (C.24)). In particular, for all $i, j \in I$ such that $U_{ij} := U_i \cap U_j \neq \emptyset$ $s_j = \psi_{ji}s_i$, where $\psi_{ji} : U_{ij} \to G$ is the transition function of the bundle. The transition functions $\{\psi_{ij}\}_{i,j \in I}$ of the bundle, associated with the open covering

$\{U_i\}_{i \in I}$, satisfy the following two conditions:

$$(a)\ \psi_{ij} = \psi_{ji}^{-1}\quad \text{and}\quad (b)\ \psi_{ij}\psi_{jk} = \psi_{ik},$$

for all $i, j, k \in I$ such that $U_{ijk} := U_i \cap U_j \cap U_k \neq \emptyset$. Given a connection $\mathcal{A}$ on M, for every $i \in I$, one can define $\mathcal{A}_i = s_i^* \mathcal{A}$, i.e., a $\mathfrak{g}$-valued one-form defined on U_i. Then, on every non-empty intersection U_{ij}

$$\mathcal{A}_j = \mathrm{Ad}_{\psi_{ji}} \circ \mathcal{A}_i + \theta_{ji}^R \tag{C.35}$$

where $\theta_{ji}^R = \psi_{ji}^* \theta^R$. Defining $C_i^{\mathcal{A}} := s_i^* C^{\mathcal{A}}$, and using (C.35) and (C.34), one can prove that

$$C_j^{\mathcal{A}} = \mathrm{Ad}_{\psi_{ji}} \circ C_i^{\mathcal{A}}, \tag{C.36}$$

on every (non-empty) intersection U_{ij}. In particular, (C.36) shows that the local forms $\{C_i^{\mathcal{A}}\}_{i \in I}$ glue together to define a global two-form $C \in \Omega_B^2(M \times_{\mathrm{Ad}} \mathfrak{g})$, i.e., a two-form on B taking values in the adjoint bundle $M \times_{\mathrm{Ad}} \mathfrak{g}$. Note that C is the unique element in $\Omega_B^2(M \times_{\mathrm{Ad}} \mathfrak{g})$ such that $\pi^* C = C^{\mathcal{A}}$. To be more precise, recall that the curvature $C^{\mathcal{A}}$ is a two-form on M, with values in $\mathfrak{g}$ such that:

(i) $C_m^{\mathcal{A}}(v, \cdot) = 0$ for every *vertical* $v \in T_m M$, i.e., the curvature is *horizontal*.
(ii) It satisfies the property of G-invariance expressed in Formula (C.27).

A form fulfilling these properties is called a *basic* form. Let $\Omega_{bas}^2(M, \mathfrak{g})$ be the vector space of the basic two-forms on M with values in $\mathfrak{g}$. Then

Proposition C.105 *There is a one-to-one correspondence between* $\Omega_{bas}^2(M, \mathfrak{g})$ *and* $\Omega_B^2(M \times_{\mathrm{Ad}} B)$, *the set of all two-forms on B with values in the adjoint bundle.*

Proof We set up the correspondence defining a map, which takes $\alpha \in \Omega_{bas}^2(M, \mathfrak{g})$ and sends it to $A \in \Omega_B^2(M \times_{\mathrm{Ad}} B)$, defined by

$$A_b(V, U) := [m, \alpha_m(v, u)], \ \forall b \in B,\ U, V \in T_b B, \tag{C.37}$$

where $m \in M$, $v, u \in T_m M$ are any elements such that $\pi(m) = b$, $\pi_{*,m}(v) = V$, and, respectively, $\pi_{*,m}(u) = U$. In (C.37), $[m, x] \in M \times_{\mathrm{Ad}} \mathfrak{g}$ denotes the class of $(m, x) \in M \times \mathfrak{g}$. The only non-trivial part of the proof is to show that this map is well defined. To this end, first suppose that $u, v, u', v' \in T_m M$ are such that $\pi_{*,m} u = U = \pi_{*,m} u'$ and $\pi_{*,m} v = V = \pi_{*,m} v'$. Then $v - v' = \xi \in \ker(\pi_{*,m})$ and $u - u' = \eta \in \ker(\pi_{*,m})$ and since

$$[m, \alpha_m(v, u)] = A_b(U, V) = [m, \alpha_m(v', u')],$$

it follows that $\alpha_m(v, u) = \alpha_m(u, v)$. Suppose now that $\pi(\tilde{m}) = b = \pi(m)$ and let $\tilde{v}, \tilde{u} \in T_{\tilde{m}} M$ such that $\pi_{*,\tilde{m}} \tilde{v} = V$ and $\pi_{*,\tilde{m}} \tilde{u} = U$. Let $g \in G$ be the unique group-element such that $\varphi_g(m) = \tilde{m}$, and assume $\tilde{u} = \varphi_{*,m}u$, $\tilde{v} = \varphi_{*,m}v$. Since $\pi \circ \varphi_g = \pi$, $\pi_{*,\tilde{m}}\tilde{\tilde{v}} = V$ and $\pi_{*,\tilde{m}}\tilde{\tilde{u}} = U$, implying that both $\tilde{v} - \tilde{\tilde{v}}$ and $\tilde{u} - \tilde{\tilde{u}}$ belong to $\ker(\pi_{*,\tilde{m}})$. In this way,

$$
\begin{aligned}
\left[\tilde{m}, \alpha_{\tilde{m}}(\tilde{v}, \tilde{u})\right] &= \left[\tilde{m}, \alpha_{\tilde{m}}(\tilde{\tilde{v}}, \tilde{\tilde{u}})\right] \\
&= \left[\varphi_g(m), \alpha_{\varphi_g(m)}\big((\varphi_g)_{*,m}v, (\varphi_g)_{*,m}u\big)\right] \\
&= \left[\varphi_g(m), (\varphi_g^*\alpha)_m(v, u)\right] \\
&\overset{(C.26)}{=} \left[\varphi_g(m), \mathrm{Ad}_g(\alpha_m(v, u))\right] \\
&= \left[m, \alpha_m(v, u)\right],
\end{aligned}
$$

which is what we needed to show.   □

Remark C.106 One can define the notion of basic k-form, for every k, and adapts the previous proof to the statement that there is a one-to-one correspondence between the $\Omega^k_{bas}(M, V)$ and $\Omega^k_B(M \times_\rho V)$, for every (finite-dimensional) representation (V, ρ) of G.   △

To prove (C.36), we are left to show that (C.35) holds true. To this end, it is worth recalling that if $f : N \times M \to Z$ is a smooth map, then $f_{*,(n_0,m_0)} : T_{n_0}N \times T_{m_0}M \to T_{f(n_0,m_0)}Z$ is given by

$$
f_{*,(n_0,m_0)}(u, v) = (f \circ i_{n_0})_{*,m_0}(u) + (f \circ i_{m_0})_{*,n_0}(v), \forall u \in T_{m_0}M, \ v \in T_{n_0}N,
\tag{C.38}
$$

where $i_{n_0} : M \to M \times N$ and $i_{m_0} : N \to M \times N$ are defined by $i_{n_0}(m) = (m, n_0)$ and, respectively, $i_{m_0}(n) = (m_0, n)$, for all $m \in M$ and $n \in N$. Let $U_{ij} \neq \emptyset$, and let $s_j = \psi_{ji}s_i$ as above and $u \in T_bU_{ij}$. To compute $(s_j)_{*,b}(u)$ in terms of the differentials of s_j and ψ_{ji}, it is convenient to think of s_j as the composition of the following maps:

$$
U_{ij} \xrightarrow{\ \Delta\ } U_{ij} \times U_{ij} \xrightarrow{\ (\psi_{ji},s_i)\ } G \times \pi^{-1}(U_{ij}) \xrightarrow{\ \varphi\ } \pi^{-1}(U_{ij})
$$

where $\Delta(b) = (b, b)$ is the diagonal map. In this way:

$$
(s_j)_{*,b}(u) = (\varphi \circ (\psi_{ji}, s_i) \circ \Delta)_{*,b}(u) = \varphi_{*,(\psi_{ji}(b),s_i(b))}\big((\psi_{ji})_{*,b}(u), (s_i)_{*,b}(u)\big),
$$

In this way, applying (C.38), one gets:

$$
(s_j)_{*,b}(u) = \big(\varphi_{\psi_{ji}(b)}\big)_{*,s_i(b)}(s_i)_{*,b}(u) + \big(\varphi_{s_i(b)}\big)_{*,\psi_{ji}(b)}(\psi_{ji})_{*,b}(u).
$$

Applying $\mathcal{A}_{s_j(b)}$ to the left-hand side of this identity, one gets:

$$
\mathcal{A}_{s_j(b)}\big((s_j)_{*,b}(u)\big) = (s_j^*\mathcal{A})_b(u) = \mathcal{A}_{j_b}(u).
$$

On the other hand, applying $\mathcal{A}_{s_j(b)}$ to the first summand of the right-hand side gives

$$
\begin{aligned}
\left\langle \mathcal{A}_{s_j(b)},\, \left(\varphi_{\psi_{ji}(b)}\right)_{*,s_i(b)} (s_i)_{*,b}(u) \right\rangle &= \left\langle \mathcal{A}_{s_j(b)},\, \left(\varphi_{\psi_{ji}(b)} \circ s_i\right)_{*,b}(u) \right\rangle \\
&= \left\langle \left(\varphi_{\psi_{ji}(b)} \circ s_i\right)^{*}_{b} \mathcal{A}_{s_j(b)},\, u \right\rangle \\
&= \left\langle \left(\varphi_{\psi_{ji}(b)} \circ s_i\right)^{*}_{b} \mathcal{A}_{\varphi_{\psi_{ji}(b)}(s_i(b))},\, u \right\rangle \\
&= \left\langle s_i^{*}{}_{b}\left(\varphi_{\psi_{ji}(b)}{}^{*}_{s_i(b)} \mathcal{A}_{\varphi_{\psi_{ji}(b)}(s_i(b))}\right),\, u \right\rangle \\
&= \left\langle \left(\mathrm{Ad}_{\psi_{ji}(b)} \circ s_i^{*}\mathcal{A}\right)_{s_i(b)},\, u \right\rangle \\
&= (\mathrm{Ad}_{\psi_{ji}} \circ \mathcal{A}_i)_b(u),
\end{aligned}
$$

while applying it to the second summand, one gets

$$
\begin{aligned}
\left\langle \mathcal{A}_{s_j(b)},\, (\varphi_{s_i(b)})_{*,\psi_{ji}(b)} (\psi_{ji})_{*,b}(u) \right\rangle &= \left\langle \mathcal{A}_{\varphi_{\psi_{ji}(b)}(s_i(b))},\, (\varphi_{s_i(b)})_{*,\psi_{ji}(b)} (\psi_{ji})_{*,b}(u) \right\rangle \\
&= \left\langle \mathcal{A}_{\varphi_{s_i(b)}(\psi_{ji}(b))},\, (\varphi_{s_i(b)})_{*,\psi_{ji}(b)} (\psi_{ji})_{*,b}(u) \right\rangle \\
&= \left\langle (\varphi_{s_i(b)}^{*} \mathcal{A})_{\psi_{ji}(b)},\, (\psi_{ji})_{*,b}(u) \right\rangle \\
&= \left(\psi_{ji}^{*}\theta^{R}\right)_{b}(u). \tag{C.39}
\end{aligned}
$$

To complete the proof, we are left to prove (C.39). More generally, we show that for all $m \in M$,

$$
\varphi_m^{*}\mathcal{A} = \theta^{R},
$$

where $\varphi_m : G \to M$ is the map $\varphi_m(g) = \varphi(g, m)$, for all $g \in G$. Let $v \in T_g G$ and $x \in \mathfrak{g}$ be the unique element such that $v = (R_g)_{*,e}x$. Then to prove (C.39), it suffices to show that $\varphi_m^{*}\mathcal{A}$ is right invariant and restricts to the identity on $\mathfrak{g}$. The latter property follows easily from the following computation:

$$
\langle (\varphi_m^{*}\mathcal{A})_e,\, x \rangle = \langle \mathcal{A}_{\varphi_m(e)},\, (\varphi_m)_{*,m}x \rangle = \langle \mathcal{A}_m,\, X_x(m) \rangle = x,
$$

for all $x \in \mathfrak{g}$. On the other hand, given $g \in G$ and $v \in T_h G$ such that $v = (R_h)_{*,e}x$, $x \in \mathfrak{g}$

$$
\begin{aligned}
\left\langle \left(R_g^{*}(\varphi_m^{*}\mathcal{A})\right)_h,\, v \right\rangle &= \left\langle \left(R_g^{*}(\varphi_m^{*}\mathcal{A})\right)_h,\, (R_h)_{*,e}x \right\rangle \\
&= \left\langle (\varphi_m^{*}\mathcal{A})_{hg},\, (R_g)_{*,h}(R_h)_{*,e}x \right\rangle \\
&= \left\langle \mathcal{A}_{\varphi_m(hg)},\, (\varphi_m)_{*,hg}(R_{hg})_{*,e}x \right\rangle
\end{aligned}
$$

$$\begin{aligned}
&= \left\langle \mathcal{A}_{\varphi_m(hg)}, (\varphi_m \circ R_{hg})_{*,e} x \right\rangle \\
&= \left\langle \mathcal{A}_{\varphi_m(hg)}, (\varphi_{\varphi_m(hg)})_{*,e} x \right\rangle \\
&= \left\langle \mathcal{A}_{\varphi_m(hg)}, X_x(\varphi_m(hg)) \right\rangle \\
&= x
\end{aligned}$$

and

$$\begin{aligned}
\left\langle (\varphi_m^* \mathcal{A})_h, v \right\rangle &= \left\langle \mathcal{A}_{\varphi_m(h)}, (\varphi_m)_{*,h} v \right\rangle \\
&= \left\langle \mathcal{A}_{\varphi_m(h)}, (\varphi_m)_{*,g}(R_h)_{*,e} x \right\rangle \\
&= \left\langle \mathcal{A}_{\varphi_m(h)}, (\varphi_m \circ R_h)_{*,e} x \right\rangle \\
&= \left\langle \mathcal{A}_{\varphi_m(h)}, (\varphi_{\varphi_m(h)})_{*,e} x \right\rangle \\
&= \left\langle \mathcal{A}_{\varphi_m(h)}, X_x(\varphi_m(h)) \right\rangle \\
&= x,
\end{aligned}$$

showing that

$$R_g^*(\varphi_m^* \mathcal{A}) = \varphi_m^* \mathcal{A}, \ \forall g \in G.$$

In the previous two computations, we used the following identity:

$$\varphi_m \circ R_h = \varphi_{\varphi_m(h)}, \ \forall h \in G, m \in M$$

which is deduced as follows:

$$(\varphi_m \circ R_h)(g) = \varphi_m(gh) = \varphi_{gh}(m) = \varphi_g(\varphi_h(m)) = \varphi_{\varphi_h(m)}(g),$$

for all $g, h \in G$ and $m \in M$.

Appendix D
Cotangent Lifts

In the first part of this appendix, we will study in some details a class of diffeomorphisms of the cotangent bundle known as *cotangent lifts*. These diffeomorphisms, known in classical mechanics under the name of *point transformations* (see [84, 250]), are obtained by *lifting* to the total space of the cotangent bundle the diffeomorphisms of the base manifold.

1 The Cotangent Lift of Diffeomorphisms

Let N be a manifold, T^*N be its cotangent bundle, and $\pi : T^*N \to N$ be the canonical projection. Denote with $\mathscr{D}(N)$ the group of diffeomorphisms of N. Given $\varphi \in \mathscr{D}(N)$, let

$$\widetilde{\varphi} : T^*N : \xrightarrow{\alpha \rightsquigarrow (\varphi^{-1})^*_{,\varphi(n)}(\alpha)} T^*N. \tag{D.1}$$

where $n = \pi(\alpha)$ and $\alpha \in T^*_n N$. Note that $(\varphi^{-1})^*_{\varphi(n)}(\alpha) \in T^*_{\varphi(n)} N$, for every (n, α) such that $n \in N$ and $\alpha \in T^*_n N$. In particular, $(\varphi^{-1})^*_{\varphi(n)} : T^*_n N \to T^*_{\varphi(n)} N$.

Lemma D.1 *Then $\widetilde{\varphi}$ is a diffeomorphism of T^*N such that $\pi \circ \widetilde{\varphi} = \varphi \circ \pi$.*

Proof The proof follows from (D.1). □

Definition D.2 The diffeomorphism $\widetilde{\varphi}$ defined in Formula (D.1) is called the *cotangent* or *total lift* of $\varphi \in \mathscr{D}(N)$.

Problem D.3 Show that the cotangent lift $\varphi \rightsquigarrow \widetilde{\varphi}$ is a group homomorphism from $\mathscr{D}(N)$ to $\mathscr{D}(T^*N)$.
(Hint: It follows from (D.1) and the chain rule) △

© The Author(s), under exclusive license to Springer Nature Switzerland AG 2026   535
A. Arsie, I. Mencattini, *Geometry of Integrable Systems*, Latin American
Mathematics Series – UFSCar subseries,
https://doi.org/10.1007/978-3-031-96282-0

The following result characterizes those diffeomorphisms of T^*N obtained by cotangent lift.

Theorem D.4 *A diffeomorphism $\psi \in \mathscr{D}(T^*N)$ preserves the canonical one-form Θ if and only if $\psi = \widetilde{\varphi}$ for some $\varphi \in \mathscr{D}(N)$.*

Proof In what follows, $\alpha \in T^*N$, i.e., $\pi(\alpha) = n$. Let $\psi = \widetilde{\varphi}$ be the cotangent lift of $\varphi \in \mathscr{D}(N)$. First we show that $(\widetilde{\varphi}^*\Theta)_\alpha = \Theta_\alpha$, for all $\alpha \in T^*N$. Let $v \in T_\alpha(T^*N)$; then

$$
\begin{aligned}
\langle(\widetilde{\varphi}^*\Theta)_\alpha, v\rangle &= \left\langle \Theta_{\left(\varphi(n),(\varphi^{-1})^*_{,\varphi(n)}(\alpha)\right)}, \widetilde{\varphi}_{*\alpha}v\right\rangle \\
&= \left\langle (\varphi^{-1})^*_{,\varphi(n)}(\alpha), (\pi \circ \widetilde{\varphi})_{*,\alpha}v\right\rangle \\
&= \left\langle (\varphi^{-1})^*_{,\varphi(n)}(\alpha), (\varphi \circ \pi)_{*,\alpha}v\right\rangle \text{ see Lemma D.1} \\
&= \left\langle (\varphi^{-1})^*_{,\varphi(n)}(\alpha), \varphi_{*,n}(\pi_{*,\alpha}v)\right\rangle \\
&= \left\langle \alpha, (\varphi^{-1})_{*,\varphi(n)}\left(\varphi_{*,n}(\pi_{*,\alpha}v)\right)\right\rangle \\
&= \left\langle \alpha, \pi_{*,\alpha}v\right\rangle \\
&= \langle \Theta_\alpha, v\rangle .
\end{aligned}
$$

We show now that the vicersa holds. Let $\psi \in \mathscr{D}(T^*N)$ such that $(\psi^*\Theta)_\alpha = \Theta_\alpha$, for all $\alpha \in T^*N$. First we show that ψ restricts to a diffeomorphism of the zero-section of T^*N (see Example C.98.

For each $n \in N$, let $0_n \in T_n^*N$ be the zero-covector. Then, given $u \in T_{\pi(\psi(0_n))}N$ and $v \in T_{\psi(0_n)}(T^*N)$ such that $\pi_{*,\psi(0_n)}(v) = u$ (v exists since π is submersive), one has

$$
\begin{aligned}
\langle\psi(0_n), u\rangle &= \left\langle \psi(0_n), \pi_{*,\psi(0_n)}v\right\rangle \\
&= \left\langle \Theta_{\psi(0_n)}, v\right\rangle \\
&= \left\langle (\psi^*\Theta)_{0_n}, (\psi^{-1})_{*,\psi(0_n)}v\right\rangle \\
&= \left\langle \Theta_{0_n}, (\psi^{-1})_{*,\psi(0_n)}v\right\rangle, \text{ since } \psi^*\Theta = \Theta \\
&= \left\langle 0_n, \pi_{*,0_n}(\psi^{-1})_{*,\psi(0_n)}v\right\rangle \\
&= 0,
\end{aligned}
$$

showing that $\psi(0_n)$ belongs to the zero-section of T^*N; more precisely, this shows that $\psi(0_n) = 0_{\pi(\psi(0_n))}$.

This proves that ψ maps the zero-section to the zero-section of T^*N. Applying the same argument ψ^{-1}, one can conclude that ψ restricts to a diffeomorphism of the zero-section. Using this remark, one can define $\varphi : N \to N$ as the map such that

$$\varphi(n) := \pi \circ \psi(0_n), \ \forall n \in N. \tag{D.2}$$

By definition, it is a smooth map, and an easy computation shows that it is invertible with inverse φ^{-1} such that $\varphi^{-1}(n) = \pi \circ \psi^{-1}(0_n)$, for all $n \in N$. In other words, Formula (D.2) defines a diffeomorphism of N. To conclude, we are left to show that:

(i) $\varphi \circ \pi(\alpha) = \pi \circ \psi(\alpha)$, for all $\alpha \in T^*N$
(ii) $\widetilde{\varphi} = \psi$, i.e., that the cotangent lift of the map φ is the diffeomorphism ψ

Point (i) follows noticing that since ψ preserves the canonical one-form Θ, it will preserve also the so-called *Euler* vector field X_Θ, defined as the vector field such that $i_{X_\Theta}\Omega = \Theta$, where $\Omega = d\Theta$. An easy computation shows that the vector field X_Θ, if written in the canonical coordinates (p, x), assumes the following form:

$$X_\Theta = \sum_{i=1}^{n} p_i \frac{\partial}{\partial p_i}, \tag{D.3}$$

The vector field in Formula (D.3) is called the *Euler* vector field. Formula (D.3) implies that the corresponding one-parameter group of diffeomorphisms has flow given by

$$T_n^*N \xrightarrow{\ \alpha \rightsquigarrow e^t \alpha\ } T_n^*N,$$

i.e., by the *dilatations* along the fibers of T^*N. Since ψ preserves Θ, , it preserves also $\Omega = d\Theta$. Since X_Θ is defined as $i_{X_\Theta}\Omega = \Theta$, X_θ is ψ-invariant, and its flow intertwines the diffeomorphism ψ, i.e.,

$$e^t \psi(\alpha) = \psi(e^t \alpha)$$

for all $\alpha \in T^*N$ and $t \in \mathbb{R}$. Using this observation, one has that

$$\varphi \circ \pi(\alpha) = \varphi \circ \pi(0_{\pi(\alpha)}) = \pi \circ \psi(\lim_{t \to -\infty} e^t \alpha) = \lim_{t \to -\infty} \pi\left(e^t \psi(\alpha)\right) = \pi \circ \psi(\alpha),$$

for all $\alpha \in T^*N$, which proves the point (i). We now show that $\widetilde{\varphi} = \psi$. Recall that the cotangent lift $\widetilde{\varphi}$ of $\varphi \in \mathscr{D}(N)$ is a diffeomorphism of the total space of the cotangent bundle, whose explicit form is given in Formula (D.1). Moreover, since ψ is a diffeomorphism, the composition $(\pi \circ \psi)_{*,\alpha} : T_\alpha(T^*N) \to T_{\varphi(n)}N$ is a surjective map, for all $\alpha \in T^*N$. Then, for every $v \in T_{\varphi(n)}N$, one can find

$u \in T_\alpha(T^*N)$ such that $v = (\pi \circ \psi)_{*,\alpha}(u)$. In this way, one has

$$
\begin{aligned}
\langle \widetilde{\varphi}(\alpha), v \rangle &= \langle \widetilde{\varphi}(\alpha), (\pi \circ \psi)_{*,\alpha}(u) \rangle \\
&= \langle (\varphi^{-1})^{*}_{,\varphi(n))}(\alpha), (\pi \circ \psi)_{*,\alpha}(u) \rangle, \text{ where } n = \pi(\alpha) \\
&= \langle \alpha, (\varphi^{-1})_{*,\varphi(n)} \circ (\pi \circ \psi)_{*,\alpha}(u) \rangle. \\
&= \langle \alpha, (\varphi^{-1} \circ \pi \circ \psi)_{*,\alpha}(u) \rangle \\
&= \langle \alpha, \pi_{*,\alpha}(u) \rangle, \text{ since for } (i) \text{ above } \pi \circ \psi = \varphi \circ \pi \\
&= \langle \Theta_\alpha, u \rangle.
\end{aligned}
$$

On the other hand,

$$
\langle \Theta_\alpha, u \rangle = \langle (\psi^*\Theta)_\alpha, u \rangle = \langle \psi(\alpha), (\pi \circ \psi)_{*,\alpha}(u) \rangle = \langle \psi(\alpha), v \rangle,
$$

for all $\alpha \in T^*N$ and $v \in T_n N$. $\qquad\qquad\qquad\qquad\qquad\qquad\qquad\qquad \square$

Remark D.5 Theorem D.4 entails that $\widetilde{\varphi}$ is a symplectomorphism of (T^*N, Ω); in fact, $\widetilde{\varphi}^* d\Theta = d(\widetilde{\varphi}^*\Theta) = d\Theta$. This observation explains why every choice of (local) coordinates on a manifold N induces canonical coordinates on its cotangent bundle endowed with its canonical symplectic form. $\qquad\qquad\qquad\qquad\qquad\qquad\triangle$

Using Theorem D.4, we show that every $X \in \mathfrak{X}(N)$ can be lifted to $\widetilde{X} \in \mathfrak{X}(T^*N)$. Let $X \in \mathfrak{X}(N)$ and let $\{\varphi_t\}_{t\in\mathbb{R}}$ the corresponding one-parameter group of diffeomorphisms defined by X.

Remark D.6 Here after, to simplify the notation, the one-parameter group of diffeomorphisms defined by X will be denoted by $\{\varphi_t\}_{t\in\mathbb{R}}$ instead by the more precise $\{\varphi_t^X\}_{t\in\mathbb{R}}$. $\qquad\qquad\qquad\qquad\qquad\qquad\qquad\qquad\qquad\triangle$

For every $t \in \mathbb{R}$ such that $\varphi_t : N \to N$ is (globally) defined, let $\widetilde{\varphi}_t : T^*N \to T^*N$ be defined by

$$
\widetilde{\varphi}_t(\alpha) = (\varphi_{-t})^{*}_{,\varphi_t(n)}\alpha,
$$

its cotangent lift, where, as usual, $n = \pi(\alpha)$. In this way, it remains defined the cotangent lift $\{\widetilde{\varphi}_t\}_{t\in\mathbb{R}}$ of the one-parameter group of diffeomorphisms $\{\varphi_t\}_{t\in\mathbb{R}}$. Taking the derivative with respect to the parameter t yields the vector field $\widetilde{X}$. In the local coordinates (p, x), the integral curves associated with the cotangent lift $\{\widetilde{\varphi}_t\}_{t\in\mathbb{R}}$ can be written as

$$
\begin{cases}
x_i(t) = x_i(\varphi_t) \\
p_i(t) = p_i(\varphi^*_{-t})
\end{cases}
$$

for $i = 1, \ldots, n$. In particular, the vector field $\widetilde{X}$ associated with the one-parameter group of diffeomorphisms $\{\widetilde{\varphi}_t\}_{t \in \mathbb{R}}$ will be written as $\widetilde{X} = X_0 + X_1$. Here $X_0 = \sum_{i=1}^{n} X_i \frac{\partial}{\partial x_i}$ and $X_i = \frac{d}{dt} x_i(\varphi_t)$, i.e., X_0 coincides with the vector field X whose one-parameter group of diffeomorphisms is $\{\varphi_t\}_{t \in \mathbb{R}}$. An explicit expression for X_1 can be found as follows:

$$p_i\left((\varphi_{-t}^*)\alpha\right) = \left\langle (\varphi_{-t}^*)\alpha, \frac{\partial}{\partial x_i} \right\rangle = \left\langle \alpha, (\varphi_{-t})_* \frac{\partial}{\partial x_i} \right\rangle = p_k(\alpha) \frac{\partial(x_k \circ \varphi_{-t})}{\partial x_i}$$

if we write $\alpha = \sum_{k=1}^{n} p_k(\alpha) dx_k$. Then it follows that

$$\frac{d}{dt} p_i\left((\varphi_{-t}^*)\alpha\right) = -p_k(\alpha) \frac{\partial X_k}{\partial x_i}.$$

So we proved the following.

Corollary D.7 *If $X \in \mathfrak{X}(N)$ and $\{\varphi_t\}_{t \in \mathbb{R}}$ are as above, then:*

(i) $\{\widetilde{\varphi}_t\}_{t \in \mathbb{R}}$ *is given by*

$$\widetilde{\varphi}_t(\alpha) = (\varphi_{-t})^*_{,\varphi_t(n)} \alpha, \ \forall \alpha \in T^*N.$$

(ii) This one-parameter group of diffeomorphism of T^*N corresponds to a vector field $\widetilde{X}$, called the cotangent lift of X, which, in (local fibered) coordinates (p, x), can be written as $\widetilde{X} = X_0 + X_1 = X + X_1$, where

$$\begin{cases} X_0 = \sum_{i=1}^{n} X_i(x) \frac{\partial}{\partial x_i} \\ X_1 = -\sum_{i,k=1}^{n} p_k \frac{\partial X_k}{\partial x_i} \frac{\partial}{\partial p_i} \end{cases} \tag{D.4}$$

Since $\{\widetilde{\varphi}_t\}_{t \in \mathbb{R}}$ is the cotangent lift of $\{\varphi_t\}_{t \in \mathbb{R}}$, each diffeomorphism $\widetilde{\varphi}_t$ of this family will satisfy the condition $\widetilde{\varphi}_t^* \Theta = \Theta$. Since this implies that $\widetilde{X}$ is an infinitesimal symmetry of the canonical one-form Θ, i.e., $\mathscr{L}_{\widetilde{X}} \Theta = 0$, $\widetilde{X}$ is a Hamiltonian vector field. In fact

Proposition D.8 *Let (M, ω) be an exact symplectic manifold (see Definition 1.14 in Chapter 1). Then, if θ is a symplectic potential of (M, ω), every $X \in \mathfrak{X}(M)$ such that $\mathscr{L}_X \theta = 0$ is Hamiltonian.*

Proof It suffices to observe that $\mathscr{L}_X \omega = d\mathscr{L}_X \theta = 0$ and that $\mathscr{L}_X \theta = i_X(d\theta) + d(i_X \theta)$. $\qquad \square$

In local coordinates, a corresponding Hamiltonian function assumes the following expression:

$$F(p, x) = \sum_{i=1}^{n} X_i(x) p_i \tag{D.5}$$

Proof From the proof of Proposition D.8, one knows that $F = i_{\tilde{X}}\Theta$. Writing the left-hand side of this identity in coordinates, one arrives to Formula (D.5).   □

Remark D.9 First we want to stress that the function defined in Formula (D.5) depends linearly on the vector field X. To underline such a dependence, we will write F_X instead of F. Then we want also to remark that the function defined in (D.5) is a smooth function on the total space of T^*N of a very special kind. In fact, it is a smooth function on (the total space of) the cotangent bundle, which is linear along the fiber of T^*M. These functions form a vector space that will be denoted by $C^\infty_{lin}(T^*N)$. Every function $F \in C^\infty_{lin}(T^*N)$ can be written locally as $F(p, x) = \sum_{i=1}^n f_i(x)p_i$.   △

Before moving forward with our discussion, we observe that

Lemma D.10 $C^\infty_{lin}(T^*N)$ *is a Poisson subalgebra of* $(C^\infty(T^*N), \{\cdot, \cdot\})$, *where* $\{\cdot, \cdot\}$ *is the Poisson bracket defined by the canonical symplectic structure* $\Omega = d\Theta$.

Proof This result follows by direct computation. If $F, G \in C^\infty_{lin}(T^*N)$, then using a local system of (fibered) coordinates (p, x), one can write $F(p, x) = \sum_{i=1}^n f_i(x)p_i$ and $G(p, x) = \sum_{i=1}^n g_i(x)p_i$, where $f_i, g_i \in C^\infty(T^*N)$, eventually defined only locally and independent on the coordinates $(p_1, \ldots, p_n)$. Using this representation, one can write

$$\{F, G\} = \sum_{i=1}^n \frac{\partial F}{\partial p_i}\frac{\partial G}{\partial x_i} - \frac{\partial G}{\partial p_i}\frac{\partial F}{\partial x_i} = \sum_{i,k=1}^n \left(f_k \frac{\partial g_i}{\partial x_k} - g_k \frac{\partial f_i}{\partial x_k} \right) p_i .$$

□

Remark D.9 implies that one can define the linear map

$$F_{(\cdot)} : \mathfrak{X}(N) \xrightarrow{\quad X=\sum_{i=1}^n X_i \frac{\partial}{\partial x_i} \rightsquigarrow F_X(p,x)=\sum_{i=1}^n X_i p_i \quad} C^\infty_{lin}(T^*N). \tag{D.6}$$

To spell out its properties, note that both $\mathfrak{X}(N)$ and $C^\infty_{lin}(T^*N)$ are endowed with a structure of Lie algebra, the first provided by the commutator between two vector fields and the second one by the (canonical) Poisson structure (see Lemma D.10). Then observe that for every $X \in \mathfrak{X}(N)$, (D.6) defines $F_X : T^*N \to \mathbb{R}$ by

$$F_X : T^*N \xrightarrow{\quad \alpha \rightsquigarrow \langle \alpha, X(\pi(\alpha)) \rangle = \sum_{i=1}^n \alpha_i X_i(n) \quad} \mathbb{R}$$

where $\alpha = \sum_{i=1}^n p_i(\alpha)dx_i = \sum_{i=1}^n \alpha_i dx_i$. Now one can prove the following.

Proposition D.11 *The linear map* $F_{(\cdot)} : \mathfrak{X}(N) \to C^\infty_{lin}(T^*N)$:

(i) *Is a homomorphism of Lie algebras, i.e.,*

$$F_{[X.Y]} = \{F_X, F_Y\}, \; \forall X, Y \in \mathfrak{X}(N) \tag{D.7}$$

(ii) Intertwines the action of the group of diffeomorphism $\mathscr{D}(N)$, i.e.,

$$F_{\varphi X} = \tilde{\varphi}^{-1} F_X, \quad \varphi \in \mathscr{D}(N), \; X \in \mathfrak{X}(N)$$

Proof

(i) It suffices to write the explicit expressions for both sides of (D.7) and then compare the results. Indeed, one has:

$$F_{[X,Y]}(p, x) = \sum_{k=1}^{n} [X, Y]_k \, p_k = \sum_{k,l=1}^{n} \left(X_l \frac{\partial Y_k}{\partial x_l} - Y_l \frac{\partial X_k}{\partial x_l} \right) p_k,$$

while, writing $F_X(p, x) = \sum_{k=1}^{n} X_k p_k$ and $F_Y(p, x) = \sum_{l=1}^{n} Y_l p_l$, one has

$$\{F_X, F_Y\}(p, x) = \sum_{k,l=1}^{n} \left\{ X_k p_k, Y_l p_l \right\} = \sum_{k,l=1}^{n} \left(X_l \frac{\partial Y_k}{\partial x_l} - Y_l \frac{\partial X_k}{\partial x_l} \right) p_k.$$

(ii) First recall that the action of $\varphi \in \mathscr{D}(N)$ on $\mathfrak{X}(N)$ and on $C^\infty(N)$ is given by the formulas

$$(\varphi X)(n) = \left(\varphi^{-1} \right)_{*, \varphi(n)} X(\varphi(n)),$$

and, respectively,

$$(\varphi f)(n) = f\left(\varphi^{-1}(n) \right),$$

for every $X \in \mathfrak{X}(N)$, $f \in C^\infty(N)$, and $n \in N$ (see Formulas 1.28). Moreover, recall that the cotangent lift defines a group homomorphism between $\mathscr{D}(N)$ and $\mathscr{D}(T^*(N))$ (see Problem D.3). Let $n = \pi(\alpha)$. Then, one has

$$\begin{aligned}
\left(\tilde{\varphi}^{-1} F_X \right)(\alpha) &= F_X\left((\varphi^{-1})^*_{,\varphi(n)}(\alpha) \right) \\
&= \left\langle (\varphi^{-1})^*_{,\varphi(n)}(\alpha), \, X(\varphi(n)) \right\rangle \\
&= \left\langle \alpha, \, (\varphi^{-1})_{*,\varphi(n)} X(\varphi(n)) \right\rangle \\
&= \langle \alpha, \, (\varphi X)(n) \rangle \\
&= F_{\varphi X}(\alpha).
\end{aligned}$$

$\square$

Problem D.12 Let $X, Y \in \mathfrak{X}(N)$ be two vector fields and $\widetilde{X}, \widetilde{Y}$ their cotangent lifts (see Formula D.4). Show that

$$\widetilde{[X, Y]} = [\widetilde{X}, \widetilde{Y}]$$

i.e., show that the (linear) map

$$\mathfrak{X}(N) \xrightarrow{\quad X=X^i \frac{\partial}{\partial x^i} \rightsquigarrow \widetilde{X}=X - p_k \frac{\partial X^k}{\partial x^i} \frac{\partial}{\partial p_i} \quad} C^\infty(T^*N)$$

is a homomorphism of Lie algebras. $\triangle$

Before moving to the next section, we would like to make one more remark. As it was shown above, cotangent lifts of diffeomorphisms of N are an important class of symplectomorphisms of (T^*N, Ω) (see Theorem D.4). There is another class of diffeomorphisms of T^*N that deserves to be mentioned, and it is defined also starting from suitable geometrical structures on N. More precisely, let $\eta \in \Omega^1(N)$ and let us define

$$t_\eta : T^*N \xrightarrow{\quad \alpha \rightsquigarrow \alpha + \eta_n \quad} T^*N, \ \forall \alpha \in T_n^*N, \text{ and } n \in N. \tag{D.8}$$

Then

Lemma D.13 *For all $\eta \in \Omega^1(N)$, the map t_η defined in Formula (D.8) is:*

(i) A diffeomorphism of the cotangent bundle T^*N
(ii) A symplectomorphism of (T^*N, Ω) if and only if η is closed

Proof The proof of part (i) of the lemma is left to the reader as an exercise. To prove part (ii), it suffices to prove that:

$$t_\eta^*\Theta = \Theta + p^*\eta, \tag{D.9}$$

where $p : T^*N \to N$ is the canonical projection. The proof of Formula (D.9) is the result of the following computation.

$$\begin{aligned}
\left\langle (t_\eta^*\Theta)_\alpha, v \right\rangle &= \left\langle \Theta_{t_\eta(\alpha)}, (t_\eta)_{*,\alpha} v \right\rangle \\
&= \left\langle \Theta_{\alpha + \eta_n}, (t_\eta)_{*,\alpha} v \right\rangle \\
&= \left\langle \alpha + \eta_n, p_{*,\alpha + \eta_n} (t_\eta)_{*,\alpha} v \right\rangle \\
&= \left\langle \alpha + \eta_n, (p \circ t_\eta)_{*,\alpha} v \right\rangle \\
&= \left\langle \alpha + \eta_n, p_{*,\alpha} v \right\rangle \text{ since } p \circ t_\eta = p
\end{aligned}$$

$$= \langle \alpha, p_{*,\alpha} v \rangle + \langle \eta_n, p_{*,\alpha} v \rangle$$

$$= \langle \Theta_\alpha, v \rangle + \langle (p^* \eta)_\alpha, v \rangle$$

$$= \langle \Theta_\alpha + (p^* \eta)_\alpha, v \rangle.$$

for all $n \in N$, $\alpha \in T_n^* N$, and $v \in T_\alpha(T^* N)$. $\qquad\square$

Remark D.14 Note that the map t_η represents a *translation* along the fibers of the cotangent bundle. When η is a closed one-form, the corresponding map t_η moves the zero-section of $T^* N$ to another Lagrangian section of the cotangent bundle of N. Furthermore, while the zero-section of $T^* N$ could be characterized as the set of all points of the cotangent bundle of N where the Liouville form Θ is identically equal to zero, one could say that, given a closed one-form η, the translation of the zero-section via the map t_η is the locus of the points of $T^* N$ where the one-form $\Theta - p^* \eta$ is identically zero. $\qquad\triangle$

2 Cotangent Lifts and Semi-direct Products

In this section, we discuss the cotangent lift of the action of a semi-direct product. To this end, we start noticing that every $f \in C^\infty(N)$ defines the diffeomorphism

$$t_f : T^* N \xrightarrow{\ \alpha \rightsquigarrow \alpha - df_n\ } T^* N, \tag{D.10}$$

where $n = \pi(\alpha)$. In particular, the map $f \rightsquigarrow t_f$ defined by (D.10) is a homomorphism between (the Abelian group) $C^\infty(N)$ and $\mathscr{D}(T^* N)$, whose kernel can be identified with $\mathbb{R}$. Note that (D.10) is a particular case of (D.8). Together with the previous map, recall that $\mathscr{D}(N)$ embeds into $\mathscr{D}(T^* N)$ via

$$\varphi \rightsquigarrow \widetilde{\varphi} \tag{D.11}$$

(see (D.1)). Observe that on $C^\infty(N)$ is defined a natural (left) action of $\mathscr{D}(N)$, $(\varphi, f) \rightsquigarrow f \circ \varphi^{-1}$ so that one can define the semi-direct product $\mathscr{D}(N) \ltimes C^\infty(N)$ whose composition law is

$$(\varphi_1, f_1)(\varphi_2, f_2) = \left(\varphi_1 \varphi_2, f_1 + f_2 \circ \varphi_1^{-1} \right) \tag{D.12}$$

With a slight abuse of notation, we will identify $C^\infty(N)$ and $\mathscr{D}(N)$ with their images under the maps (D.10) and, respectively, (D.11) into $\mathscr{D}(T^* N)$ (note however that $f \rightsquigarrow t_f$ is not injective). Under this identification, one can prove that

Lemma D.15 *The group $\mathscr{D}(N)$ is a subgroup of the normalizer of $C^\infty(N)$ in $\mathscr{D}(T^*N)$.*

Proof As usual, let $n = \pi(\alpha)$. The proof follows from the following computation:

$$
\widetilde{\varphi} t_f \widetilde{\varphi}^{-1}(\alpha) = \widetilde{\varphi}\Big((\varphi)^*_{,\varphi^{-1}(n)}(\alpha - df_{\varphi^{-1}(n)})\Big)
$$

$$
= (\varphi^{-1})^*_{,n}\Big(\varphi^*_{,\varphi^{-1}(n)}\alpha\Big) - (\varphi^{-1})^*_{,n}\Big(df_{\varphi^{-1}(n)}\Big)
$$

$$
= \alpha - (\varphi^{-1})^*_{,n} df_{\varphi^{-1}(n)}
$$

Since $(\varphi^{-1})^*_{,n} df_{\varphi^{-1}(n)} = d(f \circ \varphi^{-1})_n$ for all $n \in N$, the previous computation tells us that $\widetilde{\varphi} t_f \widetilde{\varphi}^{-1} = t_{f\circ\varphi^{-1}}$, which proves the statement. $\qquad\square$

Using the result contained in the previous lemma, one can build the semi-direct product of $\mathscr{D}(N)$ with $C^\infty(N)$ *inside* $\mathscr{D}(T^*N)$. In this case, the product between $(\widetilde{\varphi}_1, t_{f_1})$ and $(\widetilde{\varphi}_2, t_{f_2})$ is given by

$$
(\widetilde{\varphi}_1, t_{f_1})(\widetilde{\varphi}_2, t_{f_2}) = \big(\widetilde{\varphi}_1\widetilde{\varphi}_2, t_{f_1} + \widetilde{\varphi}_1 t_{f_2}\widetilde{\varphi}_1^{-1}\big) = \big(\widetilde{\varphi_1\varphi_2}, t_{f_1} + t_{f_2\circ\varphi_1^{-1}}\big)
$$

$$
= \big(\widetilde{\varphi_1\varphi_2}, t_{f_1 + f_2\circ\varphi_1^{-1}}\big) \tag{D.13}
$$

Let us now define the map $\Lambda : (\varphi, f) \rightsquigarrow (\widetilde{\varphi}, t_f)$. Then

Lemma D.16 *The map Λ is a group homomorphism between the group $\mathscr{D}(N) \ltimes C^\infty(N)$ (see Formula D.12) and the group $\mathscr{D}(T^*N)$.*

Proof In fact

$$
\Lambda((\varphi_1, f_1)(\varphi_2, f_2)) = \Lambda\big(\varphi_1\varphi_2, f_1 + f_2 \circ \varphi_1^{-1}\big) = \big(\widetilde{\varphi_1\varphi_2}, t_{f_1 + f_2\circ\varphi_1^{-1}}\big)
$$

as it follows from the definition of the map Λ and from Formula (D.12). Using now Formula, (D.13), one can write:

$$
\Lambda((\varphi_1, f_1)(\varphi_2, f_2)) = (\widetilde{\varphi}_1, t_{f_1})(\widetilde{\varphi}_2, t_{f_2}) = \Lambda(\varphi_1, f_1)\Lambda(\varphi_2, f_2).
$$

$$\square$$

Let G be a Lie group, (V, ρ) a finite dimensional G–module, $\varphi : G \times N \to N$ a (left) G-action, and $T : V \to C^\infty(N)$, $v \rightsquigarrow f_v := T(v)$ a G-equivariant linear map. Furthermore, assume that f, ρ and φ satisfy the following condition, $f_{\rho_g(v)} = f_v \circ \varphi_g^{-1}$, for all $v \in V$ and all $g \in G$. Under these assumptions, we have the following.

Proposition D.17 *The map $G \ltimes_\rho V \to \mathscr{D}(N) \ltimes C^\infty(N)$, defined by $(g, v) \rightsquigarrow (\varphi_g, f_v)$, is a group homomorphism. Composing this homomorphism with the one* *defined in Lemma D.16, one obtains a group homomorphism from $G \ltimes_\rho V$ to $\mathscr{D}(T^*N)$.*

Proof It is follows at once from the definitions of the maps involved in the statement. $\square$

One could rephrase the previous proposition saying that the semi-direct product $G \ltimes_\rho V$ acts by diffeomorphisms on T^*N. The following theorem makes this statement more precise.

Theorem D.18 *The action of $G \ltimes_\rho V$ on T^*N defined in the previous proposition is Hamiltonian with moment map $\mu : T^*N \to \mathfrak{g}^* \oplus V^*$ defined by*

$$\langle \mu(\alpha), (x, v)\rangle = F_x(\alpha) + (\pi^* f_v)(\alpha), \ \forall (x, v) \in \mathfrak{g} \oplus V,$$

where, for every $x \in \mathfrak{g}$, $F_x = F_{X_x} : T^*N \to \mathbb{R}$ is defined in (D.6) and where X_x is the fundamental vector field defined by $x \in \mathfrak{g}$ (see also Chapter 3).

Proof Let r be the Lie algebra representation associated with ρ (see Formula C.13). The infinitesimal action $G \ltimes_\rho V$ on T^*N is defined by the morphism that to every $(x, v) \in \mathfrak{g} \ltimes_r V$ associates the vector field $X_x + X_{\pi^* f_v}$. To prove the statement, it suffices to show that $\mu^\sharp : \mathfrak{g} \ltimes_r V \to C^\infty(T^*N), \{\cdot, \cdot\})$, defined by $\mu^\sharp(x, v) = F_x + \pi^* f_v$, is a Lie algebra morphism (see Section 1 in Chapter 3). This follows from a direct computation, observing that, for all $u, v \in V$, $\{\pi^* f_v, \pi^* f_u\} = 0$ since every $f \in C^\infty(N)$, $\pi^* f$ is a p-independent function. $\square$

3   Cotangent Lift of Endomorphisms

In this section, we will collect some information about the lift of endomorphisms of the tangent bundle to the cotangent bundle. Recall that an endomorphism of TQ is a smooth map $N : TQ \to TQ$ such that $N_m \in \mathrm{End}_\mathbb{K}(T_m Q)$ for all $m \in Q$. In other words, N is a smooth map between TQ and TQ, which preserves the bundle structure, i.e., $p \circ N = p$ where $p : TQ \to Q$ is the canonical projection. Furthermore, N is linear along the fibers of p. In local coordinates, $x_1, \ldots, x_n$

$$N = \sum_{i=1}^{n} N_i^j dx_i \otimes \frac{\partial}{\partial x_j},$$

where N_j^i are smooth functions locally defined on M. The endomorphisms of TM are smooth sections of $\mathrm{End}_{C^\infty(Q)}(TQ)$ and are often called tensors of type $(1, 1)$. The lift of $N \in \mathrm{End}_{C^\infty(Q)}(TQ)$ to the cotangent bundle will be the element of $\tilde{N} \in \mathrm{End}_{C^\infty(T^*Q)}(TT^*Q)$ defined by

$$i_{\tilde{N}(\xi)} \Omega = i_\xi \tau_N^* \Omega, \ \forall \xi \in \mathfrak{X}(T^*Q),$$

where $\tau_N : T^*Q \to T^*Q$ is the C^∞-map defined by

$$\langle \tau_N(\alpha), v \rangle = \langle {}^t N(\alpha), v \rangle = \langle \alpha, Nv \rangle, \ \forall \alpha \in T^*Q, \ v \in T_{\pi(\alpha)}Q. \tag{D.14}$$

Recall that the torsion of N is the tensor defined by

$$T_N(X, Y) = [NX, XY] - N([NX, Y] + [X, NY] - N[X, Y]), \ \forall X, Y \in \mathfrak{X}(Q).$$

It is possible to prove that if $T_N = 0$, then $T_{\tilde{N}} = 0$ (see [55]).

Bibliography

1. Abraham, R., and J. Marsden. 1978. *Foundations of mechanics*, vol. 75. Advanced book program. Reading, MA: The Benjamin/Cumming Publishing Company, Inc.
2. Abraham, R., J. Marsden, and T. Ratiu. 1998. *Manifolds, tensor analysis and applications*, vol. 75. Applied mathematical sciences. Berlin: Springer.
3. Adams, M., J. Harnad, and J. Hurtubise. 1993. Darboux coordinates and Liouville-Arnold integration in loop algebras. *Communications in Mathematical Physics* 155: 385–413.
4. Adams, M., J. Harnad, and J. Hurtubise. 1997. Darboux coordinates on coadjoint orbits of Lie algebras. *Letters in Mathematical Physics* 40: 41–57.
5. Adler, M. 1978/1979. On a trace functional for formal pseudo-differential operators and the symplectic structure of the Korteweg-De Vries type equations. *Inventiones mathematicae* 50: 219–248.
6. Adler, M., P. van Moerbeke, and P. Vanhaecke. 2004. *Algebraic integrability, Painlevé geometry and lie algebras*, vol. 47. A series of modern surveys in mathematics. Berlin: Springer.
7. Agrachev, A., and Y. Sachkov. 2004. *Control theory from the geometric viewpoint*, Encyclopedia of mathematical sciences control theory and optimization. Berlin: Springer.
8. Airault, M., H. McKean, and J. Moser. 1977. Rational and elliptic solutions of the Korteweg-De Vries equation and a related many-body problem. *Communications on Pure and Applied Mathematics* 30: 95–148.
9. Arnold, V. 1966 Sur la géométrie différentielle des groupes de Lie de dimension infinie et ses aplications à l'hydrodynamique des fluids parfaits. *Annales De L'Institut Fourier (Grenoble)* 16: 319–361.
10. Arnold, V. 1989. *Mathematical methods of classical mechanics*, vol. 60. Graduate texts in mathematics, GTM, 2nd ed. New York/Berlin/Heidelberg: Springer.
11. Arnold, V. 2004. *Lectures on partial differential equations*, Universititext, 1st ed. Berlin/Heidelberg: Springer.
12. Arnold, V., and B. Khesin. 1998. *Topological methods in hydrodynamics*, vol. 125. Applied mathematical sciences. New York/Berlin/Heidelberg: Springer.
13. Arnold, V. I., and A. B. Givental'. 2001. Symplectic geometry. In *Dynamical systems, IV*, vol. 4. Encyclopedia of mathematical sciences, 1–138. Berlin: Springer.
14. Arsie, A., and C. Ebenbauer. 2010. Locating omega-limit sets using height functions. *Journal of Differential Equations* 248: 2458–2469.
15. Arsie, A., and C. Ebenbauer. 2015. A Hessenberg-Jacobi isospectral flow. *Nonlinear Differential Equations and Applications (NoDEA)* 22: 87–103.

16. Arsie, A., and P. Lorenzoni. 2012. Poisson bracket on 1-forms and evolutionary partial differential equations. *Journal of Physics A: Mathematical and Theoretical* 45: 475208.

17. Arsie, A., and P. Lorenzoni. 2013. F-manifolds with eventual identities, bidifferential calculus and twisted Lenard-Magri chains. *The International Mathematics Research Notices (IMRN)* 2013 (17): 3831–3976.

18. Arsie, A., and K. Pokharel. 2015. A normalizing isospectral flow on complex hessenberg matrices. *Journal of Mathematical Analysis and Applications* 432: 787–805.

19. Audin, M. 1996. *Spinning tops*, vol. 51. Cambridge studies in advanced mathematics. Providence: Cambridge University Press.

20. Audin, M. 2004. *Torus actions on symplectic manifolds*, vol. 93. Progress in mathematics, revised ed. Basel: Birkhäuser Verlag.

21. Audin, M. 2008. *Hamiltonian systems and their integrability*, vol. 15. SMF/AMS texts and monographs. Providence: American Mathematical Society.

22. Baik, J., P. Deift, and T. Suidan. 2016. *Combinatorics and random matrix theory*, vol. 172. Graduate studies in mathematics. Providence: American Mathematical Society.

23. Bartocci, C., G. Falqui, I. Mencattini, G. Ortenzi, and M. Pedroni. 2010. On the geometric origin of the bi-Hamiltonian structure of the rational Calogero-Moser system. *International Mathematics Research Notices (IMRN)* 2010: 279–296.

24. Barucchi, G., and T. Regge. 1977. Conformal properties of a class of exactly solvable n-body problems in space dimension one. *Journal of Mathematical Physics* 18: 1149.

25. Beauville, A. 1983. Variétés kählériennes dont la première classe de chern est nulle. *Journal of Differential Geometry* 18: 755–782.

26. Ben-Zvi, D., and T. Nevins. 2008. From solitons to many-body systems. *Pure and Applied Mathematics Quarterly* 4: 319–361.

27. Ben-Zvi, D., and T. Nevins. 2008. Perverse bundles and Calogero-Moser spaces. *Compositio Mathematica* 144: 1403–1428.

28. Benenti, S. 2011. *Hamiltonian structures and generating families*, Classic reviews in mathematics and mathematical physics 1. New York: Springer.

29. Berger, C. 1955. Sur les groupes d' holonomie homogène des variétés à connexion affines et des variétés riemanniennes. *The Bulletin de la Société Mathématique de France* 83: 279–330.

30. Berndt, R. 2001. *An introduction to symplectic geometry*, vol. 26. Graduate studies in mathematics. Providence: AMS.

31. Bishop, R. L., and S. I. Goldberg. 1980. *Tensor analysis on manifolds*. New York: Dover Publications.

32. Błaszak, M. 2007. *Multi-Hamiltonian theory of dynamical systems*. Texts and monographs in physics. Berlin/Heidelberg: Springer.

33. Bloch, A., R. W. Brockett, and T. Ratiu. 1992. Completely integrable gradient flows. *Communications in Mathematical Physics* 147: 57–74.

34. Bogoyavlensky, O. 1976. On perturbations of the periodic Toda lattice. *Communications in Mathematical Physics* 51: 201–209.

35. Bolsinov, A. 1992. Compatible Poisson brackets on Lie algebras and the completeness of families of functions in involution. *Mathematics of the USSR Izvestiya* 38: 69–90.

36. Bott, R., and L. Tu. 1995. *Differential forms in algebraic topology*, vol. 82. Graduate texts in mathematics. Berlin: Springer.

37. Boualem, H., and R. Brouzet. 2021. Generically, Arnold-Liouville systems cannot be bi-Hamiltonian. *SIGMA Symmetry, Integrability and Geometry: Methods and Applications* 17: 44–45.

38. Brockett, R. 1991. Dynamical systems that sort lists, diagonalize matrices, and solve linear programming problems. *Linear Algebra and Its Applications* 146: 79–91.

39. Brouzet, R. 1990. Systèmes bihamiltonianens et complète intégrabilité en dimension 4. *C.R. Acad. Sci. Paris* 311(série I): 895–898.

40. Brouzet, R. 1991. *Géométrie des systèmes bihamiltoniens en dimension 4*, Thèse Université de Montpellier.

41. Brouzet, R. 1993. About the existence of recursion operators for completely integrable Hamiltonian systems near a Liouville torus. *Journal of Mathematical Physics* 34: 1309–1313.

42. Brouzet, R., P. Molino, and F. Turiel. 1993. Géométrie des systèmes bihamiltoniens. *Indagationes Mathematicae* 4: 269–296.

43. Burstall, F., and F. Pedit. 1994. Harmonic maps via Adler-Kostant-Symes theory. In *Harmonic maps and integrable systems*, vol. E23. Aspects of mathematics, 221–272. Wiesbaden: Vieweg+Teubner Verlag.

44. Butterfield, J. 2007. On symplectic reduction in mechanics. In *Philosophy of physics*, vol. 35. Handbook of the philosophy of science, 1–133. Amsterdam: Elsevier.

45. Calabi, E. 1979. Métriques kählériannes et fibrés holomorphes. *Ann. Sci.École Norm.Sup* 12: 269–294.

46. Calogero, F. 1971. Solution of the one-dimensional N-body problems with quadratic and/or inversely quadratic pair potentials. *Journal of Mathematical Physics* 12: 419–436. Erratum, *ibid.*, 37 (1996), 3646.

47. Candel, A., and L. Colon. 1999. *Foliations I*, vol. 23. Graduate studies in mathematics. Providence: American Mathematical Society.

48. Cannas da Silva, A. 2001. *Lectures on symplectic geometry*, vol. 1764. Lecture notes in mathematics. Berlin: Springer.

49. Caratheodory, C. 1965. *Calculus of variations and partial differential equations of the first order*. AMS, Chelsea Publishing, Providence: American Mathematical Society.

50. Cardin, F. 2015. *Elementary symplectic topology and mechanics*, vol. 16. Lecture notes of the Unione Matematica Italiana. Berlin: Springer.

51. Chu, M. T. 2008. Linear algebra algorithms as dynamical systems. *Acta Numerica* 17: 1–86.

52. Coste, A., P. Dazord, and A. Weinsten. 1987. Groupoïds sympletiques. *Publ. Dép. Math. Nouvelle Sér. A* 87: i–ii, 1–62.

53. Courant, R., and D. Hilbert. 1989. *Methods of mathematical physics, volume 2*, vol. 2. Wiley classical library. New York: Wiley.

54. Crainic, M., R. Fernandes, and I. Marcut. 2021. *Lectures on poisson geometry*, vol. 217. Graduate studies in mathematics. Providence: American Mathematical Society.

55. Crampin, M., F. Cantrijn, and W. Sarlet. 1987. Lifting geometric objects to a cotangent bundle, and the geometry of the cotangent bundle of a tangent bundle. *Journal of Geometry and Physics* 4: 469–492.

56. Crampin, M., W. Sarlet, and G. Thompson. 2000. Bi-differential calculi and bi-Hamiltonian systems. *Journal of Physics A* 33: L177–L180.

57. Crampin, M., W. Sarlet, and G. Thompson. 2000. Bi-differential calculi, bi-Hamiltonian systems and conformal Killing tensors. *Journal of Physics A* 33: 8755–8770.

58. Cushman, R. 2005. No polar coordinates. In *Geometric mechanics and symmetries*. The Peyresq lectures, vol. 306. London Mathematical Society. Lecture notes series, 211–301. Cambridge: Cambridge University Press.

59. Cushman, R., and L. Bates. 1997. *Global aspects of classical integrable systems*. Basel/Boston/Berlin: Birkhäuser Verlag.

60. Damianou, P. 2004. Master symmetries and the R-matrices for the Toda lattice. *Letters in Mathematical Physics* 20: 101–112.

61. Damianou, P. 2004. Multiple Hamiltonian structure of Bogoyavlensky-Toda lattice. *Reviews in Mathematical Physics* 16: 175–241.

62. Damianou, P., and R. Fernandes. 2008. Integrable hierarchies and the modular class. *Annales de l'Institut Fourier (Grenoble)* 58: 107–137.

63. Das, A., and S. Okubo. 1989. A systematic study of the Toda lattice. *Annals of Physics* 190: 215–232.

64. Dazord, P., and T. Delzant. 1987. Le Probleme General des Variables Action–Angles. *Journal of Differential Geometry* 26: 223–251.

65. Deift, P., L.-C. Li, T. Nanda, and C. Tomei. 1986. The Toda flow on a generic orbit is integrable. *Communications on Pure and Applied Mathematics* 39: 193–232.

66. Deift, P., L.-C. Li, and C. Tomei. 1989. Matrix factorizations and integrable systems. *Communications on Pure and Applied Mathematics* 42: 443–521.
67. Demmel, J.-W. 1997. *Applied numerical linear algebra*, vol. 242, Society for industrial and applied mathematics. Nantik, MA: SIAM.
68. Dimakis, A., and F. Müller-Hoissen. 2000. Bi-differential calculi and integrable models. *Journal of Physics A* 33: 957–974.
69. Dufour, J., and N. Zung. 2005. *Poisson structures and their normal forms*, vol. 242. Progress in mathematics. Basel/Boston/Berlin: Birkäuser.
70. Dugas, R. 1988. *A history of mechanics*, vol. 242. New York: Dover Publications.
71. Duistermaat, J. 1980. On global action-angle coordinates. *Communications on Pure and Applied Mathematics* 33: 687–706.
72. Duistermaat, J., and J. Kolk. 2000. *Lie groups*, vol. 242. Universitext. Basel/Heidelberg/New York: Springer.
73. Ebenbauer, C. 2007. A dynamical system that computes eigenvalues and diagonalizes matrices with a real spectrum. In *Proceedings of the 46th IEEE Conference on Decision and Control (CDC)*, 1704–1709. Piscataway: IEEE.
74. Ebenbauer, C., and A. Arsie. 2008. On an eigenflow equation and its Lie algebraic generalization. *Communications in Information and Systems* 8: 147–170.
75. Ebrahimi-Fard, K., A. Lundervold, I. Mencattini, and H. Z. Munthe-Kaas. 2025. Postlie algebras and isospectral flows. *Symmetry, Integrability and Geometry: Methods and Applications (SIGMA)* 11: 16p.
76. Efstathiou, K., M. Joyeux, and D. Sadovskíi. 2004. Global bending quantum numbers and the absence of monodromy in the HCN $\leftrightarrow$ CN H molecule. *Physical Review Letters* 69: 032504-1–15.
77. Eguchi, T., and A. Hanson. 1978. Asymptotically flat self-dual solutions to euclidean gravity. *Physics Letters* 74: 249–251.
78. Etingof, P. 2007. *Calogero-Moser systems and representation theory*, Zurich lectures in advanced mathematics. Zürich: European Mathematical Society (EMS).
79. Faddeev, L., and L. Takhtajan. 2013. *Hamiltonian methods in the theory of solitons*, Classics in mathematics. Reprinted edition, Berlin/Heidelberg: Springer.
80. Falqui, G., F. Magri, and M. Pedroni. 2003. The method of Poisson pairs in the theory of nonlinear PDEs. In *Direct and inverse methods in nonlinear evolution equations*, vol. 632. Lecture notes in physics, 85–136. Berlin: Springer.
81. Falqui, G., and I. Mencattini. 2017. Bi-Hamiltonian geometry and canonical spectral coordinates for the rational Calogero-Moser system. *Journal of Geometry and Physics* 118: 126–137.
82. Falqui, G., and M. Pedroni. 2003. Separation of variables for Bi-Hamiltonian systems. *Mathematical Physics Analysis and Geometry* 6: 139–179.
83. Falqui, G., and M. Pedroni. 2011. Poisson pencils, algebraic integrability and separation of variables. *Regular and Chaotic Dynamics* 16: 223–244.
84. Fasano, A., and S. Marmi. 2013. *Analytical mechanics*, vol. 242. Oxford graduate texts, Reprinted edition. Oxford, UK: Oxford University Press.
85. Fassò, F. 1999. *Notes on finite dimensional integrable systems*, Unpublished manuscript, Downloadable from the webpage of the author.
86. Faybusovich, L. 1991. Hamiltonian structure of dynamical systems which solve linear programming problems. *Physics D* 53: 217–232.
87. Fernandes, R. 1993. On the master symmetries and bi-Hamiltonian structure of the Toda lattice. *Journal of Physics A* 26: 3797–3803.
88. Fernandes, R. *Completely integrable bi-Hamiltonian systems*, Thesis Minnesota Ubliversity.
89. Fernandes, R. 1994. Completely integrable bi-Hamiltonian systems. *Journal of Dynamics and Differential Equations* 6: 53–69.
90. Fernandes, R. 2024. *Lectures on differential geometry*. Singapore: World Scientific.
91. Fiorani, E. 2009. Momentum maps, independent first integrals and integrability for central potentials. *International Journal of Geometric Methods in Modern Physics* 6: 1323–1341.

92. Flaschka, H. 1974. The Toda lattice. II inverse-scattering solution. *Progress of Theoretical Physics* 51: 703–716.
93. Flaschka, H. 1974. The Toda lattice. II: Existence of integrals. *Physical Review B* 9: 1924–1925.
94. Frölicher, A., and A. Nijenhuis. 1956. Theory of vector values differential forms, part I. *Indagationes Mathematicae* 59: 338–350.
95. Fuchssteiner, B. 1982. Lie algebra structure of degenerate Hamiltonian and bi-Hamiltonian systems. *Progress of Theoretical Physics* 68: 1082–1104.
96. Fuchssteiner, B. 1983. Master symmetries, higher order time-dependent symmetries and conserved densities of nonlinear evolution equations. *Progress of Theoretical Physics* 70: 1508–1522.
97. Geiges, H. 2008. *Introduction to contact topology*, Cambridge studies in advanced mathematics. Cambridge: Cambridge University Press.
98. Gelfand, I., and I. Dorfman. 1979. Hamiltonian operators and algebraic structures related to them. *Functional Analysis and Its Applications* 13: 248–262.
99. Gelfand, I., and I. Dorfman. 1980. Schouten bracket and Hamiltonian operators. *Functional Analysis and Its Applications* 14: 223–226.
100. Gelfand, I., and I. Zakharevich. 1991. Webs, veronese curves, and bihamiltonian systems. *Journal of Functional Analysis* 99: 150–178.
101. Gelfand, I., and I. Zakharevich. 1993. On the local geometry of bi-Hamiltonian structures. In *The Gelfand mathematical seminar, 1990–1992*, 51–112. Boston: Birkhäuser.
102. Gelfand, I., and I. Zakharevich. 2000. Webs, Lenard schemes, and the local geometry of bi-Hamiltonian Toda and Lax structures. *Selecta Mathematica, New Series* 6: 131–183.
103. Gerdjikov, V. S., G. Vilasi, and A. B. Yanovski. 2008. *Integrable Hamiltonian hierarchies – Spectral and geometric methods*, no. 748 in Lecture notes in physics. Berlin/Heidelberg: Springer.
104. Giaquinta, M., and S. Hilderbrant. 2006. *Calculus of variations, volume 1*, Grundleheren de mathematischen Wissenschaften. New York: Springer.
105. Giaquinta, M., and S. Hilderbrant. 2006. *Calculus of variations, volume 2*, Grundleheren de mathematischen Wissenschaften. New York: Springer.
106. Givental, A. 2012. Tribute to Vladimir Arnold. *Notices of the American Mathematical Society* 59: 378–399.
107. Godbillion, C. 1971. *Élements de topologie algébrique*, Colletion Méthodes. Paris: Herman.
108. Goldstein, H. 1980. *Classical mechanics*, Addison-Wesley series in physics, 2nd ed. Reading, MA: Addison-Wesley Publishing Co.
109. Görbe, T. 2016. A simple proof of Sklyanin formula for canonical spectral coordinates of the rational Calogero-Moser system. *SIGMA* 12: 5.
110. Gross, M. 1999. *Special Lagrangian fibrations II: Geometry. A survey of techniques in the study of special Lagrangian fibrations*, vol. 5. Surveys in differential geometry, 341–403. Boston: International Press. Differential geometry inspired by string theory.
111. Guest, M. 1997. *Harmonic map, loop groups, and integrable systems*, vol. 38. London mathematical society student texts. New York: Cambridge University Press.
112. Guillemin, V., and S. Sternberg. 1977. *Geometric asymptotics*, vol. 14. Mathematical surveys and monographs, revised version 1990 ed. Providence: American Mathematical Society.
113. Guillemin, V., and S. Sternberg. 1980. The moment map and collective motion. *Annals of Physics* 127: 220–253.
114. Guillemin, V., and S. Sternberg. 1990. *Symplectic techniques in physics*, 2nd ed. Cambridge: Cambridge University Press.
115. Helmke, U., and J. B. Moore. 1994. *Optimization and dynamical systems*, Communications and control engineering series. London: Springer.
116. Hénon, M. 1974. Integrals of the Toda lattice. *Physical Review B* 9: 1921–1923.
117. Hirsch, M. 1976. *Differential topology*, vol. 33. Graduate texts in mathematics, GTM, 2nd ed. New York/Berlin/Heidelberg: Springer.
118. Hitchin, N. 1992. Hyperkähler manifolds. *Astérisque* 1991/1992: 137–176.

119. Hitchin, N., A. Karlhede, Lindström, and M. U. Roček. 1987. Hyperkähler metrics and supersymmetry. *Communications in Mathematical Physics* 108: 535–589.
120. Holm, D. 2008. *Geometric mechanics, Part II rotating, translating and rolling*. London: Imperial College Press, World Scientific.
121. Huybrechts, D. 2005. *Complex geometry: An introduction*, UniversiText. Berlin/Heidelberg: Springer.
122. Huybrechts, J. 2003. Compact hyperkähler manifolds. In *Calabi-Yau manifolds and related geometries*, Universitext, 163–221. Berlin/Heidelberg/New York: Springer.
123. Iacob, A., and S. Sternberg. 1980. Coadjoint structures, solitons and integrability. In *Nonlinear evolution equations and dynamical systems*, vol. 120. Lecture notes in physics, 52–84. Berlin: Springer.
124. Ibort, A., F. Magri, and G. Marmo. 2000. Bihamiltonian structures and Stäckel separability. *Journal of Geometry and Physics* 33: 210–228.
125. Jacobi, C. 2009. *Jacobi's lectures on dynamics*, Mathematics and its applications. Providence: American Mathematical Society. Edited by A. Clebsch.
126. Joyce, D. 2000. *Compact manifolds with special holonomy*. Oxford: Oxford Science Publications, Oxford University Press.
127. Joyce, D. 2007. *Riemannian holonomy groups and calibrated geometry*, Oxford graduate texts in mathematics. Oxford: Oxford University Press.
128. Jurdjevic, V. 2009. *Geometric control theory*, Cambridge studies in advanced mathematics. Cambridge: Cambridge University Press.
129. Kalnins, E. 1986. *Separation of variables for Riemannian spaces of constant curvature*, Pitman monographs and surveys in pure and applied mathematics. Cambridge: Cambridge University Press/Wiley.
130. Kazhdan, D., B. Kostant, and S. Sternberg. 1978. Hamiltonian group actions and dynamical systems of Calogero type. *Communications on Pure and Applied Mathematics* 31: 481–507.
131. Khesin, B., and R. Wendt. 2009. *The geometry of infinite dimensional Lie groups*, vol. 51. Ergebnisse de Mathematik und ihrer Grenzgebiete. 3. Folge. A series of modern surveys in mathematics. Berlin: Springer.
132. Kirillov, A. 2004. *Lectures on the orbits method*, vol. 64. Graduate studies in mathematics. Providence: American Mathematical Society.
133. Kobayashi, S., and K. Nomizu. 1969. *Foundations of differential geometry, Vol. 1*, 2nd ed. New York: Wiley.
134. Kodama, Y. 2017. *KP solitons and the Grassmannian. Combinatorics and geometry of the two-dimensional wave patterns*. Springer briefs in mathematical physics. Providence: Springer.
135. Kodama, Y., and B. Shipman. 2008. *The finite non-periodic Toda lattice: A geometric and topological viewpoint*. arXiv:0805.1389
136. Kodama, Y., and B. Shipman. 2018. Fifty years of finite non-periodic Toda lattice: A geometrical and topological view-point. *Journal of Physics A: Mathematical and Theoretical* 51: 39.
137. Kosmann-Schwarzbach, Y. 1995. Exact Gesternhaber algebras and Lie bialgebroids. *Acta Applicandae Mathematica* 41: 153–165.
138. Kosmann-Schwarzbach, Y. 1996. The Lie bialgebroids of Poisson–Nijenhuis manifold. *Letters in Mathematical Physics* 38: 421–428.
139. Kosmann-Schwarzbach, Y. 2004. Lie bialgebras, Poisson Lie groups and dressing transformations. In *Integrability of nonlinear systems*, vol. 638. Lecture notes in physics, 107–173. Berlin: Springer.
140. Kosmann-Schwarzbach, Y. 2008. Poisson manifolds, Lie algebroids and modular classes: A survey. *SIGMA* 4: 30p.
141. Kosmann-Schwarzbach, Y., and F. Magri. 1990. Poisson–Nijenhuis structures. *Ann. Inst. Henri Poincaré, Phys. Théor.* 53: 35–81.

142. Kostant, B. 1977. Quantization and representation theory. In *Representation theory of Lie groups*, vol. 34. Lecture notes series, 287–316. London: London Mathematical Society.

143. Kostant, B. 1979. The solution to a generalized Toda lattice and representation theory. *Advances in Mathematics* 34: 195–338.

144. Krichever, I. 1978. Rational solutions of the Kadomtsev-Petviashvili equation and integrable systems of n-particles on a line. *Functional Analysis and Its Applications* 12: 59–61.

145. Krichever, I., and D. Phong. 1999. Symplectic forms in the theory of solitons. In *Integrable systems*, ed. C.-L. Terng and K. Uhlenbeck, vol. 4. Surveys in differential geometry, 239–313. Boston: International Press.

146. Krichever, I., and T. Shiota. 2013. Solitons equations and the Riemann-Schottky problem. In *Handbook of moduli Vol. II*, vol. 25. Advanced lectures in mathematics, 205–258. Sommerville, Boston, MA: International Press.

147. Landi, G., G. Marmo, and G. Vilasi. 1994. Recursion operators: Meaning and existence for completely integrable systems. *Journal of Mathematical Physics* 35: 808–815.

148. Lando, S. K., and Zvonkin, A. K. 2004. *Low-dimensional topology II*, vol. 141. Encyclopaedia of mathematical sciences. New York/Berlin/Heidelberg: Springer.

149. Laurent-Gengoux, C., E. Miranda, and P. Vanhaecke. 2010. Action-angle coordinates for integrable systems on Poisson manifolds. *International Mathematics Research Notices (IMRN)* 2011: 1839–1869.

150. Laurent-Gengoux, C., A. Pichereu, and P. Vanhaecke. 2013. *Poisson structures*, vol. 347. Grundlehren der mathematischen Wissenschaften. Heidelberg/New York/Dordrecht/London: Springer.

151. Lax, P. 1996. Outline of a theory of the KdV equation. In *Recent mathematical methods in nonlinear wave propagation (Montecatini Terme, 1994)*, vol. 1640. Lecture notes in mathematics, 70–102. Berlin: Springer.

152. Lehmann-Lejeune, J. 1966. Intégrabilité des G-strutures définies par une 1-forme 0-déformable à valeurs dans le fibré tangent. *Ann. Inst. Fourier (Grenoble)* 16: 329–387.

153. Lerman, E. *Symplectic geometry and Hamiltonian systems*, Unpublished manuscript, Downloadable from the webpage of the author.

154. Li, L.-C. 1999. Classical r-matrices and compatible Poisson structures for Lax equations on Poisson algebras. *Communications in Mathematical Physics* 203: 573–592.

155. Li, L.-C. 2023. The Toda flow on Hessenberg elements of real, split simple Lie algebras. *Physica D* 453: 133810.

156. Li, L.-C., and P. Xu. 2000. Spin Caloger-moser systems associated with simple Lie algebras. *Comptes Rendus de l'Académie des Sciences Paris* 331: 55–60.

157. L.-C. Li, and P. Xu. 2002. A class of integrable spin Calogero-Moser systems. *Communications in Mathematical Physics* 231: 257–286.

158. Libermann, P., and C. Marle. 1987. *Symplectic geometry and analytical mechanics*, vol. 35. Mathematics and its applications. Dordrecht: D. Reidel Publishing Co. Translated from the French by B. E. Schwarzbach.

159. Lorenzoni, P., and F. Magri. 2005. A cohomological construction of integrable hierarchies of hydrodynamic type. *International Mathematics Research Notices (IMRN)* 2005 (34): 2087–2100.

160. Lützen, J. 1990. *Joseph Liouville 1809–1882: Master of pure and applied mathematics*, vol. 15. Studies in the history of mathematics and physical sciences. New York: Springer.

161. Lützen, J. 1995. Interaction between mechanics and differential geometry in the 19th century. *Archive for History of Exact Sciences* 49: 1–72.

162. Mackenzie, K. 2005. *General theory of Lie groupoids and Lie algebroids*, vol. 213. London mathematical society lecture notes series. Cambridge: Cambridge University Press.

163. Magri, F. 1978. A simple model of the integrable Hamiltonian equation. *Journal of Mathematical Physics* 19: 1156–1162.

164. Magri, F. 1990. Geometry and soliton equations. In *La Mécanique Analitique de Lagrange et son Héritage*, vol. 1. Acta Academiae Scientarum Taurinensis, 181–209. Torino: Accademia delle Scienze di Torino.

165. Magri, F., P. Casati, G. Falqui, and M. Pedroni. 2004. Eight lectures on integrable systems. In *Integrability of nonlinear systems*, vol. 638. Lecture notes in physics, 209–247. Berlin: Springer.

166. Magri, F., and T. Marsico. 1996. Some development of the concept of Poisson manifold in the sense of A. Lichnerowicz. In *Gravitation, electromagnetism and geometrical structures*, Pitagora, Bologna, 207–222.

167. Magri, F., and C. Morosi. 1984. *Geometrical characterization of integrable Hamiltonian systems through the theory of poisson-nijenhuis structures*, Quaderno Università di Milano, S19.

168. Magri, F., C. Morosi, and O. Ragnisco. 1985. Reduction techniques for infinite dimensional Hamiltonian systems: Some ideas and applications. *Communications in Mathematical Physics* 99: 115–140.

169. Manakov, S. 1975. Complete integrability and stochastization of discrete dynamical systems. *Soviet Physics JETP* 40: 269–274.

170. Manakov, S. 1976. Note on the integration of Euler's equations of the dynamics of an n-dimensional rigid body. *Functional Analysis and Its Applications* 10: 328–329.

171. Marmo, G., and G. Vilasi. 1992. When do recursion operators generate new conservation laws?. *Physics Letters B* 354: 137–140.

172. Marsden, J. 1992. *Lectures on mechanics*, vol. 174. London mathematical society lecture note series. New York: Cambridge University Press.

173. Marsden, J., G. Misiołek, J.-P. Ortega, M. Perlmutter, and T. Ratiu. 2007. *Hamiltonian reduction by stages*, vol. 1913. Lecture notes in mathematics. New York: Springer.

174. Marsden, J., and T. Ratiu. 1986. Reduction of Poisson manifolds. *Letters in Mathematical Physics* 11: 161–169.

175. Marsden, J., and T. Ratiu. 1999. *Introduction to mechanics and symmetry*, vol. 17. Text in applied mathematics, 2nd ed. New York: Springer.

176. Marsden, J., and A. Weinstein. 1974. Reduction of symplectic manifolds with symmetry. *Reports on Mathematical Physics* 5: 121–130.

177. Marsico, T. 1995. *Una caratterizzazione geometrica dei sistemi che ammettono una rappresentazione di Lax estesa*, PhD thesis, Universitá degli Studi di Milano.

178. Mason, L., and N. Woodhouse. 1996. *Integrability, self-duality and twistor theory*, vol. 15. London mathematical society monograph new series. Oxford: Clarendon Press.

179. McDuff, D., and D. Salamon. 1999. *Introduction to symplectic topology*, Oxford mathematical monographs, 2nd ed. New York: Oxford University Press.

180. Meigniez, J. 2002. Submersions, fibrations and bundles. *Transactions of the American Mathematical Society* 354: 3771–3787.

181. Meyer, C. 2000. *Matrix analysis and applied linear algebra*. Philadelphia, PA: SIAM, Society for the Industrial and Applied Mathematics.

182. Meyer, K. 1973. Symmetries and integrals in mechanics. In *Dynamical systems*. Proceedings of a Symposium, University of Bahia, Salvador, 1971, 259–272. Cambridge: Academic Press.

183. Michor, P., I. Kolář, and J. Slovák. 2003. *Natural operations in differential geometry*. Berlin/Heidelberg: Springer.

184. Mischenko, A., and A. Fomenko. 1978. Generalized Liouville method of integration of Hamiltonian systems. *Functional Analysis and Its Applications* 12: 113–121.

185. Montaldi, J. 2000. Relative equilibria and conserved quantities in symmetric Hamiltonian systems. In *Peyresq lectures in nonlinear phenomena*, vol. 1, 239–280. Singapore/New Jersey/London/Hong Kong: World Scientific.

186. Monterde, J. 2004. A new proof of the existence of hierarchies of Poisson-Nijenhuis structures. *Portugaliae Mathematica* 61: 355–368.

187. Morosi, C., and L. Pizzocchero. 1996. On the Euler equation: Bi-Hamiltonian structure and integral in involution. *Letters in Mathematical Physics* 37: 117–135.

188. Moser, J. 1965. On the volume elements on a manifold. *Transactions of the American Mathematical Society* 120: 286–294.

189. Moser, J. 1975. Finitely many mass points on the line under the influence of an exponential potential- an integrable system. In *Dynamical systems, theory and applications (Recontres, Battelle Res. Inst., Seattle, Wash. 1974)*, vol. 38. Lecture notes in physics, 97–101. Berlin: Springer.

190. Moser, J. 1975. Three integrable Hamiltonian systems connected with isospectral deformations. *Advances in Mathematics* 16: 197–220.

191. Moser, J. 1978. Various aspects of integrable Hamiltonian systems. In *Proceeding C.I.M.E. conference, progress in mathematics*, Bressanone, 233–289. Basel: Birkhäuser.

192. Nakajima, H. 1999. *Lectures on Hilbert schemes of points on surfaces*, vol. 18. University lectures series. Providence: American Mathematical Society.

193. Nakamura, Y. 1993. Completely integrable gradient systems on the manifolds of Gaussian and multinomial distributions. *Japan Journal of Industrial and Applied Mathematics* 10: 179–189.

194. Nanda, T. 1985. Differential equations and the QR algorithm. *SIAM Journal on Numerical Analysis* 22: 310–321.

195. Nekhoroshev, N. 1972. Action-angle variables and their generalization. *Transactions of the Moscow Mathematical Society* 26: 181–198.

196. Nekrasov, N., and H. Braden. 2001. Instantons, Hilbert schemes and integrability. In *Integrable structures of exactly solvable two-dimensional models of quantum field theory*, vol. 35. Nato science series II: Mathematics, physics and chemistry, 35–54. Dordrecht: Springer.

197. Newell, A. 1985. *Solitons in mathematics and physics*, vol. 48. CBMS-NSF regional conference series in applied mathematics. Philadelphia, PA: SIAM.

198. Novikov, S., S. Manakov, S. Pitaevskii, and V. Zakharov. 1984. *Theory of solitons: The inverse scattering methods*. Monographs in contemporary mathematics. Berlin: Springer.

199. Nunes da Costa, J., and C. Marle. 1996. Reduction of bihamiltonian manifolds and recursion operators. In *Differential geometry and applications*, Masaryk University Brno, Czech Republic, 523–538.

200. Nunes da Costa, J., and C. Marle. 1997. Master symmetries and bi-Hamiltonian structures for the relativistic Toda lattice. *Journal of Physics A* 30: 7551–7556.

201. Oevel, W. 1987. A geometrical approach to integrable systems admitting time dependent invariants. In *Topics in soliton theory and exactly solvable nonlinear equations*, Lecture notes in mathematics, 108–124. Singapore: World Scientific.

202. Oevel, W., and M. Falck. 1986. Master symmetries for finite dimensional iintgrable systems. *Progress of Theoretical Physics* 75: 1328–1341.

203. Ohta, Y., J. Satsuma, D. Takahashi, and T. Tokihiro. 1988. An elementary introduction to Sato theory. *Progress of Theoretical Physics Supplements* 94: 210–241.

204. Olver, P. 1977. Evolution equations possessing infinitely many symmetries. *Journal of Mathematical Physics* 18: 1212–1215.

205. Ortega, J., and T. Ratiu. 2004. *Momentum maps and Hamiltonian reduction*, vol. 222. Progress in mathematics. Basel/Boston/Berlin: Birkhäuser.

206. Pedroni, M. 1995. Equivalence of the Drinfeld-Sokolov reduction to a bi-Hamiltonian reduction. *Letters in Mathematical Physics* 35: 291–302.

207. Pedroni, M. 2003. *Appunti del Corso di Aspetti Geometrici dei Sistemi Integrabili*, Unpublished manuscript, Downloadable from the webpage of the author.

208. Perelomov, A. 1990. *Integrable systems of classical mechanics and Lie algebras, vol. I*. Basel/Boston/Berlin: Birkhäuser. Translated from the Russian by A. G. Reyman.

209. Postnikov, M. 1986. *Lie groups and Lie algebras*, Lectures in geometry, SEMESTER V, Mir, Moscow.

210. Rajeev, S. 2013. *Advanced mechanics: From Euler's determinism to Arnold's chaos*. Oxford: Oxford University Press.

211. Ratiu, T. 1980. The motion of the free n-dimensional rigid body. *Indiana University Mathematics Journal* 29: 609–629.

212. Reyman, A. 1996. Poisson structures related to quantum groups. In *Quantum groups and their applications in physics*, ed. L. Castellani and J. E. Wess, Proceedings of the international school "Enrico Fermi" (Varenna 1994), 407–443. Amsterdam: IOS Press. Intern. School "Enrico Fermi" (Varenna 1994).

213. Reyman, A. G., and M. A. Semenov-Tian-Shansky. 2001. Group-theoretical methods in the theory of finite-dimensional integrable systems. In *Dynamical systems, VII*, vol. 7. Encyclopaedia of mathematical sciences, 177–225. Berlin: Springer.

214. Rutishauser, H. 1954. Ein infinitesimales Analogon zum Quotienten-Differenzen-Algorithmus. *Archiv der Mathematik* 5: 132–137.

215. Sato, M., and Y. Sato. 1983. Soliton equations as dynamical system on infinite dimensional Grassmann manifold. In *Nonlinear partial differential equations in applied sciences, Proceedings of the US-Japan Seminar, Tokyo, 1982*, vol. 81. North Holland mathematics studies, 259–271. Amsterdam/Oxford/New York: North-Holland Publishing Company.

216. Schöbel, K. 2015. *An algebraic geometric approach to separation of variables*. Berlin: Springer Spektrum.

217. Segal, G., and G. Wilson. 1985. Loop groups and equations of KdV type. *Publications Mathematiques de l'Institut des Hautes Études Scientifiques* 61: 5–65.

218. Semenov-Tian-Shansky, M. A. 2002. What is a classical R-matrix?. *Functional Analysis and Its Applications* 147: 243–348.

219. Serre, M. 1987. *Complex semisimple Lie algebras*. New York: Springer.

220. Shiota, T. 1994. Calogero-Moser hierarchy and KP hierarchy. *Journal of Mathematical Physics* 35: 5844–5849.

221. Singer, S. 2004. *Symmetry in mechanics: A gentle modern introduction*. Basel: Birkäuser.

222. Sklyanin, E. 1995. Separation of variables: New trends. *Progress of Theoretical Physics* 118: 35–60.

223. Smale, S. 1970. Topology and mechanics I. *Inventiones Mathematicae* 10: 305–331.

224. Smirnov, R. 1997. Magri-Morosi-Gel'fand-Dorfman's bi-Hamiltonian constructions in the action-angle variables. *Journal of Mathematical Physics* 38: 6444–6454.

225. Smirnov, R. 1999. The action-angle coordinates revisited: Bi-Hamiltonian systems. *Journal of Mathematical Physics* 44: 199–204.

226. Spivak, M. 1979. *A comprehensive introduction to differential geometry*, vol. I, 2nd ed. Wilmington, DE: Publish or Perish Inc.

227. Sternberg, S. 1983. *Lectures on differential geometry*, 2nd ed. New York: Chelsea Publishing Co.

228. Suris, Y. 2003. *The problem of integrable discretization: Hamiltonian approach*, vol. 219. Progress in mathematics. Basel/Boston/Berlin: Birkäuser.

229. Sutherland, B. 1972. Exact results for a quantum many-body problem in one dimension. II. *Physical Review A* 5: 1372–1376.

230. Symes, W. 1980. Hamiltonian group actions and integrable systems. *Physica D* 1: 339–374.

231. Symes, W. 1980. The QR algorithm and the scattering for the finite nonperiodic Toda lattice. *Physica D* 4: 275–280.

232. Symes, W. 1980. Systems of Toda type, inverse spectral problems, and representation theory. *Inventiones Mathematicae* 59: 13–51.

233. Tempesta, P., and G. Tondo. 2022. Higher Haantjes brackets and integrability. *Communications in Mathematical Physics* 389: 1647–1671.

234. Thomas, R. 2006. Notes on GIT and symplectic reduction for bundles and varieties. In *Essays in geometry in memory of S. S. Chern*, vol. 10. Surveys in differential Geometry, 221–273. Somerville, MA: International Press.

235. Toda, M. 1989. *Theory of non linear lattices*, no. 20 in Solid state sciences. Berlin/Heidelberg: Springer.

236. Tomei, C. 2013. The Toda lattice, old and new. *Journal of Geometric Mechanics* 5: 511–530.

237. Trusdell, C. 1977. *A first course in rational continuum mechanics: Vol. 1*, Pure and applied mathematics. London: Academic Press.

238. Turiel, F. 1980. Strutures bihamiltoniannes sur le fibré cotangent. *C. R. Acad. Sci. Paris Ser. I Math.* 315: 1085–1088.

239. Vaisman, I. 1990. *Remarks on the Lichnerowicz-Poisson cohomology. Annales de l'institut Fourier* 40: 951–963.
240. Vaisman, I. 1994. *Lectures on the geometry of Poisson manifolds*, vol. 118. Progress in mathematics. Basel/Boston/Berlin: Birkhäuser.
241. Vaisman, I. 1996. Reduction of Poisson-Nijenhuis manifolds. *Journal of Geometry and Physics* 19: 90–98.
242. Vilasi, G. 2001. *Hamiltonian dynamics*, vol. 118. Singapore/New Jersey/London/Hong Kong: World Scientific.
243. Vinogradov, A., and B. Kupershmidt. 1977. The structures of Hamiltonian mechanics. *Russian Mathematical Survey* 32: 177–243.
244. Warner, F. 1983. *Foundations of differentiable manifolds and Lie groups*, vol. 94. Graduate texts in mathematics. New York: Springer. Corrected reprint of the 1971 edition.
245. Watkins, D. S., and L. Elsner. 1990. On Rutishauser's approach to self-similar flows. *SIAM Journal on Matrix Analysis and Applications* 11: 301–311.
246. Weinstein, A. 1981. Symplectic geometry. *The Bulletin of the American Mathematical Society* 5: 1–13.
247. Weinstein, A. 1983. The local structure of Poisson manifolds. *The Journal of Differential Geometry* 18: 523–557.
248. Weinstein, A. 1985. Poisson structures and Lie algebras. In *The mathematical heritage of Élie Cartan*, vol. S131. Astérisque, SMF, 421–434. Lyon 1984.
249. Wells, R. 1983. *Differential analysis on complex manifolds*, vol. 65 of Graduate texts in mathematics, 2nd ed. New York: Springer.
250. Whittaker, E. 1988. *A treatise on the analytical dynamics of particles and rigid bodies*, Cambridge mathematical library. New York: Cambridge University Press. Reissue of the 4th edition.
251. Wilson, G. 1993. Bispectral commutative ordinary differential operators. *The Journal für die Reine und Angewandte Mathematik* 442: 177–204.
252. Wilson, G. 1998. Collisions of Calogero-Moser particles and an adelic Grassmannian. *Inventiones Mathematicae* 133: 1–41.
253. Xu, P. 1992. Poisson cohomology of regular poisson manifolds. *Annales de l'institut Fourier* 42: 967–988.
254. Yano, K., and M. Kon. 1985. *Structures on manifolds*, vol. 3. Series in pure mathematics. New York: World Scientific.
255. Zabusky, N. 1981. Computational synergetics and mathematical computation. *Journal of Computational Physics* 43: 195–249.
256. Zabusky, N. 1983. Solitons revolution. *Physics Today* 36: 95–96.
257. Zabusky, N., and M. Kruskal. 1965. Interaction of "solitons" in a collisionless plasma and the recurrence of initial states. *Physical Review Letters*, 15: 240–243.
258. Zung, N. 2018. A conceptual approach to the problem of action-angle variables. *The Archive for Rational Mechanics and Analysis* 229: 789–833.

Index

GPSR Compliance
The European Union's (EU) General Product Safety Regulation (GPSR) is a set
of rules that requires consumer products to be safe and our obligations to
ensure this.

If you have any concerns about our products, you can contact us on

ProductSafety@springernature.com

In case Publisher is established outside the EU, the EU authorized
representative is:

Springer Nature Customer Service Center GmbH
Europaplatz 3
69115 Heidelberg, Germany